1 Springer Series in
Edited by Peter Fulde

Springer Series in Solid-State Sciences

Editors: M. Cardona   P. Fulde   K. von Klitzing   H.-J. Queisser

Managing Editor: H. K. V. Lotsch

Volumes 1–49 are listed at the end of the book

50 **Multiple Diffraction of X-Rays in Crystals**
By Shih-Lin Chang

51 **Phonon Scattering in Condensed Matter**
Editors: W. Eisenmenger, K. Laßmann, and S. Döttinger

52 **Superconductivity in Magnetic and Exotic Materials**   Editors: T. Matsubara and A. Kotani

53 **Two-Dimensional Systems, Heterostructures, and Superlattices**
Editors: G. Bauer, F. Kuchar, and H. Heinrich

54 **Magnetic Excitations and Fluctuations**
Editors: S. Lovesey, U. Balucani, F. Borsa, and V. Tognetti

55 **The Theory of Magnetism II**   Thermodynamics and Statistical Mechanics   By D. C. Mattis

56 **Spin Fluctuations in Itinerant Electron Magnetism**   By T. Moriya

57 **Polycrystalline Semiconductors,** Physical Properties and Applications
Editor: G. Harbeke

58 **The Recursion Method and Its Applications**
Editors: D. Pettifor and D. Weaire

59 **Dynamical Processes and Ordering on Solid Surfaces**   Editors: A. Yoshimori and M. Tsukada

60 **Excitonic Processes in Solids**
By M. Ueta, H. Kanzaki, K. Kobayashi, Y. Toyozawa, and E. Hanamura

61 **Localization, Interaction, and Transport Phenomena**   Editors: B. Kramer, G. Bergmann, and Y. Bruynseraede

62 **Theory of Heavy Fermions and Valence Fluctuations**   Editors: T. Kasuya and T. Saso

63 **Electronic Properties of Polymers and Related Compounds**
Editors: H. Kuzmany, M. Mehring, and S. Roth

64 **Symmetries in Physics**   Group Theory Applied to Physical Problems
By W. Ludwig and C. Falter

65 **Phonons: Theory and Experiments II**
Experiments and Interpretation of Experimental Results   By P. Brüesch

66 **Phonons: Theory and Experiments III**
Phenomena Related to Phonons
By P. Brüesch

67 **Two-Dimensional Systems: Physics and New Devices**
Editors: G. Bauer, F. Kuchar, and H. Heinrich

68 **Phonon Scattering in Condensed Matter V**
Editors: A. C. Anderson and J. P. Wolfe

69 **Nonlinearity in Condensed Matter**
Editors: A. R. Bishop, D. K. Campbell, P. Kumar, and S. E. Trullinger

70 **From Hamiltonians to Phase Diagrams**   The Electronic and Statistical-Mechanical Theory of sp-Bonded Metals and Alloys   By J. Hafner

71 **High Magnetic Fields in Semiconductor Physics**
Editor: G. Landwehr

72 **One-Dimensional Conductors**
By S. Kagoshima, H. Nagasawa, and T. Sambongi

73 **Quantum Solid-State Physics**
Editors: S. V. Vonsovsky and M. I. Katsnelson

74 **Quantum Monte Carlo Methods** in Equilibrium and Nonequilibrium Systems   Editor: M. Suzuki

75 **Electronic Structure and Optical Properties of Semiconductors**
By M. L. Cohen and J. R. Chelikowsky

76 **Electronic Properties of Conjugated Polymers**
Editors: H. Kuzmany, M. Mehring, and S. Roth

77 **Fermi Surface Effects**
Editors: J. Kondo and A. Yoshimori

78 **Group Theory and Its Applications in Physics**
By T. Inui, Y. Tanabe, and Y. Onodera

79 **Elementary Excitations in Quantum Fluids**
Editors: K. Ohbayashi and M. Watabe

80 **Monte Carlo Simulation in Statistical Physics**
An Introduction
By K. Binder and D. W. Heermann

81 **Core-Level Spectroscopy in Condensed Systems**
Editors: J. Kanamori and A. Kotani

82 **Introduction to Photoemission Spectroscopy**
By S. Hüfner

83 **Physics and Technology of Submicron Structures**
Editors: H. Heinrich, G. Bauer, and F. Kuchar

84 **Beyond the Crystalline State**   An Emerging Perspective   By G. Venkataraman, D. Sahoo, and V. Balakrishnan

85 **The Fractional Quantum Hall Effect**
Properties of an Incompressible Quantum Fluid
By T. Chakraborty and P. Pietiläinen

86 **The Quantum Statistics of Dynamic Processes**
By E. Fick and G. Sauermann

87 **High Magnetic Fields in Semiconductor Physics II**
Editor: G. Landwehr

88 **Organic Superconductors**
By T. Ishiguro and K. Yamaji

89 **Strong Correlation and Superconductivity**
Editors: H. Fukuyama, S. Maekawa, and A. P. Malozemoff

90 **Early and Recent Developments in Superconductivity**
Editors: J. G. Bednorz and K. A. Müller

91 **Electronic Properties of Conjugated Polymers III**   Basic Models and Applications
Editors: H. Kuzmany, M. Mehring, and S. Roth

C. P. Slichter

Principles of Magnetic Resonance

Third Enlarged and Updated Edition

With 185 Figures

Springer-Verlag Berlin Heidelberg New York
London Paris Tokyo Hong Kong

Professor Charles P. Slichter, Ph.D.

Department of Physics, University of Illinois at Urbana-Champaign
Urbana, IL 61801, USA

Series Editors:
Professor Drs., Dres. h. c. Manuel Cardona
Professor Dr., Dr. h. c. Peter Fulde
Professor Dr., Dr. h. c. Klaus von Klitzing
Professor Dr. Hans-Joachim Queisser

Max-Planck-Institut für Festkörperforschung, Heisenbergstrasse 1
D-7000 Stuttgart 80, Fed. Rep. of Germany

Managing Editor:
Dr. Helmut K. V. Lotsch

Springer-Verlag, Tiergartenstrasse 17, D-6900 Heidelberg, Fed. Rep. of Germany

The original edition *Principles of Magnetic Resonance with Examples from Solid State Physics* was published in 1963 by Harper & Row Publishers, New York, Evanston, and London.

ISBN 3-540-50157-6 3. Aufl. Springer-Verlag Berlin Heidelberg New York
ISBN 0-387-50157-6 3rd ed. Springer-Verlag New York Berlin Heidelberg

ISBN 3-540-08476-2 2. Aufl. Springer-Verlag Berlin Heidelberg New York
ISBN 0-387-08476-2 2nd ed. Springer-Verlag New York Heidelberg Berlin

Library of Congress Cataloging-in-Publication Data. Slichter, Charles P. Principles of magnetic resonance / C. P. Slichter. – 3rd ed. p. cm. – (Springer series in solid-state sciences; 1) Bibliography: p. Includes indexes. ISBN 0-387-50157-6 (U.S.) 1. Nuclear magnetic resonance. I. Title. II. Series. QC762.S55 1989 538'.362 – dc 19 89-4117

This work is subject to copyright. All rights are reserved, whether the whole or part of the material is concerned, specifically the rights of translation, reprinting, reuse of illustrations, recitation, broadcasting, reproduction on microfilms or in other ways, and storage in data banks. Duplication of this publication or parts thereof is only permitted under the provisions of the German Copyright Law of September 9, 1965, in its version of June 24, 1985, and a copyright fee must always be paid. Violations fall under the prosecution act of the German Copyright Law.

© Springer-Verlag Berlin Heidelberg 1978 and 1990
Printed in the United States of America

The use of registered names, trademarks, etc. in this publication does not imply, even in the absence of a specific statement, that such names are exempt from the relevant protective laws and regulations and therefore free for general use.

2154/3150-543210 – Printed on acid-free paper

Preface to the Third Edition

The first edition of this book was written in 1961 when I was Morris Loeb Lecturer in Physics at Harvard. In the preface I wrote: "The problem faced by a beginner today is enormous. If he attempts to read a current article, he often finds that the first paragraph refers to an earlier paper on which the whole article is based, and with which the author naturally assumes familiarity. That reference in turn is based on another, so the hapless student finds himself in a seemingly endless retreat.

I have felt that graduate students or others beginning research in magnetic resonance needed a book which really went into the details of calculations, yet was aimed at the beginner rather than the expert."

The original goal was to treat only those topics that are essential to an understanding of the literature. Thus the goal was to be selective rather than comprehensive. With the passage of time, important new concepts were becoming so all-pervasive that I felt the need to add them. That led to the second edition, which Dr. Lotsch, Physics Editor of Springer-Verlag, encouraged me to write and which helped launch the Springer Series in Solid-State Sciences. Now, ten years later, that book (and its 1980 revised printing) is no longer available. Meanwhile, workers in magnetic resonance have continued to develop startling new insights. There are new topics which are so important that they must be included in a book that is intended to be an introductory text. The original course was one semester – the book had more topics than I taught in that amount of time. The new edition clearly would need a full year.

Throughout the history of this book, I have considered it as a textbook. Thus the aim was to explain in a rigorous but physical manner concepts which are essential for the student of magnetic resonance. Rather than giving an exhaustive treatment of any one topic, my intention has been to help prepare a student to read the literature on that topic. The main additions include an enlargement and modest rewrite of the topic of double resonance, explanations of one- and two-dimensional Fourier transform methods, of coherence transfer, of multiple quantum coherence, and of important topics related to dipolar coupling that underlie the method of spin-flip line narrowing. In the earlier editions, the chapter on the density matrix focused almost exclusively on its use in analyzing relaxation processes. The new edition explains its use in analyzing the effect of rf fields.

The first edition of this book came out shortly after the publication of Anatole Abragam's classic-to-be *Principles of Nuclear Magnetic Resonance*. There were not many other books on the market. Today there are many books on magnetic resonance, a large fraction written by some of the most illustrious founders of the field and practitioners of the art. The styles of these books are as varied as the scientific styles of their authors. As I have explored these marvellous books, I have been struck by the opportunity that exists today to get the flavor of the scientific thought of these great scientists.

When I wrote the first edition of this book at Harvard, I had no duties other than preparing and giving the lectures. The third edition has been written on evenings and weekends while I tried to carry on my normal activities. It would not have been possible without the expert, dedicated, and always cheerful support of my secretary, Ann Wells. Her association with the book has been long, since she helped type the original edition. I am also grateful for the help of Jamie Froman in the final stages.

Urbana, Illinois
June 1989   *Charles P. Slichter*

Contents

1. **Elements of Resonance** 1
   1.1 Introduction 1
   1.2 Simple Resonance Theory 2
   1.3 Absorption of Energy and Spin-Lattice Relaxation 4

2. **Basic Theory** 11
   2.1 Motion of Isolated Spins – Classical Treatment 11
   2.2 Quantum Mechanical Description of Spin in a Static Field .. 13
   2.3 Equations of Motion of the Expectation Value 17
   2.4 Effect of Alternating Magnetic Fields 20
   2.5 Exponential Operators 25
   2.6 Quantum Mechanical Treatment of a Rotating Magnetic Field . 29
   2.7 Bloch Equations 33
   2.8 Solution of the Bloch Equations for Low H_1 35
   2.9 Spin Echoes 39
   2.10 Quantum Mechanical Treatment of the Spin Echo 46
   2.11 Relationship Between Transient and Steady-State Response of a System and of the Real and Imaginary Parts of the Susceptibility 51
   2.12 Atomic Theory of Absorption and Dispersion 59

3. **Magnetic Dipolar Broadening of Rigid Lattices** 65
   3.1 Introduction 65
   3.2 Basic Interaction 66
   3.3 Method of Moments 71
   3.4 Example of the Use of Second Moments 80

4. **Magnetic Interactions of Nuclei with Electrons** 87
   4.1 Introduction 87
   4.2 Experimental Facts About Chemical Shifts 88
   4.3 Quenching of Orbital Motion 89
   4.4 Formal Theory of Chemical Shifts 92
   4.5 Computation of Current Density 96
   4.6 Electron Spin Interaction 108
   4.7 Knight Shift 113

		4.8	Single Crystal Spectra	127
		4.9	Second-Order Spin Effects – Indirect Nuclear Coupling	131
5.	Spin-Lattice Relaxation and Motional Narrowing of Resonance Lines			145
		5.1	Introduction	145
		5.2	Relaxation of a System Described by a Spin Temperature	146
		5.3	Relaxation of Nuclei in a Metal	151
		5.4	Density Matrix – General Equations	157
		5.5	The Rotating Coordinate Transformation	165
		5.6	Spin Echoes Using the Density Matrix	169
		5.7	The Response to a δ-Function	174
		5.8	The Response to a $\pi/2$ Pulse: Fourier Transform NMR	179
		5.9	The Density Matrix of a Two-Level System	186
		5.10	Density Matrix – An Introductory Example	190
		5.11	Bloch-Wangsness-Redfield Theory	199
		5.12	Example of Redfield Theory	206
		5.13	Effect of Applied Alternating Fields	215
6.	Spin Temperature in Magnetism and in Magnetic Resonance			219
		6.1	Introduction	219
		6.2	A Prediction from the Bloch Equations	220
		6.3	The Concept of Spin Temperature in the Laboratory Frame in the Absence of Alternating Magnetic Fields	221
		6.4	Adiabatic and Sudden Changes	223
		6.5	Magnetic Resonance and Saturation	231
		6.6	Redfield Theory Neglecting Lattice Coupling	234
			6.6.1 Adiabatic Demagnetization in the Rotating Frame	235
			6.6.2 Sudden Pulsing	237
		6.7	The Approach to Equilibrium for Weak H_1	239
		6.8	Conditions for Validity of the Redfield Hypothesis	241
		6.9	Spin-Lattice Effects	242
		6.10	Spin Locking, $T_{1\varrho}$, and Slow Motion	244
7.	Double Resonance			247
		7.1	What Is Double Resonance and Why Do It?	247
		7.2	Basic Elements of the Overhauser-Pound Family of Double Resonance	248
		7.3	Energy Levels and Transitions of a Model System	250
		7.4	The Overhauser Effect	254
		7.5	The Overhauser Effect in Liquids: The Nuclear Overhauser Effect	257
		7.6	Polarization by Forbidden Transitions: The Solid Effect	264
		7.7	Electron-Nuclear Double Resonance (ENDOR)	266

7.8	Bloembergen's Three-Level Maser	269
7.9	The Problem of Sensitivity	270
7.10	Cross-Relaxation Double Resonance	271
7.11	The Bloembergen-Sorokin Experiment	275
7.12	Hahn's Ingenious Concept	277
7.13	The Quantum Description	279
7.14	The Mixing Cycle and Its Equations	283
7.15	Energy and Entropy	287
7.16	The Effects of Spin-Lattice Relaxation	289
7.17	The Pines-Gibby-Waugh Method of Cross Polarization	293
7.18	Spin-Coherence Double Resonance – Introduction	295
7.19	A Model System – An Elementary Experiment: The S-Flip-Only Echo	296
7.20	Spin Decoupling	303
7.21	Spin Echo Double Resonance	311
7.22	Two-Dimensional FT Spectra – The Basic Concept	319
7.23	Two-Dimensional FT Spectra – Line Shapes	324
7.24	Formal Theoretical Apparatus I – The Time Development of the Density Matrix	325
7.25	Coherence Transfer	331
7.26	Formal Theoretical Apparatus II – The Product Operator Method	344
7.27	The Jeener Shift Correlation (COSY) Experiment	350
7.28	Magnetic Resonance Imaging	357

8.	Advanced Concepts in Pulsed Magnetic Resonance		367
	8.1	Introduction	367
	8.2	The Carr-Purcell Sequence	367
	8.3	The Phase Alternation and Meiboom-Gill Methods	369
	8.4	Refocusing Dipolar Coupling	371
	8.5	Solid Echoes	371
	8.6	The Jeener-Broekaert Sequence for Creating Dipolar Order	380
	8.7	The Magic Angle in the Rotating Frame – The Lee-Goldburg Experiment	384
	8.8	Magic Echoes	388
	8.9	Magic Angle Spinning	392
	8.10	The Relation of Spin-Flip Narrowing to Motional Narrowing	406
	8.11	The Formal Description of Spin-Flip Narrowing	409
	8.12	Observation of the Spin-Flip Narrowing	416
	8.13	Real Pulses and Sequences	421
		8.13.1 Avoiding a z-Axis Rotation	421
		8.13.2 Nonideality of Pulses	422
	8.14	Analysis of and More Uses for Pulse Sequence	423

9.	**Multiple Quantum Coherence**	431
9.1	Introduction	431
9.2	The Feasibility of Generating Multiple Quantum Coherence – Frequency Selective Pumping	434
9.3	Nonselective Excitation	444
	9.3.1 The Need for Nonselective Excitation	444
	9.3.2 Generating Multiple Quantum Coherence	445
	9.3.3 Evolution, Mixing, and Detection of Multiple Quantum Coherence	449
	9.3.4 Three or More Spins	455
	9.3.5 Selecting the Signal of a Particular Order of Coherence	463
9.4	High Orders of Coherence	470
	9.4.1 Generating a Desired Order of Coherence	471
	9.4.2 Mixing to Detect High Orders of Coherence	480
10.	**Electric Quadrupole Effects**	485
10.1	Introduction	485
10.2	Quadrupole Hamiltonian – Part 1	486
10.3	Clebsch-Gordan Coefficients, Irreducible Tensor Operators, and the Wigner-Eckart Theorem	489
10.4	Quadrupole Hamiltonian – Part 2	494
10.5	Examples at Strong and Weak Magnetic Fields	497
10.6	Computation of Field Gradients	500
11.	**Electron Spin Resonance**	503
11.1	Introduction	503
11.2	Example of Spin-Orbit Coupling and Crystalline Fields	505
11.3	Hyperfine Structure	516
11.4	Electron Spin Echoes	524
11.5	V_k Center	533
12.	**Summary**	555
	Problems	557
	Appendixes	579
A.	A Theorem About Exponential Operators	579
B.	Some Further Expressions for the Susceptibility	580
C.	Derivation of the Correlation Function for a Field That Jumps Randomly Between $\pm h_0$	584
D.	A Theorem from Perturbation Theory	585
E.	The High Temperature Approximation	589
F.	The Effects of Changing the Precession Frequency – Using NMR to Study Rate Phenomena	592

G. Diffusion in an Inhomogeneous Magnetic Field	597
H. The Equivalence of Three Quantum Mechanics Problems	601
I. Powder Patterns	605
J. Time-Dependent Hamiltonians	616
K. Correction Terms in Average Hamiltonian Theory – The Magnus Expansion	623
Selected Bibliography	629
References	639
Author Index	647
Subject Index	651

1. Elements of Resonance

1.1 Introduction

Magnetic resonance is a phenomenon found in magnetic systems that possess both magnetic moments and angular momentum. As we shall see, the term *resonance* implies that we are in tune with a natural frequency of the magnetic system, in this case corresponding to the frequency of gyroscopic precession of the magnetic moment in an external static magnetic field. Because of the analogy between the characteristic frequencies of atomic spectra, and because the magnetic resonance frequencies fall typically in the radio frequency region (for nuclear spins) or microwave frequency (for electron spins), we often use the terms *radio frequency* or *microwave spectroscopy*.

The advantage of the resonance method is that it enables one to select out of the total magnetic susceptibility, a particular contribution of interest — one that may, for example, be relatively very weak. The most spectacular example is, no doubt, the observation of the feeble nuclear paramagnetism of iron against a background of the electronic ferromagnetism. Resonance also permits the gathering of precise, highly detailed magnetic information of a type not obtainable in other ways.

One of the reasons for the impact of magnetic resonance on physics is its ability to give information about processes at the atomic level. In this book we seek to give some of the background necessary or useful to the application of magnetic resonance to the study of solids. Most of the book will be concerned with nuclear resonance, but the final chapters will focus on certain problems particularly important for electron spin resonance. Many of the principles developed in the earlier portions are, of course, equally applicable to nuclear or electron magnetic resonance. Our object is not to tell how to apply magnetic resonance to the study of solids. However, the activity in magnetic resonance has proceeded at such a vigorous pace, pouring out so many new concepts and results, that an author or lecturer faces an enormous task in the selection of material. In this book, we shall use the study of solids as a sort of ultimate goal that will help to delineate the topics for discussion and from which we shall attempt to draw most of the concrete examples of the more formal techniques.

As we remarked above, we are concerned with magnetic systems that possess angular momentum. As examples, we have electron spins, or the nuclei of atoms. A system such as a nucleus may consist of many particles coupled together so that in any given state, the nucleus possesses a total magnetic moment μ and a total angular momentum J. In fact the two vectors may be taken as parallel, so that we can write

$$\mu = \gamma J \tag{1.1}$$

where γ is a scalar called the "gyromagnetic ratio". For any given state of a nucleus, knowledge of the wave function would in principle enable us to compute both μ and J. Hence we should find that the quantity γ would vary with the state. Such calculations are beyond the scope of this book.

Of course, in the quantum theory, μ and J are treated as (vector) operators. The meaning of the concept of two operators being "parallel" is found by considering the matrix elements of the operators. Suppose we define a dimensionless angular momentum operator I by the equation:

$$J = \hbar I \;. \tag{1.2}$$

I^2 then has eigenvalues $I(I+1)$ where I is either integer or half-integer. Any component of I (for example I_z) commutes with I^2, so that we may specify simultaneously eigenvalues of both I^2 and I_z. Let us call the eigenvalues $I(I+1)$ and m, respectively. Of course m may be any of the $2I+1$ values $I, I-1, \ldots, -I$. The meaning of (1.1) is then that

$$(Im\,|\mu_{x'}|\,Im') = \gamma\hbar(Im\,|I_{x'}|\,Im') \tag{1.3}$$

where $\mu_{x'}$ and $I_{x'}$ are components of the operators μ and I along the (arbitrary) x'-direction. The validity of this equation is based on the Wigner-Eckart theorem, which we shall discuss in Chapter 10.

We shall, for the remainder of this chapter, give a very brief introduction to some of the basic facts of magnetic resonance, introducing most of the major concepts or questions that we shall explore in later chapters.

1.2 Simple Resonance Theory

We shall wish, in later chapters, to consider both quantum mechanical and classical descriptions of magnetic resonance. The classical viewpoint is particularly helpful in discussing dynamic or transient effects. For an introduction to resonance phenomena, however, we consider a simple quantum mechanical description.

The application of a magnetic field H produces an interaction energy of the nucleus of amount $-\mu \cdot H$. We have, therefore, a very simple Hamiltonian:

$$\mathcal{H} = -\mu \cdot H \;. \tag{1.4}$$

Taking the field to be H_0 along the z-direction, we find

$$\mathcal{H} = -\gamma\hbar H_0 I_z \;. \tag{1.5}$$

The eigenvalues of this Hamiltonian are simple, being only multiples ($\gamma\hbar H_0$) of the eigenvalues of I_z. Therefore the allowed energies are

$$E = -\gamma\hbar H_0 m \quad m = I, I-1, \ldots, -I \quad . \tag{1.6}$$

They are illustrated in Fig. 1.1 for the case $I = 3/2$, as is the case for the nuclei of Na or Cu. The levels are equally spaced, the distance between adjacent ones being $\gamma\hbar H_0$.

m
—3/2 ─────────
—1/2 ─────────
 1/2 ─────────
 3/2 ───────────── **Fig. 1.1.** Energy levels of (1.6)

One should hope to be able to detect the presence of such a set of energy levels by some form of spectral absorption. What is needed is to have an interaction that can cause transitions between levels. To satisfy the conservation of energy, the interaction must be time dependent and of such an angular frequency ω that

$$\hbar\omega = \Delta E \tag{1.7}$$

where ΔE is the energy difference between the initial and final nuclear Zeeman energies. Moreover, the interaction must have a nonvanishing matrix element joining the initial and final states.

The coupling most commonly used to produce magnetic resonances is an alternating magnetic field applied perpendicular to the static field. If we write the alternating field in terms of an amplitude H_x^0, we get a perturbing term in the Hamiltonian of

$$\mathcal{H}_{\text{pert}} = -\gamma\hbar H_x^0 I_x \cos\omega t \quad . \tag{1.8}$$

The operator I_x has matrix elements between states m and m', $(m'|I_x|m)$, which vanish unless $m' = m\pm 1$. Consequently the allowed transitions are between levels adjacent in energy, giving

$$\hbar\omega = \Delta E = \gamma\hbar H_0 \quad \text{or} \tag{1.9}$$

$$\omega = \gamma H_0 \quad . \quad \longrightarrow \text{resonance frequency} \tag{1.9a}$$

Note that Planck's constant has disappeared from the resonance equation. This fact suggests that the result is closely related to a classical picture. We shall see, in fact, that a classical description also gives (1.9a). By studying the two formulations (classical and quantum mechanical), one gains a great deal of added insight.

From (1.9a) we can compute the frequency needed to observe a resonance if we know the properties that determine γ. Although such calculations are of basic interest in the theory of nuclear structures, they would take us rather far afield. However, a simple classical picture will enable us to make a correct order-of-magnitude estimate of γ.

Let us compute the magnetic moment and angular momentum of a particle of mass m and charge e moving in a circular path of radius r with period T. The angular momentum is then

$$J = mvr = m\frac{2\pi r^2}{T} \quad , \tag{1.10}$$

while the magnetic moment (treating the system as a current loop of area A carrying current i) is

$$\mu = iA \quad . \tag{1.11}$$

Since $i = (e/c)(1/T)$, we get

$$\mu = \frac{e}{c}\frac{\pi r^2}{T} \quad . \tag{1.12}$$

Comparison of the expressions for μ and J therefore gives us $\gamma = e/2mc$. Besides enabling us to make an order of magnitude estimate of the expected size of γ, for our purposes the important result of this formula is that large masses have low γ's. We expect about a factor of 1,000 lower γ for nuclei than for electrons. In fact, for magnetic fields of 3,000 to 10,000 Gauss, electronic systems have a resonance at $\omega/2\pi = 10{,}000$ MHz (the 3 cm microwave region), whereas nuclear systems are typically 10 MHz (a radio frequency). Of course one can always change ω by changing H_0, but in most cases it is advantageous to use as large a magnetic field as possible, since the quanta absorbed are then larger and the resonance is correspondingly stronger.

In later sections, we shall comment somewhat more on typical experimental arrangements.

1.3 Absorption of Energy and Spin-Lattice Relaxation

We now wish to go a step further to consider what happens if we have a macroscopic sample in which we observe a resonance. For simplicity we consider a system whose nuclei possess spin $\frac{1}{2}$ (Fig. 1.2). Since there are many nuclei in our macroscopic sample, we shall specify the number in the two m states $+\frac{1}{2}$ and $-\frac{1}{2}$ by N_+ and N_-, respectively.

The total number of spins N is a constant, but application of an alternating field will cause N_+ or N_- to change as a result of the transitions induced. Let us denote the probability per second of inducing the transition of a spin with $m = +\frac{1}{2}$ to a state $m = -\frac{1}{2}$ by $W_{(+)\to(-)}$. We shall denote the reverse transition by $W_{(-)\to(+)}$. We can then write a differential equation for the change of the population N_+.

```
−1/2  ——————  N₋     Fig. 1.2. Energy levels for I = ½
       γℏH₀
+1/2  ——————  N₊
```

$$\frac{dN_+}{dt} = N_- W_{(-)\to(+)} - N_+ W_{(+)\to(-)} \quad . \tag{1.13}$$

Without as yet attempting to compute $W_{(+)\to(-)}$ or $W_{(-)\to(+)}$, we note a famous formula from time-dependent perturbation theory for the probability per second $P_{a\to b}$ that an interaction $V(t)$ induces a transition from a state (a) with energy E_a to a state (b) whose energy is E_b:

$$P_{a\to b} = \frac{2\pi}{\hbar}|(b|V|a)|^2 \delta(E_a - E_b - \hbar\omega) \quad . \tag{1.14}$$

Since $|(a|V|b)|^2 = |(b|V|a)|^2$, we note that $P_{a\to b}$ is the same as the rate $P_{b\to a}$. Such an argument describes many situations and leads to the condition $W_{(+)\to(-)} = W_{(-)\to(+)} \equiv W$.

$$\frac{dN_+}{dt} = W(N_- - N_+) \quad . \tag{1.15}$$

It is convenient to introduce the variable $n = N_+ - N_-$, the difference in population of the two levels. The two variables N_+ and N_- may be replaced by n and N, using the equations

$$N = N_+ + N_- \qquad n = N_+ - N_- \tag{1.16}$$

$$N_+ = \tfrac{1}{2}(N+n) \qquad N_- = \tfrac{1}{2}(N-n) \quad . \tag{1.16a}$$

Substitution of (1.16a) into (1.15) gives us

$$\frac{dn}{dt} = -2Wn \tag{1.17}$$

the solution of which is

$$n = n(0)e^{-2Wt} \tag{1.18}$$

where $n(0)$ is the value of n at $t = 0$. We note that if initially we have a population difference, it will eventually disappear under the action of the induced transitions.

The rate of absorption of energy dE/dt is given by computing the number of spins per second that go from the lower energy to the upper, and by subtracting the number that drop down, emitting energy in the process:

$$\frac{dE}{dt} = N_+ W\hbar\omega - N_- W\hbar\omega = \hbar\omega W n \quad . \tag{1.19}$$

Therefore, for a net absorption of energy, n must be nonzero; that is, there must be a population difference. We see that when the upper state is more highly populated than the lower, the net absorption of energy is negative – the system supplies more energy than it receives. This state of affairs is the basis of the oscillators or amplifiers known as *masers* (*m*icrowave *a*mplification by *s*timulated *e*mission of *r*adiation) or *lasers* (for *l*ight amplification).

We see that if the equations we have put down were complete, the resonant absorption of energy would eventually stop and the resonance would disappear. A more serious difficulty is seen if we assume $W = 0$ (that is, we do not apply the alternating magnetic field). Under these circumstances our equations say that $dN_+/dt = 0$. The populations cannot change. On the other hand, if we applied a static field to a piece of unmagnetized material, we should expect it to become magnetized. The preferential alignment of the nuclear moments parallel to the field corresponds to N_+ being greater than N_-. ($N_- = 0$ would represent perfect polarization, a state we should not expect to find at temperatures above absolute zero). The process of magnetization of an unmagnetized sample, therefore, requires a net number of transitions from the upper to the lower energy state. In the process, the spins give up energy – there is, so to speak, a heat transfer. Therefore there must be some other system to accept the energy. If we ask how big a population difference will eventually be found, the answer must depend upon the willingness of the other system to continue accepting energy. Speaking in thermodynamic terms, the heat flow will continue until the relative populations N_-/N_+ correspond to the temperature T of the reservoir to which the energy is given.

The final equilibrium populations N_+^0 and N_-^0 are then given by

$$\frac{N_-^0}{N_+^0} = e^{-\Delta E/kT} = e^{-\gamma \hbar H_0/kT} \quad . \tag{1.20}$$

We must postulate, therefore, that there exists a mechanism for inducing transitions between N_+ and N_-, which arises because of the coupling of the spins to some other system. Let us denote the probability per second that such a coupling will induce a spin transition upward in energy (from $+ \to -$) by $W\!\uparrow$, and the reverse process by $W\!\downarrow$. Then we have a rate equation

$$\frac{dN_+}{dt} = +N_- W\!\downarrow - N_+ W\!\uparrow \quad . \tag{1.21}$$

Let us again introduce the variables N and n; but now we no longer can assume equality of the two transition probabilities, since we know such an assumption would not give the preference for downward transitions which is necessary for the establishment of the magnetization. In fact, since in the steady-state dN_+/dt is zero, (1.21) tells us that

$$\frac{N_-^0}{N_+^0} = \frac{W\!\uparrow}{W\!\downarrow} \quad . \tag{1.22}$$

By using (1.20), we find that the ratio of $W\downarrow$ to $W\uparrow$ is not unity but rather is

$$\frac{W\downarrow}{W\uparrow} = e^{\gamma \hbar H_0 / kT} \quad . \tag{1.22a}$$

It is natural to wonder why the argument given to show the equality of $W_{(+)\to(-)}$ and $W_{(-)\to(+)}$ does not also apply here. The resolution of this paradox is that the thermal transition requires not only a coupling but also another system in an energy state that permits a transition. We can illustrate by assuming that the reservoir has only two levels whose spacing is equal to that of the nuclear system. If the nucleus and reservoir are initially in the states of Fig. 1.3a given by the crosses, conservation of energy is satisfied by simultaneous transitions indicated by the arrows. The nucleus may therefore give up energy to the lattice. On the other hand, if both systems are in the upper state (Fig. 1.3b), the simultaneous transition cannot occur because it does not conserve energy. The rate of transition of the nucleus will therefore depend not only on the matrix elements but *also* on the probability that the reservoir will be in a state that permits the transition.

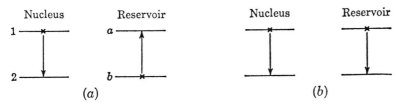

Fig. 1.3. (a) A possible transition. (b) A forbidden transition

Thus, if we label the nuclear states 1 and 2 with populations N_1 and N_2, and label the lattice states (a) and (b) with populations N_a and N_b, the number of transitions per second, such as shown in Fig. 1.3a, will be

$$\text{number/s} = N_1 N_b W_{1b \to 2a} \tag{1.23}$$

where $W_{1b\to 2a}$ is the probability per second of such a transition under the condition that the nucleus is actually in state 1 and the lattice is actually in state (b). The steady-state condition is found by equating the rate of such transitions to the rate of the inverse transition:

$$N_1 N_b W_{1b \to 2a} = N_2 N_a W_{2a \to 1b} \quad . \tag{1.24}$$

Since the quantum theory requires that $W_{1b\to 2a} = W_{2a\to 1b}$, we see that in thermal equilibrium,

$$\frac{N_1}{N_2} = \frac{N_a}{N_b} \quad . \tag{1.25}$$

That is, the nuclear levels will have the same relative populations as do those of the lattice. The nuclear population will therefore be in thermal equilibrium with that of the lattice. Note, moreover, that for this simple model, we can compute $W\uparrow$ and $W\downarrow$:

$$W\uparrow = N_a W_{2a \to 1b} \quad W\downarrow = N_b W_{1b \to 2a} = N_b W_{2a \to 1b} \tag{1.26}$$

so that $W\uparrow$ and $W\downarrow$ are seen to be unequal.

We now leave our special model and return to (1.21). By making the substitutions of (1.16a) for N_+ and N_-, we find

$$\frac{dn}{dt} = N(W\downarrow - W\uparrow) - n(W\downarrow + W\uparrow) \tag{1.27}$$

which can be rewritten as

$$\frac{dn}{dt} = \frac{n_0 - n}{T_1} \quad \text{where} \tag{1.28}$$

$$n_0 = N\left(\frac{W\downarrow - W\uparrow}{W\downarrow + W\uparrow}\right) \quad \frac{1}{T_1} = (W\downarrow + W\uparrow) \quad . \tag{1.29}$$

Since the solution of (1.28) is

$$n = n_0 + A e^{-t/T_1} \tag{1.30}$$

(where A is a constant of integration), we see that n_0 represents the thermal equilibrium population difference, and T_1 is a characteristic time associated with the approach to thermal equilibrium. T_1 is called the "spin-lattice relaxation time". For example, if we deal with a sample that is initially unmagnetized, the magnetization process is described by an exponential rise to the equilibrium:

$$n = n_0(1 - e^{-t/T_1}) \quad . \tag{1.31}$$

That is, T_1 characterizes the time needed to magnetize an unmagnetized sample.

We may now combine the two rate equations for dn/dt to find the combined transition rate due to both thermal processes and transitions induced by the applied alternating field:

$$\frac{dn}{dt} = -2Wn + \frac{n_0 - n}{T_1} \quad . \tag{1.32}$$

In the steady state, (1.32) tells us that

$$n = \frac{n_0}{1 + 2WT_1} \quad . \tag{1.33}$$

Therefore, as long as $2WT_1 \ll 1$, $n = n_0$, and the absorption of energy from the alternating field does not disturb the populations much from their thermal equilibrium values. The rate of absorption of energy dE/dt is given by

$$\frac{dE}{dt} = n\hbar\omega W = n_0\hbar\omega \frac{W}{1+2WT_1} \quad . \tag{1.34}$$

We shall see later that W is proportional to the square of the alternating magnetic field. Therefore (1.34) tells us that we can increase the power absorbed by the nuclei by increasing the amplitude of the alternating field, as long as $2WT_1 \ll 1$. However, once W is large enough so that $W \sim 1/2T_1$, this statement is no longer true. The power absorbed levels off despite an increase in W. This effect is called "saturation". Provided one has enough information to compute W (a situation often realized), one can measure T_1 by observing the saturation effect.

We have now seen several quantities that will be important in describing a magnetic resonance. The quantity T_1 will clearly be related to the microscopic details of both the nuclear system and the reservoir. We shall wish to consider what mechanisms may give rise to spin-lattice relaxation, and how to compute T_1 for any assumed mechanism. In the early work on nuclear resonance, it was feared that the spin-lattice relaxation might be so slow that a population excess might not be achieved within reasonable times. The famous Dutch physicist *C.J. Gorter*, who has made so many of the important discoveries and proposals in connection with magnetic relaxation, was the first person to look for a magnetic resonance in bulk matter [1.1]. That he failed was probably due to his bad luck in having a sample which was easily saturated because of its long T_1.

When *Purcell* et al. [1.2] first looked for a resonance of protons in paraffin, they allowed the nuclei to sit in the magnetic field H_0 for a long time before even attempting a resonance. They used a value of alternating field sufficiently low to allow them time to observe a resonance even though T_1 were many seconds. Their efforts, as with those of *Bloch* et al. [1.3], were made independently of *Gorter's*.

We have also seen that the rate of absorption is related to the transition rate W. An estimate of the size of the resonance absorption is basic to a decision about whether or not a resonance might be observed. We shall wish to consider how to calculate W. Moreover, since no resonance line is perfectly sharp, we expect that the factors governing the width of the spectral line will be of interest. Closely related is the question of what magnetic field to use in the relation $\omega = \gamma H_0$, for the nuclei are never bare. There will be magnetic fields due to electrons as well as due to other nuclei, which must be added to the external field. These fields produce effects of greatest interest, such as the splitting of the proton resonance of ethyl alcohol (CH_3CH_2OH) into three lines of relative intensities 3:2:1. They are also responsible for the fact that there is a nuclear resonance in ferromagnets even in the absence of an applied static magnetic field.

2. Basic Theory

2.1 Motion of Isolated Spins – Classical Treatment

We begin our study of the basic theory with a classical description of the motion of a spin in an external magnetic field H, assuming that H may possibly vary with time. H will produce a torque on the magnetic moment μ of amount $\mu \times H$. If we applied a magnetic field to an ordinary bar magnet, mounted with bearings so that it could turn at will, the magnet would attempt to line up along the direction of H. If H were constant in time and if the bearings were frictionless, the magnet would actually oscillate about the equilibrium direction. If the bearings were not frictionless, the oscillations would die out as the magnet gave up energy to the bearings, until eventually it would be lined up along H.

When the magnet also possesses angular momentum, the situation is modified, since it now acts like a gyroscope. As we shall see, in the event of frictionless bearings, the moment would remain at fixed angle with respect to H (providing H is constant in time), but would precess about it. The conversion of energy back and forth between potential energy and kinetic energy would not occur. It would still be true, however, that if the bearings possessed friction, the magnet would eventually become parallel to a static field H. As we shall see, the friction corresponds to relaxation processes such as T_1.

The equation of motion of the magnet is found by equating the torque with the rate of change of angular momentum J.

$$\frac{dJ}{dt} = \mu \times H \quad . \tag{2.1}$$

Since $\mu = \gamma J$, we may eliminate J, getting

$$\frac{d\mu}{dt} = \mu \times (\gamma H) \quad . \tag{2.2}$$

This equation, which holds regardless of whether or not H is time dependent, tells us that at any instant the changes in μ are perpendicular to both μ and H. Refer to Fig. 2.1 and consider the tail of the vector μ as fixed; the tip of the vector is therefore moving out of the paper. The angle θ between μ and H does not change. If H is independent of time, the vector μ therefore generates a cone.

One can proceed with the solution of (2.2) by standard methods of differential equations for various assumed time dependences of H. We shall find it

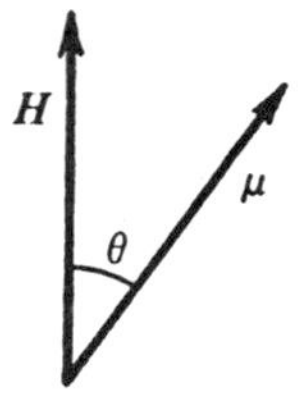

Fig. 2.1. Relation of μ to H

most useful for our future work, however, to introduce a special technique: the use of a rotating coordinate system.

Consider a vector function of time $F(t)$, which we may write in terms of its components $F_x(t)$, $F_y(t)$, $F_z(t)$, along a set of rectangular coordinates. In terms of the corresponding unit vectors i, j, and k, we have

$$F = iF_x + jF_y + kF_z \quad . \tag{2.3}$$

Ordinarily we think of i, j, and k as being constant in time, but we shall wish to be more general. Since their lengths are fixed, they can at most rotate. We shall assume they rotate with an instantaneous angular velocity Ω. Then

$$\frac{di}{dt} = \Omega \times i \quad . \tag{2.4}$$

The time derivative of F is therefore

$$\begin{aligned} \frac{dF}{dt} &= i\frac{dF_x}{dt} + F_x\frac{di}{dt} + j\frac{dF_y}{dt} + F_y\frac{dj}{dt} + k\frac{dF_z}{dt} + F_z\frac{dk}{dt} \\ &= i\frac{dF_x}{dt} + j\frac{dF_y}{dt} + k\frac{dF_z}{dt} + \Omega \times (iF_x + jF_y + kF_z) \\ &= \frac{\delta F}{\delta t} + \Omega \times F \end{aligned} \tag{2.5}$$

where we have introduced the symbol $\delta F/\delta t$, representing the time rate of change of F with respect to the coordinate system i, j, k. For example, when $\delta F/\delta t = 0$, the components of F along i, j, and k do not change in time.

By making use of (2.5), we can rewrite the equation of motion of μ in terms of a coordinate system rotating with an as yet arbitrary angular velocity Ω:

$$\frac{\delta \mu}{\delta t} + \Omega \times \mu = \mu \times \gamma H \quad \text{or} \tag{2.6}$$

$$\frac{\delta \mu}{\delta t} = \mu \times (\gamma H + \Omega) \quad . \tag{2.7}$$

Equation (2.7) tells us that the motion of μ in the rotating coordinate system obeys the same equation as in the laboratory system, *provided* we replace the actual magnetic field H by an effective field H_e:

$$H_e = H + \frac{\Omega}{\gamma} \quad . \tag{2.8}$$

We can now readily solve for the motion of μ in a static field $\mathbf{H} = \mathbf{k}H_0$ by choosing $\boldsymbol{\Omega}$ such that $\mathbf{H}_e = 0$. That is, we take $\boldsymbol{\Omega} = -\gamma H_0 \mathbf{k}$. Since in this reference frame $\delta\mu/\delta t = 0$, μ remains fixed with respect to $\mathbf{i}$, $\mathbf{j}$, and $\mathbf{k}$. The motion with respect to the laboratory is therefore that of a vector fixed in a set of axes which themselves rotate at $\boldsymbol{\Omega} = -\gamma H_0 \mathbf{k}$. In other words, μ rotates at an angular velocity $\boldsymbol{\Omega} = -\gamma H_0 \mathbf{k}$ with respect to the laboratory. The angular frequency γH_0 is called the "Larmor frequency".

We are struck by the fact that the classical precession frequency $\boldsymbol{\Omega}$ is identical in magnitude with the angular frequency needed for magnetic resonance absorption, as found by elementary quantum theory. Let us therefore look more closely at the quantum mechanical description.

2.2 Quantum Mechanical Description of Spin in a Static Field

We have seen that the quantum mechanical description of a spin in a static field gave energies in terms of the quantum number m, which was an eigenvalue of the component of spin I_z parallel to the static field H_0. The energies E_m were

$$E_m = -\gamma \hbar H_0 m \quad . \tag{2.9}$$

The corresponding eigenfunctions of the time-independent Schrödinger equation may then be denoted by $u_{I,m}$. The time-dependent solution corresponding to a particular value of m is therefore

$$\Psi_{I,m}(t) = u_{I,m} e^{-(i/\hbar)E_m t} \quad . \tag{2.10}$$

The most general time-dependent solution $\Psi(t)$ is therefore

$$\Psi(t) = \sum_{m=-I}^{+I} c_m u_{I,m} e^{-(i/\hbar)E_m t} \tag{2.11}$$

where the c_m's are complex constants. We may compute the expectation value of any observable by means of $\Psi(t)$, as we can illustrate with the x-component of magnetic moment:

$$\langle \mu_x(t) \rangle = \int \Psi^*(t) \mu_x \Psi(t) d\tau \quad .^{1} \tag{2.12}$$

We have emphasized that the expectation value of μ_x, $\langle \mu_x \rangle$ will vary in time by explicitly writing it as a function of time.

---

[1] We write a variable of integration $d\tau$ in the expression for the expectation value, in analogy to that which we would do for a spatial coordinate x, y, z or angular coordinates θ, ϕ. For spin, the notation is to be thought of as a symbolic representation of the scalar product of the two functions $\Psi(t)$ and $\mu_x \Psi(t)$

By using the fact that $\mu_x = \gamma \hbar I_x$, and that $\Psi(t)$ is given by (2.11) we find

$$\langle \mu_x(t) \rangle = \sum_{m,m'} \gamma \hbar c_{m'}^* c_m (m'|I_x|m) e^{(i/\hbar)(E_{m'}-E_m)t} \tag{2.13}$$

where

$$(m'|I_x|m) \equiv \int u_{Im'}^* I_x u_{Im} d\tau \quad , \tag{2.14}$$

is a time-independent matrix element. Expressions similar to (2.13) would hold for any operator. We denote that the expectation value will in general be time dependent, will consist of a number of terms oscillating harmonically, and that the possible frequencies

$$\frac{E_{m'} - E_m}{\hbar} \tag{2.15}$$

are just those which correspond to the frequency of absorption or emission between states m and m'. Of course it was the assumption that observable properties of any quantum system had to be given by expressions such as (2.13), which was the basis of Heisenberg and Born's formulation of the quantum theory in matrix form.

Since matrix elements $(m'|I_x|m)$ vanish unless $m' = m \pm 1$, we see that all the terms of (2.13) have an angular frequency of either $+\gamma H_0$ or $-\gamma H_0$. Their sum must also contain just γH_0. The expectation value $\langle \mu_x(t) \rangle$ therefore oscillates in time at the classical precession frequency.

It is convenient at this point to introduce the famous raising and lowering operators I^+ and I^-, defined by the equations

$$I^+ = I_x + iI_y \quad , \quad I^- = I_x - iI_y \quad . \tag{2.16}$$

We may express I_x or I_y in terms of I^+ and I^- by solving (2.16), getting

$$I_x = \frac{1}{2}(I^+ + I^-) \quad , \quad I_y = \frac{1}{2i}(I^+ - I^-) \quad . \tag{2.17}$$

The operators are called "raising" or "lowering" because of the effect they produce when they operate on a function $u_{I,m}$:

$$\begin{aligned} I^+ u_{I,m} &= \sqrt{I(I+1) - m(m+1)} u_{I,m+1} \\ I^- u_{I,m} &= \sqrt{I(I+1) - m(m-1)} u_{I,m-1} \quad . \end{aligned} \tag{2.18}$$

I^+ turns $u_{I,m}$ into a function whose m value has been raised by one unit. We see, therefore, that $(m'|I^+|m)$ vanishes unless $m' = m+1$, while $(m'|I^-|m)$ vanishes unless $m' = m-1$. Van Vleck [2.1] has characterized these as "sharper" selection rules than those of the operators I_x or I_y, which may join a state $u_{I,m}$ with either $u_{I,m+1}$ or $u_{I,m-1}$.

In order to gain further insight into the physical significance of the general expression for $\langle\mu_x(t)\rangle$, (2.13), we now consider the form it takes for a spin of $\frac{1}{2}$. By using the fact that the diagonal matrix elements of I_x vanish, we get

$$\langle\mu_x(t)\rangle = \gamma\hbar[c^*_{1/2}c_{-1/2}(\tfrac{1}{2}|I_x|-\tfrac{1}{2})e^{-i\gamma H_0 t} \\ + c^*_{-1/2}c_{1/2}(-\tfrac{1}{2}|I_x|\tfrac{1}{2})e^{i\gamma H_0 t}] \quad . \tag{2.19}$$

It is convenient to define a quantity $\omega_0 = \gamma H_0$. As we have seen, ω_0 is the angular frequency we must apply to produce resonance and is also the classical precession frequency. By utilizing the fact that $(\tfrac{1}{2}|I_x|-\tfrac{1}{2})$ is the complex conjugate of $(-\tfrac{1}{2}|I_x|\tfrac{1}{2})$, and using the symbol "Re" for "take the real part of", we get

$$\langle\mu_x(t)\rangle = 2\gamma\hbar \,\mathrm{Re}\,\{[c^*_{1/2}c_{-1/2}(\tfrac{1}{2}|I_x|-\tfrac{1}{2})e^{-i\omega_0 t}]\} \quad . \tag{2.20}$$

We evaluate the matrix element by means of (2.17) and (2.18), getting $(\tfrac{1}{2}|I_x|-\tfrac{1}{2}) = \tfrac{1}{2}$.

It is convenient at this point to express the c's in terms of two real, positive quantities a and b, and two other real quantities (which may be positive or negative) α and β:

$$c_{1/2} = a e^{i\alpha} \quad , \quad c_{-1/2} = b e^{i\beta} \quad . \tag{2.21}$$

The normalization of the wave function gives us $a^2 + b^2 = 1$. These give us

$$\langle\mu_x(t)\rangle = \gamma\hbar ab \cos(\alpha - \beta + \omega_0 t) \quad . \tag{2.22a}$$

Similarly we find

$$\langle\mu_y(t)\rangle = -\gamma\hbar ab \sin(\alpha - \beta + \omega_0 t) \tag{2.22b}$$
$$\langle\mu_z(t)\rangle = \gamma\hbar(a^2 - b^2)/2 \quad .$$

We note that both $\langle\mu_x\rangle$ and $\langle\mu_y\rangle$ oscillate in time at the Larmor frequency γH_0, but that $\langle\mu_z\rangle$ is independent of time. Moreover the maximum amplitudes of $\langle\mu_x\rangle$ and $\langle\mu_y\rangle$ are the same. If we define

$$\langle\boldsymbol{\mu}\rangle \equiv \boldsymbol{i}\langle\mu_x\rangle + \boldsymbol{j}\langle\mu_y\rangle + \boldsymbol{k}\langle\mu_z\rangle \tag{2.23}$$

and utilize the fact that $\langle\mu_x\rangle^2 + \langle\mu_y\rangle^2$ = constant, a fact readily verified from (2.23), we see that $\langle\boldsymbol{\mu}\rangle$ behaves as does a vector making a fixed angle with the z-direction, precessing in the x-y plane.

In terms of polar coordinates θ, ϕ (see Fig. 2.2), any vector $\boldsymbol{A}$ may be written as

$$A_x = A \sin\theta \cos\phi \\ A_y = A \sin\theta \sin\phi \\ A_z = A \cos\theta \quad . \tag{2.24}$$

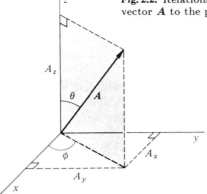

Fig. 2.2. Relationship of the components A_x, A_y, and A_z of a vector $\mathbf{A}$ to the polar angles θ, ϕ, and the magnitude A

By means of algebraic manipulation one can show that

$$\langle \mu_x \rangle = \frac{\gamma \hbar}{2} \sin \theta \cos \phi$$

$$\langle \mu_y \rangle = \frac{\gamma \hbar}{2} \sin \theta \sin \phi$$

$$\langle \mu_z \rangle = \frac{\gamma \hbar}{2} \cos \theta \qquad (2.25)$$

provided

$$\phi = \beta - \alpha - \omega_0 t$$
$$a^2 = \frac{1 + \cos \theta}{2} \quad . \qquad (2.26)$$

One may look on (2.26) as a formal change of variables, of course, but the results of (2.25) tell us that there is a simple physical significance; the expectation value of the operator μ acts as does a vector of length $\gamma\hbar/2$, whose direction is given by the spherical coordinates θ, ϕ. If the orientation is specified at any time, it can be found at future times by recognizing that it precesses at angular velocity ω_0 in the negative ϕ direction. The orientation may be specified quite arbitrarily (by specifying a or b and $\beta - \alpha$). We emphasize that an *arbitrary* orientation can be specified, since sometimes the belief is erroneously held that spins may only be found pointing either parallel or antiparallel to the quantizing field. One of the beauties of the quantum theory is that it contains features of both discreteness and continuity. In terms of the two quantum states with $m = \pm \frac{1}{2}$ we can describe an *expectation* value of magnetization which may go all the way from parallel to antiparallel, including all values in between. Thus a wave function with $a = b$ has an expectation value corresponding to a magnetization lying somewhere in the x-y plane (that is, with vanishing z-component). Just where in the plane it points is given by the complex phase $\alpha - \beta$, as well as the time at which we wish to know the orientation.

It is useful to consider briefly what we should expect for the wave function if we took a sample of many noninteracting spins which were in thermal equilibrium. There will be a wave function for each spin, but in general it will not be in one of the eigenstates ($m = +\frac{1}{2}$ or $m = -\frac{1}{2}$); rather it will be in some linear combination. For a given spin, there will be a particular set of values for a, b, α, β. The values will differ from spin to spin. For example, we have a distribution of the quantity $\alpha - \beta$ that gives the spin orientation in the x-y plane at $t = 0$. If the spins are in thermal equilibrium, the expectation value of the total magnetization must be parallel to the magnetic field. We expect, therefore, that there will be no preference for any one value of $\alpha - \beta$ over any other. That is, the spins will have a random distribution of $\alpha - \beta$. On the other hand, since the spins will be polarized to some extent, we expect to find a larger than b more often than b is larger than a. That is, the average value of a must be larger than the average value of b. Since an observable quantity can be expressed in the form of (2.13), we see that we can specify either the individual c_m's or the complex products $c_{m'}^* c_m$, which we shall label $P_{mm'}$, for convenience.

$$P_{mm'} = c_{m'}^* c_m \quad .$$

For our example:

$$P_{1/2\,1/2} = a^2 \quad , \quad P_{-1/2\,-1/2} = b^2$$
$$P_{1/2\,-1/2} = ab\,e^{i(\alpha - \beta)} \quad , \quad P_{-1/2\,1/2} = ab\,e^{i(\beta - \alpha)} \quad .$$

We may consider the $P_{mm'}$'s to be the elements of a complex matrix P. Notice that the diagonal elements ($m = m'$) give the probabilities of occupation of the various states, while the off-diagonal elements are closely related to the components of magnetic moment perpendicular to the static field. We shall make use in a subsequent section of the average of the matrix P over a statistical ensemble. The statement that in thermal equilibrium the magnetization will be parallel to the field amounts to saying that the average over the ensemble of $P_{mm'}$ for $m' \neq m$ is zero, whereas the average for $m = m'$ is the Boltzmann factor describing the probability of finding the state occupied.

(Of course, in the quantum theory, even for a number of spins with *identical* wave functions, any experiment that counts the number of spins in the various m states will find a statistical distribution not related, however, to temperature.)

2.3 Equations of Motion of the Expectation Value

The close correspondence of the classical and quantum mechanical treatments is made particularly clear by examination of a differential equation relating the time variations of the expectation values $\langle \mu_x \rangle$, $\langle \mu_y \rangle$, and $\langle \mu_z \rangle$. The equation is based on a well-known formula whose derivation we sketch.

Suppose we have a pair of wave functions $\Psi(t)$ and $\Phi(t)$, both of which are solutions of the same Schrödinger equation:

$$-\frac{\hbar}{i}\frac{\partial \Psi}{\partial t} = \mathcal{H}\Psi \qquad -\frac{\hbar}{i}\frac{\partial \Phi}{\partial t} = \mathcal{H}\Phi \quad . \tag{2.27}$$

Let us have some operator F that has no explicit time dependence. Then

$$\frac{d}{dt}\int \Phi^* F\Psi d\tau = \frac{i}{\hbar}\int \Phi^*(\mathcal{H}F - F\mathcal{H})\Psi d\tau \quad . \tag{2.28}$$

This equation is readily derived from the fact that

$$\frac{d}{dt}\int \Phi^* F\Psi d\tau = \int \frac{\partial \Phi^*}{\partial t}F\Psi d\tau + \int \Phi^* F\frac{\partial \Psi}{\partial t}d\tau \tag{2.29}$$

into which we substitute expressions for the time derivative taken from (2.27).[2]

It is convenient to write (2.28) in operator form. There is no problem with the right-hand side: It is simply $(i/\hbar)(\mathcal{H}F - F\mathcal{H})$. For the left-hand side we must define some new notation. We define the operator dF/dt by the equation

$$\int \Phi^* \frac{dF}{dt}\Psi d\tau = \frac{d}{dt}\int \Phi^* F\Psi d\tau \quad . \tag{2.30}$$

That is to say, dF/dt does *not* mean to take the derivative of F with respect to t. Such a derivative vanishes, since F does not contain the variable t. Rather dF/dt is a symbol that has the meaning of (2.30). By using dF/dt in this symbolic sense, we have

$$\frac{dF}{dt} = \frac{i}{\hbar}[\mathcal{H}, F] \tag{2.31}$$

where $[\mathcal{H}, F]$ is the usual commutator $\mathcal{H}F - F\mathcal{H}$. We may use this formalism to compute the time derivative of the expectation values of μ_x, μ_y, and μ_z. We define the x-, y-, z-axes as being fixed in space but with the z-axis coinciding at an instant with the direction of the magnetic field. (In this way we include both static and time-varying fields.) Then

$$\mathcal{H} = -\gamma \hbar H I_z \quad . \tag{2.32}$$

We shall wish to use the commutation relations for the components of angular momentum, all of which may be obtained by cyclic permutation from

$$[I_x, I_y] = iI_z \quad . \tag{2.33}$$

Then

---

[2] To prove (2.28), one must use the fact that F is an Hermitian operator. (See discussion in Sect. 2.5).

$$\begin{aligned}\frac{dI_x}{dt} &= \frac{\mathrm{i}}{\hbar}[\mathcal{H}, I_x]\\ &= -\gamma H_0 \mathrm{i}[I_z, I_x]\\ &= \gamma H_0 I_y \quad .\end{aligned} \qquad (2.34\mathrm{a})$$

Similarly,

$$\begin{aligned}\frac{dI_y}{dt} &= -\gamma H_0 I_x\\ \frac{dI_z}{dt} &= 0 \quad .\end{aligned} \qquad (2.34\mathrm{b})$$

These equations are the component equations of the vector operator equation

$$\frac{d\boldsymbol{I}}{dt} = \boldsymbol{I} \times \gamma \boldsymbol{H} \quad \text{where} \qquad (2.35)$$

$$\frac{d\boldsymbol{I}}{dt} = \boldsymbol{i}\frac{dI_x}{dt} + \boldsymbol{j}\frac{dI_y}{dt} + \boldsymbol{k}\frac{dI_z}{dt} \quad . \qquad (2.36)$$

Therefore, since $\boldsymbol{\mu} = \gamma\hbar\boldsymbol{I}$, we have the equation for the expectation value of magnetization,

$$\frac{d\langle\boldsymbol{\mu}\rangle}{dt} = \langle\boldsymbol{\mu}\rangle \times \gamma\boldsymbol{H} \qquad (2.37)$$

which is just the classical equation. In words, (2.37) tells us that the expectation value of the magnetic moment obeys the classical equation of motion. Equation (2.37) was derived for the expectation value of a magnetic moment of a single spin. If we have a group of spins with moments $\boldsymbol{\mu}_k$, for the kth spin, their total magnetic moment $\boldsymbol{\mu}$ is defined as

$$\boldsymbol{\mu} = \sum_k \boldsymbol{\mu}_k \quad . \qquad (2.38)$$

If the spins do not interact with one another, it is easy to prove that (2.37) also holds true for the expectation value of the total magnetization. Since, in practice, we measure the results of a number of spins simultaneously, the experimental measurements of magnetization measure the expectation value of the various components of magnetization. That is, the experimentally determined bulk magnetization is simply the expectation value of the total magnetic moment. Therefore the classical equation correctly describes the dynamics of the magnetization, provided the spins may be thought of as not interacting with one another.

It is important to bear in mind that (2.37) *holds true for a time-dependent H, not simply a static one.* Therefore it enables us to use a classical picture for studying the effects produced by alternating magnetic fields. We turn to that in the next section.

2.4 Effect of Alternating Magnetic Fields

The effect of an alternating magnetic field $H_x(t) = H_{x0} \cos \omega t$ is most readily analyzed by breaking it into two rotating components, each of amplitude H_1, one rotating clockwise and the other counterclockwise (see Fig. 2.3).

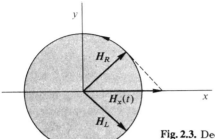

Fig. 2.3. Decomposition of a linear oscillating field into two rotating elements

We denote the rotating fields by $\boldsymbol{H}_\mathrm{R}$ and $\boldsymbol{H}_\mathrm{L}$:

$$\boldsymbol{H}_\mathrm{R} = H_1(\boldsymbol{i} \cos \omega t + \boldsymbol{j} \sin \omega t)$$
$$\boldsymbol{H}_\mathrm{L} = H_1(\boldsymbol{i} \cos \omega t - \boldsymbol{j} \sin \omega t) \quad . \tag{2.39}$$

Note that $\boldsymbol{H}_\mathrm{L}$ and $\boldsymbol{H}_\mathrm{R}$ differ simply by a replacement of ω by $-\omega$. Since one component will rotate in the same sense as the precession of the moment and the other in the opposite sense, one can show that near resonance the counterrotating component may be neglected. We shall make that approximation in what follows. Alternatively we can assume that we are finding the exact solution of a problem in which the experimental arrangement has produced a rotating field; for example, by use of two identical coils at right angles to each other and with alternating currents 90 degrees out of phase.

We shall assume we have only the field $\boldsymbol{H}_\mathrm{R}$, but this is no loss in generality because the use of a negative ω will convert it to $\boldsymbol{H}_\mathrm{L}$. In order to reserve the symbol ω for a positive quantity, we shall introduce the symbol ω_z, the component of ω along the z-axis. ω_z may therefore be positive or negative. We may, therefore, write

$$\boldsymbol{H}_1 = H_1(\boldsymbol{i} \cos \omega_z t + \boldsymbol{j} \sin \omega_z t) \tag{2.40}$$

which will give us either sense of rotation, depending on the sign of ω_z.

We now ask for the equation of motion of a spin including the effects both of $\boldsymbol{H}_1(t)$ and of the static field $\boldsymbol{H}_0 = \boldsymbol{k} H_0$.

$$\frac{d\boldsymbol{\mu}}{dt} = \boldsymbol{\mu} \times \gamma[\boldsymbol{H}_0 + \boldsymbol{H}_1(t)] \quad . \tag{2.41}$$

The time dependence of H_1 can be eliminated by using a coordinate system that rotates about the z-direction at frequency ω_z. In such a coordinate system, H_1 will be static. Since the axis of rotation coincides with the direction of H_0, H_0 will also be static. Let us take the x-axis in the rotating frame along H_1. Then (2.41) becomes

$$\frac{\delta \mu}{\delta t} = \mu \times [k(\omega_z + \gamma H_0) + i\gamma H_1] \quad . \tag{2.42a}$$

Notice that we have encountered two effects in making the transformation of (2.41) to (2.42a). The first is associated with the derivative of the rotating unit vectors and gives the term ω_z. The second is associated with expressing the vectors H_0 and H_1 in terms of their components in the rotating system and gives rise to the conversion of H_1 from a rotating to a static field. Equation (2.42a) may be rewritten to emphasize that near resonance $\omega_z + \gamma H_0 \cong 0$, by setting $\omega_z = -\omega$, where ω is now positive (we assume here that γ is positive). Then

$$\frac{\delta \mu}{\delta t} = \mu \times \gamma \left[\left(H_0 - \frac{\omega}{\gamma} \right) k + H_1 i \right]$$
$$= \mu \times H_{\text{eff}} \tag{2.42b}$$

where

$$H_{\text{eff}} = k \left(H_0 - \frac{\omega}{\gamma} \right) + H_1 i \quad .$$

Physically (2.42b) states that in the rotating frame, the moment acts as though it experienced effectively a static magnetic field H_{eff}. The moment therefore precesses in a cone of fixed angle about the direction of H_{eff} at angular frequency γH_{eff}. The situation is illustrated in Fig. 2.4 for a magnetic moment which, at $t = 0$, was oriented along the z-direction.

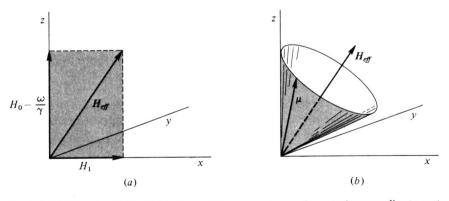

Fig. 2.4. (a) Effective field. (b) Motion of the moment μ in the rotating coordinate system

We notice that the motion of the moment is periodic. If it is initially oriented along the z-direction, it periodically returns to that direction. As it increases its angle with the z-direction, its magnetic potential energy in the laboratory reference system changes (in the laboratory system the magnetic energy with respect to H_0 is much larger than that with respect to H_1, so we customarily neglect the latter). However, all the energy it takes to tilt μ away from $\boldsymbol{H}_0$ is returned in a complete cycle of μ around the cone. There is no net absorption of energy from the alternating field but rather alternately receiving and returning of energy.

Note that if H_0 is above resonance (that is, $H_0 > \omega/\gamma$), the effective field has a positive z-component, but when H_0 lies below that resonance ($H_0 < \omega/\gamma$), the effective field has a negative z-component.

If the resonance condition is fulfilled exactly ($\omega = \gamma H_0$), the effective field is then simply iH_1. A magnetic moment that is parallel to the static field initially will then precess in the y-z plane. That is, it will precess but remaining always perpendicular to H_1. Periodically it will be lined up *opposed* to H_0. If we were to turn on H_1 for a short time (that is, apply a wave train of duration t_w), the moment would precess through an angle $\theta = \gamma H_1 t_w$. If t_w were chosen such that $\theta = \pi$, the pulse would simply invert the moment. Such a pulse is referred to in the literature as a "180 degree pulse". If $\theta = \pi/2$ (90 degree pulse), the magnetic moment is turned from the z-direction to the y-direction. Following the turn-off of H_1, the moment would then remain at rest in the rotating frame, and hence precess in the laboratory, pointing normal to the static field.

These remarks suggest a very simple method of observing magnetic resonance, illustrated in Fig. 2.5. We put a sample of material we wish to study in a coil, the axis of which is oriented perpendicular to $\boldsymbol{H}_0$. In thermal equilibrium there will be an excess of moments pointing along $\boldsymbol{H}_0$. Application of an alternating voltage to the coil produces an alternating magnetic field perpendicular to $\boldsymbol{H}_0$. By properly adjusting H_1 and t_w, we may apply a 90 degree pulse. Fol-

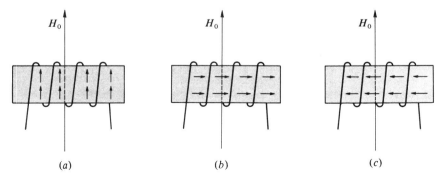

Fig. 2.5. (a) Coil containing sample. In thermal equilibrium an excess of moments is parallel to H_0. (b) and (c) Following a 90-degree pulse, the excess moments precess perpendicular to H_0

lowing the pulse, the excess magnetization will be perpendicular to H_0 and will precess at angular frequency γH_0. As a result, the moments will produce a flux through the coil which will alternate as the spins precess. The resultant induced emf may be observed.

What we have suggested so far would indicate that the induced emf would persist indefinitely, but in practice, the interactions of the spins with their surroundings cause a decay. The decay may last in liquids for many milliseconds, but in solids it is more typically $100\,\mu$s. Even during that short time, however, there are many precession periods. The technique we have described of observing the "free induction decay" (that is, decay "free" of H_1) is a commonly used technique for observing resonances. It has the great virtue of enabling one to study the resonance signal in the absence of the voltages needed to produce H_1. Since oscillators always generate noise, such a scheme may be advantageous.

One interesting application of the rotating reference frame is to prove the following theorem, which is the basis of another technique for producing resonance signals. Suppose we have a magnetic field H_0 of fixed magnitude whose direction we may vary (no other magnetic field is present). Let the magnetization M be parallel to H_0 at $t = 0$. We may describe the changing direction of H_0 by an angular velocity ω. Then the theorem states that if

$$\gamma H_0 \gg \omega$$

the magnetization M will turn with H_0, always remaining aligned along H_0 as H_0 turns.

To prove this theorem, let us assume ω to be a constant in the z-direction. We can take it perpendicular to H_0, since a component parallel to H_0 produces no effect. The relationships are shown in Fig. 2.6 at $t = 0$, with M and H_0 taken parallel to each other and pointing in the X-direction in the laboratory. If we choose a reference frame x, y, z rotating at angular velocity $\Omega_R = \omega$, H_0 appears static, but we must add an effective field Ω_R/γ. Choosing the z- and Z-axes as parallel, and x to coincide with X at $t = 0$, the effective fields and magnetization at $t = 0$ are shown in Fig. 2.7.

The effective field in the rotating frame is static and given by

$$H_{\text{eff}} = H_0 + \frac{\Omega_R}{\gamma} = H_0 + \frac{\omega}{\gamma} \quad .$$

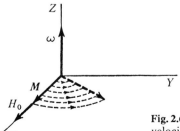

Fig. 2.6. Magnetic field H_0, magnetization M, and angular velocity ω at $t = 0$

Fig. 2.7. Magnetization M and effective field H_{eff} in the rotating coordinate system x, y, z. The magnetization will precess about the effective field in the cone of angle θ shown

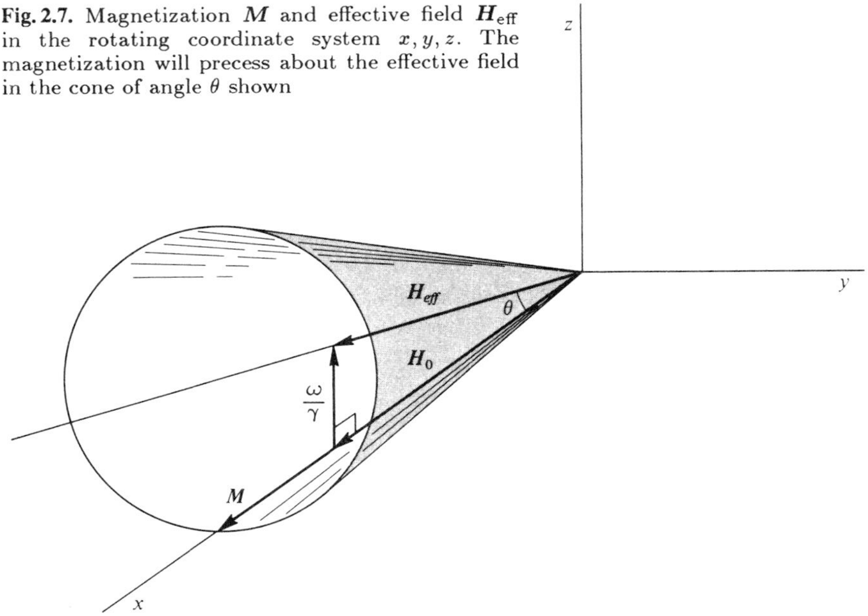

M will precess about H_{eff}, making an angle θ such that

$$\tan \theta = \frac{\omega}{\gamma H_0} \ . \tag{2.43}$$

M will therefore remain within an angle 2θ of H_0. We see that if $\omega/\gamma H_0 \ll 1$, M and H_0 remain parallel.

The fact that the magnetization follows the direction of the magnetic field when the field changes direction sufficiently slowly is described by the term *adiabatic*.

By utilizing this principle, one can turn to the case of a rotating magnetic field H_1 of frequency ω, perpendicular to a static field H_0. If one starts far below resonance, the magnetization is nearly parallel to the effective field in the rotating frame $\sqrt{H_1^2 + [(\omega/\gamma) - H_0]^2}$. As one approaches resonance, both magnitude and direction of the effective field change, but if resonance is approached sufficiently slowly, M will remain parallel to H_{eff} in the rotating frame according to the theorem we have just proved. Thus, exactly at resonance, the magnetization will lie along H_1, making a 90 degree angle with H_0 (Fig. 2.8).

If one were to continue on through the resonance, the magnetization would end up by pointing in the negative z-direction. This technique of inverting M is very useful experimentally and is called "adiabatic inversion".

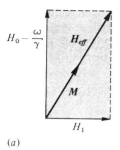 

Fig. 2.8. (a) Magnetization M and effective field H_{eff} in the rotating frame, with M parallel to H_{eff}. (b) The situation exactly at resonance, having approached resonance slowly, with M parallel to H_0 when H_0 was far above resonance

2.5 Exponential Operators

It will be useful to consider the quantum mechanical equivalent of the rotating coordinate transformation, but to do so, we shall need to employ several useful relations. We review them here for the convenience of the reader.

Suppose we have two wave functions, Φ and Ψ, that satisfy appropriate boundary conditions and have other satisfactory properties for some region of space, and suppose we have an operator F. F may be, for example, a component of spin. The operator is said to be Hermitian when

$$\int \Phi^* F \Psi d\tau = \int (F\Phi)^* d\tau \tag{2.44}$$

where the integrals are over the region of space designated. To prove that an operator is Hermitian requires some statement about the conditions Ψ and Φ are to satisfy, as well as a definition of the region. For example, if F is an operator involving derivatives, the proof that it is Hermitian may involve transforming the volume integral to a surface integral and requiring the integrand of the surface integral to vanish on the surface of the region.

Hermitian operators are important because their expectation values and eigenvalues are real. Therefore any operator that corresponds to a physically observable quantity must be Hermitian. Thus the operators I_x, I_y, and I_z are Hermitian. If they are Hermitian, it is easy to show from (2.44) that the operators $I^+ = I_x + iI_y$ and $I^- = I_x - iI_y$ are not.

In the theory of functions, it is useful to define the exponential function of the complex variable z:

$$e^z = 1 + z + \frac{z^2}{2!} + \frac{z^3}{3!} + \ldots$$

the power series converging for all z.

We define the function

$$e^F = 1 + F + \frac{F^2}{2!} + \frac{F^3}{3!} + \ldots$$

similarly, where F is now an operator. We shall be particularly interested in the function

$$e^{iF} = 1 + iF + \frac{(iF)^2}{2!} + \frac{(iF)^3}{3!} + \ldots \quad . \tag{2.45}$$

By using the series expansion, one can show that if F is Hermitian, $\exp(iF)$ is not. In fact

$$\int (e^{iF}\Phi)^* \Psi d\tau = \int \Phi^* e^{-iF} \Psi d\tau \quad . \tag{2.46}$$

The exponential function of operators obeys some of the same algebra as does the function of ordinary number, but as usual with operators, care must be taken whenever two noncommuting operators are encountered. Thus, if A and B are two operators, one can verify by means of the series expansion that

$$A e^{iB} = e^{iB} A \tag{2.47a}$$

only if A and B commute. Likewise,

$$e^{i(A+B)} = e^{iA} e^{iB} \tag{2.47b}$$

only if A and B commute.

If A and B do not commute, another useful equation may still hold. Let us define C as the commutator of A and B:

$$[A, B] = AB - BA \equiv C \quad .$$

Suppose that C commutes with both A and B:

$$[A, C] = 0 \quad , \quad [B, C] = 0 \quad .$$

Then

$$e^{(A+B)} = e^A e^B e^{-C/2} = e^{C/2} e^B e^A \quad .$$

This theorem is proved in Appendix A.

Use of the exponential function provides a particularly simple method for obtaining a formal solution of Schrödinger's equation if the Hamiltonian does not depend explicitly on time. That is, if $\Psi(t)$ is the solution of

$$-\frac{\hbar}{i} \frac{\partial \Psi(t)}{\partial t} = \mathcal{H} \Psi(t) \tag{2.48}$$

then we can express $\Psi(t)$ in terms of its value at $t = 0$, $\Psi(0)$, by the equation

$$\Psi(t) = e^{-(i/\hbar)\mathcal{H}t} \Psi(0) \quad . \tag{2.49}$$

Equation (2.49) may be verified by direct substitution into (2.48). If, for example, we consider the motion of a spin in a magnetic field so that $\mathcal{H} = -\gamma \hbar H_0 I_z$,

$$\Psi(t) = \exp\left[-(\mathrm{i}/\hbar)(-\gamma\hbar H_0 I_z)t\right]\Psi(0)$$
$$= \exp(\mathrm{i}\omega_0 t I_z)\Psi(0) \tag{2.50}$$

where $\omega_0 = \gamma H_0$.

We know that H_0 produces a rotation of the magnetic moment at angular velocity Ω given by $\Omega = -\gamma H_0 \boldsymbol{k}$. We shall call such a rotation "negative", since the component of angular velocity along the z-axis is negative. It is logical to suppose, then, that $\Psi(t)$ must correspond to the function $\Psi(0)$, referred, however, to axes rotated in the negative direction through an angle $\omega_0 t$. Thus $\exp(-\mathrm{i}I_z\phi)\Psi(0)$ should correspond to a function identical to $\Psi(0)$ referred to axes rotated through the positive angle ϕ. If we compute the expectation value or matrix elements of, for example, I_x, we find

$$\begin{aligned}\int \Psi^*(t) I_x \Psi(t)\, d\tau &= \int [\mathrm{e}^{\mathrm{i}\omega_0 t I_z}\Psi(0)]^* I_x \mathrm{e}^{\mathrm{i}\omega_0 t I_z}\Psi(0)\, d\tau \\ &= \int \Psi^*(0) \mathrm{e}^{-\mathrm{i}\omega_0 t I_z} I_x \mathrm{e}^{\mathrm{i}\omega_0 t I_z}\Psi(0)\, d\tau \\ &= \int \Psi^*(0) I_{x'}(t)\Psi(0)\, d\tau \end{aligned} \tag{2.50a}$$

where

$$I_{x'}(t) \equiv \mathrm{e}^{-\mathrm{i}\omega_0 t I_z} I_x \mathrm{e}^{\mathrm{i}\omega_0 t I_z} \quad . \tag{2.50b}$$

The last line defines the operator $I_{x'}$. We can give a simple interpretation of (2.50) as follows:

The first integral, which gives $\langle I_x(t)\rangle$, corresponds to a precessing angular momentum arising from the effect on a time-independent operator I_x of a time-dependent function $\Psi(t)$. The last integral describes the effect on a time-dependent operator $I_{x'}(t)$ of a wave function $\Psi(0)$, which is independent of time. Since the precession is in the negative sense, the first integral involves a fixed operator and a wave function fixed with respect to axes that rotate in the negative sense. Therefore the last integral must describe an operator rotating in the *positive* sense with respect to the "fixed" wave function $\Psi(0)$.

It is a simple matter to show that $I_{x'}$ is related to I_x through a rotation of axes. Let us consider

$$\mathrm{e}^{-\mathrm{i}I_z\phi} I_x \mathrm{e}^{\mathrm{i}I_z\phi} = f(\phi) \quad . \tag{2.51}$$

We wish to find $f(\phi)$, to see what meaning we can ascribe to it. Of course we could simply expand the exponentials and, using the commutation laws, try to reduce the function to something tractable. A simpler method is to show first that $f(\phi)$ satisfies a simple differential equation and then solve the equation. We have

$$\frac{df}{d\phi} = \mathrm{e}^{-\mathrm{i}I_z\phi}(-\mathrm{i}I_z I_x + \mathrm{i}I_x I_z)\mathrm{e}^{\mathrm{i}I_z\phi} \quad . \tag{2.52}$$

But, since $[I_z, I_x] = \mathrm{i}I_y$,

$$\frac{df}{d\phi} = e^{-iI_z\phi} I_y e^{iI_z\phi}. \tag{2.53}$$

Likewise

$$\begin{aligned}\frac{d^2 f}{d\phi^2} &= e^{-iI_z\phi}(-iI_z I_y + iI_y I_z)e^{iI_z\phi} \\ &= -e^{-iI_z\phi}(I_x)e^{iI_z\phi} = -f \quad \text{or} \quad \frac{d^2 f}{d\phi^2} + f = 0 \quad .\end{aligned} \tag{2.54}$$

Therefore

$$f(\phi) = A \cos \phi + B \sin \phi$$

where we must evaluate the constants of integrations. (As we shall see, the "constants" are actually operators.) Clearly, $A = f(0)$, but from (2.51), $f(0) = I_x$. Likewise, $B = f'(0) = I_y$, using (2.53). In this way we get

$$I_{x'} \equiv e^{-iI_z\phi} I_x e^{iI_z\phi} = I_x \cos \phi + I_y \sin \phi$$

$$I_{y'} \equiv e^{-iI_z\phi} I_y e^{iI_z\phi} = -I_x \sin \phi + I_y \cos \phi \tag{2.55}$$

$$I_{z'} \equiv e^{-iI_z\phi} I_z e^{iI_z\phi} = I_z \quad .$$

The quantities $I_{x'}$, $I_{y'}$, and $I_{z'}$ are clearly the components of angular momentum along a set of axes x', y', z' rotated with respect to x, y, z, as shown in Fig. 2.9. Therefore we see that we can use the exponential operator $\exp(iI_z\phi)$ to generate rotations.

It is frequently useful to work with the raising and lowering operators I^+ and I^-. Then the first two equations of (2.55) can be rewritten as

$$e^{-iI_z\phi} I^+ e^{iI_z\phi} = I^+ e^{-i\phi} \quad \text{and} \tag{2.56a}$$

$$e^{-iI_z\phi} I^- e^{iI_z\phi} = I^- e^{i\phi} \quad . \tag{2.56b}$$

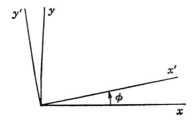

Fig. 2.9. Relation of axes x, y to x', y' and the angle ϕ

2.6 Quantum Mechanical Treatment of a Rotating Magnetic Field

We shall now use the exponential operators to perform the quantum mechanical equivalent of the classical "rotating coordinate" transformation. We shall consider a magnetic field H_1, which rotates at angular velocity ω_z, in addition to the static field $\mathbf{k}H_0$. The total field $\mathbf{H}(t)$ is then

$$\mathbf{H}(t) = \mathbf{i}H_1 \cos \omega_z t + \mathbf{j}H_1 \sin \omega_z t + \mathbf{k}H_0 \tag{2.57}$$

and the Schrödinger equation

$$-\frac{\hbar}{i}\frac{\partial \Psi}{\partial t} = -\boldsymbol{\mu}\cdot\mathbf{H}\Psi = -\gamma\hbar[H_0 I_z + H_1(I_x \cos\omega_z t + I_y \sin\omega_z t)]\Psi \quad . \tag{2.58}$$

By using (2.55) of the preceding section, we can write the Hamiltonian of (2.58) as

$$\mathcal{H} = -\gamma\hbar(H_0 I_z + H_1 e^{-i\omega_z t I_z} I_x e^{i\omega_z t I_z}) \quad . \tag{2.59}$$

We are tempted to try to "remove" the operator $\exp(i\omega_z t I_z)$ from I_x and transfer it onto Ψ, much as the reverse of the steps of (2.50) of the preceding section. Accordingly we let

$$\Psi' = e^{i\omega_z t I_z}\Psi \quad \text{or} \tag{2.60}$$

$$\Psi = e^{-i\omega_z t I_z}\Psi'.$$

The physical interpretation of (2.60) is that Ψ and Ψ' differ by a rotation of axes through an angle $\omega_z t$ (a rotating coordinate transformation).
Then, using (2.60)

$$\frac{\partial \Psi}{\partial t} = -i\omega_z I_z e^{-i\omega_z t I_z}\Psi' + e^{-i\omega_z t I_z}\frac{\partial \Psi'}{\partial t} \quad . \tag{2.61}$$

We may substitute (2.60) and (2.61) into (2.58), multiply both sides from the left by $\exp(i\omega_z t I_z)$, and obtain

$$-\frac{\hbar}{i}\frac{\partial \Psi'}{\partial t} = -[\hbar(\omega_z + \gamma H_0)I_z + \gamma\hbar H_1 I_x]\Psi' \quad . \tag{2.62}$$

In (2.62) the time dependence of $H_1(t)$ has been eliminated. In fact we recognize it as representing the coupling of the spins with an effective *static* field

$$\mathbf{k}\left(H_0 + \frac{\omega_z}{\gamma}\right) + \mathbf{i}H_1 \quad ,$$

the effective field of our classical equations. The spins are therefore quantized along the effective field in the rotating coordinate system, the energy spacing being $\gamma\hbar H_{\text{eff}}$.

The wave function Ψ' given by (2.60) is related to the function Ψ by a coordinate rotation, the "forward" motion of I_x relative to a stationary Ψ having been replaced by a stationary I_x and "backward" rotating Ψ'. As usual, resonance occurs when $\omega_z \approx -\gamma H_0$. If we define the transformed Hamiltonian $\mathcal{H}'$ by

$$\mathcal{H}' = -[(\hbar\omega_z + \gamma\hbar H_0)I_z + \gamma\hbar H_1 I_x] \tag{2.63}$$

we can formally solve (2.62):

$$\Psi'(t) = e^{-(i/\hbar)\mathcal{H}'t}\Psi'(0) \tag{2.64a}$$

whence, using (2.60),

$$\Psi(t) = e^{-i\omega_z t I_z}e^{-(i/\hbar)\mathcal{H}'t}\Psi'(0) \quad . \tag{2.64b}$$

[Note that at $t = 0$, $\Psi(0) = \Psi'(0)$].

Equation (2.64b) gives us a particularly compact way to express the solution of Schrödinger's equation when a rotating field is present.

We can illustrate the use of the wave function of (2.64b) by computing the time dependence of the expectation value of μ_z. Of course we know already what the result must be, since we have proved that the classical picture applies. Let us for simplicity assume that H_1 is applied exactly at resonance. Then, from (2.63),

$$\mathcal{H}' = -\gamma\hbar H_1 I_x \quad . \tag{2.65}$$

Then we have, using (2.64b) and (2.65),

$$\begin{aligned}\langle\mu_z(t)\rangle &= \int \Psi^*(t)\mu_z\Psi(t)d\tau \\ &= \gamma\hbar\int\left[e^{-i\omega_z t I_z}e^{i\gamma H_1 I_x t}\Psi(0)\right]^* I_z[e^{-i\omega_z t I_z}e^{i\gamma H_1 I_x t}\Psi(0)]d\tau.\end{aligned} \tag{2.66}$$

If we define ω_1,

$$\omega_1 \equiv \gamma H_1 \tag{2.67}$$

and use the fact that I_x and I_z are Hermitian, we get

$$\begin{aligned}\langle\mu_z(t)\rangle &= \gamma\hbar\int\Psi^*(0)e^{-i\omega_1 t I_x}e^{i\omega_z t I_z}I_z e^{-i\omega_z t I_z}e^{i\omega_1 t I_x}\Psi(0)d\tau \\ &= \gamma\hbar\int\Psi^*(0)e^{-i\omega_1 t I_x}I_z e^{i\omega_1 t I_x}\Psi(0)d\tau \quad .\end{aligned} \tag{2.68}$$

By using (2.55), we can write

$$e^{-i\omega_1 t I_x}I_z e^{i\omega_1 t I_x} = -I_y \sin\omega_1 t + I_z \cos\omega_1 t \quad . \tag{2.69}$$

Substituting in (2.68) we get

$$\langle \mu_z(t) \rangle = -\langle \mu_y(0) \rangle \sin \omega_1 t + \langle \mu_z(0) \rangle \cos \omega_1 t \quad . \tag{2.70}$$

If the magnetization lies along the z-axis at $t = 0$ so that $\langle \mu_y(0) \rangle = 0$, we get

$$\langle \mu_z(t) \rangle = \langle \mu_z(0) \rangle \cos \gamma H_1 t \quad . \tag{2.71}$$

Thus the z-magnetization oscillates in time at γH_1, corresponding to the precession of $\langle \mu \rangle$ about H_1 in the rotating reference frame. It is important to note that in this picture, which neglects all interactions of spins with one another or the lattice, the magnetization continues oscillating between $+\langle \mu_z(0) \rangle$ and $-\langle \mu_z(0) \rangle$ indefinitely. This behavior is very different from that which we should expect from a time-independent transition probability such as we assumed in Chapter 1. The time-independent transitions occur only if some physical process spoils the coherent precession about H_1 in the rotating reference frame.

Another approach to solving (2.62) enables us to demonstrate a very interesting and fundamental property of spin $\frac{1}{2}$ particles, their so-called spinor nature. Formally, this property describes what happens to a wave function under a rotation. One can discuss the problem by formal mathematical methods, as in group theory (see below). Here, however, we display the mathematical result by physically generating a rotation with an alternating magnetic field tuned exactly to resonance so that $(H_0 + \omega_z/\gamma) = 0$ in (2.62).

Let $u_{1/2}$ and $u_{-1/2}$ be the eigenstates of the spin operator I_z. They are independent of time. Since they form a complete set for a spin $\frac{1}{2}$ particle, we can express any function, such as ψ', as a linear combination of $u_{1/2}$ and $u_{-1/2}$ with coefficients a and b. If ψ' is itself a function of time, $\psi'(t)$, the coefficients a and b must also be functions of time:

$$\psi'(t) = a(t) u_{1/2} + b(t) u_{-1/2} \quad . \tag{2.72}$$

Substituting into (2.62) we get

$$\frac{\hbar}{i} \left(\frac{da}{dt} u_{1/2} + \frac{db}{dt} u_{-1/2} \right) = (\gamma \hbar H_1 I_x)(a u_{1/2} + b u_{-1/2}) \quad . \tag{2.73}$$

Multiplying by $u_{1/2}^*$ from the left, integrating over spin space, and utilizing the fact that $\langle \frac{1}{2} | I_x | \frac{1}{2} \rangle = \langle -\frac{1}{2} | I_x | -\frac{1}{2} \rangle = 0$, we get

$$\frac{\hbar}{i} \frac{da}{dt} = \gamma \hbar H_1 b \langle \frac{1}{2} | I_x | -\frac{1}{2} \rangle \quad . \tag{2.74}$$

Similarly we get

$$\frac{\hbar}{i} \frac{db}{dt} = \gamma \hbar H_1 a \langle -\frac{1}{2} | I_x | \frac{1}{2} \rangle \quad .$$

Utilizing the fact that $\langle \frac{1}{2} | I_x | -\frac{1}{2} \rangle = \frac{1}{2}$ and $\langle -\frac{1}{2} | I_x | \frac{1}{2} \rangle = \frac{1}{2}$ we can solve these two simultaneous differential equations to find

$$a(t) = a(0) \cos(\omega_1 t/2) + ib(0) \sin(\omega_1 t/2)$$
$$b(t) = ia(0) \sin(\omega_1 t/2) + b(0) \cos(\omega_1 t/2) \tag{2.75}$$

where

$$\omega_1 = \gamma H_1 \tag{2.76}$$

is the classical precession frequency about H_1 in the rotating frame.

The spinor property is revealed by considering a 2π pulse. We recall that such a pulse causes the expectation value of the magnetization vector to undergo a rotation about H_1 which returns it to its initial value. Denoting the pulse duration as $t_{2\pi}$, we have, then

$$\omega_1 t_{2\pi} = 2\pi \quad , \tag{2.77}$$

which gives, using (2.75),

$$a(t_{2\pi}) = -a(0) \quad , \quad b(t_{2\pi}) = -b(0) \quad . \tag{2.78}$$

These relationships show that after a 2π rotation the wave function has *not* returned to its original value but has instead changed sign. For the wave function to return to its initial value, the pulse length, t_w, must lead to a 4π rotation.

Thus, if $\omega_1 t_w = 4\pi$

$$a(t_w) = a(0) \quad , \quad b(t_w) = b(0) \quad . \tag{2.79}$$

The general property that a 2π rotation produces a sign reversal of ψ and that a 4π rotation is needed to get ψ back to its initial value is referred to as the "spinor" property of ψ. It is shared by wave functions associated with spins of $\frac{1}{2}$, $\frac{3}{2}$, $\frac{5}{2}$, etc. The existence of spinors is well known in group theory. For example, in solid-state physics the property is referred to with the term a *crystal double group* [2.2]. The wave functions of particles with spins of 0, 1, 2, etc. return to their original value under rotations of 2π.

There is something unsettling about finding that a 2π rotation does not return one to one's starting point! It is perhaps comforting in this connection to think of a Möbius strip, which is shown in Fig. 2.10. As is explained in the picture, one must go around the strip twice to reach the starting point. Thus, we have a physical manifestation or representation of (2.73).

Referring to (2.22) we see that though the wave function changes sign on a 2π rotation, the expectation values of the spin components I_x, I_y, I_z do not. The question arises whether or not the spinor nature is physically observable. The answer is yes. The first explicit demonstration was done by two groups: by *Rauch* et al. [2.3], and independently by *Werner* et al. [2.4]. Methods of providing a test had been proposed earlier by *Bernstein* [2.5] and by *Aharanov* and *Susskind* [2.6]. The essence of the idea is to produce a spatial separation of the spin-up part of the wave function from the spin-down part so that one can act on the two parts independently. Following a spatial region in which a magnetic field acts on only one component of the wave function, the two components are recombined to

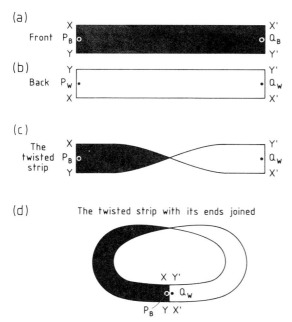

Fig. 2.10. A Möbius strip can be envisaged by starting with a strip of paper whose front (**a**) and back (**b**) sides can be distinguished by painting them black and white respectively. The paper is then twisted (**c**) and the ends joined (**d**). Suppose one then starts on the black surface at the point P_B, moves on that surface to its other end (point Q_B) at which point one crosses over to point P_W on the white surface. After going the length of the white surface, one arrives at point Q_W, adjacent to the original starting point, P_B. Thus, one has been around the strip twice to reach the starting point

study their interference. Thus, if one leaves the spin-up portion alone, it provides a fixed phase reference for the spin-down function. When one subjects the spin-down function to 2π and 4π rotations by passing it through a region of magnetic field, one finds that the interference intensity is the same for 0 and 4π rotations, but different for a 2π rotation. Ingenious NMR experiments demonstrating the spinor property have also been performed by *Stoll* and co-workers [2.7–2.9].

2.7 Bloch Equations

Both quantum mechanical and classical descriptions of the motion of noninteracting spins have in common a periodic motion of the magnetization in the rotating frame. For example, if $\gamma H_0 = \omega$ and if the magnetization is parallel to the static field at $t = 0$, the magnetization precesses around H_1 in the rotating frame, becoming alternately parallel and antiparallel to the direction of the static field. Viewed from the laboratory frame, the magnetization is continuously changing its orientation with respect to the large static field. However, the energy that must be supplied to turn the spins from parallel to antiparallel to the static field is recovered as the spins return to being parallel to the static field. Thus there is no cumulative absorption over long times but rather an alternate absorption and recovery. The situation is reminiscent of what we described in the first chapter prior to introduction of the coupling to the thermal reservoir. (We note that there the system, however, simply equalized populations, whereas our present model predicts an alternating reversal of populations. The two models must therefore be based on differing assumptions.)

Without contact to a reservoir, we have no mechanism for the establishment of the magnetization. By analogy to the equation

$$\frac{dn}{dt} = \frac{n_0 - n}{T_1} \tag{2.80}$$

and recognizing that $M_z = \gamma \hbar n/2$, we expect that it would be reasonable for M_z to be established according to the equation

$$\frac{dM_z}{dt} = \frac{M_0 - M_z}{T_1} \tag{2.81}$$

where M_0 is the thermal equilibrium magnetization. In terms of the static magnetic susceptibility χ_0 and the static magnetic field H_0, we have

$$M_0 = \chi_0 H_0 \quad . \tag{2.82}$$

We combine (2.81) with the equation for the driving of M by the torque to get

$$\frac{dM_z}{dt} = \frac{M_0 - M_z}{T_1} + \gamma(\boldsymbol{M} \times \boldsymbol{H})_z \quad . \tag{2.83}$$

Furthermore we wish to express the fact that in thermal equilibrium under a static field, the magnetization will wish to be parallel to H_0. That is, the x- and y-components must have a tendency to vanish. Thus

$$\frac{dM_x}{dt} = \gamma(\boldsymbol{M} \times \boldsymbol{H})_x - \frac{M_x}{T_2} \tag{2.84}$$

$$\frac{dM_y}{dt} = \gamma(\boldsymbol{M} \times \boldsymbol{H})_y - \frac{M_y}{T_2} \quad .$$

We have here introduced the same relaxation time T_2 for the x- and y-directions, but have implied that it is different from T_1. That the transverse rate of decay may differ from the longitudinal is reasonable if we recall that, in contrast to the longitudinal decay, the transverse decay conserves energy in the static field. Therefore there is no necessity for transfer of energy to a reservoir for the transverse decay. (This statement is not strictly true and gives rise to important effects when saturating resonances in solids, as has been described by *Redfield*. We describe *Redfield*'s theory of saturation in Chapter 6, beginning with Sect. 6.5).

On the other hand, the postulate of the particular (exponential) form of relaxation we have assumed must be viewed as being rather arbitrary. It provides a most useful postulate to describe certain important effects, but must not be taken too literally. According to (2.84), under the influence of a static field the transverse components would decay with a simple exponential. (This result is readily seen by transforming to a frame rotating at γH_0, where the effective field vanishes.)

A possible simple mechanism for T_2 for a solid in which each nucleus has nearby neighbors arises from the spread in precession rates produced by the magnetic field that one nucleus produces at another. If the nearest neighbor distance is r, we expect a typical nucleus to experience a local field $H_{\mathrm{loc}} \sim \mu/r^3$ (due to the neighbors) either aiding or opposing the static field. As a result, if all nuclei were precessing in phase at $t = 0$, they would get out of step. In a time τ such that $\gamma H_{\mathrm{loc}} \tau \cong 1$, there would be significant dephasing, and the vector sum of the moments would have thus diminished significantly. Since τ must therefore be comparable to T_2, a rough estimate for T_2 on this model is

$$T_2 = \frac{1}{\gamma H_{\mathrm{loc}}} = \frac{r^3}{\gamma^2 \hbar} \qquad (2.85)$$

often about $100\,\mu s$ for nuclei. Equations (2.83, 84) were first proposed by Felix Bloch and are commonly referred to as the "Bloch equations". Although they have some limitations, they have nevertheless played a most important role in understanding resonance phenomena, since they provide a very simple way of introducing relaxation effects.

2.8 Solution of the Bloch Equations for Low H_1

At this stage we shall be interested in the solution of the Bloch equations for low values of the alternating field, values low enough to avoid saturation. We immediately transform to the coordinate frame rotating at ω_z taking H_1 along the x-axis and denoting $H_0 + (\omega_z/\gamma)$ by h_0. Then

$$\frac{dM_z}{dt} = -\gamma M_y H_1 + \frac{M_0 - M_z}{T_1} \qquad (2.86\mathrm{a})$$

$$\frac{dM_x}{dt} = +\gamma M_y h_0 - \frac{M_x}{T_2} \qquad (2.86\mathrm{b})$$

$$\frac{dM_y}{dt} = \gamma(M_z H_1 - M_x h_0) - \frac{M_y}{T_2} \quad . \qquad (2.86\mathrm{c})$$

Since M_x and M_y must vanish as $H_1 \to 0$, we realize from (2.86a) that in a steady state, M_z differs from M_0 to order H_1^2. We therefore replace M_z by M_0 in (2.86c). The solution is further facilitated by introducing $M_+ = M_x + iM_y$. By adding (2.86b) to i times (2.86c), we get

$$\frac{dM_+}{dt} = -M_+\alpha + i\gamma M_0 H_1 \quad \text{where} \tag{2.87}$$

$$\alpha = \frac{1}{T_2} + \gamma h_0 i \quad . \tag{2.88}$$

Therefore

$$M_+ = Ae^{-\alpha t} + \frac{i\gamma M_0 H_1}{1/T_2 + i\gamma h_0} \quad . \tag{2.89}$$

If we neglect the transient term and substitute $M_0 = \chi_0 H_0$, and define $\omega_0 = \gamma H_0$, $\omega_z = -\omega$, we get

$$M_x = \chi_0(\omega_0 T_2)\frac{(\omega_0 - \omega)T_2}{1 + (\omega - \omega_0)^2 T_2^2} H_1 \tag{2.90}$$

$$M_y = \chi_0(\omega_0 T_2)\frac{1}{1 + (\omega - \omega_0)^2 T_2^2} H_1 \quad .$$

Equations (2.90) show that the magnetization is a constant in the rotating reference frame, and therefore is rotating at frequency ω in the laboratory. In a typical experimental arrangement we observe the magnetization by studying the emf it induces in a fixed coil in the laboratory. If the coil is oriented with its axis along the X-direction in the laboratory, we can calculate the emf from knowledge of the time-dependent component of magnetization M_X along the X-direction.

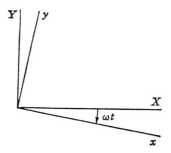

Fig. 2.11. Rotating axes x, y relative to laboratory axes X, Y

By referring to Fig. 2.11, we can relate the laboratory component M_X to the components M_x and M_y in the rotating frame. Thus

$$M_X = M_x \cos \omega t + M_y \sin \omega t \quad . \tag{2.91}$$

If we write the magnetic field as being a linear field,

$$H_X(t) = H_{X0} \cos \omega t \quad 2H_1 = H_{X0} \quad , \tag{2.92}$$

then we see that both M_x and M_y are proportional to H_{X0}, and we can write

$$M_X(t) = (\chi' \cos \omega t + \chi'' \sin \omega t) H_{X0} \quad , \tag{2.93}$$

defining the quantities χ' and χ''. By using (2.90) and (2.93), we get

$$\chi' = \frac{\chi_0}{2} \omega_0 T_2 \frac{(\omega_0 - \omega) T_2}{1 + (\omega - \omega_0)^2 T_2^2} \tag{2.93a}$$

$$\chi'' = \frac{\chi_0}{2} \omega_0 T_2 \frac{1}{1 + (\omega - \omega_0)^2 T_2^2} \quad .$$

It is convenient to regard both $M_X(t)$ and $H_X(t)$ as being the real parts of complex functions $M_X^C(t)$ and $H_X^C(t)$. Then, defining the complex susceptibility χ by

$$\chi = \chi' - i\chi'' \tag{2.94}$$

and writing

$$H_X^C(t) = H_{X0} e^{i\omega t} \tag{2.95}$$

we find

$$M_X^C(t) = \chi H_X^C(t) \quad \text{or} \tag{2.96}$$

$$M_X(t) = \text{Re} \{(\chi H_{X0} e^{i\omega t})\} \quad . \tag{2.96a}$$

Although (2.92) and (2.96a) were arrived at by considering the Bloch equations, they are in fact quite general. Any resonance is characterized by a complex susceptibility expressing the linear relationship between magnetization and applied field.

Ordinarily, if a coil of inductance L_0 is filled with a material of susceptibility χ_0, the inductance is increased to $L_0(1 + 4\pi\chi_0)$, since the flux is increased by the factor $1 + 4\pi\chi_0$ for the same current. In a similar manner the complex susceptibility produces a flux change. The flux is changed not only in magnitude but *also* in phase. By means of (2.93–96), it is easy to show that the inductance at frequency ω is modified to a new value L, given by

$$L = L_0[1 + 4\pi\chi(\omega)] \tag{2.97}$$

where $\chi(\omega) = \chi'(\omega) - i\chi''(\omega)$. It is customary in electric circuits to use the symbol j for $\sqrt{-1}$. However, in order to avoid the confusion of using two symbols for the same quantity, we use only i.[3]

---

[3] In practice, the sample never completely fills all space, and we must introduce the "filling factor" q. Its calculation depends on a knowledge of the spatial variation of the alternating field. Then (2.97) becomes

$$L = L_0[1 + 4\pi q \chi(\omega)] \quad .$$

Denoting the coil resistance in the absence of a sample as R_0, the coil impedance Z becomes

$$\begin{aligned} Z &= iL_0\omega(1 + 4\pi\chi' - i4\pi\chi'') + R_0 \\ &= iL_0\omega(1 + 4\pi\chi') + L_0\omega 4\pi\chi'' + R_0 \quad . \end{aligned} \quad (2.98)$$

The real part of the susceptibility χ' therefore changes the inductance, whereas the imaginary part, χ'', modifies the resistance. The fractional change in resistance $\Delta R/R_0$ is

$$\frac{\Delta R}{R_0} = \frac{L_0\omega}{R_0} 4\pi\chi'' = 4\pi\chi'' Q \quad (2.99)$$

where we have introduced the so-called quality factor Q, typically in a range of 50 to 100 for radio frequency coils or 1,000 to 10,000 for microwave cavities.

Assuming uniform magnetic fields occupying a volume V, the peak stored magnetic energy produced by an alternating current, whose peak value is i_0, is

$$\frac{1}{2} L_0 i_0^2 = \frac{1}{8\pi} H_{X0}^2 V \quad . \quad (2.100)$$

The average power dissipated in the nuclei $\overline{P}$ is

$$\overline{P} = \tfrac{1}{2} i_0^2 \Delta R = \tfrac{1}{2} i_0^2 L_0 \omega 4\pi\chi'' \quad . \quad (2.101)$$

By substituting from (2.100), we find

$$\overline{P} = \tfrac{1}{2} \omega H_{X0}^2 \chi'' V \quad . \quad (2.102)$$

This equation provides a simple connection between the power absorbed, χ'', and the strength of the alternating field. We shall use it as the basis of a calculation of χ'' from atomic considerations, since the power absorbed can be computed in terms of such quantities as transition probabilities. Since χ' and χ'' are always related, as we shall see shortly, a calculation of χ'' will enable us to compute χ'. Moreover, we recognize that the validity of (2.102) does not depend on the assumption of the Bloch equations.

Another useful formula can be obtained from (2.100) and (2.101), relating the average power dissipated in the coil resistance, P_c, to the strength of H_{X0}. Since P_c is half the peak power dissipated in the coil,

$$\tfrac{1}{2} R_0 i_0^2 = P_c \quad . \quad (2.103)$$

Solving for $(\tfrac{1}{2})i_0^2$ and substituting into (2.100) we get

$$\frac{H_{X0}^2}{8\pi} V = \frac{P_c Q}{\omega} \quad . \quad (2.104)$$

Utilizing the fact that $H_{X0} = 2H_1$ this can be rewritten as

$$H_1 = \sqrt{P_c Q/\nu V} \qquad (2.105)$$

where $\nu = \omega/2\pi$. Thus, if one knows the Q, the volume of the coil, and the power available from one's oscillator, one can calculate how strong an H_{X0} (or $H_1 = H_{X0}/2$) one can achieve, frequently a very useful quantity to know.

In (2.105) the units are $[H_1]$: Gauss; $[P]$: erg/s; $[\nu]$: Hz; $[V]$: cm^3. Using 1w = 10^7 erg/s, an alternative expression in mixed units is

$$H_1 = \sqrt{\frac{10PQ}{\nu V}} \qquad (2.106)$$

with $[H_1]$: Gauss; $[P]$: W; $[\nu]$: MHz; $[V]$: cm^3.

The particular functions χ' and χ'', which are solutions of the Bloch equations, are frequently encountered. They are shown in the graph of Fig. 2.12. The term *Lorentzian line* is often applied to them.

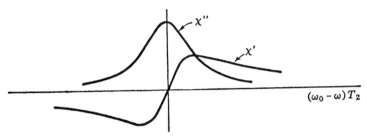

Fig. 2.12. χ' and χ'' from the Bloch equations plotted versus $x \equiv (\omega_0 - \omega)T_2$

At this time we should point out that we have computed the magnetization produced in the X-direction by an alternating field applied in the X-direction. Since the magnetization vector rotates about the Z-direction, we see that there will also be magnetization in the Y-direction. To describe such a situation, we may consider χ to be a tensor, such that

$$M^C_{\alpha'}(t) = \chi_{\alpha'\alpha} H_{\alpha 0} e^{i\omega t} \qquad \alpha = X, Y, Z \ ; \quad \alpha' = X, Y, Z \ .$$

In general we shall be interested in χ_{XX}.

2.9 Spin Echoes

Just after finishing graduate studies, *Erwin Hahn* burst on the world of science with his remarkable discovery, spin echoes [2.10]. His discovery provided the key impetus to the development of pulse methods in NMR, and must therefore be ranked among the most significant contributions to magnetic resonance. What are spin echoes and why are they so remarkable?

Suppose one applies a $\pi/2$ pulse to a group of spins to observe the free induction signal which follows turn-off of the pulse. According to the Bloch equations, the free induction signal decays exponentially with a time constant T_2. For solids, T_2 is a fraction of a millisecond, corresponding to line widths of several Gauss. For liquids, the line widths are typically much narrower, corresponding perhaps to times of several seconds. Such lines are a good deal narrower than the usual magnet homogeneity. As a result, the inhomogeneity-induced spread in precession frequency causes the spins in one portion of the sample to get out of phase with those in other portions. The free induction signal arises from the sum total of all portions of the sample. As the separate portions get out of step, the resultant signal decays. The decay time is of the order of $1/(\gamma \Delta H)$ where ΔH is the spread in static field over the sample.

Hahn made the remarkable discovery that if he applied a second $\pi/2$ pulse a time τ after the first pulse, miraculously there appeared another free induction signal at a time 2τ after the initial pulse. He named the signal the "spin echo". To produce a signal at the time of the echo, the spins must somehow have gotten back in phase. The great mystery of the spin echo was what made the spins get back in phase again? Was the echo a challenge to basic concepts of irreversibility? Was there a Maxwell demon at work producing the refocusing? *Hahn* discovered spin echoes experimentally, but was soon able to derive their existence from the Bloch equations. This solution showed that as one varied τ, the echo amplitude diminished exponentially with a time constant T_2. Thus the echo provided a way of measuring line widths much narrower than the magnet inhomogeneity. Understanding the physical basis of echo formation has led to much deeper insight into resonance phenomena in general and pulse work in particular.

The essential physical ideas of refocusing can be most easily seen by considering a pulse sequence in which the first pulse produces a rotation of $\pi/2$, the second a rotation of π. Such a sequence we denote as a $\pi/2$-π pulse sequence. It was invented by *Carr* [2.11] based on a vector model proposed by Purcell for the $\pi/2$-$\pi/2$ echo [2.10], and was first described in a famous paper by *Carr* and *Purcell* [2.12]. Consider a group of spins initially in thermal equilibrium in a static magnetic field H in the z-direction. The thermal equilibrium magnetization M_0 then lies along H as shown in Fig. 2.13a. We assume there is a spread in magnetic fields over the sample, and take the average value of field to be H_0. We first analyze what happens if we can neglect the effect of T_1 and T_2.

We apply a rotating magnetic field H_1 at $t = 0$ with frequency ω, tuned to resonance at the average field H_0. Thus

$$\omega = \gamma H_0 \quad . \tag{2.107}$$

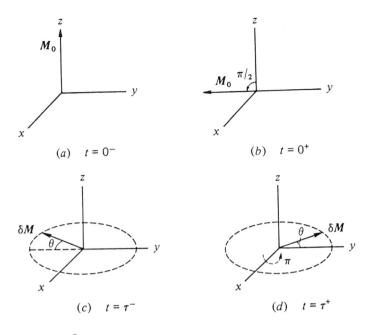
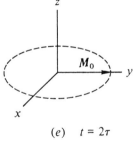

Fig. 2.13a–c. The formation of a spin echo by means of a $\pi/2$-π pulse sequence viewed in the rotating reference frame. (a) At $t = 0^-$ the magnetization, M_0 is in thermal equilibrium lying along the z-direction. (b) shows the magnetization immediately after the $\pi/2$ pulse. In (c) an element of magnetization, δM, has precessed an extra angle θ owing to the magnetic field inhomogeneity. (d) shows the effect of the π pulse on δM. In (e) we see that at time 2τ all elements of magnetization have refocused along the $+y$-direction

By proper adjustment of the pulse length t_p, we can generate a $\pi/2$ pulse. In our discussion we consider H_1 to be sufficiently strong that t_p is negligibly short. We designate the time just before or just after the initial pulse by 0^- and 0^+, respectively.

It is convenient to view the behavior of the spins in the reference frame that rotates at γH_0, with the x-axis defined as lying along H_1. Let the $\pi/2$ pulse rotate M_0 to lie along the negative y-axis (Fig. 2.13b).

This sense of rotation corresponds to a spin which has negative γ, such as an electron. For a positive γ, the rotations are positive in the left-handed sense. We give the latter case as a homework problem for the classical derivation of the echo, and treat it quantum mechanically in the next section. If there were not inhomogeneity in the static field, all spins within the sample would precess at γH_0, so that the magnetization of every portion of the sample would remain oriented along the $-y$-axis. The existence of inhomogeneity leads to a spread in precession rates and dephasing. Consider what happens during a time interval τ. In any small region of the sample, the magnetization δM will remain in the x-y plane since we are neglecting T_1 processes. But at the end of τ, the direction of δM within that plane will advance from the $-y$-direction by some angle which we call θ, given by

$$\theta = \gamma \delta H \tau \quad \text{where} \tag{2.108}$$

$$\delta H = H - H_0 \tag{2.109}$$

represents the inhomogeneity in H. The situation is shown in Fig. 2.13c. (Note that δH may be either positive or negative, so that the "advance" may be either positive or negative.)

Let us assume we can control the phase of the oscillating voltage in the second pulse so that the H_1 again lies along the $+x$-direction in the rotating frame.[4] Suppose now we apply a π pulse at $t = \tau$, again of negligible duration. We denote the time just before and just after the pulse by $t = \tau^-$ or $t = \tau^+$, respectively. The situation just after the π pulse is shown in Fig. 2.13d. Noting the orientation of δM, we immediately see that during a second time interval τ, δM will again advance through the same angle θ, which will bring it exactly along the positive y-axis at $t = 2\tau$. The argument applies to all spins, no matter what δH they experience, because the result does not depend on the angle of advance.

Though all the spins are in phase at $t = 2\tau$, they get out of phase again owing to the field inhomogeneity, so the free induction signal decays. Note that its form as a function of time during the dephasing from $t = 2\tau$ onward must be identical to the form of the decay following the initial $\pi/2$ pulse (see Fig. 2.14). The buildup of the echo signal just prior to $t = 2\tau$ is the mirror image in time of the decay after $t = 2\tau$.

It is easy to see now what will be the effect of the T_1 and T_2 terms of the Bloch equations. During the first time interval τ, the components of δM in the x-y plane will decay exponentially with T_2, and a z-component will

---

[4] With so-called coherent pulse apparatus, a highly stable oscillator generates a steady-state alternating voltage which is fed into a circuit which amplifies the rf voltage only during the time that a gate voltage is applied, thereby producing a strong rf signal coincident with the gate pulses. The output voltage is fed to the sample coil to generate the H_1. If the stable oscillator is tuned exactly to resonance, $\omega = \omega_0$, the phase of the H_1 is always the same in the ω_0 reference frame. This method of operation, together with so-called phase coherent detection of the free induction decays, was first introduced independently by *Solomon* [2.13] and by *Spokas* and *Slichter* [2.14, 15]. In [2.15], there is a description of the reasons for using phase-coherent detection. In a famous paper on nuclear relaxation in alkali metals, their colleagues, *Holcomb* and *Norberg* [2.16] introduce the use of signal averaging to improve the signal to noise ratio in pulse experiments, and point out that coherent detection is necessary to achieve the full potential of a noise integration.

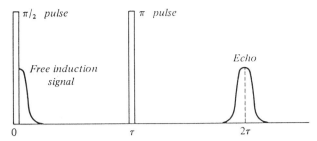

Fig. 2.14. The decay of the echo from $t = 2\tau$ onward is the same function of time as the decay of the free induction signal from $t = 0^+$ onwards. Note that the buildup before $t = 2\tau$ is the mirror image in time of the decay after 2τ. Note further that no free induction signal is produced immediately following the π pulse

develop exponentially with T_1. The π pulse will invert the z-component which has developed, so that it does not contribute to the component of δM in the x-y plane existing at $t = \tau^+$. During the next interval τ, the component δM in the x-y plane will continue to decay via the T_2 terms in the Bloch equations. As a result, the size of the magnetization producing the echo signal $M(2\tau)$ will obey

$$M(2\tau) = M_0 e^{-2\tau/T_2} \; . \tag{2.110}$$

If the second pulse is a $\pi/2$ pulse, as in *Hahn*'s original experiment, the pulse at $t = \tau$ will put some of the z-component of δM into the x-y plane. One might then wonder whether or not the echo contained an exponential depending on T_1. A detailed analysis shows that that does not happen. This result is most easily seen by pretending the magnetization developed along the z-axis arose from an independent set of spins which possess no net x-y magnetization. With respect to the new spins the situation just prior to the second pulse is like Fig. 2.13a, with δM lying along the $+z$-direction. The $\pi/2$ pulse puts the magnetization into the x-y plane. The situation at $t = 2\tau$ is identical to that of Fig. 2.13c, in which all the spins have dephased.

Thus the T_1 induced magnetization does not affect the echo. To observe the signal, one would need to refocus – which requires still another pulse. Thus with *three* pulses one could see T_1 effects.

Hahn in fact found such echoes. He observed that if he applied a third pulse at time T (hence $T - \tau$ after the second pulse), he produced an echo at $2(T - \tau)$ after the second pulse, or at $t = 2T - \tau$ after the first pulse. He found, in addition, echoes at $T + \tau$, $2T - 2\tau$, and $2T$. The various echoes induced by a third pulse are commonly referred to as "stimulated" echoes.

In liquids, the diffusional motion permits a nucleus to move between different parts of the sample where the precession rates may differ. As a result, during a spin echo the dephasing during the first interval τ may differ from the rephasing during the second interval τ, and the echo is diminished. The effect is of great practical utility as a way of measuring diffusion rates in liquids. This was discovered by *Hahn,* and reported in his first spin echo publication [2.10]. Although Hahn gives correctly the expression for the effect of diffusion on the decay of the transverse magnetization following a single pulse, there is an error in his expressions for the effect of later pulses, corrected by *Carr* and *Purcell* [2.12] in a paper which also showed how the effect of diffusion could be eliminated (Sect. 8.2). Carr and Purcell showed that diffusion led to a decay of the echo peak magnetization M, given by

$$M(2\tau) = M_0 \exp\left[-\gamma^2 \left(\frac{\partial H}{\partial z}\right)^2 \frac{2D\tau^3}{3}\right] \tag{2.111}$$

where we have assumed

$$H - H_0 = z\left(\frac{\partial H}{\partial z}\right) \tag{2.112}$$

as is the case for a magnetic field of axial symmetry about the z-axis. We derive this result in Appendix G. As a preliminary, in Appendix F we treat a simpler case in which bodily motion affects the structure of a nuclear absorption spectrum.

In an actual experiment, since H_1 is not infinite, the H_1 pulses have a nonzero duration. Let $t_{\pi/2}$ be the duration of a $\pi/2$ pulse if one were exactly at resonance ($\gamma H_1 t_{\pi/2} = \pi/2$). We call this the "$\pi/2$ pulse" in what follows even though it is not exactly so for spins which are off resonance. We then ask what influence the nonzero duration of the pulse has on the time at which the echo occurs. Indeed, one might ask whether or not the echo is formed perfectly under these circumstances. To discuss the problem, we define the times τ and τ' shown in Fig. 2.15. We define τ as the time between the end of the $\pi/2$ pulse and the start of the π pulse, τ' as the time from the end of the π pulse to the peak of the echo (assuming one is formed!).

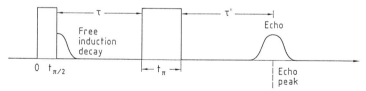

Fig. 2.15. Definition of the times τ and τ' used to discuss echo formation when one considers the effect of the finite size of H_1, hence the nonzero length of the $\pi/2$ and π pulses

To analyze this problem, let us suppose that the resonance is broadened symmetrically by a field inhomogeneity and that H_1 is tuned exactly to the center of the resonance, $\omega = \gamma H_0$. Consider an element of magnetization, δM, which is off resonance by $H - H_0 = \delta H$. Then the effective field in the rotating frame acting on this magnetization is shown in Fig. 2.16. We take the x-axis

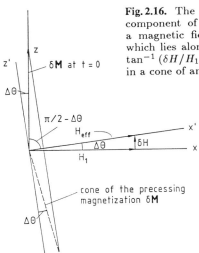

Fig. 2.16. The effect of a finite H_1 on the precession of a component of magnetization, δM, which is off resonance by a magnetic field δH. H_1 lies along the x-axis. The x'-axis, which lies along the effective field $\boldsymbol{H}_{\text{eff}}$, makes an angle $\Delta\theta = \tan^{-1}(\delta H/H_1)$ with the x-axis. Thus, ΔM precesses about $\boldsymbol{H}_{\text{eff}}$ in a cone of angle $\pi/2 - \Delta\theta$

along H_1, and the x'-axis along H_{eff}. The magnetization contribution δM will precess about $\boldsymbol{H}_{\text{eff}}$, in a cone which makes an angle $\pi/2 - \Delta\theta$ with respect to x' where $\tan\Delta\theta = \delta H/H_1$. With this as a starting point, one can then work through the details of the precession. If H_1 is much bigger than the line width, then $\Delta\theta \ll 1$, and one can reexamine the derivation given for the echo. We put the details of the derivation as a homework problem. One finds, keeping only corrections linear in $\Delta\theta$, that the "$\pi/2$ pulse" is a $\pi/2$ pulse for rotations about x'. It puts the magnetization at an angle $\Delta\theta = \delta H/H_1$ with respect to the negative y-axis. Thus, at the start of the π pulse the spin now makes an angle $\Delta\theta + \gamma\delta H\tau$ with respect to the y-axis instead of $\delta H\tau$ of (2.108). The π pulse can be shown to introduce no error in the orientation of δM in the x-y plane. Thus at a time τ' after the end of the π pulse when

$$\gamma\delta H\tau' = \Delta\theta + \gamma\delta H\tau \quad , \tag{2.113}$$

the magnetization lies along the $+y$-direction. This gives

$$\tau' = \tau + \frac{1}{\gamma H_1} \quad , \tag{2.114}$$

which is independent of δH. Thus, the δM's from all parts of the resonance line are collinear, and an echo is formed at this time. Equation (2.114) can also be written as

$$\tau' = \tau + \frac{2}{\pi} t_{\pi/2} \quad . \tag{2.115}$$

This expression was derived on the assumption that $\delta H \ll H_1$. Numerical simulations for $\delta H \gg H_1$, however, show that the correction term, $1/\gamma H_1$, remains accurate to within 10% even for lines much broader than H_1 [2.17].

2.10 Quantum Mechanical Treatment of the Spin Echo

The spin echo can also be derived quantum mechanically. Of course, since (2.37) shows that the expectation value obeys the classical equations of motion, one immediately knows that the classical derivation is true for the quantum mechanical expectation value. However, there is added insight from following the development of the actual wave function in time.

We consider, then, a sequence of events in time shown in Fig. 2.17, which shows H_1 versus time. At $t = 0$, H_1 of frequency ω is turned on along the x-axis in the reference frame rotating at $-k\omega$. It remains on until time t_1, producing a $\pi/2$ pulse. From t_1 to t_2 the spins precess freely. A second pulse is applied from t_2 to t_3, where the interval is chosen to make this a π pulse. We denote $t_2 - t_1$ as τ. Since we will be working solely in the rotating frame, we will omit from ψ and $\mathcal{H}$ the prime used earlier to designate quantities in the rotating frame.

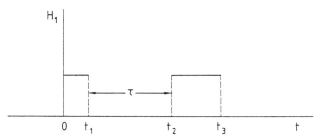

Fig. 2.17. H_1 versus time, giving the definition of the times 0, t_1, t_2, and t_3 which specify the beginning and end of the $\pi/2$ pulse ($t = 0, t_1$) and of the π pulse ($t = t_2, t_3$)

Then ψ obeys the equation

$$-\frac{\hbar}{i} \frac{\partial \psi}{\partial t} = \mathcal{H}\psi \quad \text{where} \tag{2.116}$$

$$\mathcal{H} = -\gamma\hbar(h_0 I_z + H_1 I_x) \quad \text{with} \tag{2.117}$$

$$h_0 = H_0 - \omega/\gamma \quad . \tag{2.118}$$

If γ is positive, this will produce rotations about the effective field which are left handed. If γ were itself negative, the rotations would be in the right-handed sense, as assumed in drawing Fig. 2.13.

We do not set $h_0 = 0$ since we wish to represent the fact that owing to the inhomogeneity in H_0 the typical spin is not perfectly at resonance. We introduce a distribution function $p(h_0)$ to express the fact that the number of spins with H_0 between h_0 and $h_0 + dh_0$ is

$$dN = Np(h_0)dh_0 \qquad (2.119)$$

(N is the total number of spins). See Problem 2.3 for a classical study of the effects of this distribution function.

We can simplify things by assuming that, while H_1 is on, $H_1 \gg h_0$ for all spins, so that we can approximate $\mathcal{H}$ during the pulses by

$$\mathcal{H} = -\gamma \hbar H_1 I_x \quad . \qquad (2.120a)$$

In between pulses, when $H_1 = 0$, we have

$$\mathcal{H} = -\gamma \hbar h_0 I_z \quad . \qquad (2.120b)$$

Although $\mathcal{H}$ is time dependent, its time variation occurs only at four times ($t = 0, t_1, t_2, t_3$). In between, it is independent of time, enabling us to use (2.64a) over the four time intervals 0 to t_1, t_1 to t_2, t_2 to t_3, t_3 to t. Integration of (2.116) across the discontinuities in $\mathcal{H}$ shows that $\psi(t)$ is continuous. We use this fact to join solutions across the discontinuities.

Thus, using (2.120a) and (2.64a)

$$\psi(t_1) = e^{i\gamma H_1 t_1 I_x} \psi(0) \quad , \quad \gamma H_1 t_1 = \pi/2 \quad , \qquad (2.121a)$$

$$\psi(t_2) = e^{i\gamma h_0 (t_2 - t_1) I_z} \psi(t_1) \quad , \qquad (2.121b)$$

etc.

It is now convenient to define the quantities $T(t, h_0)$ and $X(\theta)$ by

$$T(t, h_0) = e^{i\gamma h_0 t I_z} \quad . \qquad (2.122a)$$

$$X(\theta) = e^{i\theta I_x} \quad . \qquad (2.122b)$$

$T(t, h_0)$ generates the development of the wave function during those times when $H_1 = 0$ for spins which are off resonance by h_0.

As can be seen by referring to (2.55), $X(\theta)$ is the operator we need when we wish to rotate a component of spin through an angle θ in the right-handed sense about the X-axis. Thus, if I_α becomes $I_{\alpha'}$ following such a rotation,

$$I_{\alpha'} = X(\theta)^{-1} I_\alpha X(\theta) \quad . \qquad (2.123)$$

For $\theta = \pi/2$

$$\begin{aligned} X^{-1}(\pi/2) I_y X(\pi/2) &= I_z \\ X^{-1}(\pi/2) I_z X(\pi/2) &= -I_y \\ X^{-1}(\pi/2) I_x X(\pi/2) &= I_x \quad . \end{aligned} \qquad (2.124)$$

If $\theta = \pi$

$$X^{-1}(\pi)I_y X(\pi) = -I_y$$
$$X^{-1}(\pi)I_z X(\pi) = -I_z \qquad (2.125)$$
$$X^{-1}(\pi)I_x X(\pi) = I_x.$$

Let us further consider that H_1 is so large that we can neglect the time intervals t_1 and $t_3 - t_2$ compared to τ. Then, using expressions

$$\psi(t_1) = X(\pi/2)\psi(0)$$
$$\psi(t_2) = T(\tau, h_0)X(\pi/2)\psi(0) \quad , \qquad (2.126)$$

etc. we get $\psi(t)$ for times t which are after the second pulse as

$$\psi(t) = T(t - \tau, h_0)X(\pi)T(\tau, h_0)X(\pi/2)\psi(0) \qquad (2.127)$$

for a single value of h_0. Since the NMR signal arises from the transverse magnetization, we need to calculate the expectation value of I_x and I_y. We leave it as a homework problem to show that if H_1 lies along the x-axis, the expectation value of I_x is zero as in the classical picture. Instead, we here calculate $\langle I_y(t) \rangle$.

For a single spin we get

$$\langle I_y(t) \rangle = \int \psi^*(t) I_y \psi(t) d\tau_I \qquad (2.128)$$

where $d\tau_I$ stands for the "volume element" in spin space. Now, we must sum over all spins, using (2.119) to express the fact that the various spins experience different h_0's. Thus the expectation value of the total y-component of spin is

$$\langle I_{y,\text{total}}(t) \rangle = N \int p(h_0) dh_0 \int \psi^*(h_0, t) I_y \psi(h_0, t) d\tau_I \qquad (2.129)$$

where we have used a notation $\psi(h_0, t)$ which makes explicit the dependence of ψ on h_0.

We now utilize the properties of the exponential operators [see (2.50a)] to transform the integral over τ_I to get

$$\int \psi^*(h_0, t) I_y \psi(h_0, t) d\tau_I$$
$$= \int \psi^*(0) X^{-1}(\pi/2) T^{-1}(\tau, h_0) X^{-1}(\pi) T^{-1}(t - \tau, h_0) I_y T(t - \tau, h_0)$$
$$\times X(\pi) T(\tau, h_0) X(\pi/2) \psi(0) d\tau_I \quad . \qquad (2.130)$$

Utilizing the fact that $X^{-1}X = XX^{-1} = 1$, we then transform a portion of the integrand:

$$X^{-1}(\pi)T^{-1}(t-\tau,h_0)I_y T(t-\tau,h_0)X(\pi)$$
$$= \underbrace{X^{-1}(\pi)T^{-1}(t-\tau)X(\pi)}_{1} \underbrace{X^{-1}(\pi)I_y X(\pi)}_{2}$$
$$\times \underbrace{X^{-1}(\pi)T(t-\tau,h_0)X(\pi)}_{3} \quad . \tag{2.131}$$

We now have three terms identified by the three numbered braces in (2.131). We deal first with term 2, using (2.125) to get

$$X^{-1}(\pi)I_y X(\pi) = -I_y \quad .$$

We write term 3 as

$$X^{-1}(\pi)T(t-\tau,h_0)X(\pi) = X^{-1}(\pi)e^{i\gamma h_0(t-\tau)I_z}X(\pi) \quad . \tag{2.132}$$

This term has I_z in the exponent. To deal with it, we call on a theorem (given as Problem 2.4) that if R is an operator, R^{-1} its inverse, and G some other operator

$$R^{-1}(e^{iG})R = e^{i(R^{-1}GR)} \quad , \tag{2.133}$$

which is readily proved using the series expansion of the exponential and making repeated insertions of $R^{-1}R$.

Using (2.133), we then get

$$X^{-1}(\pi)\exp(ih_0(t-\tau)I_z)X(\pi)$$
$$= \exp(iX^{-1}(\pi)I_z X(\pi)h_0(t-\tau))$$
$$= \exp(-i\gamma h_0(t-\tau)I_z) \tag{2.134a}$$
$$= T^{-1}(t-\tau,h_0) \quad . \tag{2.134b}$$

We thereby get

$$\langle I_{y,\text{total}}(t)\rangle = -N\int p(h_0)dh_0 \int \psi^*(0)X^{-1}(\pi/2)T^{-1}(\tau,h_0)T(t-\tau,h_0)$$
$$\times I_y T^{-1}(t-\tau,h_0)T(\tau,h_0)X(\pi/2)\psi(0)d\tau_I \quad . \tag{2.135}$$

Now consider what happens when $t-\tau = \tau$, i.e. $t = 2\tau$. This is the time of the echo according to our previous classical calculation. At this time

$$T^{-1}(t-\tau,h_0)T(\tau,h_0) = 1 \quad \text{so that} \tag{2.136}$$

$$\langle I_{y,\text{total}}(t=2\tau)\rangle = -N\int p(h_0)dh_0$$
$$\times \int \psi^*(0)X^{-1}(\pi/2)I_y X(\pi/2)\psi(0)d\tau_I \quad . \tag{2.137}$$

The h_0 has now disappeared from the integral over $d\tau_I$, so that we can now integrate over dh_0, giving

$$\langle I_{y,\text{total}}(t = 2\tau)\rangle = -N \int \psi^*(0) X^{-1}(\pi/2) I_y X(\pi/2) \psi(0) d\tau_I \quad . \tag{2.138}$$

The integrand is just what we would have if we calculated $\langle I_{y,\text{total}}\rangle$ immediately after the first $\pi/2$ pulse. Using (2.124), in fact, we have

$$\langle I_{y,\text{total}}(t = 2\tau)\rangle = -N\langle I_z(0^-)\rangle \tag{2.139}$$

where 0^- refers to the time just before $t = 0$. If we assume the system is in thermal equilibrium before the $\pi/2$ pulse, this result states that the echo arises from the thermal equilibrium M_0. Note that the minus sign shows that the echo forms on the negative y-axis. This same result would follow from the classical argument for a nucleus with a positive γ for which the rotations about H_1 are in the left-handed sense.

Going back to (2.134) and (2.135), we see that the effect of the π pulse may be thought of as having changed the Hamiltonian after $t = \tau$ from its old value of $\mathcal{H}_{\text{old}}$ to a new value $\mathcal{H}_{\text{new}}$. In this viewpoint, the Hamiltonian from t_1 to t_2 is $\mathcal{H}_{\text{old}} = -\gamma\hbar h_0 I_z$, and then at $t = \tau$ *changes* to its negative, $\mathcal{H}_{\text{new}} = +\gamma\hbar h_0 I_z$ for later times. We replace the π pulse plus subsequent evolution under the *real* Hamiltonian with a development under a *new* or effective Hamiltonian for $t > \tau$ in which there is *no* π pulse. *The effect of the pulses is equivalent to a situation in which there are* no *pulses, but the Hamiltonian changes in time.* This concept proves very useful in magnetic resonance. It is the basis of many important pulse sequences.

In the case of the echo, the combination of $\mathcal{H}_{\text{new}}$ and $\mathcal{H}_{\text{old}}$ is such that when $t = 2\tau$, $\mathcal{H}_{\text{new}}$ has completely undone the effect of $\mathcal{H}_{\text{old}}$. Over the time interval 2τ we have eliminated the effect of magnetic field inhomogeneities from the Hamiltonian. The echo can be used to eliminate, over a particular time interval, the effect of any interaction which is equivalent to a spread in magnetic field. For example, spin-spin couplings to different nuclear species (e.g. coupling of protons to C^{13}, when C^{13} is under observation) or the effects of chemical shifts discussed in Chap. 4 all can be eliminated by use of echoes.

By using the explicit expression of (2.134a), it is easy to show that apart from a sign change the development of $\langle I_{y,\text{total}}(t)\rangle$ for $t > 2\tau$ is identical to its development immediately after the first $\pi/2$ pulse. We give this as a homework problem. This result has a useful experimental consequence. The signal immediately after the $\pi/2$ pulse gives one the Fourier transform of the line shape function $p(h_0)$, as can be seen from the results of Problem 2.3. But the signal immediately follows a strong pulse. In practice this pulse may block the signal amplifiers of the NMR apparatus, making it impossible to observe the signal until the amplifiers recover. On the other hand, the echo occurs later than a pulse by a time τ, giving the amplifiers time to recover. Thus, the echo is easier to see. It is therefore fortunate that the echo reproduces the earlier signal.

2.11 Relationship Between Transient and Steady-State Response of a System and of the Real and Imaginary Parts of the Susceptibility

Suppose, to avoid saturation, we deal with sufficiently small time-dependent magnetic fields. The magnetic system may then be considered linear. That is, the magnetization produced by the sum of two weak fields when applied together is equal to the sum of the magnetization produced by each one alone. (We shall not include the static field H_0 as one of the fields, but may find it convenient to consider small *changes* in the static field.) In a similar manner, an ordinary electric circuit is linear, since the current produced by two voltage sources simultaneously present is the sum of the currents each source would produce if the other voltage were zero.

Let us think of the magnetization $\Delta M(t)$ produced at a time t and due to a magnetic field $H(t')$ of duration $\Delta t'$ at an earlier time (see Fig. 2.18). As a result of the linearity condition we know that $\Delta M(t) \propto H(t')$. It is also $\propto \Delta t'$ as long as $\Delta t' \ll t - t'$, since two pulses slightly separate in time must produce the same effect as if they were applied simultaneously.

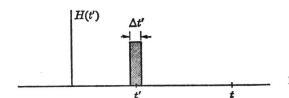

Fig. 2.18. Pulse of magnetic field

Therefore we may express the proportionality by writing

$$\Delta M(t) = m(t - t')H(t')\Delta t' \qquad (2.140)$$

where $m(t-t')$ is a "constant" for a given t and t', which, however, must depend on how long $(t - t')$ after the pulse of field we wish to know the magnetization. The total magnetization at time t is obtained by integrating (2.140) over the history of the magnetic field $H(t')$:

$$M(t) = \int_{-\infty}^{t} m(t - t')H(t')dt' \qquad (2.141)$$

Note that $m(t - t') = 0$ if $t' > t$, since the effect cannot precede the cause.

To understand just what $m(t - t')$ is, let us assume $H(t')$ is a δ-function at $t = 0$. Then the magnetization at $t > 0$ (which we shall denote by M_δ) is

$$M_\delta(t) = \int_{-\infty}^{t} m(t-t')\delta(t')dt' = m(t) \quad . \tag{2.142}$$

That is, $m(t)$ is the response to a δ-function at $t = 0$. Knowledge of $m(t)$ enables us to determine from (2.141) the magnetization resulting from a magnetic field of arbitrary time variation.

Ordinarily one determines $m(t)$ by going to the basic time-dependent differential equation which describes the behavior of the system. We can illustrate the approach by an example. Suppose we consider a system described by (2.81)

$$\frac{dM_z}{dt} = \frac{M_0 - M_z}{T_1}$$

with $M_0 = \chi_0 H_0$. We ask what the effect would be if H_0 were taken to be time dependent in magnitude though fixed in direction. We replace H_0 by $H(t)$. This situation describes the famous experiments of *C.J. Gorter* and his colleagues at Leiden [2.18].

We then have

$$\frac{dM_z}{dt} + \frac{M_z}{T_1} = \frac{\chi_0}{T_1} H(t) \quad . \tag{2.143}$$

To find the equation for $m(t)$, we recognize that $m(t)$ obeys this equation when $H(t) = \delta(t)$.

Thus,

$$\frac{dm}{dt} + \frac{m}{T_1} = \frac{\chi_0}{T_1} \delta(t) \quad . \tag{2.144}$$

We know from causality that $m(t) = 0$ for $t < 0$. We know that for $t > 0$, when the right-hand side is zero, the solution is

$$m(t) = A e^{-t/T_1} \quad . \tag{2.145}$$

To find A, we integrate (2.144) across $t = 0$ from $t = 0^-$ to $t = 0^+$:

$$\int_{t=0^-}^{t=0^+} dm + \int_{t=0^-}^{t=0^+} \frac{m}{T_1} dt = \frac{\chi_0}{T_1} \int_{t=0^-}^{t=0^+} \delta(t) dt \quad , \tag{2.146}$$

which gives

$$m(0^+) - m(0^-) = \frac{\chi_0}{T_1} \quad . \tag{2.147}$$

Since $m(0^-) = 0$, and $m(0^+) = A$ [from (2.145)], we get $A = \chi_0/T_1$, or

$$m(t) = \frac{\chi_0}{T_1} e^{-t/T_1} \quad . \tag{2.148}$$

Those readers familiar with Green's functions will recognize that $m(t)$ is a Green's function, and that what we have just done is one example of how one finds a Green's function.

If a unit step were applied at $t = 0$ (Fig. 2.19), we should have magnetization, which we shall denote as M_{step}:

$$M_{\text{step}}(t) = \int_0^t m(t - t')dt' = \int_0^t m(\tau)d\tau \quad . \tag{2.149}$$

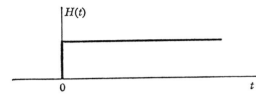

Fig. 2.19. Step function

By taking the derivative of (2.149), we find

$$m(t) = \frac{d}{dt}(M_{\text{step}}) \quad . \tag{2.150}$$

Equation (2.150) therefore shows us that knowledge of $M_{\text{step}}(t)$ enables us to compute $m(t)$.

For example, suppose we discuss the magnetization of a sample following application of a unit magnetic field in the z-direction for a system obeying the Bloch equations. We know from the Bloch equations that

$$M_z(t) = \chi_0(1 - e^{-t/T_1}) = M_{\text{step}} \quad . \tag{2.151}$$

Therefore, using (2.150),

$$m(t) = \frac{\chi_0}{T_1}e^{-t/T_1} \quad . \tag{2.152}$$

Note that in any real system, the magnetization produced by a step is bounded, so that

$$\int_0^\infty m(\tau)d\tau \tag{2.153}$$

converges.

Suppose we apply an alternating magnetic field. We shall write it as complex for simplicity:

$$H_X^C(t) = H_{X0}e^{i\omega t} \quad . \tag{2.154}$$

Then

$$M_X^C(t) = \int_{-\infty}^{t} m(t-t')H_{X0}e^{i\omega t'}dt'$$

$$= H_{X0}e^{i\omega t}\int_{-\infty}^{t} m(t-t')e^{i\omega(t'-t)}dt'$$

$$= H_{X0}e^{i\omega t}\int_{0}^{\infty} m(\tau)e^{-i\omega\tau}d\tau \quad . \tag{2.155}$$

Comparison with (2.96) shows that

$$\chi = \int_{0}^{\infty} m(\tau)e^{-i\omega\tau}d\tau$$

$$\chi' = \int_{0}^{\infty} m(\tau)\cos\omega\tau\, d\tau \tag{2.156}$$

$$\chi'' = \int_{0}^{\infty} m(\tau)\sin\omega\tau\, d\tau$$

(see footnote[5]). It is simple to show, using the integral representation of the δ-function,

$$\delta(x) = \frac{1}{2\pi}\int_{-\infty}^{+\infty} e^{ixt}dt \quad \text{that} \tag{2.157}$$

---

[5] Strictly speaking, we should turn on the alternating field adiabatically and consider the limit of slower and slower turn-on. Thus we can take

$$H_X^C(t) = H_{X0}e^{i\omega t}e^{st'} \quad , \quad s > 0 \quad .$$

As $t \to -\infty$, this function goes to zero. We compute the limit as $s \to 0$. Thus

$$M_X^C(t) = \int_{-\infty}^{t} m(t-t')H_{X0}e^{i\omega t'}e^{st'}dt'$$

$$= H_{X0}e^{i\omega t}e^{st}\int_{-\infty}^{t} m(t-t')e^{i\omega(t'-t)}e^{s(t'-t)}dt'$$

$$= H_{X0}e^{(i\omega+s)t}\int_{0}^{\infty} m(\tau)e^{-(s+i\omega)\tau}d\tau \quad \text{and}$$

$$\chi(\omega) = \lim_{s \to 0}\int m(\tau)e^{-(s+i\omega)\tau}d\tau \quad .$$

The advantage of this definition is that it has meaning for the case of a "lossless resonator" (magnetic analogue of an undamped harmonic oscillator), in which a sudden application of a field would excite a transient that would never die out.

$$m(\tau) = \frac{1}{2\pi} \int_{-\infty}^{+\infty} \chi(\omega) e^{i\omega t} d\omega \quad . \tag{2.158}$$

That is, $m(\tau)$ and $\chi(\omega)$ are Fourier transforms of each other. Knowledge of one completely determines the other. One may attempt to predict the properties of resonance lines either by analyzing the response to an alternating signal or by analyzing the transient response. *Kubo* and *Tomita* [2.19], for example, base their general theory of magnetic resonance on the transient response, calculating the response of the system to a step.

Examination of (2.156) enables us to say something about χ' and χ'' at both zero and infinite frequencies. Clearly, χ'' vanishes at $\omega = 0$, since $\sin 0$ vanishes, but χ' does not vanish at $\omega = 0$. Moreover, if $m(\tau)$ is a finite, reasonably continuous function whose total integral $\int_0^\infty m(\tau) d\tau$ is bounded, both χ' and χ'' will go to zero as $\omega \to \infty$, since the oscillations of the $\sin \omega \tau$ or $\cos \omega \tau$ will "average" the integrand to zero. Actually we may permit $m(\tau)$ to be infinite at $\tau = 0$. We can see this by thinking of $\int_0^t m(\tau) d\tau$, the response to a step. We certainly do not expect the response to a step to be discontinuous at any time other than that when the step is discontinuous ($t = 0$). Therefore $m(\tau)$ can have at most an integrable infinity at $t = 0$, since the response must be bounded. We shall represent this by a δ-function. Thus, if

$$m(\tau) = m_1(\tau) + c_1 \delta(\tau) \tag{2.159}$$

where $m_1(\tau)$ has no δ-function, we get

$$\chi'(\omega) = \int_0^\infty m_1(\tau) \cos (\omega \tau) d\tau + c_1 \quad . \tag{2.160}$$

The integral vanishes as $\omega \to \infty$, leaving us $c_1 = \chi'(\infty)$. It is therefore convenient to subtract the δ-function part from $m(\tau)$, which amounts to saying that

$$\chi(\omega) - \chi'(\infty) = \int_0^\infty m(\tau) e^{-i\omega \tau} d\tau \tag{2.161}$$

where *now* $m(\tau)$ has no δ-function part.

[Of course no physical system could have a magnetization that follows the excitation at infinite frequency. However, if one were rather making a theorem about permeability μ, $\mu(\infty)$ is *not* zero. We keep $\chi'(\infty)$ to emphasize the manner in which such a case would be treated.]

We wish now to prove a theorem relating χ' and χ'', the so-called Kramers-Kronig theorem. To do so, we wish to consider χ to be a function of a complex variable $z = x + iy$. The real part of z will be the frequency ω, but we use the symbol x for ω to make the formulas more familiar. Therefore

$$\chi(z) - \chi'(\infty) = \int_0^\infty m(\tau) e^{-iz\tau} d\tau$$

$$= \int_0^\infty m(\tau) e^{y\tau} e^{-ix\tau} d\tau \quad . \tag{2.162}$$

Since an integral is closely related to a sum, we see that $\chi(z)$ is essentially a sum of exponentials of z. Since each exponential is an analytic function of z, so is the integral, providing nothing too bizarre results from integration.

To prove that $\chi(z) - \chi'(\infty)$ is an analytic function of z, one may apply the Cauchy derivative test, which says that if

$$\chi(z) - \chi'(\infty) \equiv u + iv \tag{2.163}$$

where u and v are real, u and v must satisfy the equations

$$\frac{\partial u}{\partial x} = \frac{\partial v}{\partial y} \quad \text{and} \quad \frac{\partial v}{\partial x} = -\frac{\partial u}{\partial y} \quad . \tag{2.164}$$

From (2.162) we have

$$u = \int_0^\infty m(\tau) \cos(x\tau) e^{y\tau} d\tau \tag{2.165}$$

$$v = -\int_0^\infty m(\tau) \sin(x\tau) e^{y\tau} d\tau$$

giving

$$\frac{\partial u}{\partial x} = -\int_0^\infty m(\tau)\tau \sin(x\tau) e^{y\tau} d\tau = \frac{\partial v}{\partial y} \tag{2.166}$$

$$\frac{\partial u}{\partial y} = \int_0^\infty m(\tau)\tau \cos(x\tau) e^{y\tau} d\tau = -\frac{\partial v}{\partial x} \quad ,$$

which satisfy the Cauchy relations, *provided* it is permissible to take derivatives under the integral sign. There are a variety of circumstances under which one can do this, and we refer the reader to the discussion in *Hobson*'s book [2.20]. For our purposes, the key requirement is that the integrals in both (2.165) and (2.166) must not diverge. This prevents us in general from considering values of y that are too positive. For any reasonable $m(\tau)$ such as that of (2.152), the integrals will be convergent for $y \leq 0$, so that $\chi(z) - \chi'(\infty)$ will be analytic on the real axis and in the lower half of the complex z-plane.

Whenever we use functions $m(\tau)$ that are *not* well behaved, we shall also imply that they are to be taken as the limit of a well-behaved function. (Thus an absorption line that has zero width is physically impossible, but may be thought of as the limit of a very narrow line.)

The presence of the term $\exp(y\tau)$ tells us that

$$|\chi(z) - \chi'(\infty)| \to 0 \quad \text{as} \quad y \to -\infty \quad .$$

We already know that

$$|\chi(z) - \chi'(\infty)| \to 0 \quad \text{as} \quad x \to \pm\infty \quad .$$

Therefore $\chi(z) - \chi'(\infty)$ is a function that is analytic for $y \leq 0$ and goes to zero as $|z| \to \infty$ in the lower half of the complex plane.

Let us consider a contour integral along the path of Fig. 2.20 of the function

$$\frac{\chi(z') - \chi'(\infty)}{z' - \omega} \quad .$$

By Cauchy's integral theorem this integral vanishes, since $\chi(z)$ has no poles inside the contour.

$$\int_C \frac{\chi'(z') - \chi'(\infty)}{z' - \omega} dz' = 0 \quad . \tag{2.167}$$

Since $|\chi'(z') - \chi'(\infty)|$ goes to zero on the large circle of radius ϱ, that part of the integral gives zero contribution. There remains the contribution on the real axis plus that on the circle $z' - \omega = R\exp(i\phi)$. Thus

$$\int_{-\infty}^{\omega-R} \frac{\chi(\omega') - \chi'(\infty)}{\omega' - \omega} d\omega'$$

$$+ \int_{\pi}^{2\pi} \frac{[\chi(\omega') - \chi'(\infty)]}{Re^{i\phi}} Rie^{i\phi} d\phi + \int_{\omega+R}^{+\infty} \frac{\chi(\omega') - \chi'(\infty)}{\omega' - \omega}$$

$$= 0 = P \int_{-\infty}^{+\infty} \frac{\chi(\omega') - \chi'(\infty)}{\omega' - \omega} d\omega' + \pi i [\chi(\omega') - \chi'(\infty)] \quad , \tag{2.168}$$

where the symbol P stands for taking the principal part of the integral (that is, taking the limit of the sum of the integrals $\int_{-\infty}^{\omega-R}$ and $\int_{\omega+R}^{+\infty}$ as $R \to 0$ simultaneously in the two integrals).

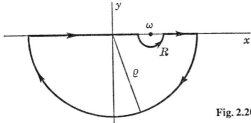

Fig. 2.20. Contour integral

Solving for the real and imaginary parts, we find

$$\chi'(\omega) - \chi'(\infty) = \frac{1}{\pi} P \int_{-\infty}^{+\infty} \frac{\chi''(\omega')}{\omega' - \omega} d\omega' \qquad (2.169)$$

$$\chi''(\omega) = -\frac{1}{\pi} P \int_{-\infty}^{+\infty} \frac{\chi'(\omega') - \chi'(\infty)}{\omega' - \omega} d\omega' \quad .$$

These are the famous Kramers-Kronig equations. Similar equations can be worked out for analogous quantities such as the dielectric constant or the electrical susceptibility.

The significance of these equations is that there are restrictions placed, for example, on the dispersion by the absorption. One cannot dream up arbitrary $\chi'(\omega)$ and $\chi''(\omega)$. To phrase alternately, we may say that knowledge of χ'' for all frequencies enables one to compute the χ' at any frequency. Note in particular that for a narrow resonance line, assuming $\chi'(\infty) = 0$, the static susceptibility χ_0 is given by

$$\chi_0 = \chi'(0) = \frac{1}{\pi} P \int_{-\infty}^{+\infty} \frac{\chi''(\omega')}{\omega'} d\omega'$$

$$= \frac{2}{\pi} \frac{1}{\omega_0} \int_{0}^{+\infty} \chi''(\omega') d\omega' \quad . \qquad (2.170)$$

The integral of $\chi''(\omega')$ is essentially the area under the absorption curve. We see that it may be computed if the static susceptibility is known.[6]

As an example, suppose

$$\chi''(\omega) = c[\delta(\omega - \Omega) - \delta(-\omega - \Omega)] \quad . \qquad (2.171)$$

The first term corresponds to absorption at frequency Ω. The second term simply makes χ'' an odd function of ω. For this function, what is $\chi'(\omega)$?

$$\chi'(\omega) - \chi'(\infty) = \frac{1}{\pi} P \int_{-\infty}^{+\infty} \frac{c[\delta(\omega' - \Omega) - \delta(-\omega' - \Omega)] d\omega'}{\omega' - \omega}$$

---

[6] Of course, if we are talking about a magnetic resonance experiment with the static field in the z-direction and the alternating field in the x-direction, we are discussing χ_{xx}. Then $\chi'(0)$ of (2.170) is $\chi'_{xx}(0)$, whereas χ_0 is usually thought of as relating the total magnetization M_0 to the field H_0, which produces it, and is thus $\chi'_{xx}(0)$. However, a small static field H_x in the x-direction simply rotates M_0, giving

$$M_x = M_0 \frac{H_x}{H_0} = \chi'_{zz}(0) H_x \quad .$$

Thus $\chi'_{xx}(0) = \chi'_{zz}(0) = \chi_0$.

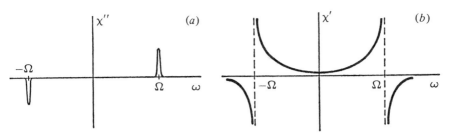

Fig. 2.21. (a) Absorption spectrum. (b) Corresponding dispersion spectrum

$$\chi'(\omega) - \chi'(\infty) = \frac{c}{\pi}\left(\frac{1}{\Omega - \omega} - \frac{1}{-\Omega - \omega}\right) = \frac{c}{\pi}\left(\frac{1}{\Omega - \omega} + \frac{1}{\Omega + \omega}\right) \quad (2.172)$$

where we have used the fact that $\delta(x) = \delta(-x)$.[7]

Of course, near resonance ($\omega \cong \Omega$), only the first term is large. The function is shown in Fig. 2.21.

2.12 Atomic Theory of Absorption and Dispersion

We shall now turn to obtaining expressions for the absorption and dispersion in terms of atomic properties such as the wave functions, matrix elements, and energy levels of the system under study. We shall compute χ'' directly and obtain χ' from the Kramers-Kronig equations.

We make the connection between the macroscopic and the microscopic properties by computing the average power $\overline{P}$ absorbed from an alternating magnetic field $H_{x0} \cos \omega t$. From (2.102) we have

$$\overline{P} = \frac{\omega}{2}\chi'' H_{x0}^2 V \quad (2.173)$$

in a volume V. It will be convenient henceforth to refer everything to a unit volume. (We shall have to remember this fact when we compute the atomic expressions in particular cases).

On the other hand, the alternating field couples to the magnetic moment μ_{xk} of the kth spin. Therefore, in our Hamiltonian we shall have a time-dependent perturbation $\mathcal{H}_{\text{pert}}$ of

---

[7] To show that $\delta(x) = \delta(-x)$ we consider the integrals $I_1 = \int_{-\infty}^{+\infty} f(x)\delta(x)dx$ and $I_2 = \int_{-\infty}^{+\infty} f(x)\delta(-x)dx$. We have immediately that $I_1 = f(0)$. To evaluate I_2, we change variable to $x' = -x$. Then we get

$$I_2 = \int_{-\infty}^{+\infty} f(-x')\delta(x')dx' = f(0) \quad .$$

Thus $I_2 = I_1$, and $\delta(x) = \delta(-x)$.

$$\mathcal{H}_{\text{pert}} = -\sum_k \mu_{xk} H_{x0} \cos \omega t$$
$$= -\mu_x H_{x0} \cos \omega t \tag{2.174}$$

where μ_x is the x-component of the total magnetic moment

$$\mu_x \equiv \sum_k \mu_{xk} \quad . \tag{2.175}$$

In the absence of the perturbation, the Hamiltonian will typically consist of the interactions of the spins with the external static field and of the coupling $\mathcal{H}_{jk}$ between spins j and k. Thus

$$\mathcal{H} = -\sum_k \mu_{zk} H_0 + \sum_{j,k} \mathcal{H}_{jk} \quad . \tag{2.176}$$

We shall denote the eigenvalues of energy of this many-spin Hamiltonian as E_a, E_b, and so on, with corresponding many-spin wave functions as $|a)$ and $|b)$. See Fig. 2.22. Because of the large number of degrees of freedom there will be a quasi-continuum of energy levels.

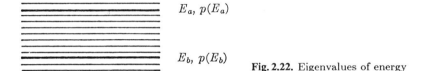

Fig. 2.22. Eigenvalues of energy

The states $|a)$ and $|b)$ are eigenstates of the Hamiltonian. The most general wave function would be a linear combination of such eigenstates:

$$\psi = \sum_a c_a |a) e^{-(\mathrm{i}/\hbar) E_a t} \tag{2.177}$$

where the c_a's are complex constants. The square of the absolute value of c_a gives the probability $p(a)$ of finding the system in the eigenstate a:

$$p(a) = |c_a|^2 \quad .$$

If the system is in thermal equilibrium, all states will be occupied to some extent, the probability of occupation $p(a)$ being given by the Boltzmann factor

$$p(E_a) = \frac{\mathrm{e}^{-E_a/kT}}{\sum_{E_c} \mathrm{e}^{-E_c/kT}} \tag{2.178}$$

where the sum E_c goes over the entire eigenvalue spectrum. The denominator is just the classical partition function Z, inserted to guarantee that the total probability of finding the system in any of the eigenstates is equal to unity; that is,

$$\sum_{E_a} p(E_a) = 1 \quad .$$

We can compute the absorption rate $\overline{P}_{ab}$, due to transitions between states a and b in terms of W_{ab}, the probability per second that a transition would be induced from a to b if the system were entirely in state a initially:

$$\overline{P}_{ab} = \hbar\omega W_{ab}[p(E_b) - p(E_a)] \quad . \tag{2.179}$$

The terms $p(E_b)$ and $p(E_a)$ come in because the states $|a\rangle$ and $|b\rangle$ are only fractionally occupied.

The calculation of the transition probability W_{ab} is well known from elementary quantum mechanics. Suppose we have a time-dependent perturbation $\mathcal{H}_{\text{pert}}$ given by

$$\mathcal{H}_{\text{pert}} = Fe^{-i\omega t} + Ge^{i\omega t} \tag{2.180}$$

where F and G are two operators. In order that $\mathcal{H}_{\text{pert}}$ will be Hermitian, F and G must be related so that for all states $|a\rangle$ or $|b\rangle$,

$$(a|F|b) = (b|G|a)^* \quad . \tag{2.181}$$

Under the action of such a perturbation we can write that W_{ab} is time independent and is given by the formula

$$W_{ab} = \frac{2\pi}{\hbar} |(a|F|b)|^2 \delta(E_a - E_b - \hbar\omega) \tag{2.182}$$

provided certain conditions are satisfied: We do not ask for details that appear on a time scale shorter than a certain characteristic time τ. It must be possible to find such a time, which will satisfy the conditions that 1) the populations change only a small amount in τ and 2) the possible states between which absorption can occur must be spread in energy continuously over a range ΔE such that $\Delta E \gg \hbar/\tau$.

These conditions are violated if the perturbation matrix element $|(a|F|b)|$ exceeds the line width, as it does when a very strong alternating field is applied. We can see this point as follows: The quantity ΔE may be taken as the line width. We have, then, that $\Delta E < |(a|F|b)|$. But under these circumstances one can show that the populations change significantly in a time of order $\hbar/|(a|F|b)|$. Thus to satisfy the condition 1 that the populations change only a small amount during τ, τ must be chosen less than $\hbar/|(a|F|b)|$. This gives us

$$|(a|F|b)| < \frac{\hbar}{\tau} \quad .$$

But, by hypothesis,

$$\Delta E < |(a|F|b)| \quad .$$

Therefore

$$\Delta E < \frac{\hbar}{\tau}$$

which violates condition 2 above. Thus it is not possible to satisfy both conditions, and the transition probability is not independent of time.

This example shows why we did not get a simple time-dependent rate process in Sect. 2.6, since for that problem, the energy levels in the absence of H_1 are perfectly sharp ($\Delta E = 0$), $|(a|F|b)| > \Delta E$.

In our formula for W_{ab} we use the δ-function. This implies that we shall eventually sum over a quasi-continuum of energy states. In writing the transition probability, it is preferable to use the δ-function form rather than the integrated form involving density of states in order to keep track of quantum numbers of individual states.

By summing over all states with $E_a > E_b$, we find

$$\overline{P} = \frac{2\pi}{\hbar} \frac{H_{x0}^2}{4} \hbar\omega \sum_{E_a > E_b} [p(E_b) - p(E_a)]|(a|\mu_x|b)|^2 \delta(E_a - E_b - \hbar\omega)$$

$$= \frac{\omega}{2} \chi'' H_{x0}^2 \quad . \tag{2.183}$$

Therefore

$$\chi''(\omega) = \pi \sum_{E_a > E_b} [p(E_b) - p(E_a)]|(a|\mu_x|b)|^2 \delta(E_a - E_b - \hbar\omega) \quad . \tag{2.184}$$

As long as $E_a > E_b$, only positive ω will give absorption because of the δ-function in (2.184). Removal of the restriction $E_a > E_b$ extends the meaning of $\chi''(\omega)$ formally to negative ω. Note that since $p(E_b) - p(E_a)$ changes sign when a and b are interchanged, $\chi''(\omega)$ is an odd function of ω, as described in the preceding section:

$$\chi''(\omega) = \pi \sum_{E_a, E_b} [p(E_b) - p(E_a)]|(a|\mu_x|b)|^2 \delta(E_a - E_b - \hbar\omega) \quad . \tag{2.185}$$

Assuming $\chi'(\infty) = 0$ for our system, we can easily compute $\chi'(\omega)$, since

$$\chi'(\omega) = \frac{1}{\pi} P \int_{-\infty}^{+\infty} \frac{\chi''(\omega')}{\omega' - \omega} d\omega' \tag{2.186}$$

$$= \pi \sum_{E_a, E_b} [p(E_b) - p(E_a)]|(a|\mu_x|b)|^2 \frac{1}{\pi} P \int_{-\infty}^{+\infty} \frac{\delta(E_a - E_b - \hbar\omega')}{\omega' - \omega} d\omega'$$

or, evaluating the integral,

$$\chi'(\omega) = \sum_{E_a, E_b} [p(E_b) - p(E_a)]|(a|\mu_x|b)|^2 \frac{1}{E_a - E_b - \hbar\omega} \quad . \tag{2.187}$$

By using the fact that a and b are dummy indices, one may also rewrite (2.187) to give

$$\chi'(\omega) = \sum_{E_a, E_b} p(E_b)|(a|\mu_x|b)|^2 \left[(E_a - E_b - \hbar\omega)^{-1} \right.$$
$$\left. + (E_a - E_b + \hbar\omega)^{-1} \right] \quad . \tag{2.187a}$$

The quanta $\hbar\omega$ correspond crudely to the energy required to invert a spin in the static field. This energy is usually much smaller than kT. For nuclear moments in strong laboratory fields ($\sim 10^4$ Gauss), T must be as low as 10^{-3} K so that $\hbar\omega$ will be as large as kT. This fact accounts for the difficulty in producing polarized nuclei. For electrons, $kT \sim \hbar\omega$ at about 1 K in a field of 10^4 Gauss. Therefore we may often approximate

$$E_a - E_b \ll kT \quad . \tag{2.188}$$

We may call this the "high-temperature approximation". By using (2.178) and (2.188), we have

$$\begin{aligned} p(E_b) - p(E_a) &= \frac{e^{-E_a/kT}[e^{(E_a-E_b)/kT} - 1]}{Z} \\ &= \frac{e^{-E_a/kT}}{Z}\left(\frac{E_a - E_b}{kT}\right) \quad . \end{aligned} \tag{2.189}$$

Substitution of (2.189) into (2.185), together with recognition that $E_a - E_b = \hbar\omega$, owing to the δ-functions, gives

$$\chi''(\omega) = \frac{\hbar\omega\pi}{kTZ} \sum_{E_a, E_b} e^{-E_a/kT} |\langle a|\mu_x|b\rangle|^2 \delta(E_a - E_b - \hbar\omega) \quad . \tag{2.190}$$

Another expression for $\chi''(\omega)$ is frequently encountered. It is the basis, for example, of *Anderson*'s theory of motional narrowing [2.21]. We discuss it in Appendix B because a proper discussion requires reference to some of the material in Chapters 3 and 5.

It is important to comment on the role of the factors $\exp(-E_a/kT)$. If one is dealing with water, for example, the proton absorption lines are found to be quite different at different temperatures. Ice, if cold enough, possesses a resonance several kilocycles broad, whereas the width of the proton resonance in liquid water is only about 1 cycle. Clearly the only difference is associated with the relative mobility of the H_2O molecule in the liquid as opposed to the solid. The position coordinates of the protons therefore play an important role in determining the resonance. Formally we should express this fact by including the kinetic and potential energies of the atoms as well as the spin energies in the Hamiltonian. Then the energies E_a and E_b contain contributions from both spin and positional coordinates. Some states $|a\rangle$ correspond to a solid, some to a liquid. The factor $\exp(-E_a/kT)$ picks out the type of "lattice" wave functions or states that are representative of the temperature, that is, whether the water molecules are in liquid, solid, or gaseous phase. Commonly the exponential factor is omitted from the expression for χ'', but the states $|a\rangle$ and $|b\rangle$ are chosen to be representative of the known state. The classic papers of *Gutowsky* and *Pake* [2.22], on the effect of hindered molecular motion on the width of resonance, use such a procedure.

Evaluation of χ'' by using (2.190) would require knowledge of the wave functions and energy levels of the system. As we shall see, we rarely have that information, but we shall be able to use (2.190) to compute the so-called moments of the absorption line. We see that the only frequencies at which strong absorption will occur must correspond to transitions among states between which the magnetic moment has large matrix elements.

3. Magnetic Dipolar Broadening of Rigid Lattices

3.1 Introduction

A number of physical phenomena may contribute to the width of a resonance line. The most prosaic is the lack of homogeneity of the applied static magnetic field. By dint of hard work and clever techniques, this source can be reduced to a few milligauss out of 10^4 Gauss, although more typically magnet homogeneities are a few tenths of a Gauss. The homogeneity depends on sample size. Typical samples have a volume between 0.1 cc to several cubic centimeters. Of course fields of ultrahigh homogeneity place severe requirements on the frequency stability of the oscillator used to generate the alternating fields. Although these matters are of great technical importance, we shall not discuss them here. If a nucleus possesses a nonvanishing electric quadrupole moment, the degeneracy of the resonance frequencies between different m-values may be lifted, giving rise to either resolved or unresolved splittings. The latter effectively broaden the resonance. The fact that T_1 processes produce an equilibrium population by balancing rates of transitions puts a limit on the lifetime of the Zeeman states, which effectively broadens the resonance lines by an energy of the order of $\hbar/T_1$.

In this chapter, however, we shall ignore all these effects and concentrate on the contribution of the magnetic dipole coupling between the various nuclei to the width of the Zeeman transition. This approximation is often excellent, particularly when the nuclei have spin $\frac{1}{2}$ (thus a vanishing quadrupole moment) and a rather long spin-lattice relaxation time.

A rough estimate of the effect of the dipolar coupling is easily made. If typical neighboring nuclei are a distance r apart and have magnetic moment μ, they produce a magnetic field H_{loc} of the order

$$H_{\text{loc}} = \frac{\mu}{r^3} \quad . \tag{3.1}$$

By using $r = 2$ Å and $\mu = 10^{-23}$ erg/Gauss (10^{-3} of a Bohr magneton), we find $H_{\text{loc}} \cong 1$ Gauss. Since this field may either aid or oppose the static field H_0, a spread in the resonance condition results, with significant absorption occurring over a range of $H_0 \sim 1$ Gauss. The resonance width on this argument is independent of H_0, but for typical laboratory fields of 10^4 Gauss, we see there is indeed a sharp resonant line. Since the width is substantially greater than the magnet inhomogeneity, it is possible to study the shape in detail without instrumental limitations.

3.2 Basic Interaction

The classical interaction energy E between two magnetic moments μ_1 and μ_2 is

$$E = \frac{\mu_1 \cdot \mu_2}{r^3} - \frac{3(\mu_1 \cdot r)(\mu_2 \cdot r)}{r^5} \tag{3.2}$$

where r is the radius vector from μ_1 to μ_2. (The expression is unchanged if r is taken as the vector from μ_2 to μ_1.) For the quantum mechanical Hamiltonian we simply take (3.2), treating μ_1 and μ_2 as operators as usual:

$$\mu_1 = \gamma_1 \hbar I_1 \quad , \quad \mu_2 = \gamma_2 \hbar I_2 \quad , \tag{3.3}$$

where we have assumed that both the gyromagnetic ratios and spins may be different. The general dipolar contribution to the Hamiltonian for N spins then becomes

$$\mathcal{H}_d = \frac{1}{2} \sum_{j=1}^{N} \sum_{k=1}^{N} \left[\frac{\mu_j \cdot \mu_k}{r_{jk}^3} - \frac{3(\mu_j \cdot r_{jk})(\mu_k \cdot r_{jk})}{r_{jk}^5} \right] \tag{3.4}$$

where the $\frac{1}{2}$ is needed, since the sums over j and k would count each pair twice, and where, of course, we exclude terms with $j = k$.

By writing μ_1 and μ_2 in component form and omitting the subscripts from r, we see from (3.2) that the dipolar Hamiltonian will contain terms such as

$$\gamma_1 \gamma_2 \hbar^2 I_{1x} I_{2x} \frac{1}{r^3} \quad , \quad \gamma_1 \gamma_2 \hbar^2 I_{1x} I_{2x} \frac{xy}{r^5} \quad . \tag{3.5}$$

If we express I_{1x} and I_{1y} in terms of the raising and lowering operators I_1^+ and I_1^-, respectively, and express the rectangular coordinates x, y, z in terms of spherical coordinates r, θ, ϕ (Fig. 3.1), we may write the Hamiltonian in a form that is particularly convenient for computing matrix elements:

$$\mathcal{H}_d = \frac{\gamma_1 \gamma_2 \hbar^2}{r^3} (A + B + C + D + E + F) \tag{3.6}$$

where

$$\begin{aligned}
A &= I_{1z} I_{2z} (1 - 3\cos^2 \theta) \\
B &= -\tfrac{1}{4}(I_1^+ I_2^- + I_1^- I_2^+)(1 - 3\cos^2 \theta) \\
C &= -\tfrac{3}{2}(I_1^+ I_{2z} + I_{1z} I_2^+) \sin\theta \cos\theta \, e^{-i\phi} \\
D &= -\tfrac{3}{2}(I_1^- I_{2z} + I_{1z} I_2^-) \sin\theta \cos\theta \, e^{i\phi} \\
E &= -\tfrac{3}{4} I_1^+ I_2^+ \sin^2 \theta \, e^{-2i\phi} \\
F &= -\tfrac{3}{4} I_1^- I_2^- \sin^2 \theta \, e^{2i\phi} \quad .
\end{aligned} \tag{3.7}$$

As we have remarked, $(\gamma_1 \gamma_2 \hbar^2)/r^3$ corresponds to the interaction of a nuclear moment with a field of about 1 Gauss, whereas the Zeeman Hamiltonian

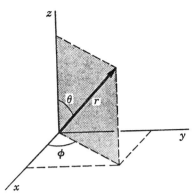

Fig. 3.1. Relationship between rectangular coordinates x, y, z (describing the position of nucleus 2 relative to nucleus 1) and the polar coordinates r, θ, ϕ

$\mathcal{H}_Z = -\gamma_1 \hbar H_0 I_{1z} - \gamma_2 \hbar H_0 I_{2z}$ corresponds to an interaction with a field of 10^4 Gauss. It is therefore appropriate to solve the Zeeman problem first and then treat the dipolar term as a small perturbation. (Actually, for two spins of $\frac{1}{2}$, an exact solution is possible.)

To see the significance of the various terms A, B, C, and so on, we shall consider a simple example of two identical moments, both of spin $\frac{1}{2}$.

The Zeeman energy and wave functions can be given in terms of the individual quantum numbers m_1 and m_2, which are the eigenvalues of I_{1z} and I_{2z}. Then the Zeeman energy is

$$E_Z = -\gamma \hbar H_0 m_1 - \gamma \hbar H_0 m_2 \quad . \tag{3.8}$$

We shall diagram the appropriate matrix elements and energy levels in Fig. 3.2. It is convenient to denote a state in which $m_1 = +\frac{1}{2}$, $m_2 = -\frac{1}{2}$ by the notation $|+-\rangle$. The two states $|+-\rangle$ and $|-+\rangle$ are degenerate, and both have $E_Z = 0$. The states $|++\rangle$ and $|--\rangle$ have, respectively, $-\hbar\omega_0$ and $+\hbar\omega_0$, where $\omega_0 = \gamma H_0$ as usual. We first ask what pairs of states are connected by the various terms in the dipolar expression. The term A, which is proportional to $I_{1z} I_{2z}$, is clearly completely diagonal: It connects $|m_1 m_2\rangle$ with $\langle m_1 m_2|$. On the other hand, B, which is proportional to $I_1^+ I_2^- + I_1^- I_2^+$, only connects $|m_1 m_2\rangle$ to states $\langle m_1 + 1, m_2 - 1|$ or $\langle m_1 - 1, m_2 + 1|$. A customary parlance is to say that B simultaneously flips one spin up and the other down. B therefore can join only the states $|+-\rangle$ and $|-+\rangle$. The states joined by A and B are shown diagrammatically in Fig. 3.3.

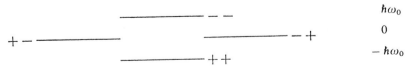

Fig. 3.2. Energy levels of two identical spins

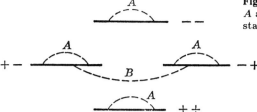

Fig. 3.3. States joined by matrix elements A and B. The dashed lines go between states that are joined

Note that B has no diagonal matrix elements for the $m_1 m_2$ representation, but it has off-diagonal elements between two states which are degenerate. The fact that off-diagonal elements join the degenerate states $|+ -)$ and $|- +)$ tells us, of course, that they are not the proper zero-order states. B therefore plays an important role in determining the proper zero-order functions. When the proper zero-order functions are determined, B turns out to have diagonal matrix elements. We shall return to this point later.

Since terms C and D each flip one spin only, they join states shown in Fig. 3.4, all of which differ by $\hbar\omega_0$ in energy. Finally E and F flip both spins up or both spins down, connecting states that differ by $2\hbar\omega_0$ (Fig. 3.5).

Fig. 3.4. States joined by the terms C and D

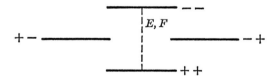

Fig. 3.5. States joined by the dipolar terms E and F

The terms C, D, E, and F therefore are off-diagonal. They produce slight admixtures of our zero-order states into the exact states. The amount of the admixture can be computed by second-order perturbation theory, using the well-known expression for the correction of the zero-order functions u_n^0 of zero-order energy E_n:

$$u_n = u_n^0 + \sum_{n'} \frac{(n'|\mathcal{H}_{\text{pert}}|n)}{E_n - E_{n'}} u_{n'}^0 \qquad (3.9)$$

where u_n is the wave function corrected for the effect of the perturbation $\mathcal{H}_{\text{pert}}$ and where, of course, the matrix elements $(n'|\mathcal{H}_{\text{pert}}|n)$ are computed between the unperturbed states of $u_{n'}^0$ and u_n^0.

By means of (3.9) we can see that the state $|++)$ will have a small admixture of $|+-)$, $|-+)$, and $|--)$. The amount of admixture will depend on $(n'|\mathcal{H}_{\text{pert}}|n)$ and $E_n - E_{n'}$. The former will be $\gamma^2\hbar^2/r^3$ multiplied by a spin matrix element. Since the spin matrix element is always of order unity, and since $H_{\text{loc}} = \gamma\hbar/r^3$, we can say $(n'|\mathcal{H}_{\text{pert}}|n) \cong \gamma\hbar H_{\text{loc}}$. On the other hand, $E_n - E_{n'} = \hbar\omega_0 = \gamma\hbar H_0$, so that

$$\left| \frac{(n'|\mathcal{H}_{\text{pert}}|n)}{E_n - E_{n'}} \right| \approx \frac{H_{\text{loc}}}{H_0} \sim 10^{-4} \tag{3.10}$$

corresponding to a very small admixture. Of course the admixture produces a second-order energy shift. As a second, and for us more important, effect, the admixture enables the alternating field to induce transitions that would otherwise be forbidden. Thus the transition from $|++)$ to $|--)$, which would be forbidden if these were exactly the states, can now take place by means of the small admixture of the states $|+-)$ and $|-+)$. (See Fig. 3.6.)

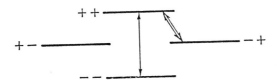

Fig. 3.6. The strong transition is indicated by the double arrow. The transition by the light arrow has non-vanishing matrix elements due to the dipole admixtures

The matrix element is smaller than the normal one, for example, between $|++)$ and $|+-)$, in the ratio H_{loc}/H_0. Therefore the intensity of the absorption, which goes as the square of the matrix element, is weaker in the ratio $(H_{\text{loc}}/H_0)^2$. The transition occurs, of course, at $\omega = 2\omega_0$. A further consequence of the admixture of states is that a transition near $\omega = 0$ can be induced. [Actually this transition is forbidden for a pair of spins, each of spin $\frac{1}{2}$, because the eigenstates of $M = m_1 + m_2 = 0$ are of different symmetry under exchange of the particle labels (the singlet and triplet states), whereas the perturbation is symmetric. If more than two spins are involved, the transition is permitted.]

The net effect of the terms C, D, E, and F is therefore to give the absorption near 0 and $2\omega_0$, shown in Fig. 3.7. The extra peaks at 0 and $2\omega_0$ are very weak and may be disregarded for our purposes. Since they are the principal effects of the terms C, D, E, and F, it will be an excellent approximation to drop C, D, E, and F from the Hamiltonian. For some of our later calculations we shall see that failure to drop these terms can lead us into erroneous results. The remaining dipolar term $A + B$ may be combined to give what we shall call $\mathcal{H}_d^0$:

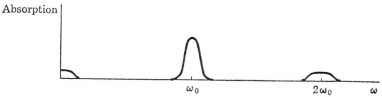

Fig. 3.7. Absorption versus frequency, including dipolar couplings. The three absorption regions have width $\sim \gamma H_{\text{loc}}$, but the intensity of the peaks at 0 and $2\omega_0$ is $\sim (H_{\text{loc}}/H_0)^2$ smaller than that at ω_0

$$\mathcal{H}_d^0 = \frac{1}{2}\frac{\gamma^2\hbar^2}{r^3}(1 - 3\cos^2\theta)(3I_{1z}I_{2z} - \mathbf{I}_1 \cdot \mathbf{I}_2) \tag{3.11}$$

and the total simplified Hamiltonian becomes

$$\mathcal{H} = \sum_k(-\gamma\hbar H_0 I_{zk}) + \frac{1}{4}\gamma^2\hbar^2\sum_{j,k}\frac{(1 - 3\cos^2\theta_{jk})}{r_{jk}^3}(3I_{jz}I_{kz} - \mathbf{I}_j \cdot \mathbf{I}_k) \quad . \tag{3.12}$$

Now the terms $\mathcal{H}_Z$ and $\mathcal{H}_d^0$ commute. [This can be seen by considering a pair of spins 1 and 2. Clearly $I_z = I_{1z} + I_{2z}$ commutes with $3I_{1z}I_{2z}$. How about the terms $\mathbf{I}_1 \cdot \mathbf{I}_2$? I_z commutes with $(\mathbf{I}_1 + \mathbf{I}_2)^2$, since $\mathbf{I}_1 + \mathbf{I}_2 = \mathbf{I}$ is the operator of the total angular momentum (any component of angular momentum commutes with the square of the total angular momentum). By writing $\mathbf{I}^2 = (\mathbf{I}_1 + \mathbf{I}_2)^2$, we have

$$\mathbf{I}^2 = \mathbf{I}_1^2 + \mathbf{I}_2^2 + 2\mathbf{I}_1 \cdot \mathbf{I}_2 \quad . \tag{3.13}$$

We see that $I_{1z} + I_{2z}$ commutes with the left side and the first two terms on the right. Therefore it must commute with $\mathbf{I}_1 \cdot \mathbf{I}_2$.] If two operators commute, we may choose an eigenfunction to be simultaneously an eigenfunction of both. Let us use α to denote the eigenvalues of $\mathcal{H}_d^0$. Then we have

$$\mathcal{H}_d^0 u_\alpha = E_\alpha u_\alpha \tag{3.14}$$

$$\mathcal{H}_Z u_M = (-\gamma\hbar H_0 M) u_M \quad \text{so that}$$

$$(\mathcal{H}_Z + \mathcal{H}_d^0)|M\alpha\rangle = (-\gamma\hbar H_0 M + E_\alpha)|M\alpha\rangle \quad . \tag{3.15}$$

These quantum numbers will prove useful later. Unfortunately all we can say about the quantum numbers α is that they exist, although we do not know them or the corresponding eigenfunctions. If $\mathcal{H}_d^0$ consisted of only the term $I_{1z}I_{2z}$, we could solve the resonance shape exactly. We could do the same if all we had was $\mathbf{I}_1 \cdot \mathbf{I}_2$. But the presence of both together, since they do not commute, spoils the two solutions. In fact, to proceed further, we are forced to go to the so-called method of moments, a clever technique due to *Van Vleck*, which enables one to compute properties of the resonance line without solving explicitly for the eigenstates and eigenvalues of energy.

3.3 Method of Moments [1]

Before outlining the method of moments, we must return to our original expression for $\chi''(\omega)$:

$$\chi''(\omega) = \frac{\pi\hbar\omega}{kTZ}\sum_{a,b}e^{-E_a/kT}|(a|\mu_x|b)|^2\delta(E_a - E_b - \hbar\omega) \quad . \tag{3.16}$$

Since we shall treat the lattice variables as parameters, the only variables coming into the problem are due to spin; that is, the quantum numbers a and b refer to spins. We shall therefore assume $E_a \ll kT$ and replace the exponentials by 1. The validity of this approximation requires some justification since the E_a's are energies of the total Hamiltonian of all N spins. If the nuclei had spin $\frac{1}{2}$, and if the only energy were the Zeeman energy, an individual spin could have energy $\pm\gamma_n\hbar H_0/2$, but the energy E_a could range between $\pm N\gamma_n\hbar H_0/2$. Since N may be very large, there are values of E_a that clearly violate the restriction $|E_a| \ll kT$. But the largest energy, $N\gamma_n\hbar H_0/2$, is realized only when all spins are parallel. The next smallest energy may be obtained by turning over one spin, it is $E_a = (N-2)\gamma\hbar H_0/2$. However, we could turn over any of the N spins, so that this energy is N-fold degenerate. We expect, in fact, that a Gaussian distribution will describe the number of states of any given energy. Typical energies E_a will therefore be of order $\sqrt{N}\gamma_n\hbar H_0/2$ which is still much larger than the Zeeman energy of a single spin. However, we know that the energy levels of a single spin give a very good prediction of the frequency of absorption, so that we suspect the $\sqrt{N}$ effect must in some sense be a red herring. That such is the case is shown in Appendix E where we demonstrate that the high temperature approximation is valid if $\gamma\hbar H_0 I \ll kT$.

With this approximation the shape of the line is then given by the factor ω and by the function $f(\omega)$ defined as

$$f(\omega) = \sum_{a,b}|(a|\mu_x|b)|^2\delta(E_a - E_b - \hbar\omega) \quad . \tag{3.17}$$

In fact, experimental determinations of $\chi''(\omega)$ enable us to compute $f(\omega)$ from (3.17), and conversely a theoretical determination of $f(\omega)$ gives us $\chi''(\omega)$. We focus therefore on $f(\omega)$. First we note that since $\chi''(\omega)$ was an odd function of ω, $f(\omega)$ is an even function. [This fact is also evident by explicit examination of $f(\omega)$.] We now define the nth moments of $f(\omega)$ by the equation

$$\langle\omega^n\rangle = \frac{\int_0^\infty \omega^n f(\omega)d\omega}{\int_0^\infty f(\omega)d\omega} \quad \text{and} \tag{3.18}$$

---
[1] See references under "Second Moment" in the Bibliography

$$\langle \Delta \omega^n \rangle = \frac{\int_0^\infty (\omega - \langle \omega \rangle)^n f(\omega) d\omega}{\int_0^\infty f(\omega) d\omega} \quad . \tag{3.19}$$

The expression (3.19) for $n = 2$ is called the "second moment". Clearly $\langle \Delta \omega^2 \rangle$ is of the order of the square of the line width, so that

$$\langle \Delta \omega^2 \rangle \cong (\gamma H_{\text{loc}})^2 \quad . \tag{3.20}$$

The two moments of (3.18) and (3.19) are closely related, as may readily be seen for $n = 2$ as follows: Expanding $(\omega - \langle \omega \rangle)^2 = \omega^2 - 2\omega \langle \omega \rangle + \langle \omega \rangle^2$, one easily shows from (3.18) and (3.19) that

$$\langle \Delta \omega^2 \rangle = \langle \omega^2 \rangle - \langle \omega \rangle^2 \quad . \tag{3.21}$$

Therefore we can compute either $\langle \Delta \omega^2 \rangle$ directly or compute it from the calculations of $\langle \omega^2 \rangle$ and $\langle \omega \rangle$. (We shall do the latter.)

To illustrate the general methods, we shall first compute $\int_0^\infty f(\omega) d\omega$, which is, of course, closely related to the area under the absorption curve. Then we shall compute $\langle \omega \rangle$ and $\langle \Delta \omega^2 \rangle$. Since $f(\omega)$ is an even function,

$$\int_0^\infty f(\omega) d\omega = \frac{1}{2} \int_{-\infty}^{+\infty} f(\omega) d\omega$$

$$= \frac{1}{2} \int_{-\infty}^{+\infty} \sum_{a,b} (a|\mu_x|b)(b|\mu_x|a) \delta(E_a - E_b - \hbar\omega) d\omega \quad . \tag{3.22}$$

The integrand picks up a contribution from the δ-function integral every time $\hbar\omega = E_a - E_b$. But for any pair of states $|a\rangle$ and $|b\rangle$, there is *some* value of ω between $-\infty$ and $+\infty$ which satisfies the condition $\hbar\omega = E_a - E_b$. (Note, if we had the integral from 0 to ∞, we should have zero from states for which $E_a - E_b$ was negative. It is for this reason that we let the integral range from $-\infty$ to $+\infty$.) Thus, changing the variable of integration from ω to $\hbar\omega$, we get

$$\int_0^\infty f(\omega) d\omega = \frac{1}{2\hbar} \sum_{a,b} (a|\mu_x|b)(b|\mu_x|a) \quad . \tag{3.23}$$

But it is a basic theorem of quantum mechanics that for any complete set $|\beta'\rangle$ and for any operators A and B,

$$\sum_{\beta'} (\beta|A|\beta')(\beta'|B|\beta'') = (\beta|AB|\beta'') \quad . \tag{3.23a}$$

So, we can rewrite (3.23) as

$$\int_0^\infty f(\omega)d\omega = \frac{1}{2\hbar}\sum_a (a|\mu_x^2|a) = \frac{1}{2\hbar}\text{Tr}\{\mu_x^2\} \tag{3.24}$$

where the symbol "Tr" stands for "trace" or sum of the diagonal matrix elements. Another important theorem tells us that when we go from one complete set of orthogonal functions $|\beta)$ to an alternative one $|\zeta)$ [$|\beta)$ can thus be expressed as a linear combination of the $|\zeta)$'s], the trace is unchanged. We may therefore choose any complete set of functions to compute the trace. In fact we shall choose a set of functions that is simply the product of the individual spin functions of quantum numbers $m_1, m_2, m_3 \ldots m_N$ for the N spins. Therefore

$$\int_0^\infty f(\omega)d\omega = \frac{1}{2\hbar}\sum_{m_1,m_2,m_3,\ldots}(m_1 m_2 m_3 \ldots |\mu_x^2|m_1 m_2 m_3 \ldots) \quad . \tag{3.25}$$

Now, since $\mu_x = \sum_j \mu_{xj}$,

$$\mu_x^2 = \sum_{j,k} \mu_{xj}\mu_{xk} \quad . \tag{3.26}$$

There are two types of terms: $j \neq k$ and $j = k$. We examine the first kind first.

Let us consider $j = 1$, $k = 2$. Then, holding $m_2, m_3, m_4 \ldots$ fixed, we can first sum over m_1. Now,

$$(m_1 m_2 m_3 \ldots |\mu_{1x}\mu_{2x}|m_1 m_2 \ldots) = (m_1|\mu_{1x}|m_1)(m_2|\mu_{2x}|m_2) \tag{3.27}$$

so that summing over m_1 gives us

$$\left[\sum_{m_1}(m_1|\mu_{1x}|m_1)\right](m_2|\mu_{2x}|m_2) \quad . \tag{3.28}$$

Now, $\sum_{m_1}(m_1|\mu_{1x}|m_1) = 0$. This may be seen by noting that when we take m_1 as eigenvalues of I_{1z}, all the diagonal elements of I_{1x} and μ_{1x} are zero. Or, alternatively, one may let m_1 be the eigenvalues of I_{1x}. But for every $+m$ value there is a corresponding negative one, so that

$$\sum_{m_1}(m_1|\mu_{1x}|m_1) = \gamma\hbar\sum_{m_1}(m_1|I_{1x}|m_1) = 0 \quad . \tag{3.29}$$

Therefore the contribution from terms $j \neq k$ vanishes. For $j = k$, we get, taking $j = 1$,

$$\frac{1}{2\hbar}\sum_{m_1 m_2}(m_1 m_2 \ldots |\mu_{1x}^2|m_1 m_2 \ldots)$$

$$= \frac{\gamma^2\hbar^2}{2\hbar}\sum_{m_1,m_2\ldots}(m_1 m_2 \ldots |I_{1x}^2|m_1 m_2 \ldots) \quad . \tag{3.30}$$

The matrix element is independent of m_2, m_3, and so on, but it is *repeated* for each combination of the other quantum numbers. Since there are $(2I+1)$ values of m_2, $(2I+1)$ values of m_3, and so on, we get the matrix for each value of m_1 repeated $(2I+1)^{N-1}$ times. On the other hand, using Tr_1 to mean a trace only over quantum numbers of spin 1, we have that $\text{Tr}_1\{\mu_{1x}^2\} = \text{Tr}_1\{\mu_{1y}^2\}$. This equation is most simply proved by first evaluating $\text{Tr}_1\{\mu_{1x}^2\}$, using eigenfunctions of I_{1x}. Then

$$\text{Tr}_1\{\mu_{1x}^2\} = \gamma^2\hbar^2 \sum_{m=-I}^{+I} m^2 \quad . \tag{3.31}$$

In a similar way, $\text{Tr}_1\{\mu_{1y}^2\}$ may be evaluated by using eigenfunctions of I_{1y}:

$$\text{Tr}_1\{\mu_{1y}^2\} = \gamma^2\hbar^2 \sum_{m=-I}^{+I} m^2 \quad . \tag{3.31a}$$

Therefore

$$\text{Tr}_1\{\mu_{1x}^2\} = \text{Tr}_1\{\mu_{1y}^2\} = \text{Tr}_1\{\mu_{1z}^2\} = \tfrac{1}{3}\text{Tr}_1\{\mu_1^2\} \quad . \tag{3.31b}$$

There are $2I+1$ diagonal matrix elements of μ_1^2, each of magnitude $\gamma^2\hbar^2 I(I+1)$. Therefore

$$\text{Tr}_1\{\mu_{1x}^2\} = \frac{\gamma^2\hbar^2 I(I+1)}{3}(2I+1) \quad .$$

Since there are N identical terms of $j = k$, finally we get as our answer

$$\int_0^\infty f(\omega)d\omega = \frac{1}{2\hbar}\gamma^2\hbar^2 \frac{I(I+1)}{3}N(2I+1)^N \quad . \tag{3.32}$$

We turn now to a calculation of the effect of the dipolar coupling on the average frequency of absorption, $\langle\omega\rangle$. The existence of such a shift implies that the local field produced by the neighbors has a preferential orientation with respect to the applied field. Since such an effect must correspond to a Lorentz local field ΔH, it must be of general order $\chi_n H_0$, where χ_n is the static nuclear susceptibility. χ_n is given by the Langevin-Debye formula: $\chi_n = N\gamma^2\hbar^2 I(I+1)/3kT$, where N is the number of nuclei per unit volume. If the distance between nearest neighbors is a, $N \cong 1/a^3$, we have, therefore, that $\Delta H \cong (\gamma\hbar/a^3)(\gamma\hbar H_0/kT) \cong H_{\text{loc}}(\gamma\hbar H_0/kT)$. Since the nuclear Zeeman energy $\gamma\hbar H_0$ is very small compared with kT, we see ΔH is very small compared with the line breadth H_{loc} and is presumably negligible. Notice that the physical significance of our expression for ΔH is that the neighbors have a slight preferential orientation parallel to the static field given by the exponent of the Boltzmann factor $(\gamma\hbar H_0/kT)$. H_{loc} has a nonzero average to this extent. Since $f(\omega)$ of (3.17) corresponds to infinite temperature, it must lead to a $\Delta H = 0$, and $\langle\omega\rangle = \omega_0$.

To compute the average frequency or first moment rigorously,

$$\langle \omega \rangle = \frac{\int_0^\infty \omega f(\omega)d\omega}{\int_0^\infty f(\omega)d\omega} \quad ,$$

is a bit more difficult than the calculation of $\int_0^\infty f(\omega)d\omega$. In (3.22) it was convenient to extend the limits of integration to go from $-\infty$ to $+\infty$. As a result, for every pair of energies E_a and E_b, there was some frequency ω such that $E_a - E_b = \hbar\omega$, regardless of whether E_a was higher or lower than E_b. We cannot do the same thing for $\langle \omega \rangle$, since

$$\int_{-\infty}^{+\infty} \omega f(\omega)d\omega = 0 \tag{3.33}$$

because the integrand is an odd function of ω. We therefore are forced to compute $\int_0^\infty \omega f(\omega)d\omega$:

$$\int_0^\infty \omega f(\omega)d\omega = \frac{1}{\hbar^2} \sum_{a,b} \int_0^{+\infty} (a|\mu_x|b)(b|\mu_x|a)(\hbar\omega)\delta(E_a - E_b - \hbar\omega)d(\hbar\omega)$$

$$= \frac{1}{\hbar^2} \sum_{E_a > E_b} (a|\mu_x|b)(b|\mu_x|a)(E_a - E_b) \quad . \tag{3.34}$$

The energies E_a and E_b are the sum of dipolar and Zeeman contributions $(-\gamma\hbar H_0 M + E_\alpha)$, as we have remarked previously. We shall assume that the dipolar energy changes are always small compared with the changes in Zeeman energy and that the latter correspond to absorption near ω_0 (our earlier discussion of the role of the terms A, B, ... F shows us this fact). Therefore, since $E_a > E_b$, we write

$$\begin{aligned} E_a &= -\gamma\hbar H_0 M + E_\alpha \\ E_b &= -\gamma\hbar H_0(M+1) + E_{\alpha'} \\ E_a - E_a &= \hbar\omega_0 + E_\alpha - E_{\alpha'} \quad . \end{aligned} \tag{3.35}$$

By using these relations, we can write (3.34) as

$$\int_0^\infty \omega f(\omega)d\omega = \frac{1}{\hbar^2} \sum_{M,\alpha,\alpha'} (M\alpha|\mu_x|M+1\alpha')$$
$$\times (M+1\alpha'|\mu_x|M\alpha)(\hbar\omega_0 + E_\alpha - E_{\alpha'}) \quad . \tag{3.36}$$

We shall first discuss the contribution of $\hbar\omega_0$ term in the parentheses. It is

$$\frac{\hbar\omega_0}{\hbar^2} \sum_{M,\alpha,\alpha'} (M\alpha|\mu_x|M+1\alpha')(M+1\alpha'|\mu_x|M\alpha) \quad . \tag{3.37}$$

Were it not for the restriction to $M + 1$, (3.37) could be converted to a trace by means of (3.23a). This restriction can be removed by using the properties of the raising and lowering operators and by noting that

$$\mu_x = \tfrac{1}{2}(\mu^+ + \mu^-) \quad .$$

Thus

$$(M+1\alpha'|\mu_x|M\alpha) = \tfrac{1}{2}(M+1\alpha'|\mu^+|M\alpha) \quad . \tag{3.38}$$

Since μ^+ connects only states M' and M in a matrix element $(M'\alpha'|\mu^+|M\alpha)$ where $M' = M + 1$, we can rewrite (3.37), summing over all values of M' as

$$\begin{aligned}
\frac{\hbar\omega_0}{4\hbar^2} \sum_{M,M',\alpha,\alpha'} & (M\alpha|\mu^-|M'\alpha')(M'\alpha'|\mu^+|M\alpha) \\
&= \frac{\omega_0}{4\hbar} \operatorname{Tr}\{\mu^-\mu^+\} \\
&= \frac{\omega_0}{4\hbar} \operatorname{Tr}(\mu_x - i\mu_y)(\mu_x + i\mu_y) \\
&= \frac{\omega_0}{4\hbar} \operatorname{Tr}\{[\mu_x^2 + \mu_y^2 + i(\mu_x\mu_y - \mu_y\mu_x)]\} \\
&= \frac{\omega_0}{2\hbar} \operatorname{Tr}\{\mu_x^2\}
\end{aligned} \tag{3.39}$$

where we have used the facts that $\operatorname{Tr}\{\mu_x^2\} = \operatorname{Tr}\{\mu_y^2\}$ and

$$\begin{aligned}
\operatorname{Tr}\{\mu_x\mu_y - \mu_y\mu_x\} &= \gamma^2\hbar^2 \operatorname{Tr}\{I_xI_y - I_yI_x\} \\
&= \gamma^2\hbar^2 i \operatorname{Tr}\{I_z\} = 0 \quad .
\end{aligned} \tag{3.40}$$

We have so far handled the $\hbar\omega_0$ term of (3.36). The technique for handling the term $E_\alpha - E_{\alpha'}$ is very simple. We know that $\mathcal{H}_d^0|M\alpha'\rangle = E_{\alpha'}|M\alpha'\rangle$. Therefore, for an operator P, we have

$$\begin{aligned}
(M'\alpha'|P\mathcal{H}_d^0|M\alpha) &= \int u_{M'\alpha'}^* P\mathcal{H}_d^0 u_{M\alpha} d\tau \\
&= \int u_{M'\alpha'}^* PE_\alpha u_{M\alpha} d\tau \\
&= E_\alpha(M'\alpha'|P|M\alpha) \quad .
\end{aligned} \tag{3.41}$$

Likewise, using the fact that $\mathcal{H}_d^0$ is Hermitian,

$$\begin{aligned}
(M'\alpha'|\mathcal{H}_d^0 P|M\alpha) &= \int u_{M'\alpha'}^* \mathcal{H}_d^0 P u_{M\alpha} d\tau \\
&= \int (\mathcal{H}_d^0 u_{M'\alpha'})^* P u_{M\alpha} d\tau
\end{aligned}$$

$$= E_{\alpha'} \int u^*_{M'\alpha'} P u_{M\alpha} d\tau$$
$$= E_{\alpha'}(M'\alpha'|P|M\alpha) \quad . \tag{3.42}$$

Therefore
$$\sum_{M,M',\alpha,\alpha'} (M\alpha|\mu^-|M'\alpha')(M'\alpha'|\mu^+|M\alpha)(E_a - E_{\alpha'})$$
$$= \sum_{M,M',\alpha,\alpha'} (M\alpha|[\mathcal{H}_d^0, \mu^-]|M'\alpha')(M'\alpha'|\mu^+|M\alpha) \tag{3.43}$$
$$= \text{Tr}\{[\mathcal{H}_d^0, \mu^-]\mu^+\} \quad .$$

A detailed evaluation of this trace shows that it vanishes. Therefore, combining the results of (3.36), (3.39), and (3.43), we get

$$\int_0^\infty \omega f(\omega) d\omega = \frac{\omega_0}{2\hbar} \text{Tr}\{\mu_x^2\} \quad . \tag{3.44}$$

But from (3.24),
$$\int_0^\infty f(\omega) d\omega = \frac{1}{2\hbar} \text{Tr}\{\mu_x^2\} \quad .$$

Therefore
$$\boxed{\langle \omega \rangle = \frac{\int_0^\infty \omega f(\omega) d\omega}{\int_0^\infty f(\omega) d\omega} = \omega_0} \quad . \tag{3.45}$$

The "average" value of the frequency is therefore unshifted by the broadening as we had expected. To get the local field correction that we mentioned in our qualitative discussion, we should, in fact, have to go back to (3.17) and include the exponential factors that we deleted in going from (3.16). [That this is true follows from the fact that the expression $\Delta H \approx H_{\text{loc}}(\gamma \hbar H_0/kT)$ depends on temperature. The only place the temperature enters is in the exponentials.]

We can compute the second moment $\langle \omega^2 \rangle$ by similar techniques:

$$\langle \omega^2 \rangle = \frac{\int_0^\infty \omega^2 f(\omega) d\omega}{\int_0^\infty f(\omega) d\omega} \quad . \tag{3.46}$$

Since we have already evaluated the denominator, all that remains is to compute the numerator:

$$\int_0^\infty \omega^2 f(\omega)d\omega = \frac{1}{2}\int_{-\infty}^{+\infty} \omega^2 f(\omega)d\omega$$

$$= \frac{1}{2}\int_{-\infty}^{+\infty} \sum_{a,b}\omega^2 (a|\mu_x|b)(b|\mu_x|a)\delta(E_a - E_b - \hbar\omega)d\omega \quad (3.47)$$

$$= \frac{1}{2\hbar^3}\sum_{a,b}(E_a - E_b)^2(a|\mu_x|b)(b|\mu_x|a) \quad .$$

By using the fact that $\mathcal{H}|a) = E_a|a)$, we see, as in (3.42) and (3.43),

$$\int_0^\infty \omega^2 f(\omega)d\omega = -\frac{1}{2\hbar^3}\sum_{a,b}(a|\mathcal{H}\mu_x - \mu_x\mathcal{H}|b)(b|\mathcal{H}\mu_x - \mu_x\mathcal{H}|a)$$

$$= -\frac{1}{2\hbar^3}\text{Tr}\{[\mathcal{H},\mu_x]^2\} \quad . \quad (3.48)$$

We can expand, using $\mathcal{H} = \mathcal{H}_Z + \mathcal{H}_d^0$ to get

$$\int_0^\infty \omega^2 f(\omega)d\omega = -\frac{1}{2\hbar^3}\text{Tr}\{[\mathcal{H}_Z,\mu_x]^2\} - \frac{2}{2\hbar^3}\text{Tr}\{[\mathcal{H}_Z,\mu_x][\mathcal{H}_d^0,\mu_x]\}$$

$$-\frac{1}{2\hbar^3}\text{Tr}\{[\mathcal{H}_d^0,\mu_x]^2\} \quad (3.49)$$

where in the "cross-term" involving $\{[\mathcal{H}_Z,\mu_x]$ and $[\mathcal{H}_d^0,\mu_x]$ we have used the basic relation true for any pair of operators A and B:

$$\text{Tr}\{AB\} = \text{Tr}\{BA\} \quad (3.50)$$

which is readily proved by applying (3.23a). If the dipolar coupling were zero only the first term on the right would survive, and of course the resonance would be a δ-function at $\omega = \omega_0$. In this case $\langle\omega^2\rangle = \omega_0^2$. Therefore we see that the first term must contribute ω_0^2 to $\langle\omega^2\rangle$. Explicit evaluation in fact verifies this result. The second, or "cross", term vanishes, since every term involves factors such as $\text{Tr}_1\{\mu_{1x}\}$. The last term, when divided by $\int_0^\infty f(\omega)d\omega$ gives

$$\frac{3}{4}\gamma^4\hbar^2 I(I+1)\left(\frac{1}{N}\right)\sum_{j,k}\frac{(1-3\cos^2\theta_{jk})^2}{r_{jk}^6} \quad . \quad (3.51)$$

Now, by (3.21),

$$\langle\Delta\omega^2\rangle = \langle\omega^2\rangle - \langle\omega\rangle^2 \quad .$$

Therefore, since $\langle\omega\rangle = \omega_0$, we have

$$\langle \Delta\omega^2 \rangle = \frac{3}{4}\gamma^4\hbar^2 I(I+1)\frac{1}{N}\sum_{j,k}\frac{(1-3\cos^2\theta_{jk})^2}{r_{jk}^6} \quad . \tag{3.52}$$

We can get a clearer understanding of (3.52) by considering an example in which all spins are located in equivalent positions, so that

$$\sum_k \frac{(1-3\cos^2\theta_{jk})^2}{r_{jk}^6}$$

is independent of j. There are then N equivalent sums, one for each value of j, giving us

$$\langle \Delta\omega^2 \rangle = \frac{3}{4}\gamma^4\hbar^2 I(I+1)\sum_k\frac{(1-3\cos^2\theta_{jk})^2}{r_{jk}^6} \quad . \tag{3.53}$$

Each term is clearly of order $(\gamma H_{\text{loc}}^k)^2$ where H_{loc}^k is the contribution of the kth spin to the local field at spin j. The important point about (3.53) is that it gives a precise meaning to the concept of a local field, which enables one to compare a precisely defined theoretical quantity with experimental values.

So far we have considered only the second moment for a case where all nuclei are identical. If more than one species is involved, we get a somewhat different answer. The basic difference is in the terms of type B in the dipolar coupling that connect states such as $|+\ -\rangle$ to $|-\ +\rangle$.

If the two states are degenerate, as in the case when the spins are identical, B makes a first-order shift in the energy. On the other hand, when the states are nondegenerate, B merely produces second order energy shifts and gives rise to weak, otherwise forbidden transitions. It is therefore appropriate to omit B when the spins are unlike.[2]

The interactions between like and unlike nuclear spins may be compared and the second moment readily obtained. If we use the symbol I for the species under observation, and S for the other species, the effective dipolar coupling between like nuclei is

$$(\mathcal{H}_d^0)_{II} = \frac{1}{4}\gamma_I^2\hbar^2\sum_{k,l}\frac{(1-3\cos^2\theta_{kl})}{r_{kl}^3}(3I_{zk}I_{zl}-\boldsymbol{I}_k\cdot\boldsymbol{I}_l) \quad . \tag{3.54}$$

In computing the second moment for like spins, the terms $\boldsymbol{I}_k\cdot\boldsymbol{I}_l$ do not contribute, since they commute with μ_x [see (3.49)]. The coupling between unlike spins is

---

[2] *Van Vleck* points out that omitting these terms for unlike spins as well as the terms C, D, E, and F for like spins is crucial in computing $(\Delta\omega^2)$. The reason is that in computing $(\Delta\omega^2)$, the rather weak satellite lines at $\omega=0$ and $\omega=2\omega_0$ correspond to a typical frequency from the center of the resonance, which is H_0/H_{loc} larger than those of the main transition. The second moment measures the square of the frequency deviation. Therefore, although the satellites are down in intensity by $(H_{\text{loc}}/H_0)^2$, they contribute an amount quite comparable to the second moment. Since we are concerned with the width of the main transition, we do not wish to include the satellites. We must exclude the terms that produce them from the Hamiltonian.

$$(\mathcal{H}_d^0)_{IS} = \gamma_I\gamma_S\hbar^2 \sum_{k,l} \frac{(1 - 3\cos^2\theta_{kl})}{r_{kl}^3} I_{zk}S_{zl} \quad . \tag{3.55}$$

Equations (3.54) and (3.55) differ primarily in the numerical factor of the zz-term, (3.55) being small by a factor of $\frac{2}{3}$. This numerical factor becomes $\frac{4}{9}$ in the second moment, giving for the final answer:

$$\langle \Delta\omega^2 \rangle_{IS} = \frac{1}{3}\gamma_I^2\gamma_S^2\hbar^2 S(S+1)\frac{1}{N}\sum_{j,k}\frac{(1 - 3\cos^2\theta_{jk})^2}{r_{jk}^6} \quad . \tag{3.56}$$

Notice that it is $S(S+1)$, not $I(I+1)$, that comes into (3.56) expressing the fact that the local magnetic field seen by nuclei I is proportional to the magnetic moment $\gamma_S\hbar\sqrt{S(S+1)}$ of the other species. The total second moment of the resonance line of spin I is given by adding the second-moment contributions of like nuclei to those of unlike nuclei.

3.4 Example of the Use of Second Moments

Since the pioneering work of *Pake* and *Gutowsky*, numerous studies of second moments have been reported. A particularly interesting example is provided by the work of *Andrew* and *Eades* [3.1] on solid benzene. By studying the various isotopic compositions in which protons were replaced by deuterons, they were able to measure the proton-proton distance between adjacent protons in the ring and to show that at temperatures above about 90 K, the benzene molecules are relatively free to reorient about the axis perpendicular to the plane of the molecule. We shall describe their work.

The three isotopic species studied by *Andrew* and *Eades* are shown in Fig. 3.8. The structure of the benzene crystal is very similar to that of a face-centered cubic crystal with the benzene molecules on the corners and face centers of the cube. However, although the sides of the unit cell are perpendicular, they are not equal in length, the a-, b-, and c-axes being, respectively, 7.44 Å, 9.65 Å, and 6.81 Å. All benzenes have their planes parallel to the cystalline b-axis. A

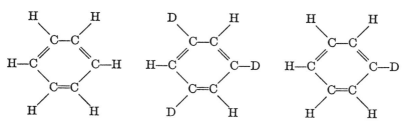

Fig. 3.8. Three species of benzene studied by *Andrew* and *Eades*

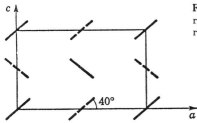

Fig. 3.9. Unit cell of the benzene crystal. Solid lines represent molecules in the $y = 0$ plane; dashed lines represent molecules $b/2$ above

rough sketch of the crystal structure is shown in Fig. 3.9, looking down the b-axis edge on to the plane of the molecules. The plane of the molecules is represented by straight lines, solid for those atoms in the $y = 0$ plane, dashed for those $b/2$ above the $y = 0$ plane. (Since the samples studied by *Andrew* and *Eades* were polycrystalline, studies of the effect of the orientation of the magnetic field relative to the crystalline axes were not possible.)

As we can see, there will be contributions to the second moment from nuclei within the same molecule and from nuclei outside the molecule. In principle, if one knew the location and orientation of all molecules, the only unknown parameter would be the distance R between adjacent protons in the ring. By using isotopic substitution, however, *Andrew* and *Eades* were able to obtain an *experimental* division of the total second moment into contributions within and outside. We can see this readily by noting that replacement of a proton by a deuteron on any given site reduces the contribution of that site to the second moment by the factor α:

$$\alpha = \frac{4}{9} \frac{\gamma_D^2 I_D(I_D + 1)}{\gamma_P^2 I_P(I_P + 1)} \tag{3.57}$$

where the subscripts P and D stand for the proton and the deuteron. By using the facts that $I_D = 1$, $I_P = \frac{1}{2}$, $(\gamma_D/2\pi) = 6.535 \times 10^2$, $(\gamma_P/2\pi) = 42.57 \times 10^2$, we have $\alpha = 0.0236$. Thus, consider S_1, the second-moment contribution from nuclei *outside* the molecule. For $C_6H_3D_3$, any given lattice position is equally likely to have a proton or a deuteron. Therefore the proton contribution to the second moment is cut by a factor of two. If all the lattice sites were occupied by deuterons, the second moment would be cut by the factor α, but since only one-half the sites are occupied by deuterons, the deuterons contribute $\alpha S_1/2$. The total second moment S_1' contributed by atoms outside the molecule is therefore

$$S_1' = \frac{S_1}{2} + \frac{\alpha S_1}{2} = \left(\frac{1+\alpha}{2}\right) S_1 \tag{3.58}$$

in which α is, of course, known.

The analysis for the contribution from atoms within the molecule proceeds in a similar way. Let S_2 and S_2' be the contribution for C_6H_6 and $C_6H_3D_3$,

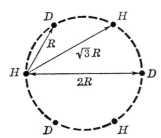

Fig. 3.10. Positions of protons and deuterons and the relative distances

respectively. S'_2 will be smaller than S_2, since the nuclei in positions 2, 4, and 6 will give only α times as big a contribution for the deuterated compound. By referring to Fig. 3.10 and recognizing the $1/r^6$ dependence of the contribution to the second moment, we see that

$$\frac{S'_2}{S_2} = \frac{\alpha\left[2 \times 1 + \left(\frac{1}{2}\right)^6\right] + 2 \times \left(\frac{1}{\sqrt{3}}\right)^6}{2 + \left(\frac{1}{2}\right)^6 + 2\left(\frac{1}{\sqrt{3}}\right)^6} = \frac{\left(1 + \frac{1}{128}\right)\alpha + \frac{1}{27}}{1 + \frac{1}{27} + \frac{1}{128}} = \delta \quad . \tag{3.59}$$

Thus we have for the second moments of $C_6H_3D_3$ and C_6H_6, respectively,

$$S' = S'_1 + S'_2 = \left(\frac{1+\alpha}{2}\right)S_1 + \delta S_2$$

$$S = S_1 + S_2 \tag{3.60}$$

where α and δ are known. Therefore, measurement of S and S' gives us S_1 and S_2 the separate contributions from outside and inside the molecule. The data for C_6H_5D provide an independent check.

On the basis of such studies *Andrew* and *Eades* determined the distance R between adjacent protons in the ring to be 2.495 ± 0.018 Å, which is consistent with a prediction of 2.473 ± 0.025 Å based on the C-C spacing as determined by x-rays, and an estimated value for the C-H bond length. Of course one can combine the x-ray and resonance data to obtain the C-H bond distance. In principle, observation of the C^{13} resonance would even permit a determination of the C-H bond length directly.

The data we have mentioned were measured at temperatures below about 90 K. A second important result of *Andrew* and *Eades* was deduced by their studies of the temperature dependence of the second moment (Fig. 3.11). The rapid drop in second moment is due to the rotation of the benzene molecules about their hexad axis. Let us discuss this effect.

The effect of rotation may be expressed very simply in terms of the angles defined in Fig. 3.12. We consider a pair of nuclei j, k fixed in a molecule, the axis of rotation of the molecule making an angle θ' with respect to the static field H_0.

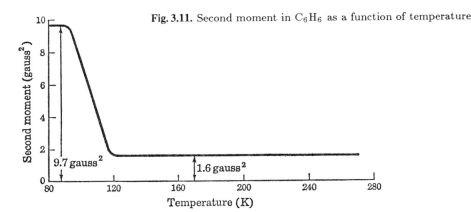

Fig. 3.11. Second moment in C_6H_6 as a function of temperature

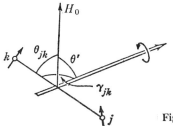

Fig. 3.12. Angles important in describing the rotation of a molecule

Let the radius vector from j to k make an angle γ_{jk} with the rotation axis. Then, as the molecule rotates, the angle θ_{jk} (between H_0 and the internuclear vector), which occurs in the factor $1 - 3\cos^2\theta_{jk}$ in the second moment, varies with time. Since the frequencies of rotation are high compared with the frequencies of interest in the resonance, it is the time average of $1 - 3\cos^2\theta_{jk}$ that affects the second moment. Assuming the motion is over a potential well of threefold or higher symmetry, this average can be shown to be independent of the details of the motion, and

$$\langle 1 - 3\cos^2\theta_{jk}\rangle_{\text{avg}} = (1 - 3\cos^2\theta')\frac{(3\cos^2\gamma_{jk} - 1)}{2} \ . \tag{3.61}$$

Equation (3.61) shows that if the axis of rotation is parallel to the internuclear axis ($\gamma_{jk} = 0$), in which case the relative position of the two nuclei is unaffected by the rotation, the angular factor is unaffected by the rotation. On the other hand, if $\gamma_{jk} = \pi/2$,

$$\langle 1 - 3\cos^2\theta_{jk}\rangle_{\text{avg}} = -\tfrac{1}{2}(1 - 3\cos^2\theta') \ . \tag{3.62}$$

In a powder sample, we find all orientations of the crystal axes with respect to H_0. For a rigid lattice, we must therefore average $(1 - 3\cos^2\theta_{jk})^2$ over the random crystal orientations. When motion sets in, we must first average $1 - 3\cos^2\theta_{jk}$ over the motion, to obtain the second moment for a given crystal orientation. Then we must average over crystal orientations.

For interacting pairs, the contribution to the second moment of the rigid lattice $\langle\Delta\omega^2\rangle_{\text{RL}}$ then goes as

$$\langle\Delta\omega^2\rangle_{\text{RL}} \propto \overline{(1 - 3\cos^2\theta_{jk})^2} \tag{3.63}$$

where the bar indicates an average over random orientations of θ_{jk}.

When rotation sets in, we have a second moment from the pair $\langle\Delta\omega^2\rangle_{\text{rot}}$ given by

$$\langle\Delta\omega^2\rangle_{\text{rot}} \propto \overline{(\langle 1 - 3\cos^2\theta_{jk}\rangle_{\text{avg}})^2} \tag{3.64}$$

where the "avg" indicates an average over rotation, and the bar indicates an average of the orientation of the rotation axis with respect to H_0. By using (3.61), we get

$$\langle\Delta\omega^2\rangle_{\text{rot}} \propto \overline{(1 - 3\cos^2\theta')^2}\left(\frac{3\cos^2\gamma_{jk} - 1}{2}\right)^2 . \tag{3.65}$$

Since the crystal axes are randomly oriented with respect to H_0, so are the *rotation* axes, specified by θ'. As a result

$$\overline{(1 - 3\cos^2\theta_{jk})^2} = \overline{(1 - 3\cos^2\theta'^2)} \quad \text{and}$$

$$\langle\Delta\omega^2\rangle_{\text{rot}} = \langle\Delta\omega^2\rangle_{\text{RL}}\left(\frac{3\cos^2\gamma_{jk} - 1)}{2}\right)^2 . \tag{3.66}$$

If $\gamma_{jk} = \pi/2$ (a pair rotating about an axis perpendicular to the internuclear axis), the contribution of the pair interaction to the second moment is reduced by a factor of 4.[3]

---

[3] The justification for averaging $1 - 3\cos^2\theta_{jk}$ over the motion *before* squaring rather than averaging $(1 - 3\cos^2\theta_{jk})^2$ may be seen also by referring to the exact expression for $\chi''(\omega)$, which was proportional to

$$\sum_{a,b} e^{-E_a/kT} (a|\mu_x|b)(b|\mu_x|a)\delta(E_a - E_b - \hbar\omega) .$$

The states $|a\rangle$ and $|b\rangle$ may be considered to involve both spin and rotational quantum numbers. But, since $E_a - E_b$ is chosen to be near the Larmor frequency, the states $|a\rangle$ and $|b\rangle$ must have the same rotational quantum numbers. Therefore, in computing the second moment, the trace will be over spin variables only, but the angular factor will be a diagonal matrix element in the "lattice" coordinates. But this means that we replace the classical $1 - 3\cos^2\theta_{jk}$ by $\int u_L^*(1 - 3\cos^2\theta_{jk})u_L d\tau$, where u_L is a lattice (in this case, rotation) state. This procedure amounts to "averaging" $1 - 3\cos^2\theta_{jk}$ over the motion prior to squaring.

For the case of benzene, *Andrew* and *Eades* found that the second moment of C_6H_6 from protons within the molecule drops from $3.10\,\text{Gauss}^2$ at low temperatures to $0.77 \pm 0.05\,\text{Gauss}^2$ at high temperatures. The assumption that the narrowing results from rotation about the hexad axis makes $\gamma_{jk} = \pi/2$, since all protons lie in a plane perpendicular to the hexad axis, and predicts that the second moment should drop to $3.10/4 = 0.78\,\text{Gauss}^2$, in excellent agreement with the observed decrease.

4. Magnetic Interactions of Nuclei with Electrons

Chemical shift:

4.1 Introduction

So far we have ignored the fact that the nuclei are surrounded by electrons with which they can interact. In this chapter we shall consider the magnetic interactions, postponing until later the consideration of the strong electrostatic effects that may be found when a nucleus possesses an electrical quadrupole moment. The magnetic coupling of the electrons to the nucleus arises from magnetic fields originating either from the motion of the electrical charges or from the magnetic moment associated with the electron spin. The former gives rise to the so-called chemical shifts; the latter, to the Knight shifts in metals and to a coupling between nuclear spins.

Both the chemical shifts and the Knight shifts have certain features in common. The total Hamiltonian of the electrons and the nuclei may be written as a sum of four terms:

$$\mathcal{H}_{nZ}(H) + \mathcal{H}_e(0) + \mathcal{H}_{eZ}(H) + \mathcal{H}_{en}$$

where $\mathcal{H}_{nZ}$ is the nuclear Zeeman coupling in the applied field H; $\mathcal{H}_e(0)$ is the Hamiltonian of the electrons (orbital and spin) in the absence of H; $\mathcal{H}_{eZ}(H)$ is the electron Zeeman energy; and $\mathcal{H}_{en}$ is the interaction between the nuclear spins and the electron orbital and spin coordinates.

If $\mathcal{H}_{en}$ were zero, the nuclear spin system would be decoupled from the electrons, and the nuclear energy levels would be solely the Zeeman levels in the applied field H. The term $\mathcal{H}_{en}$ corresponds to the extra magnetic fields the nuclei experience owing to the electrons. In a diamagnetic or paramagnetic substance, the average field a nucleus experiences owing to the electrons vanishes when H vanishes. However, since the interaction $\mathcal{H}_{eZ}(H)$ polarizes the electron system, the effect of the electron-nuclear coupling $\mathcal{H}_{en}$ is no longer zero. We may say that the nuclei experience both a direct interaction with H through $\mathcal{H}_{nZ}(H)$ and an indirect one through the interplay of $\mathcal{H}_{eZ}(H)$ and $\mathcal{H}_{en}$. The problem is very similar to the calculation of the electric field in a dielectric, in which we must add to the applied electric field the field arising from induced dipole moments in the other atoms.

Systems such as ferromagnets possess electronic magnetization even with $H = 0$. For them, the contribution of $\mathcal{H}_{en}$ is nonzero even without an applied field.

We shall consider the orbital effects first, starting with a review of the major facts about chemical shifts.

4.2 Experimental Facts About Chemical Shifts

The most famous and most quoted example of chemical shifts is ethyl alcohol, CH_3CH_2OH (see references on "Chemical Shifts" in the Bibliography). The proton resonance consists of three lines whose intensities are in the ratios $3:2:1$. If one possesses a highly homogeneous magnet, each of these lines is found to possess structure that (as we shall see) is due to effects of electron spin. The three lines are clearly due to the three "types" of protons, three in the CH_3 group, two in the CH_2 group, and one in the OH. Evidently the nuclei experience fields of local origin that are different for different molecular surroundings. A comparison of the spacing in magnetic field between the lines as a function of the frequency of the resonance apparatus shows that the splitting is proportional to the frequency. If we attribute the splitting to the fact that the nuclei must see a magnetic field ΔH in addition to the applied field H_0, we may say the resonance frequency ω obeys the equation

$$\omega = \gamma(H_0 + \Delta H) \tag{4.1}$$

where $\Delta H \propto H_0$. We may therefore define a quantity σ, which is independent of H, by the equation

$$\Delta H = -\sigma H_0 \quad . \tag{4.2}$$

If σ is positive, we must use a larger magnetic field to produce the resonance than would be necessary for the bare nucleus. Of course we never do experiments on a bare nucleus, so that what we measure are the differences in σ associated with different molecular environments. For protons, the entire range of σ's covers about one part in 10^5. For fluorine atoms, however, the range is about six parts in 10^4, two orders of magnitude larger. Because of the small size of the shifts, they are ordinarily studied in liquids where resonance lines are narrow. Since the shifts should in general depend upon the orientation of the molecule with respect to the static field, single crystal orientation studies are of interest (see Chap. 8).

As we have remarked, the chemical shifts are due to the orbital motion of electrons. It is important to contrast the orbital motion in solids or molecules with that in free atoms. We shall turn to this subject next.

4.3 Quenching of Orbital Motion

Classical electricity and magnetism tell us that a charge q moving with velocity v produces a magnetic field H at a point r' away, given by

$$H = \frac{q}{c} \frac{v \times r'}{r'^3} . \qquad (4.3)$$

If we choose rather to ask for the field at the origin of a set of coordinates due to a charge at position r, then $r' = -r$ and (4.3) becomes

$$H = \frac{q}{c} \frac{r \times v}{r^3} = \frac{q}{mc} \frac{r \times mv}{r^3} = \frac{q}{mc} \frac{L}{r^3} \qquad (4.4)$$

where L is the angular momentum of the particle about the origin. Equation (4.4) has a quantum mechanical counterpart, as we shall discuss. We see immediately, however, that for s-states, $H = 0$ at the position of a nucleus, since s-states have zero angular momentum, whereas $H \neq 0$ for p, d, and other states of nonzero angular momentum. The magnitude of H is of order

$$H \simeq \frac{\beta}{r^3} \qquad (4.5)$$

where β is the Bohr magneton (10^{-20} erg/Gauss). For fluorine the average value of $1/r^3$ for the $2p$ electrons is

$$\overline{\left(\frac{1}{r^3}\right)}_{2p} = \frac{8.9}{a_0^3} \qquad (4.6)$$

where a_0 is the Bohr radius. In other words, $\overline{(1/r^3)}$ corresponds to a typical distance of $\frac{1}{4}$ Å; the magnetic fields, to about 600,000 Gauss.

Such enormous fields would completely dominate the laboratory field H_0 for typical experiments, in contrast to the facts. (Of course, in atomic beam experiments such large couplings *can* be observed.) We must understand why the large fields of free atoms are not present in solids or molecules. The disappearance of these large fields is also closely associated with the fact that, in most substances, the atoms do not possess permanent electronic magnetic moments; that is, most substances are diamagnetic. The term *quenching of orbital angular momentum* is often applied to describe the phenomenon. Let us see how it comes about, by studying a particularly simple example.

We shall consider an atom with one electron outside closed shells in a p-state. We shall neglect spin, for convenience, although later in the book we shall return to the effect of spin in order to understand the so-called g-shifts in electron spin resonance. The three degenerate p-functions may be written in either of two ways:

$$xf(r) \quad , \quad yf(r) \quad , \quad zf(r) \quad \text{or} \qquad (4.7)$$

$$\frac{(x+iy)}{\sqrt{2}}f(r) \quad , \quad zf(r) \quad , \quad \frac{(x-iy)}{\sqrt{2}}f(r) \tag{4.8}$$

where $f(r)$ is a spherically symmetric function. The three functions of (4.8) are eigenfunctions of L_z, the z-component of angular momentum and the m-values being, from left to right, 1, 0, and -1. The wave functions of (4.7) are simply linear combinations of those of (4.8). As long as the atom is free, either set of wave functions is equally good, but if a magnetic field is applied parallel to the z-direction, the set of (4.8) must be chosen.

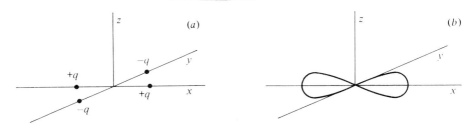

Fig. 4.1. (a) Four charges placed near an atom. The atom is assumed to be at the origin and the charges to be all equidistant from the origin. The charges $+q$ lie on the x-axis and $-q$ lie on the y-axis. (b) The wave function $xf(r)$ is largest along the x-axis for any given distance r from the origin

If now we surround the atom by a set of charges in the manner of Fig. 4.1, and for the moment assume that no static magnetic field is present, the degeneracy is lifted. The proper eigenstates are then those of (4.7), since a symmetric potential such as that of Fig. 4.1 will have vanishing matrix elements between any pair of the functions of (4.7). On the other hand, the diagonal matrix elements will be different, since the state $xf(r)$ concentrates the electron on the x-axis, near to the positive charges, whereas the state $yf(r)$ concentrates the electron near the negative charge. Clearly, $xf(r)$ will lie lowest in energy, $yf(r)$ will be highest, and $zf(r)$ will be between (unshifted in fact in the first order). The resulting energy levels are shown in Fig. 4.2.

The ground state $xf(r)$ may be written as

$$xf(r) = \frac{1}{\sqrt{2}}\left[\frac{(x+iy)}{\sqrt{2}}f(r) + \frac{(x-iy)}{\sqrt{2}}f(r)\right] \quad , \tag{4.9}$$

a linear combination of the $m = +1$ and $m = -1$ states. The states $m = +1$ and $m = -1$ correspond to electron circulation in opposite directions about the z-axis. Since they occur with *equal* weighting in (4.9), we see that $xf(r)$ corresponds to equal mixtures of the two senses of circulation, or to *zero* net circulation.

Fig. 4.2. Splitting of the three p-states by the charges of Fig. 4.1a

We can make a more precise statement by computing $\langle L_z \rangle$, the expectation value of the z-component of angular momentum. For generality, we shall make our proof for any wave function whose spatial part is real. The operator for L_z is

$$L_z = \frac{\hbar}{i}\left(x\frac{\partial}{\partial y} - y\frac{\partial}{\partial x}\right) \quad . \tag{4.10}$$

Therefore, for any wave function u_0,

$$(0|L_z|0) = \int u_0^* \frac{\hbar}{i}\left(x\frac{\partial}{\partial y} - y\frac{\partial}{\partial x}\right)u_0 \, d\tau \tag{4.11}$$

which, since u_0 is real, may be written as

$$(0|L_z|0) = \frac{\hbar}{i}\int u_0\left(x\frac{\partial}{\partial y} - y\frac{\partial}{\partial x}\right)u_0 \, d\tau \quad . \tag{4.12}$$

Since all the quantities in the integral are real, (4.12) shows that $\langle L_z \rangle$ must be pure imaginary unless the integral vanishes. But the diagonal matrix elements of any Hermitian operator are real. Therefore the integral vanishes and

$$(0|L_z|0) = 0 \quad . \tag{4.13}$$

It is clear that this proof holds for any component of angular momentum.

We say that when

$$(0|L_x|0) = (0|L_y|0) = (0|L_z|0) = 0 \tag{4.14}$$

the angular momentum is quenched.

Under what circumstances will the angular momentum be quenched? Clearly, what is needed is the possibility of choosing the eigenfunctions as real. In the absence of a magnetic field (which means that no spins are allowed to act on the orbit!), the Hamiltonian is real. If, moreover, a state is nondegenerate, its eigenfunction is always real, since it is the solution of a real differential equation (apart from an arbitrary complex constant factor, which clearly does not affect the expectation value). Therefore $(0|L_x|0) = 0$ for such a state. We conclude that whenever the crystalline electric fields leave a state nondegenerate, the orbital angular momentum of that state is quenched. The physical basis of quenching is that the external charges exert torques on the electron orbit, causing the plane of the orbit to precess. When the plane has exactly turned over, the sense of circulation is reversed. Crudely stated, the electron path has been changed from lying in a plane to being much like the path of the string in a ball of twine.

Of course application of a magnetic field will change things. We can see intuitively that a magnetic field in the z-direction will cause one sense of circulation to be favored over the other. The wave functions will readjust so that the ground state has a slight circulation in the favorable sense (the $m = -1$ state will be favored). In terms of a small quantity ε, this will make a new ground state:

$$\psi_0 = \frac{1}{\sqrt{2}}\left[(1-\varepsilon)\frac{(x+iy)}{\sqrt{2}}f(r) + (1+\varepsilon)\frac{(x-iy)}{\sqrt{2}}f(r)\right] \quad . \tag{4.15}$$

As we can see, this change results from the admixture of a small amount of the state $yf(r)$ into the ground state $xf(r)$. As we shall see, the amount ε mixed in is proportional to H_0, giving rise to a circulation that is proportional to H_0.

We turn now to a closer look at the details of chemical shifts.

4.4 Formal Theory of Chemical Shifts [1]

Chemical shifts arise because of the simultaneous interaction of a nucleus with an electron and that of the electron with the applied field H_0. A general theory has been given by *Ramsey* [4.1], but we shall present a somewhat different discussion, which breaks the calculation into two parts: 1) the determination of the electric currents produced in the molecule by the external field and 2) calculation of the magnetic field produced by these currents at the nucleus. We shall work out the theory for one electron. We start by considering the Hamiltonian of the electron. To treat the magnetic fields, we must introduce two vector potentials, A_0 and A_n, one associated with the magnetic field H_0, the other with the magnetic field H_n due to the nucleus. In terms of A_0 and A_n we have

$$H_0 = \nabla \times A_0 \quad , \quad H_n = \nabla \times A_n \quad . \tag{4.16}$$

As is well known, there is more than one vector potential that will produce a given field. Thus, if $H = \nabla \times A$, a new vector potential $A' = A + \nabla \phi$ (where ϕ is any scalar function), will give the same field, since the curl of the gradient of any function vanishes. A transformation from A to A' is called a *gauge transformation*. We must be sure that the physical results of any calculation are independent of the choice of gauge, that is, are gauge-invariant. The effect of a magnetic field is introduced into the Schrödinger equation by replacing the operator $(\hbar/i)\nabla$ by $(\hbar/i)\nabla - (q/c)A$, where q is the charge of the particle (q is positive or negative, depending on the sign of the charge of the particle).

The Hamiltonian then becomes

$$\mathcal{H} = \frac{1}{2m}\left(p - \frac{q}{c}A\right)^2 + V \tag{4.17}$$

where $p \equiv (\hbar/i)\nabla$. If one uses a different gauge, $A' = A + \nabla\phi(r)$, the new solution ψ' is related to the old one (ψ) by the (unitary) transformation

$$\psi' = \psi \exp\left[+(iq/\hbar c)\phi(r)\right] \quad . \tag{4.18}$$

If we compare the expectation values of $(\hbar/i)\nabla$, $(\psi,(\hbar/i)\nabla\psi)$, and $(\psi',(\hbar/i)\nabla\psi')$, we find that they are not equal. Therefore, since any physical observable must be independent of the choice of gauge, we see that $(\hbar/i)\nabla$ cannot be the momentum operator mv. The operator for mv, rather, is $(\hbar/i)\nabla - (q/c)A$, which is gauge-invariant. That is,

---

[1] See references under "Chemical Shifts" in the Bibliography.

$$\left(\psi', \left(\frac{\hbar}{i}\nabla - \frac{q}{c}A'\right)\psi'\right) = \left(\psi, \left(\frac{\hbar}{i}\nabla - \frac{q}{c}A\right)\psi\right) \quad . \tag{4.19}$$

In a similar manner the operator for angular momentum, $r \times mv$, is

$$r \times \left(\frac{\hbar}{i}\nabla - \frac{q}{c}A\right) \quad .$$

The distinction between mv and $p(= (\hbar/i)\nabla)$ is found in classical mechanics. Thus in terms of the Lagrangian L, the definition of the canonical momentum p_x is $\partial L/\partial \dot{x}$, whereas the x-component of linear momentum is $m\dot{x}$. When a magnetic field is present, one finds $p_x = m\dot{x} + [(q/c)A_x]$.

A quantity that will be of great importance to us is something which we shall call the current density $j(r)$. It is defined as follows:

$$j(r) = \frac{q}{2m}\frac{\hbar}{i}(\psi^*\nabla\psi - \psi\nabla\psi^*) - \frac{q^2}{mc}A\psi^*\psi \quad . \tag{4.20}$$

We note that $j(r)$ is a vector function of position, and it is real (that is, has zero imaginary part). We recognize it as being q times the quantum mechanical probability current. Explicit evaluation, first using ψ and A and then using ψ' and A', shows that $j(r)$ is gauge-invariant. Moreover, by assuming that ψ is a solution of Schrödinger's time-dependent equation, one can show that

$$\text{div}\, j + \frac{\partial \varrho}{\partial t} = 0 \quad \text{with} \quad \varrho \equiv q\psi^*\psi \quad . \tag{4.21}$$

That is, j obeys the classical equation of continuity. For stationary states, $\psi^*\psi$ is independent of time and div $j = 0$. j acts much as a classical current density. As we shall see shortly, such an interpretation is very useful in considering chemical shifts.

The Hamiltonian for our electron acted on by two magnetic fields is therefore

$$\mathcal{H} = \frac{1}{2m}\left(p - \frac{q}{c}A_0 - \frac{q}{c}A_\mathrm{n}\right)^2 + V \tag{4.22}$$

where V represents all potential energy, including that due to fields that may quench the orbital angular momentum.

It is convenient to define a quantity π by the equation

$$\pi = p - \frac{q}{c}A_0 \quad . \tag{4.23a}$$

Since both p and A_0 are Hermitian operators, so is π. Then (4.22) becomes

$$\mathcal{H} = \frac{1}{2m}\pi^2 - \frac{q}{2mc}(\pi \cdot A_\mathrm{n} + A_\mathrm{n} \cdot \pi) + \frac{q^2}{2mc^2}A_\mathrm{n}^2 + V \quad . \tag{4.23b}$$

We shall choose A_n to be

$$A_\mathrm{n} = \frac{\mu \times r}{r^3} \tag{4.24}$$

where $\boldsymbol{\mu}$ is the nuclear moment, since this vector potential generates the field of a dipole. Since μ is very small compared with electron moments, we expect to be able to treat it as an expansion parameter, and accordingly we drop the A_n^2 term in comparison with the term linear in A_n. We then have

$$\mathcal{H} = \frac{1}{2m}\pi^2 + V - \frac{q}{2mc}(\boldsymbol{\pi}\cdot\boldsymbol{A}_n + \boldsymbol{A}_n\cdot\boldsymbol{\pi}) \quad . \tag{4.25}$$

In the absence of a nuclear coupling, $(1/2m)\pi^2 + V$ is simply the Hamiltonian of the electron in the presence of the static field. We shall treat the term involving A_n as a perturbation, computing the energy by using first-order perturbation theory.

Let us consider, then, the first-order change in energy of a state whose wave function ψ is the *exact* solution of the problem of an electron acted on by the potential V and the static field $\boldsymbol{H}_0$. The energy perturbation E_{pert} is then

$$E_{\text{pert}} = -\frac{q}{2mc}\int \psi^*(\boldsymbol{\pi}\cdot\boldsymbol{A}_n + \boldsymbol{A}_n\cdot\boldsymbol{\pi})\psi\, d\tau \tag{4.26}$$

where the integration is over electron coordinates. (Actually, $\boldsymbol{A}_n$ is a function of the nuclear moment $\boldsymbol{\mu}$, which must itself be considered an operator. Thus E_{pert} will be an operator as far as the nuclear spin is concerned. We simply add it into the nuclear spin Hamiltonian.)

By using the fact that $\boldsymbol{\pi}$ is a Hermitian operator, we rewrite (4.26) as

$$E_{\text{pert}} = -\frac{q}{2mc}\int \boldsymbol{A}_n \cdot [(\boldsymbol{\pi}\psi)^*\psi + \psi^*\boldsymbol{\pi}\psi]d\tau \quad . \tag{4.27}$$

But, by using the definition of $\boldsymbol{\pi}$ given in (4.23) and of the current density in (4.20), we can write

$$\frac{q}{2m}[\psi(\boldsymbol{\pi}\psi)^* + \psi^*\boldsymbol{\pi}\psi] = \frac{q}{2m}\frac{\hbar}{i}(\psi^*\nabla\psi - \psi\nabla\psi^*) - \frac{q^2}{mc}\boldsymbol{A}_0\psi^*\psi$$
$$= \boldsymbol{j}_0(r) \quad . \tag{4.28}$$

$\boldsymbol{j}_0(r)$ is the current density flowing when the static field is on. That is, $\boldsymbol{j}_0$ is the current computed for the electron acted on by V and $\boldsymbol{H}_0$ (but not by the nucleus). Therefore

$$E_{\text{pert}} = -\frac{1}{c}\int \boldsymbol{A}_n\cdot\boldsymbol{j}_0(r)d\tau \quad . \tag{4.29}$$

[Parenthetically, this formula gives us a general expression for the change in energy δE resulting from a change in field associated with a change $\delta\boldsymbol{A}$ in vector potential, in terms of the current $\boldsymbol{j}(r)$ prior to the change $\delta\boldsymbol{A}$:

$$\delta E = -\frac{1}{c}\int \delta\boldsymbol{A}\cdot\boldsymbol{j}(r)d\tau] \quad . \tag{4.30}$$

If we now set

$$\boldsymbol{A}_n = \frac{\boldsymbol{\mu}\times\boldsymbol{r}}{r^3} \tag{4.31}$$

we get

$$E_{\text{pert}} = -\frac{1}{c} \int \frac{\mu \times r}{r^3} \cdot j_0(r) d\tau$$
$$= -\mu \cdot \left[\frac{1}{c} \int \frac{r \times j_0(r)}{r^3} d\tau \right] \quad (4.32)$$

where, as we have remarked, μ is really the operator $\gamma \hbar I$, but $j_0(r)$ is simply a vector function of position. It is important to bear in mind that $j_0(r)$ is independent of the gauge of A_0, the vector potential of the static field. Equation (4.32) is identical in form to the classical interaction of a magnetic moment μ with a current density $j_0(r)$, since the quantity in the square brackets is the field H due to the current.

There is a good deal of similarity to the expression for the magnetic moment M of the electrons:

$$M = \frac{1}{2c} \int r \times j_0(r) d\tau \quad . \quad (4.33)$$

Equation (4.32) contains the facts of the chemical shift. If we knew $j_0(r)$, we could compute the resultant field at the nucleus. We can see that there are really two parts to the theory of chemical shifts: 1) finding the current density $j_0(r)$; 2) computing the integral of (4.32) once $j_0(r)$ is known. The latter problem is entirely classical and immediately involves one in such things as multipole expansions. Thus the effect of currents on an atom distant from the nucleus in question can often be approximated by a magnetic dipole moment.

Since in general the current $j_0(r)$ flows as a result of the presence of the static field H_0, determination of $j_0(r)$ from first principles involves the solution of the quantum mechanics problem of an electron acted on by electrostatic potentials and by a static field. On the other hand, in some instances one can guess the spatial form of $j_0(r)$ and use measured magnetic susceptibilities to fix its magnitude, a technique that has been used to explain the chemical shifts of protons in various ring compounds such as benzene, in which the currents in the rings are computed to give agreement with experimental (or theoretical) magnetic moments. Alternatively, one can turn the problem around and use the measured chemical shifts to determine information about magnetic susceptibilities of atoms, molecules, or bonds. Moreover, we can see that in general the chemical shifts will be most sensitive to nearby currents because of the $1/r^3$ factor in the integral of (4.32), unless nearby currents are especially small. We shall see shortly that the small chemical shifts of protons compared, for example, to fluorine atoms results from the fact that the currents near the protons are relatively very small. In any event, (4.32) and (4.33) give concise statements of what a chemical shift or susceptibility experiment measures about the currents induced in a molecule by the external field.

4.5 Computation of Current Density

We turn now to computing $j_0(r)$. To do so, we need the wave function ψ, which describes the electron when acted on by both the electrostatic potentials and the static field. We have, then,

$$\mathcal{H}\psi = E\psi \quad \text{where} \tag{4.34}$$

$$\mathcal{H} = \frac{1}{2m}\left(p - \frac{q}{c}A_0\right)^2 + V \quad . \tag{4.35}$$

By expanding the parentheses, we find

$$\mathcal{H} = \frac{p^2}{2m} + V - \frac{q}{2mc}(p \cdot A_0 + A_0 \cdot p) + \frac{q^2}{2mc^2}A_0^2 \quad . \tag{4.36}$$

Let us assume that we know the wave functions that are the solution of the Hamiltonian $\mathcal{H}_0$ in the absence of the external field:

$$\mathcal{H}_0 = \frac{p^2}{2m} + V \quad , \quad \mathcal{H}_0\psi_n = E_n\psi_n \quad . \tag{4.37}$$

Then we can look on the terms in (4.36) that involve A_0 as perturbing the energies and wave functions. We shall compute perturbed wave functions so that we can compute the effect of the magnetic field on the current density. Of course A_0 goes to zero when H_0 vanishes, being given typically by

$$A_0 = \tfrac{1}{2}H_0 \times r \quad . \tag{4.38}$$

Although this involves a particular gauge, we can see that in any gauge, A_0 will be proportional to H_0. In the expression for the current $j_0(r)$,

$$j_0(r) = \frac{\hbar q}{2mi}(\psi^*\nabla\psi - \psi\nabla\psi^*) - \frac{q^2}{mc}A_0\psi^*\psi \quad , \tag{4.39}$$

we can compute $j_0(r)$ correctly to terms linear in H_0 as a first approximation. To do this, we need ψ and ψ^* correct to terms linear in H_0 for that part of $j_0(r)$ in the parentheses, but for the last term, we can use for ψ the unperturbed function ψ_0. Since we always have

$$\psi_0' = \psi_0 + \sum_n \frac{\langle n|\mathcal{H}_{\text{pert}}|0\rangle}{E_0 - E_n}\psi_n \quad , \tag{4.40}$$

we need keep only those parts of the perturbation that are linear in H_0, or by referring to (4.36–38), we take

$$\mathcal{H}_{\text{pert}} = -\frac{q}{2mc}(p \cdot A_0 + A_0 \cdot p) \quad . \tag{4.41}$$

Let us define

$$\varepsilon_{n0} = \frac{\langle n|\mathcal{H}_{\text{pert}}|0\rangle}{E_0 - E_n} \quad . \tag{4.42}$$

Then
$$\psi_0' = \psi_0 + \sum_n \varepsilon_{n0} \psi_n \tag{4.43}$$

which gives us for the current,

$$\begin{aligned}
\mathbf{j}_0(\mathbf{r}) &= \frac{\hbar q}{2mi}(\psi_0^* \nabla \psi_0 - \psi_0 \nabla \psi_0^*) + \sum_n \frac{\hbar q}{2mi}(\psi_0^* \nabla \psi_n - \psi_n \nabla \psi_0^*)\varepsilon_{n0} \\
&\quad + \sum_n \frac{\hbar q}{2mi}(\psi_n^* \nabla \psi_0 - \psi_0 \nabla \psi_n^*)\varepsilon_{n0}^* - \frac{q^2}{mc}\mathbf{A}_0 \psi_0^* \psi_0 \quad .
\end{aligned} \tag{4.44}$$

The term
$$\frac{\hbar}{2mi} q(\psi_0^* \nabla \psi_0 - \psi_0 \nabla \psi_0^*) \equiv \mathbf{J}(\mathbf{r}) \tag{4.45}$$

is the current that would flow when $H_0 = 0$.

When the orbital angular momentum is quenched so that ψ_0 is real, we see that $\mathbf{J}(\mathbf{r}) = 0$, and the current density vanishes at all points in the molecule in the absence of $\mathbf{H}_0$.

It is the term $\mathbf{J}(\mathbf{r})$ that gives rise, however, to the magnetic fields at a nucleus originating in the bodily rotation of the molecule, that is, the so-called spin-rotation interactions that are observed in molecular beam experiments.

It is instructive to compute $\mathbf{j}_0(\mathbf{r})$ for a *free* atom in the $m = +1$ p-state, H_0 being zero. $\mathbf{j}_0(\mathbf{r})$ then equals $\mathbf{J}(\mathbf{r})$, so that

$$\mathbf{j}_0(\mathbf{r}) = \frac{\hbar q}{2mi}(\psi_0^* \nabla \psi_0 - \psi_0 \nabla \psi_0^*) \quad . \tag{4.46}$$

But
$$\psi_0 = \left(\frac{x+iy}{\sqrt{2}}\right) f(r) \tag{4.47}$$

$$\nabla \psi_0 = \left(\frac{\mathbf{i}+i\mathbf{j}}{\sqrt{2}}\right) f(r) + \left(\frac{x+iy}{\sqrt{2}}\right) \nabla f(r)$$

giving
$$\begin{aligned}
\mathbf{j}_0(\mathbf{r}) &= \frac{\hbar q}{2m}(x\mathbf{j} - y\mathbf{i}) f^2(r) \\
&= \frac{\hbar q}{2m} \mathbf{k} \times \mathbf{r} f^2(r) \quad .
\end{aligned} \tag{4.48}$$

The current therefore flows in circles whose plane is perpendicular to the z-axis. If we define a velocity $\mathbf{v}(\mathbf{r})$ by the equation

$$\mathbf{v}(\mathbf{r}) = \frac{\mathbf{j}_0(\mathbf{r})}{q\psi^*\psi} \tag{4.49}$$

where $q\psi^*\psi$ is the charge density, we find

$$\mathbf{v}(\mathbf{r}) = \frac{\hbar}{m} \frac{\mathbf{k} \times \mathbf{r}}{(x^2 + y^2)} \tag{4.50}$$

which is tangent to a circle whose plane is perpendicular to the z-axis, so that

$$|v(r)| = \frac{\hbar}{m}\frac{1}{\sqrt{x^2+y^2}} \quad . \tag{4.51}$$

This gives a z angular momentum of

$$mv\sqrt{x^2+y^2} = \hbar \tag{4.52}$$

in accordance with our semiclassical picture of the electron in an $m = +1$ state possessing one quantum of angular momentum.

We see, therefore, the close relationship in this case of the current density, the "velocity", and our semiclassical picture of quantized orbits.

When the states ψ_0 and ψ_n may be taken as real (quenched orbital angular momentum), we have $J(r) = 0$, and

$$j_0(r) = \frac{\hbar}{2mi}q\sum_n(\varepsilon_{n0} - \varepsilon_{n0}^*)(\psi_0\nabla\psi_n - \psi_n\nabla\psi_0) - \frac{q^2}{mc}A_0\psi_0^2 \quad . \tag{4.53}$$

For (4.53) to be valid, it is actually necessary only that the *ground* state possess quenched orbital angular momentum, but for excited states, we have assumed that the real form of the wave functions has been chosen.

Let us now proceed to look at some examples. We shall consider two cases, an s-state and a p-state. It will turn out that the chemical shifts for s-states are very small but that, for p-states, the effect of the magnetic field in unquenching the orbital angular momentum plays the dominant role, giving chemical shifts two orders of magnitude larger than those typically found for s-states.

To proceed, we must now choose a particular gauge for A_0. It turns out, as we shall see, to be particularly convenient to take

$$A_0 = \tfrac{1}{2}H_0 \times r = \tfrac{1}{2}H_0 k \times r \tag{4.54}$$

although an equally correct one would be

$$A_0 = \tfrac{1}{2}H_0 \times (r - R) \tag{4.55a}$$

where R is a constant vector, or

$$A_{0z} = 0 \quad , \quad A_{0x} = H_0 y \quad , \quad A_{0y} = 0 \quad . \tag{4.55b}$$

In terms of the A_0, (4.54), we have

$$\text{div }A_0 = 0 \quad . \tag{4.56}$$

Then we have, from (4.41),

$$\mathcal{H}_\text{pert} = -\frac{q}{2mc}[A_0 \cdot p + (p \cdot A_0) + A_0 \cdot p] \tag{4.57}$$

where $(p \cdot A_0)$ means p acts solely on A_0. But since $p = (\hbar/i)\nabla$,

$$(p \cdot A_0) = \frac{\hbar}{i}(\nabla \cdot A_0) = 0 \quad . \tag{4.58}$$

Then, using (4.54), we have

$$\mathcal{H}_{\text{pert}} = -\frac{q}{2mc}(\boldsymbol{H}_0 \times \boldsymbol{r}) \cdot \boldsymbol{p}$$
$$= -\frac{q}{2mc}\boldsymbol{H}_0 \cdot (\boldsymbol{r} \times \boldsymbol{p}) \quad . \tag{4.59a}$$

We recognize that $\boldsymbol{r} \times \boldsymbol{p}$ is the operator for angular momentum in the absence of H_0. It is convenient in computing matrix elements to use the dimensionless operator $(1/i)\boldsymbol{r} \times \nabla$ for angular momentum. Denoting this by the symbol $\boldsymbol{L}$, we can write (4.59a) alternatively as

$$\mathcal{H}_{\text{pert}} = -\frac{q\hbar}{2mc}H_0 L_z \quad . \tag{4.59b}$$

Had we chosen the gauge of (4.55a), we would have

$$\mathcal{H}_{\text{pert}} = -\frac{q}{2mc}\boldsymbol{H}_0 \cdot (\boldsymbol{r} - \boldsymbol{R}) \times \boldsymbol{p}$$
$$= -\frac{q\hbar}{2mc}H_0 L_z(\boldsymbol{R}) \tag{4.59c}$$

where $L_z(\boldsymbol{R})$ is the z-component of angular momentum about the point at $\boldsymbol{R}$. The choice of gauge therefore specifies the point about which angular momentum is measured in the perturbation. It is, of course, most natural to choose $\boldsymbol{R} = 0$, corresponding to measurement of angular momentum about the nucleus, since in general the electronic wave functions are classified as linear combinations of s, p, d (and so on) functions. When the electron orbit extends over several atoms, more than one force center enters the problem. The choice of the best gauge then becomes more complicated. A closely related problem in electron spin resonance involving the g-shift is discussed in Chap. 11.

Let us now consider an s-state. Then the wave function is spherically symmetric:

$$\psi_s(\boldsymbol{r}) = \psi_s(r) \quad . \tag{4.60}$$

It is clear that, since $L_z \psi_s = 0$,

$$(n|\mathcal{H}_{\text{pert}}|\psi_s) = 0 \quad . \tag{4.61}$$

Therefore ε_{n0} is zero for all excited states, and the entire current $\boldsymbol{j}_0(r)$ comes from the last term of (4.44):

$$\boldsymbol{j}_0(r) = -\frac{q^2}{mc}\boldsymbol{A}_0 \psi_0^2 = -\frac{q^2}{2mc}H_0 \boldsymbol{k} \times \boldsymbol{r}\psi_s^2(r) \quad . \tag{4.62}$$

The current therefore flows in circles centered on the z-axis. The direction is such as to produce a magnetic moment directed opposite to H_0 so that it produces a diamagnetic moment. We see that the current direction will also produce a field opposed to H_0 at the nucleus (see Fig. 4.3).

It is interesting to note that there is a current flowing in the s-state. There must certainly be an associated angular momentum, yet we customarily think of

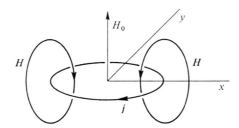

Fig. 4.3. Diamagnetic current flow in an s-state atom, and the magnetic fields produced by the current

s-states as having zero angular momentum. We are confronted with the paradox: If s-states have zero angular momentum, how can there be electronic angular momentum in a *first*-order perturbation treatment if the first-order perturbation uses the *unperturbed* wave function? The answer is that the angular momentum operator has changed from $r \times (\hbar/i)\nabla$ in the absence of a field to $r \times [(\hbar/i)\nabla - (q/c)A]$ when the field is present. By using the changed operator, the unchanged s-state has acquired angular momentum. The angular momentum is imparted to the electron by the electric field associated with turning on the magnetic field, since this electric field produces a torque about the nucleus. There is a corresponding back reaction on the magnet. We note that since A is continuously variable, we can make the angular momentum continuously variable. By using typical numbers for H_0 and r, one finds the angular momentum *much* smaller than $\hbar$. Does this fact violate the idea that angular momentum changes occur in units of $\hbar$? No, it does not, since the electron is not free but rather is coupled to the magnet. The *complete* system of magnet plus electron can only change angular momentum by $\hbar$, but the division of angular momentum between the parts of a coupled system does not have to be in integral units of $\hbar$.

We turn now to a *p*-state $xf(r)$ acted on by the crystalline field such as that discussed in Section 4.3. We duplicate the figures for the reader's convenience (Fig. 4.4).

The energy levels are then as in Fig. 4.5. Let us consider H_0 to lie along the z-direction. In contrast to the s-state, the p-state has nonvanishing matrix elements to the excited states, corresponding to the tendency of the static field to unquench the angular momentum. For this orientation of H_0, the matrix element to $zf(r)$ vanishes. That to $yf(r)$ is

$$
\begin{aligned}
(n|\mathcal{H}_{\text{pert}}|0) &= -\frac{q}{2mc}H_0\frac{\hbar}{i}\int yf(r)\left(x\frac{\partial}{\partial y} - y\frac{\partial}{\partial x}\right)xf(r)d\tau \\
&= +\frac{q}{2mc}H_0\frac{\hbar}{i}\int [yf(r)]^2 d\tau \\
&= -\frac{iq\hbar H_0}{2mc}
\end{aligned}
\tag{4.63}
$$

where we have used the fact that the function $yf(r)$ is normalized.

By using (4.63), we find

$$
\varepsilon_{n0} = \frac{(n|\mathcal{H}_{\text{pert}}|0)}{E_0 - E_n} = i\frac{q\hbar H_0}{2mc}\frac{1}{\Delta} \ .
\tag{4.64}
$$

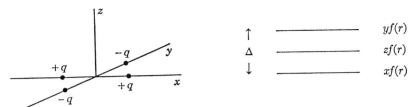

Fig. 4.4. Crystalline field due to charges $+q$ at $x = \pm a$, $y = z = 0$; $-q$ at $y = \pm a$, $x = z = 0$

Fig. 4.5. Energy levels for a crystalline field such as shown in Fig. 4.4

The term $\psi_0 \nabla \psi_n - \psi_n \nabla \psi_0$ of (4.53) is readily shown to be

$$\psi_0 \nabla \psi_n - \psi_n \nabla \psi_0 = (x\boldsymbol{j} - y\boldsymbol{i})f^2(r) \quad . \tag{4.65}$$

It is conventional that the part of (4.44) associated with the excited states is called the paramagnetic current $\boldsymbol{j}_\mathrm{P}$, since (as we shall see) it contributes a paramagnetic magnetic moment. We call the last term of (4.44) the diamagnetic current $\boldsymbol{j}_\mathrm{D}$. Then, using (4.44), (4.64) and (4.65), we get for our example

$$\boldsymbol{j}_\mathrm{P} = \frac{\hbar^2}{2m} \frac{q^2}{mc} \frac{H_0}{\Delta} \boldsymbol{k} \times \boldsymbol{r} f^2(r) \tag{4.66}$$

and by using (4.44) and (4.54),

$$\begin{aligned}\boldsymbol{j}_\mathrm{D} &= -\frac{q^2}{mc} \frac{1}{2} H_0 \boldsymbol{k} \times \boldsymbol{r} |\psi|^2 \\ &= -\frac{q^2}{2mc} H_0 (\boldsymbol{k} \times \boldsymbol{r}) x^2 f^2(r) \quad .\end{aligned} \tag{4.67}$$

It is clear that both $\boldsymbol{j}_\mathrm{P}$ and $\boldsymbol{j}_\mathrm{D}$ flow in concentric circles but in opposite directions. However, although div $\boldsymbol{j}_\mathrm{P} = 0$, the same is not true of $\boldsymbol{j}_\mathrm{D}$. Since div $\boldsymbol{j} = 0$ for a stationary state ($\boldsymbol{j} = \boldsymbol{j}_\mathrm{P} + \boldsymbol{j}_\mathrm{D}$), there is clearly a discrepancy. It may be traced to the fact that the wave functions used to derive $\boldsymbol{j}_\mathrm{P}$ and $\boldsymbol{j}_\mathrm{D}$ are not exact solutions of the crystalline field, but are, rather, only zero-order functions. The charges that give rise to the crystalline splitting will also distort the functions. For example, $xf(r)$, which points towards the positive charges, will presumably be elongated somewhat, whereas $yf(r)$ will be somewhat compressed. This will result in current flow with a radial component that will, so to speak, supply the circular currents of the diamagnetic term. However, radial currents will affect neither the chemical shift (since they produce zero field at the origin) nor the atomic magnetic moment. Therefore we shall not search for better starting functions.

The division between diamagnetic and paramagnetic currents would change if we had chosen a different gauge for $\boldsymbol{A}_0$. However, if our solution for $\boldsymbol{j}_0(\boldsymbol{r})$ were exact (to order H_0), the *total* current $\boldsymbol{j}_0(\boldsymbol{r})$ would be gauge-invariant. It is for this reason that our expression is so useful, since it holds regardless of the gauge of A_0 used to compute the currents or of the division of j_0 between paramagnetic and diamagnetic terms.

It is important to compare the relative magnitudes of j_P and j_D. From (4.66) and (4.67) we get

$$j_D = -j_P \frac{m}{\hbar^2} x^2 \Delta = -j_P \frac{\Delta}{(\hbar^2/mx^2)} \tag{4.68}$$

where $\hbar^2/mx^2$ has the units of energy (comparable to the kinetic energy of an electron whose de Broglie wavelength is x). By substituting numbers, we find

$$j_D = -j_P \frac{x^2 \Delta}{8} \tag{4.69}$$

where x is measured in angstroms and Δ in electron volts. Thus, if $\Delta = 8\,\mathrm{eV}$ (a fairly typical Δ for chemical shift problems), we see that j_P is larger for x less than 1 Å but that j_D is larger outside. As we shall see, the distance that is most important for typical chemical shifts is about 0.25 Å, so the paramagnetic current dominates; however, for computing magnetic moments, a distance of 1 Å or greater is more typical, making it difficult to assess which factor is the more important.

We can now compute the chemical shift fields H_P and H_D due to j_P and j_D:

$$\begin{aligned} H_P &= \frac{1}{c} \int \frac{\boldsymbol{r} \times \boldsymbol{j}_P}{r^3} d\tau \\ &= \frac{\hbar^2}{2m} \frac{q^2}{mc^2} \frac{H_0}{\Delta} \int \frac{\boldsymbol{r} \times (\boldsymbol{k} \times \boldsymbol{r})}{r^3} f^2(r) d\tau \quad . \end{aligned} \tag{4.70}$$

Direct evaluation shows that the x- and y-components of H_P vanish, leaving only the z-component:

$$H_P = \boldsymbol{k} \frac{\hbar^2}{2m} \frac{q^2}{mc^2} \frac{H_0}{\Delta} \int \frac{(x^2+y^2)}{r^3} f^2(r) d\tau \quad . \tag{4.71}$$

Now, for any wave function $\psi(r)$, the mean value of $1/r^3$ is given by

$$\overline{\left(\frac{1}{r^3}\right)} = \int \frac{1}{r^3} |\psi|^2 d\tau \quad , \tag{4.72}$$

so that we see

$$\int \frac{x^2 f^2(r) d\tau}{r^3} = \overline{\left(\frac{1}{r^3}\right)} = \int \frac{y^2 f^2(r) d\tau}{r^3} \quad . \tag{4.73}$$

Therefore we find

$$H_P = \boldsymbol{k} \frac{\hbar^2}{m} \frac{q^2}{mc^2} \frac{H_0}{\Delta} \overline{\left(\frac{1}{r^3}\right)} \quad . \tag{4.74}$$

We note that H_P aids the static field and is in fact proportional, in keeping with the experimental data.

The diamagnetic field, which is given by

$$H_D = -\frac{q^2}{2mc^2} H_0 \int \frac{\mathbf{r} \times (\mathbf{k} \times \mathbf{r})}{r^3} x^2 f^2(r) \quad , \tag{4.75}$$

turns out also to be in the z-direction only:

$$H_D = -\frac{kq^2}{2mc^2} H_0 \int \frac{(x^2 + y^2)}{r^3} x^2 f^2(r) \quad . \tag{4.76}$$

It is most convenient to average H_D over all orientations of H_0 with respect to the x-, y-, and z-axes. This can be shown to be equivalent to averaging H_D for H_0 parallel to the x-, y-, and z-axes, in turn giving

$$\begin{aligned} H_D &= -\frac{1}{3}\frac{q^2}{2mc^2} H_0 \int \frac{[(x^2 + y^2) + (x^2 + z^2) + (y^2 + z^2)]}{r^3} x^2 f(r) \\ &= -\frac{q^2}{3mc^2} H_0 \overline{\left(\frac{1}{r}\right)} \quad . \end{aligned} \tag{4.77}$$

Since $H_D = -\sigma_D H_0$, where σ_D is the diamagnetic contribution to the chemical shielding parameter σ, we have

$$\sigma_D = \frac{q^2}{3mc^2} \overline{\left(\frac{1}{r}\right)} \tag{4.78}$$

an expression first derived by *Lamb* [4.2] to describe the shielding of closed atomic shells.

We can likewise average H_P; however, we note here that H_P is zero when H_0 is parallel to the x-axis, since the perturbation gives zero when acting on the cylindrically symmetric function $xf(r)$. It is also convenient to assume that $yf(r)$ and $zf(r)$ are degenerate, both a distance Δ above $xf(r)$, since this corresponds to the typical case of a chemical bond. Then we have

$$H_P = \frac{2}{3}\frac{\hbar^2}{m}\frac{q^2}{mc^2}\frac{H_0}{\Delta}\overline{\left(\frac{1}{r^3}\right)} \tag{4.79}$$

and σ_P, the paramagnetic contribution to σ, is

$$\sigma_P = -\frac{2}{3}\frac{\hbar^2}{m}\frac{q^2}{mc^2}\frac{1}{\Delta}\overline{\left(\frac{1}{r^3}\right)} \quad . \tag{4.80}$$

If we take $\Delta = 4.3\,\text{eV}$, and $\overline{(1/r^3)} = 8.89/a_0^3$, where a_0 is the Bohr radius (values appropriate to the $2p$ electrons of fluorine, with the energy chosen to be appropriate for the F_2 molecule), we find $\sigma_P = -20 \times 10^{-4}$. σ_D is typically 10^{-5}. We see that this value of σ_P is quite comparable to the changes in σ observed for fluorine compounds, whereas σ_D is much too small to account for the effects. It is clear also why the range of fluorine chemical shifts is so much larger than that of protons.

Physically, the large fluorine shifts come about because the magnetic field leads to an unquenching of the angular momentum. The smaller Δ, the more effectively H_0 can "unquench".

What can we say is the cause of the s-state shift? One simple picture is to note that an s-state is a radial standing wave. Since the magnetic force is transverse to the radial motion, it produces a slow rotation quite analogous to the manner in which the Coriolis force causes the direction of a Foucault pendulum to turn.

As we have seen, for any reasonable values of Δ, the paramagnetic shielding term will completely overwhelm the diamagnetic term. What can we say about M, the electron contribution to the atomic magnetic moment?

$$M = \frac{1}{2c} \int r \times j_0 \, d\tau \quad . \tag{4.81}$$

We contrast this with the shielding field

$$H = \frac{1}{c} \int \frac{r \times j_0}{r^3} d\tau \quad . \tag{4.82}$$

Clearly, the $1/r^3$ factor makes H relatively much more sensitive to currents close to the nucleus. In fact we can quickly convert our formulas for shielding to formulas for average susceptibility χ by recognizing that only the radial averages differ. Thus the paramagnetic and diamagnetic currents contribute χ_P and χ_D, respectively, to the susceptibility χ, where

$$\chi = \chi_P + \chi_D \quad , \quad M = \chi H_0 \quad . \tag{4.83}$$

We find (averaging over all orientations)

$$\chi_P = \frac{1}{3} \frac{\hbar^2}{m} \frac{q^2}{mc^2} \frac{1}{\Delta} \quad , \quad \chi_D = -\frac{1}{6} \frac{q^2}{mc^2} \overline{r^2} \quad .$$

There are anisotropies that, for both χ_P and χ_D, are a substantial fraction of the average value. If we compare χ_P and χ_D, we have

$$\chi_D = -\chi_P \frac{\hbar^2}{m} \frac{\overline{r^2}}{2} \Delta = -\chi_P \frac{\overline{r^2}}{2} \frac{\Delta}{8} \tag{4.84}$$

where r is measured in angstroms and Δ in electron volts. In general we must expect $(\overline{r^2})^{1/2} \sim 1\,\text{Å}$, and $\Delta \sim 8\,\text{eV}$. Therefore it is clear that $\chi_D \cong -\chi_P$. We cannot decide which term is the larger without specific examination. Note in particular that the mere fact that j_P is dominant in producing the chemical shift does *not* mean that it will be the major factor in determining the atomic susceptibility. The susceptibility, since it depends on currents *far* from the origin, is much more strongly influenced by the diamagnetic currents than is the chemical shift because the diamagnetic currents are the more prominent at large distances from the origin.

It is important to bear in mind that our particular choice of gauge has made $\mathcal{H}_{\text{pert}}$ dependent on the angular momentum about the origin. For an *exact* solution, this gauge is no better than any other gauge. However, we rarely deal with exact solutions. There may then be some physical preference in choosing a gauge that puts the perturbation in terms of angular momentum measured about the most prominent force center in the problem. Since our wave functions will be approximate, the approximations will then at least be so primarily because of failure to account for the small crystalline potential rather than for the much larger atomic (central) potential.

When we are dealing with chemical shifts in molecules, we find it very hard to deal with bonds, since two force centers, one at each nucleus (for a pair bond), are important. The only simple approximation is to treat the atoms as virtually isolated, taking the excitation energies as being those of excited bonds, and including such effects as ionic character by using nonnormalized atomic wave functions. *Pople* [4.3] has discussed this problem, using a technique due to London. His results can also be obtained by using perturbation theory. A similar problem arises when calculating g-shifts, and this is discussed in Chap. 11.

When one has ring compounds such as benzene, the interatomic currents are important. Then one chooses a *molecular* force center; that is, a gauge in which L_z is the angular momentum about an axis around which the molecule has roughly cylindrical symmetry. For benzene, this is the hexad axis. About this axis only a diamagnetic current results. This current, which flows all around the benzene ring, produces chemical shifts at the proton positions. Of course, since the distance of the protons from the ring is comparable to the radius of the ring, the chemical shielding field is not accurately given by replacing the ring by a dipole.

On the other hand, when the currents $j(r)$ are well localized in an atom or in a single bond, the dimensions of which are small compared with the distance to the nucleus under study, we can represent the effect of the currents by a magnetic dipole. When we average the result over the random orientations of a molecule in a liquid, we find that the shift vanishes unless the atomic susceptibility is itself a function of the orientation of the magnetic field with respect to the molecule.

One important contribution to atomic currents, which vanishes in a liquid, is the contribution of currents in closed shells on atoms other than that containing the nucleus under study. The result in the liquid can be shown to follow simply because the current distribution in a closed shell is independent of the orientation of H_0 with respect to the molecular axes. We should emphasize that an attempt to compute the closed-shell contribution to shielding fields by *approximate* methods is dangerous, since one may find that the paramagnetic and diamagnetic contributions are large. Their algebraic sum (which is zero for an *exact* calculation in a liquid) may be nonzero unless a very accurate computation is made. It is therefore always safest 1) to judge the currents $j_0(r)$ on physical grounds, 2) always to choose gauges for each atomic current that puts $\mathcal{H}_{\text{pert}}$ as proportional to the angular momentum about the most important atomic force center, and 3) to exclude from the calculation any currents that will give an exactly zero result.

Finally, we emphasize again the fact that the a priori judgment of whether a distant atom has a paramagnetic moment or a diamagnetic moment is not possible, but rather a detailed judgment of excitation energy and mean square atomic radius is essential. Moreover, for example, a *paramagnetic* moment on another atom can produce either diamagnetic or paramagnetic shielding fields, depending on whether the anisotropic moment is largest when the internuclear axis is perpendicular or parallel to the direction of H_0.

All the expressions given involve one electron only. If there are N electrons we generalize by adding subscripts "$j = 1$ to N" to the electron position coordinate. Thus we define A_{0j} by the equation

$$\nabla_j \times A_{0j} = H_0 \quad , \quad \nabla_j \times A_{nj} = H_n \quad . \tag{4.85}$$

Typically,

$$A_{nj} = \frac{\mu \times r_j}{r_j^3} \quad , \quad A_{0j} = \tfrac{1}{2} H_0 \times (r_j - R) \tag{4.86}$$

where R is a convenient origin. We then define

$$\pi_j = \frac{\hbar}{i}\nabla_j - \frac{q}{c}A_{0j} \tag{4.87}$$

so that the Hamiltonian, including both external magnetic field and the nuclear field, is

$$\mathcal{H} = \frac{1}{2m}\sum_{j=1}^{N}\left(\pi_j - \frac{q}{c}A_{nj}\right)^2 + V \quad . \tag{4.88}$$

If we define Ψ to be the exact solution of the N electron problem in the *absence* of the nuclear coupling, it obeys the equation

$$\left(\frac{1}{2m}\sum_j \pi_j^2 + V\right)\Psi = E\Psi \quad . \tag{4.89}$$

Ψ is, of course, a function of the r_j's of all N electrons. Then we define the current associated with the jth electron as

$$J_{0j}(r_j) = \int\left[\frac{\hbar q}{2mi}(\Psi^*\nabla_j\Psi - \Psi\nabla_j\Psi^*)\right.$$
$$\left. - \frac{q}{mc}A_{0j}\Psi^*\Psi\right]d\tau_1\ldots d\tau_{j-1}d\tau_{j+1}\ldots d\tau_N \tag{4.90}$$

where the integration leaves J_{0j} a function of r_j only. In terms of (4.90), the nuclear coupling E_{pert} is then

$$E_{\text{pert}} = -\mu \cdot \frac{1}{c}\sum_j\int \frac{r_j \times J_{0j}(r_j)}{r_j^3}d\tau_j \quad . \tag{4.91}$$

The wave function Ψ, which is needed to compute the currents, is then found by using perturbation theory. Defining the functions Ψ_0, Ψ_n to be solutions of the Hamiltonian $\mathcal{H}_0$ in the absence of the external field, with eigenvalues E_0 and E_n,

$$\left(\sum_j -\frac{\hbar \nabla_j^2}{2m} + V\right)\Psi_0 = E_0 \Psi_0$$

$$\left(\sum_j -\frac{\hbar \nabla_j^2}{2m} + V\right)\Psi_n = E_n \Psi_n \quad .$$
(4.92)

We express Ψ correct to first order in H_0 as

$$\Psi = \Psi_0 + \sum_n \frac{(n|\mathcal{H}_{\text{pert}}|0)}{E_0 - E_n}\Psi_n \quad \text{where}$$
(4.93)

$$\mathcal{H}_{\text{pert}} = -\frac{q}{2mc}\sum_j(\boldsymbol{p}_j \cdot \boldsymbol{A}_{0j} + \boldsymbol{A}_{0j} \cdot \boldsymbol{p}_j) \quad ,$$
(4.94)

$\boldsymbol{p}_j$ being $(\hbar/i)\nabla_j$.

To obtain explicit solutions, one must now assume reasonable N electron wave functions for Ψ_0 and Ψ_n. Ordinarily one will choose the functions to be products of one electron functions, or perhaps pair functions to represent a covalent bond. Although j labels electrons, one can often rewrite (4.91) so that the sum over electron numbers is replaced by a sum over *orbits*, and in this way one can distinguish closed shell electrons from valence electrons.

The formalism we have discussed is useful for obtaining a physical understanding of chemical shifts. The final result can be expressed more compactly in a single formula such as has been given by *Ramsey* [4.4]. To do this, we express the magnetic field, using (4.91) as

$$\boldsymbol{H} = \frac{1}{c}\sum_j \int \frac{\boldsymbol{r}_j \times \boldsymbol{J}_{0j}(\boldsymbol{r})}{r_j^3} d\tau_j \quad .$$
(4.95)

Expressing $\boldsymbol{J}_{0j}$ by means of (4.90), Ψ by means of (4.93) and (4.94), and taking $\text{div}_j \boldsymbol{A}_{0j} = 0$, straightforward manipulations give the result:

$$\boldsymbol{H} = \frac{q^2 \hbar}{m^2 c^2}\sum_n \left\{ \frac{\left(0\left|\sum_j \frac{\boldsymbol{L}_j}{r_j^3}\right|n\right)(n|\sum_k \boldsymbol{A}_{0k} \cdot \boldsymbol{p}_k|0)}{E_n - E_0} \right.$$

$$\left. + \frac{(0|\sum_k \boldsymbol{A}_{0k} \cdot \boldsymbol{p}_k|n)\left(n\left|\sum_j \frac{\boldsymbol{L}_j}{r_j^3}\right|0\right)}{E_n - E_0} \right\}$$

$$- \frac{q^2}{mc^2}(0|\sum_j \boldsymbol{r}_j \times \boldsymbol{A}_{0j}|0) \quad .$$
(4.96)

If we further assume that $\boldsymbol{A}_{0k} = \frac{1}{2}\boldsymbol{H}_0 \times \boldsymbol{r}_k$ and that $\boldsymbol{H}_0 = H_0 \boldsymbol{k}$, we get

$$H = H_0 \frac{q^2 h^2}{2m^2c^2} \sum_n \left\{ \frac{\left(0 \left| \sum_j \frac{L_j}{r_j^3} \right| n\right)\left(n \left| \sum_k L_{zk} \right| 0\right)}{E_n - E_0} \right.$$

$$\left. + \frac{\left(0 \left| \sum_k L_{zk} \right| n\right)\left(n \left| \sum_j \frac{L_j}{r_j^3} \right| 0\right)}{E_n - E_0} \right\}$$

$$- \frac{q^2}{2mc^2} H_0 \left(0 \left| \sum_j \left[\frac{k(x_j^2 + y_j^2)}{r_j^3} - \frac{ix_j z_j}{r_j^3} - \frac{jy_j z_j}{r_j^3} \right] \right| 0\right) \quad . \tag{4.97}$$

One can proceed to evaluate this expression directly rather than to compute the current density as an explicit function of position, as we did in our examples.

4.6 Electron Spin Interaction

The coupling to electron spins produces effects using first-order perturbation theory when the electron spin moment is nonzero, as in paramagnetic or ferromagnetic materials. The Knight shifts (shift of the resonance frequency in metals relative to their positions in insultors) are an example.[2] For diamagnetic materials, one must go to second-order perturbation theory to obtain nonvanishing spin couplings. One important class of phenomena that then results is the coupling of one nucleus with another via the electrons. These couplings give rise to fine structure of resonances in liquids and to either *narrowing* or *broadening* of resonance lines in solids. For example, the indirect couplings make the pure quadrupole resonance in indium metal be about ten times broader than that computed from the direct nuclear dipolar coupling alone. However, there is no chemical shift associated with electron spin for diamagnetic substances. We shall discuss this point at the end of Sect. 4.9.

We start by discussing the form of the magnetic coupling between an electron and a nucleus. As long as the nuclear and electron moments μ_n and μ_e are far enough apart, we expect their interaction to be that of a pair of magnetic dipoles, the Hamiltonian being

$$\mathcal{H} = \frac{\mu_e \cdot \mu_n}{r^3} - \frac{3(\mu_e \cdot r)(\mu_n \cdot r)}{r^5} \tag{4.98}$$

where r is the radius vector from the nucleus to the electron. As long as the electronic wave function is a p-state, d-state, or other state of nonzero angular momentum, we expect (4.98) to be good approximation. For s-states, however, the electron wave function is nonzero at the nucleus. For these close distances, the dipole approximation is suspect. A closer examination emphasizes the troubles.

---

[2] See references to "Nuclear Magnetic Resonance in Metals" in the Bibliography.

Suppose we average $\mathcal{H}$ over an s-state electron wave function $u(r)$, as we would do to perform a first-order perturbation calculation of the coupling. There are a number of terms to (4.98), similar to the terms A, B, C, D, E, and F when computing the rigid lattice line breadth (see p. 66). Let us pick out a term A, which depends on angle and distance as $(1 - 3\cos^2\theta)/r^3$. Then, apart from a multiplicative constant, the average of such a term will be

$$\int \frac{u^2(r)}{r^3}(1 - 3\cos^2\theta)r^2 dr\, d\Omega \tag{4.99}$$

where $d\Omega$ is an element of solid angle. If we do the angular integral first, it vanishes, giving us a result of zero for (4.99).

On the other hand, if we were to integrate first over r, we would encounter trouble near $r \cong 0$, where $u^2(r) = u^2(0) \neq 0$, giving a logarithmic infinity. Since we can get either zero or infinity, depending on our method of calculation, it is clear that we cannot simply ignore the contributions when r is small.

From what we have said, it is evident that the dipole approximation has broken down. There are two effects that come in and which have been neglected. First of all, we know the nucleus has a finite size. To the extent that the nuclear magnetic moment results from the bodily rotation of the nucleus, the currents are distributed over the nuclear volume. From the electron viewpoint the spin moments of the nuclear particle are also spread over a comparable region, since the nuclear particles effectively possess much higher frequencies of motion than does the electron (the nuclear energy levels are widely spaced in energy compared with those of electrons). A second effect is that the electronic coupling to the nucleus, when computed by using a relativistic theory (the Dirac equation), shows a marked change when the electron is within a distance e^2/mc^2 of the nucleus; e^2/mc^2 is the classical radius of the electron, r_0, and is about 3×10^{-13} cm. The electron is effectively smeared out over r_0. Since the radii of nuclei are given approximately by the formula

$$r = 1.5 \times 10^{-3} A^{1/3} \text{ cm} \tag{4.100}$$

we see that the nuclear size is comparable to the electron radius r_0.

Of course, completely aside from all these remarks, the fact that the electronic potential energy is of order mc^2 near the nucleus shows us that a relativistic theory is advisable.

We shall first give a simple classical derivation of the interaction for s-states, and then we shall discuss briefly how the Dirac theory exhibits the same features. The theorems on the relation between magnetic fields and currents, which we developed in discussing chemical shifts, will show us that our result is really rigorous for the contribution to the coupling resulting from bodily motion of nuclear charges. Finally, since a volume distribution of magnetic moment (as produced by the spin moments of the nuclear particles) is equivalent to a current distribution, our result will also include the contributions of intrinsic spin. Thus, although simple, our calculation is in fact rigorous in the nonrelativistic case.

We shall represent the nucleus by a charge q going in a circular path of radius a with velocity v. This, effectively, is a current loop of current $(q/c)(1/T)$, where T is the period of the motion. We can express $\overline{H}_z$ – the magnetic field in the z-direction, due to the nucleus, averaged over the electron orbital probability density $|u(r)|^2$ – as

$$\overline{H}_z = \int H_z(\mathbf{r})|u(r)|^2 d\tau \tag{4.101}$$

where $H_z(\mathbf{r})$ is the field of the current loop. We shall take z to be normal to the loop. The other components of H can be shown to vanish when averaged, since $|u(r)|^2$ is spherically symmetric for an s-state. If we draw a sphere of radius a about the origin, we can express $H_z(\mathbf{r})$ by means of a scalar magnetic potential either for $r < a$ or $r > a$. It is straightforward to show that the contribution from regions outside the sphere vanishes from the angular integrations. If we express the scalar potential inside the sphere as a sum of products of spherical harmonics with radial functions, all contributions except the first term (the term that corresponds to a uniform field within $r < a$) vanish. We can evaluate this term simply, since it is the only one that does not vanish at $r = 0$. Therefore (4.101) may be rewritten as

$$\overline{H}_z = \int_0^a H_c u^2(r) d\tau \tag{4.102}$$

where H_c is the field at the center of the sphere. We may approximate (4.102) by recognizing that $u(r)$ varies little over the nucleus as

$$\overline{H}_z = H_c u^2(0) \frac{4\pi}{3} a^3 \quad . \tag{4.103}$$

The field at the center of the loop is simply

$$\mathbf{H}_c = \frac{q}{c} \frac{\mathbf{r} \times \mathbf{v}}{r^3} = \frac{q}{c} \frac{v}{a^2} \mathbf{k} \quad . \tag{4.104}$$

But the magnetic moment $\boldsymbol{\mu}_n$ of the nucleus is $i\pi a^2$, where i is the "current", or

$$\boldsymbol{\mu}_n = \frac{q}{c} \frac{1}{T} \pi a^2 \mathbf{k} = \mathbf{k} \frac{qav}{2c} \quad . \tag{4.105}$$

Thus

$$\mathbf{H}_c = \frac{2\boldsymbol{\mu}_n}{a^3} \quad . \tag{4.106}$$

By substituting into (4.103), we find

$$\mathbf{k}\overline{H}_z = \frac{8\pi}{3} \boldsymbol{\mu}_n u^2(0) \quad . \tag{4.107}$$

The effective interaction energy E with an electron moment $\boldsymbol{\mu}_e$ is then

$$E = -\frac{8\pi}{3} \boldsymbol{\mu}_e \cdot \boldsymbol{\mu}_n u^2(0) \quad . \tag{4.108}$$

It is convenient to express the coupling as a term in the Hamiltonian that will give this interaction. This is readily done by means of the Dirac δ-function:

$$\mathcal{H} = -\frac{8\pi}{3}\boldsymbol{\mu}_e \cdot \boldsymbol{\mu}_n \delta(\boldsymbol{r}) \tag{4.109}$$

where $\boldsymbol{r}$ is now the position of the electron relative to the nucleus. It is convenient also to re-express (4.109) in terms of the nuclear and electron spins $\boldsymbol{I}$ and $\boldsymbol{S}$. For the electron we shall use a gyromagnetic ratio γ_e, which is positive, but for the nucleus, γ_n is to have an algebraic significance, being either positive or negative. Then we have

$$\boldsymbol{\mu}_e = -\gamma_e \hbar \boldsymbol{S} \quad , \quad \boldsymbol{\mu}_n = \gamma_n \hbar \boldsymbol{I} \tag{4.110}$$

which give

$$\mathcal{H} = \frac{8\pi}{3}\gamma_e \gamma_n \hbar^2 \boldsymbol{I} \cdot \boldsymbol{S} \delta(\boldsymbol{r}) \quad . \tag{4.111}$$

We notice that in (4.108), the radius of the nuclear orbit has dropped out. Clearly, we should get the same answer for a volume distribution of circular currents. Moreover, since the smeared nuclear spin moment is equivalent to a volume distribution of current, we have also included the intrinsic spin of the nucleons if we use, for γ_n and $\boldsymbol{I}$, the experimental values. Equation (4.111) is therefore quite general. We see also that if we may *not* neglect the variations in $u(r)$ over the nucleus, the answer will be a bit different. Two isotopes that have different current distributions will then have couplings which are *not* simply in the ratio of the nuclear moments. This phenomenon is the source of the so-called hyperfine anomalies.

The treatment of the interaction by the Dirac equation is somewhat more involved. We shall sketch the important steps but leave the details to the reader. The Dirac Hamiltonian for an electron (charge $-e$) is

$$\mathcal{H} = -\boldsymbol{\alpha} \cdot (c\boldsymbol{p} + e\boldsymbol{A}) - \beta mc^2 + V \tag{4.112}$$

where $\boldsymbol{\alpha}$ and β are 4 by 4 matrices, V is the electron potential energy, and $\boldsymbol{A}$ is the vector potential. We can express $\boldsymbol{\alpha}$ and β in terms of the two by two Pauli matrices, $\boldsymbol{\sigma}$, and the two by two identity matrix $\mathbf{1}$ as

$$\boldsymbol{\alpha} = \begin{pmatrix} 0 & \sigma \\ \sigma & 0 \end{pmatrix} \quad \beta = \begin{pmatrix} 1 & 0 \\ 0 & -1 \end{pmatrix} \quad . \tag{4.113}$$

The wave functions Ψ, which are solutions of (4.112), are represented by a column matrix of four functions, but these are most conveniently expressed in terms of the functions Ψ_1 and Ψ_2, each of which is a column matrix with two elements:

$$\Psi = \begin{pmatrix} \Psi_1 \\ \Psi_2 \end{pmatrix} \quad . \tag{4.114}$$

The eigenvalues E of $\mathcal{H}$ may be written as

$$E = E' + mc^2 \tag{4.115}$$

where E' is the energy measured above mc^2, so that for a free particle at rest, $E' = 0$.

If we define

$$\boldsymbol{\pi} = c\boldsymbol{p} + e\boldsymbol{A} \tag{4.116}$$

and define ϕ as V/e, we have

$$\begin{aligned}(E' + e\phi + 2mc^2)\Psi_1 + \boldsymbol{\sigma} \cdot \boldsymbol{\pi}\Psi_2 &= 0 \\ (E' + e\phi)\Psi_2 + \boldsymbol{\sigma} \cdot \boldsymbol{\pi}\Psi_1 &= 0\end{aligned} \tag{4.117}$$

where $\phi = e/r$ is the potential due to the nucleus. As is well known, Ψ_1 is much smaller than Ψ_2 in the nonrelativistic region. One customarily calls Ψ_2 the "large component". For s-states in hydrogen, Ψ_2 is also much larger than Ψ_1, even at the nucleus. One can eliminate Ψ_1 (still with no approximations) to obtain a Hamiltonian for Ψ_2, $\mathcal{H}'$ such that

$$\mathcal{H}'\Psi_2 = E'\Psi_2 \quad . \tag{4.118}$$

By tedious manipulation one finds that

$$\begin{aligned}\mathcal{H}' = {}& \frac{1}{E' + e\phi + 2mc^2}(c^2p^2 + e^2A^2 + 2ec\boldsymbol{A} \cdot \boldsymbol{p} - iec \operatorname{div} \boldsymbol{A} \\ & + e\hbar c \boldsymbol{\sigma} \cdot \nabla \times \boldsymbol{A}) + \frac{e\hbar c}{(E' + e\phi + 2mc^2)^2} \\ & \times (ie\boldsymbol{E} \cdot \boldsymbol{A} + ic\boldsymbol{E} \cdot \boldsymbol{p} - c\boldsymbol{\sigma} \cdot \boldsymbol{E} \times \boldsymbol{p} - e\boldsymbol{\sigma} \cdot \boldsymbol{E} \times \boldsymbol{A})\end{aligned} \tag{4.119}$$

where $\boldsymbol{E}$ is the electric field due to the nucleus. For our present needs, we focus on two terms only:

$$\frac{1}{E' + e\phi + 2mc^2} e\hbar c \boldsymbol{\sigma} \cdot \nabla \times \boldsymbol{A} \quad \text{and} \tag{4.120a}$$

$$\frac{1}{(E' + e\phi + 2mc^2)^2} e^2 \hbar c \boldsymbol{\sigma} \cdot \boldsymbol{E} \times \boldsymbol{A} \quad . \tag{4.121a}$$

For our problem, the nuclear coupling is given by introducing the vector potential $\boldsymbol{A} = \boldsymbol{\mu}_\mathrm{n} \times \boldsymbol{r}/r^3$. Therefore $\nabla \times \boldsymbol{A}$ is simply the magnetic field of the nucleus computed by using the dipole approximation. As long as $e\phi \ll 2mc^2$, (4.120a) is exactly the same as (4.98) and goes as $1/r^3$. If, however, r is so small that $e\phi \sim 2mc^2$, the answer is modified. By writing $e\phi = e^2/r$, multiplying (4.120a) by r on top and bottom, using $e^2/mc^2 = r_0$ (the classical electron radius), and neglecting E', we get for (4.120a):

$$\left(\frac{2r}{2r + r_0}\right)\frac{e\hbar}{2mc}\boldsymbol{\sigma} \cdot \nabla \times \boldsymbol{A} \quad . \tag{4.120b}$$

Now we no longer have an infinity from the radial integral in computing $\boldsymbol{H}_z$, and it is clear that the angular average makes (4.120b) to be zero.

The term (4.121a) can be rewritten as

$$\left(\frac{2r}{2r+r_0}\right)^2 \frac{e^3\hbar c}{(2mc^2)^2}\boldsymbol{\sigma}\cdot\left[\frac{\boldsymbol{r}}{r^3}\times\left(\frac{\boldsymbol{\mu}_n\times\boldsymbol{r}}{r^3}\right)\right] \quad . \tag{4.121b}$$

The term in the square brackets goes as $1/r^4$. For $r \gg r_0$, (4.121b) can be shown to be of order r_0/r times (4.120b); therefore, much smaller. However, when $r \leq r_0$, the radial dependence becomes less strong, going over to the harmless $1/r_0$ near $r = 0$. The term is therefore well behaved. It also has the feature that it does not average to zero over angle of s-states. It gives the answer of (4.108) for the magnetic interaction energy.

We see that these two terms are very similar to taking a δ-function for s-states and asserting that a finite size of the electron prevents the radial catastrophe of the conventional dipolar coupling of (4.98). For computational convenience we may consider that the dipolar interaction of (4.98) should be multiplied by the function $2r/(2r + r_0)$, to provide convergence.

We shall now turn to the study of some of the important manifestations of the coupling between nuclei and electron spins, considering first the effects that are first order in the interaction and then effects that arise in second order. Further discussion of first-order effects will be found in Chap. 11 on electron spin resonance.

4.7 Knight Shift [3]

The Knight shift is named after Professor Walter Knight, who first observed the phenomenon. What he found was that the resonance frequency of Cu^{63} in metallic copper occurred at a frequency 0.23 percent higher than in diamagnetic CuCl, provided both resonances were performed at the same value of static field. Since this fractional shift is an order of magnitude larger than the chemical shifts among different diamagnetic compounds, it is reasonable to attribute it to an effect in the metal. Further studies revealed that the phenomenon was common to all metals, the principal experimental facts being four in number. By writing ω_m for the resonance frequency in the metal, ω_d for the resonance frequency in a diamagnetic reference, all at a single value of static field, there is a frequency displacement $\Delta\omega$ defined by

$$\omega_m = \omega_d + \Delta\omega \quad . \tag{4.122}$$

The four facts are

1. $\Delta\omega$ is positive (exceptional cases have been found, but we ignore them for the moment).

---

[3] See references to "Nuclear Magnetic Resonance in Metals" in the Bibliography

2. If one varies ω_d by choosing different values of static field, the *fractional shift* $\Delta\omega/\omega_d$ is unaffected.
3. The fractional shift is very nearly independent of temperature.
4. The fractional shift increases in general with increasing nuclear charge Z.

The fact that metals have a weak spin paramagnetism suggests that the shift may simply represent the pulling of the magnetic flux lines into the piece of metal. However, the susceptibilities are too small (10^{-6} cgs units/unit volume) to account for an effect of this size. As we shall see, however, the ordinary computation of internal fields in a solid that involves a spatial average of the local field is not what is wanted, since the nuclear moment occupies a very special place in the lattice – in fact a place at which the electron spends, so to speak, a large amount of time in response to the deep, attractive potential of the nuclear charge. As we shall see, the correct explanation of the Knight shift involves considering the field the nucleus experiences as a result of the interaction with conduction electrons through the s-state hyperfine coupling. If we think of the electrons in a metal as jumping rapidly from atom to atom, we see that a given nucleus experiences a magnetic coupling with many electrons. Therefore the coupling to the electron spins must be averaged over the electron spin orientations of many electrons. In the absence of an external field, there is no preferential orientation for the electron spins, and thus there is zero average magnetic coupling to the nuclei. On the other hand, the application of a static field, H_0, polarizes the electron spins, giving a nonvanishing coupling. Since the s-state interaction corresponds to the nucleus experiencing a magnetic field parallel to the electron magnetic moment,[4] and since the electron moment is preferentially parallel to H_0, the effective field at the nucleus will be increased. Since the shift in frequency is proportional to the degree of electron spin polarization, it will also be proportional to H_0 or ω_d. Moreover, since the electron polarization is temperature independent (the spin paramagnetism of a highly degenerate electron gas is independent of temperature), the shift will be temperature independent. And lastly, the Z-dependence will follow, since the wave function is larger at the position of a higher-Z nucleus, as is well known from the study of free atom hyperfine splittings. We can see from these considerations that the hyperfine coupling possesses the properties needed to explain the major facts. Let us now look into the details.

We consider a system of nuclear moments and of electrons coupled together by the hyperfine interaction. The relative weakness of the hyperfine coupling enables us to treat it by a perturbation theory in terms of the states of the electrons and the nuclear spins. We shall actually be able to avoid specifying the nuclear states, since we shall show that the effect of the interaction is simply to add an effective magnetic field parallel to the applied field. However, it is

---

[4] The dependence of the interaction on the relative orientation of nuclear and electron moments is most readily seen by replacing the nucleus by a current loop.

necessary to specify the electron wave function, which is, of course, a formidable task from a rigorous viewpoint, one that has not in fact been carried out because electrons couple to each other so strongly via the long range Coulomb interaction. Therefore we are forced to an approximation. We shall consider the electrons as being noninteracting – or at least only weakly so. Bohm and Pines have shown that this approximation has considerable theoretical justification. For a review of the Bohm-Pines theory, see [4.5]. By means of a canonical transformation, they show that the principal effect of the Coulomb interaction is to give rise to a set of collective modes of oscillation, the plasma modes. The basic frequency of excitation of the plasma is so high that we may ordinarily consider the system to be in the ground plasma state. There still remain individual particle motions. The residual interaction between particles, however, is very weak and falls off nearly exponentially with distance. For low energy processes that do not excite the plasma modes, we may therefore treat the electrons as weakly interacting.

We shall describe the system, therefore, with a Hamiltonian:

$$\mathcal{H} = \mathcal{H}_e + \mathcal{H}_n + \mathcal{H}_{en} \tag{4.123}$$

where $\mathcal{H}_e$ describes a group of weakly interacting electrons, $\mathcal{H}_n$ is the nuclear Hamiltonian and includes the Zeeman energy of the nuclei in the static field H_0 as well as the magnetic dipolar coupling among the nuclei, and where $\mathcal{H}_{en}$ is the magnetic interaction between the nuclei and the electron spins. We omit the coupling of the nuclei to the electron orbital motion because it gives effects comparable to the chemical shifts. (Of course the electrons in the metal are free, so that the orbital effect is a bit different from that in an insulator.) It can be shown that the conventional dipolar coupling between nuclear and electron spins, (4.98), contributes nothing in a cubic metal. For noncubic metals it gives rise to Knight shifts which depend on the orientation of H_0 with respect to the crystalline axes. Since resonance in metals is usually performed on powders (to permit adequate penetration of the alternating field into the material, a problem that we may call the "skin depth" problem), the anisotropy manifests itself through a line broadening. In the interest of simplicity, we shall confine our attention to the δ-function coupling:

$$\mathcal{H}_{en} = \frac{8\pi}{3}\gamma_e\gamma_n\hbar^2 \sum_{j,l} \boldsymbol{I}_j \cdot \boldsymbol{S}_l \delta(\boldsymbol{r}_l - \boldsymbol{R}_j) \tag{4.124}$$

where $\boldsymbol{r}_l$ is the radius vector to the position of the lth electron, and $\boldsymbol{R}_j$ that to the position of the jth nucleus.

Equation (4.124) can be rewritten in a highly useful manner. When we discuss electric charges in quantum mechanics we introduce a charge density operator $\varrho^{(op)}(\boldsymbol{r})$ given by

$$\varrho^{(op)}(\boldsymbol{r}) = \sum_l q_l \delta(\boldsymbol{r}_l - \boldsymbol{r}) \tag{4.125}$$

where q_l is the charge of the lth particle, and the sum goes over the particles.

In a similar way we introduce a spin magnetization density operator of electron magnetization

$$M^{(\mathrm{op})}(\boldsymbol{r}) = \sum_l -\gamma_e \hbar S_l \delta(\boldsymbol{r}_l - \boldsymbol{r}) \quad . \tag{4.126}$$

The expectation value of this operator is the classical spin magnetization density. Using this definition, we get

$$\mathcal{H}_{\mathrm{en}} = -\sum_j \frac{8\pi}{3} M^{(\mathrm{op})}(\boldsymbol{R}_j) \cdot (\gamma_n \hbar \boldsymbol{I}_j) \quad . \tag{4.127}$$

Since we treat the nuclei and electrons as only weakly interacting, we may write the complete wave function ψ as a product of the (many particle) wave functions ψ_e and ψ_n of the electrons and nuclei:

$$\psi = \psi_e \psi_n \quad . \tag{4.128}$$

(Of course this wave function would be exact if $\mathcal{H}_{\mathrm{en}}$ were zero.) We shall then perform a perturbation calculation of the energy E_{en}:

$$E_{\mathrm{en}} = \int \psi^* \mathcal{H}_{\mathrm{en}} \psi \, d\tau_e d\tau_n \tag{4.129}$$

where $d\tau_e$ and $d\tau_n$ indicate integration over electron and nuclear coordinates (spatial and spin). Of course we shall wish to see the effect of (4.129) on *transitions* of the nuclear system from one nuclear state ψ_n to another $\psi_{n'}$. Since the transitions are in the nuclear system, they leave the electron state ψ_e unchanged. In computing the energy of the nuclear transition $E_{\mathrm{en}} - E_{\mathrm{en}'}$, we have to compute both E_{en} and $E_{\mathrm{en}'}$. Both energies involve the integral over the electron coordinates. It is convenient for us to postpone a specification of the nuclear states and to compute, therefore, the electronic integral

$$\mathcal{H}'_{\mathrm{en}} = \int \psi_e^* \mathcal{H}_{\mathrm{en}} \psi_e \, d\tau_e \tag{4.130}$$

which is simply the first step in computing (4.129) on the assumption of a product function, $\psi = \psi_e \psi_n$ or $\psi' = \psi_e \psi_{n'}$. We denote (4.130) by $\mathcal{H}'_{\mathrm{en}}$ to emphasize that the nuclear coordinates still appear as operators.

We can re-express (4.130) using (4.127) to obtain

$$\mathcal{H}'_{\mathrm{en}} = -\sum_j \frac{8\pi}{3} M(\boldsymbol{R}_j) \cdot \gamma_n \hbar \boldsymbol{I}_j \quad , \tag{4.131}$$

where now $M(\boldsymbol{R}_j)$ is the classical spin magnetization density at the position of the jth nucleus.

The function ψ_e will itself be a simple product of one electron functions if we assume that the electrons do not interact among themselves – or at least interact weakly. For the individual electrons we shall take the so-called Bloch functions. We remind ourselves of what these are: If the electrons were thought of as moving in a one-dimensional box of length a (Fig. 4.6), the position coordinate

Fig. 4.6. Metal represented by a box, with a potential depth V_0

being x, the wave functions would be $\sin kx$ or $\cos kx$, where only those values of k are allowed that satisfy the proper boundary conditions at $x = 0$ and $x = a$. In order to describe a situation in which a current can flow, it is customary to consider instead solutions $\exp(ikx)$, where now the allowed values of k are those that make the wave function the same at $x = a$ as it is at $x = 0$. The periodic boundary conditions for a three-dimensional box give solutions of the form

$$\psi = e^{i\boldsymbol{k}\cdot\boldsymbol{r}} \quad . \tag{4.132}$$

These solutions are modified in a very simple way to take account of the fact that the real potential is very deep in the vicinity of the nuclei. The wave functions, called *Bloch functions*, are then of the form

$$\psi_{\boldsymbol{k}} = u_{\boldsymbol{k}}(\boldsymbol{r})e^{i\boldsymbol{k}\cdot\boldsymbol{r}} \quad . \tag{4.133}$$

That is, there is still a quantity $\boldsymbol{k}$, the allowed values of which are given by requiring periodicity of the wave function on the walls of a box, but the plane wave $\exp(i\boldsymbol{k}\cdot\boldsymbol{r})$ is multiplied by a modulating function $u_{\boldsymbol{k}}(\boldsymbol{r})$, which is a function possessing the lattice periodicity. A typical $u_{\boldsymbol{k}}(\boldsymbol{r})$ peaks up strongly near a nucleus. The fact that we explicitly label u with a subscript $\boldsymbol{k}$ points out that u will in general vary with $\boldsymbol{k}$.

We shall need to add a spin coordinate as well, giving us finally a function

$$\psi_{\boldsymbol{k}s} = u_{\boldsymbol{k}}e^{i\boldsymbol{k}\cdot\boldsymbol{r}}\psi_s \tag{4.134}$$

where ψ_s is a spin function. The wave function for the N electrons, ψ_e, will then be a product of $\psi_{\boldsymbol{k}s}$'s, properly antisymmetrized to take account of the Pauli exclusion principle. We can do this readily in terms of the permutation operator P [4.6]:

$$\psi_e = \frac{1}{\sqrt{N!}} \sum_P (-1)^P P \psi_{\boldsymbol{k}s}(1)\psi_{\boldsymbol{k}'s'}(2)\psi_{\boldsymbol{k}''s''}(3)\ldots\psi_{\boldsymbol{k}^N s^N}(N) \tag{4.135}$$

where the symbol $(-1)^P$ means to take a plus or minus sign, depending on whether or not the permutation involves an even or an odd number of interchanges. The factor $1/\sqrt{N!}$ is, of course, simply normalization.

Let us compute $\mathcal{H}'_{\text{en}j}$, the contribution to (4.130) of the jth nuclear spin, and choose the origin of coordinates at that nuclear site ($\boldsymbol{R}_j = 0$). Then we have

$$\mathcal{H}'_{\text{en}j} = \frac{8\pi}{3}\gamma_e\gamma_n\hbar^2 \boldsymbol{I}_j \cdot \int \psi_e^* \sum_l \boldsymbol{S}_l\delta(\boldsymbol{r}_l)\psi_e d\tau_e \quad . \tag{4.136}$$

Since the operator $\boldsymbol{S}_l\delta(\boldsymbol{r}_l)$ involves only one electron, there are not contributions to (4.136) from terms in which electrons are exchanged, and therefore we get

$$\frac{8\pi}{3}\gamma_e\gamma_n\hbar^2 \boldsymbol{I}_j \cdot \sum_l \int [\psi_{\boldsymbol{k}s}(1)\psi_{\boldsymbol{k}'s'}(2)\ldots]^* \boldsymbol{S}_l$$
$$\times \delta(\boldsymbol{r}_l)[\psi_{\boldsymbol{k}s}(1)\psi_{\boldsymbol{k}'s'}(2)\ldots] d\tau_1 d\tau_2 \ldots \quad . \tag{4.137}$$

We now assume the electrons are quantized along the z-direction by the external static field H_0. Then the only contribution to (4.137) comes from S_{zl}. (We could alternatively have kept just that part proportional to I_{zj}, assuming the nuclear spins to be quantized along H_0, the result being the same.) We can write the results of (4.137) as

$$\frac{8\pi}{3}\gamma_e\gamma_n\hbar^2 I_{zj} \sum_{\boldsymbol{k},s} |u_{\boldsymbol{k},s}(0)|^2 m_s p(\boldsymbol{k},s) \tag{4.138}$$

where the sum is over all values of $\boldsymbol{k}$, s, and where $p(\boldsymbol{k},s)$ is a factor that is 1 if $\boldsymbol{k}$, s are occupied by an electron, zero otherwise. The factor m_s is the m value of the state $\psi_{\boldsymbol{k},s}$; hence it is $+\frac{1}{2}$ or $-\frac{1}{2}$, and of course $u_{\boldsymbol{k},s}(0)$ is the wave function evaluated at the position of nucleus j.

If we average this expression over a set of occupations $p(\boldsymbol{k},s)$, which are representative of the temperature of the electrons, we can write for the effective interaction with jth nucleus,

$$\frac{8\pi}{3}\gamma_e\gamma_n\hbar^2 I_{zj} \sum_{\boldsymbol{k},s} |u_{\boldsymbol{k},s}(0)|^2 m_s f(\boldsymbol{k},s) \tag{4.139}$$

where $f(\boldsymbol{k},s)$ is the Fermi function. For the electrons at absolute zero, $f(\boldsymbol{k},s)$ is 1 for all $\boldsymbol{k}$, s, which makes the total (that is, spin plus spatial) electron energy less than the Fermi energy E_F and is zero for energies greater than E_F. At temperatures above absolute zero, $f(\boldsymbol{k},s)$ is modified within about kT of E_F (Fig. 4.7). Of course $f(E)$ is

$$f(E) = \frac{1}{1 + \exp\left[(E - E_F)/kT\right]} \quad . \tag{4.140}$$

The notation $f(\boldsymbol{k},s)$ means $f(E)$, of course, where E is the energy of an electron with wave vector $\boldsymbol{k}$ and spin coordinate s. Typically, $E = E_{\boldsymbol{k}} + E_{\text{spin}}$, where E_{spin} is the energy associated with the spin orientation and $E_{\boldsymbol{k}}$ is the sum of the kinetic and potential energies of an electron of wave vector $\boldsymbol{k}$. (We shall call $E_{\boldsymbol{k}}$ the

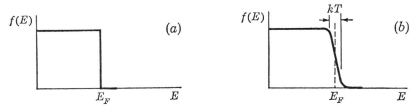

Fig. 4.7. (a) Fermi function $f(E)$ at absolute zero. (b) Fermi function at a temperature T above absolute zero

translational energy of the electron). For example, E_spin is the Zeeman energy of the electron spin in the static field H_0. There may be other contributions to E_spin, however, from the electrostatic coupling between electrons, which depends on their relative spin orientation. If we consider a typical term in the sum of (4.139) corresponding to a single value of $\boldsymbol{k}$, there are two values of m_s, giving us

$$\frac{8\pi}{3}\gamma_n\hbar I_{zj}[\gamma_e\hbar(\tfrac{1}{2})f(\boldsymbol{k},\tfrac{1}{2}) + \gamma_e\hbar(-\tfrac{1}{2})f(\boldsymbol{k},-\tfrac{1}{2})]|u_{\boldsymbol{k}}(0)|^2 \quad . \tag{4.141}$$

As we can see, the quantity in the brackets is (apart from a minus sign, since $\boldsymbol{\mu}_e = -\gamma_e\hbar\boldsymbol{S}$) the average contribution of state $\boldsymbol{k}$ to the z-component of electron magnetization of the sample. We shall denote it by $\overline{\mu_{z,\boldsymbol{k}}}$. The *total* z-magnetization of the electrons, $\overline{\mu_z}$, is then

$$\overline{\mu_z} = \sum_{\boldsymbol{k}} \overline{\mu_{z\boldsymbol{k}}} \quad . \tag{4.142}$$

If we define the total spin susceptibility of the electrons, χ_e^s, by

$$\overline{\mu_z} = \chi_e^s H_0 \tag{4.143}$$

and define a quantity $\chi_{\boldsymbol{k}}^s$ by

$$\overline{\mu_{z\boldsymbol{k}}} = \chi_{\boldsymbol{k}}^s H_0 \tag{4.144}$$

then (4.142) is equivalent to

$$\chi_e^s = \sum_{\boldsymbol{k}} \chi_{\boldsymbol{k}}^s \quad . \tag{4.145}$$

We can therefore write (4.141) as

$$-\frac{8\pi}{3}\gamma_n\hbar I_{zj}|u_{\boldsymbol{k}}(0)|^2 \chi_{\boldsymbol{k}}^s H_0 \quad , \tag{4.146}$$

so that the total effective interaction for spin j is

$$-\frac{8\pi}{3}\gamma_n\hbar I_{zj}\sum_{\boldsymbol{k}}|u_{\boldsymbol{k}}(0)|^2 \chi_{\boldsymbol{k}}^s H_0 \quad . \tag{4.147}$$

Our problem now is to evaluate the summation. It would be simple to make the assumption of completely free electrons, but it is actually no harder to treat the case of a material with a more complicated band structure. We shall do the latter, since the resulting expression will enable us to make some important comparisons between experiment and theory.

We assume, therefore, that the energy of the electrons, apart from spin effects, is determined by its $\boldsymbol{k}$ vector. For free electrons, the dependence would be

$$E_{\boldsymbol{k}} = \frac{\hbar^2 k^2}{2m} \tag{4.148}$$

so that in k-space, all electrons on a sphere of given k would have the same energy. That is, the points of constant energy form a surface, in this case a sphere

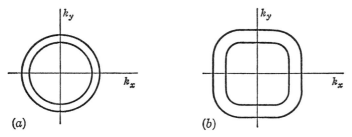

Fig. 4.8a,b. Intersection of two surfaces of constant energy with the $k_z = 0$ plane. (a) Circular section of a free electron. (b) A less symmetric section for a hypothetical "real" substance

(Fig. 4.8). In general the effect of the lattice potential is to distort the surfaces from spheres. We shall assume that the function $|u_{\bm{k}}(0)|^2 \chi^{\text{s}}_{\bm{k}}$ varies slowly as one moves the point $\bm{k}$ in k-space from one allowed $\bm{k}$-value to the next, so that we can define a density function to describe the number of allowed $\bm{k}$-values in any region. Let us define $g(E_{\bm{k}}, A)dE_{\bm{k}}dA$ as the number of allowed $\bm{k}$-values lying within a certain region of k-space, defined as follows: It is a small cylinder lying between the energy surfaces $E_{\bm{k}}$ and $E_{\bm{k}} + dE_{\bm{k}}$ (Fig. 4.9). Its surface area on the top or bottom surface is an element of dA of the constant energy surface. We denote the particular coordinates on the surface also by the symbol A in $g(E, A)$. The *total* number of states dN between $E_{\bm{k}}$ and $E_{\bm{k}} + dE_{\bm{k}}$ is found by summing the contributions over the entire surface:

$$dN = dE_{\bm{k}} \int_{E_{\bm{k}}=\text{const}} g(E_{\bm{k}}, A)dA \equiv \varrho(E_{\bm{k}})dE_{\bm{k}} \quad . \tag{4.149}$$

We can use these functions to evaluate the summation by replacing it with an integral:

$$\sum_{\bm{k}} |u_{\bm{k}}(0)|^2 \chi^{\text{s}}_{\bm{k}} = \int |u_{\bm{k}}(0)|^2 \chi^{\text{s}}_{\bm{k}} g(E_{\bm{k}}, A)dE_{\bm{k}}dA \quad . \tag{4.150}$$

Now $\chi^{\text{s}}_{\bm{k}}$ depends on the Fermi functions $f(\bm{k}, \frac{1}{2})$ and $f(\bm{k}, -\frac{1}{2})$ and thus on the energy $E_{\bm{k}}$ and on the difference in energy of a spin in state $\bm{k}$ with spin up versus that with spin down ($\gamma_{\text{e}}\hbar H_0$ for free electrons). Therefore $\chi^{\text{s}}_{\bm{k}}$ would be the same

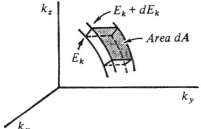

Fig. 4.9. Volume in k-space associated with $dE\,dA$

for any states k having the same value of translational energy E_k. Even when we allow the electrostatic coupling between the electrons to affect the energy to turn over a spin, it may be reasonable to assume that this modification depends at most on E_k. We therefore assume that χ_k^s is a function only of the energy E_k:

$$\chi_k^s = \chi_k^s(E_k) \quad . \tag{4.151}$$

We can therefore rewrite (4.150) as

$$\sum_k |u_k(0)|^2 \chi_k^s = \int |u_k(0)|^2 \chi^s(E_k) g(E_k, A) dA dE_k \quad . \tag{4.152}$$

If we have any function F of E, we define its average value over the surface of constant translational energy E_k, $\langle F(k) \rangle_{E_k}$, as

$$\langle F(k) \rangle_{E_k} = \frac{\int F(k) g(E_k, A) dA}{\int g(E_k, A) dA} = \frac{1}{\varrho(E_k)} \int F(k) g(E_k, A) dA \quad . \tag{4.153}$$

Then we can set the integral over dA in (4.152) equal to

$$\int |u_k(0)|^2 g(E_k, A) dA = \varrho(E_k) \langle |u_k(0)|^2 \rangle_{E_k} \tag{4.154}$$

giving

$$\sum_k |u_k(0)|^2 \chi_k^s = \int \langle |u_k(0)|^2 \rangle_{E_k} \chi^s(E_k) \varrho(E_k) dE_k \quad . \tag{4.155}$$

Now $\chi^s(E_k)$ is zero for all values of E_k that are not rather near to the Fermi energy, since for small values of E_k the two spin states are 100 percent populated, whereas for large E_k, *neither* spin state is occupied. $\chi^s(E_k)$ must look much like Fig. 4.10. $\chi^s(E_k)$ will be nonzero for a region of width about kT around the Fermi energy E_F.

We are therefore justified in assuming $\langle |u_k(0)|^2 \rangle_{E_k}$ varies sufficiently slowly to be evaluated at the Fermi energy and taken outside of the integral,

$$\sum_k |u_k(0)|^2 \chi_k^s = \langle |u_k(0)|^2 \rangle_{E_F} \int \chi^s(E_k) \varrho(E_k) dE_k \quad . \tag{4.156}$$

The integral remaining in (4.156) is readily evaluated in terms of (4.145), since

$$\begin{aligned}\chi_e^s &= \sum_k \chi_k^s = \int \chi_k^s g(E_k, A) dE_k dA \\ &= \int \chi^s(E_k) g(E_k, A) dE_k dA\end{aligned} \tag{4.157a}$$

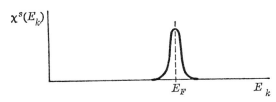

Fig. 4.10
Function $\chi^s(E_k)$ versus E_k

which, performing the integral over A, becomes

$$\chi_e^s = \int \chi^s(E_k)\varrho(E_k)dE_k \quad . \tag{4.157b}$$

Therefore, using (4.147), (4.156), and (4.157), we may say that the interaction with jth nuclear spin is

$$-\gamma_n \hbar I_{zj}\left[\frac{8\pi}{3}\langle|u_{\boldsymbol{k}}(0)|^2\rangle_{E_F} \chi_e^s H_0\right] \quad . \tag{4.158}$$

This is entirely equivalent to the interaction with an *extra* magnetic field ΔH, which aids the applied field H_0, and is given in magnitude by the equation

$$\frac{\Delta H}{H_0} = \frac{8\pi}{3}\langle|u_{\boldsymbol{k}}(0)|^2\rangle_{E_F} \chi_e^s \quad . \tag{4.159}$$

We see that this formula has all the correct properties to explain the experimental results.

1) It predicts that a higher frequency is needed for the metal than in the diamagnetic reference.

2) The fractional shift is independent of ω.

3) Since both $\langle|u_{\boldsymbol{k}}(0)|^2\rangle_{E_F}$ and χ^s are independent of temperature, $\Delta H/H_0$ is as well. Since the larger-Z atoms will have a larger value of $\langle|u_{\boldsymbol{k}}(0)|^2\rangle_{E_F}$ corresponding to the pulling in of their wave function by the larger nuclear charge, the increase of Knight shift with Z is explained.

The Knight shift formula can be checked if one can measure independently $\Delta H/H_0$, $\langle|u_{\boldsymbol{k}}(0)|^2\rangle_{E_F}$ and χ_e^s. There is only one case for which all three quantities are known (Li metal). The spin susceptibility has been measured by *Schumacher* [4.7] by a method we will describe shortly. *Ryter* [4.8] has measured $\langle|u_{\boldsymbol{k}}(0)|^2\rangle_{E_F}$ by measuring the shift of the electron resonance by the nuclear moments. This shift, ΔH_e, is given by

$$\frac{\Delta H_e}{H_0} = \frac{8\pi}{3}\langle|u_{\boldsymbol{k}}(0)|^2\rangle_{E_F} \chi_n^s \tag{4.160}$$

where χ_n^s is the nuclear susceptibility of the Li^7 nuclei. Denoting the number of nuclei per unit volume by N, we have

$$\chi_n^s = \frac{N\gamma_n^2\hbar^2 I(I+1)}{3kT} \quad . \tag{4.161}$$

Thus, since χ_n^s is known, measurement of ΔH_e gives $\langle|u_{\boldsymbol{k}}(0)|^2\rangle_{E_F}^2$. In order to enhance the size of the shift, *Ryter* polarized the nuclei, using the so-called Overhauser effect (see Chap. 7). It is necessary then to modify the formulas slightly, but the principle remains the same.

For comparison with experiment it is convenient to compute $\langle|u_{\boldsymbol{k}}(0)|^2\rangle_{E_F}$ using wave functions normalized to the atomic volume, which we shall call

P_F. We shall use P_A to denote the wave function density at the nucleus for a free atom. It is then convenient to discuss the ratio $P_\mathrm{F}/P_\mathrm{A}$ for Li and Na. A comparison of *Ryter*'s values, theoretical values, and values deduced from combining *Schumacher*'s measurement of $\chi_\mathrm{e}^\mathrm{s}$ with measured Knight shifts is given in Table 4.1. There is excellent agreement among all three values. For Na, *Kohn* and *Kjeldaas* [4.9, 10] find $P_\mathrm{F}/P_\mathrm{A} = 0.80 \pm 0.03$. We can combine this value with the measured Knight shift to obtain a quasi-experimental value of $\chi_\mathrm{e}^\mathrm{s}$. Before presenting these results, we shall describe *Schumacher*'s direct measurement of $\chi_\mathrm{e}^\mathrm{s}$.

Table 4.1

	$P_\mathrm{F}/P_\mathrm{A}$ in Li
Kohn and *Kjeldaas* theoretical	0.49 ± 0.05
Experimental ($\chi_\mathrm{e}^\mathrm{s}$ plus Knight shift)	0.45 ± 0.03
Ryter (experimental)	0.442 ± 0.015

The fundamental problem in measuring $\chi_\mathrm{e}^\mathrm{s}$ is how to distinguish it from the other contributions to the total susceptibility, which (in a metal) are quite comparable in size. The method employed by *Schumacher* is to isolate the spin contribution by use of magnetic resonance. From the Kramers-Kronig relations we have for the electron spin susceptibility $\chi_\mathrm{e}^\mathrm{s}$:

$$\chi_\mathrm{e}^\mathrm{s} = \frac{2}{\pi} \int_0^\infty \frac{\chi_\mathrm{e}'' d\omega}{\omega} \tag{4.162}$$

where χ_e'' is the imaginary part of the conduction electron spin susceptibility. For a sufficiently narrow resonance, we may neglect the variation in ω across the absorption line, taking it out of the integral. We can then change the integration from one over frequency to one over field, using $\omega = \gamma H$:

$$\chi_\mathrm{e}^\mathrm{s} = \frac{2}{\pi} \frac{1}{\omega_0} \int_0^\infty \chi_\mathrm{e}'' d\omega = \frac{2}{\pi} \frac{\gamma_\mathrm{e}}{\omega_0} \int_0^\infty \chi_\mathrm{e}'' dH \quad . \tag{4.163}$$

An absolute measurement of the area under the resonance curve will therefore enable one to determine $\chi_\mathrm{e}^\mathrm{s}$. (Actually, the approximation of a narrow resonance is not well fulfilled in *Schumacher*'s case; however, the final formula can be shown still to be correct. For a discussion of these points the reader is referred to *Schumacher*'s paper [4.7]. We may also note that the resonant absorption of energy at the cyclotron frequency occurs degenerate with that of the spin. However, the rapid electron collisions broaden it so much as to render it unobservable. One may be confident that it is only χ_e'' that is being measured.)

Absolute measurements of absorption are always very difficult. *Schumacher* circumvented them by making use of the nuclear resonance of the Li^7 or Na^{23}

nuclei in the same sample for which he measured the conduction electron resonance. For the nuclei, one has a spin susceptibility χ_n^s, given by (4.161) and for which

$$\chi_n^s = \frac{2}{\pi} \frac{\gamma_n}{\omega_0} \int_0^\infty \chi_n'' dH \quad . \tag{4.164}$$

By choosing $\omega_0/2\pi \cong 10$ Mc/s, *Schumacher* could observe either the electron or nuclear resonances simply by changing H_0, the remainder of the apparatus being left unchanged. The nuclear resonance occurred at about 10,000 Gauss, whereas the electron resonance was at only a few Gauss. Then, if we denote the area under the electron or nuclear resonances by A_e or A_n, respectively, we have

$$\frac{\chi_e^s}{\chi_n^s} = \frac{\gamma_e}{\gamma_n} \frac{A_e}{A_n} \quad . \tag{4.165}$$

Since χ_n^s can be computed, we are thus able to determine χ_e^s. Note that we can measure the "area" in any units we wish (such as square centimeters on the face of an oscilloscope) as long as they are the same for both resonances. We do not even need to know how much sample we have, since it is the same for both electrons and nuclei.

The experimental values obtained are listed in Table 4.2 together with various theoretical values. The first column of experimental and theoretical numbers [4.5] gives theoretical values based on noninteracting electrons, but an effective mass has been introduced to take account of the lattice potential. The effective masses are computed by Harvey Brooks, using the quantum defect method. The second column shows the theoretical values obtained by *Sampson* and *Seitz*, who took into account the electron-electron coupling by means of an interpolation formula of Wigner. The next column shows theoretical values due to *Pines,* based upon the Bohm-Pines collective description. Next we give values obtained by using the Knight shift and the *Kohn-Kjeldaas* theoretical values of P_F/P_A. The last column shows *Schumacher*'s results. (See [4.11] for subsequent results.)

Table 4.2 χ_e^s (all values are 10^6 cgs volume units)

	Free electrons	Sampson and Seitz	Bohm-Pines	Knight shift and theoretical P_F/P_A	Schumacher
Li	1.17	2.92	1.87	1.85 ± 0.20	2.08 ± 0.10
Na	0.64	1.21	0.85	0.83 ± 0.03	0.95 ± 0.10

We point out that if *Ryter*'s value of P_F/P_A is used, the Knight shift value is raised, providing excellent agreement with that of *Schumacher*.

The Knight shift calculation is closely related to the problem of the nuclear resonance in samples in which the electron magnetization does not vanish when

$H_0 = 0$, as with ferromagnets or antiferromagnets. Let us discuss the case of ferromagnetism briefly.[5]

Once again the electron-nuclear interaction will consist of the sum of the conventional dipolar coupling $\mathcal{H}_d$ and the s-state interaction $\mathcal{H}_s$:

$$\mathcal{H}_{en} = \mathcal{H}_d + \mathcal{H}_s \quad . \tag{4.166}$$

We shall average this over the electron wave function ψ_e, as with the Knight shift, to get an effective nuclear Hamiltonian $\mathcal{H}'_{en}$, which will contain the nuclear spins as operators:

$$\mathcal{H}'_{en} = \int_G \psi_e^*(\mathcal{H}_d + \mathcal{H}_s)\psi_e d\tau_e \tag{4.167}$$

where $d\tau_e$ stands for integration over all electron coordinates (spin and spatial), and where the symbol G signifies that the spatial integration goes over the entire physical volume of the sample. By breaking G into the atomic cells $G_1, G_2, \ldots G_N$ of the N atoms of the crystal, we may formally interpret the contribution of $\mathcal{H}_d$ to (4.167) as arising from the summation of the dipolar fields of the electrons on the various atoms. The wave function ψ_e will provide a detailed picture of the spatial distribution of the electron magnetization in each atomic cell. Evaluation of this term is identical to computing the local field due to a volume distribution of electron magnetization. Suppose the magnetization is uniform throughout the sample. If the lattice has cubic symmetry, $\mathcal{H}_d$ will contribute an effective field given by the Lorentz local field:

$$\frac{4\pi}{3}\boldsymbol{M} - \overleftrightarrow{\alpha} \cdot \boldsymbol{M} \tag{4.168}$$

where $\boldsymbol{M}$ is the magnetic dipole moment per unit volume and $\overleftrightarrow{\alpha}$ is a demagnetizing factor (in general a tensor) that expresses the effect of the "magnetic poles" on the outer surface of the sample. For example $\overleftrightarrow{\alpha} = 4\pi/3(\boldsymbol{ii} + \boldsymbol{jj} + \boldsymbol{kk})$ for a sphere.

The s-state term may be interpreted as contributing a magnetic field $\boldsymbol{H}_{sj}$ at the jth nucleus.

$$\boldsymbol{H}_{sj} = -\frac{8\pi}{3}\gamma_e\hbar \sum_l \int \psi_e^* S_l \delta(\boldsymbol{r}_l - \boldsymbol{R}_j)\psi_e d\tau_e \quad . \tag{4.169}$$

If we took the wave function to be a product of one electron states $|\beta\rangle$ of a set of quantum numbers β, we would have, then,

$$\boldsymbol{H}_{sj} = -\frac{8\pi}{3}\gamma_e\hbar \sum_{\beta \text{ occupied}} \langle\beta|S\delta(\boldsymbol{r} - \boldsymbol{R}_j)|\beta\rangle \tag{4.170}$$

---

[5] See the references listed under "Nuclear Resonance in Ferromagnets and Antiferromagnets" in the Bibliography.

where "occupied" means that we include in the sum only those states $|\beta)$ containing an electron and we have omitted the subscript l from S and r. By using the fact that $\boldsymbol{\mu}_e = -\gamma_e \hbar S$, we have

$$H_{sj} = \frac{8\pi}{3} \sum_{\beta \text{ occupied}} (\beta|\boldsymbol{\mu}_e \delta(r-R_j)|\beta) \quad . \tag{4.171}$$

Since the matrix element involves coordinates of only one electron, the various values of l now appear as the values of β that are occupied. In a substance such as iron, we may think of some values of β as corresponding to closed shells, some to the $3d$ band, and some to the $4s$ band. We shall discuss these contributions shortly.

The contribution of the term $\mathcal{H}_d$ is somewhat different for a ferromagnet than for a paramagnet. For the latter, the magnetization is uniform in both magnitude and direction for ellipsoidal samples, and the simple demagnetizing arguments follow. For a ferromagnet the magnetization within a domain is uniform, but the various domains have differing magnetization vectors. Thus, for a soft ferromagnet in zero applied field, the magnetization averaged over a volume large compared with the domain size is zero. The density of magnetic poles on the outer surface therefore vanishes. Within the body of the ferromagnet, div $M = 0$ even at domain boundaries. If, then, we calculate the dipolar contribution to the magnetic field at a nucleus, we may proceed as follows.

We draw a small sphere about the nucleus, of radius small enough to lie within one domain. We compute the field due to magnetization on atoms within the sphere by a direct sum. The atoms outside the sphere are treated in the continuum approximation. For cubic symmetry, the atoms within the sphere give zero total contribution. The atoms outside the sphere contribute as a result of the surface pole density on the inner sphere and the outer sample surface. The former is the contribution $4\pi M/3$, where M is the magnetization within the domain containing the nucleus. The latter contributes $- \overleftrightarrow{\alpha} \cdot M'$, where M' is the magnetization averaged over a volume large compared with a domain size. The total field seen by the jth nucleus H_{Tj} is therefore given by

$$H_{Tj} = H_0 + \frac{4\pi}{3} M - \overleftrightarrow{\alpha} \cdot M' + H_{sj} \tag{4.172}$$

where H_0 is an externally applied field. Although H_0 and M' vanish in zero applied field, H_{Tj} does not. Therefore we have a "zero field" resonance. Such a resonance was first observed by *Gossard* and *Portis* [4.12] in the face-centered cubic form of cobalt. Using the Co^{59} resonance, the measured $H_{sj} = 213,400$ Gauss. In iron, H_{sj} is 330,000 Gauss. H_{sj} has also been observed by means of the Mössbauer effect. It was discovered there that application of a static field H_0 *lowered* the resonance frequency, showing the H_{sj} points opposed to the magnetization M.

The contribution from the $3d$ and $4s$ shells in iron is expected by *Marshall* [4.13] to give field of 100,000 to 200,000 Gauss *parallel* to the local magnetization. Therefore the inner electrons must give a field of about 400,000 Gauss opposed to the local magnetization [4.14].

This phenomenon, called *core polarization*, was actually already known from electron magnetic resonance of paramagnetic ions for which the $4s$ electrons are missing. In principle the $3d$ electrons are incapable of giving an isotropic hyperfine coupling, since d-states vanish at the nucleus. However, the d-electrons are coupled to inner shell electrons electrostatically, the coupling for an inner electron of spin parallel to the d-electron spin being different from that of an electron whose spin is opposed to that of the d-electron. Consequently the spatial part of two wave functions such as the $3s$ are different for the two spin states. The spin magnetization of the two electrons does *not* add to zero at all points of the electron cloud. We can see from (4.171) that if the $3s$ electron densities at the nucleus differ, there will be a nonzero contribution from the $3s$ electrons to H_{sj}, even though their spins are opposed.

4.8 Single Crystal Spectra

We saw in Sect. 4.5 that the chemical shift for a p-state such as $xf(r)$ depended on the orientation of the static field $\boldsymbol{H}_0$, with respect to the bond axes. If $\boldsymbol{H}_0$ was either parallel or perpendicular to the bond direction, an extra field was induced, acting on the nucleus, which was parallel to $\boldsymbol{H}_0$. For an arbitrary orientation, one can resolve $\boldsymbol{H}_0$ into x-, y-, and z-components. H_{0x}, H_{0y}, H_{0z}, in general getting an interaction of external field with nuclear spin involving bilinear products such as $H_{0x} I_y$ etx. *Bloembergen* and *Rowland* [4.15] showed theoretically and experimentally that the same thing is true for a Knight shift if one includes the conventional dipolar coupling (4.98) of the nucleus with the spins of the conduction electrons in addition to the Fermi contact term. Although the conventional dipolar term does not give rise to a net shift of the resonance if one averages over all orientations of $\boldsymbol{H}_0$ with respect to the crystal axes, as for a powder sample or as in a liquid metal, it nevertheless gives an orientation dependence to the Knight shift.

To deal with these anisotropies, one expresses the bilinear form with coupling coefficients which are components of a tensor interaction. Then, writing the Knight shift K and the chemical shift σ as tensors using a dyadic notation where

$$\overleftrightarrow{\sigma} = i\sigma_{xx}i + i\sigma_{xy}j + i\sigma_{xz}k + j\sigma_{yx}i + j\sigma_{yy}j + j\sigma_{yz}k$$
$$+ k\sigma_{zx}i + k\sigma_{zy}j + k\sigma_{zz}k \tag{4.173}$$

and similarly for $\overleftrightarrow{K}$, we get a Hamiltonian

$$\mathcal{H} = -\gamma\hbar \boldsymbol{H}_0 \cdot (\overleftrightarrow{1} - \overleftrightarrow{\sigma} + \overleftrightarrow{K}) \cdot \boldsymbol{I} \tag{4.174}$$

where $\overleftrightarrow{1}$ is the identity dyadic

$$\overleftrightarrow{1} = ii + jj + kk \quad . \tag{4.175}$$

Explicit evaluation of $\sigma_{\alpha\beta}$ and $K_{\alpha\beta}$ ($\alpha = x,y,z$; $\beta = x,y,z$) shows that both tensors are symmetric

$$\sigma_{xy} = \sigma_{yx} \quad , \quad \text{etc.} \quad , \tag{4.176}$$

so that one can find principal axes. Denoting these axes by X, Y, Z with unit vectors i_p, j_p, and k_p, we have

$$\overleftrightarrow{\sigma} = i_\text{p}\sigma_{XX}i_\text{p} + j_\text{p}\sigma_{YY}j_\text{p} + k_\text{p}\sigma_{ZZ}k_\text{p} \tag{4.177}$$

and similarly for $\overleftrightarrow{K}$. In general, $\overleftrightarrow{K}$ and $\overleftrightarrow{\sigma}$ do not have to have the same principal axes, but we will assume for simplicity that they do.

Then if $\boldsymbol{H}_0$ lies along a principal axis the resonance frequency is

$$\begin{aligned} \omega_X &= \gamma H_0(1 - \sigma_{XX} + K_{XX}) \equiv \omega_a + \omega_0 \\ \omega_Y &= \gamma H_0(1 - \sigma_{YY} + K_{YY}) \equiv \omega_b + \omega_0 \\ \omega_Z &= \gamma H_0(1 - \sigma_{ZZ} + K_{ZZ}) \equiv \omega_c + \omega_0 \end{aligned} \tag{4.178}$$

where

$$\omega_0 = \gamma H_0 \quad . \tag{4.179}$$

If $\boldsymbol{H}_0$ lies in a more general direction we can express (4.174) by defining an effective field $\boldsymbol{H}_\text{eff}$

$$\mathcal{H} = -\gamma\hbar \boldsymbol{H}_\text{eff} \cdot \boldsymbol{I} \quad \text{where} \tag{4.180}$$

$$\begin{aligned} H_{\text{eff}\,X} &= (1 - \sigma_{XX} + K_{XX})H_{0X} \\ H_{\text{eff}\,Y} &= (1 - \sigma_{YY} + K_{YY})H_{0Y} \\ H_{\text{eff}\,Z} &= (1 - \sigma_{ZZ} + K_{ZZ})H_{0Z} \quad . \end{aligned} \tag{4.181}$$

Then the resonance frequency ω is given by

$$\omega = \gamma H_\text{eff} \quad \text{where} \tag{4.182a}$$

$$H_\text{eff} = \sqrt{H_{\text{eff}\,X}^2 + H_{\text{eff}\,Y}^2 + H_{\text{eff}\,Z}^2} \quad . \tag{4.182b}$$

Defining the three direction cosines α_X, α_Y, and α_Z by

$$\begin{aligned} \alpha_X &= H_{0X}/H_0 \\ \alpha_Y &= H_{0Y}/H_0 \\ \alpha_Z &= H_{0Z}/H_0 \quad , \end{aligned} \tag{4.183}$$

we get from (4.180) and (4.181)

$$\omega = \gamma H_0 \left[(1 - \sigma_{XX} + K_{XX})^2 \alpha_X^2 + (1 - \sigma_{YY} + K_{YY})^2 \alpha_Y^2 \right.$$

$$+ (1 - \sigma_{ZZ} + K_{ZZ})^2 \alpha_Z^2 \Big]^{1/2} \quad . \tag{4.184}$$

Since the shift parameters σ_{XX}, K_{XX}, etc. are in general very small compared to 1, and since $\alpha_X^2 + \alpha_Y^2 + \alpha_Z^2 = 1$, we can rewrite (4.183) to get

$$\begin{aligned}\omega &= \gamma H_0[1 + (K_{XX} - \sigma_{XX})\alpha_X^2 + (K_{YY} - \sigma_{YY})\alpha_Y^2 + (K_{ZZ} - \sigma_{ZZ})\alpha_Z^2]\\ &= \omega_0 + \omega_a \alpha_X^2 + \omega_b \alpha_Y^2 + \omega_c \alpha_Z^2 \quad .\end{aligned} \tag{4.185}$$

The approximation that the shifts are small could in fact be introduced in the original Hamiltonian. Then we write $\mathcal{H}$ as a large part $\mathcal{H}_0$ plus a perturbation $\mathcal{H}_\mathrm{p}$

$$\mathcal{H} = \mathcal{H}_0 + \mathcal{H}_\mathrm{p} \quad . \tag{4.186}$$

Then with $\boldsymbol{H}_0 = \boldsymbol{k} H_0$, where $\boldsymbol{k}$ is a unit vector along the laboratory z-direction, we get

$$\mathcal{H}_0 = -\gamma \hbar H_0 I_z \tag{4.187a}$$

$$\mathcal{H}_\mathrm{p} = +\gamma \hbar H_0 \boldsymbol{k} \cdot (\overleftrightarrow{\sigma} - \overleftrightarrow{K}) \cdot \boldsymbol{I} \quad . \tag{4.187b}$$

But (4.187a) shows that I_z commutes with the large part of the Hamiltonian, so we keep only those terms in $\mathcal{H}_\mathrm{p}$ which are diagonal in I_z. Thus

$$\mathcal{H}_\mathrm{p} = \gamma \hbar H_0 \boldsymbol{k} \cdot (\overleftrightarrow{\sigma} - \overleftrightarrow{K}) \cdot \boldsymbol{k} I_z \quad . \tag{4.188}$$

Utilizing (4.183) we have

$$\boldsymbol{k} \cdot \boldsymbol{i}_\mathrm{p} = \alpha_X \quad , \quad \boldsymbol{k} \cdot \boldsymbol{j}_\mathrm{p} = \alpha_Y \quad , \quad \boldsymbol{k} \cdot \boldsymbol{k}_\mathrm{p} = \alpha_Z \tag{4.189}$$

giving us

$$\mathcal{H}_\mathrm{p} = \gamma \hbar H_0 I_z [\alpha_X^2 (\sigma_{XX} - K_{XX}) + \alpha_Y^2 (\sigma_{YY} - K_{YY}) \\ + \alpha_Z^2 (\sigma_{ZZ} - K_{ZZ})] \tag{4.190}$$

or

$$\mathcal{H}_\mathrm{p} = -\gamma \hbar H_0 [1 + \alpha_X^2 (K_{XX} - \sigma_{XX}) + \alpha_Y^2 (K_{YY} - \sigma_{YY}) \\ + \alpha_Z^2 (K_{ZZ} - \sigma_{ZZ})] I_z \quad . \tag{4.191}$$

It is often convenient to express the orientation of $\boldsymbol{H}_0$ in terms of spherical coordinates (θ, ϕ), so that

$$\begin{aligned}\alpha_X &= \sin \theta \cos \phi \\ \alpha_Y &= \sin \theta \sin \phi \\ \alpha_Z &= \cos \theta \quad .\end{aligned} \tag{4.192}$$

Then

$$\omega = \omega_0 + \omega_a \sin^2 \theta \cos^2 \phi + \omega_b \sin^2 \theta \sin^2 \phi + \omega_c \cos^2 \theta \tag{4.193}$$

and

$$\mathcal{H} = -\gamma\hbar H_0[1 + (K_{XX} - \sigma_{XX})\sin^2\theta\cos^2\phi$$
$$+ (K_{YY} - \sigma_{YY})\sin^2\theta\sin^2\phi$$
$$+ (K_{ZZ} - \sigma_{ZZ})\cos^2\theta]I_z \quad . \tag{4.194}$$

This result can be expressed in another manner which is useful when discussing the techniques of magic angle spinning which we take up in Chapter 8. Utilizing the trigonometric identities

$$\cos^2\phi = \tfrac{1}{2}(1 + \cos 2\phi)$$
$$\sin^2\phi = \tfrac{1}{2}(1 - \cos 2\phi) \quad , \tag{4.195}$$

we find

$$\omega = \omega_0 + \frac{1}{3}(\omega_a + \omega_b + \omega_c) + \left(\frac{2\omega_c - \omega_a - \omega_b}{3}\right)\left(\frac{3\cos^2\theta - 1}{2}\right)$$
$$+ \left(\frac{\omega_a - \omega_b}{2}\right)\sin^2\theta\cos 2\phi \quad . \tag{4.196}$$

Defining

$$\overline{\sigma} = \tfrac{1}{3}(\sigma_{XX} + \sigma_{YY} + \sigma_{ZZ})$$
$$\sigma_{\text{LO}} = \frac{2\sigma_{ZZ} - \sigma_{XX} - \sigma_{YY}}{3} = (\sigma_{ZZ} - \overline{\sigma}) \tag{4.197}$$

(LO for "longitudinal")

$$\sigma_{\text{TR}} \equiv \sigma_{XX} - \sigma_{YY} = (\sigma_{XX} - \overline{\sigma}) - (\sigma_{YY} - \overline{\sigma})$$

(TR for "transverse") and correspondingly for the components of K, we get

$$\mathcal{H} = -\gamma\hbar H_0\left[1 + (\overline{K} - \overline{\sigma}) + (K_{\text{LO}} - \sigma_{\text{LO}})\left(\frac{3\cos^2\theta - 1}{2}\right)\right.$$
$$\left.+ \left(\frac{K_{\text{TR}} - \sigma_{\text{TR}}}{2}\right)\sin^2\theta\cos 2\phi\right]I_z \quad . \tag{4.198}$$

The two angular functions are linear combinations of the spherical harmonics Y_{2m}. Therefore, they average to zero over a sphere. Since in a liquid there is generally rapid tumbling motion, such a spherical average is appropriate, giving

$$\omega = \omega_0 + \tfrac{1}{3}(\omega_a + \omega_b + \omega_c)$$
$$= \gamma H_0\left[\left(1 + \frac{K_{XX} + K_{YY} + K_{ZZ}}{3}\right) - \left(\frac{\sigma_{XX} + \sigma_{YY} + \sigma_{ZZ}}{3}\right)\right] \quad . \tag{4.199}$$

For a system such as a bond $xf(r)$ which possesses axial symmetry, it is convenient to take ω_c to lie along the axis, so that $\omega_a = \omega_b$, giving

$$\omega = \omega_0 + \frac{1}{3}(\omega_a + \omega_b + \omega_c) + \left(\frac{2\omega_c - \omega_a - \omega_b}{3}\right)\left(\frac{3\cos^2\theta - 1}{2}\right) \quad . \tag{4.200}$$

Often in working with solids one has a sample consisting of many small crystallites in the form of a powder. Then the orientation of the crystallites is random with respect to H_0. The anisotropic nature of the chemical and Knight shifts then gives rise to a line broadening and to spectra which have quite distinct shapes depending on the relative values of ω_a, ω_b, and ω_c. Such absorption spectra are called "powder patterns" and are discussed in Appendix I.

4.9 Second-Order Spin Effects – Indirect Nuclear Coupling

We have discussed the role of electron spin coupling to nuclei in paramagnetic or ferromagnetic materials. Since, in a diamagnetic substance, the total spin of the electrons vanishes, the nuclei experience zero coupling to the electron spins in first order. Effects are found if one considers the coupling in second order,[6] however. The coupling is manifested through an apparent coupling of nuclei among themselves, the so-called indirect coupling.

The indirect coupling was discovered independently by *Hahn* and *Maxwell* [4.16] and by *Gutowsky* et al. [4.17]. The phenomena they observed are illustrated by the case of PF$_3$, a molecule in which all nuclei have spin $\frac{1}{2}$. In liquid PF$_3$, the rapid tumbling narrows the line. It is found that both the P^{31} and F^{19} resonances consist of several lines, as illustrated in Fig. 4.11. Since all the fluorine nuclei are chemically equivalent, the splittings cannot be due to chemical shifts. (Furthermore there is only one phosphorus atom per molecule but four phosphorus frequencies.) The fact that the individual lines themselves are narrow shows that the motion is sufficiently rapid to narrow the direct dipolar coupling. Moreover, the splittings are found to be independent both of temperature and static field. The number and relative intensity of lines are as though each nuclear species experienced a magnetic field proportional to the z-component of the total spin of the other species. It was found that $\delta\omega_F = \delta\omega_P$ (see Fig. 4.11), where $\delta\omega_F$ and $\delta\omega_P$ are the frequency separations of adjacent lines in the phosphorus and fluorine spectra, respectively. These facts indicated the coupling was somehow related to the nuclear magnetic moments.

The original explanation proposed was that one nucleus induced currents in the electron cloud, which then coupled to the other nucleus. In a simple picture, the induced currents are represented by an induced electron magnetic moment.

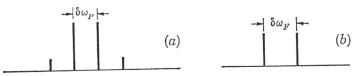

Fig. 4.11. (a) The P^{31} resonance in PF$_3$. The lines are equally spaced an amount $\delta\omega_P$, and the intensities are 1:3:3:1. (b) The F^{19} resonance in PF$_3$

---

[6] See references to "$I_1 \cdot I_2$ Coupling" in the Bibliography.

If this moment were *isotropic* (as one changes the orientation of the molecule with respect to the nuclear moment), the coupling to a second nucleus would average to zero in a liquid, owing to the rapid random tumbling of the molecules. However, as we observed in connection with the chemical shifts, the induced moment is in general not isotropic. We can estimate the size of the coupling between the two nuclei from second-order perturbation theory. The first nucleus exerts a magnetic field $\gamma_1 \hbar \overline{(1/r^3)}$ where $\overline{(1/r^3)}$ is the average of the inverse cube of the distance between the electron and the first nucleus that partially unquenches the orbital angular momentum, producing a fractional admixture of excited state of $\gamma_1 \gamma_e \hbar^2 \overline{(1/r^3)}/\Delta E$, where ΔE is the energy to the excited state. A complete unquenching would produce a magnetic field at the second nucleus of $\gamma_e \hbar/R^3$, where R is the distance between the nuclei (we treat the electron orbital magnetization as equivalent to a magnetic dipole). Therefore the order of magnitude of the nuclear-nuclear interaction energy E_{12} on this model is

$$E_{12} \cong \frac{\gamma_1 \gamma_e \hbar^2 \overline{(1/r^3)}}{\Delta E} \frac{\gamma_e \hbar}{R^3} \gamma_2 \hbar \quad . \tag{4.201}$$

This formula fails by an order of magnitude or more in accounting for the facts. However, it has the virtue of making it seem reasonable that the splittings in PF$_3$ were an order of magnitude larger than those in PH$_3$, since clearly this mechanism is closely related to chemical shifts that are always smaller for hydrogen than for fluorine.

As was pointed out by *Hahn* and *Maxwell*, and *Gutowsky* and *McCall*, any mechanism such as we have described, and which will lead to a result that is bilinear in the two nuclear moments, must take a very simple form. Since the interaction is averaged over all molecular orientations, it can depend only on the relative orientation of the nuclei; hence it must be of the form

$$A_{12} \boldsymbol{\mu}_1 \cdot \boldsymbol{\mu}_2 \tag{4.202}$$

where A_{12} is independent of temperature and field. These workers also pointed out that this particular form would also explain the puzzling fact that, for example, there were apparently no splittings of fluorines by fluorines in PF$_3$. We shall not give the proof here but physically the explanation is based on the idea that the interaction energy, (4.202), which depends on the relative orientation of spins, is unchanged if *both* spins are rotated through the same angle. For equivalent nuclei such as the three fluorines in PF$_3$, one cannot rotate one fluorine spin without rotating the others by an equal amount, since the alternating and static fields are identical at all three fluorines. Therefore the coupling between equivalent nuclei does not affect the resonance frequency.

Ramsey and *Purcell* [4.18] proposed another mechanism utilizing the electron spins, which was substantially larger because, as we shall explain, it allowed the *two* nuclei to interact with nearby electrons, in contrast with the orbital mechanism where only one nucleus is on the same atom as the electron that is polarized. We may schematize their mechanism as shown in Fig. 4.12.

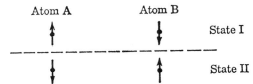

Fig. 4.12. Because of the bonds between atoms A and B, the electron wave function is formed by an equal mixture of the state I, in which the electron moment on atom A points up (that on B, down), and state II, in which the spin orientation is reversed

In the absence of a nuclear moment the electron bond will consist of an equal mixture of the states I and II, shown in Fig. 4.12. If now we put a nucleus on atom A with its magnetic moment pointing up, state I will be slightly favored over state II. The electronic spin magnetic moment of atom A will have a slight polarization up; that on atom B, down. Therefore a nucleus on atom B will find a nonzero field owing to its own electron. Since this field would reverse if the nucleus on atom A were reversed, an effective nuclear-nuclear coupling results. We can easily estimate the size of the coupling. The fractional excess of state I over state II will be

$$\frac{\frac{8\pi}{3}\gamma_1\gamma_e\hbar^2|u(0)|_A^2}{\Delta E} = \frac{\text{hyperfine energy}}{\text{electrostatic energy}} \qquad (4.203)$$

where $|u(0)|_A^2$ is the wave-function density of the electron at atom A, and ΔE is the energy to an appropriate excited state. The coupling of the electrons on atom B to nucleus 2 are thus given by the product of the electron spin coupling if the electron spin is in one orientation only, $(8\pi/3)\gamma_1\gamma_e\hbar^2|u(0)|_B^2$, times the excess fraction of the time the electron is in the favored orientation. Thus the coupling is

$$\frac{\left(\frac{8\pi}{3}\gamma_1\gamma_e\hbar^2|u(0)|_A^2\right)\left(\frac{8\pi}{3}\gamma_1\gamma_e\hbar^2|u(0)|_B^2\right)}{\Delta E} \qquad (4.204)$$

This coupling turns out to have the correct order of magnitude. If the electron functions do not contain an *s*-part, we should instead use the ordinary dipolar coupling between the electron and nuclear spins.

The extension of these ideas to solids was made independently by *Bloembergen* and *Rowland* [4.19] and by *Ruderman* and *Kittel* [4.20]. We shall discuss the situation for metals, confining our attention to the coupling via the *s*-state hyperfine coupling. Since the metal is not diamagnetic, we should concern ourselves with the possibility of a first-order effect — related, therefore, to the Knight shift. This mechanism of coupling was originally proposed by *Fröhlich* and *Nabarro* [4.21]. As *Yosida* [4.22] has explained, however, the Fröhlich-Nabarro effect is actually included in the second-order calculation. We shall discuss the physical reason shortly, but for the present, we shall simply ignore any first-order effect. *Van Vleck* [4.23] also discussed this effect.

The effect of the magnetic moment of a nucleus at any lattice site is to make that site a region favorable for an electron of parallel magnetic moment but unfavorable for an electron of antiparallel moment. In order to take advantage of the magnetic interaction, an electron of parallel moment will distort its wave function to be larger in the vicinity of the nucleus. The distortion is brought about by mixing in other states k of the same spin orientation. As we shall see, the result is as though only states above the Fermi surface were added. The wave functions of the Bloch states are added so as to be in phase with the unperturbed function (Fig. 4.13) at the nucleus in order to interfere constructively at that point, but because of the spread in wavelengths, they rapidly get out of step as one moves away from the nucleus.

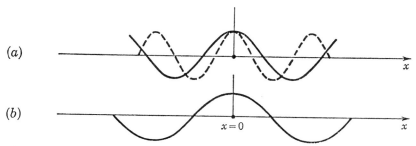

Fig. 4.13a,b. The unperturbed function and two of the higher states mixed in. The nucleus is at $x = 0$, which guarantees that the waves be in phase at $x = 0$. Note that the admixed waves beat with one another. (a) Two of the waves mixed in by the perturbation. (b) Unperturbed function

As a result of the beats between the unperturbed and perturbed functions, the original uniform distribution of spin-up charge density (we neglect the variations due to the lattice charge) is changed to have an oscillatory behavior, which dies out as one goes away from the nucleus. The characteristic length describing the attenuation is the wavelength of electrons at the Fermi surface. The resulting charge density of electrons whose moment is parallel to the nucleus is shown in Fig. 4.14.

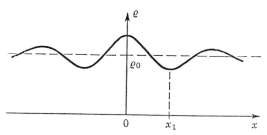

Fig. 4.14. The charge density of electrons whose magnetic moments are parallel to the nuclear moment. The nucleus is located at $x = 0$. ϱ_0 is the charge density in the absence of a nuclear moment. At $x = x_1$, the electron charge is deficient, so that the net electron moment there is opposed to the nuclear moment

We now turn to the actual calculation of these effects. For simplicity we shall calculate the interaction between the two nuclei directly rather than compute the changes in the spatial distribution of electron spin. However, the oscillatory nature of the charge will be apparent from the answer.

We consider, therefore, an electron-nuclear coupling $\mathcal{H}_{en}$ involving only two nuclei of spins I_1 and I_2 and for simplicity, treat only the effect of the s-state coupling. We have, then,

$$\mathcal{H}_{en} = \frac{8\pi}{3}\left[\gamma_1\gamma_e\hbar^2 I_1 \cdot \sum_l S_l\delta(r_l - R_1) + \gamma_2\gamma_e\hbar^2 I_2 \cdot \sum_l S_l\delta(r_l - R_2)\right]$$
$$= \mathcal{H}_1 + \mathcal{H}_2 \qquad (4.205)$$

where we have allowed the nuclei to be different by using two values of gyromagnetic ratio γ_1 and γ_2 and spins I_1 and I_2.

We must take into account the exclusion principle for the electrons. Two methods are available. We could use (4.205) to find perturbed one-electron functions and then fill these functions in accordance with the exclusion principle. Or, we could do perturbation theory in which we used the many electron functions as the unperturbed states. It is this latter procedure that we shall utilize, since it emphasizes that basically we are dealing with a many-electron problem and that the states we use are but one approximation.

Let us therefore consider a many-electron state $|0\rangle$ with energy E_0 and excited states $|n\rangle$ with energy E_n, and compute the second-order energy shift due to $\mathcal{H}_{en}$. We shall as usual assume the total wave function of the system to be a product of the electron and nuclear functions. Denoting the latter by ψ_α with energy E_α, where α represents the nuclear spin quantum numbers, we shall wish to compute the second-order energy shift of states $|0\rangle\psi_\alpha$.

Therefore, for the second-order shift, we have $\Delta E_{0\alpha}^{(2)}$ of the state $|0\rangle\psi_\alpha$:

$$\Delta E_{0\alpha}^{(2)} = \sum_{n,\alpha'} \frac{(0\alpha|\mathcal{H}_{en}|n\alpha')(n\alpha'|\mathcal{H}_{en}|0\alpha)}{(E_0 + E_\alpha) - (E_n + E_{\alpha'})} . \qquad (4.206)$$

Since the electronic energy differences are generally much greater than the differences in nuclear energy, we can neglect E_α and $E_{\alpha'}$ in the denominator.[7]

Writing $\mathcal{H}_{en}$ as $\mathcal{H}_1 + \mathcal{H}_2$, we find

$$\Delta E_{0\alpha}^{(2)} = \sum_{n,\alpha'} \frac{1}{E_0 - E_n}[(0\alpha|\mathcal{H}_1|n\alpha')(n\alpha'|\mathcal{H}_1|0\alpha)$$
$$+ (0\alpha|\mathcal{H}_2|n\alpha')(n\alpha'|\mathcal{H}_2|0\alpha) + (0\alpha|\mathcal{H}_1|n\alpha')(n\alpha'|\mathcal{H}_2|0\alpha)$$
$$+ (0\alpha|\mathcal{H}_2|n\alpha')(n\alpha'|\mathcal{H}_1|0\alpha)] . \qquad (4.207)$$

---

[7] One can think of this as being, therefore, basically a calculation of the coupling between nuclei in zero external field. However, it turns out that the field dependence is very small. This results from the fact that the states E_n have a continuous distribution in energy starting from E_0. Neglecting $E_\alpha - E'_\alpha$ does not seriously perturb the spectra of excited states or the matrix elements. A further discussion of these points is given immediately following (4.223).

The first two terms in the brackets represent the changes in energy we should have were one or the other of the nuclei the only one. The last two terms represent the extra energy when both are simultaneously present, and are thus an interaction energy. Since it is only the interaction that we wish to calculate, we shall consider the last two terms only. We have then

$$\Delta E_{0\alpha}^{(2)} = \sum_{n,\alpha'} \frac{(0\alpha|\mathcal{H}_1|n\alpha')(n\alpha'|\mathcal{H}_2|0\alpha)}{E_0 - E_n} + \text{complex conjugate} \quad . \quad (4.208)$$

Now, from the form of $\mathcal{H}_1$ and $\mathcal{H}_2$, we can write them as

$$\mathcal{H}_1 = \boldsymbol{I}_1 \cdot \boldsymbol{G}_1 = \sum_{\beta=x,y,z} I_{1\beta} G_{1\beta}$$
$$\mathcal{H}_2 = \boldsymbol{I}_2 \cdot \boldsymbol{G}_2 = \sum_{\beta=x,y,z} I_{2\beta} G_{2\beta} \quad (4.209)$$

where G_1 and G_2 do not involve the nuclear spin coordinates. By using these relations,

$$\Delta E_{0\alpha}^{(2)} = \sum_{\beta,\beta'} \sum_n \frac{(0|G_{1\beta}|n)(n|G_{2\beta'}|0)}{E_0 - E_n} \sum_{\alpha'} (\alpha|I_{1\beta}|\alpha')(\alpha'|I_{2\beta'}|\alpha) + \text{c.c.}$$
$$= \sum_{\beta,\beta'} \sum_n \frac{(0|G_{1\beta}|n)(n|G_{2\beta'}|0)}{E_0 - E_n} (\alpha|I_{1\beta} I_{2\beta'}|\alpha) + \text{c.c.} \quad . \quad (4.210)$$

To evaluate the energy of (4.210), we should need to specify the nuclear states $|\alpha\rangle$. Just what they would be would depend on the total nuclear Hamiltonian, which would include such things as the coupling of the nuclei to the external static field H_0, the dipolar coupling between nuclei, and so forth. It is convenient to note that whatever the states $|\alpha\rangle$ may be, the energy $\Delta E_{0\alpha}^{(2)}$ is just what we should find as the first-order perturbation contribution of an extra term in the nuclear Hamiltonian $\mathcal{H}_{\text{eff}}$ given by

$$\mathcal{H}_{\text{eff}} = \sum_{\beta,\beta'} I_{1\beta} I_{2\beta'} \left[\sum_n \frac{(0|G_{1\beta}|n)(n|G_{2\beta'}|0)}{E_0 - E_n} \right] + \text{c.c.} \quad . \quad (4.211)$$

By expressing the couplings G_1 and G_2 explicitly, we obtain

$$\mathcal{H}_{\text{eff}} = C \sum_n \boldsymbol{I}_1 \cdot \frac{(0|\sum_l \boldsymbol{S}_l \delta(\boldsymbol{r}_l - \boldsymbol{R}_1)|n)(n|\sum_l \boldsymbol{S}_l \delta(\boldsymbol{r}_l - \boldsymbol{R}_2)|0)}{(E_0 - E_n)} \cdot \boldsymbol{I}_2 + \text{c.c.} \quad (4.212)$$

where

$$C = \frac{64\pi^2}{9} \gamma_1 \gamma_2 \gamma_e^2 \hbar^4 \quad .$$

We shall now take the states $|0\rangle$ to be products of Bloch functions. Denoting the product of a Bloch function and a spin function by letters A, and so on,

$$|0) = \frac{1}{\sqrt{N!}} \sum_P (-1)^P P[A(1)B(2)C(3)\ldots]$$
$$|n) = \frac{1}{\sqrt{N!}} \sum_P (-1)^P P[A'(1)B'(2)C'(3)\ldots] \quad .$$
(4.213)

Of course the permutation causes any function to vanish if any two functions such as A and B are identical. Consider the matrix element of a perturbation V, which is symmetric among all the electrons; that is, it is unchanged by interchanging the electron numbering:

$$(n|V|0) = \frac{1}{\sqrt{N!}} \sum_{P,P'} (-1)^{P+P'}$$
$$\times \int P'[A'(1)B'(2)\ldots]^* V P[A(1)B(2)C(3)\ldots] d\tau \quad . \quad (4.214)$$

Since V is symmetric, it is only the *relative* ordering of the states that counts. We can express this fact by defining the permutation P''. P'' is the permutation that, following P, gives the same ordering of electrons as P' alone. That is, $P''P = P'$. By making this substitution into (4.214), we get

$$(n|V|0) = \frac{1}{N!} \sum_{P,P''} (-1)^{(2P+P'')} \int P''P[A'(1)B'(2)\ldots]^*$$
$$\times V P[A(1)B(2)C(3)\ldots] d\tau$$
$$= \frac{1}{N!} \sum_{P,P''} (-1)^{P''} \int P''P[A'(1)B'(2)\ldots]^* V P[A(1)B(2)\ldots] d\tau$$
$$= \sum_{P''} (-1)^{P''} \int P''[A'(1)B'(2)\ldots]^* V [A(1)B(2)\ldots] d\tau \quad (4.215)$$

where the last step follows because V is unchanged by relabelling the electrons. Let us now consider V to be a sum of one-electron operators:

$$V = \sum_l V(l) \quad (4.216)$$

where $V(l)$ depends only on the coordinates of the lth particle. Consider, for example, the contribution from $l = 1$:

$$(n|V_1|0) = \sum_{P''} (-1)^{P''} \int P''[A'(1)B'(2)\ldots]^* V(1)[A(1)B(2)C(3)\ldots] d\tau \quad .$$
(4.217)

This will vanish unless the state $|n)$ contains $B(2)C(3)\ldots$. Let us therefore write such an excited state as

$$|n) = \frac{1}{\sqrt{N!}} \sum_P (-1)^P P[A'(1)B(2)C(3)\ldots] \quad . \quad (4.218)$$

This makes

$$(n|V_1|0) = \sum_{P''}(-1)^{P''}\int P''[A'(1)B(2)C(3)\ldots]^*$$
$$\times V(1)[A(1)B(2)C(3)\ldots]d\tau$$
$$= 0 \quad \text{if } A' \text{ is identical to any of the functions } B, C, D, \text{ etc.}$$
$$= (A'|V(1)|A) \quad \text{otherwise} \quad . \tag{4.219}$$

It is clear that the various values of l in (4.216) will simply pick out the various states A, B, C in $|0\rangle$, and the sum over excited states $|n\rangle$ will pick out the states A', B', and so on, which are *not* occupied in $|0\rangle$. We can therefore write (4.212) as

$$\mathcal{H}_{\text{eff}} = C \sum_{\substack{k,s \text{ occupied} \\ k',s' \text{ unoccupied}}} I_1$$
$$\times \frac{(ks|S\delta(r-R_1)|k's')(k's'|S\delta(r-R_2)ks)}{E_{ks} - E_{k's'}} \cdot I_2 + \text{c.c.} \tag{4.220}$$

where we have replaced $E_0 - E_n$ by $E_{ks} - E_{k's'}$, since the states E_0 and E_n differ in energy solely by the transfer of one electron from the state $|ks\rangle$ to $|k',s'\rangle$. The terms "k,s occupied", "k',s' unoccupied", refer to whether or not these states are occupied by electrons in the wave function $|0\rangle$.

If we define the functions $p(k,s)$ by

$$p(k,s) = 1 \quad \text{if } k,s \text{ is occupied in state } |0\rangle$$
$$= 0 \quad \text{if } k,s \text{ is unoccupied in state } |0\rangle \tag{4.221}$$

we can easily remove the restrictions on k,s from the summation:

$$\mathcal{H}_{\text{eff}} = C \sum_{k,s;k',s'} I_1 \cdot \frac{(ks|S\delta(r-R_1)|k's')(k's'|S\delta(r-R_2)|ks)}{E_{ks} - E_{k's'}}$$
$$\times \frac{p(k,s)[1 - p(k',s')]}{E_{ks} - E_{k's'}} \cdot I_2 + \text{c.c.} \tag{4.222}$$

In order to express the variation of $\mathcal{H}_{\text{eff}}$ with the temperature of the electrons, we must average $\mathcal{H}_{\text{eff}}$ over an ensemble. This will simply replace $p(k,s)$ by $f(k,s)$, the Fermi function. We have then,

$$\mathcal{H}_{\text{eff}} = C \sum_{k,s;k',s'} I_1 \cdot \frac{(ks|S\delta(r-R_1)|k's')(k's'|S\delta(r-R_2)|ks)}{E_{ks} - E_{k's'}}$$
$$\times \frac{f(k,s)[1 - f(k',s')]}{E_{ks} - E_{k's'}} \cdot I_2 + \text{c.c.}$$
$$= C \sum_{k,s;k',s'} I_1 \cdot (s|S|s')(s'|S|s) \cdot I_2$$
$$\times \frac{(k|\delta(r-R_1)|k')(k'|\delta(r-R_2)|k)f(k,s)[1-f(k',s')]}{E_{ks} - E_{k's'}} + \text{c.c.} \quad . \tag{4.223}$$

Now, the Fermi functions as well as the energies E_{ks} depend on the energy associated with the electron spin quantum number. For example, the electron spin Zeeman energy changes with s. However, the Fermi levels of the spin-up and spin-down distributions coincide. Thus, at absolute zero, there is a continuous range of E_{ks} up to the Fermi energy and a continuous range of $E_{k's'}$ from the Fermi energy on up. The matrix elements of the δ-functions vary slowly with energy. Therefore the variation of $\mathcal{H}_{\text{eff}}$ with the electron spin energy is very small, and we may forget about the energy of the electron spin coordinate, writing to a good approximation (4.223) in the form it would have in zero magnetic field:

$$\mathcal{H}_{\text{eff}} = C \sum_{kk';s,s'} \boldsymbol{I}_1 \cdot (s|\boldsymbol{S}|s')(s'|\boldsymbol{S}|s) \cdot \boldsymbol{I}_2$$
$$\times \frac{(k|\delta(r-R_1)|k')(k'|\delta(r-R_2)|k)f(k)[1-f(k')]}{E_k - E_{k'}} + \text{c.c.} \quad (4.224)$$

We can now perform the sums over s and s':

$$\sum_{s,s'} \boldsymbol{I}_1 \cdot (s|\boldsymbol{S}|s')(s'|\boldsymbol{S}|s) \cdot \boldsymbol{I}_2 = \sum_{\substack{\beta,\beta'=x,y,z \\ s,s'}} I_{1\beta}(s|S_\beta|s')(s'|S_{\beta'}|s)I_{2\beta'}$$
$$= \sum_{\beta,\beta'} I_{1\beta} I_{2\beta'} \operatorname{Tr}\{S_\beta S_{\beta'}\} \quad . \quad (4.225)$$

But, as we saw in Chap. 3,

$$\operatorname{Tr}\{S_\beta S_{\beta'}\} = \frac{1}{3}S(S+1)(2S+1)\delta_{\beta\beta'}$$
$$= \frac{\delta_{\beta\beta'}}{2} \quad \text{since} \quad S = \frac{1}{2} \quad , \quad (4.226)$$

which gives, finally,

$$\mathcal{H}_{\text{eff}} = \boldsymbol{I}_1 \cdot \boldsymbol{I}_2 \frac{C}{2} \left\{ \sum_{kk'} \frac{(k|\delta(r-R_1)|k')(k'|\delta(r-R_2)|k)f(k)[1-f(k')]}{E_k - E_{k'}} \right.$$
$$\left. + \text{c.c.} \right\} \quad (4.227)$$
$$= A_{12} \boldsymbol{I}_1 \cdot \boldsymbol{I}_2 \quad ,$$

where A_{12} is a constant independent of spin. We now evaluate the matrix elements in terms of the Bloch functions:

$$\psi_k = u_k(r) e^{i k \cdot r} \quad (4.228)$$

where, as before, $u_k(r)$ has the periodicity of the lattice. We have, then,

$$(k'|\delta(r-R_2)|k) = u_{k'}^*(R_2) u_k(R_2) e^{i(k-k') \cdot R_2} \quad (4.229)$$

so that,

$$(k|\delta(r - R_1)|k')(k'|\delta(r - R_2)|k)$$
$$= u^*_{k'}(R_2)u_{k'}(R_1)u^*_k(R_1)u_k(R_2)e^{i(k-k')\cdot(R_2-R_1)} \quad . \tag{4.230}$$

If we assume that R_1 and R_2 are equivalent sites (as, for example, in a simple metal), and define

$$R_{12} = R_2 - R_1 \quad , \tag{4.231}$$

we have

$$\mathcal{H}_{\text{eff}} = \frac{C}{2}I_1 \cdot I_2 \sum_{k,k'} \frac{|u_{k'}(0)|^2|u_k(0)|^2 2\cos[(k-k')\cdot R_{12}]}{E_k - E_{k'}} f(k)[1 - f(k')]$$

$$= I_1 \cdot I_2 \frac{64}{9}\pi^2\gamma_e^2\gamma_1\gamma_2\hbar^4$$
$$\times \sum_{k,k'} \frac{|u_{k'}(0)|^2|u_k(0)|^2 \cos[(k-k')\cdot R_{12}]f(k)[1-f(k')]}{E_k - E_{k'}} \quad . \tag{4.232}$$

It is not possible to evaluate the summation without either some further approximations or some explicit information on the k dependence of wave functions and energy. If one assumes spherical energy surfaces, an effective mass m^*, and that $|u_{k'}(0)|^2$ and $|u_k(0)|^2$ may be replaced by a value appropriate to k and k' near the Fermi energy, one can evaluate the sums. In these terms,

$$E_k = \frac{\hbar^2}{2m^*}k^2 \quad . \tag{4.233}$$

This gives us

$$\mathcal{H}_{\text{eff}} = I_1 \cdot I_2 \frac{64}{9}\pi^2\gamma_e^2\gamma_1\gamma_2\hbar^4|u_{k_F}(0)|^4$$
$$\times \frac{2m^*}{\hbar^2}\sum_{k,k'} \frac{\cos[(k-k')\cdot R_{12}]}{k^2 - k'^2} f(k)[1 - f(k')] \quad . \tag{4.234}$$

The number dN of states in k-space within a solid angle $d\Omega$ and between two spherical shells of radii k and dk is

$$dN = \frac{d\Omega}{4\pi}\frac{k^2 dk}{2\pi^2} \quad . \tag{4.235}$$

Denoting the angle between k and R_{12} as θ, that between k' and R_{12} as θ' (see Fig. 4.15), and using R for $|R_{12}|$, we have

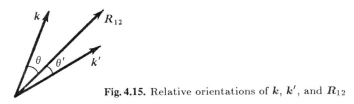

Fig. 4.15. Relative orientations of k, k', and R_{12}

$$\sum_{k,k'} \frac{\cos\left[(k-k')\cdot R_{12}\right]f(k)[1-f(k')]}{k^2-k'^2}$$

$$= \left(\frac{1}{2\pi}\right)^6 \iint \frac{[\cos(kR\cos\theta)\cos(k'R\cos\theta')}{k^2-k'^2}$$

$$+ \frac{\sin(kR\cos\theta)\sin(k'R\cos\theta')]}{k^2-k'^2}$$

$$\times f(k)[1-f(k')]k^2 k'^2 d\Omega\, d\Omega'\, dk\, dk' \quad . \tag{4.236}$$

The integrals over $d\Omega = -2\pi d(\cos\theta)$ and $d\Omega' = -2\pi d(\cos\theta')$ are readily performed to give, for the summation of (4.236),

$$\frac{4}{(2\pi)^4}\frac{1}{R^2} \iint \frac{\sin kR \sin k'R}{k^2-k'^2} kk' f(k)[1-f(k')] dk\, dk' \quad . \tag{4.237}$$

This integral may be evaluated at absolute zero by noting that the limits on k' can go from 0 to ∞, not just k_F to ∞, since if $k' < k_F$ for each $k = k_1$, $k' = k_2$, there is a $k = k_2$, $k' = k_1$ which has the opposite sign for the integrand. The range of k' is then extended to cover $-\infty$ to $+\infty$, and the integral is evaluated by a contour integral. The infinity at $k = k'$ is avoided by taking a principal part. The final result is

$$\mathcal{H}_{\text{eff}} = -\frac{2}{9\pi}\gamma_e^2 \gamma_1 \gamma_2 \hbar^2 m^* |u_{k_F}(0)|^4$$

$$\times \frac{(\sin 2k_F R - 2k_F R \cos 2k_F R)}{R^4} I_1 \cdot I_2 \quad . \tag{4.238}$$

We note that this expression indeed corresponds to an oscillatory behavior as one varies R. For large distances the coupling goes as

$$+\frac{\cos(2k_F R)}{R^3} \quad . \tag{4.239}$$

We see that the dependence on $|u(0)|^4$ will cause the coupling to be large for large-Z atoms. In fact the coupling for the heavier elements is substantially larger than the direct dipolar coupling.

Had we used the conventional dipolar form of coupling between the nuclear and electron spins appropriate to "non" s-states, we should have obtained

$$\mathcal{H}_{\text{eff}} = \left[I_1 \cdot I_2 - \frac{3(I_1 \cdot R_{12})(I_2 \cdot R_{12})}{R_{12}^2}\right] B_{12} \tag{4.240}$$

where B_{12} is a complicated function. It vanishes if there is no "non" s-character to the wave function as seen by a single nucleus; for large distances, B_{12} typically falls off as $1/R_{12}^3$. For a discussion of B_{12} the reader is referred to the paper by *Bloembergen* and *Rowland* [4.19]. In the case of molecules such as PF_3, *Gutowsky* et al. [4.17] show that the assumption of p-type wave functions on *both* P and F together with the "non" s-state coupling gives a coupling of the form

$$A_{12}\mathbf{I}_1 \cdot \mathbf{I}_2 \tag{4.241}$$

when averaged over all orientations of the molecule in the external field. The coupling does not vanish when averaged over molecular orientations, since the induced spin magnetization itself depends on the orientation of the molecule with respect to the nuclear spins.

Couplings such as (4.240) have the same spin dependence as that of the direct dipolar coupling. To emphasize this similarity, the coupling is often referred to as the "pseudo-dipolar" coupling. On the other hand, a coupling such as $A_{12}\mathbf{I}_1 \cdot \mathbf{I}_2$ has the same form as the electrostatic exchange coupling. Since in our case the physical origin is not an exchange integral, the term is referred to as the "pseudo-exchange" coupling.

The effect of pseudo-exchange or pseudo-dipolar coupling on the width and shape of resonance lines can be analyzed by simply adding these terms to the dipolar terms. The number of situations that arise are very numerous. In liquids, the pseudo-dipolar coupling averages to zero, but the pseudo-exchange does not, giving rise to resolved splittings.

For solids, both terms have effects. The pseudo-exchange term, since it commutes with $I_x = I_{1x} + I_{2x}$, has no effect on the second moment of the resonance but does increase the fourth moment. As *Van Vleck* discusses, this must mean that the *central* portion of the resonance is narrowed, but the wings are enhanced (see Fig. 4.16), since the fourth moment is more affected by the wings than is the second moment. The fact that the central portion appears sharper gives rise to the terms *exchange narrowing* or *pseudo-exchange narrowing*. Where a real exchange interaction exists, as in electron resonances, the exchange narrowing may be very dramatic.

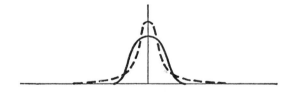

Fig. 4.16. Solid curve, the resonance shape with vanishing pseudo-exchange coupling. The dashed curve shows the shape when pseudo-exchange is included

If the two nuclei are *not* identical, one can approximate the pseudo-exchange coupling as $A_{12}I_{1z}I_{2z}$. Since this interaction does not commute with $I_{1x} + I_{2x}$, it increases the second moment. The resonance curve then appears broadened, and one speaks of "exchange broadening". If there is a quadrupole interaction that makes the various m-states unequally spaced, the exchange coupling can lead to a broadening, even when the nuclei are identical.

We conclude by remarking on the Fröhlich-Nabarro effect that the presence of one nucleus causes the electrons to repopulate their spin states, producing a magnetic field at other nuclei in much the same way as the static field produces a Knight shift. *Yosida* has analyzed the problem by performing a spatial Fourier analysis of the electron-nuclear interaction. Denoting the wave vector of a Fourier component by q, Yosida points out that the Fröhlich-Nabarro effect is the $q = 0$ term (infinite wavelength). In the second-order perturbation, a component $q \neq 0$ joins two electron states k and k' that satisfy the relation

$$k' = k + q \quad . \tag{4.242}$$

Since in a proper second-order calculation the excited and ground states must differ, we see from (4.242) that we must exclude $q = 0$. *Yosida* treats $q = 0$ by first-order theory, but $q \neq 0$ by second order, adding the results. The answer he obtains in this way is identical to that of *Bloembergen-Rowland* and *Ruderman-Kittel*, taking a principal part as mentioned on p. 141. The simplest way to see that the answers will be the same is to consider instead the Knight shifts.

There are two ways one can calculate the Knight shift. The first method is to assume a uniform static field and to compute the first-order nuclear-electron coupling of the *polarized* electron state. The second method is more complicated. We assume a static field but one that is oscillating spatially with wave vector q. For such a spatial oscillatory field there is no net spin polarization, since the field points up in some regions of space as much as it points down in others.

The usual first-order interaction vanishes. If one now goes to second-order perturbation in which one matrix element is the electron-nuclear coupling, the other, the electron-applied field interaction, a nonzero result is found. That is, the static field induces a spatially varying spin polarization. Let us choose the maximum of the static field to be at the position of the nucleus. When q gets very small (long wavelength), we expect that the result must be the same as if the field were strictly uniform. Therefore the limit of the second-order answer as $q \to 0$ must be the usual Knight shift. This result can in fact be verified.

In evaluating (4.238) using principal parts, one takes the limit as $k' \to k$. This procedure, as with the Knight shift, includes the first-order perturbation repopulation contribution, or the Fröhlich-Nabarro effect.

Before concluding this chapter, we should consider the role of electron spin in the chemical shift of diamagnetic substances. In the absence of an applied magnetic field, diamagnetic substances are characterized by a total electron-spin quantum number S of zero. The application of a field H_0 in, say, the z-direction adds a term $\mathcal{H}_{SZ}$, the spin Zeeman interaction, to the Hamiltonian:

$$\mathcal{H}_{SZ} = \gamma_e \hbar H_0 \sum_{j=1}^{n} S_{zj} = \gamma_e \hbar H_0 S_z \tag{4.243}$$

where j labels the electrons, and where

$$S_z \equiv \sum_{j=1}^{N} S_{zj} \quad . \tag{4.244}$$

Since the ground-state wave function $|0\rangle$ is a spin zero function, we have that

$$S_z |0\rangle = 0 \tag{4.245}$$

so that all matrix elements of $\mathcal{H}_{SZ}$ to excited states $|n\rangle$ vanish:

$$(n|\mathcal{H}_{SZ}|0) = \gamma_e \hbar H_0 (n|S_z|0) = 0 \quad . \tag{4.246}$$

The ground state is therefore strictly decoupled from all other states as far as the spin Zeeman coupling is concerned. The applied field is therefore unable to induce any net spin, and there is no phenomenon analogous to the unquenching of the orbital angular momentum.

The fact that the spins actually couple to a magnetic field makes this result seem strange. Intuitively we except that, given a strong enough magnetic field, the spins *must* be polarized. The paradox is resolved by considering an example, the hydrogen molecule. The ground state is the singlet bonding state, but there is a triplet antibonding state. In the presence of an applied field, the states split, as shown in Fig. 4.17.

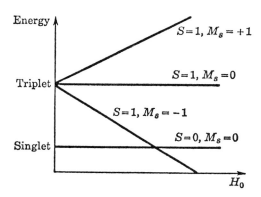

Fig. 4.17. Effect of the applied field H_0 on the singlet- and triplet-spin states of a hydrogen molecule. If H_0 were large enough, a triplet state would be lowest and the ground state would possess a magnetic moment

As we can see, for large enough H_0 the $S = 1$, $M_S = -1$ state crosses the $S = 0$, $M_S = 0$ state. The ground state is then a triplet state, corresponding to a spin polarization. However, since the singlet-triplet splitting in zero field is several electron volts, the crossing of levels could never be produced by an attainable laboratory field.

Further insight is obtained by considering the effect of a hypothetical mixing of a triplet state into the ground state. If the result is to induce a net spin polarization in the positive z-direction on one atom, it induces an equal and opposite spin polarization in the negative z-direction on the other atom. Clearly there is zero net spin Zeeman interaction with an applied field in the z-direction. Since such a spin polarization gives no net lowering of energy, it is not in fact induced.

Note, however, that if the two atoms are dissimilar, there may be a different induced orbital moment which, through the spin-orbit coupling, could then induce such a spin polarization. Thus, in a molecule such as HI, the iodine orbital magnetization could induce a spin polarization into the bond, giving a spin contribution to the chemical shift on the hydrogen as well as on the iodine.

5. Spin-Lattice Relaxation and Motional Narrowing of Resonance Lines

5.1 Introduction

We turn now to a discussion of how the nuclei arrive at their thermal equilibrium magnetization via the process of spin-lattice relaxation. We shall find it convenient to discuss two techniques for computing T_1.

The first method is appropriate when the coupling of the nuclei with one another is much stronger than with the lattice. In this case an attempt to compute the population changes of an individual nucleus due to the coupling to the lattice is complicated by the presence of a much stronger coupling of the nuclei among themselves. The first method makes the assumption that the strong coupling simply establishes a common temperature for the spins and that the lattice coupling causes this temperature to change. There is a close analogy with the process of heat transfer between a gas and the walls of its container, in which the role of the collisions within the gas is to maintain a thermal equilibrium among the gas molecules. In the collision of molecules with the wall, we consider the molecules to have the velocity distribution appropriate to thermal equilibrium. As we shall see the first method leads to a formula for T_1 that is particularly convenient when the lattice is readily described in quantum mechanical terms. For example, relaxation in a metal involves the transfer of energy to the conduction electrons, which are readily thought of in terms of Bloch functions and the exclusion principle.

The second method is that of the so-called density matrix. Although this is a completely general method, it finds its greatest utility for systems in which the lattice is naturally described classically and in which the resonance width is substantially narrowed by the motion of the nuclei. Moreover, when motion takes place, the relaxation time T_2, which describes coupling between the nuclei, becomes long, and it may be a very poor approximation to assume that a spin temperature is achieved rapidly as compared to T_1. Thus the second method is useful when the first one fails. Because of their large mass, the motion of the nuclei is often given very well in classical terms. In fact an attempt to describe the motion of molecules in a liquid quantum mechanically would be quite cumbersome. Consequently the density matrix method is well suited to discussing cases in which motional narrowing takes place. An added feature is that both T_1 and T_2 processes (relaxation of I_z and I_x or I_y) can be treated by the density matrix method, provided there is motional narrowing.

The density matrix method is very closely related to the conventional time-dependent perturbation theory. Actually the two are entirely equivalent. The density matrix method, however, gives results in a particularly useful form. It is ideal for treating problems in which phase coherence is important, and in fact the density matrix or a mathematical equivalent is necessary to treat such problems. In any event, much of the formalism of this chapter applies equally well to systems other than spins. For example, dielectric relaxation can be treated by these methods.

As we see, the two approaches complement one another, one applying to the broad resonances of a rigid lattice and the other being most useful when the resonance has been narrowed by nuclear motion.

5.2 Relaxation of a System Described by a Spin Temperature [1]

A system with a set of energies E_a, E_b, and so on, which is in thermal equilibrium with a reservoir of temperature T, occupies the levels with probabilities $p(E_a)$, $p(E_b)$, and so forth, which are given by

$$\frac{p(E_a)}{p(E_b)} = \frac{e^{-E_a/kT}}{e^{-E_b/kT}} \quad \text{so that, since} \tag{5.1}$$

$$\sum_{E_a} p(E_a) = 1 \tag{5.2}$$

we have

$$p(E_a) = \frac{e^{-E_a/kT}}{\sum_c e^{-E_c/kT}} = \frac{e^{-E_a/kT}}{Z} \quad \text{where} \tag{5.3}$$

$$Z = \sum_c e^{-E_c/kT} \tag{5.4}$$

is the partition function or "sum of states".

These equations may actually have two interpretations, which we might illustrate by considering N identical spins. The first interpretation considers the spins as isolated from one another. The "system" consists then of a single spin, and the energies E_a represent the possible energies of the single spin. The second interpretation considers the system to be formed by all N spins. In this case, E_a represents the total energy of all N spins. We shall find it convenient to use both interpretations. We should perhaps note that the first interpretation is correct only if the spins can be considered to obey Maxwell-Boltzmann statistics, but the second interpretation holds true whether or not the individual particles obey Maxwell-Boltzmann, Fermi-Dirac, or Bose-Einstein statistics. The statistics

---

[1] See references to "Spin Temperature" in the Bibliography

would enter only when we tried to express the wave function of the total system in terms of the wave functions of individual spins.

We shall say that any system whose population obeys (5.1) is described by a temperature T, even when the system is not in equilibrium with a reservoir. Equation (5.3) enables us to make a simple plot to schematize the populations. We illustrate in Fig. 5.1 a case based on our first interpretation: we consider the population of the various energy states of a single spin of $I = \frac{3}{2}$ acted on by a static magnetic field.

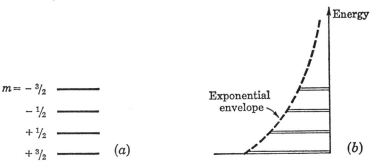

Fig. 5.1. (a) Energy levels of a spin 3/2 nucleus. (b) Bar graph of population versus energy. The lengths of the bars are determined by the exponential envelope

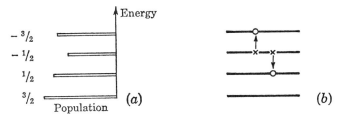

Fig. 5.2. (a) Population distribution not describable by a temperature. (b) A possible transition of a pair of spins from the states designated by crosses to those designated by circles

A system such as shown in Fig. 5.2 clearly does not correspond to thermal equilibrium, since the bar graph envelope is not an exponential.

In Fig. 5.2b we indicate a transition that could take place, conserving the total energy of the spins. Two spins designated by crosses couple, inducing transitions to the states designated by circles, one spin going up in energy and the other down. (Such a transition would be induced by the $I_1^+ I_2^-$ terms of the dipolar coupling.) The number of transitions per second from cross to circle, $dN/dt)_{x \to 0}$, will be the product of the probabilities of finding two spins in the initial state times the probability of transition $W_{x \to 0}$ if the spins are in the initial state. Thus,

$$\left.\frac{dN}{dt}\right)_{x \to 0} = p_{-1/2} p_{-1/2} W_{x \to 0} \quad . \tag{5.5}$$

The inverse reaction from the circle to the cross will have a rate $dN/dt)_{0 \to x}$, given by

$$\left.\frac{dN}{dt}\right)_{0 \to x} = p_{-3/2} p_{1/2} W_{0 \to x} \quad . \tag{5.6}$$

If we equate these rates, we guarantee equilibrium. This is the assumption that equilibrium is obtained by detailed balance. Since $W_{0 \to x} = W_{x \to 0}$, we find

$$p_{-1/2} p_{-1/2} = p_{-3/2} p_{1/2} \quad \text{or} \quad \frac{p_{-3/2}}{p_{-1/2}} = \frac{p_{-1/2}}{p_{+1/2}} \quad . \tag{5.7}$$

But this is just the condition of thermal equilibrium among the states, since they are equally spaced in energy.

We see, therefore, that thermal equilibrium is reached by processes such as we have indicated in Fig. 5.2b. The typical rate for such a process is of the order of the inverse of the rigid lattice line breadth, or between 10 to 100 μs for typical nuclei. Therefore, if T_1 is milliseconds to seconds, we should consider the nuclear populations to be given by a Boltzmann distribution.

We shall now proceed to consider the relaxation of a system of nuclear spins whose Hamiltonian $\mathcal{H}$ has eigenvalues E_n, and in which the fractional occupation of state n is p_n. (*Thus n designates a state of the total system, rather than the energy of a single spin*). Normalization requires that

$$\sum_n p_n = 1 \quad . \tag{5.8}$$

The average energy of the system, $\overline{E}$, is then

$$\overline{E} = \sum_n p_n E_n \quad . \tag{5.9}$$

We shall assume further that the energies E_n are measured from a reference such that

$$\sum_n E_n = \operatorname{Tr} \mathcal{H} = 0 \quad , \tag{5.10}$$

a condition that is fulfilled for both the Zeeman and dipolar energies.

To compute the relaxation, we shall consider changes in the average energy. If we define $\beta = 1/kT$ to represent the spin temperature, we have that

$$\frac{d\overline{E}}{dt} = \frac{d\overline{E}}{d\beta} \frac{d\beta}{dt} \quad . \tag{5.11}$$

But since $\overline{E} = \sum_n p_n E_n$, we have also that

$$\frac{d\overline{E}}{dt} = \frac{d}{dt} \sum_n p_n E_n = \sum_n E_n \frac{dp_n}{dt} \quad . \tag{5.12}$$

We shall assume that the p_n's obey simple linear rate equations. Introducing W_{mn} as the probability per second that the lattice induces a transition of the system from m to n if the system is in state m, the rate equation is

$$\frac{dp_n}{dt} = \sum_m (p_m W_{mn} - p_n W_{nm}) \quad . \tag{5.13}$$

This equation is frequently called the "master" equation.

By substituting into (5.12) we have, then, that

$$\frac{d\overline{E}}{dt} = \sum_{m,n} (p_m W_{mn} - p_n W_{nm}) E_n$$

$$= \frac{1}{2} \sum_{m,n} (p_m W_{mn} - p_n W_{nm})(E_n - E_m) \tag{5.14}$$

where the second form is introduced because it treats the labels m and n more symmetrically. By equating (5.11) and (5.14), we obtain a differential equation for the changes in spin temperature. We have two problems: (1) finding $d\overline{E}/d\beta$ and (2) seeing what becomes of (5.14) when we introduce the requirement that at all times a spin temperature apply.

We turn first to evaluating $d\overline{E}/d\beta$:

$$p_n = \frac{e^{-\beta E_n}}{Z} \quad \text{and} \tag{5.15}$$

$$\frac{d\overline{E}}{d\beta} = \frac{d}{d\beta} \sum_n p_n(\beta) E_n \quad . \tag{5.16}$$

We first seek an approximate expression for Z. Once again, assuming the temperature to be high enough so that $\beta E_n \ll 1$ for the majority of states, we expand $\exp(-\beta E_n)$ in a power series and keep only the leading terms:

$$Z = \sum_n \left(1 - \beta E_n + \frac{\beta^2 E_n^2}{2!} + \dots \right) \quad . \tag{5.17}$$

Approximating such a power series expansion by the leading terms clearly has validity if $|E_n| \ll kT$ for the significant energies. However, the approximation proves legitimate under less stringent conditions using an argument similar to that in Appendix E both here and in (5.19–26).

When we utilize (5.10), the second term on the right of (5.17) vanishes. We then neglect the β^2 term, and Z becomes equal to the total number of states. Since this is also Z for infinite temperature Z_∞, we may say that

$$Z = Z_\infty \quad . \tag{5.18}$$

When we utilize this fact and (5.15), (5.16) becomes

$$\begin{aligned}
\frac{d\overline{E}}{d\beta} &= -\frac{1}{Z_\infty} \sum_n E_n^2 \mathrm{e}^{-\beta E_n} \\
&= -\frac{1}{Z_\infty} \sum_n E_n^2 (1 - \beta E_n + \ldots) \\
&\cong -\frac{1}{Z_\infty} \sum_n E_n^2
\end{aligned} \qquad (5.19)$$

again in the high-temperature limit. Thus

$$\frac{d\overline{E}}{dt} = -\frac{d\beta}{dt} \frac{\sum_n E_n^2}{Z_\infty} . \qquad (5.20)$$

We now turn to evaluation of (5.14). Since the system is always describable by a temperature, we have

$$p_n = p_m \mathrm{e}^{(E_m - E_n)\beta} . \qquad (5.21)$$

We shall furthermore assume that when the system is in thermal equilibrium with the lattice, the transitions between every pair of levels are in equilibrium. This is the so-called principle of detailed balance. Denoting by p_n^L the value of p_n when the spins are in thermal equilibrium with the lattice, the principle of detailed balance says that

$$p_m^L W_{mn} = p_n^L W_{nm} \quad \text{or that} \qquad (5.22)$$

$$W_{mn} = W_{nm} \frac{p_n^L}{p_m^L} = W_{nm} \mathrm{e}^{(E_m - E_n)\beta_L} \qquad (5.23)$$

where $\beta_L = 1/(kT_L)$.

By substituting (5.21) and (5.23) into (5.14), we find

$$\frac{d\overline{E}}{dt} = \frac{1}{2} \sum_{m,n} p_m W_{mn} [1 - \mathrm{e}^{(E_m - E_n)(\beta - \beta_L)}](E_n - E_m) . \qquad (5.24)$$

Now, expanding the exponential, we find

$$\frac{d\overline{E}}{dt} \cong \frac{1}{2} \sum_{m,n} p_m W_{mn} (E_n - E_m)^2 (\beta - \beta_L) . \qquad (5.25)$$

Now

$$p_m = \frac{\mathrm{e}^{-\beta E_m}}{Z_\infty} \cong \frac{1 - \beta E_m + (\beta^2 E_m^2/2!)}{Z_\infty} \cong \frac{1}{Z_\infty} . \qquad (5.26)$$

Thus, combining (5.26) with (5.25) and equating the resultant $d\overline{E}/dt$ to that of (5.20), we find

$$\frac{d\beta}{dt} = (\beta_L - \beta) \left[\frac{1}{2} \frac{\sum_{m,n} W_{mn}(E_m - E_n)^2}{\sum_n E_n^2} \right] = \frac{\beta_L - \beta}{T_1} \qquad (5.27)$$

where

$$\frac{1}{T_1} = \frac{1}{2} \frac{\sum\limits_{m,n} W_{mn}(E_m - E_n)^2}{\sum\limits_{n} E_n^2} \ . \tag{5.28}$$

Equation (5.28) was first derived by *Gorter* on the assumption that $|\beta - \beta_\mathrm{L}| \ll \beta$ [5.1]. As we can see, this restriction is not necessary.

The great advantage of (5.28) is that, by postulating a temperature, it has taken into account the spin-spin couplings. The rate equations, (5.14), would by themselves imply that there are multiple time constants that describe the spin-lattice relaxation, but the assumption of a temperature forces the whole system to relax with a single exponential.

We may get added insight by viewing our states n as being *nearly* exact solutions of the nuclear spin Hamiltonian, between which (since they are not *exact* states) transitions take place rapidly to guarantee a thermal equilibrium, but between which the lattice also makes much slower transitions. After each lattice transition, which disturbs the nuclear distribution, the nuclei readjust among their approximate levels so that the lattice once again finds the spins distributed according to a temperature when it induces the next spin transition. Our formalism implies that treating the states n as being exact makes a negligible difference in the answer.

5.3 Relaxation of Nuclei in a Metal

We now turn to an example of the application of (5.28). We shall consider the relaxation of nuclei in a metal by their coupling to the spin magnetic moments of the conduction electrons. This is the dominant relaxation mechanism.

In a T_1 process, the nucleus undergoes a transition in which it either absorbs or gives up energy. In order to conserve energy, the lattice must undergo a compensating change. For coupling to the conduction electrons, we may think of the nuclear transition as involving a simultaneous electron transition from some state of wave vector k and spin orientation s, to a state k', s'. We may think of this as a scattering problem. Denoting the initial and final nuclear quantum numbers as m and n, respectively, we have that the number of transitions per second from the initial state of nucleus and electron $|mks\rangle$ to the final state $|nk's'\rangle$, $W_{mks,nk's'}$, is

$$W_{mks,nk's'} = \frac{2\pi}{\hbar} |\langle mks|V|nk's'\rangle|^2 \delta(E_m + E_{ks} - E_n - E_{k's'}) \tag{5.29}$$

where V is the interaction that provides the scattering, and where (5.29) assumes that there is an electron in $|ks\rangle$ and there is *none* in $|k's'\rangle$. The total probability per second of nuclear transitions is obtained by adding up the $W_{mks,nk's'}$'s for all initial and final electron states. We have

$$W_{mn} = \sum_{\substack{ks \text{ occupied} \\ k's' \text{ unoccupied}}} W_{mks,nk's'} \quad . \tag{5.30}$$

The sum over "ks occupied" is, of course, equivalent to summing over electrons. We can remove the restrictions on ks and $k's'$ by introducing the quantity p_{ks}, which is defined to be unity if ks is occupied; zero, otherwise. This gives us

$$W_{mn} = \sum_{ks;k's'} W_{mks,nk's'} p_{ks}(1 - p_{k's'}) \quad . \tag{5.31}$$

By averaging (5.31) over an ensemble of electron systems, we simply replace p_{ks} by the Fermi function $f(E_{ks})$, which we abbreviate as $f(\boldsymbol{k},s)$:

$$W_{mn} = \sum_{ks;k's'} W_{mks,nk's'} f(\boldsymbol{k},s)[1 - f(\boldsymbol{k}',s')] \quad . \tag{5.32}$$

We must now express $W_{mks,nk's'}$ explicitly. To do so, we must specify the interaction V. For metals with a substantial s-character to the wave function at the Fermi surface, the dominant contribution to V comes from the s-state coupling between the nuclear and electron spins:

$$V = \frac{8\pi}{3} \gamma_e \gamma_n \hbar^2 \boldsymbol{I} \cdot \boldsymbol{S} \delta(\boldsymbol{r}) \tag{5.33}$$

where we have chosen the nucleus $\boldsymbol{I}$ to be at the origin. For the electron wave function, we shall take a product of a spin function and a Bloch function $u_{\boldsymbol{k}}(\boldsymbol{r})\exp(\mathrm{i}\boldsymbol{k}\cdot\boldsymbol{r})$. Therefore the initial wave function is

$$|m\boldsymbol{k}s) = |m)|s) u_{\boldsymbol{k}}(\boldsymbol{r}) \mathrm{e}^{\mathrm{i}\boldsymbol{k}\cdot\boldsymbol{r}} \quad . \tag{5.34}$$

It is a simple matter to compute the matrix element of (5.29):

$$(m\boldsymbol{k}s|V|n\boldsymbol{k}'s') = \frac{8\pi}{3} \gamma_e \gamma_n \hbar^2 (m|\boldsymbol{I}|n) \cdot (s|\boldsymbol{S}|s') u_{\boldsymbol{k}}^*(0) u_{\boldsymbol{k}'}(0) \tag{5.35}$$

which gives us

$$\begin{aligned} W_{mks,nk's'} = \frac{2\pi}{\hbar} \frac{64\pi^2}{9} \gamma_e^2 \gamma_n^2 \hbar^4 \sum_{\alpha,\alpha'=x,y,z} & (m|I_\alpha|n)(n|I_{\alpha'}|m) \\ \times\, & (s|S_\alpha|s')(s'|S_{\alpha'}|s)|u_{\boldsymbol{k}}(0)|^2 |u_{\boldsymbol{k}'}(0)|^2 \\ \times\, & \delta(E_m + E_{\boldsymbol{k}s} - E_n - E_{\boldsymbol{k}'s'}) \quad . \end{aligned} \tag{5.36}$$

We can substitute this expression into (5.32) to compute W_{mn}. We are once again faced with a summation over $\boldsymbol{k}$ and $\boldsymbol{k}'$ of a slowly varying function. As before, we replace the summation by an integral, using the density of states $g(E_{\boldsymbol{k}},A)$ introduced in Sect. 4.7. This gives us

$$W_{mn} = \frac{2\pi}{\hbar} \frac{64\pi^2}{9} \gamma_e^2 \gamma_n^2 \hbar^4 \sum_{\alpha,\alpha's,s'} (m|I_\alpha|n)(n|I_{\alpha'}|m)(s|S_\alpha|s')(s'|S_{\alpha'}|s)$$

$$\times \int |u_{\boldsymbol{k}}(0)|^2 |u_{\boldsymbol{k}'}(0)|^2 f(\boldsymbol{k},s)[1-f(\boldsymbol{k}',s')]g(E_{\boldsymbol{k}},A)$$
$$\times g(E_{\boldsymbol{k}'},A')\delta(E_m - E_n + E_{\boldsymbol{k}s} - E_{\boldsymbol{k}'s'})dE_{\boldsymbol{k}}dAdE_{\boldsymbol{k}'}dA' . \quad (5.37)$$

We integrate first over dA and dA', utilizing the relations of (4.153, 154) to perform the integrals, and introducing again the average of $|u_{\boldsymbol{k}}(0)|^2$ over the energy surface, $E_{\boldsymbol{k}}$, $\langle |u_{\boldsymbol{k}}(0)|^2 \rangle_{E_{\boldsymbol{k}}}$. We also assume that the energy $E_{\boldsymbol{k}s}$ appearing in the Fermi functions remains constant on a surface of constant $E_{\boldsymbol{k}}$, an assumption that would be fulfilled unless the spin energy depended on the location on the surface $E_{\boldsymbol{k}}$. The integration over $dE_{\boldsymbol{k}'}$ is then easy because of the delta function. We have that

$$E_{\boldsymbol{k}s} + E_m = E_{\boldsymbol{k}'s'} + E_n \quad (5.38)$$

and, assuming $E_{\boldsymbol{k}s} = E_{\boldsymbol{k}} + E_s$, we get for W_{mn}

$$W_{mn} = \frac{2\pi}{\hbar} \frac{64\pi^2}{9} \gamma_e^2 \gamma_n^2 \hbar^4 \sum_{\alpha,\alpha' s, s'} (m|I_\alpha|n)(n|I_{\alpha'}|m)(s|S_\alpha|s')(s'|S_{\alpha'}|s)$$
$$\times \int \langle |u_{\boldsymbol{k}}(0)|^2 \rangle_{E_{\boldsymbol{k}}} \langle |u_{\boldsymbol{k}'}(0)|^2 \rangle_{E_{\boldsymbol{k}'}} f(E_{\boldsymbol{k}s})[1 - f(E_{\boldsymbol{k}'s'} + E_m - E_n)]$$
$$\times \varrho(E_{\boldsymbol{k}})\varrho(E_{\boldsymbol{k}'})dE_{\boldsymbol{k}} \quad (5.39)$$

$$E_{\boldsymbol{k}'} = E_{\boldsymbol{k}} + E_s - E_{s'} + E_m - E_n \quad . \quad (5.40)$$

Since $E_m - E_n$, the nuclear energy change, is very small compared with kT, the Fermi function $f(E_{\boldsymbol{k}s} + E_m - E_n)$ may be replaced by $f(E_{\boldsymbol{k}s})$. Actually, doing so makes $W_{mn} = W_{nm}$. It is, in fact, at this point that the slight difference between W_{mn} and W_{nm} arises. It is the slight difference, we recall, that gives rise to the establishment of the thermal equilibrium nuclear population. We are allowed to neglect the difference here in computing W_{mn}, since we have already included the effect in (5.23).

Moreover, since both $\varrho(E_{\boldsymbol{k}'})$ and $\langle |u_{\boldsymbol{k}'}(0)|^2 \rangle_{E_{\boldsymbol{k}'}}$ are slowly varying functions of $E_{\boldsymbol{k}'}$, we may set them equal to their values when $E_{\boldsymbol{k}} = E_{\boldsymbol{k}'}$. In fact we shall evaluate $\varrho(E_{\boldsymbol{k}'})$, and so on at $E_{\boldsymbol{k}s}$. This gives us for the integral in (5.39):

$$\int_0^\infty \langle |u_{\boldsymbol{k}}(0)|^2 \rangle_E^2 \varrho^2(E) f(E)[1 - f(E)]dE \quad (5.41)$$

where we have set the lower limits as zero, since the only contributions to the integral come from the region near the Fermi surface, $E = E_F$.

Since (5.41) is independent of the spin quantum numbers s and s', we may now evaluate the spin sum of (5.39):

$$\sum_{s,s'}(s|S_\alpha|s')(s'|S_{\alpha'}|s) = \sum_s (s|S_\alpha S_{\alpha'}|s)$$
$$= \text{Tr}\{S_\alpha S_{\alpha'}\} = \delta_{\alpha\alpha'}\tfrac{1}{3}S(S+1)(2S+1)$$
$$= \delta_{\alpha\alpha'}/2 \quad \text{since} \quad S = \tfrac{1}{2} \quad . \quad (5.42)$$

This gives us

$$W_{mn} = \frac{64}{9}\pi^3\hbar^3\gamma_e^2\gamma_n^2 \sum_\alpha (m|I_\alpha|n)(n|I_\alpha|m)$$
$$\times \int \langle|u_k(0)|^2\rangle_E^2 \varrho^2(E) f(E)[1-f(E)]dE \quad . \tag{5.43}$$

Now

$$f(E)[1-f(E)] = -kT\frac{\partial f}{\partial E} \tag{5.44}$$

which follows directly from the fact that

$$f(E) = \frac{1}{e^{(E-E_F)/kT}+1}$$

and $f(E)[1-f(E)]$ peaks up very strongly (Fig. 5.3) when $E \cong E_F$.

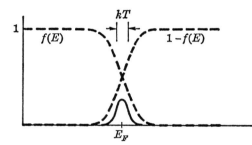

Fig. 5.3. Functions $f(E)$, $1-f(E)$, and $f(E)[1-f(E)]$. The solid line shows $f(E)[1-f(E)]$

Since $f(0) = 1$ and $f(\infty) = 0$, and since $f(E)[1-f(E)]$ peaks up only within a width kT, it is also closely related to a δ-function when in an integral of other functions that vary slowly over kT.

$$f(E)[1-f(E)] = kT\delta(E - E_F) \tag{5.45}$$

(see footnote[2]). By utilizing this fact, we have, finally,

---

[2] Equation (5.45) can be derived simply. Let $G(E)$ be a slowly varying function of energy. Then, utilizing the fact that $f(E)[1-f(E)]$ is nonvanishing within only kT about E_F, we can expand $G(E)$ in a power series about E_F:

$$G(E) = G(E_F) + (E - E_F)\frac{dG}{dE}\bigg|_{E_F} + \frac{(E-E_F)^2}{2!}\frac{d^2G}{dE^2}\bigg|_{E_F} + \cdots \quad .$$

Thus

$$\int_0^\infty G(E)f(E)[1-f(E)]dE = G(E_F)\int_0^\infty f(E)[1-f(E)]dE$$
$$+ \frac{dG}{dE}\bigg|_{E_F}\int_0^\infty (E-E_F)f(1-f)dE + \frac{d^2G}{dE^2}\bigg|_E \int_0^\infty (E-E_F)^2 f(1-f)dE \ldots \quad .$$

$$W_{mn} = \frac{64}{9}\pi^3 \hbar^3 \gamma_e^2 \gamma_n^2 \langle |u_{\boldsymbol{k}}(0)|^2 \rangle_{E_F}^2 \varrho^2(E_F) kT \sum_\alpha |(m|I_\alpha|n)|^2 \quad . \tag{5.46}$$

We note that W_{mn} is proportional to the temperature T. This fact has a simple physical interpretation. When the nuclei undergo a transition, they give an energy to the electrons that is very small compared with kT. Most of the electrons are unable to take part in the relaxation because they have no empty states nearby in energy into which they can make a transition. It is only those electrons in the tail of the distribution that are important. Their number is proportional to kT.

We can write (5.46) as

$$W_{mn} = a_{00} \sum_\alpha |(m|I_\alpha|n)|^2 \quad , \tag{5.47}$$

defining the quantity a_{00}, which is independent of the nuclear states n and m. If we have more than one nucleus, it can be shown [5.2] that W_{mn} is given by a sum over the N nuclei, labeled by i or j, $(i,j = 1$ to $N)$:

$$W_{mn} = \sum_{i,j} a_{ij} \sum_\alpha (m|I_{i\alpha}|n)(n|I_{j\alpha}|m) \tag{5.48}$$

where the coefficient $i = j$ is a_{00}, and where a_{ij} for $i \neq j$ falls off rapidly with the distance apart of the nuclei i and j. These terms arise because the electron wave function extends over many nuclei, so that more than one nucleus may scatter the electron from a given initial to a given final state.

By returning to (5.47) and employing our formula for T_1, we have

$$\begin{aligned}\frac{1}{T_1} &= a_{00} \frac{1}{2} \frac{\sum_{m,n,\alpha} (m|I_\alpha|n)(n|I_\alpha|m)(E_m - E_n)^2}{\sum_m E_m^2} \\ &= -\frac{a_{00}}{2} \frac{\sum_{m,n,\alpha} (m|[\mathcal{H},I_\alpha]|n)(n|[\mathcal{H},I_\alpha]|m)}{\sum_m E_m^2} \\ &= -\frac{a_{00}}{2} \frac{\sum_{\alpha=x,y,z} \mathrm{Tr}\{[\mathcal{H},I_\alpha]^2\}}{\mathrm{Tr}\{\mathcal{H}^2\}} \quad . \end{aligned} \tag{5.49}$$

A similar expression is found by using (5.48). The important point is that we do not need to solve for the explicit eigenstates and eigenvalues, but only evaluate the traces in a convenient representation.

The first term, using (5.44), gives $G(E_F)kT$. The second term vanishes, since the integrand is an odd function of $E - E_F$, and the third term gives a contribution proportional to $(kT)^3$, as seen by changing the integrand from E to $x \equiv E/kT$. If we neglect the third and higher terms, the answer is just what we should have if we replaced $f(1-f)$ by $kT\delta(E - E_F)$. The corrections are generally of order $(kT/E_F)^2$ smaller, the exact form depending on the functional dependence of G on the energy E.

For our problem of a single spin, the quantum numbers m and n would label the $2I+1$ eigenstates of I_z. Then, using the fact that

$$\mathcal{H} = -\gamma \hbar H_0 I_z \quad \text{and that} \tag{5.50}$$

$$[I_z, I_x] = iI_y \quad \text{etc.} \tag{5.51}$$

we find

$$\sum_\alpha \text{Tr}\{[\mathcal{H}, I_\alpha]^2\} = -\gamma_n^2 \hbar^2 H_0^2 \, \text{Tr}\{[I_x^2 + I_y^2]\} \tag{5.52}$$

$$\sum_\alpha \text{Tr}\{\mathcal{H}^2\} = \gamma_n^2 \hbar^2 H_0^2 \, \text{Tr}\{I_z^2\}$$

so that, since $\text{Tr}\{I_x^2\} = \text{Tr}\{I_y^2\} = \text{Tr}\{I_z^2\}$,

$$\frac{\sum_\alpha \text{Tr}\{[\mathcal{H}, I_\alpha]^2}{\text{Tr}\{\mathcal{H}^2\}} = -2 \tag{5.53}$$

$$\frac{1}{T_1} = a_{00} = \frac{64}{9}\pi^3 \hbar^3 \gamma_e^2 \gamma_n^2 \langle |u_k(0)|\rangle^2_{E_F} \varrho^2(E_F) kT \quad . \tag{5.54}$$

The quantity $\langle |u_k(0)|\rangle^2_{E_F}$ appearing in the expression also occurred in the expression for the Knight shift, $\Delta H/H$:

$$\frac{\Delta H}{H} = \frac{8\pi}{3} \langle |u_k(0)|\rangle^2_{E_F} \chi_e^s \quad . \tag{5.55}$$

We can therefore use (5.55) to evaluate $\langle |u_k(0)|\rangle^2_{E_F}$, giving

$$T_1 \left(\frac{\Delta H}{H}\right)^2 = \left[\frac{\chi_e^s}{\varrho(E_F)}\right]^2 \frac{1}{\pi kT} \frac{1}{\gamma_n^2 \gamma_e^2 \hbar^3} \quad . \tag{5.56}$$

For a Fermi gas of noninteracting spins, one can show that χ_e^s given by

$$\chi_0^s = \frac{\gamma_e^2 \hbar^2}{2} \varrho_0(E_F) \tag{5.57}$$

where we have put subscripts "0" on χ^s and $\varrho(E_F)$ to label them as appropriate to noninteracting electrons. In this approximation one has

$$T_1 \left(\frac{\Delta H}{H}\right)^2 = \frac{\hbar}{4\pi kT} \frac{\gamma_e^2}{\gamma_n^2} \quad . \tag{5.58}$$

Equation (5.58) is commonly called the "Korringa relation", after *Dr. J. Korringa* who first published it [5.3]. It provides a very convenient way to use measured Knight shifts to predict spin-lattice relaxation times. A more accurate expression is obtained from (5.56) and (5.57):

$$T_1 \left(\frac{\Delta H}{H} \right)^2 = \frac{\hbar}{4\pi k T} \frac{\gamma_e^2}{\gamma_n^2} \left[\frac{\chi_e^s}{\chi_0^s} \frac{\varrho_0(E_F)}{\varrho(E_F)} \right]^2 \quad . \tag{5.59}$$

The T appearing in the Korringa relation represents only one contribution to the relaxation time – that due to the coupling of nuclei to the magnetic moment of s-state electrons. One expects that the experimental T_1 should, if anything, be shorter. It is therefore interesting to examine a table given by *Pines* [5.4]. In it he lists experimental T_1's, those computed from (5.58) (the Korringa relation) and those computed from (5.59), using *Pines'* theoretical values of χ_e^s/χ_0^s and $\varrho_0(E_F)/\varrho(E_F)$.

Table 5.1. Experimental and theoretical T_1's (all times in ms)

	T_1 (Experimental)	T_1 (Korringa)	T_1 (Pines)
Li	150 ± 5	88	232
Na	15.9 ± 0.3	10.3	18.1
Rb^{95}	2.75 ± 0.2	2.1	2.94
Cu	3.0 ± 0.6	2.3	4.0
Al	6.3 ± 0.1	5.1	6.5

We note that the Korringa T_1's are all *shorter* than the experimental ones. The discrepancy cannot be removed by appealing to other relaxation processes, since if we included them, the theoretical T_1 would be even shorter than those computed by the Korringa relation, and the discrepancy would be still greater. On the other hand, the Pines' values, based on inclusion of the electron-electron couplings, make the predicted values *longer* than the experimental. The discrepancy between the Pines' values and the experimental is perhaps a measure of the importance of relaxation processes we have not computed.

5.4 Density Matrix – General Equations

As we have remarked, the concept of a spin temperature is not always valid. We turn now to discussion of a method of attack that is very useful when the spin temperature concept breaks down – the technique of the density matrix. An exceptionally good discussion of the density matrix is that of *Tolman* [5.5].

The method has the further advantage of giving one a discussion of both T_1 and T_2 processes in a natural way. It is ideally suited to treating problems in which the resonance is narrowed by the bodily motion of the nuclei. It is also applicable to broad line spectra, where it can in fact be used for an alternate derivation of our equation for T_1 of Sect. 5.3. As we shall see, the method is simply a variant of the usual time-dependent perturbation theory, but one that is in a particularly useful form.

We begin by considering a system described by a wave function ψ, at some instant of time, and ask for the expectation value $\langle M_x \rangle$ of some operator such as the x-component of magnetization, M_x. We have, then,

$$\langle M_x \rangle = (\psi, M_x \psi) \quad . \tag{5.60}$$

Suppose we now expand in a complete set of orthonormal functions u_n, which are independent of time:

$$\psi = \sum_n c_n u_n \quad . \tag{5.61}$$

If ψ varies in time, so must the c_n's. In terms of the functions u_n, we have

$$\langle M_x \rangle = \sum_{n,m} c_m^* c_n (m|M_x|n) \quad . \tag{5.62}$$

If we change the wave function, $\langle M_x \rangle$ will differ because the coefficients $c_m^* c_n$ will differ, but the matrix elements $(m|M_x|n)$ will remain the same. Correspondingly, for a given ψ, the effect of calculating expectation values of different operators is found in the different matrix elements, but the coefficients $c_m^* c_n$ remain the same. We can conveniently arrange the coefficients $c_n c_m^*$ to form a matrix. We note that, to compute any observable, we can specify either all the c_n's or all the products $c_n c_m^*$. However, since we always wish the c's in the form of products, to calculate observable properties of the system, we find knowledge of the products more useful than knowledge of the individual c's.

It is convenient to think of the matrix $c_n c_m^*$ as being the representation of an operator P, the operator being defined by its matrix elements:

$$(n|P|m) = c_n c_m^* \quad . \tag{5.63}$$

In terms of (5.63) we have, then,

$$\langle M_x \rangle = \sum_{n,m} (n|P|m)(m|M_x|n) \quad . \tag{5.64}$$

The result of the operator P acting on a function u_m may be written as

$$P u_m = \sum_n a_n u_n \tag{5.65}$$

since the u_n's form a complete set. As usual, we find the a_n's by multiplying both sides from the left by u_n^* and integrating:

$$a_n = \int u_n^* P u_m d\tau = (n|P|m) \quad \text{so that} \tag{5.66}$$

$$P u_m = \sum_n u_n (n|P|m) \quad . \tag{5.67}$$

Likewise we have

$$M_x u_n = \sum_m u_m (m|M_x|n) \quad . \tag{5.68}$$

Therefore
$$PM_x u_n = \sum_m P u_m (m|M_x|n)$$
$$= \sum_{m,n'} u_{n'}(n'|P|m)(m|M_x|n) \tag{5.69}$$
so that
$$(n'|PM_x|n) = \sum_m (n'|P|m)(m|M_x|n) . \tag{5.70}$$

By using (5.62), we have that
$$\langle M_x \rangle = \sum_{m,n} (n|P|m)(m|M_x|n)$$
$$= \sum_n (n|PM_x|n)$$
$$= \text{Tr}\{PM_x\} = \text{Tr}\{M_x P\} . \tag{5.71}$$

Also we note that P is an Hermitian operator. This we prove by noting that the definition of an Hermitian operator P is
$$\int u_n^* P u_m d\tau \equiv \int (Pu_n)^* u_m d\tau = \left(\int u_m^* P u_n d\tau \right)^* \tag{5.72}$$
or
$$(n|P|m) = (m|P|n)^* . \tag{5.73}$$
But
$$(n|P|m) = c_n c_m^* \quad , \quad (m|P|n) = c_m c_n^* \tag{5.74}$$
so that (5.73) is satisfied.

Often we shall be concerned with problems in which we wish to compute the average expectation value of an ensemble of systems. The matrix elements $c_n c_m^*$ will then vary from system to system to the extent that they have differing wave functions, but the matrix elements $(m|M_x|n)$ will be the same. If we use a bar to denote an ensemble average, we have, then,
$$\overline{\langle M_x \rangle} = \sum_{n,m} \overline{c_n c_m^*} (m|M_x|n) . \tag{5.75}$$

The quantities $\overline{c_n c_m^*}$ form a matrix, and it is this matrix that we call the "density matrix". We shall consider it to be the matrix of an operator ϱ, defined by the equation
$$(n|\varrho|m) = \overline{c_n c_m^*} = \overline{(n|P|m)} . \tag{5.76}$$

Since P is an Hermitian operator, it is clear that ϱ is as well. Equation (5.64) becomes, then,
$$\overline{\langle M_x \rangle} = \sum_{n,m} (n|\varrho|m)(m|M_x|n) = \text{Tr}\{\varrho M_x\} = \text{Tr}\{M_x \varrho\} . \tag{5.77}$$

For the future, we shall omit the bar indicating an ensemble average to simplify the notation, but of course we realize that whenever the symbol ϱ is used, an ensemble average is intended.

Of course the wave function ψ, describing whatever system we are considering, will develop in time. Since the u_n's are independent of time, the coefficients c_n must carry the time dependence. It is straightforward to find the differential equation they obey in terms of Hamiltonian $\mathcal{H}$ of the system, since

$$-\frac{\hbar}{i}\frac{\partial \psi}{\partial t} = \mathcal{H}\psi \qquad (5.78)$$

which gives, using (5.61),

$$-\frac{\hbar}{i}\sum_n \frac{dc_n}{dt}u_n = \sum_n c_n \mathcal{H} u_n \quad .$$

We can pick out the equation for one particular coefficient, c_k, by multiplying both sides by u_k^* and integrating:

$$-\frac{\hbar}{i}\frac{dc_k}{dt} = \sum_n c_n (k|\mathcal{H}|n) \quad . \qquad (5.79)$$

This equation is the well-known starting point for time-dependent perturbation theory. We can use (5.79) to find a differential equation for the matrix elements of the operator P, since

$$\begin{aligned}
\frac{d}{dt}(k|P|m) &= \frac{d}{dt}(c_k c_m^*) \\
&= c_k \frac{dc_m^*}{dt} + \frac{dc_k}{dt} c_m^* \\
&= \frac{i}{\hbar}\sum_n [c_k c_n^* (n|\mathcal{H}|m) - (k|\mathcal{H}|n) c_n c_m^*] \\
&= \frac{i}{\hbar}(k|P\mathcal{H} - \mathcal{H}P|m)
\end{aligned} \qquad (5.80)$$

where we have used (5.70) for the last step. We can write (5.80) in operator form as

$$\frac{dP}{dt} = \frac{i}{\hbar}[P, \mathcal{H}] \quad . \qquad (5.81)$$

This equation looks very similar to that of (2.31) for the time derivative of an observable, except for the sign change.

If we perform an ensemble average of the various steps of (5.80), assuming $\mathcal{H}$ to be identical for all members of the ensemble, we find a differential equation for the density matrix ϱ. Since the averaging simply replaces P by ϱ, the equation for ϱ is

$$\frac{d\varrho}{dt} = \frac{i}{\hbar}[\varrho, \mathcal{H}] \quad . \qquad (5.82)$$

The density matrix is the quantum mechanical equivalent of the classical density ϱ of points in phase space, and (5.82) is the quantum mechanical form of Liouville's theorem describing the time rate of change of density at a fixed point in phase space.

In the event that $\mathcal{H}$ is independent of time, we may obtain a formal solution of (5.82):

$$\varrho(t) = e^{-(i/\hbar)\mathcal{H}t}\varrho(0)e^{(i/\hbar)\mathcal{H}t} \quad . \tag{5.83}$$

In terms of functions u_n, which are *eigenfunctions* of the Hamiltonian $\mathcal{H}$, we have, for example,

$$\begin{aligned}(k|\varrho(t)|m) &= \int u_k^* e^{-(i/\hbar)\mathcal{H}t}\varrho(0)e^{(i/\hbar)\mathcal{H}t} u_m d\tau \\ &= \int \left(e^{(i/\hbar)\mathcal{H}t} u_k\right)^* \varrho(0)e^{(i/\hbar)\mathcal{H}t} u_m d\tau \quad .\end{aligned} \tag{5.84}$$

By utilizing the fact that $\mathcal{H}u_m = E_m u_m$, and using the power series expansion of the exponential operator, we get

$$(k|\varrho(t)|m) = e^{(i/\hbar)(E_m - E_k)t}(k|\varrho(0)|m) \tag{5.85}$$

for the time-dependent matrix element in terms of the matrix element of ϱ at $t = 0$.

So far we have talked about the density matrix without ever exhibiting explicitly an operator for ϱ. For the sake of concreteness, we shall do so now. We shall take an example of a spin system in thermal equilibrium at a temperature T. We shall take as our basis states, u_n, the eigenstates of the Hamiltonian of the problem, $\mathcal{H}_0$. The populations of the eigenstates are then given by the Boltzmann factors, giving for the diagonal elements of ϱ:

$$\overline{c_m c_m^*} = \frac{e^{-E_m/kT}}{Z} \tag{5.86}$$

where, as usual,

$$Z = \sum_n e^{-E_n/kT} \quad .$$

If we write

$$c_n = |c_n|e^{i\alpha_n}$$

we have that

$$\overline{c_m c_n^*} = \overline{|c_m||c_n|e^{i(\alpha_m - \alpha_n)}} \quad . \tag{5.87}$$

It is customary in statistical mechanics to assume that the phases α_n are statistically independent of the amplitudes $|c_n|$ and that, moreover, α_m or α_n have all values with equal probability. This hypothesis, called the "hypothesis of random phases", causes all the off-diagonal elements of (5.87) to vanish. If, for example, we were to compute the average magnetization perpendicular to the static field for a group of noninteracting spins, as we did in (2.88), the vanishing of the off-diagonal elements of ϱ would make the transverse components of magnetization vanish, as they must, of course, for a system to be in thermal equilibrium. More generally, we see from (5.85) that the off-diagonal elements of ϱ oscillate harmonically in time. If they do not vanish, we expect that there will be some observable property of the system which will oscillate in time according to (5.75). But we should then not have a true thermal equilibrium, since for thermal equilibrium we mean that all properties are independent of time. Therefore we must assume that all the off-diagonal elements vanish. Note, however, from (5.85) (which applies to the situation in which the basis functions are eigenfunctions of the Hamiltonian) that if the off-diagonal elements vanish at any one time, they vanish for all time.

We have, therefore,

$$(n|\varrho|m) = (\delta_{nm}/Z)e^{-E_n/kT} \quad . \tag{5.88}$$

It is worth noting that the operator for ϱ is on a different footing from most other operators such as that for momentum. In the absence of a magnetic field, the latter is always $\hbar\nabla/i$. For a given representation the density matrix may, however, be specified quite arbitrarily, subject only to the conditions that it be Hermitian, that its diagonal elements be greater than or equal to zero, and that they sum to unity. There is therefore no operator known a priori. However, in certain instances the matrix elements $(n|\varrho|m)$ can be obtained very simply from a specific operator for ϱ. When this is possible, we can use operator methods to calculate properties of the system. We now ask what operator will give the matrix elements of (5.88) bearing in mind that the u_n's, and so on, are eigenfunctions of $\mathcal{H}_0$.

Using the fact that

$$e^{-\mathcal{H}_0/kT} u_m = e^{-E_m/kT} u_m \tag{5.89}$$

(which can be proved from the expansion of the exponentials), we can see readily that the explicit form of ϱ is

$$\varrho = \frac{1}{Z} e^{-\mathcal{H}_0/kT} \quad . \tag{5.90}$$

We can use this expression now to compute the average value of any physical property. Thus suppose we have an ensemble of single spins with spin I, acted on by a static external field. Then $\mathcal{H}_0$ is the Hamiltonian of a single spin:

$$\mathcal{H}_0 = -\gamma_n \hbar H_0 I_z \quad . \tag{5.91}$$

We shall illustrate the use of the density matrix to compute the average value of the z-magnetization $\overline{\langle M_z \rangle}$. It is

$$\overline{\langle M_z \rangle} = \text{Tr}\{M_z \varrho\} = \frac{1}{Z} \text{Tr}\{M_z e^{-\mathcal{H}_0/kT}\} \quad . \tag{5.92}$$

In the high-temperature approximation we can expand the exponential, keeping only the first terms. By utilizing the fact that Tr $\{M_z\} = 0$, we have

$$\overline{\langle M_z \rangle} = \frac{1}{Z}\text{Tr}\left\{M_z\left(1 - \frac{\mathcal{H}_0}{kT} + \dots\right)\right\}$$
$$\cong \frac{1}{Z}\text{Tr}\left\{\left(\frac{\gamma_n^2 \hbar^2 H_0 I_z^2}{kT}\right)\right\} \quad . \tag{5.93}$$

Now, in the high-temperature limit, $Z = 2I+1$. Since Tr $\{I_z^2\} = \frac{1}{3}I(I+1)(2I+1)$, we get

$$\overline{\langle M_z \rangle} = \frac{\gamma_n^2 \hbar^2 I(I+1)}{3kT} H_0 \tag{5.94}$$

which we recognize as Curie's law for the magnetization. The density matrix gives, therefore, a convenient and compact way of computing thermal equilibrium properties of a system.

One situation commonly encountered is that of a Hamiltonian consisting of a large time-independent interaction $\mathcal{H}_0$, and a much smaller but time-dependent term $\mathcal{H}_1(t)$. The equation of motion of the density matrix is then

$$\frac{d\varrho}{dt} = \frac{i}{\hbar}[\varrho, \mathcal{H}_0 + \mathcal{H}_1] \quad . \tag{5.95}$$

If $\mathcal{H}_1$ were zero, the solution of (5.95) would be

$$\varrho(t) = e^{-(i/\hbar)\mathcal{H}_0 t} \varrho(0) e^{(i/\hbar)\mathcal{H}_0 t} \quad . \tag{5.96}$$

Let us then define a quantity ϱ^* (the star does not mean complex conjugate) by the equation

$$\varrho(t) = e^{-(i/\hbar)\mathcal{H}_0 t} \varrho^*(t) e^{(i/\hbar)\mathcal{H}_0 t}. \tag{5.97}$$

If, in fact, $\mathcal{H}_1$ were zero, comparison of (5.96) and (5.97) would show that ϱ^* would be a constant. (Note, moreover, that at $t = 0$, ϱ^* and ϱ are identical.) For small $\mathcal{H}_1$, then, we should expect ϱ^* to change slowly in time. Substituting (5.97) into the left side of (5.95) gives us the differential equation obeyed by ϱ^*:

$$-\frac{i}{\hbar}[\mathcal{H}_0, \varrho] + e^{-(i/\hbar)\mathcal{H}_0 t}\frac{d\varrho^*}{dt} e^{(i/\hbar)\mathcal{H}_0 t} = \frac{i}{\hbar}[\varrho, \mathcal{H}_0 + \mathcal{H}_1] \quad . \tag{5.98}$$

We note that the commutator of ϱ with $\mathcal{H}_0$ can now be removed from both sides. Then, multiplying from the left by $\exp((i/\hbar)\mathcal{H}_0 t)$ and from the right by $\exp(-(i/\hbar)\mathcal{H}_0 t)$, and defining

$$\mathcal{H}_1^* = e^{i\mathcal{H}_0 t/\hbar}\mathcal{H}_1 e^{-(i/\hbar)\mathcal{H}_0 t} \tag{5.99}$$

we get, from (5.98),

$$\frac{d\varrho^*}{dt} = \frac{i}{\hbar}[\varrho^*, \mathcal{H}_1^*(t)] \quad . \tag{5.100}$$

Equation (5.100) shows us, as we have already remarked, that the operator ϱ^* would be constant in time if the perturbation $\mathcal{H}_1$ were set equal to zero.

The transformation of the operator $\mathcal{H}_1$ given by (5.99) is a canonical transformation, and the new representation is termed the *interaction representation*. The relationship of ϱ and ϱ^* is illustrated by considering the expansion of the wave function Ψ in a form

$$\Psi = \sum_n a_n e^{-(i/\hbar)E_n t} u_n \tag{5.101}$$

instead of

$$\Psi = \sum_n c_n u_n \tag{5.102}$$

where the u_n's and E_n's are the eigenfunctions and eigenvalues of the Hamiltonian $\mathcal{H}_0$. In the absence of $\mathcal{H}_1$, the a_n's would then be constant in time. We shall show that the matrix $\overline{a_n a_m^*}$ is simply $(n|\varrho^*|m)$. We note first that replacing $\varrho(0)$ by ϱ^* in (5.83–85) gives that

$$(n|\varrho^*|m) = e^{(i/\hbar)(E_n - E_m)t}(n|\varrho|m) \quad . \tag{5.103}$$

Since (5.101) and (5.102) give the same Ψ, we must have

$$c_n = a_n e^{-(i/\hbar)E_n t} \quad , \quad \text{so that} \tag{5.104}$$

$$a_n a_m^* = c_n c_m^* e^{(i/\hbar)(E_n - E_m)t} . \tag{5.105}$$

Comparison with (5.103) shows that

$$\overline{a_n a_m^*} = (n|\varrho^*|m) \quad \text{Q.E.D.} \tag{5.106}$$

There is likewise a simple relationship between $(n|\mathcal{H}_1^*|m)$ and $(n|\mathcal{H}_1|m)$. By an argument quite identical to that of (5.83–85) we have

$$(n|\mathcal{H}_1^*|m) = \int u_n^* e^{(i/\hbar)\mathcal{H}_0 t} \mathcal{H}_1 e^{-(i/\hbar)\mathcal{H}_0 t} u_m d\tau$$

$$= e^{(i/\hbar)(E_n - E_m)t}(n|\mathcal{H}_1|m) \quad . \tag{5.107}$$

Now we proceed to solve the equation of motion for ϱ^*, (5.100). By integrating from $t = 0$, we have

$$\varrho^*(t) = \varrho^*(0) + \frac{i}{\hbar} \int_0^t [\varrho^*(t'), \mathcal{H}_1^*(t')] dt' \quad . \tag{5.108}$$

This has not as yet produced a solution, since $\varrho^*(t')$ in the integral is unknown. We can make an approximate solution by replacing $\varrho^*(t')$ by $\varrho^*(0)$, its value at $t = 0$. This gives us

$$\varrho^*(t) = \varrho^*(0) + \frac{i}{\hbar} \int_0^t [\varrho^*(0), \mathcal{H}_1^*(t')] dt' \quad . \tag{5.109}$$

We can make a closer approximation by an iteration procedure, using (5.109) to get a better value of $\varrho^*(t')$, to put into the integrand of (5.108). Thus we find

$$\varrho^*(t) = \varrho^*(0) + \frac{i}{\hbar} \int_0^t \left[\left\{ \varrho^*(0) + \frac{i}{\hbar} \int_0^{t'} [\varrho^*(0), \mathcal{H}_1^*(t'')] dt'' \right\}, \mathcal{H}_1^*(t') \right] dt'$$

$$= \varrho^*(0) + \frac{i}{\hbar} \int_0^t [\varrho^*(0), \mathcal{H}_1^*(t')] dt'$$

$$+ \left(\frac{i}{\hbar}\right)^2 \int_0^t \int_0^{t'} [[\varrho^*(0), \mathcal{H}_1^*(t'')], \mathcal{H}_1^*(t')] dt' dt'' \quad . \tag{5.110}$$

We could continue this iteration procedure. Since each iteration adds a term one power higher in the perturbation $\mathcal{H}_1^*$, the successive iterations are seen to consist of higher and higher perturbation expansions in the interaction $\mathcal{H}_1$. For our purposes, we shall not go higher than the second. Actually, we shall find it most convenient to calculate the derivative of ϱ^*. Taking the derivative of (5.110) gives

$$\frac{d\varrho^*(t)}{dt} = \frac{i}{\hbar}[\varrho^*(0), \mathcal{H}_1^*(t)] + \left(\frac{i}{\hbar}\right)^2 \int_0^t [[\varrho^*(0), \mathcal{H}_1^*(t')], \mathcal{H}_1^*(t)] dt' \quad . \tag{5.111}$$

It is important to note that (5.111) is entirely equivalent to ordinary time-dependent perturbation theory carried to second order. However, instead of solving for the behaviors of a_n and a_m, we are solving for the behavior of the products $a_n a_m^*$ which are more directly useful for calculating expectation values.

5.5 The Rotating Coordinate Transformation

We saw in Chap. 2 that it is often convenient to go to the rotating frame when a system is acted on by alternating magnetic fields. We explored both a classical treatment and a quantum mechanical treatment. In the latter we transformed the Schrödinger equation. We now examine how to transform the differential equation for the density matrix.

Let us consider first the case of an isolated spin acted on only by the static field H_0 and a rotating field $\boldsymbol{H}_1(t)$ given by

$$\boldsymbol{H}_1(t) = H_1(t)(\boldsymbol{i} \cos \omega t - \boldsymbol{j} \sin \omega t) \tag{5.112}$$

where we have already recognized the sense of rotation for a positive γ in the negative sign in front of $\boldsymbol{j}$.

In the laboratory reference frame

$$\frac{d\varrho}{dt} = \frac{i}{\hbar}(\varrho \mathcal{H} - \mathcal{H} \varrho) \tag{5.113}$$

$$\mathcal{H} = -\gamma\hbar[H_0 I_z + H_1(I_x \cos \omega t - I_y \sin \omega t)] \quad . \tag{5.114}$$

Utilizing (2.55) we express this as

$$\mathcal{H} = -\gamma\hbar[H_0 I_z + H_1 e^{+i\omega t I_z} I_x e^{-i\omega t I_z}] \quad . \tag{5.115}$$

Defining an operator R as

$$R \equiv e^{i\omega t I_z} \tag{5.116}$$

we have

$$\mathcal{H} = -\gamma\hbar(H_0 I_z + H_1 R I_x R^{-1}) \quad \text{so that} \tag{5.117}$$

$$\frac{d\varrho}{dt} = i\varrho\gamma(H_0 I_z + H_1 R I_x R^{-1}) - i\gamma(H_0 I_z + H_1 R I_x R^{-1})\varrho \quad . \tag{5.118}$$

Following the same reasoning as in Chap. 2, we now seek to eliminate the "R" operators by defining a new variable ϱ_R

$$\varrho_R \equiv R^{-1}\varrho R \quad \text{or} \tag{5.119}$$

$$\varrho = R\varrho_R R^{-1} \quad . \tag{5.120}$$

Substituting (5.120) into (5.118), we get

$$i(\omega I_z \varrho - \varrho\omega I_z) + R\frac{d\varrho_R}{dt}R^{-1}$$
$$= -i\gamma H_0(\varrho I_z - I_z \varrho) - i\gamma H_1(\varrho R I_x R^{-1} - R I_x R^{-1}\varrho) \quad . \tag{5.121}$$

Multiplying from the left by R^{-1}, from the right by R, utilizing the fact that I_z and R commute, and employing (5.119) and (5.120), we get

$$\frac{d\varrho_R}{dt} = \frac{i}{\hbar}(\varrho_R \mathcal{H}_{\text{eff}} - \mathcal{H}_{\text{eff}}\varrho_R) \quad \text{where} \tag{5.122}$$

$$\mathcal{H}_{\text{eff}} \equiv -\hbar(\gamma H_0 - \omega)I_z - \gamma\hbar H_1 I_x \tag{5.123}$$

is our old friend, the Hamiltonian of (2.63), which expresses the interaction of the spin with the effective magnetic field in the rotating reference frame.

Let us now allow there to be more than one spin, allow the spins to interact with each other by the dipole-dipole coupling or by indirect spin-spin coupling through bonding electrons, and include Knight shifts and chemical shifts as well.

Then, labeling individual spins by k or j, we have the Hamiltonian in the laboratory frame as

$$\mathcal{H} = -\sum_k \gamma\hbar H_0 I_{zk}(1 - \sigma_k) + \frac{1}{2}\sum_{k,j} \mathcal{H}_{kj}$$
$$- \gamma\hbar H_1\left(\sum_k I_{xk} \cos \omega t - \sum I_{yk} \sin \omega t\right) \tag{5.124}$$

where σ_k expresses the chemical or Knight shifts. We omit a chemical shift correction from H_1 since $H_1 \ll H_0$ always. Defining

$$I_x \equiv \sum_k I_{xk} \quad , \quad I_y \equiv \sum_k I_{yk} \quad , \quad I_z \equiv \sum_k I_{zk} \quad , \tag{5.125}$$

we now express $\mathcal{H}$ as a sum

$$\mathcal{H} = \mathcal{H}_Z + \mathcal{H}_P + \mathcal{H}_1 \quad \text{with} \tag{5.126}$$

$$\mathcal{H}_Z = -\gamma \hbar H_0 I_z$$

$$\mathcal{H}_P = \sum_k \gamma \hbar H_0 \sigma_k I_{zk} + \frac{1}{2} \sum_{k,j} \mathcal{H}_{jk} \tag{5.127}$$

$$\mathcal{H}_1 = -\gamma \hbar H_1 e^{i\omega t I_z} I_x e^{-i\omega t I_z} \quad .$$

Following the reasoning of Chap. 3, we keep only that portion of the spin-spin coupling which commutes with $\mathcal{H}_Z$. Thus, we express $\mathcal{H}_{ik}$ as a sum of terms like the A, B, C, ..., F of (3.7), keeping only $\mathcal{H}_{jk}^0$, the terms which commute with I_z:

$$[\mathcal{H}_{jk}^0, I_z] = 0 \quad . \tag{5.128}$$

Thus we take as our total Hamiltonian $\mathcal{H}'$ in this perturbation sense

$$\mathcal{H}' = -\gamma \hbar H_0 I_z + \sum_k \gamma \hbar H_0 \sigma_k I_{zk}$$
$$+ \frac{1}{2} \sum_{j,k} \mathcal{H}_{jk}^0 - \gamma \hbar H_1 R I_x R^{-1} \quad . \tag{5.129}$$

We can now repeat the derivation of (5.122) above, noting that our new Hamiltonian differs from that of (5.114) by addition of the terms

$$\sum_k \gamma \hbar H_0 \sigma_k I_{zk} + \frac{1}{2} \sum_{jk} \mathcal{H}_{ij}^0 \quad , \tag{5.130}$$

all of which commute with R. (It is interesting to note that this expression through its σ_k *seems* to require that we know the absolute chemical shift relative to a bare nucleus, i.e. a nucleus stripped of all its electrons. That viewpoint is wrong. All one needs do is pick a convenient reference compound for which one defines the chemical shift to be zero. γ is then defined as being the ratio of the angular frequency to the static field for resonance for that reference compound. We discuss this topic further at the end of this section).

Thus we readily find

$$\frac{d\varrho_R}{dt} = \frac{i}{\hbar}(\varrho_R \mathcal{H}'_{\text{eff}} - \mathcal{H}'_{\text{eff}} \varrho_R) \quad \text{where} \tag{5.131}$$

$$\mathcal{H}'_{\text{eff}} = -\gamma\hbar(H_0 - \omega/\gamma)I_z - \gamma\hbar H_1 I_x$$
$$+ \sum_k \gamma\hbar H_0 \sigma_k I_{zk} + \frac{1}{2}\sum_{k,j} \mathcal{H}^0_{kj} \quad . \tag{5.132}$$

This new effective Hamiltonian reminds us by the presence of the terms $\mathcal{H}^0_{jk}$ that if spins interact with one another in the laboratory frame, they do also in the rotating frame. It also reminds us from the terms involving σ_k that when there is more than one chemical shift, one cannot be simultaneously exactly on resonance for all nuclei.

Since $\mathcal{H}'_{\text{eff}}$ is independent of time, we can solve formally for $\varrho_R(t)$ in terms of its value at an earlier time ($t = 0$) by

$$\varrho_R(t) = e^{-i\mathcal{H}'_{\text{eff}}t/\hbar} \varrho_R(0) e^{i\mathcal{H}'_{\text{eff}}t/\hbar} \quad . \tag{5.133}$$

It is important to keep in mind that in general the interaction representation differs from the rotating coordinate transformation. Both may be used with a large time-independent Hamiltonian $\mathcal{H}_0$ and a time-dependent $\mathcal{H}_1(t)$.

ϱ^* (ϱ expressed in the interaction representation) is related to ϱ by

$$\varrho^*(t) = e^{i\mathcal{H}_0 t/\hbar} \varrho(t) e^{-i\mathcal{H}_0 t/\hbar} \tag{5.134}$$

whereas ϱ_R is related to ϱ by

$$\varrho_R(t) = e^{-i\omega t I_z} \varrho(t) e^{i\omega t I_z} \quad . \tag{5.135}$$

If we have the Hamiltonian of (5.114),

$$\varrho^*(t) = e^{-i\omega_0 t I_z} \varrho(0) e^{i\omega_0 t I_z} \quad . \tag{5.136}$$

These *do* become the same if we are exactly at resonance ($\omega_0 = \omega$), but otherwise would differ. If the *only* time dependence in the problem is from H_1, the rotating coordinate transformation eliminates it.

The division of the Hamiltonian of (5.127) into $\mathcal{H}_Z$ and $\mathcal{H}_P$ is not the only division one might think to make. For example, if one defines an average chemical shift for the system of N nuclei, $\overline{\sigma}$, by the equation

$$\overline{\sigma} = \frac{1}{N} \sum_k \sigma_k \tag{5.137}$$

then one might prefer to have the Zeeman energy be the average value including the chemical shifts. We can do this by making a division into an $\mathcal{H}'_Z$ and $\mathcal{H}'_P$ defined as

$$\mathcal{H}'_Z = -\gamma\hbar H_0(1 - \overline{\sigma}) I_z \tag{5.138a}$$

$$\mathcal{H}'_P = \sum_k \gamma\hbar H_0(\sigma_k - \overline{\sigma}) I_{zk} + \frac{1}{2}\sum_{j,k} \mathcal{H}_{ij} \quad . \tag{5.138b}$$

Note, however, that (5.138a) can be converted to (5.127) if we define a new γ, called γ',

$$\gamma' = \gamma(1 - \overline{\sigma}) \tag{5.139}$$

and a new zero of chemical shifts

$$\sigma'_k = \sigma_k - \overline{\sigma} \tag{5.140}$$

which makes

$$\mathcal{H}'_Z = -\gamma' \hbar H_0 I_z \tag{5.141a}$$

$$\mathcal{H}'_P = \sum_k \gamma' \hbar H_0 \sigma'_k I_{zk} \quad , \tag{5.141b}$$

where in getting the second equation we drop terms such as $\overline{\sigma}\sigma_k$.

Clearly the form of (5.141) is the same as that of (5.127). We have simply redefined γ to correspond to a different zero of chemical shifts.

While (5.138) guarantees that the Zeeman term falls in the midst of the resonance lines, it seems to imply through (5.137) that we know the individual σ_k's, and go through some evaluation to get $\overline{\sigma}$. What is important to realize is that there is some arbitrariness in the division between $\mathcal{H}_Z$ and $\mathcal{H}_P$ which really corresponds to the arbitrariness in the choice of a zero of chemical shift.

5.6 Spin Echoes Using the Density Matrix

Now that we have obtained the differential equation for the density matrix in the laboratory frame, in the interaction representation, and in the rotating frame, we turn to methods of solution. The rest of this chapter is devoted to that task. We will first discuss several problems involving pulse techniques, then we turn to use of the density matrix to discuss relaxation. As topics involving pulse techniques we shall first examine how to use the density matrix to derive the spin echo, giving us a chance to use this formalism to examine a problem whose physics we have now seen both classically and quantum mechanically (using wave functions). We then turn to some new physics which underlies one of the major techniques employed in NMR today, Fourier transform methods. To do this we first set the stage by deriving a result important to linear response theory, the theory underlying Fourier transform NMR. This result is the response of a system to a δ-function excitation. Armed with that result, we next go on to explain what is meant by Fourier transform NMR, and then prove the fundamental theorem which specifies what it measures. We then further develop a concrete appreciation of the density matrix by exploring what happens to the actual matrix elements under the action of an alternating magnetic field.

The most general approach to solving the equation is to pick an appropriate family of wave functions as a basis set, then convert the operator equation to a set of coupled differential equations between the matrix elements of ϱ and $\mathcal{H}$. Thus if we pick a basis set $|\alpha\rangle$, where α stands for the set of eigenvalues which distinguish the states, we convert the equation

$$\frac{d\varrho}{dt} = \frac{i}{\hbar}(\varrho\mathcal{H} - \mathcal{H}\varrho) \quad \text{to} \tag{5.142}$$

$$\frac{d}{dt}(\alpha|\varrho|\alpha') = \frac{i}{\hbar}\sum_{\beta}((\alpha|\varrho|\beta)(\beta|\mathcal{H}|\alpha') - (\alpha|\mathcal{H}|\beta)(\beta|\varrho|\alpha')) \tag{5.143}$$

In many cases there are then straightforward methods of solution. Once one has obtained the $(\alpha|\varrho|\alpha')$'s, one can then calculate the expectation values of observables using (5.77),

$$\langle M_x \rangle = \text{Tr}\{M_x \varrho\} = \sum_{\alpha,\alpha'}(\alpha'|M_x|\alpha)(\alpha|\varrho|\alpha') \quad . \tag{5.144}$$

The weakness of this approach is that it may fail to reveal important physical insights available from other methods.

The "other methods" are based on manipulating operators, and are analogous to the quantum mechanical derivation of the echo carried out in Chap. 2. We now turn to several illustrations of the use of the operator approach. Later we return to use of (5.143).

We begin with a derivation of the spin echo of a group of noninteracting spins using a $(\pi/2, \pi)$ pulse sequence. We deal with this problem by first transforming to the rotating frame so that our basic equation is given by (5.123)

$$\frac{d\varrho_R}{dt} = \frac{i}{\hbar}(\varrho_R \mathcal{H}_{\text{eff}} - \mathcal{H}_{\text{eff}} \varrho_R) \quad .$$

As in Chap. 2, we assume that the static field is inhomogeneous with a distribution function $p(h_0)dh_0$ for the N spins, where dN, the number of spins whose resonance lies between h_0 and $h_0 + dh_0$, is given by (2.119):

$$dN = Np(h_0)dh_0 \quad ,$$

and where h_0 is given by (2.118):

$$h_0 = H_0 - \omega/\gamma \quad .$$

We take the distribution function to be symmetrical about a center value ($h_0 = 0$) and we assume ω is to be at resonance when $h_0 = 0$.

Since we have already treated this problem in Chap. 2 using wave functions, we will employ the same notation (Fig. 5.4). At $t = 0^-$, the density matrix has its initial value $\varrho_R(0^-)$ which represents thermal equilibrium in the laboratory frame. At $t = 0$, H_1 is turned on. Although there is a discontinuity in $\mathcal{H}_{\text{eff}}$ at $t = 0$, the right-hand side of (5.123) remains finite. Therefore, integrating (5.123) with respect to time from $t = 0^-$ to $t = 0^+$, we find

$$\varrho_R(0^+) = \varrho_R(0^-) \quad . \tag{5.145}$$

A similar argument holds of the other discontinuities. During the time interval from $t = 0$ to $t = t_1$ or from t_2 to t_3, the Hamiltonian is given by (2.117)

$$\mathcal{H}_{\text{eff}} = -\gamma\hbar(h_0 I_z + H_1 I_x) \quad .$$

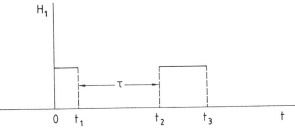

Fig. 5.4. H_1 vs time, giving the definition of the times 0, t_1, t_2, and t_3 which specify the beginning and end of the $\pi/2$ pulse $(t = 0, t_1)$ and of the π pulse $(t = t_2, t_3)$

As in Chap. 2, we approximate this as

$$\mathcal{H}_{\text{eff}} = -\gamma \hbar H_1 I_x \quad . \tag{5.146}$$

Then, utilizing (5.83), we write

$$\begin{aligned}\varrho_R(t_1) &= e^{i\omega_1 t_1 I_x} \varrho_R(0^+) e^{-i\omega_1 t_1 I_x} \\ &= e^{i\omega_1 t_1 I_x} \varrho_R(0^-) e^{-i\omega_1 t_1 I_x}\end{aligned} \tag{5.147}$$

where

$$\omega_1 = \gamma H_1 \quad . \tag{5.148}$$

Likewise, during the interval t_1 to t_2, where H_1 is zero, we have

$$\varrho_R(t_2) = e^{i\gamma h_0 (t_2 - t_1) I_z} \varrho_R(t_1) e^{-i\gamma h_0 (t_2 - t_1) I_z} \quad . \tag{5.149}$$

Defining [as in (2.122)]

$$X(\theta) = e^{i\theta I_x} \tag{5.150a}$$

$$T(t, h_0) = e^{i\gamma h_0 t I_z} \quad , \tag{5.150b}$$

selecting $\omega_1 t_1 = \pi/2$ and $\omega_1 (t_3 - t_2) = \pi$ for the $\pi/2$ and π pulses, and choosing times t later than t_3, we get $\varrho_R(t, h_0)$ for those spins off resonance by h_0 as

$$\begin{aligned}\varrho_R(t, h_0) = {}&T(t - \tau, h_0) X(\pi) T(\tau, h_0) X(\pi/2) \varrho_R(0^-) \\ &\times X^{-1}(\pi/2) T^{-1}(\tau, h_0) X^{-1}(\pi) T^{-1}(t - \tau, h_0) \quad .\end{aligned} \tag{5.151}$$

Following our procedure from Chap. 2, we insert $X(\pi) X^{-1}(\pi) = 1$ to the left of $T(t - \tau, h_0)$ and utilize (2.134) to get

$$X(\pi) X^{-1}(\pi) T(t - \tau, h_0) X(\pi) = X(\pi) T^{-1}(t - \tau, h_0) \quad . \tag{5.152}$$

Thus

$$\begin{aligned}\varrho_R(t, h_0) = {}&X(\pi) T^{-1}(t - \tau, h_0) T(\tau, h_0) X(\pi/2) \varrho_R(0^-) \\ &\times X^{-1}(\pi/2) T^{-1}(\tau, h_0) T(t - \tau, h_0) X^{-1}(\pi) \quad .\end{aligned} \tag{5.153}$$

Thus, when $t - \tau = \tau$, or $t = 2\tau$,

$$\varrho_R(2\tau, h_0) = X(\pi)X(\pi/2)\varrho_R(0^-)X^{-1}(\pi/2)X^{-1}(\pi) \qquad (5.154a)$$

$$= X(3\pi/2)\varrho_R(0^-)X^{-1}(3\pi/2) \qquad (5.154b)$$

$$= X(-\pi/2)\varrho_R(0^-)X^{-1}(-\pi/2) \quad , \qquad (5.154c)$$

where the last step utilizes the fact that $X(2\pi) = X(0)$ when X operates on ϱ [contrast this with (2.78)]. Equation (5.154) shows that when $t = 2\tau$, ϱ is independent of h_0, hence we have an echo. Moreover, $\varrho_R(2\tau)$ is the *same* as we would have if instead of applying two pulses spaced apart by τ, we had applied a single pulse which produced a $-\pi/2$ rotation.

We now proceed to calculate the magnetization signal at the echo peak. Since $\varrho_R(0^-)$ corresponds to thermal equilibrium in the laboratory frame, we have, using (5.119) and (5.90),

$$\begin{aligned}\varrho_R(0^-) &= \mathrm{e}^{-\mathrm{i}\omega t I_z}\varrho(0^-)\mathrm{e}^{\mathrm{i}\omega t I_z} \\ &= \frac{1}{Z}\mathrm{e}^{-\mathrm{i}\omega t I_z}\mathrm{e}^{-\mathcal{H}_0/kT}\mathrm{e}^{\mathrm{i}\omega t I_z} \quad .\end{aligned} \qquad (5.155)$$

Since $\mathcal{H}_0 = -\gamma\hbar H_0 I_z$, we get

$$\varrho_R(0^-) = \frac{1}{Z}\mathrm{e}^{\gamma\hbar H_0 I_z/kT} \quad . \qquad (5.156)$$

In the high temperature approximation, $Z = 2I + 1$, giving

$$\varrho_R(0^-) = \frac{1}{(2I+1)}\left(1 + \frac{\gamma\hbar H_0 I_z}{kT} + \ldots\right) \quad . \qquad (5.157)$$

(Note that since the spins do not interact, we can work with the Hamiltonian of one spin. However, we eventually include the effect of all N spins when we integrate over h_0. This situation is to be contrasted with that of Sect. 5.7 where the Hamiltonian is that of N coupled spins).

We have operators $M_x(= \gamma\hbar I_x)$ and M_y corresponding to magnetization along the x-axis (the H_1 direction) and the y-axis in the rotating frame given by

$$\begin{aligned}\langle M_x(t)\rangle &= \gamma\hbar\,\mathrm{Tr}\,\{\varrho_R(t)I_x\} \\ \langle M_y(t)\rangle &= \gamma\hbar\,\mathrm{Tr}\,\{\varrho_R(t)I_y\} \quad .\end{aligned} \qquad (5.158)$$

Combining we get

$$\langle M_x(t)\rangle + \mathrm{i}\langle M_y(t)\rangle = \langle M^+(t)\rangle = \gamma\hbar\,\mathrm{Tr}\,\{\varrho_R I^+\} \quad . \qquad (5.159)$$

Since both $\langle M_x(t)\rangle$ and $\langle M_y(t)\rangle$ are real, we can get them both readily by taking the real and the imaginary parts of (5.159). We utilize (5.154), (5.156), and (5.157) to calculate $\langle M_x(2\tau)\rangle$ and $\langle M_y(2\tau)\rangle$

$$\begin{aligned}\langle M^+(2\tau)\rangle &= N\int p(h_0)dh_0\gamma\hbar\,\mathrm{Tr}\,\{I^+\varrho(2\tau, h_0)\} \\ &= \frac{\gamma\hbar N}{(2I+1)}\int p(h_0)dh_0\,\mathrm{Tr}\,\{G\}\end{aligned} \qquad (5.160a)$$

where

$$\text{Tr}\{G\} = \text{Tr}\left\{I^+X(-\pi/2)\underbrace{\left(1+\underbrace{\frac{\gamma\hbar H_0 I_z}{kT}}_{b}\right)}_{a}X^{-1}(-\pi/2)\right\} . \quad (5.160b)$$

In the trace, the term labeled a is readily shown to be zero:

$$\text{Tr}\{I^+X(-\pi/2)X^{-1}(-\pi/2)\} = \text{Tr}\{I^+\} = 0 . \quad (5.161)$$

The physical significance of this result is that at infinite temperature, where only the a term remains, the thermal equilibrium magnetization vanishes. Therefore, no pulse sequence can produce a signal. We will encounter a similar situation in all pulse sequences starting from thermal equilibrium.

The b term gives

$$\langle M^+(2\tau)\rangle = \frac{(\gamma\hbar)^2 N}{(2I+1)}\frac{H_0}{kT}\text{Tr}\{I^+X(-\pi/2)I_z X^{-1}(-\pi/2)\} . \quad (5.162)$$

Now $X(-\pi/2) = X^{-1}(\pi/2)$. Hence, using (2.124) $[X^{-1}(\pi/2)I_z X(\pi/2) = -I_y]$ we get

$$\langle M^+(2\tau)\rangle = \frac{(\gamma\hbar)^2 N}{(2I+1)}\frac{H_0}{kT}\text{Tr}\{I^+(-I_y)\} . \quad (5.163)$$

Since $\text{Tr}\{I_x I_y\} = 0$ and $\text{Tr}\{I_y^2\} = (2I+1)I(I+1)/3$, we get

$$\langle M^+(2\tau)\rangle = -\frac{N\gamma^2\hbar^2 I(I+1)}{3kT}H_0 i = -i\chi_0 H_0 . \quad (5.164)$$

The $-i$ shows that at $t = 2\tau$ the magnetization has zero component along the x-direction (i.e. H_1), that its magnitude is the same as the thermal equilibrium magnetization, $\chi_0 H_0$, and that it points along the negative y-direction.

The approach we have described can be extended readily to deal with arbitrary sequences of pulses. Let us consider that at $t = 0^-$ the system is in thermal equilibrium with density matrix $\varrho_R(0^-) = \varrho(0^-)$, as in (5.152). Starting at $t = 0$ the Hamiltonian in the rotating frame takes on a value $\mathcal{H}_A$ for a time interval t_A, then jumps suddenly to $\mathcal{H}_B$ for a time t_B, and so on. In the example above

$$\begin{aligned}
t_A &= t_1 & \mathcal{H}_A &= -\gamma\hbar(h_0 I_z + H_1 I_x) \\
t_B &= t_2 - t_1 & \mathcal{H}_B &= -\gamma\hbar h_0 I_z \\
t_C &= t_3 - t_2 & \mathcal{H}_C &= -\gamma\hbar(h_0 I_z + H_1 I_x) \\
t_D &= t - t_3 & \mathcal{H}_D &= -\gamma\hbar h_0 I_z .
\end{aligned} \quad (5.165)$$

Corresponding to t_A and $\mathcal{H}_A$, there is a time development operator T_A given by

$$T_A = \exp(-i\mathcal{H}_A t_A/\hbar) . \quad (5.166)$$

Similar equations hold for the succeeding intervals. Thus, we get

$$\varrho_R(t_Z) = T_Z T_Y \ldots T_B T_A \varrho(0^-) T_A^{-1} T_B^{-1} \ldots T_Y^{-1} T_Z^{-1} . \quad (5.167)$$

We chose to neglect all terms in the Hamiltonian but the term involving H_1 when H_1 is on. That approximation is often useful (as illustrated above) but would not be allowable if one were trying to understand what happens when H_1 is comparable to the line width. The general formula (5.167) does not necessarily include this approximation, but the approximation is readily introduced if desired.

5.7 The Response to a δ-Function

Now that we have looked at a familiar problem with the density matrix formalism, we turn to a problem we have not treated quantum mechanically before, the response of the system to a δ-function. This topic is important in part because, as we saw in Chap. 2, if we know this response we can use it to calculate both χ' and χ''. But it is also important because it underlies the theory of Fourier transform NMR which we take up in the next section. We therefore start by recalling (2.156):

$$\chi' = \int_0^\infty m(\tau) \cos(\omega\tau) d\tau \quad , \quad \chi'' = \int_0^\infty m(\tau) \sin(\omega\tau) d\tau \tag{2.156}$$

where $m(\tau)$ is the response at time τ to a delta-function applied at $\tau = 0$.

We now turn to the calculation of $m(\tau)$. To do this we must first settle on the Hamiltonian. We take a Hamiltonian which has substantial practical utility, that of a group of N identical spins, coupled to each other with dipolar, pseudodipolar, and pseudoexchange interactions, experiencing chemical and Knight shifts. A more general Hamiltonian would include different nuclear species, quadrupole coupling, and relaxation phenomena to a thermal reservoir.

With the approximation above we will then have a Hamiltonian

$$\mathcal{H} = -\sum_k \gamma\hbar H_0 I_{zk}(1-\sigma_k) + \frac{1}{2}\sum_{i,j}\mathcal{H}_{ij} - \sum_k \gamma\hbar C\delta(t)I_x \quad , \tag{5.168}$$

which is similar to that of (5.123) except for the replacement of the H_1 driving term with a magnetic field $C\delta(t)$. The parameter C is used to define the size of the driving term. This is a Hamiltonian in the laboratory frame.

In (5.168), the δ-function field has been applied along the x-direction in the laboratory. It will produce magnetization in general with components along the laboratory x-, y-, and z-directions. Thus, in general such a drive produces a magnetization in the α'-direction for a δ-function in the α-direction. This fact requires that we label $m(\tau)$ by subscripts α and α' just as we label the χ by $\chi_{\alpha'\alpha}$ on p. 58. Thus, let $m_{\alpha'\alpha}(\tau)$ be the magnetization in the α'-direction produced by a δ-function field applied in the α-direction. Equation (2.93a) gives χ'_{xx} and χ''_{xx}.

The meaning of the constant C can be seen by considering the integral of the driving field $C\delta(t)$:

$$\int_{0^-}^{0^+} C\delta(t)dt = C \quad . \tag{5.169}$$

But, by definition, $m(t)$ is the response when this integral is unity, i.e. when $C = 1$, or, alternatively, if $M(\tau)$ is the response when $C \neq 1$,

$$m(\tau) = M(\tau)/C \quad . \tag{5.170}$$

While in principle one can set $C = 1$, in the process one conceals some units, and thus complicates the use of dimensional arguments to check the correctness of formulas. We will therefore keep C, realizing that eventually we will utilize (5.170). Indeed, to keep the units explicit, we will replace C by a product $H\tau_0$:

$$H\tau_0 \equiv C \tag{5.171}$$

where one can think of H as the height and τ_0 the duration of a rectangular pulse which, in the limit that τ_0 goes to zero, represents the δ-function.

To find $m_{xx}(\tau)$ we need to solve the density matrix with the Hamiltonian of (5.168), with the initial condition that $\varrho(0^-)$ represents thermal equilibrium.

At $t = 0$, the driving term dominates, since the δ-function is the biggest term. At all other times, only the Zeeman and spin-spin terms are present. Thus, if $t > 0$, we have a time-independent Hamiltonian. We will treat it in the same spirit as the Hamiltonian of (5.126) in which we recognize that the spin-spin coupling terms are perturbations, so that it is appropriate to drop the terms in the spin-spin coupling which do not commute with the Zeeman coupling. We thus have as our Hamiltonian

$$\mathcal{H} = -\gamma\hbar H_0 I_z + \sum_k \gamma\hbar H_0 \sigma_{zk} I_{zk} + \frac{1}{2}\sum_{j,k} \mathcal{H}^0_{jk} - \gamma\hbar I_x H\tau_0\delta(t) \quad , \tag{5.172}$$

where $\mathcal{H}^0_{jk}$ is the portion of $\mathcal{H}_{jk}$ which commutes with the mean Zeeman energy

$$[I_z, \mathcal{H}^0_{ij}] = 0 \quad . \tag{5.173}$$

It is convenient, as in (5.126), to define three quantities

$$\mathcal{H}_Z \equiv -\gamma\hbar H_0 I_z$$

$$\mathcal{H}_P \equiv \sum_k \gamma\hbar H_0 \sigma_{zk} I_{zk} + \frac{1}{2}\sum_{j,k} \mathcal{H}^0_{jk} \tag{5.174}$$

$$\mathcal{H}_\delta = -\gamma\hbar I_x H\tau_0\delta(t) \quad .$$

Note from (5.174) that $\mathcal{H}_Z$ and $\mathcal{H}_P$ commute

$$[\mathcal{H}_P, \mathcal{H}_Z] = 0 \quad . \tag{5.175}$$

$\mathcal{H}_Z$ is the Zeeman energy, apart from chemical and Knight shift differences. $\mathcal{H}_P$ is a much smaller term ("P" for perturbation) which gives rise to splittings and line widths. For $t > 0$, since we have then $\mathcal{H}_\delta = 0$, we have a time-independent Hamiltonian $\mathcal{H}_Z + \mathcal{H}_P$. Then, using (5.95), for $t > 0$,

$$\varrho(t) = \exp(-i(\mathcal{H}_Z + \mathcal{H}_P)t/\hbar)\varrho(0^+)\exp(i(\mathcal{H}_Z + \mathcal{H}_P)t/\hbar) \quad . \tag{5.176}$$

To find $\varrho(0^+)$ we must solve for the time development under the action of the δ-function. Since we seek the linear response, it is appropriate to treat $\mathcal{H}_\delta$ as a perturbation, keeping only the first-order term. We therefore turn to (5.111), using $\mathcal{H}_\delta$ for $\mathcal{H}_1$, keeping only the first term on the right. In this equation

$$\varrho^*(t) = \exp(-i(\mathcal{H}_Z + \mathcal{H}_P)t/\hbar)\varrho(t)\exp(i(\mathcal{H}_Z + \mathcal{H}_P)t/\hbar)$$

$$\mathcal{H}_\delta^*(t) = \exp(-i(\mathcal{H}_Z + \mathcal{H}_P)t/\hbar)\mathcal{H}_\delta(t)\exp(i(\mathcal{H}_Z + \mathcal{H}_P)t/\hbar) \quad . \tag{5.177}$$

Then

$$\frac{d\varrho^*(t)}{dt} = \frac{i}{\hbar}[\varrho^*(0^-)\mathcal{H}_\delta^*(t) - \mathcal{H}_\delta^*(t)\varrho^*(0^-)] \quad . \tag{5.178}$$

Integrating from $t = 0^-$ to $t = 0^+$, we get

$$\varrho^*(0^+) - \varrho^*(0^-) = \frac{i}{\hbar} \int_{0^-}^{0^+} [T(t)\varrho(0^-)\mathcal{H}_\delta(t)T^{-1}(t)$$

with
$$\qquad - T(t)\mathcal{H}_\delta(t)\varrho(0^-)T^{-1}(t)]dt \tag{5.179}$$

$$T \equiv \exp(-i(\mathcal{H}_Z + \mathcal{H}_P)t/\hbar) \tag{5.180a}$$

and (for future use)

$$T_Z \equiv \exp(-i\mathcal{H}_Z t/\hbar) \quad , \quad T_P \equiv \exp(-i\mathcal{H}_P t/\hbar) \quad . \tag{5.180b}$$

But over the zero time interval we can neglect the time dependence of everything except the δ-function giving

$$\varrho^*(0^+) = \varrho^*(0^-) - \frac{i}{\hbar}\gamma\hbar H\tau_0[\varrho(0^-)I_x - I_x\varrho(0^-)] \tag{5.181}$$

and indeed

$$\varrho^*(0^+) = \varrho(0^+) \quad , \quad \varrho^*(0^-) = \varrho(0^-) \quad . \tag{5.182}$$

Now $\varrho(0^-)$ corresponds to thermal equilibrium, hence is given by

$$\varrho(0^-) = \frac{1}{Z}\exp[-(\mathcal{H}_Z + \mathcal{H}_P)/kT] \cong \frac{1}{Z}\exp(-\mathcal{H}_Z/kT) \quad . \tag{5.183}$$

In the high temperature approximation this gives

$$\varrho(0^-) = \frac{1}{Z(\infty)}\left(1 + \frac{\gamma\hbar H_0 I_z}{kT}\right) \quad . \tag{5.184}$$

Following the argument after (5.158), we drop the first term in the parenthesis since it contributes zero magnetization, getting

$$\varrho(0^-) = \frac{1}{Z(\infty)} \frac{\gamma\hbar H_0}{kT} I_z \quad, \qquad \text{therefore} \tag{5.185}$$

$$\varrho(0^+) = \frac{1}{Z(\infty)} \frac{\gamma\hbar H_0}{kT}[I_z - i\gamma H\tau_0(I_z I_x - I_x I_z)]$$

$$= \frac{1}{Z(\infty)} \frac{\gamma\hbar H_0}{kT}(I_z + \gamma H\tau_0 I_y) \quad. \tag{5.186}$$

The first term, linear in I_z, corresponds to a thermal equilibrium magnetization. Therefore, when put in (5.176), it will not lead to any transverse magnetization.

To get $m_{xx}(t)$, therefore, we keep only the term involving I_y, utilizing (5.170)

$$(H\tau_0) m_{xx}(t) = \langle M_x(t) \rangle$$

$$= \gamma\hbar \, \text{Tr}\,\{I_x \varrho(t)\}$$

$$= \gamma\hbar \, \text{Tr}\left\{ I_x T(t) \frac{\gamma\hbar H_0}{Z(\infty) kT} \gamma (H\tau_0) I_y T^{-1}(t) \right\} \tag{5.187}$$

or

$$m_{xx}(t) = \frac{\gamma^2 \hbar^2 H_0}{Z(\infty) kT} \gamma \, \text{Tr}\,\{I_x T(t) I_y T^{-1}(t)\} \quad. \tag{5.188}$$

A useful variant on (5.188) can be obtained utilizing the fact that $\mathcal{H}_Z$ and $\mathcal{H}_P$ commute, and expressing T_Z as $\exp(i\omega_0 t I_z)$, with $\omega_0 = \gamma H_0$.
Then

$$\text{Tr}\,\{I_x T I_y T^{-1}\}$$

$$= \text{Tr}\,\{T_Z^{-1} I_x T_Z T_P I_y T_P^{-1}\}$$

$$= \text{Tr}\,\{e^{-i\omega_0 t I_z} I_x e^{i\omega_0 t I_z} T_P I_y T_P^{-1}\}$$

$$= \text{Tr}\,\{I_x T_P I_y T_P^{-1}\} \cos \omega_0 t + \text{Tr}\,\{I_y T_P I_y T_P^{-1}\} \sin \omega_0 t \quad. \tag{5.189}$$

This equation is closely related to (B.18) in Appendix B. At $t = 0^+$, $T_P = T_P^{-1} = 1$ so that the coefficient of the $\cos \omega_0 t$ term vanishes ($\text{Tr}\,\{I_x I_y\} = 0$), and the coefficient of the $\sin \omega_0 t$ term is $\text{Tr}\,\{I_y^2\}$, which is nonzero. For times short compared to $\hbar/|\mathcal{H}_P|$, where by $|\mathcal{H}_P|$ we refer to the magnitude of a typical nonzero matrix element of $\mathcal{H}_P$, we can still approximate T_P as 1. Consequently, the magnetization $m_{xx}(t)$ will start as $\sin \omega_0 t$, a signal oscillating at the Larmor frequency. Since $\mathcal{H}_P$ gives rise to spectral width, over times of the order of the inverse of the spectral width, T_P will differ significantly from 1, and we may expect both traces to contribute. In Appendix B, Eq. (B.18), we show that if $\mathcal{H}_P$ contains only spin-spin terms and γ is chosen such that the σ_k's are all zero, the first trace always vanishes. The effect of the operators T_P is to modulate the coefficients of the oscillations at ω_0 over times of the order of the inverse of the spectral frequency width, $\Delta\omega$.

Thus, for times short compared to the inverse of the spectral frequency width, $\Delta\omega$, we may write

$$m_{xx}(t) = \gamma \underbrace{\frac{\gamma^2 \hbar^2 \operatorname{Tr}\{I_y^2\}}{Z(\infty)kT}}_{a} H_0 \sin \omega_0 t \quad \text{with} \quad t \ll 1/\Delta\omega \quad . \tag{5.190}$$

It is straightforward to show that the term a for a many-spin system is the static susceptibility χ_0. Hence

$$m_{xx}(t) = \chi_0 \omega_0 \sin \omega_0 t \quad \text{for} \quad t \ll 1/\Delta\omega \quad . \tag{5.191}$$

This result has a simple classical significance (Fig. 5.5).

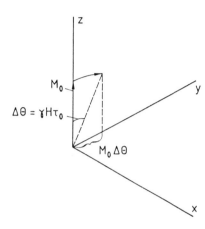

Fig. 5.5. A magnetization, $M_0 = \chi_0 H_0$, initially along the z-direction is rotated through a small angle $\Delta\theta = \gamma H \tau_0$ by a field H, of duration τ_0, along the laboratory x-direction. The resulting magnetization has a component $M_0 \Delta\theta$ along the laboratory y-direction. At later times, it precesses about z, producing a nonzero component along the x-axis which oscillates as $\sin \omega_0 t$

Considering the δ-function to be a field in the x-direction of strength $H \gg H_0$, on for a time $\tau_0 \ll 1/\omega_0$, we see that the field tilts the thermal equilibrium magnetization $\chi_0 H_0$ to have a component in the positive laboratory y-direction.

Since we are seeking a response (i.e. an x-y magnetization plane) linear in the driving term, H, we keep the angle of rotation, $\Delta\theta$, small

$$\Delta\theta = \gamma H \tau_0 \quad , \tag{5.192}$$

giving

$$M_y = \chi_0 H_0 (\gamma H \tau_0) = H \tau_0 \chi_0 \omega_0 \quad . \tag{5.193}$$

This magnetization precesses at ω_0, in the left-handed sense, giving a $\sin \omega_0 t$ dependence to M_x. Thus, utilizing (5.170), $m_{xx}(t)$ will be given as

$$m_{xx}(t) = \chi_0 \omega_0 \sin \omega_0 t \tag{5.194}$$

for times short compared to the dephasing effects arising from chemical and Knight shift differences and spin-spin couplings.

We see, therefore, that (5.188) and (5.189) have simple physical significance. In fact, we can usefully rewrite them as

$$m_{xx}(t) = \frac{\chi_0 \omega_0}{\text{Tr}\{I_y^2\}} \Big[\cos(\omega_0 t) \text{Tr}\{I_x T_P I_y T_P^{-1}\}$$
$$+ \sin(\omega_0 t) \text{Tr}\{I_y T_P I_y T_P^{-1}\}\Big] \quad . \tag{5.195}$$

While (5.195) looks relatively simple, involving just a few traces, its actual evaluation may be enormously difficult since the equation contains all the content of the theory of line shapes! Nevertheless, the equation gives a compact statement for attack by the various formal methods which have been developed to deal with the line-shape problem.

In principle, this $m_{xx}(t)$ will yield both the absorption and dispersion spectra utilizing (2.156)

$$\chi'_{xx}(\omega) = \int_0^\infty m_{xx}(\tau) \cos(\omega\tau) d\tau$$

$$\chi''_{xx}(\omega) = \int_0^\infty m_{xx}(\tau) \sin(\omega\tau) d\tau \quad . \tag{2.156}$$

5.8 The Response to a $\pi/2$ Pulse: Fourier Transform NMR

We turn now to the topic of Fourier transform NMR, a method of doing magnetic resonance which is today all pervasive in the field of high resolution liquid spectra, and which forms the basis for one of the most important other developments in NMR, so-called two-dimensional Fourier transform NMR (which we take up in Chap. 7).

In 1957, *Lowe* and *Norberg* [5.6] discovered an important theoretical result which was confirmed experimentally by them in conjunction with *Bruce* [5.7]. *Lowe* and *Norberg* showed theoretically that in solids with dipolar broadening, the Fourier transform of the free induction decay following a $\pi/2$ pulse gave the shape of the absorption line. They verified their result experimentally with a single crystal of CaF_2 by comparing their experimental free induction decays with the Fourier transform of *Bruce*'s steady-state absorption data taken on the same crystal.

In 1966, *Ernst* and *Anderson* [5.8] pointed out that there were substantial experimental advantages to the use of pulses over steady-state methods when dealing with complex spectra. (We discuss advantages later in this section.) Their method, which consists of first recording the free-induction signal, then taking its Fourier transform, is referred to as Fourier transform NMR. The method allows one to obtain both the absorption spectrum and the dispersion spectrum. They show that the cosine transform of the free induction decay gives one whereas the sine transform gives the other. (See also the book by *Ernst* et al. [5.9].)

These remarks are reminiscent of linear response theory discussed in Sect. 2.11 where we found that χ' and χ'' are given by cosine and sine transforms of the function $m(\tau)$, (2.156):

$$\chi' = \int_0^\infty m(\tau) \cos(\omega\tau) d\tau \quad , \quad \chi'' = \int_0^\infty m(\tau) \sin(\omega\tau) d\tau \tag{5.196}$$

where $m(\tau)$ is the response of the system to a δ-function which we treated in Sect. 5.6.

We therefore demonstrate that in the linear response regime at high temperatures, a $\pi/2$ pulse (in fact, any pulse) is equivalent to a δ-function excitation, hence that the transforms of the free induction signal give χ' and χ''. Let us consider, then, the Hamiltonian of (5.124) describing a group of N coupled spins, allowing there to be both chemical and Knight shifts. We then have a Hamiltonian $\mathcal{H}'$ given by keeping only that part of the spin-spin coupling which commutes with the Zeeman coupling as in (5.172)

$$\mathcal{H}' = \mathcal{H}_Z + \mathcal{H}_P + \mathcal{H}_1 \quad \text{with} \tag{5.197}$$

$$\mathcal{H}_Z = -\gamma\hbar H_0 I_z$$

$$\mathcal{H}_P = \sum_k \gamma\hbar H_0 \sigma_k + \frac{1}{2}\sum_{k,j} \mathcal{H}_{kj}^0 \tag{5.198}$$

$$\mathcal{H}_1 = -\gamma\hbar H_1 R I_x R^{-1} \quad \text{with}$$

$$[\mathcal{H}_{kj}^0, I_z] = 0 \tag{5.199}$$

$$R \equiv e^{i\omega t I_z} \tag{5.200}$$

expressing the fact that the spin is acted on by a rotating magnetic field which is to generate a $\pi/2$ pulse. During the pulse, we go to the rotating frame where ϱ_R is described by (5.131):

$$\frac{d\varrho_R}{dt} = \frac{i}{\hbar}(\varrho_R \mathcal{H}'_{\text{eff}} - \mathcal{H}'_{\text{eff}} \varrho_R) \tag{5.201}$$

with $\mathcal{H}'_{\text{eff}}$ given by (5.132) and ϱ_R by (5.116) and (5.119). While the H_1 is on, we take it to be so large that we can neglect everything else in $\mathcal{H}'$ giving

$$\mathcal{H}' = -\gamma\hbar H_1 I_x \quad . \tag{5.202}$$

If the H_1 is turned on suddenly at $t = 0$ and turned off at $t = t_1$, we have, then [as in (5.147)],

$$\begin{aligned}\varrho_R(t_1) &= e^{+i\omega_1 t_1 I_x} \varrho_R(0^+) e^{-i\omega_1 t_1 I_x} \\ &= e^{+i\omega_1 t_1 I_x} \varrho_R(0^-) e^{-i\omega_1 t_1 I_x}\end{aligned} \tag{5.203}$$

with $\omega_1 = \gamma H_1$. Taking $\varrho_R(0^-)$ to represent thermal equilibrium, we have, utilizing (5.155–157) generalized for the N-body Hamiltonian as in (5.184),

$$\varrho_R(0^-) = \varrho(0^-) = \frac{1}{Z(\infty)}\left(1 + \frac{\gamma\hbar H_0 I_z}{kT}\right) \tag{5.204}$$

whence, as in (5.150a) and (5.151)

$$\varrho_R(t_1) = \frac{1}{Z(\infty)}\left(1 + \frac{\gamma\hbar H_0}{kT}X(\omega_1 t_1)I_z X^{-1}(\omega_1 t_1)\right) \tag{5.205}$$

where

$$X(\theta) \equiv e^{+i\theta I_x} \; . \tag{5.206}$$

Now, utilizing (2.55),

$$X(\theta)I_z X^{-1}(\theta) = I_y \sin\theta + I_z \cos\theta \tag{5.207}$$

so that

$$\varrho_R(t_1) = \frac{1}{Z(\infty)}\left(1 + \frac{\gamma\hbar H_0}{kT}[I_y \sin\omega_1 t_1 + I_z \cos\omega_1 t_1]\right) \; . \tag{5.208}$$

The first term on the right will not contribute to the magnetization, following the argument of (5.160) and (5.161) (i.e. it is all that remains as $T \to \infty$).

Moreover, the term involving I_z is, apart from a constant of proportionality, the same as the temperature-dependent contribution to a ϱ_R corresponding to thermal equilibrium. Thus, it contributes nothing to transverse magnetization. [The validity of neglecting the terms involving 1 and I_z can be verified by direct calculation of $\langle M_x(t)\rangle$ and $\langle M_y(t)\rangle$.] We are left, then, with

$$\varrho_R(t_1) = \frac{1}{Z(\infty)}\frac{\gamma\hbar H_0}{kT}\sin(\omega_1 t_1)I_y \; . \tag{5.209}$$

Therefore, utilizing (5.120),

$$\begin{aligned}\varrho(t_1) &= R\varrho_R(t_1)R^{-1}\\ &= \frac{1}{Z(\infty)}\frac{\gamma\hbar H_0}{kT}\sin(\omega_1 t_1)[R(t_1)I_y R^{-1}(t_1)]\end{aligned} \tag{5.210}$$

where R is given by (5.200):

$$R(t_1) = e^{i\omega t_1 I_z} \; . \tag{5.211}$$

To get $\varrho(t)$ for $t > t_1$, we introduce the operators T, T_Z, and T_P defined as in (5.180):

$$T_Z(t) = \exp(-i\mathcal{H}_Z t/\hbar) = \exp(i\omega_0 t I_z)$$

$$T_P(t) = \exp(-i\mathcal{H}_P t/\hbar)$$

$$T(t) = \exp(-i(\mathcal{H}_P + \mathcal{H}_Z)t/\hbar) = T_Z T_P = T_P T_Z \; . \tag{5.212}$$

(Do not confuse the operator $T(t)$ with the temperature T!)

Thus, since $\mathcal{H}_Z + \mathcal{H}_P$ are independent of time, we get from (5.83)

$$\begin{aligned}\varrho(t) &= T(t-t_1)\varrho(t_1)T^{-1}(t-t_1) \\ &= \frac{\gamma\hbar H_0}{Z(\infty)kT}\sin(\omega_1 t_1)T(t-t_1)R(t_1)I_y R^{-1}(t_1)T^{-1}(t-t_1)\end{aligned} \quad (5.213)$$

Now

$$\begin{aligned}&T(t-t_1)R(t_1) \\ &= \exp\left[i\omega_0(t-t_1)I_z\right]\exp(-i\mathcal{H}_P(t-t_1)/\hbar)\exp(i\omega t_1 I_z) \\ &= \exp(i\omega_0 t I_z)\exp(-i\mathcal{H}_P t/\hbar)\exp(i\mathcal{H}_P t_1/\hbar)\exp\left[i(\omega-\omega_0)t_1 I_z\right] \\ &= T(t)T_P^{-1}(t_1)\exp(i(\omega-\omega_0)t_1 I_z) \quad .\end{aligned} \quad (5.214)$$

If, now, we neglect the effect of $\mathcal{H}_P$ and $(\omega-\omega_0)t_1$ during t_1 we can approximate

$$T_P^{-1}(t_1)e^{i(\omega-\omega_0)t_1 I_z} \cong 1 \quad (5.215)$$

getting

$$T(t-t_1)R(t_1) = T(t) \quad . \quad (5.216)$$

Therefore, the expectation value of the transverse magnetization $\langle M_x \rangle$, following (5.158), is

$$\begin{aligned}\langle M_x(t)\rangle &= \gamma\hbar\,\text{Tr}\{\varrho(t)I_x\} \\ &= \frac{\gamma^2\hbar^2 H_0}{Z(\infty)kT}\sin(\omega_1 t_1)\,\text{Tr}\{I_x T I_y T^{-1}\} \quad .\end{aligned} \quad (5.217)$$

This result is *identical* to the result of (5.187) except that $\sin\omega_1 t_1$ replaces $\gamma H \tau_0$. We introduce the notation

$$\langle M_x(t)\rangle_\theta = \frac{\gamma^2\hbar^2 H_0}{Z(\infty)kT}\sin(\theta)\,\text{Tr}\{I_x T I_y T^{-1}\} \quad , \quad (5.218)$$

the magnetization following a pulse with

$$\omega_1 t_1 \equiv \theta \quad . \quad (5.219)$$

Then we can write

$$\langle M_x(t)\rangle_\theta = m_{xx}(t)\frac{\sin\theta}{\gamma H \tau_0} \quad . \quad (5.220)$$

Now, $\langle M_x(t)\rangle_\theta$ is just the signal we would observe from a pulse in which $\gamma H_1 t_1 = \theta$, produced by a rotating field which at $t=0$ lay along the x-axis according to the convention of (5.114). This is also the rotating field generated by a linearly polarized field along the x-axis given by

$$H_x(t) = H_{x0}\cos\omega t \quad (5.221)$$

where, following (2.83),

$$H_{x0} = 2H_1 \quad . \quad (5.222)$$

Utilizing (5.196) and (5.219) with $H\tau_0 = 1$, we have then

$$\chi'(\omega) = \frac{\gamma}{\sin\theta} \int_0^\infty \langle M_x(\tau)\rangle_\theta \cos(\omega\tau) d\tau \tag{5.223a}$$

$$\chi''(\omega) = \frac{\gamma}{\sin\theta} \int_0^\infty \langle M_x(\tau)\rangle_\theta \sin(\omega\tau) d\tau \quad . \tag{5.223b}$$

Equation (5.223b) includes the result of *Lowe* and *Norberg* (they did not include chemical or Knight shifts). Equation (5.223) includes the result of *Ernst* and *Anderson* that the cosine and sine transforms yield respectively the dispersion and the absorption. [It is useful to keep in mind that the terms "cosine" and "sine" imply a particular phase convention. Ours is defined by (5.221).]

The validity of (5.218) rests on the approximation that during the pulse we can neglect the effect of $\mathcal{H}_P$ and $(\omega - \omega_0)t_1$. Since $\mathcal{H}_P$ leads to a spectral width, $\Delta\omega$, this approximation is equivalent to

$$t_1\Delta\omega \ll 1 \quad \text{and} \quad (\omega - \omega_0)t_1 \ll 1 \quad \text{or} \tag{5.224a}$$

$$\gamma H_1 \gg \Delta\omega \quad \text{and} \quad \gamma H_1 \gg |\omega - \omega_0| \quad . \tag{5.224b}$$

The classical derivation of the δ-function response given in (5.192–194) involved assuming that a magnetic field H satisfy

$$H \gg H_0 \quad , \tag{5.225}$$

where the field H lasts for a time τ_0. This condition signifies that the magnetization is rotated during a time less than a Larmor period in H_0. This appears to be a much stricter condition than (5.224b). The reason one can use an $H_1 \ll H_0$ is that H_1 is rotating at a frequency ω which is nearly the same frequency as the free precession frequency ω_0. Thus, while H_1 is on, the total phase angle about the H_0 axis which the spin advances is ωt_1. If a δ-function had kicked the spin at $t = 0$, the subsequent precession angle in the same time t_1 would have been $\omega_0 t_1$. But since $|(\omega_0 - \omega)t_1|$ and $\Delta\omega t_1 \ll 1$ we can take $\omega_0 t_1 = \omega t_1$. Thus, the H_1 builds up the transverse magnetization over the time t_1, but does not in the process introduce a phase error relative to a "zero time", or δ-function, creation of transverse magnetization followed by free precession for t_1.

We note also that this argument explains why there is nothing magical about exciting the free induction decay with a $\theta = \pi/2$ pulse. Equation (5.221) shows that any angle of θ will do. [The formula (5.221) appears to blow up for $\sin\theta = 0$, which simply reflects the vanishing of $\langle M_x(t)\rangle_\theta$ for that case].

There are several very important advantages to the Fourier transform approach. The first is that since one is recording the NMR signal when the H_1 is zero, one eliminates noise from the oscillator which drives the H_1. For steady-state apparatus, the oscillator noise is reduced by bridge balance, but typically such an approach reduces noise which arises from amplitude modulations of H_1 but does not reduce noise which arises from frequency modulation.

If one is dealing with a complex spectrum which has many absorption lines well separated from one another, as in many high resolution spectra of liquids, there is an additional advantage to the Fourier transform method. A steady-state search necessarily spends much of its time looking *between* the resonance lines of such a spectrum. The search procedure is thus inefficient. By pulse excitation, all the spins are induced to broadcast their presence at their characteristic frequencies, thus one immediately gains in data collection by eliminating the dead time of a steady-state search. Of course, one must use a broader band system, since the steady-state system need only have the band width necessary to examine a single line whereas the pulse system needs a band width large enough to include the whole spectrum. However, since the noise voltage goes as the square root of the band width, one still is better off with the pulsed approach.

An important point is that the Fourier transform method is the basis for the enormously powerful two-dimensional Fourier transform techniques which have revolutionized NMR.

It is useful at this point to make several remarks about instrumentation. The signal $M_x(\tau)$ consists in general of a sum of oscillating signals (corresponding to each resonance frequency which has been excited) which decay as a result of line broadening and relaxation phenomena. In general, all of the spread in oscillations is small compared to the Larmor frequency. To record these frequencies takes very high frequency response. It is customary to use a technique called "mixing", which in essence translates the frequency of oscillation while preserving phase relationships. We will not go into the details of how mixers work, but essentially they work as shown in Fig. 5.6. The mixer has three pairs of terminals, one at which the signal is applied, one to which a reference voltage is applied, and one which is the output, the "mixed" signal.

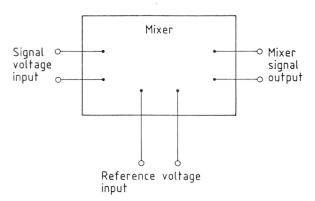

Fig. 5.6. Schematic diagram of a mixer. The radiofrequency magnetic resonance signal is fed into the signal voltage input terminals; a reference voltage whose phase can be adjusted is fed into the reference voltage input. The voltage at the mixer signal output goes to recording equipment (oscilloscopes or multichannel digital recorders)

If
$$V_{\text{sig}}(t) = V_S \cos(\omega_S t + \phi)$$
$$= V_S(\cos\phi \cos\omega_S t - \sin\phi \sin\omega_S t) \quad (5.226a)$$

$$V_{\text{ref}}(t) = V_R \cos(\omega_R t + \theta) \quad . \quad (5.226b)$$

Then,
$$V_{\text{out}} = V_S \cos[(\omega_S - \omega_R)t + \phi - \theta] \quad . \quad (5.227)$$

In expressing the reference, we have introduced a phase angle θ to represent the fact that the phase of the reference can be set by the experimenter.

Suppose $\theta = 0$, then
$$V_{\text{out}} = V_S \cos[(\omega_S - \omega_R)t + \phi] \quad . \quad (5.228)$$

At first sight, this expression appears to be equivalent to (5.226a). However, there is an ambiguity since the cosine is unchanged when the sign of its argument is changed. Thus, one does not know whether $\omega_R < \omega_S$ or $\omega_S < \omega_R$ from recording (5.227) alone. To resolve the ambiguity, we record a second signal with $\theta = -\pi/2$. If $\theta = -\pi/2$

$$V_{\text{out}} = V_S \cos[(\omega_S - \omega_R)t + (\phi + \pi/2)] \quad (5.229a)$$

$$= V_S \sin[(\omega_S - \omega_R)t + \phi] \quad . \quad (5.229b)$$

Together, (5.228) and (5.229) enable us to determine both the sign of $(\omega_S - \omega_R)$, and the phase angle ϕ.

Recording both such signals is called *quadrature detection*. It requires two mixers, whose reference signals differ by $\pi/2$ in phase, and the sets of recording apparatus to record the output signals from both mixers.

In the example above we took the reference phases to be "0" and "$-\pi/2$". To apply (5.196) we must recognize that there is an absolute meaning to the phase since we took the rf driving magnetic field to vary as $\cos\omega t$ in our definition of χ' and χ'' [see, for example, (2.92), (2.95) and (5.198)]. It is best, then, to consider that the two reference phases are

$$\theta = \theta_1 \quad \text{and} \quad (5.230)$$

$$\theta = \theta_1 - \pi/2$$

where θ_1 is unknown until one goes through some procedure to set it. Using θ_1, we then have from the first mixer

$$V_{\text{out}} = V_S \cos[(\omega_S - \omega_R)t + \phi - \theta_1] \quad (5.231a)$$

and from the second mixer

$$V_{\text{out}} = V_S \sin[(\omega_S - \omega_R)t + \phi - \theta_1] \quad . \quad (5.231b)$$

These two outputs together enable one to determine ω_S and $\phi - \theta_1$. If one has a procedure to determine θ_1 in absolute terms so that one can set it to zero, the cos output gives the dispersion signal and the sin output gives the absorption signal.

5.9 The Density Matrix of a Two-Level System

We have seen examples of the use of operator methods in conjunction with the density matrix to calculate the results of a sequence of pulses. These methods are extremely powerful and help us see what is happening in physical terms. However, there is also great insight to be found in actually following what is happening to the individual matrix elements of the density matrix for an example which we already understand thoroughly, the effect of an alternating field on a spin $\frac{1}{2}$ system. [The spirit is that of (5.143).] The Hamiltonian in the laboratory system for this problem is then

$$\mathcal{H} = -\gamma\hbar H_0 I_z - \gamma\hbar H_1 (I_x \cos \omega t - I_y \sin \omega t) \quad , \tag{5.232}$$

where we assume a positive γ and a rotating field.

As in (5.115), we write

$$\mathcal{H} = -\gamma\hbar (H_0 I_z + \mathrm{e}^{\mathrm{i}\omega t I_z} I_x \mathrm{e}^{-\mathrm{i}\omega t I_z}) \quad . \tag{5.233}$$

Defining, as in (5.116),

$$R = \mathrm{e}^{\mathrm{i}\omega t I_z} \quad , \tag{5.234}$$

we then transform to the rotating coordinate frame to have (5.122)

$$\frac{d\varrho_R}{dt} = \frac{\mathrm{i}}{\hbar}(\varrho_R \mathcal{H}_{\mathrm{eff}} - \mathcal{H}_{\mathrm{eff}} \varrho_R) \quad \text{with} \tag{5.235}$$

$$\mathcal{H}_{\mathrm{eff}} = -\gamma\hbar[(H_0 - \omega/\gamma)I_z + H_1 I_x] \quad \text{and} \tag{5.236a}$$

$$\varrho_R = R^{-1} \varrho R \quad . \tag{5.236b}$$

We take $H_0 = \omega/\gamma$, exact resonance, giving

$$\mathcal{H}_{\mathrm{eff}} = -\hbar \omega_1 I_x \quad \text{with} \tag{5.237}$$

$$\omega_1 \equiv \gamma H_1 \quad . \tag{5.238}$$

The energy levels and eigenstates in the laboratory system in the absence of H_1 are those of the Zeeman Hamiltonian

$$\mathcal{H}_Z = -\gamma\hbar H_0 I_z \quad , \tag{5.239}$$

hence are given by eigenfunctions and eigenvalues of I_z:

$$I_z |\mu\rangle = \mu |\mu\rangle \quad \mu = \pm \tfrac{1}{2} \tag{5.240}$$

$$E_\mu = (\mu|\mathcal{H}_Z|\mu) = -\gamma\hbar H_0\mu = -\hbar\omega_0\mu \quad . \tag{5.241}$$

We use these as basis states to express ϱ, (5.235), in matrix form:

$$\frac{d}{dt}(\mu|\varrho_R|\mu') = \frac{i}{\hbar}\sum_{\mu''}((\mu|\varrho_R|\mu'')(\mu''|\mathcal{H}_{\text{eff}}|\mu')$$
$$- (\mu|\mathcal{H}_{\text{eff}}|\mu'')(\mu''|\varrho_R|\mu')) \quad .$$

A somewhat more compact notation is useful, defined as

$$(\tfrac{1}{2}|\varrho_R| - \tfrac{1}{2}) = \varrho_{+-} \tag{5.242}$$

and so forth for the other matrix elements.

Recognizing that

$$(\mu|I_x|\mu') = 0 \tag{5.243}$$

unless $\mu \neq \mu'$, we then get four equations such as

$$\frac{d\varrho_{++}}{dt} = -i\omega_1\left[\varrho_{+-}(-|I_x|+) - (+|I_x|-)\varrho_{-+}\right] \quad . \tag{5.244}$$

But $(+|I_x|-) = \tfrac{1}{2}$, so that

$$\frac{d\varrho_{++}}{dt} = -\frac{i\omega_1}{2}(\varrho_{+-} - \varrho_{-+}) \tag{5.245a}$$

$$\frac{d\varrho_{--}}{dt} = -\frac{i\omega_1}{2}(\varrho_{-+} - \varrho_{+-}) \tag{5.245b}$$

$$\frac{d\varrho_{+-}}{dt} = -\frac{i\omega_1}{2}(\varrho_{++} - \varrho_{--}) \tag{5.245c}$$

$$\frac{d\varrho_{-+}}{dt} = -\frac{i\omega_1}{2}(\varrho_{--} - \varrho_{++}) \quad . \tag{5.245d}$$

We note immediately that if $H_1 = 0$, all elements of ϱ_R become independent of time. They change with time only while the pulse is on. Adding (5.245a) and (5.245b) we get

$$\frac{d\varrho_{++}}{dt} + \frac{d\varrho_{--}}{dt} = 0 \quad , \quad \text{or} \tag{5.246}$$

$$\varrho_{++}(t) + \varrho_{--}(t) = \text{const}$$
$$= \varrho_{++}(0) + \varrho_{--}(0) \quad , \tag{5.247}$$

a statement of the conservation of the total probability of finding the spin in either the up or down spin states.

In a similar way, adding (5.245c) and (5.245d) gives

$$\varrho_{+-}(t) + \varrho_{-+}(t) = \text{const}$$
$$= \varrho_{+-}(0) + \varrho_{-+}(0) \quad . \tag{5.248}$$

If we take the differences instead of the sums of the two pairs of equations, we get

$$\frac{d}{dt}(\varrho_{++} - \varrho_{--}) = -i\omega_1(\varrho_{+-} - \varrho_{-+})$$

$$\frac{d}{dt}(\varrho_{+-} - \varrho_{-+}) = -i\omega_1(\varrho_{++} - \varrho_{--}) \quad .$$
(5.249)

These equations state that the off-diagonal matrix elements are "fed" by the action of H_1 on the population difference $\varrho_{++} - \varrho_{--}$. By taking the time derivatives of (5.249), we get

$$\frac{d^2}{dt^2}(\varrho_{++} - \varrho_{--}) + \omega_1^2(\varrho_{++} - \varrho_{--}) = 0 \quad \text{and}$$

$$\frac{d^2}{dt^2}(\varrho_{+-} - \varrho_{-+}) + \omega_1^2(\varrho_{+-} - \varrho_{-+}) = 0 \quad .$$
(5.250)

Therefore,

$$\varrho_{++}(t) - \varrho_{--}(t) = A \cos \omega_1 t + B \sin \omega_1 t$$

$$\varrho_{+-}(t) - \varrho_{-+}(t) = C \cos \omega_1 t + D \sin \omega_1 t \quad .$$
(5.251)

A and C are obtained by evaluating the left-hand sides at $t = 0$. B and D are found by evaluating the derivatives of the left-hand sides at $t = 0$, utilizing (5.249). The result is

$$\varrho_{++}(t) - \varrho_{--}(t) = [\varrho_{++}(0) - \varrho_{--}(0)] \cos \omega_1 t$$
$$\qquad - i[\varrho_{+-}(0) - \varrho_{-+}(0)] \sin \omega_1 t$$

$$\varrho_{+-}(t) - \varrho_{-+}(t) = [\varrho_{+-}(0) - \varrho_{-+}(0)] \cos \omega_1 t$$
$$\qquad - i[\varrho_{++}(0) - \varrho_{--}(0)] \sin \omega_1 t \quad .$$
(5.252)

Combining (5.252) with (5.247) and (5.248) gives

$$\varrho_{++}(t) = \frac{1}{2} + \frac{1}{2}[\varrho_{++}(0) - \varrho_{--}(0)] \cos \omega_1 t$$
$$\qquad - \frac{i}{2}[\varrho_{+-}(0) - \varrho_{-+}(0)] \sin \omega_1 t$$
(5.253)

and

$$\varrho_{-+}(t) = \frac{1}{2}[\varrho_{+-}(0) + \varrho_{-+}(0)] + \frac{1}{2}[\varrho_{-+}(0) - \varrho_{+-}(0)] \cos \omega_1 t$$
$$\qquad + \frac{i}{2}[\varrho_{++}(0) - \varrho_{--}(0)] \sin \omega_1 t \quad .$$
(5.254)

Suppose, now, we wish to calculate the expectation values of the x-, y-, and z-components of magnetization in the rotating frame. We utilize (5.159)

$$\langle M^+(t) \rangle = \gamma \hbar \, \text{Tr} \{ \varrho_R(t) I^+ \}$$
(5.255a)

$$\langle M_z(t)\rangle = \gamma\hbar\,\text{Tr}\{\varrho_R(t)I_z\} \quad . \tag{5.255b}$$

Expressing these in matrix form, we get

$$\langle M^+(t)\rangle = \gamma\hbar\sum_{\mu,\mu'}(\mu|\varrho_R(t)|\mu')(\mu'|I^+|\mu) \quad \text{and}$$
$$\langle M_z(t)\rangle = \gamma\hbar\sum_{\mu,\mu'}(\mu|\varrho_R(t)|\mu')(\mu'|I_z|\mu) \quad , \tag{5.256}$$

which give, on utilizing the explicit matrix elements of I^+ and I_z,

$$\langle M^+(t)\rangle = \gamma\hbar\varrho_{-+}(t) \tag{5.257a}$$

$$\langle M_z(t)\rangle = \frac{\gamma\hbar}{2}[\varrho_{++}(t) - \varrho_{--}(t)] \quad . \tag{5.257b}$$

These equations show immediately that the existence of transverse magnetization depends on the existence of nonzero off-diagonal elements of ϱ_R. Moreover, the z-magnetization is proportional to the difference in the diagonal elements of ϱ_R, hence it will vanish unless $\varrho_{++}(t) \neq \varrho_{--}(t)$.

The significance of (5.252) and (5.253) is clarified by considering $\varrho(0)$ to correspond to thermal equilibrium, in which case the off-diagonal terms $\varrho_{+-}(0)$ and $\varrho_{-+}(0)$ vanish. Note, incidentally that (5.73) requires that $\varrho_{+-}(t)$ and $\varrho_{-+}(t)$ are complex conjugates of one another.

Then we get

$$\varrho_{++}(t) - \varrho_{--}(t) = [\varrho_{++}(0) - \varrho_{--}(0)]\cos\omega_1 t \tag{5.258a}$$

and

$$\varrho_{-+}(t) = \frac{\text{i}}{2}[\varrho_{++}(0) - \varrho_{--}(0)]\sin\omega_1 t \quad . \tag{5.258b}$$

If we let the H_1 stay on for a time t_1, we get, then, for all later times t that

$$\begin{aligned}\langle M^+(t)\rangle &= \text{i}\frac{\gamma\hbar}{2}[\varrho_{++}(0) - \varrho_{--}(0)]\sin\omega_1 t_1 \\ &= \text{i}\langle M_z(0)\rangle\sin\omega_1 t_1\end{aligned} \tag{5.259a}$$

and

$$\begin{aligned}\langle M_z(t)\rangle &= \frac{\gamma\hbar}{2}[\varrho_{++}(0) - \varrho_{--}(0)]\cos\omega_1 t_1 \\ &= \langle M_z(0)\rangle\cos\omega_1 t_1 \quad .\end{aligned} \tag{5.259b}$$

Thus following a $\pi/2$ pulse ($\omega_1 t_1 = \pi/2$)

$$\begin{aligned}\langle M_x(t)\rangle &= 0 \\ \langle M_y(t)\rangle &= \langle M_z(0)\rangle \\ \langle M_z(t)\rangle &= 0 \quad ,\end{aligned} \tag{5.260}$$

which agrees with the classical picture of rotation about H_1 by $\pi/2$ in the left-handed sense.

Following a π pulse ($\omega_1 t_1 = \pi$)

$$\begin{aligned} \langle M_x(t)\rangle = \langle M_y(t)\rangle &= 0 \\ \langle M_z(t)\rangle &= -\langle M_z(0)\rangle \quad . \end{aligned} \tag{5.261}$$

Again, the result agrees with the classical result.

Equation (5.257a) tells us useful facts about the off-diagonal elements of ϱ_R:

a) ϱ_{+-} must be nonzero for there to be transverse magnetization.
b) ϱ_{+-} must be pure real for the transverse magnetizations to lie along the x-axis.
c) ϱ_{+-} must be pure imaginary for the transverse magnetization to lie along the y-axis.

Since use of an H_1 which lies along the x-axis will produce components of magnetization perpendicular to the x-axis (assuming thermal equilibrium at $t = 0$), it will necessarily produce a purely imaginary ϱ_{+-} no matter what angle, $\omega_1 t_1$, is chosen. To produce magnetization along the x-axis, H_1 must have a component normal to the x-axis. Such a magnetic field will produce a ϱ_{+-} with a nonvanishing real component.

In general, if ϱ_{-+} is given by

$$\varrho_{-+} = Ae^{+i\phi} \quad \text{(and hence} \quad \varrho_{+-} = Ae^{-i\phi}) \tag{5.262}$$

where A is real,

$$\langle M^+(t)\rangle = \gamma\hbar A(\cos\phi + i\sin\phi) \quad \text{so that} \tag{5.263}$$

$$\langle M_x(t)\rangle = \gamma\hbar A\cos\phi \tag{5.264a}$$

$$\langle M_y(t)\rangle = \gamma\hbar A\sin\phi \quad . \tag{5.264b}$$

Thus ϕ gives the phase angle of the x-y magnetization in the x-y plane.

5.10 Density Matrix – An Introductory Example

Since the formal equations of the density matrix get rather involved, it is a good idea at this point to consider an example that will make the notation more concrete. We shall calculate the probability per second of transitions from a state k to a state m. We consider that only state k is occupied at $t = 0$. This assumption is not necessary, but it has the advantage, that, with it, $(d/dt)(m|\varrho|m)$ is directly the probability per second of a transition.

Thus initially all c_n's are zero except c_k. As a result, at $t = 0$, the only nonvanishing element of the density matrix is $(k|\varrho|k)$, which is equal to 1. By using (5.103), we see that

$$(n|\varrho^*(0)|m) = (n|\varrho(0)|m)$$
$$= 0 \quad \text{unless} \quad n = m = k \tag{5.265}$$
$$(k|\varrho^*(0)|k) = (k|\varrho(0)|k) = 1 \quad .$$

Then, taking the mm matrix element of (5.111), and using (5.103) also, we have

$$\frac{d}{dt}(m|\varrho^*(t)|m) = \frac{d}{dt}(m|\varrho|m)$$
$$= \frac{i}{\hbar}\sum_n [\underbrace{(m|\varrho^*(0)|n)(n|\mathcal{H}_1^*|m)}_{A} - \underbrace{(m|\mathcal{H}_1^*|n)(n|\varrho^*(0)|m)]}_{B}$$
$$+ \left(\frac{i}{\hbar}\right)^2 \int_0^t (m|[\underbrace{\varrho^*(0)\mathcal{H}_1^*(t')\mathcal{H}_1^*(t)}_{C} - \underbrace{\mathcal{H}_1^*(t')\varrho^*(0)\mathcal{H}_1^*(t)}_{D}$$
$$- \underbrace{\mathcal{H}_1^*(t)\varrho^*(0)\mathcal{H}_1^*(t')}_{E} + \underbrace{\mathcal{H}_1^*(t)\mathcal{H}_1^*(t')\varrho^*(0)]}_{F}|m)dt' \; . \tag{5.266}$$

We consider first the terms A and B. Since $m \neq k$ (the transition is between two distinct states), A and B both vanish according to (5.265). To treat terms such as C, we write

$$(m|\varrho^*(0)\mathcal{H}_1^*(t')\mathcal{H}_1^*(t)|m) = \sum_{m'}(m|\varrho^*(0)|m')(m'|\mathcal{H}_1^*(t')\mathcal{H}_1^*(t)|m) \quad .$$

Since
$$(m|\varrho^*(0)|m') = 0 \quad (m \neq k) \quad ,$$

the terms C and F vanish, leaving

$$\frac{d}{dt}(m|\varrho|m) = \frac{1}{\hbar^2}\int_0^t [(m|\mathcal{H}_1^*(t')|k)(k|\mathcal{H}_1^*(t)|m)$$
$$+ (m|\mathcal{H}_1^*(t)|k)(k|\mathcal{H}_1^*(t')|m)]dt' \quad . \tag{5.267}$$

We convert the matrix elements in the integrand by (5.107):

$$(m|\mathcal{H}_1^*(t)|n) = e^{(i/\hbar)(E_m - E_n)t}(m|\mathcal{H}_1(t)|n) \quad . \tag{5.268}$$

We shall also adopt a convenient abbreviation. We shall define the quantum numbers m, n, and so on, in terms of the corresponding energies, measured in radians per second:

$$\frac{E_m}{\hbar} \equiv m \qquad \frac{E_k}{\hbar} \equiv k \quad , \quad \text{etc.} \tag{5.269}$$

Then we have, using (5.268) and (5.269),

$$\frac{d}{dt}(m|\varrho|m) = \frac{1}{\hbar^2} \int_0^t [(m|\mathcal{H}_1(t')|k)(k|\mathcal{H}_1(t)|m)e^{i(m-k)(t'-t)}$$
$$+ (m|\mathcal{H}_1(t)|k)(k|\mathcal{H}_1(t')|m)e^{i(m-k)(t-t')}]dt' \quad . \tag{5.270}$$

So far, our equations have been quite general as to the nature of the perturbation $\mathcal{H}_1(t)$. For example, it could vary sinusoidally in time. We shall assume for our example, however, that $\mathcal{H}_1(t)$ varies randomly in time. By this we mean that we shall consider a number of ensembles of systems of identical $\mathcal{H}_0$ and identical $\varrho(0)$. However, we shall consider that $\mathcal{H}_1(t)$ varies from ensemble to ensemble, with properties described below. We shall therefore perform an average of the ensembles. We denote this average by a bar. Then

$$\frac{d}{dt}\overline{(m|\varrho|m)} = \frac{1}{\hbar^2} \int_0^t [\overline{(m|\mathcal{H}_1(t')|k)(k|\mathcal{H}_1(t)|m)}e^{i(m-k)(t'-t)}$$
$$+ \overline{(m|\mathcal{H}_1(t)|k)(k|\mathcal{H}_1(t')|m)}e^{i(m-k)(t-t')}]dt' \quad . \tag{5.271}$$

As an example, suppose $\mathcal{H}_1(t)$ were the dipolar coupling between nuclear moments in a liquid. It might vary in time, owing to the thermal motion of nuclei in the liquid. These motions would in general be different among various parts of the fluid, all at the same temperature. We shall assume, moreover, that the ensemble average, such as

$$\overline{(m|\mathcal{H}_1(t')|k)(k|\mathcal{H}_1(t)|m)} \tag{5.272}$$

depends on t and t' *only* through their difference τ, defined by

$$t - t' = \tau \quad . \tag{5.273}$$

That is, we assume

$$\overline{(m|\mathcal{H}_1(t-\tau)|k)(k|\mathcal{H}_1(t)|m)} \tag{5.274}$$

is independent of t but is a function both of τ and of the pair of levels, m and k, with which we are concerned. The fact that (5.274) is independent of t is expressed by the statement that the perturbation is stationary. (A more complex case could be handled, but it would make everything much more complicated.) The dependence on τ, m, and k leads us to define a quantity $G_{mk}(\tau)$ by the equation

$$G_{mk}(\tau) = \overline{(m|\mathcal{H}_1(t-\tau)|k)(k|\mathcal{H}_1(t)|m)} \quad . \tag{5.275}$$

Since $\mathcal{H}_1(t)$ is a stationary perturbation, we have that

$$G_{mk}(\tau) = \overline{(m|\mathcal{H}_1(t)|k)(k|\mathcal{H}_1(t+\tau)|m)}$$
$$= \overline{(k|\mathcal{H}_1(t+\tau)|m)(m|\mathcal{H}_1(t)|k)}$$
$$= G_{km}(-\tau) \quad . \tag{5.276}$$

The function $G_{mk}(\tau)$ is called the "correlation function" of $\mathcal{H}_1(t)$, since it tells how $\mathcal{H}_1$ at one time is correlated to its value at a later time. For a typical perturbation we have

$$\overline{(k|\mathcal{H}_1(t)|m)} = 0 \quad . \tag{5.277}$$

Thus, if $\mathcal{H}_1(t)$ and $\mathcal{H}_1(t+\tau)$ were unrelated, we could average the two terms in the product separately, getting

$$G_{mk}(\tau) = \overline{(k|\mathcal{H}_1(t)|m)}\,\overline{(m|\mathcal{H}_1(t+\tau)|k)} = 0 \quad . \tag{5.278}$$

However, we have, for $\tau = 0$,

$$G_{mk}(0) = \overline{|(k|\mathcal{H}_1(t)|m)|^2} \geq 0 \quad . \tag{5.279}$$

For typical physical systems, the perturbation $\mathcal{H}_1(t)$ varies in time, owing to some physical movement. For times less than some critical time τ_c, called the "correlation time", the motion may be considered negligible, so that $\mathcal{H}_1(t) \cong \mathcal{H}_1(t+\tau)$. For $\tau > \tau_c$, the values of $\mathcal{H}_1(t+\tau)$ become progressively less correlated to $\mathcal{H}_1(t)$ as τ is lengthened, so that G_{km} goes to zero. Thus $G_{mk}(\tau)$ has a maximum at $\tau = 0$, and falls off for $|\tau| > \tau_c$, as in Fig. 5.7. A function $\mathcal{H}_1(t)$ with the above properties will be called a "stationary random function of time".

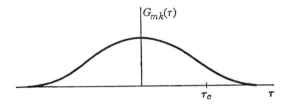

Fig. 5.7. Function $G_{mk}(\tau)$ for a typical physical system

Bearing in mind these properties, we now rewrite (5.271) in terms of $G_{mk}(\tau)$:

$$\frac{d}{dt}(m|\varrho|m) = \frac{1}{\hbar^2} \int_0^t [G_{mk}(\tau)e^{-i(m-k)\tau} + G_{mk}(-\tau)e^{i(m-k)\tau}]d\tau$$

$$= \frac{1}{\hbar^2} \int_{-t}^{t} G_{mk}(\tau)e^{-i(m-k)\tau}d\tau \quad . \tag{5.280}$$

This equation tells us the rate of change of $(m|\varrho|m)$ at a variable time t. We now note that if $t \gg \tau_c$, the limits of integration may be taken as $\pm\infty$, so that $(d/dt)(m|\varrho|m)$ becomes independent of time. There is a range of $0 < t < \tau_c$ for which the transition rate is not constant. And, of course, if $(m|\varrho|m)$ becomes comparable to unity, we do not expect our peturbation approximation to hold, since the initial population $(k|\varrho|k)$ would necessarily be strongly depleted.

We shall now confine our attention to times longer than τ_c, assuming that for them $(m|\varrho|m)$ has not grown unduly. Then we have

$$\frac{d}{dt}(m|\varrho|m) = \frac{1}{\hbar^2} \int_{-\infty}^{+\infty} G_{mk}(\tau) e^{-i(m-k)\tau} d\tau = W_{km} \quad (5.281)$$

where W_{km} is the probability per second of a transition from state k to m.

Equation (5.281) is closely related to the well-known result from time-dependent perturbation theory:

$$W_{km} = \frac{2\pi}{\hbar} |(k|V|m)|^2 \varrho(E_f)$$

where $(k|V|m)$ is the matrix element of the interaction between states k and m, and $\varrho(E_f)$ is the density of final states in energy. In fact, by utilizing (5.275), we see that W_{km} of (5.275) involves a product of two matrix elements of the perturbation. In the present case the energy levels are sharp, but the perturbation is spread in frequency, whereas the usual time-dependent perturbation is monochromatic but the energy levels are smeared. In view of the similarity to the usual time-dependent theory, we are not surprised that the terms A and B of (5.266), which involve only one matrix element of the perturbation, vanish.

The integral of (5.281) is reminiscent of a Fourier transform. Let us then define a quantity $J_{mk}(\omega)$ by the equation

$$J_{mk}(\omega) = \int_{-\infty}^{+\infty} G_{mk}(\tau) e^{-i\omega\tau} d\tau \quad (5.282)$$

with the inverse relation

$$G_{mk}(\tau) = \frac{1}{2\pi} \int_{-\infty}^{+\infty} J_{mk}(\omega) e^{i\omega\tau} d\omega \quad . \quad (5.283)$$

$J_{mk}(\omega)$ may be thought of as the spectral density of the interaction matrix G_{mk}. We expect that J_{mk} will therefore contain frequencies up to the order of $1/\tau_c$ (see Fig. 5.8). In terms of J_{mk}, we have

$$W_{km} = \frac{J_{mk}(m-k)}{\hbar^2} \quad . \quad (5.281a)$$

Now, in a typical case, the interaction matrix element $(m|\mathcal{H}_1(t)|k)$ runs over a set of values as time goes on. As one varies something such as the temperature, the *rate* at which $(m|\mathcal{H}_1(t)|k)$ changes may speed up or slow down (τ_c changes),

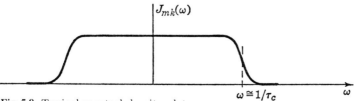

Fig. 5.8. Typical spectral density plot

but the set of values covered remains unchanged. As a physical example, the dipole-dipole coupling of a pair of nuclei depends on their relative positions. If the nuclei diffuse relative to each other, the coupling takes on different values. The possible values are independent of the rate of diffusion, since they depend only on the radius vector from one nucleus to the other and on the spatial orientation of the moments. However, the duration of any magnitude of interaction will depend on the rate of diffusion. Then we have that

$$\overline{|(m|\mathcal{H}_1(t)|k)|^2} = G_{mk}(0) \tag{5.284}$$

is independent of τ_c. But, from (5.283)

$$G_{mk}(0) = \frac{1}{2\pi} \int_{-\infty}^{+\infty} J_{mk}(\omega)d\omega \tag{5.284a}$$

which tells us that the area of the spectral density curve remains *fixed* as τ_c varies. A set of curves of $J_{mk}(\omega)$ for three different τ_c's is shown in Fig. 5.9. A simple consequence of the fact that the area remains fixed as τ_c varies is found by considering Fig. 5.9. We note that if the frequency difference $m - k$ were equal to the value ω_1, the spectral density $J_{mk}(\omega_1)$ of the curve of medium τ_c would be the greatest of the three at ω_1. Consequently, as τ_c is varied, the transition probability W_{km} has a maximum. The maximum will occur when $(m - k)\tau_c \cong 1$, since it is for this value of τ_c that the spectrum extends up to $(m - k)$ without extending so far as to be diminished.

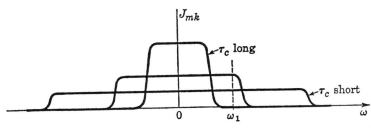

Fig. 5.9. $J_{mk}(\omega)$ for three values of correlation time, illustrating the variations that keep the area under the curves constant as τ_c changes. $J_{mk}(\omega_1)$ is a maximum for the curve of medium τ_c

In the event that $\tau_c \ll 1/(m - k)$, $J_{mk}(\omega)$ extends far above the frequency of the transition. It is then often a good approximation to say that

$$J_{mk}(m - k) \cong J_{mk}(0) \quad . \tag{5.285}$$

Let us define a quantity α such that

$$2\left(\frac{\alpha}{\tau_c}\right) J_{mk}(0) = \int_{-\infty}^{+\infty} J_{mk}(\omega)d\omega \tag{5.286}$$

where $2(\alpha/\tau_c)$ is the width of a rectangular $J_{mk}(\omega)$ whose height is $J_{mk}(0)$ and whose area is the same as that of the real $J_{mk}(\omega)$. From our remarks, $\alpha \cong 1$. Then we have, combining (5.284), (5.284a), and (5.286),

$$J_{mk}(0) = \frac{\tau_c}{2\alpha} 2\pi \overline{|(m|\mathcal{H}_1(t)|k)|^2} \quad . \tag{5.287}$$

By utilizing (5.281a) and (5.285), then, we have

$$W_{km} = \frac{\overline{|(m|\mathcal{H}_1(t)|k)|^2}}{\hbar^2} \frac{\pi}{\alpha} \tau_c \quad , \tag{5.288}$$

which holds for $\tau_c \ll 1/(m-k)$.

Since one can often estimate the mean squared interaction as well as the correlation time, (5.288) provides a very simple method of computing the transition probability in the limit of short correlation time (rapid motion). As a side remark, we note that for a flat spectrum $J_{mk}(\omega)$, (5.288) will hold approximately for all $\tau_c \leq 1/(m-k)$. We can use it to estimate roughly the *maximum* rate of transition that $\mathcal{H}_1(t)$ is capable of producing under the most favorable circumstances, by noting that this occurs when $\tau_c \cong 1/(m-k)$. We get the approximate relation, then,

$$W_{km})_{\max} \cong \frac{\overline{|(m|\mathcal{H}_1(t)|k)|^2}}{\hbar^2} \frac{\pi}{\alpha(m-k)} \quad . \tag{5.289}$$

Our remarks so far apply to many forms of interaction $\mathcal{H}_1(t)$. In order to be more concrete, we now specialize, to consider a particular form for the interaction $\mathcal{H}_1(t)$. We shall assume that it consists of a fluctuating magnetic field with x-, y-, and z-components that couple to the components of the nuclear moment:

$$\mathcal{H}_1(t) = -\gamma_n \hbar [H_x(t) I_x + H_y(t) I_y + H_z(t) I_z]$$
$$= -\gamma_n \hbar \sum_{q=x,y,z} H_q(t) I_q \quad . \tag{5.290}$$

Then

$$G_{mk}(\tau) = \overline{(m|\mathcal{H}_1(t)|k)(k|\mathcal{H}_1(t+\tau)|m)}$$
$$= \gamma_n^2 \hbar^2 \sum_{q,q'} (m|I_q|k)(k|I_{q'}|m) \overline{H_q(t) H_{q'}(t+\tau)} \tag{5.291}$$

where we have recognized that it is $H_q(t)$ that varies from one member of the ensemble to another. For simplicity we shall suppose that the x-, y-, and z-components of field fluctuate independently. This means in effect that knowledge of H_x at some time is not sufficient to predict H_y at that time. With this assumption, we get in (5.291) only terms for which $q = q'$. Let us define

$$j_{mk}^q(\omega) = \gamma_n^2 \hbar^2 |(m|I_q|k)|^2 \int_{-\infty}^{+\infty} \overline{H_q(t) H_q(t+\tau)} e^{-i\omega\tau} d\tau \quad . \tag{5.292}$$

Then we have

$$W_{km} = \frac{1}{\hbar^2} \sum_q j^q_{mk}(m-k) \quad . \tag{5.293}$$

Evaluation of $j^q_{mk}(\omega)$ requires information on the physical basis of the field fluctuation. Moreover, even when one possesses that information, the mathematical problem of finding the correlation function may still be too difficult to solve. One can then often assume a function that, on general physical grounds, has about the correct behavior. For certain simple cases one can actually compute the correlation function. One such case arises when the field $H_q(\tau)$ takes on either of two values and makes transitions from one value to the other at a rate that is independent of the time since the preceding transition. Assuming the field takes on values of $\pm h_q$, we show in Appendix C that

$$\overline{H_q(t)H_q(t+\tau)} = h_q^2 e^{-|\tau|/\tau_0} \tag{5.294}$$

where τ_0 is time defined by (5.295) in terms of the probability per second, W, that $H_q(\tau)$ will jump from $+h_q$ to $-h_q$:

$$\frac{1}{\tau_0} = 2W \quad . \tag{5.295}$$

We assume for our example that this time is the same for all three components of the field. Clearly τ_0 may be taken as the correlation time.[3] Substituting this form into (5.292) we find

$$j^q_{mk}(\omega) = \gamma_n^2 \hbar^2 |(k|I_q|m)|^2 h_q^2 \frac{2\tau_0}{1+\omega^2\tau_0^2} \quad . \tag{5.296}$$

The transition probability W_{km} is then

$$W_{km} = [\sum \gamma_n^2 h_q^2 |(m|I_q|k)|^2] \frac{2\tau_0}{1+(m-k)^2\tau_0^2} \quad . \tag{5.297}$$

It is interesting to apply this formula to compute the T_1 for the case of spin $\frac{1}{2}$. For this case, as we saw in Chapter 1,

$$\frac{1}{T_1} = 2W_{1/2,-1/2} \quad . \tag{5.298}$$

Assuming a strong static field in the z-direction, the matrix elements between the levels are

$$\begin{aligned}|(\tfrac{1}{2}|I_x|-\tfrac{1}{2})|^2 &= \tfrac{1}{4} \\ |(\tfrac{1}{2}|I_y|-\tfrac{1}{2})|^2 &= \tfrac{1}{4} \\ |(\tfrac{1}{2}|I_z|-\tfrac{1}{2})|^2 &= 0 \quad .\end{aligned} \tag{5.299}$$

---

[3] Note that although we have not given a precise definition of the correlation time, once the correlation function is specified, there is always a precisely defined parameter (in this case τ_0) that enters the problem and characterizes the time scale.

We shall assume $h_x = h_y = h_z$, so that

$$h_x^2 = \tfrac{1}{3} h_0^2 \tag{5.300}$$

where $h_0^2 = h_x^2 + h_y^2 + h_z^2$. Then, since $m - k = \omega_0$, the Larmor frequency

$$\frac{1}{T_1} = 2\gamma_n^2 \frac{h_0^2}{3} \frac{\tau_0}{1+\omega_0^2 \tau_0^2} \ . \tag{5.301}$$

This function is plotted in Fig. 5.10. We see that it indeed has the sort of variation we had predicted earlier from our general arguments based on the "constant area" under the curve of $J_{mk}(\omega)$. We note that the minimum comes when $\omega_0 \tau_0 = 1$.

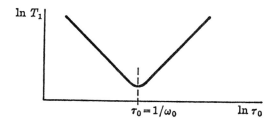

Fig. 5.10. Variation of T_1 with the correlation time τ_0

By utilizing this fact, we can calculate the minimum value of T_1 from (5.301) and obtain, in agreement with (5.289)

$$\left.\frac{1}{T_1}\right)_{\min} = \frac{1}{3} \frac{\gamma_n^2 h_0^2}{\omega_0} \ . \tag{5.302}$$

Now, when τ_0 gets very long, the field H_z simply gives rise to a static line broadening (in this case, since only two discrete values can occur, the static line consists of a pair of lines at $\pm h_z$) and h_0^2 is simply related to this static width. Of course, ω_0 is known for a typical experiment. Equation (5.302) therefore tells us the remarkable fact that knowledge of the rigid lattice line breadth, plus the resonant frequency, enables us to compute the most effective relaxation possible from allowing the line broadening interaction to fluctuate. In general the correlation time is changed by changing the temperature of the sample. Although (5.302) enables us to predict the minimum value of T_1, it cannot tell us at what temperature the minimum will occur unless, of course, we *know* the dependence of τ_0 on temperature. We note in addition that the measurement of T_1 as a function of temperature can provide information about the temperature variation of whatever physical process produces the fluctuations.

5.11 Bloch-Wangsness-Redfield Theory

We turn now to a more general treatment of the density matrix, following the ideas of *Redfield* [5.10], which are closely related to a treatment of relaxation due to *Wangsness* and *Bloch* [5.11]. All of the basic ideas are anticipated physically in the basic work of *Bloembergen* et al. [5.12]. The development of the basic equation of *Redfield*'s theory is a generalization of the treatment given in the previous section for computing transition probabilities. *Redfield* shows that the elements of the density matrix obey a set of linear differential equations of the following form: [4]

$$\frac{d}{dt}\varrho^*_{\alpha\alpha'} = \sum_{\beta,\beta'} R_{\alpha\alpha',\beta\beta'} e^{i(\alpha-\alpha'-\beta+\beta')t} \varrho^*_{\beta\beta'} \tag{5.303}$$

where $R_{\alpha\alpha',\beta\beta'}$ is constant in time. In this equation the time-dependent exponential has the property of making any terms unimportant unless $\alpha - \alpha' = \beta - \beta'$. Therefore we could write (5.303) as

$$\frac{d}{dt}\varrho^*_{\alpha\alpha'} = {\sum}' R_{\alpha\alpha',\beta\beta'} \varrho^*_{\beta\beta'} \tag{5.304}$$

where the prime on the summation indicates that we keep only those terms for which $\alpha - \alpha' = \beta - \beta'$. The diagonal part of this equation (that is, the part we should have if we kept $\alpha = \alpha'$, $\beta = \beta'$ only) has the same form as the "master" equation, (5.13). The conditions under which (5.303) or (5.304) hold are given in terms of $R_{\alpha\alpha'\beta\beta'}$, τ_c, and a time interval Δt. Δt defines a sort of "coarse graining". That is, we shall assume that we never try to follow the details of ϱ^* for time intervals *less* than Δt. We must be able to choose such a time, subject to the simultaneous conditions that

$$\Delta t \gg \tau_c \quad \text{and} \tag{5.305}$$

$$\frac{1}{R_{\alpha\alpha',\beta\beta'}} \gg \Delta t \quad . \tag{5.306}$$

Equation (5.305) will permit us to set the limits of certain integrals as $\pm \infty$, as we did in (5.281). Equation (5.306) will guarantee that, during Δt, the density matrix does not change too drastically, so that our perturbation expansion has validity. Since the $R_{\alpha\alpha',\beta\beta'}$'s are comparable to the inverse of the relaxation times, these conditions are equivalent to saying that T_1 and T_2 are much longer than τ_c. These are also the conditions of motional narrowing. The reader will pehaps note that these are the very conditions stated in Section 2.11, which must hold if simple time-independent transition probabilities are to hold. What we are doing here is to generalize the usual time-dependent perturbation theory so that it includes the coherence effects associated with phase factors in the wave function.

---

[4] For compactness in writing, we use the notation $\varrho_{\alpha\alpha'}$, for the matrix element $(\alpha|\varrho|\alpha')$.

The beauty of (5.304) is that it provides us with a simple set of linear differential equations among the elements of the density matrix, which in principle we can always solve. They will lead to a set of "normal modes". We note that there is a great deal of similarity to the rate equation describing population changes. Moreover, expressions for the $R_{\alpha\alpha'\beta\beta'}$'s are given by *Redfield's* theory, so that we have the relaxation times given in terms of the atomic properties.

Before outlining the derivation of (5.303) and (5.304), we remark that there are two ways of utilizing (5.303) or (5.304). In the first method, one solves for the behavior of each separate element of the density matrix and then computes the time dependence of the physical variables of interest (such as the x-components of magnetization M_x) by the fundamental equation

$$\overline{\langle M_x \rangle} = \sum_{\alpha,\alpha'} \varrho_{\alpha\alpha'}(\alpha'|M_x|\alpha) \quad . \tag{5.307}$$

The second method involves seeking a differential equation for $\overline{\langle M_x \rangle}$. One does this by writing

$$\frac{d}{dt}\overline{\langle M_x \rangle} = \frac{d}{dt} \sum_{\alpha,\alpha'} \varrho_{\alpha\alpha'}(\alpha'|M_x|\alpha)$$

$$= \sum_{\alpha,\alpha'} \left(\frac{d\varrho_{\alpha\alpha'}}{dt}\right)(\alpha'|M_x|\alpha) \quad . \tag{5.308}$$

We then use (5.303) to express the time derivative $(d/dt)\varrho_{\alpha\alpha'}$. In fact, since

$$\varrho^*_{\alpha\alpha'} = e^{i(\alpha-\alpha')t}\varrho_{\alpha\alpha'} \tag{5.309}$$

we have

$$\frac{d\varrho^*_{\alpha\alpha'}}{dt} = i[\alpha - \alpha']\varrho^*_{\alpha\alpha'} + e^{i(\alpha-\alpha')t}\frac{d\varrho_{\alpha\alpha'}}{dt} \quad . \tag{5.310}$$

This relation enables us to transform (5.303). By substituting into (5.303) and utilizing (5.309), we find

$$\frac{d\varrho_{\alpha\alpha'}}{dt} = i(\alpha' - \alpha)\varrho_{\alpha\alpha'} + \sum_{\beta,\beta'} R_{\alpha\alpha',\beta\beta'}\varrho_{\beta\beta'}$$

$$= \frac{i}{\hbar}[\varrho, \mathcal{H}_0]_{\alpha\alpha'} + \sum_{\beta,\beta'} R_{\alpha\alpha',\beta\beta'}\varrho_{\beta\beta'} \quad . \tag{5.311}$$

We then substitute this expression into (5.308), obtaining

$$\frac{\overline{d\langle M_x \rangle}}{dt} = \sum_{\alpha,\alpha',\beta,\beta'} \left(\frac{i}{\hbar}[\varrho, \mathcal{H}_0]_{\alpha\alpha'} + R_{\alpha\alpha',\beta\beta'}\varrho_{\beta\beta'}\right)(\alpha'|M_x|\alpha) \quad . \tag{5.312}$$

Although it is not obvious from (5.312), under some circumstances the right-hand side is proportional to a linear combination of $\overline{\langle M_x(t) \rangle}$, $\overline{\langle M_y(t) \rangle}$, and $\overline{\langle M_z(t) \rangle}$, giving us a set of differential equations similar to those of Bloch. If these equations are fewer in number than are the equations of the elements of the density matrix, their solution may be considerably simpler than the original set. This trick will work when the relaxation mechanism and operators are such that the expectation values of the operators pick out only a small number of the possible normal modes. We shall illustrate this use of (5.312) shortly. First, however, we turn to a description of the derivation of *Redfield*'s fundamental equation.

Our starting point is the basic equation for the time derivative of ϱ^*, (5.111):

$$\frac{d\varrho^*}{dt} = \frac{i}{\hbar}[\varrho^*(0), \mathcal{H}_1^*(t)] + \left(\frac{i}{\hbar}\right)^2 \int_0^t [[\varrho^*(0), \mathcal{H}_1^*(t')], \mathcal{H}_1^*(t)]dt' \quad . \tag{5.313}$$

We compute the $\alpha\alpha'$ matrix element. There will be contributions from both terms on the right. We consider first the contribution of $[\varrho^*(0), \mathcal{H}_1^*(t)]$:

$$(\alpha|[\varrho^*(0), \mathcal{H}_1^*(t)]|\alpha') = \sum_\beta (\alpha|\varrho^*(0)|\beta)(\beta|\mathcal{H}_1^*(t)|\alpha')$$
$$- (\alpha|\mathcal{H}_1^*(t)|\beta)(\beta|\varrho^*(0)|\alpha') \quad . \tag{5.314}$$

We shall now introduce the idea of an ensemble of ensembles whose density matrix coincides at $t = 0$ but whose perturbations $\mathcal{H}_1(t)$ are different. (We are therefore *not* allowing an applied alternating field to be present. That is, we are describing relaxation in the absence of an alternating field. One could add the effect of the alternating field readily. We discuss it in the next section.) We shall assume that the ensemble average of $\mathcal{H}_1(t)$ vanishes. This amounts to our assuming that $\mathcal{H}_1(t)$ does not produce an average frequency shift.[5]

Let us discuss this point. In general we expect

$$\mathcal{H}_1(t) = \sum H_q(t) K^q \tag{5.315}$$

where K^q is a function of the spin coordinates, and $H_q(t)$ is independent of spin. For example, we saw that $\mathcal{H}_1(t)$ had this form if it consisted of the coupling of a fluctuating magnetic field with the x-, y-, or z-components of spin. In that case q took on three values corresponding to the three components x, y, and z. If $\mathcal{H}_1(t)$ represented a dipole-dipole interaction of two spins, there would be six values of q corresponding to the terms, A, B, ..., F into which we broke the dipolar coupling in Chapter 3.

Since we are dealing with stationary perturbations, the ensemble average of $\mathcal{H}_1^*(t)$ is equivalent to a time average. In general we assume that the time average of $H_q(t)$ vanishes, causing $\mathcal{H}_1^*(t)$ to have a vanishing ensemble average. As a consequence we shall set

$$\overline{(\alpha|\mathcal{H}_1^*(t)|\beta)} = 0 \tag{5.316}$$

---

[5] If there *is* a shift, we can *include* the average shift in $\mathcal{H}_0$, redefining $\mathcal{H}_1(t)$ to give a zero shift.

where the bar indicates an ensemble average. This means, we have remarked, that we cannot let $\mathcal{H}_1(t)$ be a time-dependent driving field such as that applied to observe a resonance.

On the basis of (5.316) the first term on the right of (5.313) vanishes when averaged over an ensemble.

We proceed in a similar way to compute the $\alpha\alpha'$ matrix element of the second term on the right of (5.313). By utilizing the fact that

$$(\beta|\mathcal{H}_1^*(t)|\beta') = e^{i(\beta-\beta')t}(\beta|\mathcal{H}_1(t)|\beta') \tag{5.317}$$

and defining

$$\tau = t - t' \tag{5.318}$$

we find

$$\begin{aligned}
\frac{d\varrho_{\alpha\alpha'}^*}{dt} = \frac{1}{\hbar^2} \sum_{\beta,\beta'} \int_0^t &[(\alpha|\mathcal{H}_1(t-\tau)|\beta)(\beta'|\mathcal{H}_1(t)|\alpha') \\
&\times e^{-i(\alpha-\beta)\tau} e^{i(\alpha-\beta+\beta'-\alpha')t} \varrho_{\beta\beta'}^* \\
&+ (\alpha|\mathcal{H}_1(t)|\beta)(\beta'|\mathcal{H}_1(t-\tau)|\alpha') e^{i(\alpha'-\beta')\tau} e^{i(\alpha-\beta+\beta'-\alpha')t} \varrho_{\beta\beta'}^* \\
&- \varrho_{\alpha\beta}^*(\beta|\mathcal{H}_1(t-\tau)|\beta')(\beta'|\mathcal{H}_1(t)|\alpha') e^{i(\beta'-\beta)\tau} e^{i(\beta-\alpha')t} \\
&- (\alpha|\mathcal{H}_1(t)|\beta)(\beta|\mathcal{H}_1(t-\tau)|\beta') e^{i(\beta'-\beta)\tau} e^{i(\alpha-\beta')t} \varrho_{\beta'\alpha'}^*] d\tau \quad.
\end{aligned} \tag{5.319}$$

We now perform an average over ensembles of differing $\mathcal{H}_1(t)$, obtaining such terms as

$$\frac{1}{\hbar^2} \int_0^t \overline{(\alpha|\mathcal{H}_1(t-\tau)|\beta)(\beta'|\mathcal{H}_1(t)|\alpha')} e^{-i(\alpha-\beta)\tau} e^{i(\alpha-\beta+\beta'-\alpha')t} d\tau \quad. \tag{5.320}$$

We assume that the average,

$$\overline{(\alpha|\mathcal{H}_1(t-\tau)|\beta)(\beta'|\mathcal{H}_1(t)|\alpha')} \quad, \tag{5.321}$$

is independent of t and goes to zero when τ exceeds some critical value τ_c. In this case we can consider times t greater than τ_c, permitting us to extend the upper limits of integration to $\tau = +\infty$.

We now define the correlation function $G_{\alpha\beta\alpha'\beta'}(\tau)$ as

$$G_{\alpha\beta\alpha'\beta'}(\tau) \equiv \overline{(\alpha|\mathcal{H}_1(t)|\beta)(\beta'|\mathcal{H}_1(t+\tau)|\alpha')} \quad. \tag{5.322}$$

By utilizing (5.315) we have that

$$G_{\alpha\beta\alpha'\beta'}(\tau) = \sum_{q,q'} (\alpha|K^q|\beta)(\beta'|K^{q'}|\alpha') \overline{H_q(t) H_{q'}(t+\tau)} \quad. \tag{5.323}$$

We then define the spectral density $L_{qq'}(\omega)$ of the interaction as

$$L_{qq'}(\omega) \equiv \int_0^\infty \overline{H_q(t)H_{q'}(t+\tau)} e^{-i\omega\tau} d\tau \quad . \tag{5.324}$$

By utilizing the fact that $\overline{H_q(t)H_{q'}(t+\tau)}$ is real[6] and is an even function of τ, it is convenient to define the real and imaginary parts $L_{qq'}(\omega)$:

$$\text{Re}\{L_{qq'}(\omega)\} = \frac{1}{2} \int_{-\infty}^{+\infty} \overline{H_q(t)H_{q'}(t+\tau)} \cos(\omega\tau) d\tau \equiv k_{qq'}$$

$$\text{Im}\{L_{qq'}(\omega)\} = -\int_0^\infty \overline{H_q(t)H_{q'}(t+\tau)} \sin(\omega\tau) d\tau \quad . \tag{5.325}$$

Since the only important contributions to (5.319) come from terms that satisfy the condition $\alpha - \alpha' = \beta - \beta'$, we can combine the first two terms on the right of (5.319) as

$$\frac{1}{\hbar^2}\left(\sum_{\beta,\beta'}\sum_{q,q'}\{(\alpha|K^q|\beta)(\beta'|K^{q'}|\alpha')[k_{qq'}(\alpha-\beta)+k_{qq'}(\alpha'-\beta')]\right.$$
$$\left. \times e^{i(\alpha-\beta+\beta'-\alpha')t}\varrho^*_{\beta\beta'}\}\right) \quad . \tag{5.326}$$

The last two terms of (5.319) are

$$\frac{1}{\hbar^2} \sum_{\beta,\beta',q,q'} \{\varrho^*_{\alpha\beta}[(\beta|K^q|\beta')(\beta'|K^{q'}|\alpha')L_{qq'}(\beta-\beta')]e^{i(\beta-\alpha')t}$$
$$+ \varrho^*_{\beta'\alpha'}[(\beta|K^q|\beta')(\alpha|K^{q'}|\beta)L_{qq'}(\beta-\beta')]e^{i(\alpha-\beta')t}\} \quad . \tag{5.327}$$

The imaginary part of $L_{qq'}$ can be shown to give rise to a frequency shift corresponding to the second-order frequency shift of a static interaction.

We shall neglect this effect, keeping only terms proportional to Re$\{L_{qq'}\}$ because they give the relaxation. Therefore we replace the $L_{qq'}$'s by the $k_{qq'}$'s.

In analogy to our earlier discussion we now define the spectral densities $J_{\alpha\alpha'\beta\beta'}(\omega)$ as

$$J_{\alpha\alpha'\beta\beta'}(\omega) = \int_{-\infty}^{+\infty} \overline{(\alpha|\mathcal{H}_1(t)|\alpha')(\beta'|\mathcal{H}_1(t+\tau)|\beta)} e^{-i\omega\tau} d\tau \quad . \tag{5.328}$$

Then, combining (5.326), (5.327), and (5.328), we have

$$\frac{d\varrho^*_{\alpha\alpha'}}{dt} = \sum_{\beta,\beta'} R_{\alpha\alpha'\beta\beta'} e^{i(\alpha-\alpha'-\beta+\beta')t}\varrho^*_{\beta\beta'}(0) \quad , \tag{5.329}$$

where $R_{\alpha\alpha',\beta\beta'}$ is given as

---

[6] As long as the K_q's are taken as Hermitian operators, the H_q's are real. If one chooses the K_q's as non-Hermitian, the H_q's become complex, but the only $L_{qq'}$'s that are nonzero then involve q and q''s which make $\overline{H_q(t)H_{q'}(t+\tau)}$ real.

$$R_{\alpha\alpha',\beta\beta'} = \frac{1}{2\hbar^2}[J_{\alpha\beta\alpha'\beta'}(\alpha'-\beta') + J_{\alpha\beta\alpha'\beta'}(\alpha-\beta)$$
$$-\delta_{\alpha'\beta'}\sum_\gamma J_{\gamma\beta\gamma\alpha}(\gamma-\beta) - \delta_{\alpha\beta}\sum_\gamma J_{\gamma\alpha'\gamma\beta'}(\gamma-\beta')] \quad .$$
(5.330)

Equation (5.329) relates $d\varrho^*/dt$ at time $t > \tau_c$ to ϱ^* at $t = 0$. It is the first term in a power-series expansion. In order for the convergence to be good, it must imply that $\varrho^*_{\beta\beta'}$ at time t not be vastly different from this value at $t = 0$. This implies that we can find a range of times t such that $t \gg \tau_c$, but still $\varrho^*_{\beta\beta'}(t) \cong \varrho^*_{\beta\beta'}(0)$. This latter condition implies that

$$\frac{1}{R_{\alpha\alpha',\beta\beta'}} \gg t \quad . \tag{5.331}$$

The important trick now is to note that if (5.331) holds true, we can replace $\varrho^*_{\beta\beta'}(0)$ by $\varrho^*_{\beta\beta'}(t)$ on the right side of (5.329). By this step, we convert (5.329) into a differential equation for ϱ^*, which will enable us to find ϱ^* by "integration" at times so much later than $t = 0$ that $\varrho^*_{\beta\beta'}(t)$ will no longer be nearly its value at $t = 0$. The resulting equation is (5.303).

The physical significance of our conditions is now seen to be that we never ask for information over time intervals comparable to τ_c, and that in this time interval the density matrix must not change too much. In practice, this implies that

$$T_1, \ T_2 \gg \tau_c \quad . \tag{5.332}$$

As we shall see in greater detail, the condition $\tau_c \ll T_2$ is just that for which the resonance lines are "narrowed" by the "motion" that produces the fluctuations in $\mathcal{H}_1(t)$.

Because

$$R_{\alpha\alpha,\beta\beta} = R_{\beta\beta,\alpha\alpha} \tag{5.333}$$

(that is, the transition probability from α to β is equal to that from β to α), the solution of the Redfield equation is an *equal* distribution among all states. This situation corresponds to an infinite temperature. Clearly the equations do not describe the approach to an equilibrium at a finite temperature. The reason is immediately apparent – our equation involves the spin variables only, making no mention of a thermal bath. The bath coordinates are needed to enable the spins to "know" the temperature.

A rigorous method of correcting for the bath is to consider that the density matrix of (5.313) is for the total system of bath and spins. Since in the absence of $\mathcal{H}_1$ the spins and lattice are decoupled, we may take the density matrix to consist of a product of that for the spins, σ, and that for the lattice, ϱ^L. We take for our basic Hamiltonian $\mathcal{H}_0$ the *sum* of the lattice and the spin Hamiltonians (which, of course, commute). $\mathcal{H}_1$ commutes with neither and induces simultaneous transitions in the lattice and the spin system. Then we have

$$\varrho^* = \sigma^* \varrho^{L*} \quad . \tag{5.334}$$

Introducing spin quantum numbers s and s', and lattice quantum numbers f and f', we replace α by sf, and so on. Then we assume that the lattice remains in thermal equilibrium despite the spin relaxation:

$$\varrho^L_{ff'} = \delta_{ff'} \frac{e^{-\hbar f/kT}}{\sum_{f''} e^{-\hbar f''/kT}} \quad . \tag{5.335}$$

We then find the differential equation for

$$\frac{d}{dt}(\varrho^{L*}_{ff'} \sigma^*_{ss'}) = \delta_{ff'} \varrho^{L*}_{ff'} \frac{d}{dt} \sigma^*_{ss'} \tag{5.336}$$

and sum over f. The result, in the high temperature limit, is simply to give a modified version of *Redfield*'s equation, with density matrix for spin σ replaced by the difference between σ and its value for thermal equilibrium at the lattice temperature $\sigma(T)$.

We therefore simply assert that for an interaction in which the lattice couples to the spins via an interaction $\mathcal{H}_1$ (which, to the spins, is time dependent),[7] the role of the lattice is to modify the *Redfield* equation to be

$$\frac{d\sigma^*_{\alpha\alpha'}}{dt} = \sum_{\beta,\beta'} R_{\alpha\alpha',\beta\beta'} e^{+i(\alpha-\alpha'-\beta+\beta')t} [\sigma^*_{\beta\beta'} - \sigma^*_{\beta\beta'}(T)] \tag{5.337}$$

where $\alpha, \alpha', \beta, \beta'$ stand for spin quantum numbers, and where $\sigma_{\beta\beta'}(T)$ is the thermal equilibrium value of $\sigma_{\beta\beta'}$:

$$\sigma_{\beta\beta'}(T) = \delta_{\beta\beta'} \frac{e^{-\hbar\beta/kT}}{\sum_{\beta''} e^{-\hbar\beta''/kT}} \quad . \tag{5.338}$$

That (5.337) should hold true is not surprising in view of our remarks in Chapter 1 concerning the approach to thermal equilibrium. We note here, however, that our remarks apply not only to the level populations (the diagonal elements of σ) but also to the off-diagonal elements.

---

[7] $\mathcal{H}_1$ involves both spin and lattice coordinates. If we treat the lattice quantum mechanically, the lattice variables are operators, and $\mathcal{H}_1$ does not involve the time explicitly. If we treat the lattice classically, $\mathcal{H}_1$ involves the time explicity. That this must be so is evident, since the coupling must be time dependent to induce spin transitions between spin states of different energy. However, it is time independent when the lattice makes a simultaneous transition that just absorbs the spin energy.

5.12 Example of Redfield Theory

We turn now to an example to illustrate both the method of *Redfield* and some simple physical consequences. The example we choose is that of an ensemble of spins which do not couple to one another but which couple to an external fluctuating field, different at each spin. The external field possesses x-, y-, and z-components. This example possesses many of the features of a system of spins with dipolar coupling. However, it is substantially simpler to treat; moreover, it can be solved exactly in the limit of very short correlation time. For the case of dipolar coupling, the fluctuations of the dipole field arise from bodily motion of the nuclei, as when self-diffusion occurs. The correlation time corresponds to the mean time a given pair of nuclei are near each other before diffusing away. Our simple model gives the main qualitative features of the dipolar coupling if we simply consider the correlation time to correspond to that for diffusion. In particular, then, our model will exhibit the important phenomenon of motional narrowing, which has been so beautifully explained in the original work of *Bloembergen* et al.

Before plunging into the analysis, we can remark on certain simple features that will emerge. At the end of this section we develop these simple arguments further, showing how to use them for more quantitative results.

We may distinguish between the effects of the x-, y-, and z-components of the fluctuating field. A component H_z causes the precession rate to be faster or slower. It, so to speak, causes a spread in precessions. It will evidently *not* contribute to the spin-lattice relaxation because that requires changes in the component of magnetization parallel to H_0, but it will contribute to the decay of the transverse magnetization even if the fluctuations are so slow as to be effectively static. In fact, as we shall see, it is H_z that contributes to the rigid-lattice line breadth. The phenomenon of motional narrowing corresponds to a sort of averaging out of the H_z effect when the fluctuations become sufficiently rapid.

The x- and y-components of fluctuating field are most simply viewed from the reference frame rotating with the precession. Components fluctuating at the precession frequency in the laboratory frame can produce quasi-static components in the rotating frame perpendicular to the static field. They can cause changes in components of magnetization, either parallel or perpendicular to the static field. The former is a T_1 process; the latter, a T_2 process. Clearly the two processes are intimately related, since the magnetization vector of an individual spin is of fixed length. The transverse components of fluctuating magnetic fields will be most effective when their Fourier spectrum is rich at the Larmor frequency. For either very slow or very rapid motion, the spectral density at the Larmor frequency is low, but for motions whose correlations time τ is of order $1/\omega_0$, the density is at a maximum. The contribution of H_x and H_y to the longitudinal and transverse relaxation rates therefore has a maximum as τ is changed.

Let us consider, then, an interaction $\mathcal{H}_1(t)$ given by

$$\mathcal{H}_1(t) = -\gamma_n \hbar \sum_q H_q(t) I_q \qquad (5.339)$$

where $q = x, y, z$, and

$$\mathcal{H}_0 = -\gamma_n \hbar H_0 I_z = -\hbar \omega_0 I_z \tag{5.340}$$

where ω_0 is the Larmor frequency. We characterize the eigenstates by α, the eigenvalues of (5.340). These are ω_0 times the usual m-values of the operator I_z. (Here $m = I, I-1, \ldots -I$). We shall continue to use the notation $\alpha, \alpha', \beta, \beta'$, however, rather than m, in order to keep the equations similar to those we have just developed. The matrix elements $(\alpha|\mathcal{H}_1(t)|\alpha')$ are

$$(\alpha|\mathcal{H}_1(t)|\alpha') = -\gamma_n \hbar \sum_q H_q(t)(\alpha|I_q|\alpha') \quad . \tag{5.341}$$

Then the spectral density functions $J_{\alpha\beta\alpha'\beta'}(\omega)$ are

$$\frac{1}{2\hbar^2} J_{\alpha\beta\alpha'\beta'}(\omega) = \frac{1}{2\hbar^2} \int_{-\infty}^{+\infty} \overline{(\alpha|\mathcal{H}_1(t)|\beta)(\beta'|\mathcal{H}_1(t+\tau)|\alpha')} e^{-i\omega\tau} d\tau$$

$$= \frac{\gamma_n^2}{2} \sum_{q,q'} (\alpha|I_q|\beta)(\beta'|I_{q'}|\alpha') \int_{-\infty}^{+\infty} \overline{H_q(t) H_{q'}(t+\tau)} e^{-i\omega\tau} d\tau \quad . \tag{5.342}$$

We now use the symbol $k_{qq'}(\omega)$ introduced in the preceding section as

$$k_{qq'}(\omega) = \frac{1}{2} \int_{-\infty}^{+\infty} \overline{H_q(t) H_{q'}(t+\tau)} e^{-i\omega\tau} d\tau \quad . \tag{5.343}$$

Clearly the fluctuation effects, correlation time, and so on are all associated with the $k_{qq'}$'s. For simplicity let us assume that the fluctuations of the three components of field are independent. That is, we shall assume

$$\overline{H_q(t) H_{q'}(t+\tau)} = 0 \quad \text{if} \quad q \neq q' \quad . \tag{5.344}$$

For example, (5.344) will hold true if, for any value of the component H_q, the values of $H_{q'}$ occur with equal probability as $|H_{q'}|$ or $-|H_{q'}|$. We note that $k_{qq}(\omega)$ gives the spectral density at frequency ω of the q-component of the fluctuating field. With the assumption of (5.344) we have, then,

$$\frac{1}{2\hbar^2} J_{\alpha\beta\alpha'\beta'} = \gamma_n^2 \sum_q (\alpha|I_q|\beta)(\beta'|I_q|\alpha') k_{qq}(\omega) \quad . \tag{5.345}$$

We now seek to find the effect of relaxation on the x-, y-, and z-components of the spins. To do this, we utilize the second technique described in the preceding section, that of finding a differential equation for the expectation value of the spin components. Let us therefore ask for $(d/dt)\langle I_r \rangle$, $r = x, y$, or z. By using (5.312), we find

$$\frac{d\langle I_r \rangle}{dt} = \sum_{\alpha,\alpha'} \frac{i}{\hbar} [\varrho, \mathcal{H}_0]_{\alpha\alpha'} (\alpha'|I_r|\alpha) + \sum_{\alpha,\alpha',\beta,\beta'} R_{\alpha\alpha',\beta\beta'} \varrho_{\beta\beta'} (\alpha'|I_r|\alpha) \quad . \tag{5.346}$$

The first term on the right, involving $\mathcal{H}_0$, can be handled readily:

$$\sum_{\alpha,\alpha'} \frac{i}{\hbar}[\varrho, \mathcal{H}_0]_{\alpha\alpha'}(\alpha'|I_r|\alpha) = \frac{i}{\hbar}\text{Tr}\{(\varrho\mathcal{H}_0 - \mathcal{H}_0\varrho)I_r\}$$

$$= \frac{i}{\hbar}\text{Tr}\{\varrho\mathcal{H}_0 I_r - \varrho I_r \mathcal{H}_0\}$$

$$= \frac{i}{\hbar}\text{Tr}\{\varrho[\mathcal{H}_0, I_r]\}$$

$$= -i\gamma_n H_0 \text{Tr}\{\varrho[I_z, I_r]\} \quad . \tag{5.347}$$

If $r = z$, this term vanishes. If $r = x$, we have

$$-i\gamma_n H_0 \text{Tr}\{\varrho[I_z, I_x]\} = -i\gamma_n H_0 \text{Tr}\{iI_y\varrho\}$$

$$= +\gamma_n H_0 \langle I_y \rangle \quad . \tag{5.348}$$

If $r = y$, we get $-\gamma_n H_0 \langle I_x \rangle$. Thus we have

$$\sum_{\alpha,\alpha'} \frac{i}{\hbar}[\varrho, \mathcal{H}_0]_{\alpha\alpha'}(\alpha'|I_r|\alpha) = \gamma_n(\langle \boldsymbol{I} \rangle \times \boldsymbol{H}_0)_r \tag{5.349}$$

which is the driving term of the Bloch equations describing the torque due to the external field. The second term on the right of (5.346) involves the relaxation terms:

$$\sum_{\alpha,\alpha',\beta,\beta'} R_{\alpha\alpha',\beta\beta'} \varrho_{\beta\beta'}(\alpha'|I_r|\alpha) \quad . \tag{5.350}$$

As we have seen, $R_{\alpha\alpha',\beta\beta'}$ is itself the sum of four terms [see (5.330)]. We shall discuss the first term, $J_{\alpha\beta\alpha'\beta'}(\alpha' - \beta')$. By using (5.345), we find

$$\frac{1}{2\hbar^2} \sum_{\alpha,\alpha',\beta,\beta'} J_{\alpha\beta\alpha'\beta'}(\alpha' - \beta')\varrho_{\beta\beta'}(\alpha'|I_r|\alpha)$$

$$= \gamma_n^2 \sum_{\alpha,\alpha',\beta,\beta',q} (\alpha|I_q|\beta)(\beta'|I_q|\alpha')(\beta|\varrho|\beta')(\alpha'|I_r|\alpha)k_{qq}(\alpha' - \beta')$$

$$= \gamma_n^2 \sum_{\alpha',\beta',q} (\beta'|I_q|\alpha')(\alpha'|I_r I_q \varrho|\beta')k_{qq}(\alpha' - \beta') \tag{5.351}$$

where the last step follows from the basic properties of orthogonality and completeness of the wave functions $|\alpha)$, and so on. We are able to "collapse" the indices α and β, but we cannot do the same trick for α' and β' because they occur not only in the matrix elements but also in the k_{qq}'s.

In a similar way one can obtain expressions for the other three terms in $R_{\alpha\alpha',\beta\beta'}$, getting finally

$$\sum_{\alpha,\alpha',\beta,\beta'} R_{\alpha\alpha',\beta\beta'} \varrho_{\beta\beta'}(\alpha'|I_r|\alpha)$$

$$= \gamma_n^2 \sum_{\alpha,\beta,q} (\beta|I_q|\alpha)(\alpha|I_r I_q - I_q I_r)\varrho|\beta)k_{qq}(\alpha - \beta)$$

$$+ \gamma_n^2 \sum_{\alpha,\beta,q} (\beta|I_q|\alpha)(\alpha|\varrho(I_q I_r - I_r I_q)|\beta) k_{qq}(\beta - \alpha)$$

$$= \gamma_n^2 \sum_{\alpha,\beta,q} (\beta|I_q|\alpha)(\alpha|[[I_r, I_q], \varrho]|\beta) k_{qq}(\beta - \alpha) \tag{5.352}$$

where, in the last step, we have utilized the fact that $k_{qq}(\omega)$ is an even function of ω.

To proceed further, we must now specify whether r is x, y, or z. First let us consider $r = z$. Then, since I_r will commute with I_z, we get nothing from $q = z$ in the last line of (5.352). Since matrix elements of I_x vanish except for $\Delta m = \pm 1$, the only states (α and β) that are joined by I_q for $q = x$ have $|\beta - \alpha| = \omega_0$, the Larmor frequency. Since $[I_z, I_x] = iI_y$ and $[I_x, I_y] = iI_z$, we have, then,

$$\gamma_n^2 \sum_{\alpha,\beta}(\beta|I_x|\alpha)(\alpha|[[I_z, I_x]\varrho]|\beta) k_{xx}(\alpha - \beta)$$

$$= \gamma_n^2 [\sum_{\alpha,\beta}(\beta|I_x|\alpha)(\alpha|iI_y\varrho - i\varrho I_y|\beta)] k_{xx}(\omega_0)$$

$$= i\gamma_n^2 k_{xx}(\omega_0) \operatorname{Tr} \{I_x I_y \varrho - I_x \varrho I_y\}$$

$$= i\gamma_n^2 k_{xx}(\omega_0) \operatorname{Tr} \{(I_x I_y - I_y I_x)\varrho\}$$

$$= -\gamma_n^2 k_{xx}(\omega_0) \operatorname{Tr} \{I_z \varrho\}$$

$$= -\gamma_n^2 k_{xx}(\omega_0)\langle I_z \rangle \quad . \tag{5.352a}$$

The term $q = y$ gives, in a similar manner,

$$-\gamma_n^2 k_{yy}(\omega_0)\langle I_z \rangle \quad . \tag{5.353}$$

All told, then,

$$\sum_{\alpha,\alpha',\beta,\beta'} R_{\alpha\alpha',\beta\beta'} \varrho_{\beta\beta'}(\alpha'|I_z|\alpha) = -\gamma_n^2[k_{xx}(\omega_0) + k_{yy}(\omega_0)]\langle I_z \rangle \quad . \tag{5.354}$$

By combining (5.346), (5.349), and (5.354), we have

$$\frac{d\langle I_z \rangle}{dt} = \gamma(\langle \mathbf{I} \rangle \times \mathbf{H}_0)_z - \gamma_n^2[k_{xx}(\omega_0) + k_{yy}(\omega_0)]\langle I_z \rangle \quad . \tag{5.355}$$

This equation relaxes toward $\langle I_z \rangle = 0$ rather than the thermal equilibrium value I_0. To remedy the situation, we should replace ϱ by $\varrho - \varrho(T)$, as discussed in the preceding section. This result simply makes $\langle I_z \rangle$ relax toward the thermal equilibrium value I_0:

$$\frac{d\langle I_z \rangle}{dt} = \gamma(\langle \mathbf{I} \rangle \times \mathbf{H}_0)_z - \gamma_n^2[k_{xx}(\omega_0) + k_{yy}(\omega_0)](\langle I_z \rangle - I_0) \quad . \tag{5.356}$$

This is clearly one of the Bloch equations, with T_1 given by the expression

$$\frac{1}{T_1} = \gamma_n^2[k_{xx}(\omega_0) + k_{yy}(\omega_0)] \quad . \tag{5.357}$$

We can proceed in a similar way to find the relaxation of the x-component. For it, the value $q = x$ contributes nothing, but $q = y$ or z does. The situation for $q = y$ is similar to the one we have just discussed, connecting states α and β which differ by ω_0. On the other hand, when $q = z$, the states α and β are the *same* (I_z is diagonal), so that $\alpha - \beta = 0$. The spectral density of H_z at zero frequency enters. Therefore we find

$$\sum_{\alpha,\alpha',\beta,\beta'} R_{\alpha\alpha',\beta\beta'} \varrho_{\beta\beta'}(\alpha'|I_x|\alpha) = -\gamma_n^2 [k_{yy}(\omega_0) + k_{zz}(0)]\langle I_x \rangle \tag{5.358}$$

which gives for the derivative of $\langle I_x \rangle$:

$$\frac{d\langle I_x \rangle}{dt} = \gamma_n(\langle \boldsymbol{I} \rangle \times \boldsymbol{H}_0)_x - \gamma_n^2[k_{yy}(\omega_0) + k_{zz}(0)]\langle I_x \rangle \quad . \tag{5.359}$$

For this equation there is no effect of replacing ϱ by $\varrho - \varrho(T)$, since in thermal equilibrium $\langle I_x \rangle = 0$. Equation (5.359) and a similar one for $\langle I_y \rangle$ look very much like the transverse Bloch equation describing a T_2 process, except that the T_2 for $\langle I_x \rangle$ differs from that for $\langle I_y \rangle$. Labeling these as T_{2x} and T_{2y} we get

$$\frac{1}{T_{2x}} = \gamma_n^2 [k_{yy}(\omega_0) + k_{zz}(0)] \tag{5.360a}$$

$$\frac{1}{T_{2y}} = \gamma_n^2 [k_{xx}(\omega_0) + k_{zz}(0)] \quad . \tag{5.360b}$$

If we did not have the driving term from $\boldsymbol{H}_0$, these equations would cause $\langle I_x \rangle$ and $\langle I_y \rangle$ to decay with the rates T_{2x} and T_{2y} respectively. However, in general the precession rate ω_0 is much faster than $1/T_{2x}$ or $1/T_{2y}$. Thus, we must average the T_2 effects over the precessional motion. That is, the amplitude of the transverse magnetization will decay slowly compared to the precession frequency.

It is straightforward to determine this rate. We define the transverse magnetization $M_\perp$ by the equation

$$M_\perp = iM_x + jM_y \quad \text{so that} \tag{5.361}$$

$$M_\perp^2 = M_x^2 + M_y^2 \quad .$$

Then

$$\frac{dM_\perp^2}{dt} = 2M_\perp \frac{dM_\perp}{dt} \tag{5.362a}$$

$$= 2M_x \frac{dM_x}{dt} + 2M_y \frac{dM_y}{dt} \quad . \tag{5.362b}$$

From (5.359) and (5.360), we get

$$\frac{dM_x}{dt} = \omega_0 M_x - \frac{M_x}{T_{2x}} \quad , \quad \frac{dM_y}{dt} = -\omega_0 M_x - \frac{M_y}{T_{2y}} \quad . \tag{5.363}$$

Using (5.363) and (5.362b), we get

$$M_\perp \frac{dM_\perp}{dt} = M_x \omega_0 M_y - \frac{M_x^2}{T_{2x}} - M_y \omega_0 M_x - \frac{M_y^2}{T_{2y}}$$

$$= -\frac{M_x^2}{T_{2x}} - \frac{M_y^2}{T_{2y}} \quad . \tag{5.364}$$

We now introduce the precessional motion by replacing M_x and M_y

$$\begin{aligned} M_x &= M_\perp \cos(\omega_0 t + \phi) \\ M_y &= -M_\perp \sin(\omega_0 t + \phi) \end{aligned} \tag{5.365}$$

in (5.364), getting

$$M_\perp \frac{dM_\perp}{dt} = -M_\perp^2 \left(\frac{\cos^2(\omega_0 t + \phi)}{T_{2x}} + \frac{\sin^2(\omega_0 t + \phi)}{T_{2y}} \right) \quad . \tag{5.366}$$

Averaging this equation over one period, we replace $\cos^2(\omega_0 t + \phi)$ and $\sin^2(\omega_0 t + \phi)$ by $\frac{1}{2}$ getting

$$\frac{dM_\perp}{dt} = -M_\perp \frac{1}{2} \left(\frac{1}{T_{2x}} + \frac{1}{T_{2y}} \right) = -\frac{M_\perp}{T_2} \quad . \tag{5.367}$$

$$\frac{1}{T_2} = \frac{1}{2} \left(\frac{1}{T_{2x}} + \frac{1}{T_{2y}} \right) \quad . \tag{5.368}$$

Combining (5.360) with (5.368), we see that

$$\frac{1}{T_2} = \frac{1}{2T_1} + \gamma_n^2 k_{zz}(0) \quad . \tag{5.369}$$

Our relaxation mechanism therefore leads to the Bloch equations. Of course one cannot expect that in general the Bloch equations follow from an arbitrary $\mathcal{H}_1(t)$, and one would have to study each case to see whether or not the Bloch equations resulted.

To proceed further, we need to know something about the spectral densities of the x-, y-, and z-components of the fluctuating field. We shall once again assume a simple exponential correlation function, with the same correlation time τ_0, for $q = x, y$, and z:

$$\overline{H_q(t)H_q(t+\tau)} = \overline{H_q^2} \exp(-|\tau|/\tau_0) \quad , \tag{5.370}$$

which gives

$$k_{qq}(\omega) = \overline{H_q^2} \frac{\tau_0}{1 + \omega^2 \tau_0^2} \quad . \tag{5.371}$$

In terms of (5.371) we have, then,

$$\frac{1}{T_1} = \gamma_n^2(\overline{H_x^2} + \overline{H_y^2})\frac{\tau_0}{1+\omega_0^2\tau_0^2}$$

$$\frac{1}{T_{2x}} = \gamma_n^2\left(\overline{H_z^2} + \overline{H_y^2}\frac{\tau_0}{1+\omega_0^2\tau_0^2}\right) \quad (5.372)$$

$$\frac{1}{T_{2y}} = \gamma_n^2\left(\overline{H_z^2}\tau_0 + \overline{H_x^2}\frac{\tau_0}{1+\omega_0^2\tau_0^2}\right) \quad ,$$

from which we get

$$\frac{1}{T_2} = \gamma_n^2\overline{H_z^2}\tau_0 + \frac{1}{2T_1}$$

$$= \gamma_n^2\left(\overline{H_z^2}\tau_0 + \frac{1}{2}(\overline{H_x^2} + \overline{H_y^2})\frac{\tau_0}{1+\omega_0^2\tau_0^2}\right) \quad . \quad (5.373)$$

We note first of all that T_1 goes through a minimum as a function of τ_0 when $\omega_0\tau_0 = 1$. The fluctuating fields that determine T_1 are the x- and y-components at the Larmor frequency. If we view the problem from the rotating frame (that is, one rotating at the Larmor frequency), these results are reasonable, since the T_1 corresponds to a change in the z-magnetization. Such a change is brought about by "static" fields in either the x- or y-directions in the rotating frame, since in the rotating frame the effective field is zero (H_1, of course, is zero). But "static" x- or y-fields in the rotating frame oscillate at ω_0 in the laboratory frame.

On the other hand, the decay of the x-magnetization must arise from the "static" y- or z-fields in the rotating frame. Since the z-axes of the laboratory and rotating frames coincide, for the z-field it is the static laboratory component that counts, but for the y-field, it is the laboratory component at the Larmor frequency that is important. We note that in the limit of very rapid motion ($\omega_0\tau_0 \ll 1$), and assuming an isotropic fluctuating field

$$\overline{H_x^2} = \overline{H_y^2} = \overline{H_z^2} \quad , \quad (5.374)$$

T_1 and T_2 are equal. Physically, for our model, this result signifies that for a very short correlation time, the spectral density of the fluctuating field is "white" to frequencies far above the Larmor frequency, so that the x-, y-, and z-directions in the *rotating* reference frame see equivalent fluctuating fields.

The two terms in the expression for T_2 have simple physical meanings. One term depends on H_z. It represents the dephasing of the spins due to the spread in precession rates arising from the fact that H_z can aid or oppose H_0. This term can be readily derived by a simple argument, which we give below. The second term, as we shall explain, results from broadening of the energy levels due to the finite lifetime a spin is in a given energy state.

Let us now turn to a simple derivation of the first term of the equation for T_2. We assume the field as a value $|H_z|$ for a time τ; then it jumps randomly to $\pm|H_z|$. Such a change in field in practice arises because a nucleus moves relative to its neighbors by diffusion. In the time τ, a spin will precess an *extra* phase angle $\delta\phi$ over its normal precession:

$$\delta\phi = \pm\gamma_n |H_z|\tau \quad . \tag{5.375}$$

After n such intervals, the mean square dephasing $\overline{\Delta\phi^2}$ will be

$$\overline{\Delta\phi^2} = n\delta\phi^2 = n\gamma_n^2 H_z^2 \tau^2 \quad . \tag{5.376}$$

The number of intervals n in a time t is simply

$$n = \frac{t}{\tau} \quad . \tag{5.377}$$

If we take as T_2 the time for a group of spins in phase at $t = 0$ to get about one radian out of step, we find

$$1 = \frac{T_2}{\tau}\gamma_n^2 H_z^2 \tau^2 \quad \text{or} \tag{5.378}$$

$$\frac{1}{T_2} = \gamma_n^2 H_z^2 \tau \quad . \tag{5.379}$$

We note that the *shorter* τ (that is, the more *rapid* the motion[8]), the narrower the resonance. This phenomenon is therefore called *motional narrowing*. We see that the motion narrows the resonance because it allows a given spin to sample many fields H_z, some of which cause it to advance in phase; others, to be retarded. The dephasing takes place, then, by a random walk of small steps, each one much less than a radian.

In contrast, when there is no motion, a given spin experiences a constant local field. It precesses either faster or slower than the average, and the dephasing of a group of spins arises from the inexorable accumulation of positive or negative phase.

The contrast with "collision broadening" of spectral lines is great. In that case the *phase* of the oscillation is changed by each collision. Since the frequency is unperturbed between collisions, there is no loss in phase memory except during a collision. Since each collision gives a loss in phase memory, a more rapid collision rate produces a shorter phase memory and a broader line. With motional narrowing, there is no phase change when H_z is changing from one value to another because the change is very rapid, but there is a phase change during the time a given value of H_z persists. More rapid motion diminishes the loss in phase memory in each interval.

We have considered just one term of our expression for T_2. The other term, which involves H_y^2, clearly has the same dependence on τ_0 as does the spin-lattice relaxation. We interpret it as the broadening of the line due to the finite life of a spin in any eigenstate as a result of the spin-lattice relaxation. The lifetime

---
[8] Of course the word "motion" here refers to translation of the position of the nucleus, not to the change in spin orientation.

is finite because a field in the y-direction can change the z-magnetization. We should estimate the order of magnitude of the lifetime broadening to be

$$\Delta E = \frac{\hbar}{T_1} \quad \text{or} \tag{5.380}$$

$$\Delta \omega = \frac{\Delta E}{\hbar} = \frac{1}{T_1} \;.$$

Assuming isotropic fluctuating fields, we see that our example actually gives

$$\frac{1}{T_2} = \frac{1}{T_{2'}} + \frac{1}{2T_1} \tag{5.381}$$

where $T_{2'}$ is the broadening due to the spread in the z-field. The quantity $1/T_{2'}$, is often called the *secular broadening*; the term $1/2T_1$ is called the *nonsecular* or *lifetime broadening*. More generally, we replaced $1/2T_1$ by $1/T_{1'}$, where $T_{1'}$ (which gives the nonsecular broadening) is related to T_1.

As we have remarked, if we consider the secular broadening, we note that as τ_0 decreases, T_2 increases, or the line narrows. Of course, we have seen that as one increases τ_0 (slows the motion), the validity of the Redfield equations ceases when $T_2 \cong \tau_0$. For longer τ_0's, we cannot apply the Redfield equation. The longest τ_0 for which the Redfield theory can apply, then, is $\tau_0 = T_2$, or

$$\frac{1}{\tau_0} = \frac{1}{T_2} = \gamma_n^2 \overline{H_z^2} \tau_0 \quad \text{or} \tag{5.382}$$

$$\gamma_n |H_z| \tau_0 = 1 \;. \tag{5.383}$$

As we can see from our simple model, this is just the value of τ_0 at which a typical spin gets one radian out of phase *before* there is a jump. For *longer* τ_0's, the spins are dephased before there is a chance for a jump. That is, they do not dephase by a random walk. The line breadth is then independent of the jump rate, giving one the temperature-independent, rigid-lattice line breadth.

The two contributions to the line breadth (secular and nonsecular) are plotted in Fig. 5.11.

If one analyzes the relaxation via other mechanisms, the same general features are found. The fact that more than one transition may be induced will make important the spectral densities at frequencies other than 0 and ω_0. Often, $2\omega_0$

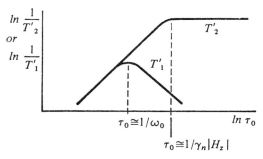

Fig. 5.11. Secular (T_2') and nonsecular (T_1') broadening versus τ_0. For the example in the text, $T_1' = 2T_1$

comes in. For example, when the relaxation arises from the coupling of one nucleus with another by means of their magnetic dipole moments, the E and F terms of Chapter 3, which involve the product of two raising or two lowering operators, connect states differing in energy by $2\hbar\omega_0$.

Our formulas show us that the measurement of T_1 and T_2 will enable us to determine τ_0. When the fluctuations in the interaction $\mathcal{H}_1(t)$ arise from bodily motion that varies with temperature, we can use resonance to study the temperature variation of τ_0. Often one has a "barrier" to motion and an activation energy E such that

$$\tau_0 = \tau_\infty e^{E/kT} \quad , \tag{5.384}$$

where τ_∞ is the value of τ_0 for infinite temperature. The temperature variation of T_1 or T_2 gives one a convenient measure of E and τ_∞. The narrowing studies of *Andrew* and *Eades* [5.13] performed on molecular crystals, provide one such example. Another one is the work of *Holcomb* and *Norberg* [5.14] on self-diffusion in the alkali metals, and subsequently, *Seymour* [5.15] and *Spokas* [5.16] on aluminium. Here one has the interesting fact that, using resonance, these workers could measure the self-diffusion rate of both Li and Al (for which there is no radioactive isotope for use in the conventional tracer technique).

5.13 Effect of Applied Alternating Fields

So far we have excluded applied alternating fields from the time-dependent coupling $\mathcal{H}_1(t)$. Let us now assume that such fields are present, giving an extra term, $\mathcal{H}_2(t)$, in the Hamiltonian. We can include its effect in a straightforward way, as has been shown by Bloch. Introduction of $\mathcal{H}_2(t)$ simply replaces $\mathcal{H}_1^*(t)$ by $\mathcal{H}_1^*(t) + \mathcal{H}_2^*(t)$ in (5.313):

$$\frac{d\varrho^*}{dt} = \frac{i}{\hbar}[\varrho^*(0), \mathcal{H}_1^*(t) + \mathcal{H}_2^*(t)]$$
$$+ \left(\frac{i}{\hbar}\right)^2 \int_0^t [[\varrho^*(0), (\mathcal{H}_1^*(t') + \mathcal{H}_2^*(t'))], (\mathcal{H}_1^*(t) + \mathcal{H}_2^*(t))]dt' \tag{5.385}$$

The effect of $\mathcal{H}_1^*$ in the first term on the right vanished when averaged over an ensemble. For that reason we were forced to consider the second term. In general the contribution of $\mathcal{H}_2^*(t)$ to the first term does *not* vanish, since $\mathcal{H}_2^*(t)$ is identical for all members of the ensemble. If $\mathcal{H}_2^*(t)$ is not too strong, we expect that the first-order term is all that is necessary, and therefore we neglect the role of $\mathcal{H}_2^*$ in the integral. Physically, this approximation amounts to our saying that

$$\frac{d\varrho^*}{dt} = \left.\frac{d\varrho^*}{dt}\right)_{\mathcal{H}_2} + \left.\frac{d\varrho^*}{dt}\right)_{\text{relax}} \tag{5.386}$$

where $d\varrho^*/dt)_{\mathcal{H}_2}$ is the rate of change of ϱ^* due solely to $\mathcal{H}_2$, and $d\varrho^*/dt)_{\text{relax}}$ is the rate of change of ϱ^* we should have if $\mathcal{H}_2$ were zero. What we are neglecting are, therefore, nonlinear effects in the interactions.

Under what circumstances do we expect this approximation to work? The answer is that neither perturbation must change ϱ^* too much during the time t (which is the upper limit of the integral), for the $\mathcal{H}_2^*$ terms in the integral represent the fact that $\mathcal{H}_1^*$ is acting on a ϱ^*, which is not in fact $\varrho^*(0)$ but must be corrected for the driving by $\mathcal{H}_2^*$. Since we wish to choose t as about τ_c, this requirement means that

$$\frac{|(\alpha|\mathcal{H}_2|\alpha')|\tau_c}{\hbar} \ll 1 \quad . \tag{5.387}$$

If $\mathcal{H}_2$ is too large to satisfy (5.387), we should then attempt to solve first for the combined effect of $\mathcal{H}_0$ and $\mathcal{H}_2$, using perturbation theory for $\mathcal{H}_1$. For example, we may use a rotating coordinate transformation rather than a transformation to the interaction representation, thereby converting $\mathcal{H}_2^*(t)$ to a static interaction, which could then be removed by a second transformation to the interaction representation of the effective field.

It is interesting to note that there is in fact a very close similarity between the interaction representation and the usual rotating coordinate transformation that renders H_1 a static field. Both are transformations to rotating coordinate systems. The interaction representation is a transformation to a system rotating at the Larmor frequency, whereas the usual transformation goes to a coordinate system rotating with H_1.

For simplicity we shall assume that (5.387) is satisfied. Note that since T_1 and T_2 are much longer than τ_c for all our equations to be valid, (5.387) can still be easily satisfied even under conditions of saturation.

There is one further consequence of the addition of a term $\mathcal{H}_2(t)$. We have remarked that when we treat the lattice classically, the density matrix relaxes to its value at infinite temperature rather than to the thermal equilibrium value $\varrho(T)$:

$$\varrho(T) = \frac{1}{Z(T)} e^{-\mathcal{H}_0/kT} \quad , \tag{5.388}$$

where $Z(T)$ is the sum of states. When the change of $\mathcal{H}_2$ is small during the time τ_c, which characterizes the "lattice" motion, we expect that $\mathcal{H}_2$ looks like a "static" coupling to the lattice and that we should assume that the system relaxes at each instant toward the instantaneous density matrix:

$$\varrho(T,t) = \frac{1}{Z(T,t)} \exp\left(-(\mathcal{H}_0 + \mathcal{H}_2(t))/kT\right) \quad . \tag{5.389}$$

This equation can be verified by treating the lattice quantum mechanically. When τ_c is long compared with the period of $\mathcal{H}_2$, (5.388) applies.

Under circumstances where the Bloch equations hold, the short τ_c leads often to $T_1 = T_2$, and the Bloch equations become

$$\frac{d\boldsymbol{M}}{dt} = \gamma \boldsymbol{M} \times \boldsymbol{H}(t) + \frac{\boldsymbol{M}_0 - \boldsymbol{M}}{T_1} \quad \text{where} \tag{5.390}$$

$$\boldsymbol{M}_0 = \chi_0 \boldsymbol{H}(t) \tag{5.391}$$

and $H(t)$ is the instantaneous applied field. Explicit solution shows that the solution of (5.390) differs significantly from the normal Bloch equations (assuming $T_1 = T_2$, but $M_0 = \chi_0 M_0$) only when the line width is comparable to the resonance frequency.

We can combine (5.386) with (5.311) to obtain the complete differential equation for the density matrix, including an applied alternating field:

$$\frac{d\varrho_{\alpha\alpha'}}{dt} = \frac{i}{\hbar}(E_{\alpha'} - E_\alpha)\varrho_{\alpha\alpha'} + \frac{i}{\hbar}\sum_{\alpha''}[\varrho_{\alpha\alpha''}(\alpha''|\mathcal{H}_2(t)|\alpha')$$
$$- (\alpha|\mathcal{H}_2(t)|\alpha'')\varrho_{\alpha''\alpha'}] + \sum_{\beta,\beta'} R_{\alpha\alpha',\beta\beta'}[\varrho_{\beta\beta'} - \varrho_{\beta\beta'}(T)] \quad (5.392)$$

where, for $\varrho_{\beta\beta'}(T)$, we use either (5.388) or (5.389), depending on the circumstances.

In order to make (5.392) more concrete, let us suppose we have a two-level system. Thus we may have a spin $\frac{1}{2}$ particle quantized by a static field in the z-direction. We label the states 1 and 2 and find that the density matrix has four elements ϱ_{11}, ϱ_{22}, ϱ_{12}, ϱ_{21}.

In the relaxation terms $R_{\alpha\alpha',\beta\beta'}$, the only terms of importance (as we have seen) involve $\alpha - \alpha' = \beta - \beta'$. The only terms that count are therefore

$$\begin{aligned} R_{11,22} &= R_{22,11} \equiv 1/\tau_1 \\ R_{12,12} &= R_{21,21} \equiv -1/\tau_2 \end{aligned} \tag{5.393}$$

By assuming $\mathcal{H}_2$ joins states 1 and 2 only, and denoting $(1|\mathcal{H}_2(t)|2)$ by $\mathcal{H}_{12}(t)$, we find

$$\frac{d\varrho_{11}}{dt} = -\frac{d\varrho_{22}}{dt}$$
$$= \frac{\varrho_{22} - \varrho_{11} - [\varrho_{22}(T) - \varrho_{11}(T)]}{\tau_1} + \frac{i}{\hbar}(\varrho_{12}\mathcal{H}_{21}(t) - \mathcal{H}_{12}(t)\varrho_{21}) \tag{5.394}$$

and

$$\frac{d\varrho_{12}}{dt} = -\frac{\varrho_{12}}{\tau_2} + \frac{i}{\hbar}(E_2 - E_1)\varrho_{12} + \frac{i}{\hbar}(\varrho_{11} - \varrho_{22})\mathcal{H}_{12}(t) \quad . \tag{5.395}$$

If E_2 is larger than E_1 and $\mathcal{H}_{12}$ oscillates at frequency ω, we can solve (5.394) and (5.395) in steady state by assuming that

$$\begin{aligned} \varrho_{12} &= r_{12}e^{i\omega t}, & \varrho_{21} &= r_{21}e^{-i\omega t}, \\ \varrho_{11} &= r_{11}, & \varrho_{22} &= r_{22} \end{aligned} \tag{5.396}$$

where the $r_{\alpha\alpha'}$'s are complex constants. The details of the solution are left as a problem, but the form of the answer is identical to that of the Bloch equations.

If we write

$$\mathcal{H}_2(t) = V \cos \omega t \tag{5.397}$$

where V is an operator, and define ω_0 by the relation $E_2 - E_1 \equiv \hbar\omega_0$, we have in the limit of small V (that is, no saturation)

$$r_{11} = \varrho_{11}(T) = \frac{e^{-E_1/kT}}{e^{-E_1/kT} + e^{-E_2/kT}} \cong \frac{1}{2}e^{-E_1/kT}$$

$$\begin{aligned} r_{12} &= \frac{i}{2\hbar} \frac{V_{12}\tau_2}{1 + i(\omega - \omega_0)\tau_2}[\varrho_{11}(T) - \varrho_{22}(T)] \\ &\cong \frac{i\tau_2}{1 + i(\omega - \omega_0)\tau_2} \frac{V_{12}\omega_0}{4kT} \quad . \end{aligned} \tag{5.398}$$

We note that r_{12} differs from zero only near to resonance and that τ_2 characterizes the width of frequency over which r_{12} is nonzero. If the states 1 and 2 are the two Zeeman states of a spin $\frac{1}{2}$ nucleus and a static magnetic field parallel to the z-direction, then the transverse magnetization M_x has matrix elements only between states 1 and 2, the diagonal elements being zero. Therefore

$$\begin{aligned} \langle M_x(t) \rangle &= r_{12}e^{i\omega t}(2|M_x|1) + r_{21}e^{-i\omega t}(1|M_x|2) \\ &= 2\operatorname{Re}\{r_{12}e^{i\omega t}(2|M_x|1)\} \quad . \end{aligned} \tag{5.399}$$

Taking

$$V = -M_x H_{x0} \tag{5.400}$$

and recalling that χ is defined as

$$\langle M_x(t) \rangle = \operatorname{Re}\{\chi H_{x0}e^{i\omega t}\} \quad , \tag{5.401}$$

we see that

$$\chi(\omega) = \frac{i\tau_2}{1 + i(\omega - \omega_0)\tau_2} \frac{\omega_0 |(1|M_x|2)|^2}{2kT} \quad . \tag{5.402}$$

Now, using the fact that $I = \frac{1}{2}$, we have

$$\chi(\omega) = \frac{i\tau_2}{1 + i(\omega - \omega_0)\tau_2} \frac{\omega_0}{2} \frac{\gamma^2 \hbar^2 I(I+1)}{3kT} \quad , \tag{5.403}$$

which agrees with the expression for the Bloch equation derived in Chapter 2.

We note that we could determine both τ_1 and τ_2 from first principles by computing $R_{11,22}$ and $R_{12,12}$. Alternatively we could simply treat τ_1 and τ_2 as phenomenological constants to be given by experiment.

If there are more than two levels to a system, the solution may be carried out analogously by simply setting all off-diagonal elements of $\varrho_{\alpha\alpha'}$ equal to zero, except those near resonance with the alternating frequency ($E_\alpha - E_{\alpha'} \cong \hbar\omega$).

6. Spin Temperature in Magnetism and in Magnetic Resonance

6.1 Introduction

In Chapter 5 we employed the concept of spin temperature to discuss relaxation. The idea of spin temperature was introduced by *Casimir* and *du Pre* [6.1] to give a thermodynamic treatment of the experiments of *Gorter* and his students on paramagnetic relaxation. It was *Van Vleck* [6.2] who first employed the concept for a detailed statistical mechanical calculation of the relaxation times of paramagnetic ions. Both in this case, and also in his general statistical mechanical treatment of static properties of paramagnetic atoms [6.3], he recognized and emphasized the fact that expansion of the partition function Z in powers of $1/T$ enabled one to calculate Z without the necessity of solving for the energies and eigenfunctions of the Hamiltonian. *Waller* evidently was the first person to use this property [6.4]. From the partition function, one can compute all the static properties of the system, such as the specific heat, the entropy, the magnetization, and the energy. For example, the average energy of a system, $\overline{E}$, at a temperature T is given by

$$\overline{E} = kT^2 \frac{\partial}{\partial T} \ln Z \quad . \tag{6.1}$$

In 1955, *Redfield* [6.5] showed that the conventional theory of saturation did not properly account for the experimental facts of nuclear resonance in solids. In one of the most important papers ever written on magnetic resonance, he showed that the conventional approach in essence violated the second law of thermodynamics. He went on to show that saturation in solids can be described simply by applying the concept of spin temperature to the reference frame that rotates in step with the alternating field H_1. To understand his ideas, one needs to understand certain concepts which predate the discovery of magnetic resonance – ideas such as adiabatic demagnetization.

In this chapter we begin by describing a simple experiment which displays the failing of the pre-Redfield theory of magnetic resonance. Then we turn to a discussion of the use of spin temperature in nonresonance cases to build background for the application of these same ideas in the rotating reference frame. We then discuss the Redfield theory of saturation in solids.

6.2 A Prediction from the Bloch Equations

Let us consider a simple resonance experiment with a rotating magnetic field of angular frequency ω, transverse to the static field H_0, tuned exactly to resonance

$$\omega = \gamma H_0 \quad . \tag{6.2}$$

We discuss it by means of the Bloch equations.

It is convenient to transform to a reference frame rotating at ω with H_1 along the x-axis as is done in Sect. 2.8. Exactly at resonance, the Bloch equations become

$$\frac{dM_z}{dt} = -\gamma M_y H_1 + \frac{M_0 - M_z}{T_1} \tag{6.3a}$$

$$\frac{dM_x}{dt} = -\frac{M_x}{T_2} \tag{6.3b}$$

$$\frac{dM_y}{dt} = \gamma M_z H_1 - \frac{M_y}{T_2} \quad . \tag{6.3c}$$

Suppose we now orient M along H_1 so that at $t = 0$ $M_x = M_0$, $M_y = M_z = 0$. From (6.3b) we see that M_x will decay to zero in a time T_2. The low H_1 steady-state solution of the Bloch equations shows that they describe a Lorentzian line with a frequency width

$$\Delta\omega = \frac{1}{T_2} \quad . \tag{6.4}$$

For solids typically

$$\Delta\omega \cong \gamma H_{\text{neighbor}} \cong \frac{\gamma\mu}{a^3} \cong \frac{\gamma^2}{a^3}\sqrt{I(I+1)}\sqrt{Z} \tag{6.5}$$

where H_{neighbor} is the nuclear magnetic dipole field due to neighbors, and a is the distance to the Z nearest neighbor. For typical solids $\Delta\omega$ is of order of a few tens of kilocycles (e.g. $\Delta\omega \cong 2\pi \times 10\,\text{kc}$ for Al metal).

Further examination of (6.3a) and (6.3c) shows that they do not involve M_x and that if M_y and M_z are initially zero they would remain zero were it not for the term involving T_1. If $T_1 \gg T_2$, therefore, and for times up to about T_1 after M has been oriented along H_1, we still have $M_y = M_z = 0$.

Therefore these equations predict that when M is aligned along H_1, it will decay to zero in a time T_2, typically of order 10^{-4} to 10^{-5} sec.

Experimentally this prediction (decay in T_2 when M is along H) is found to be correct for liquids, but it is *not* correct for solids. Rather, for solids it is found that as long as H_1 is turned on and is sufficiently strong, the decay rate of M_x is much more like T_1 than a time T_2 which characterizes the line width. *Redfield* first stated this fact on the basis of his steady-state experiments, but he did not actually perform the experiment we have described. That experiment was first performed by *Holton* et al. [6.6]. Why do the Bloch equations fail for solids but not for liquids? *Redfield* has given an explanation. We will discuss the conditions for validity of the Bloch equations later in this chapter, after we have discussed some important background material on spin temperature in cases where there is no alternating field H_1 present.

6.3 The Concept of Spin Temperature in the Laboratory Frame in the Absence of Alternating Magnetic Fields

Let us now turn to a discussion of the application of the concept of spin temperature to magnetic experiments not involving resonance. A typical system we might consider is a group of N spins of spin I, gyromagnetic ratio γ, acted on by an external field H_0, and coupled together by a magnetic dipolar interaction represented by a dipolar Hamiltonian $\mathcal{H}_d$. We denote the Zeeman Hamiltonian by $\mathcal{H}_Z$. The solutions of the Schrödinger equation are then wave functions ψ_n of energy E_n of the total system.

$$\mathcal{H}\psi_n = (\mathcal{H}_Z + \mathcal{H}_d)\psi_n = E_n \psi_n \quad . \tag{6.6}$$

Unfortunately, (6.6) is exceedingly difficult to solve, depending as it does on the coordinates of 10^{22} spins.

One can assume, however, that if the spin system is in thermal equilibrium with a reservoir of temperature θ, the various states n of the total system would be occupied with fractional probabilities p_n given by the Boltzmann factor

$$p_n = \frac{1}{Z} e^{-E_n/k\theta} \tag{6.7}$$

where Z is the partition function

$$Z = \sum_n e^{-E_n/k\theta} \quad . \tag{6.8}$$

Quantities such as the average energy $\overline{E}$ and magnetization $\overline{M}_z$ would then be given by

$$\overline{E} = \sum_n p_n E_n \tag{6.9a}$$

$$\overline{M}_z = \sum_n \gamma \hbar (n|I_z|n) p_n \quad . \tag{6.9b}$$

As we have remarked, *Van Vleck* recognized that expressions such as (6.9) could be evaluated without solving the Schrödinger equation because they could be expressed as traces. For example we can express the partition function as a trace as follows.

$$Z = \sum_n e^{-E_n/k\theta} = \sum_n (n|e^{-\mathcal{H}/k\theta}|n) = \text{Tr}\{e^{-\mathcal{H}/k\theta}\} \quad . \tag{6.10}$$

Since the trace is independent of the particular representation used to evaluate it, we can use a convenient representation. For example, we could evaluate (6.10) in principle by using as basis functions the eigenfunctions of the z-component of spin I_{zk} of all the individual nuclei. However, to do so, we need to take one more step, expansion of Z in a power series.

It is often valid for nuclei and for electrons to use the high temperature approximation. We expand the exponential in a power series, keeping only the leading terms. Then the sums are easy to do.

$$Z = \text{Tr}\left\{1 - \frac{\mathcal{H}}{k\theta} + \frac{1}{2}\frac{\mathcal{H}^2}{k^2\theta^2} - \cdots\right\} = (2I+1)^N + \frac{1}{2k^2\theta^2}\text{Tr}\{\mathcal{H}^2\} + \cdots \quad (6.11)$$

where we have used the fact that $\text{Tr}\{\mathcal{H}\} = 0$, as can be readily verified for both $\mathcal{H}_Z$ and $\mathcal{H}_d$.

Using these methods one finds

$$\overline{E} = -\frac{C(H_0^2 + H_L^2)}{\theta} \quad \text{where} \quad (6.12)$$

$$C = \frac{N\gamma^2\hbar^2 I(I+1)}{3k} \quad (6.13)$$

is the Curie constant, and H_L is a quantity we call the local field, which is of the order of the field one nucleus produces at a neighbor (several Gauss) and is defined by

$$CH_L^2 = \frac{1}{k(2I+1)^N}\text{Tr}\{\mathcal{H}_d^2\} \quad . \quad (6.14)$$

Since the trace in (6.14) can be computed, H_L may be considered to be precisely known. One finds

$$H_L^2 = \gamma^2\hbar^2 I(I+1)\sum_j (1/r_{jk})^6 \quad . \quad (6.15)$$

One can compute the magnetization M and finds it obeys Curie's law

$$\boldsymbol{M} = \frac{C\boldsymbol{H}_0}{\theta} \quad . \quad (6.16)$$

Note that this is a vector equation, so that $\boldsymbol{M}$ and $\boldsymbol{H}_0$ are parallel.

A moment's reflection shows that (6.16) is truly remarkable. It states that the vectors $\boldsymbol{M}$ and $\boldsymbol{H}_0$ are parallel no matter what the size of H_0 as long as the high temperature approximation is valid. Suppose H_0 is small, comparable to the local field a nucleus experiences from its neighbors. One might then suppose that nuclei would tend to line up along the direction of the local field, not along H_0. It would seem reasonable to suppose that the degree of polarization one could achieve per unit of applied field would be less when $H_0 \ll H_L$ than when $H_0 \gg H_L$. Equation (6.16) shows that this intuitive argument is incorrect – the degree of polarization per unit of applied field is independent of the size of H_0 relative to the local field. Not only is this true of the magnitude of M, but also of the direction as well.

Another useful property of (6.16) to note is that when $H_0 = 0$, $M = 0$. Thus, suppose H_0 were turned to zero so suddenly that M does not have time to change. Immediately after $H_0 = 0$ we have a case where $M \neq 0$, $H_0 = 0$. But if there were a temperature, (6.16) shows M must be zero. Therefore we can use (6.16) to conclude that at this instant of time the system is not describable by a temperature.

Another quantity of great utility is the entropy σ. We know from statistical mechanics that the entropy measures the degree of order in a system. In a reversible process in which there is no heat flow into or out of a system, the entropy of that system remains constant.

$$\sigma = \frac{\overline{E} + k\theta \ln Z}{\theta} \quad . \tag{6.17}$$

Evaluating Z and $\overline{E}$ we get

$$\sigma = Nk \ln(2I+1) - \frac{C}{2} \frac{(H_0^2 + H_L^2)}{\theta^2} \quad . \tag{6.18}$$

6.4 Adiabatic and Sudden Changes

The significance of these results is more fully realized by considering the behavior of the spin system when the applied field H_0 is made a function of time. For simplicity, let us assume the spin system is thermally isolated from the outside world and that it may or may not be in thermal equilibrium. The first of these assumptions is satisfied if the experiments we conduct are performed on a time scale short compared to the spin-lattice relaxation time. The assumption implies that the Hamiltonian of the system includes only variables internal to the system since spin-lattice relaxation results from terms involving variables both internal and external to the system as illustrated by (5.33). If the system is in internal thermal equilibrium, a spin temperature applies, so that (6.18) holds. We can then consider three cases.

a) *The Hamiltonian is independent of time* (the applied field is static). In this case, the average energy is constant in time, whether or not the system is describable by a spin temperature. If the system has many parts which are strongly coupled together, but is not initially in a state of internal equilibrium, we expect that irreversible processes within the system will eventually bring the system into an internal equilibrium describable by a spin temperature θ. During that process, the energy is conserved since a time-independent Hamiltonian corresponds to a system on which there are no applied forces. Proof of the constancy of the energy is left as a homework problem (Problem 6.5).

b) *The Hamiltonian changes slowly in time.* The criterion of slowness is that at all times internal transfers of energy shall be fast enough so that the system is always describable by a single temperature θ. Under this circumstance, the changes are reversible, and the entropy of the system remains constant.

c) *The Hamiltonian changes discontinuously in time.* Such a change could occur if H_0 can be changed quickly. The term "discontinuous" means that the change is so fast that the various spins making up the system do not change direction during the change.

Let us now investigate these various cases more fully.

1) *Time-Independent Hamiltonian*

Consider a system not in thermal equilibrium initially. Let us suppose that the parts of the system are coupled together. We then expect that eventually the system will achieve an internal equilibrium described by a "final" temperature θ_f. If we know the energy of the system at $t = 0$ (call it E_0), we can compute the temperature θ_f as follows, making use of the fact that for a time-independent Hamiltonian the energy is conserved.

Utilizing (6.12) which relates the energy $\overline{E}$ to the temperature, and applying conservation of energy, we get that

$$\overline{E} = E_0 \quad \text{or} \tag{6.19}$$

$$\theta_f = -\frac{C(H_0^2 + H_L^2)}{E_0} . \tag{6.20}$$

We can compute the magnetization M_f which finally results after internal thermal equilibrium is reached by means of Curie's law

$$M_f = \frac{CH_0}{\theta_f} \tag{6.21}$$

where we take θ_f from (6.20).

During the process of establishing internal thermal equilibrium, the entropy is not constant since irreversible processes are taking place. But eventually the entropy σ_f is given by the thermal equilibrium value for a temperature θ_f

$$\sigma_f = Nk \ln(2I + 1) - \frac{C(H_0^2 + H_L^2)}{2\theta_f^2} . \tag{6.22}$$

If the Hamiltonian can be divided into two parts which commute, the energy of each part is separately conserved. The system then cannot be expected to reach an equilibrium described by a single temperature, but rather is expected to reach an equilibrium in which each commuting part is described by a temperature. We encounter this situation when we apply these ideas to the rotating frame later in the chapter. In some cases the subsystem may not have enough complexity to make us confident it will eventually be describable by a temperature, but even so we often blithely proceed to assume that a temperature is achieved.

2) *Slow or Adiabatic Changes*

We keep in mind that to be adiabatic, a change in H_0 must satisfy two conditions. The first is that there should be no heat flow into or out of the spin system. The condition will be satisfied if we make the changes on a time scale rapid compared to the spin-lattice relaxation time T_1. Frequently one has T_1's of seconds, and by cooling may achieve T_1's of hours. Such long times are practically infinite.

The second condition we must satisfy is that after each small change in H_0 we must allow a new temperature to be reached before making another small change. This condition is typically that we must change H_0 slowly on a time scale defined by the precession period of nuclei in the local field of neighbors ($1/\gamma H_L$). This time scale is a few tenths of a millisecond. Between these time intervals there is a readily achievable range for which H_0 can be changed adiabatically.

For an adiabatic change we have a constant entropy. Thus from (6.18) we get that $(H_0^2 + H_L^2)/\theta^2$ remains constant. If we start in an initial field H_i at temperature θ_i and change the field adiabatically to a final value H_f, the temperature θ_f is given by the relation

$$\frac{H_i^2 + H_L^2}{\theta_i^2} = \frac{H_f^2 + H_L^2}{\theta_f^2} \quad . \tag{6.23}$$

A famous result, cooling by adiabatic demagnetization, can be seen in (6.23) by taking $H_i^2 \gg H_L^2$, and $H_f^2 \ll H_L^2$ corresponding to what happens when a sample initially in a strong magnetic field has that field turned to zero. Then we find

$$\frac{\theta_f}{\theta_i} = \frac{H_L}{H_i} \ll 1 \quad . \tag{6.24}$$

Thus, the final temperature of the spins is much colder than the initial temperature. If initially the spins were in thermal equilibrium with a thermal bath (such as the lattice cooled to liquid helium temperatures), the final temperature would be a good deal less than the bath temperature. Note that the bigger H_i and the smaller H_L, the greater the cooling. To reduce H_L, it is common to dilute the magnetic atoms.

Figure 6.1 shows θ_f versus H_f for an initial field H_i much larger than the local field, and for an arbitrary initial temperature θ_i. Note that as one lowers the

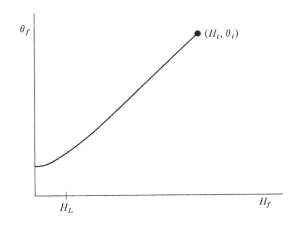

Fig. 6.1. The final temperature θ_f reached at a final magnetic field H_f for adiabatic changes in the applied field from an initial value H_i at temperature θ_i. Note that θ_f does not change much with H_f for $H_f < H_L$

field from its initial value, the temperature drops until the applied field becomes comparable to the local field. Further decreases in applied field do not then produce much lowering of the temperature.

The magnetization can be computed from Curie's law if the temperature is obtained from (6.23). For the case that there is an initial magnetization M_i in an initial field H_i much larger than H_L, the final magnetization M_f is then easily shown to be

$$M_f = M_i \frac{H_f}{\sqrt{H_f^2 + H_L^2}} . \qquad (6.25)$$

This result is shown in Fig. 6.2. Note that during the cooling process, M_f remains at the initial value M_i until H_f gets comparable to the local field H_L, because for large values of H, θ_f is proportional to H_f.

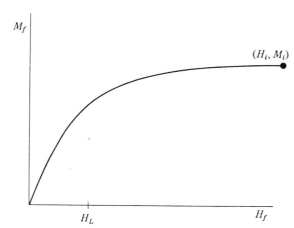

Fig. 6.2. Magnetization M_f versus applied field H_f for adiabatic changes in H_f. It is assumed that $H_i^2 \gg H_L^2$. Note that M_f is independent of H_f for $H_f^2 \gg H_L^2$, and goes to zero when H_f goes to zero

This equation is plotted in Fig. 6.2. Note that if we lower the field until it is zero, $M_f = 0$. However, since the field changes are *reversible* (and in fact at all times keep the entropy constant), we can recover M_i by raising H_f from zero back up to its initial value H_i. The recovery of M_f in such a process does *not* involve spin-lattice relaxation since, as we have postulated, everything is done on a time scale short compared to T_1! This curve is often a surprise to those of us who first learned magnetism by studying magnetic resonance, since we learned that one needs T_1 to produce magnetization from an unmagnetized sample. Actually, if one uses Curie's law and (6.23), one sees that if one starts in zero field with spins which are in thermal equilibrium with the lattice ($\theta = \theta_l$), the mere act of adiabatically turning on the static field to $H_0 \gg H_L$ will produce a magnetization. The size is

$$M = CH_L/\theta_l = (M_0 H_L/H_0) \qquad (6.26)$$

where M_0 is the magnetization one gets when the spin-lattice relaxation produces thermal equilibrium between the spins and lattice in the field H_0. If $H_0 \gg H_\mathrm{L}$, the magnetization is small, but nevertheless it is not zero.

One of the remarkable features of (6.25) or Fig. 6.2 is that when $H_\mathrm{f} = 0$, $M_\mathrm{f} = 0$. As we have remarked, this is a very general consequence of Curie's law no matter what the temperature. (Note, however, that the derivation of Curie's law utilizes the high temperature approximation. At low enough temperatures this approximation breaks down, allowing spontaneous magnetization to be a possible zero field state). What is remarkable is that the degree of order of the spin system is just the same when $M = 0$ as when $M = M_\mathrm{i}$. Although it is clear that a system with net magnetization is ordered, how can there be order when $M = 0$? The answer to this paradox is that even when $H_0 = 0$, spins still experience magnetic fields owing to the presence of their neighbors. A typical spin will point either with or against the local field it experiences. For a highly ordered system, there will be a substantial excess pointing with the local field rather than against it. Since the local fields at different nuclear sites have random orientation (we here rule out highly ordered spin arrangements such as in a ferromagnet), there is no resultant *macroscopic* magnetization resulting from the alignment along the microscopic local fields.

3) *Sudden Switching*

We have considered a process which is reversible. Now we turn to one where things happen suddenly, resulting in irreversibility. Suppose we describe the system by a wave function ψ. The time-dependent Schrödinger equation is

$$-\frac{\hbar}{i}\frac{\partial \psi}{\partial t} = \mathcal{H}(t)\psi$$
$$= -\boldsymbol{M} \cdot \boldsymbol{H}(t)\psi + \mathcal{H}_\mathrm{d}\psi \quad (6.27)$$

where $\boldsymbol{M}$ is the total magnetic moment operator and $\mathcal{H}_\mathrm{d}$ the dipolar coupling. The time dependence of $\mathcal{H}$ arises because the applied field $\boldsymbol{H}$ is time dependent. For the case of sudden switching, we take $\boldsymbol{H}$ as independent of time except for $t = 0$, at which time it jumps discontinuously from one value to another. Denoting by 0^- and 0^+ times just before and just after $t = 0$, we then have

$$\psi(0^+) - \psi(0^-) = -\int_{0^-}^{0^+}\frac{i}{\hbar}\mathcal{H}(t)\psi(t)dt = 0 \quad \text{or} \quad (6.28)$$

$$\psi(0^+) = \psi(0^-) \quad (6.29)$$

since $\mathcal{H}(t)$, though discontinuous, is nevertheless never infinite. Thus we have that the wave function just before the switch is identical to its value just after.

We can utilize (6.29) to see that a sudden change in $\boldsymbol{H}$ produces a change in the expectation value of the energy $\overline{E}$. The expectation value of energy at any time t, $\overline{E}(t)$, is

$$\overline{E}(t) = \langle \psi(t), \mathcal{H}(t)\psi(t) \rangle \quad . \quad (6.30)$$

Thus

$$\overline{E}(0^-) = \langle \psi(0^-), \mathcal{H}(0^-)\psi(0^-)\rangle$$
$$= -\langle M(0^-)\rangle \cdot H(0^-) + \langle \mathcal{H}_d(0^-)\rangle \tag{6.31}$$

where $\langle M(0^-)\rangle$ and $\langle \mathcal{H}_d(0^-)\rangle$ are the expectation values of magnetization and dipolar energy at $t = 0^-$, defined as

$$\langle M(0^-)\rangle \equiv \langle \psi(0^-), M\psi(0^-)\rangle$$
$$\langle \mathcal{H}_d(0^-)\rangle \equiv \langle \psi(0^-), \mathcal{H}_d \psi(0^-)\rangle \quad . \tag{6.32}$$

Likewise

$$\overline{E}(0^+) = -\langle M(0^+)\rangle \cdot H(0^+) + \langle \mathcal{H}_d(0^+)\rangle \quad . \tag{6.33}$$

The fact that $\psi(0^-) = \psi(0^+)$, however, means that the expectation values of both magnetization and dipolar energy are the same at $t = 0^+$ as at $t = 0^-$. This result expresses the physical fact that all spins point the same direction at $t = 0^+$ as they did at $t = 0^-$.

Thus we can write

$$\overline{E}(0^+) = -\langle M(0^-)\rangle \cdot H(0^+) + \langle \mathcal{H}_d(0^-)\rangle \quad . \tag{6.34}$$

This equation is very useful since in general the system just before the sudden change is assumed to be in internal thermal equilibrium with temperature θ_i. Thus we can compute the expectation values by the methods of Sect. 6.3.

So far we have not employed the density matrix notation, simply to make this chapter accessible to those readers who have not yet become familiar with it. We can write our results compactly by recognizing that (6.29) is equivalent to saying that the density matrix ϱ obeys the relation

$$\varrho(0^+) = \varrho(0^-) \quad . \tag{6.35}$$

Then

$$\langle M(0^+)\rangle = \mathrm{Tr}\{\varrho(0^+)M\} = \mathrm{Tr}\{\varrho(0^-)M\} = \langle M(0^-)\rangle \quad . \tag{6.36}$$

Assuming thermal equilibrium at $t = 0^-$ at a temperature θ_i we get

$$\varrho(0^-) = \frac{\exp(-\mathcal{H}(0^-)/k\theta_i)}{Z(0^-)} \quad . \tag{6.37}$$

Therefore

$$\langle \mathcal{H}_d(0^-)\rangle = \mathrm{Tr}\{\varrho(0^-)\mathcal{H}_d\}$$
$$= \frac{\mathrm{Tr}\{\mathcal{H}_d \exp(-\mathcal{H}(0^-)/k\theta_i)\}}{Z(0^-)} \quad . \tag{6.38}$$

Manipulation, using the high temperature approximation plus the definition (6.14), gives

$$\langle \mathcal{H}_d(0^-)\rangle = -\frac{CH_L^2}{\theta_i} \tag{6.39}$$

whence we get

$$\overline{E}(0^+) = -C\frac{\boldsymbol{H}(0^-)\cdot\boldsymbol{H}(0^+)}{\theta_i} - \frac{CH_L^2}{\theta_i} \quad . \tag{6.40}$$

As discussed earlier, immediately after switching the field, the spin system is in general not in internal thermal equilibrium even though it was in thermal equilibrium at $t = 0^-$. If we wait a long enough time, we expect a temperature to be achieved. Call that time t_f and the temperature θ_f. We can compute θ_f by recognizing that for $t > 0$ the Hamiltonian is time independent, hence the energy is conserved. Therefore,

$$\overline{E}(t_f) = \overline{E}(0^+) \quad . \tag{6.41}$$

But

$$\overline{E}(t_f) = -C\frac{[H^2(0^+) + H_L^2]}{\theta_f}$$

and $\overline{E}(0^+)$ is given by (6.39). Thus

$$\theta_f = \theta_i \frac{H(0^+)^2 + H_L^2}{\boldsymbol{H}(0^-)\cdot\boldsymbol{H}(0^+) + H_L^2} \quad . \tag{6.42}$$

The significance of (6.42) can be seen by considering a particular example in which the applied magnetic field is turned suddenly from its initial value, $\boldsymbol{H}_0$, to zero at $t = 0$. Then $\boldsymbol{H}(0^-) = \boldsymbol{H}_0$, $\boldsymbol{H}(0^+) = 0$, giving

$$\theta_f = \theta_i \quad . \tag{6.43}$$

We contrast this with the result of turning the field slowly to zero

$$\theta_f = \theta_i \frac{H_L}{H_0} \quad . \tag{6.44}$$

The sudden turn-off leaves the temperature unchanged, the slow turn-off produces cooling.

The results of sudden changes are summarized in Fig. 6.3, for the case described above in which H_0 is turned suddenly to zero at $t = 0$.

Adiabatic and sudden changes in H_0 have been very useful in magnetic resonance. One of the first uses of adiabatic changes was to measure the magnetic field dependence of the spin-lattice relaxation time. In 1948, *Turner* et al. [6.7] were studying the spin-lattice relaxation time T_1 of protons in insulators. The relaxation times were, in some instances, many minutes long. To observe the dependence of T_1 on static field, they used a field cycling technique in which they observed the resonance at a high field, but allowed the spins to relax in lower fields to which they cycled between their observations. *Pound* [6.8] discovered that the nuclear relaxation times in a crystal of LiF were so long that he could take his sample out of the apparatus into the earth's magnetic field and then return it with only a small loss in magnetization. With *Ramsey* [6.9] he demonstrated that if this sample were removed from the strong field to a small static field (40 Gauss

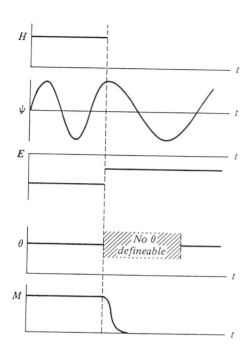

Fig. 6.3. Magnetic field (H), wave function (ψ), energy ($\overline{E}$), spin temperature (θ), and magnetization (M) as functions of time for the case of a magnetic field turned suddenly to zero

or less), subjected to an audio-frequency magnetic field, and then returned to the original magnetic field, he could detect at the strong field the resonant absorption of energy by the spin system when in the weak field.

The experiment we have just described is closely related to the famous experiment on negative temperatures by *Pound* and *Purcell* [6.10]. They too started with the LiF sample in a strong field, but removed the sample to a solenoid whose field they suddenly reversed, producing a situation in which the magnetization is antiparallel to the static field. They then returned the sample to the strong field where they observed the relaxation of the magnetization component along the static field. They point out that such a circumstance, with the *upper* energy Zeeman levels more populated than the lower, corresponds to a negative Zeeman temperature. Any system with an upper bound on its energy levels can in principle have a negative temperature. In addition to the original paper by *Pound* and *Purcell*, we call the reader's attention to the wonderful account of this experiment and discussion of the negative temperature concept to be found in *Van Vleck'* s [6.11] lecture on the concept of temperature in magnetism. (Professor Van Vleck cautions that there are some incorrect numbers in his paper.) This important experiment in which a population inversion was produced is the forerunner of the maser and the laser.

Demagnetizing to zero field is the basis for the experiments of *Hebel* and *Slichter* [6.12] and *Redfield* and *Anderson* [6.13, 14] to measure the nuclear relaxation in superconductors. The problem here is that since superconductors exclude the magnetic field, it is hard to observe a magnetic resonance in a metal in the superconducting state. One starts the field cycle with a magnetic field of sufficient strength to suppress the superconductivity. When one turns the field to zero, one achieves two effects: (1) the nuclear spins are cooled and (2) the sample becomes superconducting. With the field zero, the nuclear spin temperature relaxed towards the lattice temperature so that when the magnet was turned on again after a variable time τ, the metal returned to the normal state and the spins warmed to a hotter temperature than they had before the cycle began. The increase in temperature is measured by a swift pass though resonance. By varying τ, one could deduce the zero-field spin-lattice relaxation time. The NMR results showed that as one lowered the sample temperature below the superconducting transition temperature, T_c, the nuclear relaxation rate was at first faster than it would have been if the metal were normal, then at lower temperature it became slower. The temperature dependence of the nuclear relaxation rate is dramatically different from the temperature dependence of ultrasonic absorption, as was shown by *Morse* and *Bohm* [6.15]. They found that ultrasonic absorption dropped rapidly relative to its value in the normal metal on cooling below T_c. The contrast between the behavior of two low energy scattering processes, nuclear relaxation and sound absorption, is difficult to understand in a one-electron theory of metals, but finds a natural explanation in the Bardeen, Cooper, and Schrieffer (BCS) theory of superconductivity. These experiments constitute a direct verification of the concept of electron pairs, which is the basis of the BCS theory. (See, for example, *Leon Cooper*'s Nobel Prize lecture [6.16].)

Anderson [6.17] working with *Redfield,* combined field cycling with the application of an audio-frequency magnetic field applied while the static field was zero to heat the spins to plot out the zero-field absorption characteristics of a spin system. Thus they used field cycling to give them the sensitivity of resonance in a strong field to monitor the effects they produced in zero field.

Abragam and *Proctor* [6.18] did further studies establishing the validity of the spin temperature concept, again using LiF. An important result of their experiments was their observation of the transfer of energy between the two spin systems (Li^7 and F^{19}) even in a static field which greatly exceeded the dipolar fields exerted by the nuclei on their neighbors. We discuss these experiments further in Sect. 7.10.

6.5 Magnetic Resonance and Saturation

The analysis of magnetic resonance by *Bloembergen* et al. using standard perturbation theory is given rather compactly in (1.32) – the differential equation for the population difference, n, between the two energy levels of a system of spin $\frac{1}{2}$ particles,

$$\frac{dn}{dt} = -2W(\omega)n + \frac{n_0 - n}{T_1} \quad . \tag{6.45}$$

$W(\omega)$ is the probability/second that a spin will be flipped by the radio-frequency field H_1. Standard perturbation theory shows

$$W(\omega) = \tfrac{\pi}{2}\gamma^2 H_1^2 g(\omega) \tag{6.46}$$

where $g(\omega)$ is a function normalized to unit area having the same dependence on frequency as the absorption line – that is, it expresses the fact that the frequency of H_1 must be close to resonance for H_1 to induce transitions. n_0 is the thermal equilibrium population difference, and T_1 the spin-lattice relaxation time.

It is always possible, at least conceptually, to consider T_1 infinite, in which case (6.45) is especially simple to solve.

$$n = Ae^{-2Wt} \tag{6.47}$$

where A is a constant of integration.

It is well to recall the conditions for the validity of (6.46). They are two:

i) The perturbation matrix elements inducing transitions must be small compared to the width of the final state energy levels. This means $H_1 \ll H_L$.
ii) The wave function must not change much. We note, however, that (6.47) predicts that $n \to 0$ as $t \to \infty$.

To satisfy condition (ii) we expect that we must consider times less than $1/W$. Thus, though it is always easy to meet condition (i) by making H_1 small, no matter how weak H_1, if we wait long enough we violate (ii). [Note we are requiring here also that $1/W \ll T_1$, otherwise the T_1 term would rescue condition (ii) even for times long compared to $1/W$].

We have the interesting problem, therefore, that we do not know how to integrate the equations of motion beyond a time for which n is almost its value at $t = 0$. (See Fig. 6.4).

The solution of this problem was found by *Redfield* [6.19] in a truly remarkable paper, the more so when one recognizes that it was his first work on magnetic resonance. In it he shows that the Bloch equations, when applied to a solid, violate the second law of thermodynamics. The essence of his approach is to note that a *resonant* time-dependent perturbation, no matter how weak,

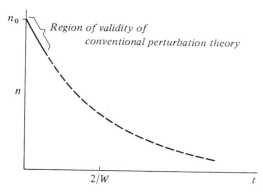

Fig. 6.4. Conventional saturation theory predicts that the population n goes to zero exponentially with time t. However, the assumptions on which it is based are valid only for the initial part of the curve where $n \cong n_0$

will eventually produce large effects. Whenever a perturbation of small size can produce a big effect, it is dangerous to treat it lightly. He therefore eliminates the time dependence by transforming to a reference system in which the Hamiltonian is essentially time independent. The residual time dependence is not of the dangerous variety. For such a transformed system, the energy is conserved. Moreover, the system is highly complex, consisting of many interacting spins. One can thus predict that after a sufficiently long time the system will be found in a state of internal equilibrium. That is, it will be in one of its most probable states. Phrased alternatively, the energy states will be occupied according to a Boltzmann distribution at some temperature θ.

We consider the Hamiltonian $\mathcal{H}$, given by

$$\mathcal{H} = \mathcal{H}_Z(t) + \mathcal{H}_d \tag{6.48}$$

where $\mathcal{H}_Z(t)$ is the Zeeman interaction with the static field H_0 and the rotating field of amplitude H_1 and angular frequency ω rotating in the sense of the nuclear precession. The rotating field makes $\mathcal{H}_Z$ time dependent. We are, of course, considering T_1 infinite. Utilizing the methods of Sect. 2.6, we transform to the rotating reference frame, getting a transformed Hamiltonian $\mathcal{H}'$:

$$\begin{aligned}\mathcal{H}' &= -\gamma\hbar[(H_0 - \omega/\gamma)I_z + H_1 I_x] + \mathcal{H}_d^0 \\ &\quad + \text{term oscillating at} \pm\omega, \pm 2\omega \quad .\end{aligned} \tag{6.49}$$

In deriving (6.49) we have used relations such as

$$-I_x \sin\phi + I_y \cos\phi = e^{-iI_z\phi} I_y e^{iI_z\phi} \quad \text{and} \tag{6.50}$$

$$e^{-iI_z\omega_z t} e^{iI_z\omega_z t} = 1 \tag{6.51}$$

to transform products such as

$$e^{iI_z\omega_z t} I_x I_y e^{-iI_z\omega_z t} \tag{6.52}$$

to expressions involving $\sin\omega_z t$ and $\cos\omega_z t$. The term $\mathcal{H}_d^0$ is that part of the dipolar coupling which commutes with I_z. Physically, it is the part of $\mathcal{H}_d$ which is unchanged by rotation about the z-axis. (The two statements are of course equivalent since one can consider I_z as generating rotations). It is the sum of the terms A and B discussed in Sect. 3.2.

The form of $\mathcal{H}_d^0$ is

$$\mathcal{H}_d^0 = \frac{\gamma^2\hbar^2}{4} \sum_{j,k} \frac{(1 - 3\cos^2\theta_{jk})}{r_{jk}^3}(3I_{zj}I_{zk} - \boldsymbol{I}_j\cdot\boldsymbol{I}_k) \quad , \tag{6.53}$$

where θ_{jk} and r_{jk} are coordinates of nucleus j with respect to nucleus k.

The term in the square brackets in (6.49) may be considered as the coupling of the spins to an effective static field $\boldsymbol{H}_e$ as we noted in Chap. 2.

$$\boldsymbol{H}_e = \boldsymbol{k}(H_0 - \omega/\gamma) + \boldsymbol{i}H_1 \quad . \tag{6.54}$$

In the absence of the time-dependent terms, the energy levels of $\mathcal{H}'$ would be split by H_e and the dipolar couplings, so that we expect typical splittings to be

$$\Delta E \approx \gamma\hbar\sqrt{H_e^2 + H_L^2} \tag{6.55}$$

where the square root is a convenient way of including the two limiting cases of $H_e \gg H_L$ and $H_e \ll H_L$.

The time-dependent terms will connect states of order $\gamma\hbar H_0$ apart in energy. Unless $H_0^2 \approx H_L^2 + H_e^2$, a very low resonance field indeed, the time-dependent terms are far from resonance and can be neglected since they are unable to produce transitions. They are not dangerous. We thus obtain a Hamiltonian which we call $\mathcal{H}$, omitting primes for simplicity of notation.

$$\mathcal{H} = -\gamma\hbar \boldsymbol{I} \cdot \boldsymbol{H}_e + \mathcal{H}_d^0 = \mathcal{H}_Z + \mathcal{H}_d^0 \quad . \tag{6.56}$$

Now, in the absence of H_1, $\mathcal{H}_Z$ and $\mathcal{H}_d^0$ commute, since the fact that $[I_z, \mathcal{H}_d^0] = 0$ was the definition of $\mathcal{H}_d^0$. Under this circumstance $\mathcal{H}_Z$ and $\mathcal{H}_d^0$ would separately be constants of the motion. However, if $H_1 \neq 0$, $[\mathcal{H}_Z, \mathcal{H}_d^0] \neq 0$, and the Zeeman and dipolar systems can then exchange energy. Since $\mathcal{H}$ is independent of time, the total energy is conserved. Moreover, the system is very complex. *Redfield* therefore postulates that no matter what the state of the system at $t = 0$, a long time later it will be in a state of internal equilibrium described by a Boltzmann distribution. In other words, there will eventually be a temperature θ which can be assigned to the spins. We can thus say that the density matrix ϱ is

$$\varrho = \frac{e^{-\mathcal{H}/k\theta}}{Z} \tag{6.57}$$

with $Z = \text{Tr}\{\exp(-\mathcal{H}/k\theta)\}$, where $\mathcal{H}$ is the effective Hamiltonian of (6.56). Of course, we expect that after a long enough time *Redfield*'s hypothesis would be fulfilled (unless there were some hidden selection rule which we have overlooked, such as the fact that $\mathcal{H}_Z$ and $\mathcal{H}_d^0$ are perfectly isolated from one another if H_1 is zero). But the really important question becomes, how long does it take to reach equilibrium? The answer to this question clearly depends on the size of H_1, since H_1 is needed to prevent isolation of the dipolar and Zeeman systems. We return to the question later, but for the present consider that the time is short enough to make the establishment of a temperature practical.

6.6 Redfield Theory Neglecting Lattice Coupling

The significance of (6.56) can be appreciated by calculating again the energy E, entropy σ, and magnetization M. We find easily that

$$\overline{E} = -C\frac{(H_e^2 + H'^2_L)}{\theta} \tag{6.58}$$

$$M = C\frac{H_e}{\theta} \tag{6.59}$$

$$\sigma = Nk\ln(2I+1) - \frac{C}{2}\frac{(H_e^2 + H_L'^2)}{\theta^2} \tag{6.60}$$

where C is the Curie constant, and where

$$CH_L'^2 = \frac{1}{k(2I+1)^N}\text{Tr}\{(\mathcal{H}_d^0)^2\} \quad . \tag{6.61}$$

Evaluation of the trace of (6.61) gives, for a system with only one species,

$$H_L'^2 = \tfrac{1}{3}\langle\Delta H^2\rangle \tag{6.62}$$

where $\langle\Delta H^2\rangle$ is the second moment of the resonance line. Following our earlier convention of omitting primes, we shall now use H_L for H_L', using the prime only when we wish to distinguish between the local field in the laboratory reference frame and the rotating reference frame.

It is important to notice that the Redfield assumption leads to Curie's law and that the vector nature of the law shows that the nuclear magnetization is always parallel to the effective field when the system is describable by a temperature in the rotating frame. Thus, if one is exactly at resonance, where $H_e = iH_1$, the magnetization is perpendicular to the static field.

Since the form of (6.58), (6.59) and (6.60) is identical to that of the corresponding equations in the laboratory frame, most of the equations of Sect. 6.4 can be immediately applied to the rotating frame by replacing H_0 with H_e and H_L^2 with $H_L'^2$.

6.6.1 Adiabatic Demagnetization in the Rotating Frame

An adiabatic demagnetization in the rotating frame can be performed readily, as was demonstrated by *Holton* [6.20]. Suppose that initially $H_1 = 0$, and the sample is magnetized to its thermal equilibrium value kM_0. Let us shift H_0 far above the resonance value ω/γ, and turn on H_1. (We assume $H_0 - \omega/\gamma$ to be much bigger than H_L and H_1.) We now have an effective field which is virtually parallel to M. Next we change H_0 to approach resonance at a rate sufficiently slow to satisfy the criterion for a reversible change. (We are here assuming T_1 to be infinite, which we achieve in practice by performing all experiments in a time shorter than T_1).

We then have that

$$M = M_0\frac{H_e}{\sqrt{H_e^2 + H_L^2}} \quad . \tag{6.63}$$

Notice that M is parallel to H_e, as is required by Curie's law. Thus, as one approaches resonance, M changes direction, always pointing along H_e. In general, the magnitude of M also changes, unless $H_e^2 \gg H_L^2$. We can experimentally measure M by suddenly turning off H_1, leaving M to precess freely about

H_0. The induced voltage immediately after turn-off is proportional to M_x. (Use of a phase sensitive detector enables one to measure M_x and M_y, but for these experiments $M_y = 0$). One can measure M_z by noting that though M_x decays to zero within a time of order $1/\gamma H_L$, after turning off H_1, M_z does not change. One can thus wait till M_x has decayed, and then apply a $\pi/2$ pulse which rotates M_z into the x-y plane for inspection. The theoretical values of M_x and M_z are

$$M_x = M\frac{H_1}{H_e} = M_0 \frac{H_1}{\sqrt{H_e^2 + H_L^2}} \qquad (6.64)$$

$$M_z = M\frac{h_0}{H_e} = M_0 \frac{h_0}{\sqrt{H_e^2 + H_L^2}} \qquad (6.65)$$

where h_0 is the component of the effective field in the z-direction:

$$h_0 \equiv (H_0 - \omega/\gamma) \quad . \qquad (6.66)$$

Notice that if one does an adiabatic demagnetization exactly to resonance, the value of magnetization

$$M_x = M_0 \frac{H_1}{\sqrt{H_1^2 + H_L^2}} \qquad (6.67)$$

will persist indefinitely as long as H_1 remains on (actually, when relaxation to the lattice is included, it decays, but on a time scale typically of order T_1). We contrast this prediction with our earlier conclusion from the Bloch equations that M_x decays to zero in T_2, where $T_2 \approx 1/\gamma H_L$. The fact that M does not shrink as long as H_e is kept constant, and that M precesses in step with H_1 is often described by the graphic term "spin locking". If H_1^2 is comparable to H_L^2, M_x will be less than M_0. However, the "loss" of magnetization is recoverable. Were one to go back off resonance, M would grow back to M_0 when $H_e^2 \gg H_L^2$. Figure 6.5 shows (6.67).

When *Redfield* proposed his theory, the fact that the spins were locked to H_e was one of the most surprising results. It is, of course, nothing but the rotating frame equivalent of the statement that the magnetization in the usual laboratory frame adiabatic demagnetization has a one-to-one correspondence with H_0, as expressed in (6.25).

Note that if one pulses on H_0 when $(H_0 - \omega/\gamma) \gg H_1$ and H_L, and then changes H_0 so that one passes through resonance, continuing until one is far on the other side of the resonance ($H_0 - \omega/\gamma$ negative), one will have turned M antiparallel to H_0. Moreover, although near to resonance, one might have $M < M_0$ (if $H_1 < H_L$); by the time one is far from resonance one would have $M \cong M_0$. This experiment provides a simple means of turning over the magnetization.

One can see from (6.64) that if one demagnetizes exactly to resonance, the magnetization will be the full M_0 provided $H_1 \gg H_L$. The resonance signal is then as big as can be achieved using a $\pi/2$ pulse. If $H_1 \lesssim H_L$, one will not

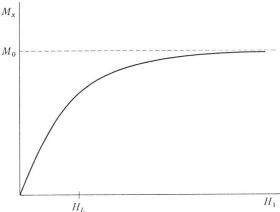

Fig. 6.5. The transverse magnetization M_x produced in a demagnetization experiment in the rotating frame, as a function of the strength of alternating field H_1 employed. H_1 is assumed to be turned on when H_0 is well off resonance, with the sample initially at its thermal equilibrium magnetization M_0

achieve the full magnetization at resonance. Were one to reduce H_1 slowly to zero after arriving at resonance, M would shrink to zero. In this manner all of the order represented by M_0 prior to demagnetization would have been put into the dipolar system, that is, into alignment of spins along their local fields. One could, at a later time, slowly turn H_1 back on again, thereby recovering the magnetization.

6.6.2 Sudden Pulsing

A situation frequently encountered in experiment and interesting to contrast with adiabatic demagnetization is the effect of a sudden change in $\boldsymbol{H}_e$. We treat an especially simple case, that of suddenly turning on $\boldsymbol{H}_1$ at time $t = 0$. We assume that before we turn on H_1, we are off resonance by an amount h_0, with the system initially magnetized to M_0 along H_0. We utilize the ideas of sudden changes discussed in Sect. 6.4.

The sudden turn-on of H_1 is so fast that the system has the same wave function or density matrix just after turn-on as it had before. The dipolar energy E_d which depends on the relative orientation of spins is thus the same at $t = 0^+$ as at $t = 0^-$. Moreover, since the dipolar Hamiltonian $\mathcal{H}_d^0$ is the same in both laboratory and rotating frame,

$$\overline{E}_d = -\frac{CH_L^2}{\theta_1} = -M_0 \frac{H_L^2}{H_0} \tag{6.68}$$

where θ_1 is the lattice temperature. The Zeeman energy is

$$\overline{E}_Z = -\boldsymbol{M} \cdot \boldsymbol{H}_e = -M_0 h_0 \quad . \tag{6.69}$$

The total energy, $\overline{E} = \overline{E}_Z + \overline{E}_d$, is then

$$\overline{E} = -M_0(h_0 + H_L^2/H_0) \cong -M_0 h_0 \quad . \tag{6.70}$$

A long time later a spin temperature will be established together with a magnetization M parallel to H_e giving

$$\overline{E} = -C\left(\frac{H_e^2 + H_L^2}{\theta}\right) = -M\left(\frac{H_e^2 + H_L^2}{H_e}\right) \quad . \tag{6.71}$$

But the total energy is conserved once H_1 has been turned on, so that, equating (6.70) and (6.71), we get

$$M = M_0 \frac{h_0 H_e}{H_e^2 + H_L^2} \quad , \tag{6.72a}$$

$$M_x = M_0 \frac{H_1 h_0}{H_e^2 + H_L^2} \quad , \tag{6.72b}$$

$$M_z = M_0 \frac{h_0^2}{H_e^2 + H_L^2} \quad . \tag{6.72c}$$

This equation shows that exactly at resonance, M would vanish. The null is very sharp, M_x varying linearly with h_0, so that observation of the null provides a precise method of observing exact resonance.

It is interesting to contrast these results with conventional saturation theory. For a spin $\frac{1}{2}$ system, M_z is related to the population difference n by the equation

$$M_z = \frac{\gamma \hbar n}{2} \quad . \tag{6.73}$$

Saturation theory says that the equilibrium population difference, assuming infinite T_1, is $n = 0$, hence $M_z = 0$. Equation (6.72c), on the other hand, says M_z will be zero only when saturation is performed exactly at resonance ($h_0 = 0$). Conventional saturation theory assumes the transverse magnetization vanishes as well as M_z. Equation (6.72b) shows that in general, M_x is large, and may in fact be larger than M_z.

If $H_e^2 \gg H_L^2$, (6.72) has a simple geometrical meaning. After H_1 is pulsed on, M precesses about H_e. The component of M parallel to H_e cannot decay without energy exchange to the lattice, but the component perpendicular can decay since the local field gives a spread in precession frequencies. Thus, after several times $1/\gamma H_L$, M will be parallel to H_e, and will have a magnitude given by the projection of the initial M_0 on H_e.

6.7 The Approach to Equilibrium for Weak H_1

We saw in Sect. 6.5 that standard perturbation theory predicted that following the turn-on of a weak H_1 the population difference n would go to zero for long times, although we recognized that we could not rigorously apply perturbation theory to times greater than $1/W$. The requirement of a weak H_1 was necessary in order that perturbation theory be valid for at least short times. Of course, since M_z is proportional to n, this implies M_z would go to zero. In Sect. 6.6, however, we saw that M_z would, under these conditions, go to an equilibrium value

$$M_z)_{\text{equil}} = M_0 \frac{h_0^2}{h_0^2 + H_L^2} \tag{6.74}$$

where we have assumed $H_1^2 \ll h_0^2$ and $H_1^2 \ll H_L^2$ (although we note that the equilibrium expressions in Sect. 6.6 were not limited to weak H_1). *It is therefore clear, as we suspected, that for long times, perturbation theory does not give correct predictions.* For short time intervals, however, it must be correct. Recalling the proportionality between M_z and n, we know that

$$\frac{dM_z}{dt} = -2W(\omega)M_z \tag{6.75}$$

for times short compared to $1/W(\omega)$. How can we describe M_z for longer times?

The solution to this problem was worked out by *Provotorov* [6.21] in an elegant paper utilizing powerful techniques. Rather than outlining his analysis, we will give an alternative derivation of his result.

We note that in the absence of H_1, the Zeeman interaction in the rotating frame is just

$$\mathcal{H}_Z = -\gamma\hbar h_0 I_z \quad . \tag{6.76}$$

Let us make the assumption that we can assign a temperature θ_Z to this Zeeman Hamiltonian, and θ_d to the dipolar $\mathcal{H}_d^0$. This assumption may not be rigorously correct, but it is at least simple, and corresponds to the facts at the time H_1 is turned on. Immediately before turning on H_1 the dipolar system is at the lattice temperature θ_l, since $\mathcal{H}_d^0$ is the same in the rotating or laboratory frame. In the laboratory frame we have

$$M_0 = \frac{CH_0}{\theta_l} \quad . \tag{6.77}$$

But in the rotating frame

$$M_0 = \frac{Ch_0}{\theta_Z} \quad . \tag{6.78}$$

Inasmuch as $H_0 \gg h_0$, $\theta_Z \ll \theta_l$. Thus the Zeeman temperature in the rotating frame, θ_Z, is very cold compared to the dipolar temperature θ_d. Turning on H_1 couples the two reservoirs and they approach the final equilibrium value θ given by the analysis of Sect. 6.6. The coupling of H_1 produces transitions between the energy levels of the Zeeman and dipolar systems which we assume are governed by simple rate equations for the population of the various states. (This assumption is similar to our postulates of Sect. 5.2. It is quite common in all cross-relaxation calculations. *Provotorov* makes it implicit in his work when he evaluates the relaxation times.) Since there are many states, a large number of coupled rate equations similar to (5.13) result.

As has been shown by *Schumacher* [6.22], when the two systems are characterized by temperatures, the many equations reduce to two coupled linear rate equations, one for $(1/\theta_Z)$, the other for $(1/\theta_d)$ much as the many equations represented by (5.13) reduce to a single rate equation (5.27). But the conservation of energy gives a relationship between θ_Z and θ_d:

$$-\frac{Ch_0^2}{\theta_Z} - \frac{CH_L^2}{\theta_d} = \text{const} \quad . \tag{6.79}$$

Equation (6.79) is a first integral of the coupled equations, so that one of the resultant time constants is infinite, and a *single* exponential results. Since $M_z \propto 1/\theta_Z$, this means M_z relaxes according to a *single* exponential towards its equilibrium value. Using the fact that $M_z = M_0$ initially, we get an equation for M_z as a function of time,

$$M_z - M_z)_{\text{equil}} = (M_0 - M_z)_{\text{equil}})e^{-t/\tau} \quad . \tag{6.80}$$

The only unknown in this equation is τ. We can, however, easily calculate it as follows. Taking the derivative of (6.80), evaluating it at $t = 0$, and comparing with (6.75) (which must be valid initially where perturbation theory is correct), we get

$$\frac{1}{\tau} = 2W(\omega)\frac{M_0}{M_0 - M_z)_{\text{equil}}} \quad . \tag{6.81}$$

Using (6.72c) for $M_z)_{\text{equil}}$ we get

$$\frac{1}{\tau} = 2W(\omega)\frac{h_0^2 + H_L^2}{H_L^2} = \pi\gamma^2 H_1^2 \left(\frac{h_0^2 + H_L^2}{H_L^2}\right)g(\omega) \quad . \tag{6.82}$$

This result is the same as that first found by *Provotorov*, as indeed it must be since we have made the same approximations as he. The complete time development of the magnetization is therefore

$$M_z = \frac{M_0}{h_0^2 + H_L^2}\left[h_0^2 + H_L^2 \exp\left(-\pi\gamma^2 H_1^2 \frac{(h_0^2 + H_L^2)}{H_L^2}g(\omega)t\right)\right] \quad . \tag{6.83}$$

This expression is remarkable since it involves the successful integration of the equations of motion well beyond the time $(1/W)$ for which perturbation theory is usually valid (see Fig. 6.6).

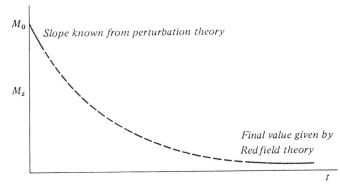

Fig. 6.6. M_z versus t during saturation as given by the argument in the text. The equation for M_z as a function of time is found by joining the initial region, where dM_z/dt is known from perturbation theory, to the equilibrium value, given by Redfield theory, using *Schumacher'*s observation that the approach to equilibrium involves a single exponential

6.8 Conditions for Validity of the Redfield Hypothesis

We have noted that the concept of spin temperature in the rotating frame has as a basic requirement neglect of certain time-dependent terms in the Hamiltonian transformed to the rotating frame. This means, in essence $H_0 \gg \sqrt{H_e^2 + H_L^2}$. On the basis of the previous section we can now add a second requirement. We saw that in the absence of H_1, the equality of dipolar and Zeeman temperatures in the laboratory reference frame implied inequality in the rotating frame. Spin-lattice relaxation will attempt to bring the two temperatures together at the lattice temperature in laboratory frame in a time T_1. On the other hand, the presence of H_1 attempts to equalize them in rotating frame. The two tendencies are in conflict. The strength of the tendency is indicated by the corresponding relaxation times (a short relaxation time means a correspondingly strong tendency for the associated equilibrium). Thus a spin temperature will be established in the rotating frame only if τ is much less than T_1. Thus we find that a spin temperature is established in the rotating frame and that we must employ *Redfield'*s approach provided

$$\frac{T_1}{\tau} \gg 1 \quad \text{or}$$

$$\pi\gamma^2 H_1^2 \left(\frac{h_0^2 + H_L^2}{H_L^2} \right) g(\omega) T_1 \gg 1 \quad . \tag{6.84}$$

This is almost exactly the conventional condition for saturation. Note that the longer T_1, the smaller the H_1 which will satisfy (6.84). In particular, frequently (6.84) is satisfied even when $H_1 \ll H_L$.

6.9 Spin-Lattice Effects

So far we have considered the case of magnetic resonance on a time scale short compared to the spin-lattice relaxation time. In many instances one performs transient experiments which satisfy this condition. However, an equally important case arises when one does experiments on a time scale *long* compared to T_1 as when one employs steady-state apparatus. One can still satisfy the criterion for validity of the Redfield theory given by (6.84), but the spin temperature θ in the rotating frame is now determined by the coupling to the lattice. As we shall see, this statement by no means implies that $\theta = \theta_l$, but only that it is a *function* of θ_l. Fortunately it is very simple to generalize our previous treatment to include the lattice. As a matter of fact, in *Redfield*'s famous paper he considered just this case, but we have put off consideration of the lattice relaxation until this stage to simplify the discussion.

In general, when the spin system exchanges energy with the lattice, the internal equilibrium of the spin system in the rotating frame is momentarily disturbed. The basic assumption we make is that the cross-relaxation between the dipolar and Zeeman systems is so rapid compared to T_1 that, following an exchange of energy of the spins with the lattice, a new spin temperature is rapidly established. Thus, the lattice always finds the spin system described by a temperature in the rotating frame. We consider that there are three basic relaxation equations for the classical magnetizations M_x[1] and M_z and for the expectation value of the dipolar energy $\langle \mathcal{H}_d^0 \rangle$, which we write down phenomenologically in a form chosen to assure that in the absence of H_1 the spin system would reach thermal equilibrim with the lattice.

$$\frac{\partial M_z}{\partial t} = \frac{M_0 - M_z}{T_a} \tag{6.85}$$

$$\frac{\partial M_x}{\partial t} = -\frac{M_x}{T_b} \tag{6.86}$$

$$\frac{\partial \langle \mathcal{H}_d^0 \rangle}{\partial t} = \frac{\langle \mathcal{H}_d^0 \rangle_l - \langle \mathcal{H}_d^0 \rangle}{T_c} \tag{6.87}$$

where T_a and T_b and T_c are relaxation times corresponding to exchange of energy with the lattice, and where $\langle \mathcal{H}_d^0 \rangle_l$ is the value of $\langle \mathcal{H}_d^0 \rangle$ when the spin temperature is equal to the lattice temperature. We have used partial derivative signs in these equations to emphasize that they represent the changes in these quantities induced by lattice coupling only. Thus, although T_a is the usual T_1, T_b is *not* the usual T_2 ($\approx 1/\gamma H_L$). T_b is generally of order T_1, a much longer time. A good analogy to these equations is to think of a Boltzmann equation in

---

[1] We do not worry about M_y since its relaxation does not change the energy in the rotating frame (in essence we assume $M_y = 0$).

statistical mechanics for which there are a number of collision terms, each of which changes the distribution function f and for each one of which one could compute $\partial f/\partial t$. Thus, if the Zeeman energy $-\boldsymbol{M}\cdot\boldsymbol{H}_e$ and the dipolar energy $\langle\mathcal{H}_d^0\rangle$ were not in thermal equilibrium with one another, the magnetization would grow (or shrink) until equilibrium was reached. That process clearly produces a contribution to the three time derivatives (6.85–87) which we have not included in the equations. The lattice coupling, represented by T_a, T_b, and T_c, must be expected to push the spin system out of thermal equilibrium in the rotating frame, but the internal couplings of the spin system counteract that trend by producing energy exchanges within the spin system. Changes within the spin system must conserve the total energy of the spins. If then we consider the rate of change of the expectation value of energy, $\overline{E}$, we can ignore any changes which simply redistribute energy within the spin system, and keep only changes associated with the lattice.

Using the fact that

$$\overline{E} = -M_z h_0 - M_x H_1 + \langle\mathcal{H}_d^0\rangle \tag{6.88}$$

we can find the rate at which the lattice coupling changes $\overline{E}$ by taking the time derivative of (6.88):

$$\frac{d\overline{E}}{dt} = -h_0\frac{\partial M_z}{\partial t} - H_1\frac{\partial M_x}{\partial t} + \frac{\partial}{\partial t}\langle\mathcal{H}_d^0\rangle. \tag{6.89}$$

The only contributions to the time derivatives we need consider are those given by (6.85), (6.86), and (6.87) since other contributions do not change the total energy $\overline{E}$. Employing (6.85–87), assuming that M always lies along H_e, and using (6.58) and (6.59), we readily find an equation for the magnitude of M

$$\frac{dM}{dt} = \frac{1}{T_{1\varrho}}(M_{\text{eq}} - M) \tag{6.90a}$$

and for the spin temperature θ

$$\frac{d}{dt}\left(\frac{1}{\theta}\right) = \frac{1}{T_{1\varrho}}\left[\left(\frac{1}{\theta_{\text{eq}}}\right) - \frac{1}{\theta}\right] \tag{6.90b}$$

where we have introduced a notation $T_{1\varrho}$ and where

$$M_{\text{eq}} = \frac{M_0 H_{\text{eff}}(h_0/T_a)}{(h_0^2/T_a) + (H_1^2/T_b) + (H_L^2/T_c)} \tag{6.91a}$$

$$\frac{1}{T_{1\varrho}} = \frac{1}{h_0^2 + H_1^2 + H_L^2}\left(\frac{h_0^2}{T_a} + \frac{H_1^2}{T_b} + \frac{H_L^2}{T_c}\right). \tag{6.91b}$$

[We have here neglected the term $\langle\mathcal{H}_d^0\rangle_l$ of (6.87)].

Note in particular that $M_{\text{eq}} = 0$ exactly at resonance, is positive when $h_0 > 0$ (i.e., M is parallel to H_e), but is negative for $h_0 < 0$ (that is, M is antiparallel to H_e). The last case corresponds to a negative spin temperature in the rotating frame. Equation (6.91a) shows that the equilibrium θ is far from the lattice temperature θ_l and may even be of the opposite sign. Since M_0 is inversely proportional to θ_l, it is still true that θ_l determines θ, even though they are quite different. The negative temperature one sometimes finds is a simple manifestation of the fact that M_z always tends to be positive whether h_0 is positive or negative. In fact, one can say that the equilibrium is reached as follows: The strong internal coupling of the spin system (which guarantees a spin temperature) keeps M along H_e, since Curie's law is a vector law. The lattice is attempting (a) to make the z-component of M be M_0, but (b) the x-component be zero. (a) would make M bigger than M_0 so that its projection on the z-axis is M_0, whereas (b) would make M be zero. The lattice is thus fighting itself since (a) and (b) are inconsistent. The equilibrium value of (6.91a) results.

6.10 Spin Locking, $T_{1\varrho}$, and Slow Motion

As discussed above, and shown in Fig. 6.7, a spin temperature in the rotating frame is established in a time τ from an arbitrary initial condition without exchange of energy with the lattice. But this is only a quasi-equilibrium value since, over the subsequent time $T_{1\varrho}$, the spin temperature changes as energy is exchanged with the lattice to drive M to M_{equil} of (6.91a) (see Fig. 6.7). During this process M lies along H_{eff}. Thus if one starts at resonance having oriented M along H_1, M will not decay in a time characterized by the inverse of the

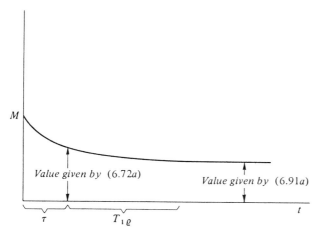

Fig. 6.7. The hierarchy of times following sudden turn-on of H_1. During time τ, a spin temperature is established in the rotating frame, the value of which is determined by the expectation value of the energy, E, immediately after turning on H_1.

Over a longer time $T_{1\varrho}$, the spin temperature in the rotating frame changes, changing M correspondingly in response to energy transfer with the lattice.

Both processes require *Redfield'*s spin temperature hypothesis to analyze since we have assumed $\tau \ll T_{1\varrho}$

line width, but rather with the time $T_{1\varrho}$ which requires energy exchange with the lattice. In principle, by going to low enough temperatures one should be able to make $T_{1\varrho}$ as long as one pleases. If one does this, magnetization along $\boldsymbol{H}_{\mathrm{eff}}$ in the rotating frame, following at time τ to establish a spin temperature, will remain without decay for as long as one has chosen to make $T_{1\varrho}$. This time may be seconds in metals or even hours in insulators.

Even though we can lock the magnetization along $\boldsymbol{H}_{\mathrm{eff}}$ for a time $T_{1\varrho}$ when H_1 is on, if we *remove H_1 suddenly, M will decay to zero in a time of order of the inverse line width*. That is, the Bloch equations with the usual meaning of T_2 give a rough qualitative description of what happens.

In contemplating the Redfield theory, it is helpful to go back to the situation of the laboratory frame without alternating fields. There we see that starting with a magnetized system we can turn H_0 to zero slowly and later turn it back up to its original value. When $H_0 = 0$, $M = 0$, but the full M is recovered when H_0 is turned back on, all without exchange of energy with the lattice. In zero field the order is manifested by the preferential alignment of nuclear moments along the direction of the local fields of their neighbors.

The existence of order in the local fields is the basis of a technique [6.13] for observing motions which are much too slow to be seen as a T_1 minimum or a line-narrowing. Consider a nucleus #1 with a neighbor nucleus #2. Suppose now that #2 makes a sudden jump, as in the process of diffusion. The duration of the jump is perhaps 10^{-12} to 10^{-13}, very fast compared to nuclear precession frequencies. Thus the local field at #1 arising from #2 changes suddenly in both magnitude and direction. The orientation of the spin of nucleus #1 is thus somewhat randomized relative to the local field. If the mean time a given nucleus sits between jumps is τ_m, the alignment of nuclei in the local fields of neighbors, that is the ordered state, can persist only a time of τ_m. Thus to carry out a full demagnetization and remagnetization cycle with full recovery of the initial magnetization, we must remain in zero field for a time less than τ_m. Of course even were there *no* jumping, any T_1 process would change the entropy of the spin system, so that in any event we must complete the cycle in a time less than T_1. We can conclude

i) The demagnetization-remagnetization cycle can be used to monitor the zero field T_1.
ii) We can detect jumping when $\tau_\mathrm{m} < T_1$.

We can apply these same concepts to adiabatic demagnetization experiments performed in the rotating reference frame, as was done by *Slichter* and *Ailion* [6.23, 24] and by *Look* and *Lowe* [6.25]. The analysis is quite straight forward. Since jumping causes sudden changes in the dipolar energy, it gives a contribution to $1/T_\mathrm{c}$ which is

$$\frac{1}{T_{\mathrm{c\ jumping}}} = \frac{2}{\tau_\mathrm{m}}(1-p) \tag{6.92}$$

where p is a quantity expressing the fact that the local field after a jump is not completely random relative to its value before the jump. It can be calculated. Its value depends on the jump mechanism (vacancy, interstitial, etc.). The 2 arises because the local field is a coupling between pairs of nuclei, either of which can jump.

Thus, simple measurements of $T_{1\varrho}$ enable one to measure atomic motion or molecular reorientation when τ_m is comparable to $T_{1\varrho}$ arising from other causes.

7. Double Resonance

7.1 What Is Double Resonance and Why Do It?

One of the most important developments beyond the original concept of magnetic resonance is so-called double resonance in which, as the name suggests, one excites one resonant transition of a system while simultaneously monitoring a different transition. There are many reasons for doing double resonance. The goals include polarizing nuclei, enhancing sensitivity, simplifying spectra, unravelling complex spectra, and generating coherent radiation (e.g. masers and lasers). The field has been characterized by a rich inventiveness which seems to continue unabated to the present day, and which makes the task of understanding all that has been done rather overwhelming. It is useful, therefore, to find some means of classifying the work into broad areas which involve related concepts. We shall employ three broad categories.

The first category of double resonance makes use of spin-lattice relaxation mechanisms. We call it the *Pound-Overhauser double resonance* after two of the important pioneers. The method involves a family of energy levels whose populations are ordinarily held in thermal equilibrium by thermal relaxation processes. If one saturates one of the transitions (1.34) one so-to-speak clamps their populations together (i.e. forces them to be equal). The thermal relaxation processes then repopulate all the levels, producing unusual population differences which may possess useful properties (for example, the upper of two energy levels may have a *larger* population than the lower, or a population difference which is normally small may become large). Included in this category are methods of dynamic polarization of nuclei (the Overhauser effect and solid effect), electron-nuclear double resonance (ENDOR), and masers and lasers.

A second category depends on cross-relaxation phenomena, hence we call it *cross-relaxation double resonance*. The fundamental concept is that if two spin systems can exchange energy (i.e. cross-relax), one can detect the absorption of energy by a resonant alternating field tuned to one spin system by use of a second alternating field to monitor the temperature of a second spin system. Various experiments involving cycling the "static" magnetic fall in this category, as does the Hartmann-Hahn method which is the basis for the technique referred to today as "CP" (meaning "cross polarization") introduced by Pines, Gibby, and Waugh for enhancing the sensitivity of C^{13} spectroscopy. In general these techniques require that some appropriate cross-relaxation process exist and that it be much faster than certain spin-lattice relaxation processes which would otherwise destroy the effect.

The third category depends in general on the existence of spin-spin couplings which in many cases must not be unduly obscured by either spin-lattice relaxation or cross-relaxation. We shall therefore call it *spin coherence double resonance* because it depends on the ability of spins to precess coherently for a sufficient time to reveal the spin-spin splittings. Typically, one here makes use of the fact that when two nuclei are coupled, changing the spin orientation of one nucleus changes the precession frequency of the nuclei to which it is coupled, so that the second nucleus can reveal in this way when the first nucleus is being subjected to a resonant alternating magnetic field.

Among the examples of this category are topics known as spin decoupling, spin tickling, spin echo double resonance (SEDOR), coherence transfer, and two-dimensional Fourier transform NMR (2D-FT NMR). The last of these is fundamental to many important areas of resonance ranging from NMR imaging to determining the structure of complex biomolecules.

Our purpose in introducing three classifications is pedagogical. Other scientists might pick different classifications. Of course, most schemes of classification are not perfect. For example, the nuclear Overhauser effect leads to effects in 2D-FT NMR which are exceedingly useful and important. Moreover, 2D-FT NMR involving only one nuclear species is not a double resonance experiment (only one oscillator is used) but it can be conceptually viewed as one in which the ability of a large H_1 to excite all the nuclei in a spectrum obviates the necessity of having a separate oscillator for each NMR line. It is not our intention in picking three classifications to imply that there is only one idea involved in each category. Indeed, within the groupings we employ there are numerous important innovations. For example, it was many years after the first spin decoupling experiments were performed that 2D Fourier transform spectroscopy was invented.

7.2 Basic Elements of the Overhauser-Pound Family of Double Resonance

The first double resonance experiment was carried out by *Pound* [7.1] on the Na^{23} nuclear resonance in $NaNO_3$. His aim was to prove that the spin lattice relaxation mechanism was via time-dependent electric field gradients. He computed the thermally induced transition probabilities between the various spin states based on this mechanism. In the $NaNO_3$ crystal, the Na resonance is split by an axially symmetrical electric field gradient into three distinct absorption lines corresponding to transitions between the m values of $\frac{3}{2}$ to $\frac{1}{2}$, $\frac{1}{2}$ to $-\frac{1}{2}$, and $-\frac{1}{2}$ to $-\frac{3}{2}$. He predicted and observed that saturating any one transition (e.g., the $\frac{3}{2}$ to $\frac{1}{2}$) would produce level populations of all the states which differ from thermal equilibrium. Thus when he saturated the $\frac{3}{2}$ to $\frac{1}{2}$ transition, the intensity of the $\frac{1}{2}$ to $-\frac{1}{2}$ transition increased by a factor of $\frac{5}{3}$ times its normal intensity. We will discuss below why such intensity changes occur in connection with the Overhauser effect. By means of a series of experiments saturating the various transitions while observing the others, he verified that the spin-lattice relaxation was via electric quadrupole coupling.

The second double resonance experiment performed was by *Carver* [7.2]. *Overhauser,* while still a graduate student at Berkeley, had predicted [7.3] that if one saturated the conduction electron spin resonance in a metal, the nuclear spins would be polarized 1000-fold more strongly than their normal polarization in the absence of electron saturation. (For an excellent summary of dynamic nuclear polarization by one of the important pioneers, see [7.4].) Crudely speaking, *Overhauser* predicted a polarization of the nuclei which they would have if the electron spin Boltzmann factor were used in place of the nuclear spin Boltzmann factor. It may be hard for readers today to appreciate the deep scepticism with which *Overhauser*'s proposal was greeted by the resonance community, because today there are many schemes for dynamic polarization, and the principle forms the basis for many other important techniques. However, in 1953 there was a widespread (though short-lived) belief that *Overhauser*'s scheme must violate the second law of thermodynamics. *Carver*'s experiment was not only the first demonstration of the dynamic polarization of nuclei, but perhaps more importantly showed that *Overhauser*'s revolutionary thought was correct. His concept reoriented the thinking of resonators. While his concept stimulated the subsequent invention of other schemes of dynamic polarization, perhaps more important was the stimulus it provided to exploration of other novel effects of pumping transitions and doing double resonance.

Figure 7.1 shows the first dynamic polarization of nuclei, the Li nuclei in lithium metal. To perform this experiment it is necessary to have an H_1 of 5 to 10 Gauss to saturate the conduction electron spin resonance. One also wants a size of metal particle small compared to the skin depth at the electron frequency so that all the nuclei are polarized by the electron saturation. Accordingly, *Carver* used a solenoid to generate a static field of about 30 Gauss, which put the electron spin resonance at 84 MHz, and the lithium nuclear resonance at a frequency of 50 kHz. At that, low frequency, the nuclear resonance was too weak to be observed directly (Fig. 7.1a), but popped up from the noise instantly when the electron saturating oscillator was turned on. *Carver* calibrated the degree of polarization by observing the proton resonance in mineral oil (Fig. 7.1c). He went on to show that dynamic polarization did not require a metal by working with the proton resonance of liquid ammonia in which sodium had been dissolved. The sodium atoms in such solutions ionize, giving free electrons whose resonance is readily saturated.

The principles underlying the Overhauser effect are in fact identical to those on which *Pound* based his experiment. The Overhauser effect provided a strong impetus to the development of double resonance methods, in part because *Over-*

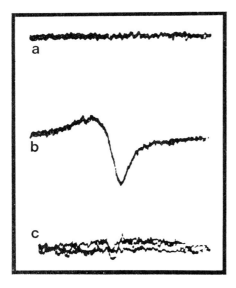

Fig. 7.1. Demonstration of the Overhauser nuclear polarization effect on Li^7 nuclei in metallic lithium by *Carver*. The oscilloscope picture shows nuclear absorption plotted vertically versus magnetic field. The magnetic field excursion is about 0.2 Gauss. The top line shows the normal Li^7 nuclear resonance (lost in noise at the 50 kHz frequency of the NMR apparatus). The middle line shows the Li^7 nuclear resonance enhanced by saturating the electron spin resonance. The experimental conditions of (a) and (b) differ only in turning on the electron spin saturating oscillator. The bottom line shows the proton resonance from a glycerine sample containing eight times as many protons under the same experimental conditions, from which one concludes the Li^7 nuclear polarization was increased by a factor of 100

hauser had made an ingenious and daring prediction which many talented physicists judged to be wrong, and in part because he addressed a topic of interest to a community of scientists much broader than resonators. Although *Pound* did not recognize that if his concept were applied to other systems it could lead to large nuclear polarizations, there can be little doubt that even without *Overhauser'* s contribution, *Pound'* s invention of double resonance would have caused a burgeoning of the field.

7.3 Energy Levels and Transitions of a Model System

To understand the principles of the many applications, one can consider a very simple system consisting of a nucleus with spin $I = \frac{1}{2}$ coupled to an electron of spin $S = \frac{1}{2}$, acted on by an external static magnetic field H_0. The Hamiltonian for this system is

$$\mathcal{H} = \gamma_e \hbar H_0 S_z + A \mathbf{I} \cdot \mathbf{S} - \gamma_n \hbar H_0 I_z \tag{7.1}$$

where we have used subscripts e and n to denote electrons and nuclei, and where we have taken the form of electron-nuclear coupling appropriate for s-states.

We assume that $\gamma_e \hbar H_0 \gg A$ (the strong field approximation). Of course $\gamma_e \gg |\gamma_n|$. These assumptions make S_z nearly commute with $\mathcal{H}$; hence m_S, the eigenvalue of S_z, is a good quantum number. Only the term $A I_z S_z$ of the electron-nuclear coupling gives diagonal terms, so the Hamiltonian is effectively

$$\mathcal{H} = \gamma_e \hbar H_0 S_z + A I_z S_z - \gamma_n \hbar H_0 I_z \quad . \tag{7.2}$$

Hence we take m_I, the eigenvalues of I_z, as another good quantum number. The energy eigenvalues are then

$$E = \gamma_e \hbar H_0 m_S + A m_I m_S - \gamma_n \hbar H_0 m_I$$
$$m_S = \pm \tfrac{1}{2}, \quad m_I = \pm \tfrac{1}{2} \ . \tag{7.3}$$

It is convenient to label states and wave functions by a convention

$$\psi = |\varepsilon \mu\rangle \tag{7.4}$$

where $\varepsilon = 2m_S$, $\mu = 2m_I$. A state with $m_S = +\tfrac{1}{2}$, $m_I = -\tfrac{1}{2}$ is written as $|+ -\rangle$. The selection rule for transitions induced by an applied alternating field is $\Delta m_S = \pm 1$, $\Delta m_I = 0$, or $\Delta m_S = 0$, $\Delta m_I = \pm 1$. The first corresponds to an electron spin resonance, the second to nuclear resonance. (See Sect. 11.3 for a detailed discussion).

Their resonance frequencies, ω_e and ω_n, are

$$\omega_e = \gamma_e H_0 + \frac{A}{\hbar} m_I \tag{7.5a}$$

$$\omega_n = \gamma_n H_0 - \frac{A}{\hbar} m_S \ . \tag{7.5b}$$

There are thus four allowed transitions. They are shown in Fig. 7.2. The relative size of $|A|$ and $(\gamma_n \hbar H)$ influences the appearance of the energy level diagram slightly as is shown in Fig. 7.3. If the nucleus and electron under consideration are far apart, typically $|A| < |\gamma_n \hbar H_0|$ (Fig. 7.3a). If the nucleus and electron are close together, usually $|A| > |\gamma_n \hbar H_0|$ (Fig. 7.3b). The former case is encountered for a typical nucleus in a solid which has a low concentration of paramagnetic centers. The latter case is encountered for nuclei of paramagnetic atoms, or for nuclei which are very close neighbors of a paramagnetic center.

For the rest of our discussion we adopt the convention of drawing figures which look like Fig. 7.2a, though we do not mean to imply thereby anything about the relative size of $|A|$ and $|\gamma_n \hbar H_0|$.

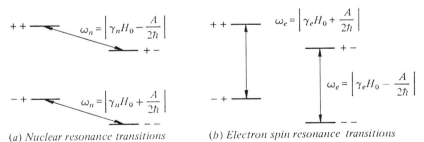

(a) Nuclear resonance transitions   (b) Electron spin resonance transitions

Fig. 7.2a,b. The energy level diagram and allowed transitions of a system consisting of an electron of spin $S = \tfrac{1}{2}$ coupled to a nucleus of spin $I = \tfrac{1}{2}$ acted on by an external static magnetic field H_0. The figure assumes γ_n is negative

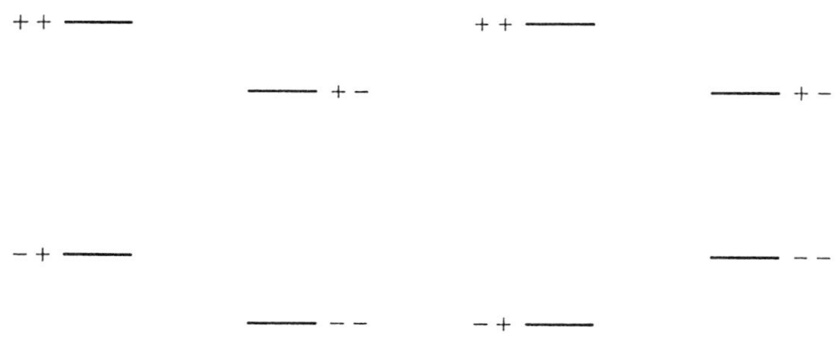

```
 + +  ─────                         + +  ─────


              ─────  + −                           ─────  + −




 − +  ─────                                                ─────  − −

                    ─────  − −
                                   − +  ─────
```
(a) *Distant nucleus*   (b) *Nearby nucleus*

Fig. 7.3a,b. The effect of the size of the electron-nuclear coupling $A/\hbar$ relative to the nuclear resonance frequency ω_n on the appearance of the energy level diagrams. For nuclei far from the electron, usually $|A| < |\gamma_n \hbar H_0|$. For nuclei near to the electron, frequently $|A| > |\gamma_n \hbar H_0|$. The figure assumes a negative γ_n and A positive. What would it look like for positive γ_n?

We have assumed a particularly simple form of electron spin-nuclear spin coupling, that which arises from the Fermi contact hyperfine expression. The most general form of spin-spin coupling would be obtained by adding also the dipolar coupling of Sect. 3.2 between the electron spin and the nuclear spin. Then, as is discussed in Chapter 11, the Hamiltonian would be

$$\mathcal{H} = \gamma_e \hbar H_0 S_z + A_{x'x'} S_{x'} I_{x'} + A_{y'y'} I_{y'} S_{y'} + A_{zz'} I_{z'} S_{z'} - \gamma_n \hbar H_0 I_n \quad (7.6)$$

where the axes x', y', z' are a set of principal axes.

Solution of the more general Hamiltonian of (7.6) still gives an energy level diagram which looks much like either Fig. 7.2a or 7.2b as long as $\gamma_e \hbar H_0 \gg |A_{x'x'}|$, $|A_{y'y'}|$, and $|A_{z'z'}|$. In this approximation, m_S, the eigenvalue of S_z, is still a good number, but m_I, the eigenvalue of I_z, is not necessarily so. Thus the lowest order wave functions ψ_i (where i distinguishes the four states) may not be

$$\psi_i = |m_S m_I) \quad (7.7)$$

but may instead be linear combinations of such states.

$$\psi_i = \sum_{m_S, m_I} c_{i m_S m_I} |m_S m_I) \quad . \quad (7.8)$$

If we keep the Hamiltonian (7.2) but solve it more exactly, we find that the states are of the form of (7.8) rather than of (7.7). Since there are still only four energy levels, the notation $|\varepsilon \mu)$ [i.e. $|++\rangle$, $|-+\rangle$, $|+-\rangle$, $|--\rangle$] still has validity since it suffices to distinguish four levels.

However, while

$$S_z |\varepsilon \mu) \cong \tfrac{1}{2} \varepsilon |\varepsilon \mu) \quad (7.9)$$

in general

$$I_z |\varepsilon \mu) \neq \tfrac{1}{2} \mu |\varepsilon \mu) \quad . \quad (7.10)$$

The result of using the more general form of function of (7.8), whether as a result of solving the simpler Hamiltonian of (7.1) more exactly, or as a result of solving the more general (7.7), is that, on application of an applied alternating magnetic field, transitions other than those shown in Fig. 7.2 become possible. We adopt the convention of calling transitions other than the four in Fig. 7.2 "forbidden transitions".

In the absence of applied alternating fields, populations of the energy levels of the combined spin system are given by the Boltzmann factors when the system is in thermal equilibrium. As discussed in Chapter 1, the achievement of thermal equilibrium can be thought of as resulting from transitions induced by the coupling to a thermal reservoir in which the thermally induced transition probability $W_{\varepsilon\eta,\varepsilon'\eta'}$ from state $|\varepsilon\eta\rangle$ to state $|\varepsilon'\eta'\rangle$ is related to the rate $W_{\varepsilon'\eta',\varepsilon\eta}$ by

$$\frac{W_{\varepsilon\eta,\varepsilon'\eta'}}{W_{\varepsilon'\eta',\varepsilon\eta}} = \frac{p_{\varepsilon'\eta'}}{p_{\varepsilon\eta}} \qquad (7.11)$$

where $p_{\varepsilon\eta}$ is the thermal equilibrium probability of occupation of state $|\varepsilon\eta\rangle$. Or, using the Boltzmann relation,

$$\frac{W_{\varepsilon\eta,\varepsilon'\eta'}}{W_{\varepsilon'\eta',\varepsilon\eta}} = \exp\left[(E_{\varepsilon\eta} - E_{\varepsilon'\eta'})/kT\right] \quad . \qquad (7.12)$$

It becomes convenient at this point to switch to a more compact notation. Since there are only four states, we label them 1, 2, 3, or 4 as shown in Fig. 7.4. The notation W_{ij} then corresponds to the thermally induced transition rate from state i to state j. Transitions W_{ij} in which the electron spin is flipped but not the nuclear spin are shown in Fig. 7.5.

As an aid to memory, transitions between 1 and 2 or between 3 and 4 are electron transitions, transitions between 1 and 3 or 2 and 4 are nuclear transitions. Also, level 2 lies below level 1, level 4 lies below level 3 in Fig. 7.4.

$\psi_1 = |++\rangle$ ———

——— $\psi_3 = |+-\rangle$

$\psi_2 = |-+\rangle$ ———

——— $\psi_4 = |--\rangle$

Fig. 7.4. Definition of the four states 1, 2, 3, and 4 in terms of the earlier notation $|\varepsilon\eta\rangle$

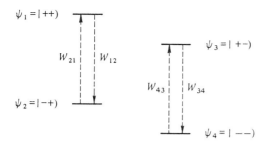

Fig. 7.5. Some thermally induced flipping rates which involve the transition of the electron spin. The direction of the arrow indicates the direction of the transition corresponding to the rates W_{21}, W_{12}, etc.

7.4 The Overhauser Effect

Though *Overhauser*'s original proposal pertained to polarization of nuclei in a metal, the principle can be illustrated by considering the model system discussed in the previous section. We take the simplest Hamiltonian, (7.1), with the solutions of (7.3).

We assume that the principal relaxation mechanisms are those shown in Fig. 7.6 which involve electron spin relaxation (W_{12}, W_{21}, W_{34}, W_{43}) and a combined nucleus-electron spin flip (W_{23}, W_{32}) such as one obtains from the Fermi contact interaction in a metal as explained in Sect. 5.3. An applied alternating field induces transitions of the electrons between levels 1 ($|++\rangle$) and 2 ($|-+\rangle$) at a rate W_e. W_e corresponds to an electron spin resonance. The probabilities of occupation of levels i, p_i, then obey the following differential equations similar to (5.13):

$$\frac{dp_1}{dt} = p_2 W_{21} - p_1 W_{12} + (p_2 - p_1) W_e \tag{7.13a}$$

$$\frac{dp_2}{dt} = p_1 W_{12} - p_2 W_{21} + p_3 W_{32} - p_2 W_{23} + (p_1 - p_2) W_e \tag{7.13b}$$

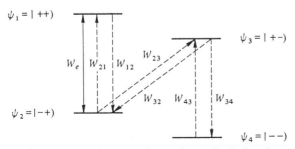

Fig. 7.6. The Overhauser effect. Thermally induced transitions W_{ij} shown attempt to maintain thermal equilibrium. An applied alternating field induces electron spin transitions at a rate W_e between the states 1 and 2

$$\frac{dp_3}{dt} = p_2 W_{23} - p_3 W_{32} + p_4 W_{43} - p_3 W_{34} \tag{7.13c}$$

$$\frac{dp_4}{dt} = p_3 W_{34} - p_4 W_{43} \quad . \tag{7.13d}$$

Since the probabilities of occupation must add to one, we have that

$$p_1 + p_2 + p_3 + p_4 = 1 \quad . \tag{7.14}$$

We seek a steady-state solution, so we set the left side of (7.13a–d) zero. We also assume W_e can be made sufficiently large to produce complete saturation (i.e., we take the limit as W_e goes to infinity). Therefore

$$p_1 = p_2 \quad . \tag{7.15}$$

Equation (7.13d) then gives us

$$p_3 = p_4 \frac{W_{43}}{W_{34}} \tag{7.16}$$

which is the normal thermal equilibrium population ratio for this pair of states.

Use of (7.13d) in (7.13c) gives

$$p_3 = p_2 \frac{W_{23}}{W_{32}} \tag{7.17}$$

which is the normal thermal equilibrium population of states 2 and 3.

Thus for the family of transitions shown, while the saturation will change all the populations, it only affects the population *ratio* of the pair of states between which W_e acts. This result is a special property of the assumption that there are no W_{ij}'s which *also* couple state 1 to states 3 or 4.

Since the ratios p_3/p_4 and p_2/p_3 are both thermal equilibrium, it follows that the ratio of p_4 to p_2 is also thermal equilibrium.

For a pair of levels in thermal equilibrium

$$p_j = p_i e^{(E_i - E_j)/kT} \tag{7.18a}$$

$$\equiv p_i B_{ij} \tag{7.18b}$$

which defines the quantity B_{ij}, the Boltzmann ratio (whence the symbol B). Note the convention on the order of the symbols E_i, E_j of (7.18a) and the subscripts i, j on B_{ij}. We therefore have

$$p_1 = p_2 \quad , \quad p_3 = p_2 B_{23} \quad , \quad p_4 = p_2 B_{24} \quad . \tag{7.19}$$

Whence

$$p_1 = p_2 = \frac{1}{2 + B_{23} + B_{24}} \quad , \quad p_3 = \frac{B_{23}}{2 + B_{23} + B_{24}} \quad ,$$

$$p_4 = \frac{B_{24}}{2 + B_{23} + B_{24}} \quad . \tag{7.20}$$

We have stated that the Overhauser effect produces a nuclear polarization. Let us compute the nuclear polarization.

In a state $|\varepsilon\eta\rangle$, the nuclear spin expectation value is $\eta/2$ ($\eta = +1$ or -1). Therefore, the average expectation value of nuclear spin I_z, $\langle I_z \rangle$ is

$$\langle I_z \rangle = \sum_i p_i \langle i | I_z | i \rangle \tag{7.21a}$$

$$= \tfrac{1}{2}(p_1 + p_2 - p_3 - p_4) \tag{7.21b}$$

$$= \frac{1}{2}\frac{2 - B_{23} - B_{24}}{2 + B_{23} + B_{24}} . \tag{7.21c}$$

To appreciate the significance of this expression, we evaluate it in the high temperature approximation:

$$B_{ij} \cong 1 + \frac{E_i - E_j}{kT}$$

$$\langle I_z \rangle = \frac{1}{2}\frac{(E_3 - E_2) + (E_4 - E_2)}{4kT} . \tag{7.22a}$$

Now $E_3 - E_2 = \gamma_e \hbar H_0 + \gamma_n \hbar H_0$

$$E_4 - E_2 = \frac{A}{2} + \gamma_n \hbar H_0 \quad \text{so that} \tag{7.22b}$$

$$\langle I_z \rangle = \frac{1}{2}\frac{\gamma_e \hbar H_0 + (A/2) + 2\gamma_n \hbar H_0}{4kT} . \tag{7.23a}$$

Using the approximation that $\gamma_e \hbar H_0$ is much the largest term, we get

$$\langle I_z \rangle \cong \frac{1}{2}\frac{\gamma_e \hbar H_0}{4kT} . \tag{7.23b}$$

If the electron spin resonance were not being saturated, it is easy to show that the thermal equilibrium $\langle I_z \rangle$

$$\langle I_z \rangle_{\text{therm}} = \frac{1}{2}\frac{B_{21} + 1 - B_{23} - B_{24}}{B_{21} + 1 + B_{23} + B_{24}} \tag{7.24}$$

which, in the high temperature approximation, gives

$$\langle I_z \rangle_{\text{therm}} = \frac{1}{2}\frac{(E_2 - E_1) + (E_3 - E_2) + (E_4 - E_2)}{4kT} . \tag{7.25}$$

Since $E_2 - E_1 = -\gamma_e \hbar H_0 - A/2$, this gives

$$\langle I_z \rangle_{\text{therm}} = \frac{1}{2}\frac{\gamma_n \hbar H_0}{2kT} . \tag{7.26}$$

This expression is the same as would be found for the nucleus if acted upon by the external field only.[1]

Comparing (7.23b) with (7.26), we see that electron saturation has increased $\langle I_z \rangle$ by the ratio

$$\frac{\langle I_z \rangle}{\langle I_z \rangle_{\text{therm}}} = \frac{\gamma_e}{2\gamma_n} \qquad (7.27)$$

which means the nucleus is polarized as though its magnetic moment were comparable to the much larger moment of an electron! If both electron transitions are saturated simultaneously

$$\frac{\langle I_z \rangle}{\langle I_z \rangle_{\text{therm}}} = \frac{\gamma_e}{\gamma_n} \quad . \qquad (7.28)$$

Equation (7.28) is the result *Overhauser* originally predicted for a metal. In a metal, one electron couples to many nuclei, so there is only a single electron spin resonance, not a resolved pair as in our example.

7.5 The Overhauser Effect in Liquids: The Nuclear Overhauser Effect

The essential feature of the Overhauser polarization of nuclei is that there be a dominant nuclear relaxation process which requires a simultaneous nuclear spin flip and electron spin flip. In metals, the strong role of the Fermi contact term leads to relaxation processes in which the nuclear and the electron spins flip in opposite directions through terms such as I^+S^- or I^-S^+. Even with a conventional dipole-dipole coupling there are terms [the terms B, E, and F of (3.7)] which have this property of correlated flips if they can lead to relaxation. In liquids, the translational and molecular rotational degrees of freedom introduce a time dependence to the dipolar coupling which enables these terms to produce relaxation. Therefore, paramagnetic ions in liquids should lead to an Overhauser effect. As we have remarked, *Carver* and *Slichter* [7.2] demonstrated this result using solutions of Na dissolved in liquid ammonia. The Na ions give up their valence electron, producing isolated electron spins which relax the H nuclei of

---

[1] One is at first surprised that the coupling to the electron does not come into $\langle I_z \rangle_{\text{therm}}$ since one can view the energy levels as though there were an effective magnetic field H_{eff} acting on the nucleus given by

$$H_{\text{eff}} = H_0 - \frac{Am_S}{\gamma_n \hbar} \quad .$$

Since m_S occurs with virtually equal probability as $+\frac{1}{2}$ and $-\frac{1}{2}$, and since in the high temperature approximation the magnetization is proportional to the effective field, the term $Am_S/\gamma_n \hbar$ averages to zero over an ensemble of electron-nuclear systems.

the ammonia. A similar effect should arise with two nuclei, as was recognized by *Holcomb* [7.5], *Bloch* [7.6], and *Solomon* [7.7], and was demonstrated by *Solomon* in a pioneering double resonance experiment in HF utilizing the H^1 and F^{19} resonances.

In Sect. 3.4 we saw that the dipolar coupling between nuclei in a solid could be used to obtain information about structure of molecules or solids. In liquids the dipolar coupling is averaged out owing to the rapid translational and rotational motion. Nevertheless, the dipolar coupling still plays an important role in determining the relaxation times. It has turned out that the nuclear Overhauser effect (both static and transient) gives important information about which resonance lines correspond to nuclei that are physically close to one another in a molecule. It is thus playing an important role in structure determinations of complex molecules, especially when combined with two-dimensional Fourier transform methods. We therefore seek to understand the Overhauser effect in liquids in greater detail.

In order to see more clearly how the liquid degrees of freedom make an Overhauser effect possible, we examine a concrete example, the nuclear Overhauser effect of a molecule containing two spin $\frac{1}{2}$ nuclei. An example is HF, the system studied by *Solomon*. We shall demonstrate that saturating the resonance of one nucleus (the S species) produces a polarization of the other species (the I spins) and then show that if one disturbs one species from thermal equilibrium with a pulse, the effect is revealed in the transient behavior of the other species. It is, then, convenient to write the Hamiltonian as

$$\mathcal{H} = -\gamma_I \hbar H_0 I_z - \gamma_S \hbar H_0 S_z + \mathcal{H}_{12}(t) \quad , \tag{7.29}$$

where $\mathcal{H}_{12}(t)$ is the dipolar coupling between the two nuclei, augmented, perhaps, by any pseudoexchange or pseudodipolar coupling (Sect. 4.9), and where the explicit inclusion of t emphasizes that in liquids the dipolar coupling is in general time dependent since the radius vector from one nucleus to its neighbor within the molecule is continually changing direction. There is also coupling of the nuclei in one molecule with those in another. We omit these effects since we are concerned with demonstrating principles. Our calculation would therefore apply rigorously only to a case in which the molecules are present in low concentration in a liquid which does not otherwise contain nuclear spins. Since the Hamiltonian commutes with both I_z and S_z, we can label its eigenstates by m_I and m_S, the eigenvalues of I_z and S_z.

The energy levels then become

$$E = -\gamma_I \hbar H_0 m_I - \gamma_S \hbar H_0 m_S + A m_I m_S \quad , \tag{7.30}$$

where A is the pseudoexchange coefficient, the only term of $\mathcal{H}_{12}$ which does not average to zero under the molecular tumbling. If $A \neq 0$, the resonances of the I-nuclei or of the S-nuclei are doublets. We shall for simplicity assume that A is zero, so that both the I resonance and the S resonance consist of single resonance lines. Figure 7.7 shows the energy levels, labeled by $m_I m_S$, and also

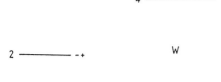

Fig. 7.7. The energy levels $|m_S m_I\rangle$ of a pair of nuclei (positive γ's). An applied alternating field induces the transition 1 to 2 and 3 to 4 in which m_I remains fixed, but m_S changes. The transition rate W is defined in the text

numbered 1, 2, 3, and 4. We assume that we are irradiating the S transition, producing a transition probability per second of W between states 1 and 2 and between 3 and 4. To find what this does to the I-spin polarization, we calculate $\langle I_z \rangle$ given by

$$\langle I_z \rangle = \sum_{i=1}^{4} \langle i|I_z|i\rangle p_i = \tfrac{1}{2}(p_1 + p_2 - p_3 - p_4) \quad . \tag{7.31a}$$

In a similar way

$$\langle S_z \rangle = \tfrac{1}{2}(p_1 + p_3 - p_2 - p_4) \quad . \tag{7.31b}$$

To find the p_i's, we must solve the four rate equations for the populations p_i ($i = 1$ to 4)

$$\frac{dp_4}{dt} = p_1 W_{14} + p_2 W_{24} + p_3 W_{34} - p_4(W_{41} + W_{42} + W_{43})$$
$$+ (p_3 - p_4)W \tag{7.32a}$$

$$\frac{dp_3}{dt} = p_1 W_{13} + p_2 W_{23} + p_4 W_{43} - p_3(W_{31} + W_{32} + W_{34})$$
$$+ (p_4 - p_3)W \quad , \tag{7.32b}$$

etc. Adding these two equations, gives

$$\frac{d(p_3 + p_4)}{dt} = p_1(W_{14} + W_{13}) + p_2(W_{24} + W_{23}) - p_3(W_{31} + W_{32})$$
$$- p_4(W_{41} + W_{42}) \quad . \tag{7.33}$$

In the steady state (where all dp_i/dt's vanish), (7.32) shows that as W is increased, p_3 and p_4 become progressively more nearly equal. In a similar manner, p_1 and p_2 become progressively more nearly equal. We therefore set $p_1 = p_2$ and $p_3 = p_4$, obtaining from (7.33)

$$p_4 = p_1 \left(\frac{W_{14} + W_{13} + W_{24} + W_{23}}{W_{41} + W_{31} + W_{42} + W_{32}} \right) = p_1 \frac{F}{G} \quad , \tag{7.34}$$

which defines F and G.

Now, utilizing (7.12) and (7.18)

$$W_{14} = W_{41}B_{14} = W_{41}\exp\left[(E_1 - E_4)/kT\right] \tag{7.35}$$

and so on for W_{13}, W_{24}, and W_{23}. Since $\sum p_i = 1$, $p_1 = p_2$, $p_3 = p_4$, and $p_4 = p_1(F/G)$, we get

$$p_1 + p_4 = \tfrac{1}{2} = p_1(1 + F/G) \quad , \quad \text{hence} \tag{7.36a}$$

$$p_1 = \frac{1}{2(1 + F/G)} \quad . \tag{7.36b}$$

Therefore

$$p_1 - p_4 = \frac{1}{2}\left(\frac{1 - (F/G)}{1 + (F/G)}\right) \quad . \tag{7.37}$$

Utilizing $p_1 = p_2$ and $p_3 = p_4$, (7.31) becomes

$$\langle I_z \rangle = (p_1 - p_4) = \frac{1}{2}\frac{G - F}{G + F} \quad . \tag{7.38}$$

Utilizing (7.34) and (7.35) we can get an exact expression for F and G. However, in the case of a liquid, the high temperature approximation is valid even for electrons (perhaps liquid He is an exception!) so that

$$B_{ij} = 1 + (E_i - E_j)/kT \tag{7.39}$$

giving

$$\langle I_z \rangle = \frac{1}{2kT}\left\{\frac{W_{41}(E_4 - E_1) + W_{31}(E_3 - E_1)}{2(W_{41} + W_{31} + W_{42} + W_{32})}\right.$$
$$\left. + \frac{W_{42}(E_4 - E_2) + W_{32}(E_3 - E_2)}{2(W_{41} + W_{31} + W_{42} + W_{32})}\right\} \quad . \tag{7.40}$$

Examination of Fig. 7.7 shows that W_{41} is the transition from $|--\rangle$ to $|++\rangle$ in which both spins flip up, W_{31} is the transition from $|+-\rangle$ to $|++\rangle$ in which the I-spin flips up, W_{42} is the transition from $|--\rangle$ to $|-+\rangle$, also a transition in which the I-spin flips up. Finally W_{32} is the transition from $|+-\rangle$ to $|-+\rangle$ in which the S-spin flips down and the I-spin flips up.

We shall define a new notation which makes these processes more explicit by defining U_M where M is the total change in $m_I + m_S$ in the transition:

$$U_0 \equiv W_{32} \quad , \quad U_1 \equiv W_{42} \quad , \quad U_2 \equiv W_{41} \quad . \tag{7.41}$$

As we shall see, $W_{42} = W_{31}$, so that

$$U_1(\omega) = W_{31}(\omega) \quad . \tag{7.42}$$

To calculate these transition rates we need several equations. From (5.281)

$$W_{km} = \frac{1}{\hbar^2} \int_{-\infty}^{+\infty} G_{mk}(\tau) \exp[-i(m-k)\tau d\tau] \quad \text{with}$$

$$G_{mk}(\tau) \equiv \overline{(m|\mathcal{H}_1(t)|k)(k|\mathcal{H}_1(t+\tau)|m)} \quad .$$

Taking

$$G_{mk}(\tau) = G_{mk}(0)\exp(-|\tau|/\tau_c)$$

we get from (5.297)

$$W_{km} = \frac{G_{mk}(0)}{\hbar^2} \frac{2\tau_c}{1+(m-k)^2\tau_c^2} \quad .$$

For dipolar coupling, the $\mathcal{H}_1(t)$'s represent the various terms A, B, C, etc. from (3.7). Defining

$$A_0 = \gamma_I \gamma_S \hbar^2 / r^3 \tag{7.43}$$

and recalling that

$$B = -\tfrac{1}{4}(S^-I^+ + S^+I^-)(1 - 3\cos^2\theta)$$

$$C = -\tfrac{3}{2}(S_z I^+ + S^+ I_z)\sin\theta\cos\theta e^{-i\phi}$$

$$E = -\tfrac{3}{4}(S^+ I^+)\sin^2\theta e^{-2i\phi}$$

we get that

$$U_0 = \frac{A_0^2}{\hbar^2}\left(\frac{1}{16}\right)\langle(1-3\cos^2\theta)^2\rangle_{4\pi}(+ - |I^+S^-| - +)(- + |I^-S^+| + -) \tag{7.44}$$

where $\langle(1-3\cos^2\theta)^2\rangle_{4\pi}$ means the average over the 4π solid angle of $(1 - 3\cos^2\theta)^2$. Thus

$$\langle(1-3\cos^2\theta)^2\rangle_{4\pi} = \frac{1}{4\pi}\int(1-3\cos^2\theta)^2 \sin\theta\, d\theta\, d\phi = 4/5 \quad . \tag{7.45}$$

Likewise

$$\langle \sin^2\theta \cos^2\theta \rangle_{4\pi} = 2/15 \tag{7.46}$$

$$\langle \sin^4\theta \rangle_{4\pi} = 8/15.$$

In this manner we get

$$U_0 = \frac{A_0^2}{\hbar^2}\frac{1}{10}\frac{\tau_c}{1+(\omega_S-\omega_I)^2\tau_c^2} \quad , \quad 2U_1 = \frac{A_0^2}{\hbar^2}\frac{3}{10}\frac{\tau_c}{1+\omega_I^2\tau_c^2} \quad ,$$

$$U_2 = \frac{A_0^2}{\hbar^2}\frac{3}{5}\frac{\tau_c}{1+(\omega_I+\omega_S)^2\tau_c^2} \quad . \tag{7.47}$$

The resulting polarization of the I-spins is thus, using (7.40), (7.41), and (7.47),

$$\langle I_z \rangle = \frac{\hbar}{4kT} \frac{U_2(\omega_I + \omega_S) + 2U_1\omega_I + U_0(\omega_I - \omega_S)}{U_2 + 2U_1 + U_0}$$

$$= \frac{\hbar}{4kT} \frac{\frac{3}{5}\frac{(\omega_I+\omega_S)}{1+(\omega_I+\omega_S)^2\tau_c^2} + \frac{3}{10}\frac{\omega_I}{1+\omega_I^2\tau_c^2} + \frac{1}{10}\frac{(\omega_I-\omega_S)}{1+(\omega_S-\omega_I)^2\tau_c^2}}{\frac{3}{5}\frac{1}{1+(\omega_I+\omega_S)^2\tau_c^2} + \frac{3}{10}\frac{1}{1+\omega_I^2\tau_c^2} + \frac{1}{10}\frac{1}{1+(\omega_S-\omega_I)^2\tau_c^2}} \quad . \quad (7.48)$$

Suppose $|\omega_S| \gg \omega_I$ (as for S being an electron spin, in which case ω_S is negative) and that $\omega_S^2\tau_c^2 \ll 1$. Then

$$\langle I_z \rangle = \frac{\hbar\omega_S}{4kT} \frac{(3/5 - 1/10)}{(3/5 + 1/10)} = \frac{5}{7}\frac{\hbar\omega_S}{4kT} \quad , \quad (7.49)$$

which is *negative* for electrons.

On the other hand, if there were a strong Fermi contact term, so that $U_0 \gg U_1$ or U_2, as with the conventional Overhauser effect,

$$\langle I_z \rangle = -\frac{\hbar\omega_S}{4kT} \quad , \quad (7.50)$$

which for negative ω_S is positive. Therefore, dipolar relaxation produces an Overhauser effect of the opposite sign to the conventional Overhauser effect since for dipolar coupling $U_2 \gg U_0$.

We have carried out the solution for the steady-state polarization of the I-spins. If one is doing pulsed experiments, as is frequently the case for double resonance experiments or for two-dimensional Fourier transform experiments, the pulses disturb the I-S spin system from thermal equilibrium, following which the thermal processes bring the system back to thermal equilibrium. One then speaks of a transient nuclear Overhauser effect. Thus, in general, one wishes to find the time dependence of the observables $\langle I_z(t) \rangle$ or $\langle S_z(t) \rangle$ for some sort of initial conditions. *Solomon,* in [7.7], calculates the transient response and demonstrates what happens in a set of classic experiments. To do the calculation, one needs to solve for the time dependence of the p_i's, starting with equations such as (7.32) with $W = 0$. Thus we have

$$\frac{dp_4}{dt} = p_1 W_{14} + p_2 W_{24} + p_3 W_{34} - p_4(W_{41} + W_{42} + W_{43}) \quad . \quad (7.51)$$

Writing $W_{14} = W_{41}B_{14}$ and taking

$$B_{14} = 1 + \varepsilon_{14} \quad \text{where} \quad (7.52)$$

$$\varepsilon_{14} = \frac{E_1 - E_4}{kT} \quad (7.53)$$

we get

$$\frac{dp_4}{dt} = W_{41}(p_1 - p_4) + W_{42}(p_2 - p_4) + W_{43}(p_3 - p_4)$$
$$+ W_{41}p_1\varepsilon_{14} + W_{42}p_2\varepsilon_{24} + W_{43}p_3\varepsilon_{34} \quad . \quad (7.54)$$

Now, in thermal equilibrium we have from $p_j(T) = p_i(T)B_{ij}$ that

$$p_j(T) - p_i(T) = p_i(T)(B_{ij} - 1) = p_i(T)\varepsilon_{ij} \quad . \tag{7.55}$$

Therefore, in (7.54) we recognize that since $p_i \cong p_i(T) \cong \frac{1}{4}$

$$W_{41}p_1\varepsilon_{14} \cong W_{41}[p_4(T) - p_1(T)] \quad . \tag{7.56}$$

Substituting into (7.54) we get

$$\begin{aligned}\frac{dp_4}{dt} &= W_{41}\{p_1 - p_4 - [p_1(T) - p_4(T)]\} \\ &+ W_{42}\{p_2 - p_4 - [p_2(T) - p_4(T)]\} \\ &+ W_{43}\{p_3 - p_4 - [p_3(T) - p_4(T)]\}\end{aligned} \tag{7.57}$$

with similar equations for dp_i/dt for $i = 1, 2$, and 3.

We saw in Chap. 1 that the fact that $W_{ij} \neq W_{ji}$ is important in producing a thermal equilibrium population. That is the physical significance of (7.56). Clearly the family of equations represented by (7.54) describes a system which will relax to the thermal equilibrium populations $p_j(T)$. We can then replace the W_{ij}'s by the U_m's, taking

$$\begin{aligned}W_{32} &\cong W_{23} = U_0 \\ W_{42} &\cong W_{24} = U_1 = W_{31} \cong W_{13} \\ W_{31} &\cong W_{13} = U_2\end{aligned} \tag{7.58a}$$

and introducing U_1' by the relations

$$W_{12} \cong W_{21} \equiv U_1' = W_{34} \cong W_{43} \quad . \tag{7.58b}$$

U_1' differs from U_1 (7.47) by the substitution of ω_S for ω_I. Making use of these relationships, one can now show that

$$\begin{aligned}\frac{d}{dt}(p_1 + p_2 - p_3 - p_4) &= 2U_1[(p_4 - p_2) + (p_3 - p_1)] \\ &+ 2U_0(p_3 - p_2) + 2U_2(p_4 - p_1) - 2U_1[p_1(T) - p_2(T) + p_3(T) - p_1(T)] \\ &- 2U_0[p_3(T) - p_2(T)] - 2U_1[p_4(T) - p_1(T)] \quad .\end{aligned} \tag{7.59}$$

From (7.31) one can show that

$$p_1 - p_4 = \langle I_z \rangle + \langle S_z \rangle \quad , \quad p_2 - p_3 = \langle I_z \rangle - \langle S_z \rangle \quad . \tag{7.60}$$

Therefore (7.59) can be rewritten utilizing I_0 and S_0 to represent the thermal equilibrium values of $\langle I_z \rangle$ and $\langle S_z \rangle$ as

$$\frac{d\langle I_z \rangle}{dt} = [I_0 - \langle I_z \rangle](U_0 + 2U_1 + U_2) + [S_0 - \langle S_z \rangle](U_2 - U_0) \quad . \tag{7.61a}$$

In a similar manner, we get

$$\frac{d\langle S_z \rangle}{dt} = [I_0 - \langle I_z \rangle](U_2 - U_0) + (S_0 - \langle S_z \rangle)(U_0 + 2U_1' + U_2) \quad . \tag{7.61b}$$

These two coupled linear differential equations have solutions consisting of real exponentials. They may be viewed as a normal modes problem with imaginary frequencies, i.e. real exponentials. There are thus two time constants. These equations show that as long as $\langle S_z \rangle = S_0$ and $\langle I_z \rangle = I_0$, both $d\langle S_z \rangle/dt$ and $d\langle I_z \rangle/dt$ remain zero. However, if one disturbs either spin system from thermal equilibrium, *both* spin systems will respond. Thus, if one makes $\langle S_z \rangle = 0$ by applying a $\pi/2$ pulse to the S-spins, not only will $d\langle S_z \rangle/dt$ be different from zero, but also $d\langle I_z \rangle/dt$ will be nonzero.

In general, then, when one applies pulses to a system of coupled spins, all spin populations will be found to respond as a result of the relaxation. We pose the detailed solutions of these rate equations as a homework problem.

7.6 Polarization by Forbidden Transitions: The Solid Effect

In order for the Overhauser effect to work, the nuclear relaxation process (such as W_{13}, W_{24}) cannot be allowed to short circuit relaxation in which both an electron and a nucleus flip, such as W_{23} or W_{14}. It is not always possible to meet those conditions.

The condition which one can, in general, be quite sure will hold is that pure electron spin-flip processes (such as W_{12} or W_{34}) are much the fastest W_{ij}'s because electrons couple more strongly to the lattice than do nuclei. *Jeffries* [7.8] and independently *Abragam* et al. [7.9] recognized that the so-called forbidden transitions were not *strictly* forbidden in many useful cases, and that one could use them to good advantage in achieving polarization. In fact, *Erb* et al. [7.10] independently discovered the effect experimentally. Using this scheme, *Jeffries* and his colleagues at Berkeley, and experimentalists at Saclay collaborating with *Abragam*, have obtained proton polarizations of over 70%, and have made a rich variety of applications. Invention of this technique was another major step forward. The phenomenon is often referred to as the solid effect.

There are two possible forbidden transitions. They are shown in Figs. 7.8 and 7.9. The transition of Fig. 7.8 can be induced by an alternating field *parallel* to the static field when the simple isotropic electron-nuclear coupling of (7.1) is solved to the next higher order to include the effect of the term $A(I_x S_x + I_y S_y)$ in admixing $|+-\rangle$ with $|-+\rangle$.

The transitions of both Figs. 7.8 and 7.9 can be induced by alternating fields perpendicular to H_0 when the more general Hamiltonian, (7.6), is solved to adequate precision. For example, the dipole-dipole coupling between a nucleus a distrance r from the electron makes the transition matrix elements of Fig. 7.8 and 7.9 of order $\gamma_e \hbar / r^3 H_0$ times the matrix element of Fig. 7.6 in which only an electron is flipped. The ratio also depends on the angle the static field makes with the axis connecting the nucleus and the electron. (The effect of the dipolar coupling is expressible more precisely in terms of the contribution of the dipole-dipole coupling components $A_{x'x'}$, $A_{y'y'}$, and $A_{z'z'}$).

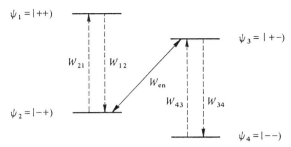

Fig. 7.8. The use of a forbidden transition which flips both the electron and nucleus simultaneously, W_{en}, to produce nuclear polarization. It is assumed that transitions which involve an electron spin-flip only are the only significant thermal processes

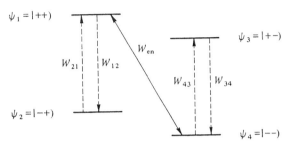

Fig. 7.9. A forbidden transition alternative to that of Fig. 7.8, which produces nuclear polarization of the opposite sign

Even though the transition probability W_{en} may be small, it is frequently possible to achieve strong enough alternating magnetic fields to make it larger than the thermal transition rates W_{ij} in which a nucleus flips, so that the transition connected by W_{en} produces effective population equalization.

Let us analyze the case of Fig. 7.9 for which the transition between states $1[\psi_1 = |++\rangle]$ and $4[(\psi_4 = |--\rangle)]$ is saturated. Since we assume the only thermal transitions of consequence are those shown, we immediately can write down

$$p_1 = p_4 \tag{7.62a}$$

$$p_2 = p_1 B_{12} \tag{7.62b}$$

$$p_3 = p_4 B_{43} \quad \text{so that} \tag{7.62c}$$

$$p_1 = p_4 = \frac{1}{2 + B_{12} + B_{43}} \tag{7.63a}$$

$$p_2 = \frac{B_{12}}{2 + B_{12} + B_{43}} \tag{7.63b}$$

$$p_3 = \frac{B_{43}}{2 + B_{12} + B_{43}} \quad . \tag{7.63c}$$

Using (7.21b),

$$\langle I_z \rangle = \frac{1}{2}(p_1 + p_2 - p_3 - p_4) = \frac{1}{2}\frac{B_{12} - B_{43}}{2 + B_{12} + B_{43}} \quad . \tag{7.64}$$

To appreciate the meaning of this expression we look at the high temperature limit (evaluation of the expression for all temperatures is given as a homework problem).

$$\langle I_z \rangle = \frac{1}{2}\frac{\gamma_e \hbar H_0}{2kT} \tag{7.65}$$

so that the enhancement over the normal polarization is

$$\frac{\langle I_z \rangle}{\langle I_z \rangle_{\text{therm}}} = \frac{\gamma_e}{\gamma_n} \quad , \tag{7.66}$$

the full Overhauser effect.

7.7 Electron-Nuclear Double Resonance (ENDOR)

A double resonance experiment of great historical importance was performed by *George Feher* [7.11]. He named it electron-nuclear double resonance (ENDOR). The purpose of this technique is to resolve otherwise unresolvable resonance lines. The concept involves in essence observing a nuclear resonance through its effect on an electron spin resonance. *Feher* was studying the electron spin resonance of electrons bound to donor atoms in silicon. 5 % of the silicon nuclei, the isotope Si^{29} with spin $\frac{1}{2}$, possess a nuclear magnetic moment. As a result of the large radius of the orbit, the electron spin couples magnetically to many Si^{29} nuclei. Since the Si^{29} nucleus can have two spin orientations, and since the Si^{29} hyperfine fields depend on where in the donor orbit the nucleus sits, there are many different local fields Si^{29} nuclei can produce at the electron.

The essential situation is shown by considering a Hamiltonian of the one electron interacting with N Si^{29} spins. (For simplicity, to avoid the needed discussion of the statistical effect of an isotopic abundance, let us suppose all Si sites had Si^{29} atoms.) We take it to be of the form

$$\mathcal{H} = \gamma_e \hbar H_0 S_z + \boldsymbol{S} \cdot \sum_{i=1}^{N} A_i \boldsymbol{I}_i - \gamma_n \hbar H_0 I_z \tag{7.67}$$

where A_i is the hyperfine coupling of the ith nucleus (for simplicity we assume a simple Fermi contact form of coupling rather than the more general tensor form). The A_i's are determined by the crystallographic location of the ith nucleus relative to the donor atom. The solution is a straightforward generalization of that of Sect. 7.3, using the N quantum numbers m_i, the eigenvalues of I_{zi},

$$E = \gamma_e \hbar H_0 m_S + \sum_i A_i m_i m_S - \sum_i \gamma_n \hbar H_0 m_i \quad . \tag{7.68}$$

Electron resonance occurs when $\Delta m_S = \pm 1$, $\Delta m_i = 0$ for all i. The frequency ω_e is

$$\omega_e = \gamma_e H_0 + \sum_i \frac{A_i}{\hbar} m_i \quad . \tag{7.69}$$

Since there are many possible values of A_i, and since each m_i can be $+\frac{1}{2}$ or $-\frac{1}{2}$, (7.69) describes many frequencies, in fact virtually a continuum, rather than a set of resolved lines such as in the case of an electron interacting with only a single nucleus. For a sample consisting of many donors, there are 2^N various ways of assigning the m_i's.

The problem of resolving the hyperfine lines may be likened to that of a man with several telephones on his desk, all of which ring at the same time. If he tries to answer them all, he hears a jumble of conversations as all the callers speak to him at once. Of course his callers have no problem — they hear only one voice, though he hears several.

Feher recognized that each Si^{29} nucleus experiences the hyperfine field of only one electron. Thus the *nuclear* resonances are sharp. Each Si^{29} site gives rise to two resonance transitions, corresponding to whether the electron hyperfine field aids or opposes the applied magnetic field:

$$\omega_{ni} = \gamma_n H_0 - A_i m_S \quad . \tag{7.70}$$

The nuclear transition frequencies of Si^{29} nuclei located at each distinct crystallographic location relative to the donor atom give rise to a distinct pair of lines. For many sites, the A_i is sufficiently large that these lines are well resolved from the other transitions corresponding to other sites.

Recognition that nuclear resonance would give resolved lines even though electron spin resonance did not was a great insight on *Feher*'s part. But, however, the nuclear resonance signal would be weak since the donor concentration ($\sim 10^{17}$ per cm^3) is typically so low. *Feher* conceived of the clever method of detecting the nuclear resonance by its effect on the electron spin resonance.

Consider the situation in Fig. 7.10. We show 4 levels only, which represent the four energies we get if we hold all m_i's fixed except for those of the jth nucleus:

$$E = (\gamma_e \hbar H_0 + \sum_{i \neq j} A_i m_i) m_S + A_j m_j m_S - \gamma_n \hbar H_0 m_j \quad . \tag{7.71}$$

Clearly we can define an effective static field acting on the electron of

$$H_{\text{eff}} = H_0 + \sum_{i \neq j} \frac{A_i}{\gamma_e \hbar} m_i \tag{7.72}$$

for this case.

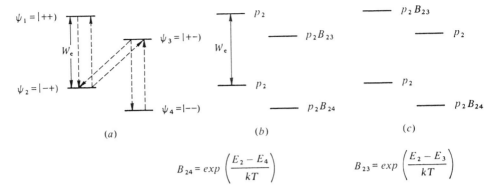

Fig. 7.10a–c. *Feher*'s scheme for observing nuclear resonance by monitoring its effect on the electron spin resonance. Figure 7.10a shows the states of (7.71) for the electron and the jth nucleus, with the electron transition being shown between states 1 and 2, and thermal relaxation processes such as in the Overhauser effect. Figure 7.10b shows the populations produced by saturating the electron transition. Figure 7.10c shows the effect of an adiabatic passage through the resonant frequency of states 1 and 3, with W_e turned off. When W_e is turned back on, it finds the electron transition between states 1 and 2 momentarily unsaturated

We assume for concreteness that we have the same thermal relaxation processes as with the Overhauser effect (Sect. 7.4), and saturate the same transition. The resulting populations are shown in Fig. 7.10b. Suppose we momentarily switched off W_e, switched on the nuclear oscillator (W_n) close to the transition frequency between states 1 and 3, and did an adiabatic passage with ω_n through the nuclear resonance line. The passage will exchange the populations between states 1 and 3, producing the populations shown in Fig. 7.10c. If one then turns on W_e, initially the electron resonance is no longer fully saturated, so the electron spin resonance has a changed signal height which decays as the transition saturates. Thus, the nuclear resonance produces an effect on the electron spin resonance.

We have described a particular method of doing ENDOR. There are many variations. The technique has been exceedingly important in mapping wave functions of a variety of paramagnetic centers, and thus is one of the principal ways one knows the structure of many important point imperfections in solids. (For an excellent review of the use of ESR and ENDOR to study structure of point imperfections, see [7.12].) It is of great importance for structural determinations of biological molecules. For his discovery of ENDOR and applications, *Feher* was awarded the Buckley Prize in Solid State Physics of the American Physical Society.

7.8 Bloembergen's Three-Level Maser

A maser (microwave amplification by stimulated emission of radiation) works on the principle that irradiation of an absorption line associated with transition between two energy levels will lead to a net emission of energy when the upper level is more highly populated than the lower one. *Townes* [7.13], and [7.14], and independently *Prokhorov* and *Basov* [7.15], recognized that if such a situation could be achieved, it could be used as the basis of a new type of oscillator. *Townes* achieved population inversion with ammonia gas by physically separating the molecules in the upper energy level from the lower energy level by means of properly shaped electric fields, in essence a molecular beam apparatus. The frequency of the oscillator fell in the microwave region.

Shortly thereafter, *Bloembergen* recognized that one could produce the energy level inversion by use of the Overhauser-Pound principle of level repopulation through saturation. It required a system with three or more energy levels (Fig. 7.11).

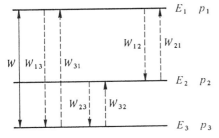

Fig. 7.11. *Bloembergen*'s three-level maser. The thermal relaxation is shown by the dashed arrows, the induced transition by the two-headed solid arrow.

If the populations p_1 and p_3 are equalized by saturation, p_2 reaches a value which causes an inverted population either of p_1 relative to p_2 or of p_2 relative to p_3, depending on the relative strength of thermal rates between states 1 and 2 versus 2 and 3

If, for simplicity, we can neglect the thermal relaxation between states 2 and 3, then

$$p_1 = p_3 \tag{7.73a}$$

$$p_2 = p_1 B_{12} \quad \text{so that} \tag{7.73b}$$

$$p_1 = p_3 = \frac{1}{2 + B_{12}} \tag{7.74a}$$

$$p_2 = \frac{B_{12}}{2 + B_{12}} \tag{7.74b}$$

so that $p_2/p_3 = B_{12}$.
Since

$$B_{12} = e^{(E_1 - E_2)/kT} > 1 \tag{7.75}$$

we have that
$$\frac{p_2}{p_3} > 1 \tag{7.76}$$
making the upper level more highly populated than the lower, the condition for maser operation at the frequency $(E_1 - E_2)/\hbar$.

Bloembergen pointed out that the theoretical situation he described could be achieved with paramagnetic ions, as was shortly demonstrated [7.16].

Bloembergen's concept has been of enormous value. Level inversion by pumping another transition is the principle on which all lasers operate. The pumping may be done optically, as when a light source is used, or by the equivalent of pumping (which generates nonequilibrium populations between a pair of levels) in chemical reactions (chemical lasers) or electric discharges (electric discharge laser).

Prokhorov and *Basov* [7.15] independently developed many of the same ideas.

7.9 The Problem of Sensitivity

Sooner or later all resonators want to observe a resonance which is too weak to be seen. Resonances may be weak because the number of spins is low, for example since the isotopic species is rare, or the nuclei occupy special positions such as being surface atoms on a crystal or neighbors of an imperfection. The resonances may be weak because the nuclear γ is small.

The development of superconducting solenoids with their high magnetic fields has helped enormously in increasing sensitivity of NMR equipment because the quanta absorbed in resonance are thereby increased. Reduction of noise by narrow banding also helps. The practical problem used to be that a narrow bandwidth was achieved by using long integrating time constants, and those in turn meant it took a long time to sweep across the resonance line. By the time one was near the end of the sweep, the apparatus was likely to have drifted, giving distorted line shapes or spurious signals from drift in the apparatus baseline. The advent of the multichannel digital signal averager has dramatically changed all that by permitting cumulative addition of a number of rapid sweeps. If the apparatus drifts, one stops sweeping, readjusts the equipment, and continues sweeping. Moreover one sweeps only as long as is needed to achieve the desired signal-to-noise ratio. The signal accumulated is proportional to the number of sweeps N. Since the noise is random and both positive and negative, its amplitude increases as $\sqrt{N}$, giving an overall increase in signal-to-noise ratio proportional to $N/\sqrt{N} = \sqrt{N}$.

As a practical matter, this means one must increase the averaging time fourfold to gain a mere factor of 2 in signal-to-noise. This increase may not matter if it means going from 1 min to 4 min, but if one is already averaging for

1 h, going to 4 h is a hard way to improve signal-to-noise! These facts illustrate that when one is using averaging there is a big premium on setting variables such as H_1, modulation amplitude, sweep excursion, etc., to the optimum. A corollary is when looking for an unknown, be willing to set the equipment to give a maximum signal even though you may thereby distort the resonance line (for example, from using too large a modulation, or from partially saturating). After the resonance has been found you can focus attention on it with parameters adjusted to avoid distortion.

7.10 Cross-Relaxation Double Resonance

No matter what one does to improve one's apparatus, one eventually reaches the limit of the current state of experimental art. What then if the signal is still too weak to see? We turn now to the use of double resonance, assuming there are two resonances which can be excited, one the "weak" one too difficult to see directly, the other a "strong" one observable by conventional means. For example, the weak one might result from a low abundance species with spin S, whereas the strong one might result from an abundant species with spin I. (For simplicity, we use the term "rare" to imply the species whose resonance is weak. Since a low γ could also make the resonance weak, "low γ" can be substituted for "rare" in most cases.)

If an H_1 is applied at resonance to the rare species, those nuclei absorb energy and their spin temperature rises. If the rare and abundant spins could exchange energy, the abundant spins would thereby get hotter, and their signal amplitude would diminish. If the abundant species is thermally isolated from the outside world (i.e., has a long T_1), this temperature rise can be made quite large and thus readily observable merely by sitting on the weak resonance long enough. By going to low temperatures, T_1 of many systems can be made exceedingly long. Thus the crucial question becomes how well can the two species exchange energy.

The mixing of the two spin systems was studied by *Abragam* and *Proctor* [7.17]. They employed the Li^7 and F^{19} resonances in LiF. Their experiments were part of their studies of the fundamentals of spin temperature and of adiabatic demagnetization. Working in a field of several thousand gauss to observe the resonances, they prepared the system in some nonequilibrium state in the strong field (for example, they inverted the F^{19} magnetization), removed the sample to a lower static field, allowed the Li^7 and F^{19} spins to mix, then returned the sample to the strong field for inspection of the Li^7 and F^{19} resonances. In this way they found that in a field of 75 Gauss the spins come to a common spin temperature in a mixing time of 6 s, that the time was unobservably short at 30 Gauss, and longer than the T_1's (several minutes) above 100 Gauss.

A detailed study of cross-relaxation was made by *Bloembergen* et al. [7.18] and by *Pershan* [7.19] for both electron spin systems and nuclear spin systems. They show that the crucial problem is the failure of the Zeeman energy to match when two different nuclear species undergo mutual spin flips. The mismatch in Zeeman energy must be made up by the dipolar coupling between the spins. The simplest sort of process arises when two nuclei have nearly the same γ. For example, H^1 and F^{19} have γ's which differ by 5%. Owing to the existence of terms such as S^+I^- or S^-I^+ in the dipolar coupling [the B term of the dipolar Hamiltonian of (3.7)], the dipolar coupling couples states in which the proton spin flips up (down) and the fluorine spin flips down (up). If we could consider the γ's as being identical, we would then have a situation such as shown in Fig. 7.12. Because the individual energy levels match, the initial state of the two spins, indicated by the two ×'s, has the same energy as the final state indicated by the two o's. A system started in the × state will undergo a transition to the o state by means of the S^+I^- part of the B term in the dipolar coupling. For H^1 and F^{19}, the fact that the γ's differ by 5% means that in a strong magnetic field such as 10 kGauss, the Zeeman energy difference between the × and the o states would correspond to the energy of F^{19} or H^1 in a field of 500 Gauss. Such an energy mismatch would prevent the transition unless there were some other energy reservoir whose energy could change to make up the difference. A possible candidate is the dipolar energy reservoir. For a pair of spins, it has a typical value of $\gamma_I \gamma_S \hbar^2 / r^3$, which, expressed in units of magnetic field, is only a few gauss for reasonable values of r. Thus, it cannot make up for an energy mismatch of 500 Gauss. If, however, the static field were much lower, the mismatch would be correspondingly reduced, and the mutual flips might become possible!

Returning to LiF in a field of 10 kGauss, the Li^7 resonance occurs at 16.547 MHz and the F^{19} resonance occurs at 40.055 MHz, a ratio of 2.420. Clearly the mismatch here is even worse than for H^1 and F^{19}. What Bloembergen and his colleagues recognized [7.18, 19] was that 2.4 is close to 2.0, hence a process in which two Li^7 nuclei flip up and one F^{19} nucleus flips down comes much closer to satisfying energy conservation than a process in which only one Li^7 flips. Now, Li^7 has a spin of $\frac{3}{2}$, but for the sake of argument we are going to pretend it has a spin $\frac{1}{2}$, since the explanation is then simpler. To estimate the rate, one must utilize a formula for transition probabilities such as (2.182) (with

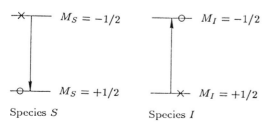

Fig. 7.12. A mutual spin-flip from the (×) to the (o) state exchanges energy between the I-spins and the S-spins. Shown vertically are the allowed energy levels. For this example the energy spacing is assumed to be the same for the two species, and both nuclei have spin $\frac{1}{2}$

$\omega = 0$). One therefore looks for a matrix element which connects the initial state (two Li^7 spins up, one F^{19} spin down) to the final state (two Li^7 spins down, one F^{19} spin up). If we denote the Li^7 by S and the F^{19} by I, and use subscripts i, j, k to distinguish individual nuclei, a matrix element of the dipolar term B, $(a|S_i^+ I_k^-|b)$, between states a and b flips only *one* Li^7 spin. To describe two spin flips we must go to second-order perturbation to get the effective matrix element $(a|\mathcal{H}_{\text{eff}}|b)$ joining the initial and final states:

$$(a|\mathcal{H}_{\text{eff}}|b) = \sum_c \frac{(a|\mathcal{H}_d|c)(c|\mathcal{H}_d|b)}{E_b - E_c} , \qquad (7.77)$$

where one matrix element might involve $S_i^+ I_{zj}$ and the other $S_k^+ I_j^-$. There are other possibilities, for example $S_i^+ I_k^-$ combined with $S_{zi} S_j^+$. The energy difference $E_b - E_c$ depends on what the initial and intermediate states are, but is of the order of magnitude of the Zeeman energies involved.

In general, if one considers a process with n_F fluorine spin flips down and n_{Li} Li spin flips up, with $\hbar\omega_{\text{Li}}$ or $\hbar\omega_F$ energy change per Li or per F spin, the energy mismatch ΔE between initial and final states is

$$\Delta E = n_F \hbar\omega_F - n_{\text{Li}} \hbar\omega_{\text{Li}} \quad \text{or} \qquad (7.78)$$

$$\frac{\Delta E}{\hbar\omega_F} = n_F - n_{\text{Li}} \frac{\hbar\omega_{\text{Li}}}{\hbar\omega_F} . \qquad (7.79)$$

For
$$n_F = 1 , \quad n_{\text{Li}} = 2 \quad \frac{\Delta E}{\hbar\omega_F} = 0.17$$

but for
$$n_F = 2 , \quad n_{\text{Li}} = 5 \quad \frac{\Delta E}{\hbar\omega_F} = 0.066 .$$

Thus, the more spins we allow to flip simultaneously, the closer we can make the energies of initial and final states match. However, more spins flipping requires more dipolar terms acting simultaneously and thus requires third- or higher-order perturbation expressions. The first-order matrix elements are of order H_L, where H_L is some sort of dipolar field of one nucleus at a neighbor. The second-order expression (7.77) is then of order H_L^2/H_0, where H_0 is the static field. Each increase in order of perturbation contributes a factor H_L/H_0. Thus, since $H_L \ll H_0$, one expects the perturbation order which works to be the lowest one for which $\Delta E \cong \sqrt{n}\hbar\omega_L$, where ω_L is the local field in frequency units, and n the total number of spins which flip.

Since the energy mismatch ΔE is proportional to H_0, and the cross-relaxation matrix elements are inversely proportional to a power of H_0, one can see that the cross-relaxation rate should be a steep function of H_0, as is illustrated in Fig. 7.13 showing Pershan's measurements of the H_0 dependence of the cross-relaxation rate in LiF.

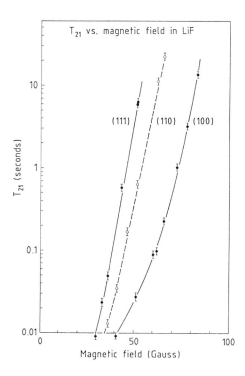

Fig. 7.13. Data of *Pershan* for the cross-relaxation time T_{21} for Li^7-F^{19} in LiF versus static magnetic field for three crystal orientations (see [7.19])

Clearly the use of cross-relaxation as part of double resonance leads in a natural way to cycling the static field between low values for cross-relaxation and high values for observation.

Anderson [7.20], working with Redfield, combined field cycling with the application of an audio-frequency magnetic field applied while the static field was zero to heat the spins to plot out the zero-field absorption characteristics of a spin system. Thus they used field cycling to give them the sensitivity of resonance in a strong field to monitor the effects they produced in zero field.

Redfield [7.21], *Fernelius* [7.22], *Slusher* and *Hahn* [7.23], *Minier* [7.24], and others utilized field cycling to observe quadrupole splittings of nuclei near to foreign atoms. If one has a sample which is a powder (e.g. a metal sample where one employs a powder to overcome problems of the skin effect), an advantage of demagnetizing to zero field is that the quadrupole splittings are then identical in all the individual crystallites. There is therefore no longer any powder broadening of the NMR transitions [7.21–24]. To observe very large quadrupole splittings in powders, there is a problem of coupling the energy absorbed by the quadruple transition into the rest of the spin system. It is necessary then to adjust the amplitude of the "audio" frequency alternating magnetic field according to principles of Hahn and Hartmann discussed later in this chapter.

Weitekamp et al. [7.25] have employed field cycling to zero field together with Fourier transform NMR to simplify dipolar spectra of powdered samples.

When one has a single crystal in which some nuclei have a quadrupole splitting and others do not, there are magnetic fields other than zero in which the energy level splittings of the two systems become equal. Cross-relaxation is then rapid. *Edmonds* [7.26] has reviewed nuclear quadrupole double resonance.

7.11 The Bloembergen-Sorokin Experiment

An important advance in cross-relaxation double resonance was discovered by *Bloembergen* and *Sorokin* in their studies of the cesium halides [7.27], especially CsBr. In fact, they found virtually all the essential elements of what has come to be known as the Hartmann-Hahn method which we discuss later, though it seems evident that the full significance of the results of *Bloembergen* and *Sorokin* was not appreciated owing to the fact that they discussed the problem entirely in the Cs^{133} rotating frame instead of using the doubly rotating frame which we take up in the next section.

We give a brief account here of what they found, then return in the next section to discuss their results a bit further, using the concepts of *Hartmann* and *Hahn* [7.28] and of *Lurie* and *Slichter* [7.29] concerning spin temperature in the doubly rotating frame.

We shall discuss two important discoveries made by *Bloembergen* and *Sorokin*. The first has to do with spin-locking (Sect. 6.6.1). We saw in Chap. 6 that if the H_1 is sufficiently strong (6.84) one must use *Redfield*'s ideas of spin temperature in the rotating reference frame. Then, if one puts the magnetization along the H_1 in the rotating frame, it will persist for times as long as those for which the spin system may be considered to be thermally isolated from the surroundings. In their sample of CsBr, *Bloembergen* and *Sorokin* found that the Cs^{133} spin-lattice relaxation time was 10–20 min at room temperature, ten thousand times longer than the Br T_1's. Thus they expected to be able to maintain (i.e. "lock") the Cs^{133} magnetization lined up along the $(H_1)_{Cs}$ for many minutes. Yet, when they attempted to observe the Cs^{133} while sweeping through the line using an H_1 for which Redfield theory should apply, they found no signal even when they took only 15 s to go through the line. They found that the explanation was that the short Br T_1 causes the Cs-Br dipolar coupling to fluctuate in time, thereby providing a relaxation mechanism for $T_{1\varrho}$ Eq. (6.91b). Since $T_{1\varrho}$ specifies how long the spins can be locked, the Br T_1 thereby limits the Cs^{133} spin-locking. A simple view of their discovery is that the Br T_1 contributes to the relaxation rates T_b and T_c of (6.87) and (6.91b). The short relaxation time of the Br results from the modulation of the Br electric quadrupole interaction by means of lattice vibrations.

Thus, we see that if one nuclear species has a dipole coupling to a second species with a fast spin-lattice relaxation time, $T_{1\varrho}$ of the first species will be shortened, and one can detect the existence of the second species by comparing, for the first species, T_1 with $T_{1\varrho}$.

Their second discovery was that if they utilized two oscillators, one tuned to the Cs^{133} resonance, the other tuned near the Br^{79} or Br^{81} resonance, they could utilize the rapid spin-lattice relaxation time of the bromine to polarize the Cs^{133}. This effect they labeled the transverse Overhauser effect. In their experiment, they applied a strong rf field $H_1)_{Cs}$ tuned exactly to the Cs resonance, and a rather weak rf field $H_1)_{Br}$ tuned to a frequency ν' which differed from the Br resonant frequency ν_{Br} by an amount given by

$$\nu' = \nu_{Br} \pm \gamma_{Cs} H_1)_{Cs}/2\pi \quad . \tag{7.80}$$

As we explain rigorously in the next section, this condition can be understood by considering the effective fields $H_{\text{eff}})_{Br}$ and $H_{\text{eff}})_{Cs}$ in their respective rotating frames. In the bromine frame,

$$\boldsymbol{H}_{\text{eff}})_{Br} = \boldsymbol{k}\frac{2\pi(\nu' - \nu_{Br})}{\gamma_{Br}} + i H_1)_{Br} \quad . \tag{7.81}$$

Since $H_1)_{Br}$ is small,

$$\boldsymbol{H}_{\text{eff}})_{Br} \cong \boldsymbol{k} 2\pi(\nu' - \nu_{Br})/\gamma_{Br} \quad . \tag{7.82}$$

Likewise,

$$\boldsymbol{H}_{\text{eff}})_{Cs} = i H_1)_{Cs}$$

since the Cs is tuned exactly to resonance. The Bloembergen-Sorokin condition is a special case of the equation

$$\gamma_{Br}|\boldsymbol{H}_{\text{eff}}|_{Br} = \gamma_{Cs}|\boldsymbol{H}_{\text{eff}}|_{Cs} \quad , \tag{7.83}$$

which has come to be known as the Hartmann-Hahn condition. As we shall see, when this condition is satisfied, the two spin systems can come to a common spin temperature in the doubly rotating frame. Equation (7.83) is a condition for rapid cross-relaxation in the doubly rotating frame.

Since *Bloembergen* and *Sorokin* analyzed their experiment by means of the single rotating frame of the Cs^{133}, they did not have spin systems which came to a common temperature in that frame of reference. They discussed the problem in terms of an Overhauser effect (indeed they called it the transverse Overhauser effect). They showed that when (7.80) was satisfied, the Cs magnetization M_{Cs} and the Br magnetization M_{Br} satisfied the relation

$$M_{Cs} = \pm (\gamma_{Br} M_{Br})/\gamma_{Cs} \quad . \tag{7.84}$$

In Sect. 7.16 we shall see how we can get their result in a simple manner by combining *Redfield*'s results on the calculation of M_{eq}, (6.91a), with the idea of spin temperature in the doubly rotating frame.

7.12 Hahn's Ingenious Concept

The problem of matching the Zeeman splittings of two different species could be solved if by magic one could apply one magnetic field to the I-spins, a second magnetic field to the S-spins. How can one do this to spins which are neighbors on the atomic scale?

The magical solution was found by the Wizard of Resonance, *Erwin Hahn* and demonstrated by the Wizard and his Sorcerer's Apprentice *Sven Hartmann* [7.28]. Hahn recognized that alternating magnetic fields have a negligible effect on nuclei unless the frequency of alternation is close to the precession frequency. Thus if he applied two alternating fields $H_1)_I$ and $H_1)_S$ at frequencies ω_I and ω_S, respectively, the frequencies being chosen to satisfy the respective resonance conditions,

$$\omega_I = \gamma_I H_0 \quad , \quad \omega_S = \gamma_S H_0 \quad , \tag{7.85}$$

$H_1)_I$ would have negligible effect on the S-spins and $H_1)_S$ would have negligible effect on the I-spins. Each species could then be viewed in its own rotating reference frame. Figure 7.14 shows the two rotating reference frames. Note that the z-axis is the same in both frames. It is in fact the direction of H_0 in the laboratory. In Fig. 7.14 the magnetization vectors M_I and M_S are also shown. Note that in a general situation they do not lie along either the z-axis or the respective H_1's, though they might if special ways of preparing the system were used. For the case where an M is not parallel to its H_1, it will precess about the H_1 in the rotating frame with precession frequency Ω_I or Ω_S given by

$$\Omega_I = \gamma_I H_1)_I \quad , \quad \Omega_S = \gamma_S H_1)_S \quad . \tag{7.86}$$

These equations suggest immediately that we can make the precession frequencies match by adjusting the ratio of the H_1's to satisfy the relation

$$\frac{H_1)_I}{H_1)_S} = \frac{\gamma_S}{\gamma_I} \quad . \tag{7.87}$$

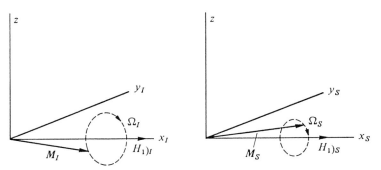

Fig. 7.14. The rotating reference frames show the axes x_I and x_S along the respective rf fields $H_1)_I$ and $H_1)_S$. Note that the z-axis is in common

This condition is known as the Hahn condition. It corresponds to making the Zeeman splittings of the I-spins quantized along $H_1)_I$ in the I-spin rotating frame equal to the Zeeman splitting of the S-spins quantized along $H_1)_S$ in the S-spin rotating frame.

Satisfying the Hahn condition matches the Zeeman splittings, but in reference systems which are exotic to say the least. Our real objective in producing the matching is to permit the two spin systems to couple. How can we use the exotic matching to couple the systems? We know that the spin systems are coupled via the dipolar interaction. For two spins of unlike species that portion which broadens the line is the A term of the dipolar coupling $\mathcal{H}_{dA}$

$$\mathcal{H}_{dA} = \frac{\gamma_I \gamma_S}{r_{IS}^3} \hbar^2 I_z S_z (1 - 3 \cos^2 \theta_{IS}) \quad . \tag{7.88}$$

Since the z-axis in the rotating frame is the same as in the laboratory, this coupling is unaffected by the transformation of spin variables to the rotating frame. (Note: Sometimes confusion arises as to what variables are transformed. In quantum mechanics one can formally introduce operators which transform only I, only S, or only r, or any combination. Since r_{IS} and θ_{IS} are unaffected by rotations of spatial coordinates about the z-axis. (7.88) is unaffected whether or not r is transformed. As we show below, the usual approach transforms only I and S in order to remove the time dependence of the H_1 couplings to the respective spins.)

Taking the classical view, if M_S is precessing around $H_1)_S$ in the S-spin rotating frame, M_{zS} will be oscillating sinusoidally at $\Omega_S = \gamma_S H_1)_S$. This produces, via $\mathcal{H}_{dA}$ (7.88), a time-dependent coupling to M_{zI}. But M_{zI} is the component of M_I transverse to $H_1)_I$, hence the coupling produces the same effect as applying an alternating field along the z-direction of frequency Ω_S. When the driving frequency Ω_S matches the I-spin resonance frequency Ω_I, the S-spins cause the I-spins to absorb energy, and M_I to nutate away from the direction of $H_1)_I$.

We have described two spin systems which are coupled. We know with coupled systems that when the natural frequencies coincide, resonant transfer of energy results and we need to worry about the back reaction. Thus for two coupled pendulums, one at rest initially, the other set in motion, after a while the first reaches a maximum amplitude, with the second one at rest. Then the energy exchange reverses, the pendulum which was driven now drives, the pendulum which drove now is driven.

If we had only a pair of spins, a similar situation would occur. However, typically there are many coupled I-spins, many S-spins, perhaps coupled, perhaps not, but in different neighbor environments in any case. For such a case it becomes useful to assign a temperature θ_I to the I-spins in their rotating frame and a temperature θ_S to the S-spins in their rotating frame. Under these circumstances, M_I always points along $H_1)_I$ and M_S always points along $H_1)_S$, but as energy is exchanged, the sizes of M_I and M_S change. In equilibrium $\theta_I = \theta_S$. Causing θ_S to increase produces a heating of θ_I.

We see, therefore, that there is a way of coupling two different spin systems so that they can exchange energy. We need now to do three things:

1. set the qualitative arguments on a firm quantum mechanical basis;
2. describe the experimental steps involved in putting *Hahn*'s idea to use;
3. analyze the expected results.

7.13 The Quantum Description

We start by writing the Hamiltonian of the system in the laboratory frame.

$$\mathcal{H} = \mathcal{H}_{ZI}(t) + \mathcal{H}_{ZS}(t) + (\mathcal{H}_d)_{II} + (\mathcal{H}_d)_{SS} + (\mathcal{H}_d)_{IS} \quad , \tag{7.89}$$

where $\mathcal{H}_{ZI}(t)$ is the Zeeman energy of the I-spins. It includes both a static interaction with field $H_0\mathbf{k}$ and a time-dependent interaction with the two alternating fields. It is most convenient to consider that rotating fields have been applied, instead of linearly polarized alternating fields. We have, then,

$$\begin{aligned}\mathcal{H}_{ZI}(t) = &- \gamma_I \hbar \mathbf{I} \cdot [\mathbf{k} H_0 + \mathbf{i}(H_1)_I \cos \omega_{zI} t + \mathbf{j}(H_1)_I \sin \omega_{zI} t \\ &+ \mathbf{i}(H_1)_S \cos \omega_{zS} t + \mathbf{j}(H_1)_S \sin \omega_{zS} t] \quad ,\end{aligned} \tag{7.90}$$

where ω_{zI} and ω_{zS} may be positive or negative, to represent either sense of rotation, and where

$$\mathbf{I} = \sum_j \mathbf{I}_j \tag{7.90a}$$

is the total spin vector of the I-spins.

The terms $(\mathcal{H}_d)_{IS}$, etc., represent the magnetic dipolar coupling of the I-spins with the S-spins, and so forth.

We now wish to transform to a rotating reference system. In so doing, we are following *Redfield* as discussed in Chapter 6. However, our problem is somewhat different from his since we have two rotating fields. We therefore wish to transform in such a way that we view the I-spins and S-spins in their respective reference frames. The transformation is readily accomplished by introducing the unitary operator T defined as

$$T = \exp(i\omega_{zI} I_z t)\exp(i\omega_{zS} S_z t) \quad , \quad \text{where} \tag{7.91}$$

$$I_z = \sum_j I_{zj} \quad , \quad S_z = \sum_k S_{zk} \quad , \tag{7.92}$$

are the total z-components of angular momentum of the two spin species.

We define a new wave function ψ' by the equation

$$\psi' = T\psi \quad . \tag{7.93a}$$

then substituting $T^{-1}\psi'$ for ψ, Schrödinger's equation becomes

$$-\frac{\hbar}{i}\frac{\partial \psi'}{\partial t} = \mathcal{H}'\psi' \quad , \tag{7.93b}$$

where $\mathcal{H}'$ is a transformed Hamiltonian.

Explicit evaluation of $\mathcal{H}'$ using the techniques of Chapter 2 gives

$$\begin{aligned}\mathcal{H}' = &-\gamma_I \hbar[(H_0 + \omega_{zI}/\gamma_I)I_z + (H_1)_I I_x] \\ &-\gamma_S \hbar[(H_0 + \omega_{zS}/\gamma_S)S_z + (H_1)_S S_x] + \mathcal{H}_{dII}^0 + \mathcal{H}_{dIS}^0 + \mathcal{H}_{dSS}^0 \\ &+ \text{time-dependent terms we ignore} \quad .\end{aligned} \tag{7.94}$$

The terms $\mathcal{H}_{dII}^0$, etc., represent that part of the dipolar coupling $\mathcal{H}_{dII}$ that commutes with the Zeeman interaction between the spins and the static laboratory field H_0. These terms are usually called the "secular part" of the dipolar interaction. We write them out explicitly below.

The time-dependent terms are of two sorts. One variety arises from the nonsecular parts of the dipolar coupling. They oscillate at frequencies ω_{zI}, ω_{zS}, or $\omega_{zI} \pm \omega_{zS}$. The second sort arises from couplings of the I-spins to $(H_1)_S$ and the S-spins to $(H_1)_I$. These oscillate at a frequency $(\omega_{zI} - \omega_{zS})$. Since ω_{zI}, ω_{zS}, and $(\omega_{zI} \pm \omega_{zS})$ are all far from any of the energy level spacings in the rotating frame, they can be neglected. One must remember, however, that it is conceivable that should there be a quadrupolar interaction added, and should the two nuclei have similar γ's (as do, for example, Cu^{63} and Cu^{65}), the frequency $(\omega_{zI} - \omega_{zS})$ might, in fact, be close to a possible transition.

If one chooses

$$\omega_{zI} = -\gamma_I H_{00} \quad , \quad \omega_{zS} = -\gamma_S H_{00} \quad , \tag{7.95}$$

where H_{00} is a particular value of the static field H_0 around which we later wish to make variations, then the two nuclei are exactly at resonance. Note that the two ω's are negative if the γ's are positive, representing the fact that nuclei of positive γ rotate in the "negative" sense about H_0. Making use of (7.95), defining h_0 by $H_0 - H_{00} = h_0$, and neglecting the time-dependent terms of $\mathcal{H}'$, we have

$$\begin{aligned}\mathcal{H}' = &-\gamma_I \hbar[h_0 I_z + (H_1)_I I_x] - \gamma_S \hbar[h_0 S_z + (H_1)_S S_x] \\ &+ \mathcal{H}_{dII}^0 + \mathcal{H}_{dIS}^0 + \mathcal{H}_{dSS}^0 \quad .\end{aligned} \tag{7.96}$$

It is convenient to define the Zeeman energies $\mathcal{H}_{ZI}$ and $\mathcal{H}_{ZS}$ by the equations

$$\mathcal{H}_{ZI} = -\gamma_I \hbar[h_0 I_z + (H_1)_I I_x] \quad , \quad \text{etc.} \quad . \tag{7.97}$$

The dipolar terms are

$$\begin{aligned}\mathcal{H}_{dII}^0 &= \frac{\gamma_I^2 \hbar^2}{4} \sum_{j,k}(3I_{zj}I_{zk} - \mathbf{I}_j \cdot \mathbf{I}_k)\left(\frac{1 - 3\cos^2\theta_{jk}}{r_{jk}^3}\right) \\ \mathcal{H}_{dIS}^0 &= \gamma_I \gamma_S \hbar^2 \sum_{k,p} I_{zk} S_{zp} \frac{(1 - 3\cos^2\theta_{kp})}{r_{kp}^3}\end{aligned} \tag{7.98}$$

and similarly for $\mathcal{H}^0_{dSS}$. To these may be added the pseudo-dipolar and pseudo-exchange couplings when their size is large enough to be important.

It is the term $\mathcal{H}^0_{dIS}$ which gives rise to the effects observed by *Bloembergen* and *Sorokin* [7.27] in their studies of CsBr. They found that the rapid bromine spin-lattice relaxation could communicate itself to the Cs nuclei through this term when the Cs nuclei were quantized along their own H_1.

We can view the various terms of (7.96) as energy reservoirs of Zeeman or dipolar energy. Since the various terms do not commute, they can exchange energy. Such processes may be termed cross-relaxation in the double-rotating reference frame. The rate of cross-relaxation will depend on how the energy levels of the different terms match, on the heat capacities, and on the strength of coupling as measured by the failure of terms to commute with one another. Thus, we note that the Zeeman term of the I-spins $\mathcal{H}_{ZI}$ commutes with the Zeeman energy of the S-spins $\mathcal{H}_{ZS}$. However, as long as $(H_1)_I \neq 0$, $\mathcal{H}_{ZI}$ does not commute with either $\mathcal{H}^0_{dII}$ or $\mathcal{H}^0_{dIS}$, and we can transfer energy between $\mathcal{H}_{ZI}$ and $\mathcal{H}^0_{dII}$ or $\mathcal{H}^0_{dIS}$. Moreover, $\mathcal{H}^0_{dIS}$ provides a coupling mechanism to transfer energy between $\mathcal{H}_{ZI}$ and $\mathcal{H}_{ZS}$ [provided $(H_1)_S \neq 0$].

All these remarks lead one, following *Redfield,* to assume that if one waits long enough, the various parts of (7.96) will come to an equilibrium in which the system can be described by a common temperature θ. For some purposes, it may also be possible and convenient to assume that various parts may come to common temperatures faster than the whole system achieves a single temperature. This is the viewpoint *Lurie* [7.29] adopts to calculate some cross-relaxation times.

We therefore, make the assumption that when the system has achieved a common temperature it is described by a density matrix ϱ given as

$$\varrho = \frac{\exp(-\mathcal{H}'/k\theta)}{\mathrm{Tr}\{\exp(-\mathcal{H}'/k\theta)\}} \tag{7.99}$$

where $\mathcal{H}'$ is given in (7.96). In terms of ϱ we can calculate the average energy E and the average magnetization vector $\langle M_I \rangle$ in the high temperature approximation

$$\begin{aligned} E &= \mathrm{Tr}\{\varrho\mathcal{H}'\} \\ &= -\frac{C_I[(H_1)_I^2 + h_0^2 + H_L^2] + C_S[(H_1)_S^2 + h_0^2]}{\theta} \end{aligned} \tag{7.100a}$$

$$\langle M_I \rangle = \mathrm{Tr}\{\varrho\gamma_I\hbar I\} = \frac{C_I(H_{\mathrm{eff}})_I}{\theta} \tag{7.100b}$$

where C_I and C_S are the Curie constants given in terms of the number of I- or S-spins per unit volume N_I or N_S, and Boltzmann's constant k by

$$C_I = \frac{\gamma_I^2\hbar^2 I(I+1)N_I}{3k} \quad, \quad \mathrm{etc.} \quad, \tag{7.101}$$

and where H_L^2 is defined by the equation

$$-\frac{C_I H_L^2}{\theta} = \text{Tr}\{\varrho(\mathcal{H}_{dII}^0 + \mathcal{H}_{dIS}^0 + \mathcal{H}_{dSS}^0)\}. \tag{7.102}$$

Evaluating the trace gives

$$H_L^2 = \frac{1}{3}\langle \Delta^2 H \rangle_{II} + \langle \Delta^2 H \rangle_{IS} + \frac{1}{3}\frac{\gamma_S^2 N_S S(S+1)}{\gamma_I^2 N_I I(I+1)}\langle \Delta^2 H \rangle_{SS}, \tag{7.103}$$

where $\langle \Delta^2 H \rangle_{\alpha\beta}$ is the contribution (in Gauss) of the β-spins to the second moment of the α-spin resonance line. H_L has the dimensions of a magnetic field. Although we call it the "local field", it should not be confused with the Lorentz local field. Actually, H_L is introduced simply to enable us to factor C_I out of various equations. The fact that the dipolar energy is $(-C_I H_L^2/\theta)$ makes it appear superficially that we have taken into account only the I-spins in calculating the dipolar energy. However, that such is not the case is seen by examining (7.102) and (7.103) which exhibit explicitly the dipolar contribution of the S species to the expression for H_L. $C_I H_L^2$ measures the total dipolar contribution to the spin specific heat. Note that although the term "local field" sounds vague, H_L can in fact be calculated *exactly* and is to be considered throughout as a precisely predicted quantity. The only exception to the statement is found when pseudo-dipolar coupling becomes prominent as in higher atomic number elements. In that case one would need to know the magnitude of the pseudo-dipolar coupling terms to make quantitative predictions.

We can observe $\langle M_I \rangle$ from our oscilloscope photographs of the initial height of the free induction decay following turn-off of $(H_1)_I$.

One further expression is needed. It is the expression for the magnetization M_I found after the demagnetization of I's. That is, suppose $(H_1)_S = (H_1)_I = 0$, and that $M_I = kM_{I0}$ where

$$M_{I0} = C_I H_0/\theta_l \tag{7.104}$$

is the thermal equilibrium magnetization of the I-spins at the lattice temperature θ_l. With $h_0 \gg H_L$, we switch on $(H_L)_I$, and slowly reduce h_0 to zero. We then end up, according to (7.100) with

$$\langle M_I \rangle = iM_{I0}\frac{(H_1)_I}{[(H_1)_I^2 + H_L^2]^{1/2}}. \tag{7.105}$$

Note that if we were to change $(H_1)_I$ slowly, $\langle M_I \rangle$ would follow $(H_1)_I$, in accord with (7.105).

7.14 The Mixing Cycle and Its Equations

There are two general ways of doing double resonance. The first was proposed by *Hartmann* and *Hahn* [7.28]. The second, which is a variant of the first, was demonstrated by *Lurie* [7.29].

As will become apparent, the two different techniques have simple analogies in thermodynamics. Consider two bodies connected by a rod to provide thermal contact. One body, of small heat capacity, represents the low abundance S-spins; the other, of large heat capacity, represents the I-spins. *Hartmann* and *Hahn*'s experiments are analogous to heating the large object by holding the small one at constant elevated temperature. The rate of heating depends on the thermal conductivity of the rod and the heat capacity of the large object (the I-spin system), but is independent of the heat capacity of the small object (the S-spin system) since we never let its temperature change. A theoretical prediction of the rate of heating of the large object would require knowledge of its heat capacity and of the thermal conductivity of the rod. In resonance language, that means we must calculate a cross-relaxation time. This cannot be done exactly.

Lurie's experiment is analogous to breaking the thermal contact between the rod and the small object, heating the small object to a known temperature, disconnecting the heater, and reconnecting the rod. After a sufficiently long time, the entire system of large object, small object, and rod, comes to a common temperature. Since the S system has a relatively small heat capacity, the final temperature is not much different from the initial temperature of the large object. However, we can repeat the cycle. In fact, if we do the thermal mixing N times, the heating of the I system is as great as it would be for a single mixing with an S system whose heat capacity is N times larger than it actually is. Since N may be made very large, a significant effect can be achieved even when the S-spins have a very small relative heat capacity.

Calculation of the temperature rise requires knowledge only of the heat capacities of the parts. It is not even necessary for the heat capacity of the rod to be small since we can easily include its effect. Calculation of the heat capacities of the spin systems is simple and can be done exactly. We therefore have a simple, exact theory to compare with experiment.

As *Hahn* and *Hartmann*'s analysis shows, the effective thermal conductivity of the rod depends on the size of the two rotating fields. The Hahn condition provides the fastest mixing or largest thermal conductivity. The heat capacity of the spin system is determined in large measure by the strength of the H_1's. We can therefore vary the heat capacities experimentally although we must remember that when the H_1 ratio does not satisfy the Hahn condition, it may take a longer time for a uniform temperature to be reached. The dipolar coupling between the two different species provides the thermal contact or "rod". As we have remarked, we can easily calculate its heat capacity. Likewise, there is a contribution to the heat capacity from the dipolar coupling of the I-spins among themselves and the S-spins among themselves. All these effects can be rigorously and simply included. In the process, *Lurie* demonstrated that it is not necessary for the H_1's to be large compared to the local fields and further demonstrated the coupling in cases where $(H_1)_I$ has been turned to zero.

We now turn to an analysis of the Lurie experiment. We shall assume throughout that spin-lattice relaxation can be neglected during the times of the experiment. Spin-lattice processes can be included readily, but one must be careful in so doing to include the sort of transverse Overhauser effects described by *Bloembergen* and *Sorokin* [7.27].

We begin by the demagnetization process. This brings $\langle M_I \rangle$ down along the x_I-axis, its magnitude being given by (7.105). Let us call this magnetization $\langle M_I \rangle_i$. During this process, since $(H_1)_S$ is zero, $\mathcal{H}_{ZS}$ commutes exactly with the rest of the Hamiltonian. So, likewise, does $\gamma_S \hbar S_z$. Therefore, $\langle M_S \rangle$ remains unaffected, and points along the static laboratory field H_0. The rest of $\mathcal{H}'$ is at a common temperature θ_i which we can compute from (7.100b) and (7.105):

$$\theta_i = C_I (H_1)_I / \langle M_I \rangle_i \quad . \tag{7.106}$$

This temperature is, of course, very much lower than the lattice temperature θ_l.

We now turn on $(H_1)_S$ suddenly. In such a rapid process, the state of the system does not change. The dipolar energy and the Zeeman energy of the I-spins are therefore unchanged. The S-spin Zeeman energy E_S

$$E_S = -\langle M_S \rangle \cdot (H_{\text{eff}})_S = 0 \quad , \tag{7.107}$$

since $\langle M_S \rangle$ and $(H_{\text{eff}})_S$ are perpendicular. The total energy of the system E_i is therefore

$$E_i = -\frac{C_I[(H_1)_I^2 + H_L^2]}{\theta_i} \quad . \tag{7.108}$$

After a sufficiently long time, the S-spins Zeeman energy comes into thermal equilibrium with the rest of the system at a common final temperature θ_f. The final energy E_f is then

$$E_f = -\frac{C_I[(H_1)_I^2 + H_L^2] + C_S(H_1)_S^2}{\theta_f} \quad . \tag{7.109}$$

But since the total system is isolated and its Hamiltonian independent of time, its energy cannot change. Therefore,

$$E_i = E_f \quad . \tag{7.110}$$

This gives us that

$$\frac{\theta_i}{\theta_f} = \frac{C_I[(H_1)_I^2 + H_L^2]}{C_I[(H_1)_I^2 + H_L^2] + C_S(H_1)_S^2} = \frac{1}{1+\varepsilon} \quad , \tag{7.111}$$

where

$$\varepsilon \equiv \frac{C_S(H_1)_S^2}{C_I[(H_1)_I^2 + H_L^2]} . \tag{7.112}$$

In the process the magnitude of $\langle M_I \rangle$ drops from its initial value $(M_I)_i$ to a final value $(M_I)_f$ which, in view of Curie's law, is

$$(M_I)_f/(M_I)_i = 1/(1 + \varepsilon) . \tag{7.113}$$

We now suddenly turn off $(H_1)_S$. Once again the system immediately after the change has the same wave function as it did just before. The expectation value of $\mathcal{H}_{ZI}$ and of the dipolar energies is thus unchanged, but that of $\mathcal{H}_{ZS}$ is zero since $(H_{\text{eff}})_S = 0$. The total energy E'_f is therefore

$$E'_f = -\frac{C_I[(H_1)_I^2 + H_L^2]}{\theta_f} . \tag{7.114}$$

Immediately after the turn-off of $(H_1)_S$, $\langle M_S \rangle$ is nonzero. Therefore, we see that we do not have thermal equilibrium. If we wait for a sufficiently long time, the entire system will come to a common temperature θ_{ff}. In the process, $\langle M_S \rangle$ decays to zero. This is an irreversible decay. We shall in fact calculate the entropy increase below.

When the system has reached the final temperature θ_{ff}, the energy is E_{ff}. Using (7.100a), we have

$$E_{\text{ff}} = -C_I[(H_1)_I^2 + H_L^2]/\theta_{\text{ff}} . \tag{7.115}$$

But $E_{\text{ff}} = E'_f$ since the spin system is isolated from the outside world and has a Hamiltonian independent of time. Therefore, using (7.114) and (7.115)

$$\theta_f = \theta_{\text{ff}} . \tag{7.116}$$

Using Curie's law we see that following the turn-off of $(H_1)_S$, $\langle M_I \rangle$ does not change.

For one complete on-off cycle, therefore, we can argue that $\langle M_I \rangle$ is reduced by the factor $1/(1 + \varepsilon)$.

We can repeat the argument for another on-off cycle. The magnetization $M_I(N)$ after N cycles is thus given in terms of its value $M_I(0)$ prior to the first cycle by

$$M_I(N)/M_I(0) = [1/(1 + \varepsilon)]^N . \tag{7.117}$$

When $\varepsilon \ll 1$, as in our experiments, we can write this as

$$M_I(N)/M(0) = e^{-N\varepsilon} ,$$

where ε is given in (7.112).

Double resonance is possible even when $(H_1)_I \ll H_L$. Experimentally we accomplish this by performing adiabatic reduction of $(H_1)_I$ after M_I has been brought along $(H_1)_I$ in the rotating frame. After $(H_1)_I \simeq 0$, $M_I \simeq 0$ from Curie's law, however, we have retained the order in the I-spin system, the order now being with respect to the local field [7.30]. $(H_1)_S$ is now cycled on and off N times. After the Nth cycle, $(H_1)_I$ is adiabatically returned to its original value. The resulting M_I is then observed by rapidly turning off $(H_1)_I$ and observing the free induction decay.

The analysis for the case when $(H_1)_I \ll H_L$ is essentially the same as given above; however, now, the term $C_I(H_1)_I^2/\theta$ no longer appears in (7.108) and (7.109), and ε reduces to

$$\varepsilon = C_S(H_1)_S^2/C_I H_L^2 \quad . \tag{7.118}$$

Lurie studied Li metal for which the 93 % abundant isotope Li^7 gives the strong or I-spin resonance; the 7 % abundant isotope Li^6 is the weak or S-spin system. To observe the effect of the double resonance, he measured the amplitude of the Li^7 free induction decay after N mixing cycles.

Figure 7.15 displays *Lurie'*s data demonstrating the effect of N on the destruction of the Li^7 signal. The solid line has no adjustable parameters apart from the $N = 0$ intercept.

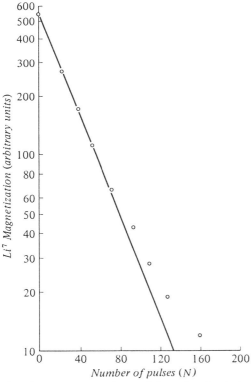

Fig. 7.15. *Lurie'*s experimental data showing $\ln(M_7)$ versus N at 1.5 K. $(H_1)_7 = 2.14\,\text{G}$, $(H_1)_6 = 5.4\,\text{G}$, nearly satisfying the Hahn condition. $t_{\text{on}} = t_{\text{off}} = 4\,\text{ms}$. The solid line is calculated using (7.117) and (7.118)

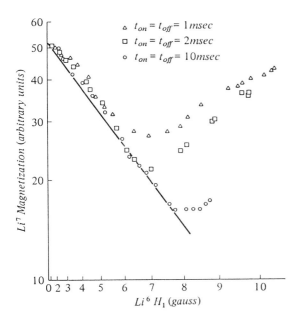

Fig. 7.16. $\ln(M_7)$ versus $(H_1)_6^2$ at 1.5 K for $N \to 22$, $t_{\rm on} = t_{\rm off} = 1, 2$, and 10 ms, $(H_1)_7 = 2.30$ G. For this value of $(H_1)_7$, $(H_1)_6 = 6.1$ G satisfies the Hahn condition. The solid line is calculated from (7.112) and (7.117)

Figure 7.16 shows how the spin temperature idea works even when the Hahn condition is not perfectly satisfied. Note that the solid line has no adjustable parameters. The deviations for larger $H_1)_6$ and shorter $t_{\rm on}$ and $t_{\rm off}$ arise because there is not enough time for a common spin temperature to be established.

In this example a very large destruction is produced. Alternatively, one can conclude that a much smaller fraction of Li^6 nuclei could produce an observable destruction. In this case one begins to worry about the long distance that energy must diffuse from the hot Li^6 spins to couple to the distant Li^7's. This problem has been investigated experimentally by several authors [7.23, 31, 32].

7.15 Energy and Entropy

It is interesting to follow the changes in energy and entropy of the spins that take place during the double resonance. The essence of the experiment is the heating of the I-spins brought about by the contact with the hot S-spins. There is a net flow of energy into the system as a result of work done on the S-spins. The destruction of the I magnetization corresponds to an irreversible loss of order, that is, an entropy increase. The energy of the system is, of course, the expectation value of the Hamiltonian of (7.96). Basic theorems of quantum (and classical) mechanics tell us that the total energy remains constant as long as $\mathcal{H}$ does not explicitly depend on time. Rearrangements of energy within the total system, even when $\mathcal{H}$ is independent of time, we identify as a heat flow within the system. Changes of the total energy due to variation of an external parameter we call work on or by the spin system.

We can follow the cycle by considering work and heat transfer in the rotating frame. Consider one complete cycle of $(H_1)_S$ on and off. We start by turning on $(H_1)_S$ suddenly. Bearing in mind that the S Zeeman energy is

$$\langle \mathcal{H}_{ZS} \rangle = -\langle M_S \rangle \cdot (H_1)_S \quad , \tag{7.119}$$

and that during the sudden turn-on the dipolar energy does not have time to change, we see that it takes no work to turn on $(H_1)_S$ since $\langle M_S \rangle$ is initially zero. The establishment of the S magnetization causes $\langle \mathcal{H}_{ZS} \rangle$ to go from zero to a negative value. That is, there is a heat flow from the S-spin Zeeman reservoir to the rest of the spin system. That this is the direction of heat flow is reasonable since the initial zero $\langle M_S \rangle$ in the presence of a nonzero $(H_1)_S$ can be viewed as saying that $\mathcal{H}_{ZS}$ has an infinite temperature. Associated with this heat flow between systems at different temperatures there must be an entropy increase.

Following establishment of $\langle M_S \rangle$, we turn off $(H_1)_S$. Using (7.119) we can see that we must do positive work on $\mathcal{H}_{ZS}$ in the process. [Note that during the turn-on or turn-off, which takes place suddenly, there is no time for heat flow, so that we can compute the work done solely from the changes in $\langle \mathcal{H}_{ZS} \rangle$ given by (7.119).] Following turn-off, $\langle M_S \rangle$ decays irreversibly to zero. Again there must be an entropy increase associated with the irreversibility. We are now ready to repeat the cycle. Note that we have done a net amount of work on the spin system, and that there has been an irreversible loss in order.

Had we turned on $(H_1)_S$ in the next cycle *before* $\langle M_S \rangle$ had been able to decay, the S-spins would have done positive work back on us. In fact, had $\langle M_S \rangle$ not decayed at all, we would have gotten back as much work as we put in when we turned off $(H_1)_S$. We would not then have done any net work in a cycle. Moreover, apart from the effects at the original turn-on, there would have been no irreversible loss in magnetization of either spin system. We must allow a sufficiently long time for the irreversible features to occur.

The entropy σ of the system can be calculated starting from the basic equation

$$\sigma = \frac{E + k\theta \ln Z}{\theta} \tag{7.120}$$

where Z is the partition function. Making use of the high-temperature approximation, we can evaluate (7.120) to get

$$\sigma \cong k\left[\ln(\text{Tr}\, 1) - \frac{1}{2k^2\theta^2} \frac{\text{Tr}\{\mathcal{H}^2\}}{\text{Tr}\, 1}\right] \quad , \tag{7.121}$$

where Tr 1 means the total number of states, and equals

$$\text{Tr}\, 1 = (2I + 1)^{N_I}(2S + 1)^{N_S} \quad . \tag{7.122}$$

Evaluating the $\text{Tr}\{\mathcal{H}^2\}$, as in (7.100a) gives (with $H_0 = 0$)

$$\sigma = k[N_I \ln(2I+1) + N_S \ln(2S+1)]$$
$$- \frac{C_I[(H_1)_I^2 + H_L^2] + C_S(H_1)_S^2}{2\theta^2} \quad . \tag{7.123}$$

[Note that (7.105), describing the adiabatic demagnetization, follows from use of (7.100b), Curie's law, together with the requirement that the entropy given by (7.123) remain constant.] Since we have worked out the temperature at each part of the cycle, we can use (7.123), together with the approximation that ε is small, to find that in one complete cycle starting at temperature θ there is a total increase in entropy $\sigma_f - \sigma_i$ of

$$\sigma_f - \sigma_i = \frac{C_S(H_1)_S^2}{\theta^2} \quad . \tag{7.124}$$

Half of the increase occurs following the turn-on, the other half following the turn-off at $(H_1)_S$. We note that the larger $(H_1)_S$, the larger the change of entropy per cycle. Since the existence of M_I is a sign of order, we see that a large $(H_1)_S$ leads to a large destruction of M_I, as was, in fact, already expressed by (7.118).

7.16 The Effects of Spin-Lattice Relaxation

In Sect. 7.11, we saw that *Bloembergen* and *Sorokin* observed two striking effects in their pioneering double resonance experiment. As we have mentioned, they gave a full explanation of their results. We turn now to an alternative description of their experiment to show how one can think of their experiments in terms of two concepts: (1) spin temperature in the doubly rotating frame and (2) Redfield theory of the equilibrium spin temperature reached in the rotating frame when spin-lattice relaxation is included. Thus, we show that their experiment can be viewed as a combination of *Redfield's* experiments on saturation with the Hartmann-Hahn and Lurie-Slichter experiments.

As we explained in Sect. 7.11, *Bloembergen* and *Sorokin* observed a species, I, (Cs^{133} in their case) which had a long T_1. The $(H_1)_I$ they used was such that one should consider the I-spins to be described by a spin temperature in their rotating frame. They observed that for the I-spins, $T_{1\varrho}$ was determined by the T_1 of the S-spins, leading to a circumstance that $(T_{1\varrho})_I \ll (T_1)_I$.

They observed further that if they were on resonance for the I-spins

$$\gamma_I H_0 = \omega_I \tag{7.125}$$

but off resonance for the S-spins such that

$$|\gamma_S H_0 - \omega_S| = \gamma_I (H_1)_I \quad \text{with} \tag{7.126a}$$

$$\gamma_S^2 (H_1)_S^2 \ll (\gamma_S H_0 - \omega_S)^2 \quad , \tag{7.126b}$$

they produced an I magnetization, M_I, given by

$$M_I = \pm M_I(\theta_l, H_0)\frac{\gamma_S}{\gamma_I} \quad , \tag{7.127}$$

where $M_I(\theta_l, H_0)$ is the I magnetization when the I-spins are in thermal equilibrium in a static field H_0 with a lattice at temperature θ_l.

To understand how $(T_1)_S$ can determine $T_{1\varrho}$ for a case in which $(H_1)_S = 0$, we follow the Bloembergen-Sorokin analysis. We consider a case in which $(H_1)_I^2 \gg H_L^2$ where H_L is some sort of local field. Then the I-spins, if exactly at resonance, are quantized along the I-spin x-axis, the direction of $(H_1)_I$, and the S-spins are quantized along the z-axis, the direction of H_0. The term $\mathcal{H}_{dIS}$, given by (3.55), involves products such as $I_{zk}S_{zp}$, which have matrix elements which are diagonal in S_{zp}, but are off-diagonal in I_{xp}. However, owing to the spin-lattice relaxation time of the S-spins, the diagonal matrix elements of S_{zp} are functions of time with a correlation time of $(T_1)_S$. Defining

$$C_{kp} = \frac{\gamma_I \gamma_S \hbar^2}{r_{kp}^3}(1 - 3\cos^2\theta_{kp}) \quad , \tag{7.128}$$

Bloembergen and *Sorokin* show that

$$\frac{1}{(T_{1\varrho})_I} = \frac{1}{\hbar^2}\sum_{k,p} C_{kp}^2 \frac{S(S+1)}{3} \frac{(T_1)_S}{1 + \omega_{1I}^2(T_1)_S^2} \quad . \tag{7.129}$$

The reader should compare this expression for the relaxation of $\langle I_x \rangle$ along $(H_1)_x$ with the expression for T_1 of (5.372) which applies to the relaxation of $\langle I_z \rangle$ along H_0. Physically, the fact that the z-component of the S-spins is fluctuating produces a time-dependent magnetic field on the I-spins along the z-direction. A field of this polarization is transverse to the I-spin quantization direction, and can therefore relax the I-spin magnetization along its "static" field in the rotating reference frame. The factor $T_1/(1 + \omega_{1I}^2 T_{1S}^2)$ gives the spectral density at the frequency ω_{1I} which can produce flips of the I-spins quantized along $(H_1)_I$.

If $(H_1)_S$ is turned on to a level sufficiently strong that the S-spins obey the Redfield spin-temperature condition in their rotating frame (6.84), then the magnitude of the S-spin magnetization, M_S, and of the I-spin magnetization, M_I, will be given by an argument such as is given in Sect. 6.9. The calculation we described above for $(T_{1\varrho})_I$ can be viewed as a calculation of T_{bI} defined by a modified (6.86):

$$\frac{\partial M_{xI}}{\partial t} = -\frac{M_{xI}}{T_{bI}} \quad \text{with} \tag{7.130}$$

$$\frac{1}{T_{bI}} = \frac{1}{\hbar^2}\sum_{k,p} C_{kp}^2 \frac{S(S+1)}{3} \frac{(T_1)_S}{1 + \omega_{1I}^2(T_1)_S^2} \quad . \tag{7.131}$$

The z-component of M_I will not be relaxed directly by the time variation of I-S dipolar coupling since that coupling commutes with I_z. Thus in the equation

$$\frac{\partial M_{zI}}{\partial t} = \frac{M_{0I} - M_{zI}}{(T_a)_I} \tag{7.132}$$

we can set

$$T_{aI} = (T_1)_I \quad . \tag{7.133}$$

For the S-spins we will have, in analogy to (7.132),

$$\frac{\partial M_{zS}}{\partial t} = \frac{M_{0S} - M_{zS}}{T_{aS}} \quad \text{with} \tag{7.134}$$

$$T_{aS} = (T_1)_S \quad , \quad \text{and} \tag{7.135}$$

$$\frac{\partial M_{xS}}{\partial t} = -\frac{M_{xS}}{T_{bS}} \tag{7.136}$$

where $T_{bS} \sim (T_1)_S$.

We will have three dipolar relaxation equations similar to (6.87):

$$\frac{\partial}{\partial t}\langle \mathcal{H}_{dII}\rangle = -\frac{\langle \mathcal{H}_{dII}\rangle}{T_{cII}} \tag{7.137a}$$

$$\frac{\partial}{\partial t}\langle \mathcal{H}_{dIS}\rangle = -\frac{\langle \mathcal{H}_{dIS}\rangle}{T_{cIS}} \tag{7.137b}$$

$$\frac{\partial}{\partial t}\langle \mathcal{H}_{dSS}\rangle = -\frac{\langle \mathcal{H}_{dIS}\rangle}{T_{cIS}} \tag{7.137c}$$

where we have neglected the terms like $\langle \mathcal{H}_d^0\rangle$ of (6.87).

If we assume that $(T_1)_S \ll (T_1)_I$ the lattice will flip individual S-spins much more rapidly than individual I-spins, so we can neglect any rates dependent on I-spin flips.

Therefore

$$T_{cIS} \approx T_{cSS} \tag{7.138}$$

and both will be much shorter than T_{cII}. Likewise

$$T_{bs} \ll T_{bI} \quad . \tag{7.139}$$

In analogy with (6.91) we will then find for the Bloembergen-Sorokin case

$$\overline{E} = -(h_0)M_{zS} - (H_1)_I M_{xI} + \langle \mathcal{H}_{dII}^0 \rangle \\ + \langle \mathcal{H}_{dIS}^0 \rangle + \langle \mathcal{H}_{dSS}^0 \rangle \quad . \tag{7.140}$$

Then

$$\frac{\partial \overline{E}}{\partial t} = -(h_0)_S \frac{\partial M_{zS}}{\partial t} - (H_1)_I \frac{\partial M_{xI}}{\partial t} \\ + \frac{\partial}{\partial t}\langle \mathcal{H}_{dII}^0\rangle + \frac{\partial}{\partial t}\langle \mathcal{H}_{dIS}^0\rangle + \frac{\partial}{\partial t}\langle \mathcal{H}_{dSS}^0\rangle \quad . \tag{7.141}$$

Keeping only the fast terms which relax at rates comparable to $(T_1)_S$, we get

$$\frac{\partial \overline{E}}{\partial t} = -(h_0)_S \frac{(M_{0S} - M_{zS})}{T_{aS}} - \frac{\langle \mathcal{H}^0_{dIS} \rangle}{T_{cIS}} - \frac{\langle \mathcal{H}^0_{dSS} \rangle}{T_{cSS}} \quad . \tag{7.142}$$

We now assume the system has a common spin temperature θ so that

$$M_{zS} = \frac{C_S (h_0)_S}{\theta} \tag{7.143a}$$

$$M_{xI} = \frac{C_I (H_1)_I}{\theta} \tag{7.143b}$$

$$\langle \mathcal{H}^0_{dIS} \rangle = -\frac{C_S H^2_{IS}}{\theta} \tag{7.144a}$$

$$\langle \mathcal{H}^0_{dSS} \rangle = -\frac{C_S H^2_{SS}}{\theta} \tag{7.144b}$$

where H^2_{IS} and H^2_{SS} are defined by (7.144).

To find the equilibrium spin temperature, we set $d\overline{E}/dt = 0$ and substitute (7.143) and (7.144) into (7.142), getting

$$\frac{C_S}{\theta} = \frac{M_{0S}(h_0)_S}{T_{aS}} \bigg/ \left(\frac{(h_0)^2_S}{T_{aS}} + \frac{H^2_{LIS}}{T_{cIS}} + \frac{H^2_{LSS}}{T_{cSS}} \right) \quad . \tag{7.145}$$

If one is off resonance $(h_0)_S$ by a sufficiently large amount compared to the local fields, one can keep only the term involving $(h_0)_S$ in (7.145), getting

$$\frac{C_S}{\theta} = \frac{M_{0S}}{(h_0)_S} \quad . \tag{7.146}$$

But, since

$$M_S = \frac{C_S (h_0)_S}{\theta} \quad , \tag{7.147a}$$

$$M_S = M_{0S} \quad . \tag{7.147b}$$

Recognizing that

$$M_{0S} = \frac{C_S H_0}{\theta_l} \tag{7.148}$$

we get, combining (7.146) and (7.147),

$$M_I = \frac{C_I (H_1)_I}{\theta} = \frac{C_I (H_1)_I}{C_S (h_0)_S} \frac{C_S H_0}{\theta_l} = \frac{(H_1)_I}{(h_0)_S} M_{0I} \quad . \tag{7.149}$$

But from (7.145)

$$\gamma_I (H_1)_I = \gamma_S |(h_0)_S|$$

so that we get the Bloembergen-Sorokin result that

$$M_I = \frac{\gamma_S}{\gamma_I} \frac{|(h_0)_S|}{(h_0)_S} M_{0I} \quad . \tag{7.150}$$

The ratio $|(h_0)_S|/(h_0)_S$ is either $+1$ or -1 depending on whether ω_S lies below or above resonance respectively. The two possible signs reflect the fact that θ is positive in the former case [M_S parallel to $(H_{\text{eff}})_S$] and negative in the latter [M_S antiparallel to $(H_{\text{eff}})_S$].

The size of M_I is larger than its thermal equilibrium value in the ratio γ_S/γ_I. This result is just the result of the Overhauser effect, hence the name "transverse Overhauser effect".

In considering this result, we see that it follows very directly from the realization that if the S-spins are far from resonance

$$M_S = M_{0S} \quad , \tag{7.151}$$

and that the I-spins and S-spins are at a common spin temperature in the doubly rotating frame.

We could achieve the same result by an appropriate adiabatic demagnetization experiment. We turn on $(H_1)_I$ and we turn on an $(H_1)_S$ which satisfies the Hartmann-Hahn condition at a frequency ω_S far from resonance. We then sweep ω_S onto resonance. The two spin systems will then mix. However, since they have comparable heat capacity, the common spin temperature they reach will be hotter than the temperature the S-spins would reach if the spins did not mix. So we turn off $(H_1)_S$, go off resonance, magnetize the S-spins again, and repeat cooling the I-spins several times more. In this way, we can bring the I-spins asymptotically to the value given by (7.150). However, we can only hold this result for a time $(T_{1\varrho})_I$ given by (7.129). The Bloembergen-Sorokin method gives one a steady-state M_I along $(H_1)_I$.

Note that if one sits with the S-spins exactly at resonance [$(h_0)_S = 0$], (7.145) and (7.146a) show that the equilibrium values of M_S and M_I would vanish even if one satisfied the Hartmann-Hahn condition. Thus, to use the fast relaxation of the S-spins to generate a strong I-spin magnetization, the S-spins must be far from resonance (above or below).

Equation (7.145) can be viewed as giving the spin temperature of the S-spins. If the Hartmann-Hahn condition is satisfied, the I-spins will rapidly come to this spin temperature. However, if one is far from the Hartmann-Hahn condition, the I-spins will not reach this temperature, as is shown from the experiment of *Lurie* et al. (Fig. 7.16).

7.17 The Pines-Gibby-Waugh Method of Cross Polarization

The methods of *Hartmann* and *Hahn* or *Lurie* et al. involve detection of a rare species by observing its effect on an abundant species. *Pines, Gibby,* and *Waugh* [7.33] pointed out that under some circumstances it is preferable to perform the experiment in a manner to observe the rare species. Specifically, suppose C^{13} is the rare species (its natural abundance is 1.1%), and H^1 is the abundant species, as when one is working with solid hydrocarbons. Each nonequivalent C^{13} site will have its own chemical shift. One would like to record each C^{13} chemical shift. We have seen (Sect. 5.8) that the Fourier transform method is then a very efficient way of collecting data, since a single C^{13} $\pi/2$ pulse will excite all the C^{13} lines simultaneously. If one observes the H^1 resonance, one can only measure the individual C^{13} resonance lines using a point-by-point search. However, if one brings the H^1 magnetization along $(H_1)_H$, the hydrogen H_1, one can rapidly polarize the C^{13} nuclei by turning on their H_1, $(H_1)_C$, to the Hartmann-Hahn value. Within a cross-relaxation time, the C^{13} will be polarized. Since there are many more hydrogen atoms than C^{13} nuclei, in this process the H^1 spin temperature will hardly change. Therefore

$$M_C = \frac{C_C(H_1)_C}{\theta} \quad . \tag{7.152}$$

But

$$\theta = \frac{C_H(H_1)_H}{M_H} \quad \text{and} \tag{7.153a}$$

$$\gamma_C(H_1)_C = \gamma_H(H_1)_H \quad , \quad \text{so that} \tag{7.153b}$$

$$M_C = M_H \frac{C_C}{C_H}\left(\frac{\gamma_H}{\gamma_C}\right) \quad . \tag{7.154}$$

Suppose, now, that initially M_H has its thermal equilibrium value in H_0

$$M_H = M_H(\theta_l) = \frac{C_H H_0}{\theta_l} \quad . \tag{7.155}$$

The corresponding $M_C(\theta_l)$ is given by

$$M_C(\theta_l) = C_C \frac{H_0}{\theta_l} \quad . \tag{7.156}$$

Utilizing these equations in (7.154), we find

$$M_C = M_C(\theta_l) \frac{\gamma_H}{\gamma_C} \quad . \tag{7.157}$$

This is the Bloembergen-Sorokin enhancement, characteristic of an Overhauser effect.

Having polarized the C^{13}'s, one turns off their H_1, and records the free induction decay. Meanwhile, one leaves on the proton H_1 for two reasons: (1) to maintain the proton magnetization [spin-locked by $(H_1)_H$] and (2) to provide so-called H^1-C^{13} spin decoupling. This last topic we take up later in the chapter, where we see that the presence of the strong H^1 acting on the protons effectively wipes out the H^1 splittings of the C^{13} resonance. By this means the C^{13} resonance consists of a family of lines each of which goes with a given C^{13} chemical shift, but without splittings arising from the C^{13}-H^1 spin-spin coupling.

We now turn on $(H_1)_C$ again, repolarizing the C^{13}'s along their H_1. We again turn off $(H_1)_C$ and record the C^{13} free induction decay. Each repetition of this cycle will heat the H^1 nuclei in accord with (7.112) and (7.117), so that the C^{13} signal gets progressively smaller until it becomes necessary to turn off $(H_1)_H$ to allow the proton spin-lattice relaxation time to replenish the H^1 magnetization.

We turn now to a few remarks about the nature of the rare spin spectra. If we recall that in general the spin-temperature approach requires that we be working with solids, we should distinguish single crystals from powders. In a single crystal a given type of C^{13} chemical site might have several different bond orientations with respect to H_0. For example, in a benzene molecule (C_6H_6), H_0 might lie along one CH bond, in which case it would also lie along one other in the molecule, but would make a 60° angle with the other four bonds. Thus we would have two C^{13} lines, with one twice the intensity of the other. If the sample is a powder, the C^{13} spectrum of this bond would be a smear. By magic-angle spinning (discussed in Sect. 8.9) one could narrow this pattern to a single narrow line at the average chemical shift position of C^{13}'s in that molecular position.

7.18 Spin-Coherence Double Resonance – Introduction

The Overhauser-Pound and the cross-relaxation methods of double resonance are both dependent on the existence of relaxation processes of sufficient vigor. In the case of Overhauser-Pound schemes it is necessary that appropriate spin-lattice mechanisms are strong. In the case of cross-relaxation double resonance, cross-relaxation times must be sufficiently rapid (relative to spin-lattice relaxation) for the method to work. We now turn to the third family of double-resonance schemes. For these schemes to work, there must be a coupling between the two spin systems which manifests itself in splittings of the spectral lines which would exist in the absence of the coupling. We observe the double resonance by producing some form of modulation of this coupling. Crudely speaking, we observe the effect on splittings of the I-spin resonance produced by exciting the S-spin resonance. If $\delta\omega$ is a typical such spin-spin splitting, the essential condition for viability of this third family is that the effects of the splitting not be obscured by relaxation processes. If one thinks of $\delta\omega$ as some sort of beat frequency, the period of the beats is $1/\delta\omega$. The requirement is then physically that the coherence of this beat not be interrupted by relaxation. If T stands for the relaxation time (spin-spin, or spin-lattice) which could interrupt the coherence, observation of the coherence requires that $T \gg 1/\delta\omega$. The requirement of long relaxation times shows that spin-coherence double resonance tends to work under circumstances where the other methods do not.

7.19 A Model System – An Elementary Experiment: The S-Flip-Only Echo

To begin the discussion of spin-coherence double resonance we are going to imagine ourselves to be working in resonance before the invention of all the varied double resonance methods which exist today. We suppose that we have observed a H^1 resonance which is split into a doublet by coupling to another nucleus (for example a C^{13}). Since the C^{13} affects the H^1 spectrum, we ask ourselves is there not some way we can use the H^1 resonance to detect when the C^{13} is brought to resonance?

We first describe the model system, a pair of coupled spin $\frac{1}{2}$ nuclei I and S with a Hamiltonian similar to that of (7.29)

$$\mathcal{H} = -\gamma_I \hbar H_0 I_z - \gamma_S \hbar H_0 S_z + \mathcal{H}_{12}(t) \quad .$$

Initially we shall ignore relaxation effects, considering $\mathcal{H}_{12}$ to be independent of time. For simplicity we take it to be

$$\mathcal{H}_{12} = A\mathbf{I}\cdot\mathbf{S} \quad . \tag{7.158}$$

This is the form of the indirect spin-spin coupling in liquids. If I and S represent different nuclear species, the terms $I_x S_x$ and $I_y S_y$ are nonsecular. Then we approximate

$$\mathcal{H}_{12} = AI_z S_z \quad . \tag{7.159}$$

This is also the form which the dipolar coupling between two nonequivalent spins takes if we keep only the secular part, see (3.55). For indirect spin coupling in liquids, it is conventional to substitute J for A, and to speak of the J coupling. Since we wish to include dipolar coupling between unlike nuclei, we utilize A as the coupling constant.

If I and S are the same species but have different chemical shifts, we can represent the chemical shift difference by defining

$$\gamma_I = \gamma(1-\sigma_I) \quad , \quad \gamma_S = \gamma(1-\sigma_S) \quad , \tag{7.160}$$

where γ is characteristic of the species. If then

$$|(\gamma_I - \gamma_S)H_0| \gg A \tag{7.161}$$

we can approximate $\mathcal{H}_{12}$ by (7.159), but if the chemical shift differences are comparable to A or less than A, we must keep the complete expression $A\mathbf{I}\cdot\mathbf{S}$ for the coupling in a liquid. For a dipolar coupling in a solid, we would need to include both the A and B terms of (3.7). At this point we make a diversion to discuss notation. In the literature it is customary to refer to the case where (7.161) is satisfied as an "AX" case, whereas the case where the full expression (7.159) is needed as an "AB" case (do not confuse the A of AX with the A of $A\mathbf{I}\cdot\mathbf{S}$!). Each letter refers to a given chemical shift or resonance frequency.

Letters next to each other in the alphabet have chemical shifts comparable to their spin-spin couplings. Letters not adjacent have big differences in resonance frequency compared to their spin-spin coupling, either because the chemical shifts are quite different, or because the nuclei are different species (H^1 versus C^{13}). An A_2 system has two nuclei with the same chemical shift. An AB_2 or an AX_2 system has three nuclei. In the former case the chemical shift differences between the identical pair are not large compared to the A-B spin-spin coupling, in the latter case (AX_2) they are large. A notation AMX means three nuclei with big chemical shift differences. Note that if two nuclei are different species, they would be called an AX system.

We shall treat an AX system, hence take

$$\mathcal{H} = -\gamma_I \hbar H_0 I_z - \gamma_S \hbar H_0 S_z + A I_z S_z \quad . \tag{7.162}$$

The energy levels are

$$E = -\hbar\omega_{0I} m_I - \hbar\omega_{0S} m_S + A m_I m_S \quad , \tag{7.163}$$

where

$$\omega_{0I} \equiv \gamma_I H_0 \tag{7.164a}$$

$$\omega_{0S} \equiv \gamma_S H_0 \quad . \tag{7.164b}$$

If we apply alternating fields at frequencies ω_I or ω_S (near to the resonance frequencies of the I-spins or the S-spins, respectively), the resonance condition is

$$\omega_I = \omega_{0I} - a m_S \tag{7.165a}$$

$$\omega_S = \omega_{0S} - a m_I \quad \text{where} \tag{7.165b}$$

$$a \equiv A/\hbar \tag{7.166}$$

Since $m_S = \pm \frac{1}{2}$, the I resonance consists of two lines, spaced apart in angular frequency by a, as does the S resonance. Thus, through the couplings the presence of the S manifests itself in the I resonance. We now ask the question: *Is there some way we can use this manifestation to enable us to employ the I resonance as a means of detecting when we are producing a resonance with the S-spins?* In fact, in 1954, *Virginia Royden* [7.34] was interested in just this question in order to measure precisely the ratio of the γ's of H^1 to C^{13}. We describe her experiment and that of *Bloom* and *Shoolery* [7.35] based on the theory of *Bloch* [7.36] in Sect. 7.20.

One simple concept immediately comes to mind. Suppose the S-spins were polarized so that they were all in a given state m_S, say $m_S = +\frac{1}{2}$. Then the I spectrum would occur at $\omega_I = \omega_{0I} - a/2$. We now start searching for the S resonance by applying π pulses to them of angular frequency ω_S. We slowly sweep ω_S. After each S-spin pulse, we inspect the I resonance by inspecting

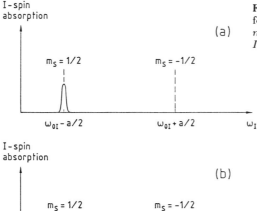

Fig. 7.17. I-spin absorption spectrum for S-spins 100% polarized with **(a)** $m_S = +\frac{1}{2}$ and **(b)** $m_S = -\frac{1}{2}$ versus I-spin search frequency ω_I

the I absorption spectrum. It will be unchanged *unless* ω_S was tuned to the S resonance. In that case, we change m_S from $+\frac{1}{2}$ to $-\frac{1}{2}$ and the I resonance shifts from $\omega_I = \omega_{0I} - a/2$ to $\omega_I = \omega_{0I} + a/2$. The situation is illustrated in Fig. 7.17.

Simple and beautiful as this scheme may seem, it alas cannot be used exactly as described without first getting the S-spins polarized, a very difficult task indeed. In fact, at normal applied field H_0 and temperatures θ, the condition $\gamma_S \hbar H_0 / k\theta \ll 1$ holds, so that to a good approximation one has an equal number of spins $m_S = +\frac{1}{2}$ as $m_S = -\frac{1}{2}$.

We can reanalyze the experiment very simply by saying that half of the I-spins have neighbors with $m_S = +\frac{1}{2}$, the other half have $m_S = -\frac{1}{2}$. Let us call these two groups (1) and (2) respectively. Figure 7.18 shows the "before" and "after" spectra for the two groups produced by the π pulse on the S-spins. Group (1) will have a resonance at $\omega_I = \omega_{0I} - a/2$ before the S-spins are flipped, and a resonance at $\omega_I = \omega_{0I} + a/2$ after. Group (2) will have a "before" resonance at $\omega_I = \omega_{0I} + a/2$, which will be shifted to $\omega_I = \omega_{0I} - a/2$ after the S-spins are flipped. We observe of course the spectrum of all the I-spins, which is the sum of the spectra of the two groups (1) and (2). As we see from Fig. 7.18e,f, the total spectrum is the same before and after. It looks as though our scheme has failed.

Fortunately, though the simplest version of our concept does not work, it can be modified very simply in any of several methods to be made to work. The essential point is that any one I-spin belongs to either group (1) or group (2). Since *either* group (1) or group (2) is affected by flipping the S-spins, both group (1) and group (2) can tell the difference between whether or not the S-spins are flipped. We must do an experiment which utilizes the memory that spins have

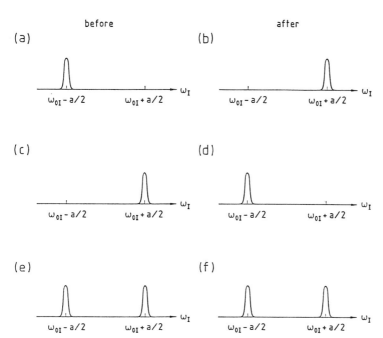

Fig. 7.18. The I-spin absorption before (left column) and after (right column) applying a π pulse to the S-spins, for the group (1) I-spins (**a, b**), the group (2) I-spins (**c, d**), and the total of all I-spins (**e, f**). Note that though the spectrum of *either* group (1) or group (2) alone is changed by flipping the S-spins, the *total* intensity pattern (**e, f**) is unaffected

of their prior history. An experiment which displays a spin's memory is easy to concoct. At $t = 0$ we apply a $\pi/2$ pulse to the I-spins with a sufficiently strong $(H_1)_I$ to flip the spins in both the I absorption lines. We view the I-spins in the reference frame rotating at ω_{0I}. In this frame, we will have two magnetization vectors of equal length, one precessing at angular frequency $a/2$, the other at $-a/2$. If we assume that immediately after the $\pi/2$ pulse the I magnetization lies along the y-axis [the situation if $(H_1)_I$ were along the x-axis], the magnetization will obey

$$\langle M_I^+(t)\rangle = \frac{C_0 i}{2}[\cos(at/2) + i\sin(at/2)]$$

$$+ \frac{C_0 i}{2}[\cos(at/2) - i\sin(at/2)] \quad (7.167\text{a})$$

$$= \frac{C_0 i}{2}(e^{iat/2} + e^{-iat/2}) \quad \text{where} \quad (7.167\text{b})$$

$$C_0 = \langle M_I(\theta_l)\rangle \quad (7.168)$$

gives the thermal equilibrium magnetization of the I-spins for the lattice temperature θ_l.

Then, at a time t_π we apply a π pulse to the S-spins. The group (1) spins will instantly change their precession frequency from $\omega_{0I} - a/2$ to $\omega_{0I} + a/2$ and the group (2) spins from $\omega_{0I} + a/2$ to $\omega_{0I} - a/2$. Thus, both groups will reverse their precession directions in the rotating frame. (Readers who are not content with the present informal treatment can look ahead at Sect. 7.24 for a formal justification of the relationships we use here.)

We can utilize the complex notation to express this rotation reversal. If a vector is represented by a complex number A with initial components A_x and A_y obeying

$$A_x = A_0 \cos \phi \quad , \quad A_y = A_0 \sin \phi \tag{7.169}$$

where A_0 is the magnitude of the vector, then $A = A_x + iA_y = A_0 \exp(i\phi)$.

If this vector is rotated at angular velocity ω, it becomes at a time t later

$$A(t) = A e^{i\omega t} = A_0 e^{i\phi} e^{i\omega t} \quad . \tag{7.170}$$

Therefore, we can apply this relationship to the two components of $\langle M_I^+(t) \rangle$ given in (7.167b) to get the magnetization at a time $t - t_\pi$ after the S-spin π pulse as

$$\langle M_I^+(t) \rangle = \frac{C_0 i}{2}(e^{iat_\pi/2} e^{-ia(t-t_\pi)/2} + e^{-iat_\pi/2} e^{ia(t-t_\pi)/2}) \quad . \tag{7.171}$$

Thus

$$\langle M_{Iy}(t) \rangle = C_0 \cos (at_\pi/2 - a(t - t_\pi)/2) \quad . \tag{7.172}$$

At time $t - t_\pi = t_\pi$ ($t = 2t_\pi$), $\langle M_{Iy}(t) \rangle$ will once again be C_0. In other words, we have produced an echo of the I-spins by flipping the S-spins. The S-spins so-to-speak cause the I-spin "field inhomogeneity" (a two-valued function). By flipping the S-spins we reverse this field inhomogeneity.

The theoretical time development of $\langle M_{Iy}(t) \rangle$ is shown in Fig. 7.19. An experimental demonstration of this phenomenon by Pennington et al. from the author's laboratory for the P^{31} resonance of liquid $(C_2H_5O)_2PHO$ is shown in Fig. 7.20.

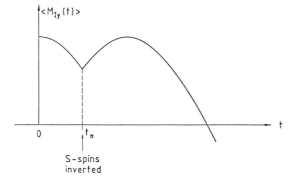

Fig. 7.19. The y-component of the I-spin magnetization versus time, t. At $t = t_\pi$, the S-spins are inverted by a π pulse

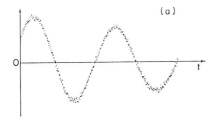

(a)

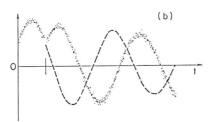

(b)

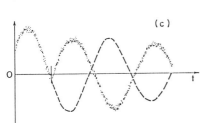

(c)

Fig. 7.20a–c. P^{31} NMR signal versus time for diethyl phosphite $[(C_2H_5O)_2PHO]$. (a) The precession of the P^{31} nuclei due to their coupling to the H^1 nuclei; (b) and (c) the precession of the P^{31} nuclei with the H^1 nuclei inverted at the indicated times. The dashed lines in (b) and (c) show what the precession would have been had there been no H^1 inversion. (Data taken by Charles Pennington with assistance from Jean-Philippe Ansermet, Dale Durand, and David Zax.)

A useful way of viewing this experiment is to plot the phases $\phi_1(t)$ and $\phi_2(t)$ of the two groups of I-spins where we define their phases at $t = 0$ as being given by (7.169). From (7.171), they are

$$\phi_1(t) = \begin{cases} (\pi/2) + at/2 & \text{for } t < t_\pi \\ (\pi/2) + at_\pi/2 - a(t - t_\pi)/2 & \text{for } t > t_\pi \end{cases}$$

$$\phi_2(t) = \begin{cases} (\pi/2) - at/2 & \text{for } t < t_\pi \\ (\pi/2) - at_{\pi/2} + a(t - t_\pi)/2 & \text{for } t > t_\pi \end{cases} \quad (7.173)$$

These are shown in Fig. 7.21.

If there were more than one pair of spins I-S, each pair with its characteristic chemical shift, the situation would be more complicated, but the same principles would apply. Let us label each I-S pair by k. ($k = 1$ to N if there are N distinct pairs.) We shall assume the I-S coupling is only within a given pair. Then we have N Hamiltonians of the form

$$\mathcal{H}_k = -\hbar\omega_{0Ik}I_{zk} - \hbar\omega_{0Sk}S_{zk} + A_k I_{zk} S_{zk} \quad , \quad (7.174)$$

so that, defining $a_k = A_k/\hbar$,

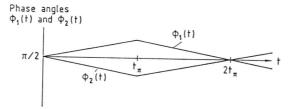

Fig. 7.21. The phase $\phi_1(t)$ and $\phi_2(t)$ of the two groups of spins (1) and (2) versus time. At t_π, the S-spins are inverted. At time $2t_\pi$, $\phi_1(t)$ and $\phi_2(t)$ are equal, corresponding to the two components of I magnetization being collinear, the peak of the "echo"

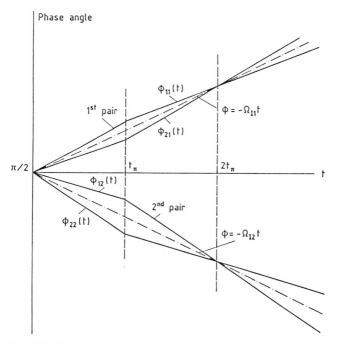

Fig. 7.22. The phase angles $\phi_{1k}(t)$ and $\phi_{2k}(t)$ versus time, t, for a case with two I-S pairs ($k = 1$ and 2). At $t = t_\pi$ the S-spins are flipped by π. At $t = 2t_\pi$, $\phi_{1k}(t) = \phi_{2k}(t)$ for *both* spin groups, indicating that both components of I magnetization for a given pair are collinear. Note, however, that since the two pairs have different chemical shifts, the I magnetizations of the different pairs are out of phase with each other

$$\langle M_I^+(t)\rangle_{\text{lab}} = \frac{C_0 i}{2} \sum_k (\exp[-i(\omega_{0Ik} - a_k/2)t]$$
$$+ \exp[-i(\omega_{0Ik} + a_k/2)t] \tag{7.175}$$

in the laboratory. Relative to some frame rotating at ω_I we have magnetization $\langle M_I^+(t)\rangle_{\omega_I}$. Defining

$$\Omega_{Ik} \equiv \omega_{0Ik} - \omega_I \quad , \tag{7.176}$$

$$\langle M_I^+(t) \rangle_{\omega_I} = \frac{C_0 i}{2} \sum_k (\exp[-i(\Omega_{Ik} - a_k/2)t]$$
$$+ \exp[-i(\Omega_{Ik} + a_k/2)t] \quad . \tag{7.177}$$

This general expression as well as the effect of a π pulse applied to the S-spins at time t_π can be represented by plotting $\phi_{1k}(t)$ and $\phi_{2k}(t)$ given by

$$\phi_{1k}(t) = (\pi/2) - (\Omega_{Ik} - a_k/2)t \quad \text{for} \quad t < t_\pi \tag{7.178}$$
$$= (\pi/2) - (\Omega_{Ik} - a_k/2)t_\pi - (\Omega_{Ik} + a_k/2)(t - t_\pi) \quad \text{for} \quad t > t_\pi \quad .$$

A similar expression holds for $\phi_2(k)$ except that the sign of a_k is reversed. In Fig. 7.22 we plot the ϕ's for a case of two I-S groups ($N = 2$). It is convenient to give a name to the experiment we have been analyzing. We shall call it the *S-flip-only echo*.

7.20 Spin Decoupling

The ideas we have just developed lead in a natural way to the concept of spin decoupling, one of the earliest goals of double resonance experiments. We will follow a pedagogical rather than a historical order.

The goal of decoupling is to simplify spectra. A typical NMR spectrum consists of many lines arising from the combined effect of chemical shifts and spin-spin couplings. Decoupling is a process which effectively eliminates the spin-spin couplings. It is very important both for high resolution spectra of liquids and spectra of solids, for example to eliminate the effect of proton spins on C^{13} spectra. We shall discuss removing the spin-spin coupling between different nuclear species such as H^1 and C^{13}. The basic idea on which all decoupling schemes work can be understood physically as follows. The existence of two lines in the I-spin spectrum corresponds to the fact that the S-spins have two orientations, up and down. If we can cause the S-spins to flip back and forth between the up and down orientations sufficiently rapidly, we should achieve an effect much like the motional narrowing of resonance lines. That is, an I-spin will precess at a time-averaged frequency rather than at one or the other of two discrete frequencies.

To study decoupling, we are going to pick a scheme which is easy to analyze. We will look at the effect on the I-spins of applying a sequence of π pulses to the S-spins. We shall see that as the time between S-spin π pulses gets shorter and shorter, the I-spin spectrum goes from two distinct lines separated by a in angular frequency, to a single line at the average frequency.

There are two general approaches we could take to showing this result. One would be to try to calculate what the I-spin absorption spectrum $\chi''(\omega)$ looks like as we flip the S-spins at a progressively more rapid rate. Or, we could look at the Fourier transform of the absorption line, which we do by looking at the time development of the I-spin magnetization following a $\pi/2$ pulse applied to the I-spins. In the absence of S-spin pulses, the I-spin magnetization will consist of two components, one oscillating at $\omega_{0I} + a/2$, the other at $\omega_{0I} - a/2$. So, if we choose $\omega_I = \omega_{0I}$, in the ω_I reference frame the components will precess at angular frequencies $+a/2$ and $-a/2$. If the system were perfectly decoupled, a would become effectively zero, so both components are at rest in the ω_{0I} reference frame, and the transverse I-spin magnetization will be a constant in time. (We are not including relaxation effects in our Hamiltonian. If we did, the I-spin transverse magnetization would decay rather than remaining a constant in time.)

We shall follow the Fourier transform approach for which, in fact, we have already set the stage in Sect. 7.19. There we analyzed what happens if we apply a single π pulse to the S-spins at a time $t = t_\pi$.

Let us then look back at Fig. 7.19 and at (7.171), which show that at time $t = 2t_\pi$ the magnetization $\langle M_I^+(t) \rangle$ has returned to its value at time $t = 0^+$, immediately after the $\pi/2$ pulse. Though the magnetization at the time of the π pulse is smaller by the factor $\cos(at_\pi/2)$, the π pulse has produced an echo at $t = 2t_\pi$. If we now think of ourselves as having a fresh start at $t = 2t_\pi$, we realize we can repeat the echo by applying a second π pulse to the S-spins. If we apply it at a time t'_π, hence $t'_\pi - 2t_\pi$ after the first echo, we will *again* form an echo. It will occur at a time $2(t'_\pi - 2t_\pi)$ after the first echo, or at a time t'' given by

$$t'' = t'_\pi + (t'_\pi - 2t_\pi) \quad , \tag{7.179}$$

see Fig. 7.23a. If we chose t'_π to be given by

$$t'_\pi = 3t_\pi \quad \text{then} \tag{7.180}$$

$$t'' = 4t_\pi \tag{7.181}$$

so that the time delay of the second echo after the first is identical to the time delay of the first echo after the initial signal. We can repeat this process again and again. The situation is shown in Fig. 7.23.

The phase angles $\phi_1(t)$ and $\phi_2(t)$ of Fig. 7.21 will then appear as shown in Fig. 7.24a. If we choose a shorter time for t_π, as shown in Fig. 7.24b, the maximum excursions of $\phi_1(t)$ and $\phi_2(t)$ away from $\pi/2$ will be less, so the magnetization versus time will appear like the solid curves in Fig. 7.25 rather than the dashed curves. Clearly, $\langle M_{Iy}(t) \rangle$ will be a periodic function as long as there is no relaxation, with period, T, equal to $2t_\pi$. We can think of $\langle M_{Iy}(y) \rangle$ as consisting of a constant with a superimposed periodic ripple. Figure 7.25 shows that the shorter t_π, the smaller the ripple, and the more $\langle M_{Iy}(t) \rangle$ approaches a constant in the reference frame rotating at ω_{0I}, hence behaves as though a were zero. Thus, we have "decoupled" the S-spins from the I-spins.

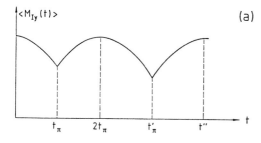
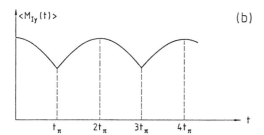

Fig. 7.23
(a) The magnetization $\langle M_{Iy}(t) \rangle$ versus t for the case of a π pulse applied to the S-spins at $t = t_\pi$, producing a refocusing at $t = 2t_\pi$, followed by another π pulse at $t = t'_\pi$, with a second refocusing at time t''. (b) The effect on $\langle M_{Iy}(t) \rangle$ of applying a pair of π pulses to the S-spins at times t_π and $t = 3t_\pi$, with resulting echoes at $t = 2t_\pi$ and $4t_\pi$

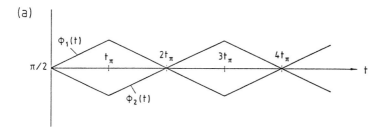
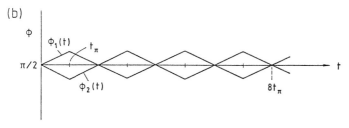

Fig. 7.24. (a) The phase angles $\phi_1(t)$ and $\phi_2(t)$ of the two components of I-spin magnetization. The π pulses applied to the S-spins at $t = t_\pi$ and $3t_\pi$ cause ϕ_1 and ϕ_2 to become equal at $t = 2t_\pi$ and $4t_\pi$, corresponding to the maxima in the curves of $\langle M_{Iy}(t) \rangle$ shown in Fig. 7.23b. (b) The effect on $\phi_1(t)$ and $\phi_2(t)$ of shortening t_π to half its value in (a) above. The echoes $[\phi_1(t) = \phi_2(t)]$ occur more frequently, and the maximum difference between $\phi_1(t)$ and $\phi_2(t)$, which occurs at $t = t_\pi$, $3t_\pi$, $5t_\pi$, etc., is reduced

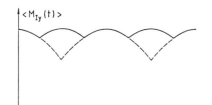

Fig. 7.25. The effect on $\langle M_{Iy}(t)\rangle$ of halving the time t_π. The solid-dashed curve corresponds to t_π of Fig. 7.24a, the solid-only curve corresponds to Fig. 7.24b. Note that $\langle M_{Iy}(t)\rangle$ can be viewed as a steady value with a superposed ripple, that the ripple is periodic with period $2t_\pi$, and that the amplitude of the ripple diminishes as t_π is shortened

We can express the existence of the ripple mathematically if we wish, by recognizing from Fig. 7.25 that since $\langle M_{Iy}(t)\rangle$ is a periodic function with period $2t_\pi$ it can be expressed as a Fourier series

$$\langle M_{Iy}(t)\rangle = \sum_{n=0}^{\infty} A_n \cos(2\pi nt/T) \quad \text{with} \tag{7.182}$$

$$T = 2t_\pi \quad . \tag{7.183}$$

The coefficient of the constant term, A_0, then turns out to be

$$A_0 = \langle M_I(\theta_l)\rangle \frac{\sin(at_\pi/2)}{at_\pi/2} \quad , \tag{7.184a}$$

which shows that as $at_\pi/2$ shrinks, A_0 approaches $\langle M_I(\theta_l)\rangle$.

The higher coefficients are

$$A_n = 2\langle M_I(\theta_l)\rangle \frac{\sin(at_\pi/2)}{at_\pi/2}(-1)^{n+1}\frac{(at_\pi/2)^2}{(\pi n)^2 - (at_\pi/2)^2} \quad , \tag{7.184b}$$

which become much smaller than the constant term as $at_\pi \to 0$. Note that if $at_\pi/2$ is very large, so that there are many oscillations in the I-spin rotating frame between S-spin flips, the coefficients A_n are very small for small n. However, for $\pi n = at_\pi/2$, the coefficients peak. (For larger values of n they decrease again.) Substituting into (7.182) this gives a time dependence for such values of n of

$$A_n \cos(2\pi nt/2t_\pi) = A_n \cos(at/2) \quad . \tag{7.185}$$

This value corresponds to the I-spins precessing at $+a/2$ and $-a/2$ in the rotating frame, the situation without decoupling. The transition from the split lines to the strongly decoupled lines will therefore look much like Fig. F.3 in Appendix F. Strongly decoupled spins require that there be no peak in the A_n's at frequencies $\approx \pm a/2$, hence that in (7.185) there be no term for $n \geq 1$ which has a denominator close to zero. This condition requires $(at_\pi/2) < \pi$.

If we took a Fourier transform of (7.182), we would get a constant term in the reference frame rotating at ω_{0I}, and side bands at frequencies $\pm(n/2t_\pi)$, where n is an integer. The amplitude of the side bands diminishes relative to the constant term as one shortens t_π.

We can summarize for the case that we flip the S-spins sufficiently frequently: following a $\pi/2$ pulse applied to the I-spins, they precess as though the coupling to the S-spins were turned off. This method of decoupling was in fact proposed by *Freeman* et al. in 1979 [7.37]. They call it "spin-flip" decoupling. In their paper, they discuss its advantages over other methods, as well as discussing interesting variants.

Another method of producing such a zero time-average z-component is to apply a constant H_1 at the S-spin resonance which will cause the S-spin to precess about the H_1. As a result, the z-component of S oscillates sinusoidally at angular frequency $\gamma_S(H_1)_S$. This method is historically the first method employed by *Royden* [7.34] and by *Bloom* and *Shoolery* [7.35].

To analyze it, we take the Hamiltonian to be

$$\mathcal{H} = -\gamma_I \hbar H_0 I_z - \gamma_S \hbar H_0 S_z + A I_z S_z \\ - \gamma_S \hbar (H_1)_S (S_x \cos \omega_S t - S_y \sin \omega_S t) \quad . \tag{7.186}$$

We have included the Zeeman interactions of the two spins, their spin-spin coupling, and the interaction of a rotating magnetic field $(H_1)_S$ with the S-spins. We have *not* included the interaction of $(H_1)_S$ with the I-spins since we assume that ω_S is close to the S-spin resonance, but far from the I-spin resonance.

We now view the problem in a double reference frame which is the lab frame for the I-spins and one rotating at ω_S for the S-spins. That is, if the original Hamiltonian $\mathcal{H}$ and wave function ψ obey

$$-\frac{\hbar}{i} \frac{\partial \psi}{\partial t} = \mathcal{H}\psi \quad , \tag{7.187}$$

where $\mathcal{H}$ is given by (7.186), we define a transformed wave function ψ' and transformed Hamiltonian $\mathcal{H}'$ by

$$\psi' = e^{-i\omega_S t S_z} \psi \quad , \tag{7.188}$$

which, when substituted into (7.187), gives

$$-\frac{\hbar}{i} \frac{\partial \psi'}{\partial t} = \mathcal{H}'\psi' \quad \text{with} \tag{7.189}$$

$$\mathcal{H}' = -\gamma_I \hbar H_0 I_z - \gamma_S \hbar \{[H_0 - (\omega_S/\gamma_S)]S_z + (H_1)_S S_x\} \\ + A I_z S_z \quad . \tag{7.190}$$

Defining

$$(h_0)_S \equiv H_0 - (\omega_S/\gamma_S) \quad \text{and} \tag{7.191a}$$

$$(\boldsymbol{H}_{\text{eff}})_S = \boldsymbol{i}(H_1)_S + \boldsymbol{k}(h_0)_S \tag{7.191b}$$

we get

$$\mathcal{H} = -\gamma_I \hbar H_0 I_z - \gamma \hbar (\boldsymbol{H}_{\text{eff}})_S \cdot \boldsymbol{S} + A I_z S_z \quad . \tag{7.192}$$

This Hamiltonian can in fact be solved exactly, but we will leave the exact solution as a homework problem. Instead, to solve this equation, we assume that we have applied a strong decoupling field so that

$$\gamma\hbar(H_{\text{eff}})_S \gg A \tag{7.193}$$

and treat the term AI_zS_z as a perturbation. It is then convenient to define axes (x', y', z') such that z' lies along $(\boldsymbol{H}_{\text{eff}})_S$, and the y- and y'-axes are coincident (Fig. 7.26).

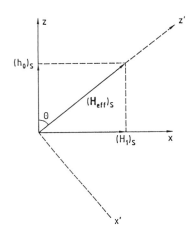

Fig. 7.26. The effective field acting on the S-spins in their rotating frame as a result of $(H_1)_S$ and of being somewhat off-resonance for the S-spins $[(h_0)_S \neq 0]$

This gives

$$S_z = S_{z'}\cos\theta - S_{x'}\sin\theta \quad \text{with} \tag{7.194a}$$

$$\cos\theta = \frac{(h_0)_S}{\sqrt{(H_1)_S^2 + (h_0)_S^2}} \ . \tag{7.194b}$$

The unperturbed Hamiltonian $\mathcal{H}_0'$, is then

$$\mathcal{H}_0' = -\gamma_I\hbar H_0 I_z - \gamma_S(H_{\text{eff}})_S S_{z'} \tag{7.195}$$

with eigenstates $|m_I m_{S'}\rangle$ which are eigenfunctions of I_z and $S_{z'}$. The perturbation AI_zS_z then becomes

$$AI_zS_z = AI_z(S_{z'}\cos\theta - S_{x'}\sin\theta) \ . \tag{7.196}$$

But the term in $S_{x'}$ has zero diagonal element in the $|m_I m_{S'}\rangle$ representation, giving for the energy eigenvalues

$$E'_{m_I m_{S'}} = -\gamma_I\hbar H_0 m_I - \gamma_S\hbar(H_{\text{eff}})_S m_{S'} + Am_I m_{S'}\cos\theta \ . \tag{7.197}$$

A weak rf field tuned close to the I-spin resonance can then induce transitions with a selection rule given by

$$(m_I m_{S'} | I_x | m'_I m'_{S'}) = (m_I | I_x | m'_I) \delta_{m_{S'} m'_{S'}} \quad , \tag{7.198}$$

so that the energy changes ΔE are

$$\Delta E(m_{S'}) = \gamma_I \hbar H_0 - A m_{S'} \cos\theta$$
$$= \gamma_I \hbar H_0 - A m_{S'} \frac{(h_0)_S}{\sqrt{(h_0)_S^2 + (H_1)_S^2}} \quad . \tag{7.199}$$

Corresponding to the two values of $m_{S'} = \pm \frac{1}{2}$, there are thus two I-spin resonance lines whose spacing is a function of how far ω_S is tuned away from the resonance frequency ω_{0S}. At exact S-spin resonance, $(h_0)_S = 0$, the I-spin resonance consists of a single line. Figure 7.27 shows how $\Delta E(m_{S'}) - \gamma_I \hbar H_0$ goes with $(h_0)_S/(H_1)_S$, demonstrating the collapse of the splitting as one tunes the S-spins to resonance.

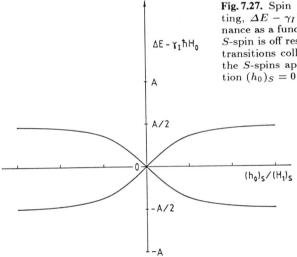

Fig. 7.27. Spin decoupling: the energy splitting, $\Delta E - \gamma_I \hbar H_0$, seen in the I-spin resonance as a function of $(h_0)_S$ the amount, the S-spin is off resonance, showing that the two transitions collapse to a single transition as the S-spins approach their resonance condition $(h_0)_S = 0$

The exact solution of (7.193) introduces another feature, the fact that the direction of the effective magnetic field acting on the S-spins depends on m_I, the orientation of the I-spins. This same problem arises in electron spin resonance and is treated in Chap. 11, (11.83–88). There it is shown that as a result the frequency of transition for $\Delta m_I = \pm 1$ is modified, and in addition transitions normally forbidden in which $m_{S'}$ changes become possible.

Decoupling by applying a steady alternating field was, as remarked previously, demonstrated by *Royden* [7.34] and by *Bloom* and *Shoolery* [7.35] based on ideas of *Bloch* [7.36] presented in an invited talk at an American Physical Society meeting. In Fig. 7.28 we show data from *Bloom* and *Shoolery*, demonstrating experimentally the collapse of the splitting as the S-spin oscillator approaches resonance.

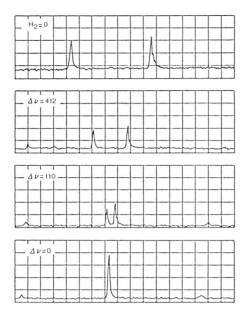

Fig. 7.28. Demonstration by *Bloom* and *Shoolery* of the effect on the F^{19} resonance of varying the P^{31} resonance frequency for an $(H_1)_P$ of 800 Hz. $\Delta\nu$ is the amount (in Hz) that the P^{31} is off-resonance. As $\Delta\nu$ approaches zero, the splitting goes to zero. "H_2" in the top line denotes $(H_1)_P$

As is evident from Figs. 7.27 and 28, the collapse of the splittings is only perfect when the S-spins are exactly on resonance, and the size of the splitting grows linearly with the S-spin offset for small offset. As a result, it takes a large $(H_1)_S$ to deal with a situation in which there are several S-spin chemical shifts present in the spectrum. *Anderson* and *Nelson* [7.38] and *Freeman* and *Anderson* [7.39] employed frequency modulation of $(H_1)_S$ to improve the decoupling over a broader range of $(h_0)_S$.

Ernst proposed a method of broad-band decoupling [7.40] in which he used a noise signal to modulate an rf carrier, for example, by switching the rf phase between two values (0 and π) at random time intervals. Figure 7.29, reproduced from his paper, shows a comparison of noise modulation with an unmodulated $(H_1)_S$ as a function of the S-spin offset for the case of $CHFCl_2$ in which he observed the F^{19} spectrum while decoupling the protons. Clearly noise modulation produces a dramatic improvement in the ability to decouple when not perfectly at resonance. This result follows from the fact that with a sufficiently broad noise spectrum, the residual splitting varies quadratically with offset instead of linearly as in Fig. 7.27. Some other decoupling schemes for broad-band decoupling are a coherent phase alternation method described by *Grutzner* and *Santini* [7.41], in which a 50% duty cycle square wave is used to phase modulate $(H_1)_S$. *Basus* et al. [7.42] discuss the advantages of adding a swept frequency for ω_S to the coherent phase shifting. The approach of Freeman, Kempsell, and Levitt can be further improved by the use of so-called composite pulses. Invented by *Levitt* and *Freeman* [7.43], they consist of a closely spaced group of pulses to achieve a desired rotation (such as a π pulse)

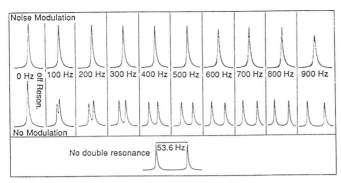

Fig. 7.29. Comparison of the decoupling with random noise and with a coherent rf frequency. The F^{19} spectrum of $CHFCl_2$ is observed while the proton frequency region is strongly irradiated. The spectra are taken for several different frequency offsets $\Delta\omega = \omega_S - \omega_2$, where ω_S is the proton resonance frequency and ω_2 is the carrier frequency of the strong rf field. The rf amplitude is identical throughout all the spectra and is $\gamma_S H_2/2\pi = 1370\,\text{Hz}$. The random noise is obtained by a binary phase modulation of the carrier with a 1023-bit shift register sequence with a shift frequency of 3 KHz. (From [7.40])

which is insensitive to some parameter such as the frequency offset of $(H_1)_S$. For example, instead of an $X(\pi)$ pulse, they suggest and analyze an $X(\pi/2)$ followed immediately by a $Y(\pi)$ followed immediately by another $X(\pi/2)$. The theory of composite pulses has been reviewed recently by *Levitt* [7.44] and by *Shaka* and *Keeler* [7.45]. *Freeman* has given an excellent review of the status of broad-band decoupling [7.46].

7.21 Spin Echo Double Resonance

We turn now to a slight variant of the basic double resonance experiment described in Sect. 7.20. The technique was first demonstrated by *Kaplan* and *Hahn* [7.47] and by *Emshwiller* et al. [7.48], and is called spin echo double resonance (SEDOR). This technique is important because the observation of a SEDOR proves that the two nuclei involved are physically near each other. Consider an ordinary spin echo with the echo formed at 2τ. During the first time interval τ, the spins dephase, but during the second interval τ they rephase. Hahn pointed out that if a nucleus of spin I has a neighboring spin belonging to a different species, an S-spin, the neighbor produces a local field which may aid or oppose the applied field, broadening the resonance much like a magnet inhomogeneity or the existence of several chemical shifts. Of course, magnet inhomogeneities do not affect the echo amplitude since the dephasing effect of the inhomogeneity during the first interval τ is exactly undone by a rephasing during the second interval τ. He noted, however, that if one added a second oscillator to flip the S-spin with a π pulse at the time the I-spin is given its π pulse, the sign of the field the S-spin produced at the I-spin would be opposite in the two time intervals τ. Thus, if a neighbor dephased the I-spin during the first interval, it

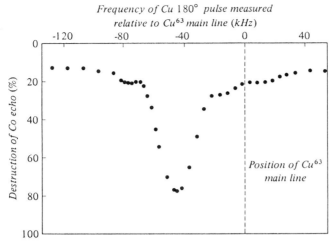

Fig. 7.30. Boyce's observation of SEDOR in a Cu powder containing 0.54 at. % Co near the Cu^{63} resonance frequence of pure Cu at 9907 Gauss and 1.5 K, showing the satellite due to the $\frac{1}{2} \leftrightarrow -\frac{1}{2}$ transition of the first neighbor. The parameters used are $\tau = 250\,\mu s$, $H_{1S} \simeq 5$ Gauss for a π pulse of the Cu $\frac{1}{2} \leftrightarrow -\frac{1}{2}$ transition. 100 echoes were averaged for each point and the dots are larger than the scatter and drift. Note that the much more abundant Cu^{63} nuclei a long distance from the Co, the so-called Cu main line, do not show up in the Co SEDOR

would continue to dephase during the second interval, producing a smaller echo at 2τ.

An example of a resonance detected in this manner is shown in Fig. 7.30. *Boyce* [7.49–51] studied a dilute alloy of Co in Cu. He wished to observe the nuclear resonance of Cu nuclei which are near neighbors of the Co nuclei. For such a dilute alloy, the neighbor resonances are weak in amplitude, frequently hidden under the tail of the resonance of Cu nuclei which are distant from impurities, the so-called "Cu main lines". Boyce was able to reveal the "hidden" resonance of the first neighbor by doing a spin echo double resonance in which he observed the effect on the Co spin-echo amplitude of a π pulse applied over a sequence of frequencies close to the Cu main line. Since distant nuclei do not produce much field at the Co, the "main line" has negligible effect on the Co echo height. But the first neighbor has a big effect. In this instance the use of double resonance can be seen as a way of selecting pairs of nuclei which are close in space.

Makowka et al. [7.52] utilized Pt^{195}-C^{13} SEDOR to detect the surface layer of Pt nuclei for small metal particles of Pt on whose surface they had adsorbed a monolayer of CO enriched to 90 % in C^{13}. The metal particles, which were tens of angstroms in diameter, were supported on Al_2O_3, a typical supported catalyst. Makowka et al. observed Pt^{195} spin echoes. Figure 7.31 shows their data. The straight Pt^{195} spin echo gives the line shape of the Pt nuclei in the small metal particles. This line is over 3 kGauss wide! To observe the Pt^{195} resonance of

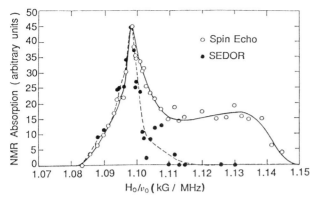

Fig. 7.31. Measurement of *Makowka* et al. [7.52] of the spin-echo (o) and SEDOR (•) line shapes of Pt^{195} for a sample of small particles of Pt metal supported on Al_2O_3 whose surface is coated with $C^{13}O$ molecules. The metal particles have diameters of a few tens of Å. The Pt^{195} spin echo gives the total line shape of all the Pt^{195} in the particle. The SEDOR data, involving Pt^{195}-$C^{13}O$ double resonance, and an add-subtract method, give the NMR line shape of the Pt^{195} nuclei in close proximity to the CO molecules, i.e. the surface layer of Pt atoms

the surface Pt atoms only, they employed an add-subtract technique in which on alternate spin echo cycles they applied a C^{13} π pulse coincident with the Pt^{195} π pulse. For those Pt nuclei far from the C^{13}, the echo was unaffected by the C^{13} π pulse, whereas for the surface atoms, the Pt^{195} echo was diminished by the C^{13} pulse. Thus, subtracting the Pt^{195} echoes when the C^{13} pulse was applied from those when it was not, the signal from the Pt^{195} nuclei not bonded to C^{13} nuclei vanishes.

To analyze the SEDOR signal, we note that the precession angle, θ, of the I-spins off resonance by $(h_0)_I$ during an interval τ is

$$\theta = (\gamma(h_0)_I - am_S)\tau \quad , \tag{7.200}$$

where $(h_0)_I$ represents the extent to which the particular I-spin is off resonance due either to magnetic field inhomogeneities or chemical shifts. Suppose, then, that at $t = 0$ we apply a $\pi/2$ pulse with $(H_1)_I$ along the I-spin x-axis in their rotating frame. This puts the corresponding magnetization $M((h_0)_I)$ along the $+y$-axis (Fig. 7.32a). At time τ^- just before the I and S π pulses, θ has the value θ_1 (Fig. 7.32b):

$$\theta_1 = \theta_0 \mp a\tau/2 \quad \text{where} \tag{7.201}$$

$$\theta_0 = \gamma_I (h_0)_I \tau \quad .$$

The I-spin π pulse reflects $M((h_0)_I)$ about the x-axis, producing the situation shown in Fig. 7.32c.

During the next time interval, τ, θ advances an angle θ_2 (Fig. 7.32d) given by

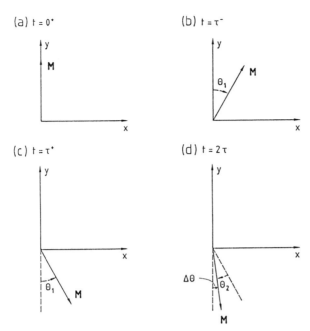

Fig. 7.32a–d. The I-spin magnetization for spin echo double resonance (SEDOR) seen in the I-spin rotating frame. M is the magnetization component off-resonance by $(h_0)_I$ for a particular value of m_S. At $t = 0$, $(H_1)_I$ is applied along the x-axis, producing a $\pi/2$ rotation of M onto the y-axis (a). During the time τ, M precesses through a net angle θ_1 given by (7.201) (b). At $t = \tau$, π pulses are applied along x_I and x_S to both the I-spins and the S-spins, so that at $t = \tau^+$, M is at the position shown in (c). During the next interval, τ, M precesses through the angle θ_2 given by (7.202), producing the situation of (d)

$$\theta_2 = \theta_0 \pm a\tau/2 \quad , \tag{7.202}$$

where the upper signs of (7.201) and (7.202) go together, as do the lower signs. Thus $M((h_0)_I)$ makes an angle $\Delta\theta$ (Fig. 7.32d) given by

$$\Delta\theta = \theta_1 - \theta_2 = \mp a\tau \quad . \tag{7.203}$$

Since $\Delta\theta$ is independent of θ_0 or $(h_0)_I$, it is independent of magnetic field inhomogeneity or of the existence of multiple chemical shifts. Since m_S is equally likely to be either $+\frac{1}{2}$ or $-\frac{1}{2}$, two resultant magnetization vectors, equal in length to $M_{0I}/2$ form, making angles $\Delta\theta$ of $a\tau$ or $-a\tau$ with the negative y-axis. Thus

$$\langle M_y(2\tau) \rangle = -M_{0I} \cos a\tau \quad . \tag{7.204}$$

It is useful for future reference to derive this result using the density matrix formalism. But first we demonstrate a useful theorem.

While the (H_1)'s are on, we customarily neglect the spin-spin coupling, giving in the doubly rotating frame a Hamiltonian

$$\mathcal{H}(I,S) = -\gamma_I\hbar[(h_0)_I I_z + (H_1)_I I_x]$$
$$-\gamma_S\hbar[(h_0)_S S_z + (H_1)_S S_x] \tag{7.205a}$$

$$\equiv \mathcal{H}(I) + \mathcal{H}(S) \quad , \tag{7.205b}$$

where by $\mathcal{H}(I)$ [$\mathcal{H}(S)$] we mean a Hamiltonian which is a function only of the spin components of the I-spins [S-spins]. If this Hamiltonian acts from a time t_1 to a time t_2, we have

$$\psi(t_2) = \exp\left(-\frac{i}{\hbar}\mathcal{H}(I,S)(t_2 - t_1)\right)\psi(t_1) \quad . \tag{7.206}$$

Since $\mathcal{H}(I)$ and $\mathcal{H}(S)$ commute, we can write this as

$$\begin{aligned}\psi(t_2) &= \exp\left(-\frac{i}{\hbar}\mathcal{H}(I)(t_2-t_1)\right)\exp\left(-\frac{i}{\hbar}\mathcal{H}(S)(t_2-t_1)\right)\psi(t_1)\\ &= \exp\left(-\frac{i}{\hbar}\mathcal{H}(S)(t_2-t_1)\right)\exp\left(-\frac{i}{\hbar}\mathcal{H}(I)(t_2-t_1)\right)\psi(t_1)\end{aligned} \tag{7.207}$$

The significance of (7.207) is that in analyzing the effect of the rf pulses, at a given time, we can treat the effects of $(H_1)_I$ independently of those of $(H_1)_S$. Thus, we can treat first the I-spins, then the S-spins, or vice versa. We do not have to treat both spins at once. Of course, the conclusion of this theorem is only valid during the time we can neglect the spin-spin coupling.

We now employ this theorem to analyze SEDOR using the density matrix. We consider a sequence in which the I-spin and S-spin pulses are defined in the doubly rotating frame. At $t = 0^-$ the spins are in thermal equilibrium. At $t = 0$ we apply a $\pi/2$ pulse with $(H_1)_I$ along the I-spin x-axis. We denote it as an $X_I(\pi/2)$ pulse. We wait a time τ, then apply simultaneously $X_I(\pi)$ and $X_S(\pi)$ pulses. We then follow the density matrix $\varrho(t)$ and the magnetization at later time, t, such that $t > \tau$, looking for an echo at $t - \tau = \tau$ or $t = 2\tau$. We denote this sequence with the notation

$$X_I(\pi/2)\ldots\tau\ldots[X_I(\pi), X_S(\pi)]\ldots(t-\tau) \quad .$$

We shall assume the I-spins are off resonance by $(h_0)_I$ and the S-spins by $(h_0)_S$, to allow for chemical shifts, and use the notations of (7.176)

$$\Omega_I = \gamma_I(h_0)_I - \omega_I \quad , \qquad \Omega_S = \gamma_S(h_0)_I - \omega_S \quad . \tag{7.208}$$

Then,

$$\langle I^+(t)\rangle = \text{Tr}\left\{I^+\varrho(t)\right\} \quad . \tag{7.209}$$

To find $\varrho(t)$, we must first know its value at $\varrho(0^-)$, corresponding to thermal equilibrium. We thus take

$$\varrho(0^-) = \frac{1}{Z}\left(1 + \frac{\hbar\omega_{0I}I_z + \hbar\omega_{0S}S_z}{kT}\right) \tag{7.210}$$

with Z the partition function. In calculating $\langle I^+(t)\rangle$ we need keep only the term

$$\varrho(0^-) = \frac{\hbar\omega_{0I}}{ZkT}I_z \quad . \tag{7.211}$$

We thus have

$$\begin{aligned}\langle I^+(t)\rangle = \frac{\hbar\omega_{0I}}{ZkT}\text{Tr}\,\{&I^+\exp\left(i(\Omega_I I_z + \Omega_S S_z - aI_z S_z)(t-\tau)\right)X_I(\pi)X_S(\pi)\\ &\times\exp\left(i(\Omega_I I_z + \Omega_S S_z - aI_z S_z)\tau\right)X_I(\pi/2)I_z X_I^{-1}(\pi/2)\\ &\times\exp\left(-i(\Omega_I I_z + \Omega_S S_z - aI_z S_z)\tau\right)X_S^{-1}(\pi)X_I^{-1}(\pi)\\ &\times\exp\left(-i(\Omega_I I_z + \Omega_S S_z - aI_z S_z)(t-\tau)\right)\} \quad . \end{aligned} \tag{7.212}$$

Utilizing that

$$X_I(\pi/2)I_z X_I^{-1}(\pi/2) = I_y \tag{7.213}$$

and inserting

$$X_S^{-1}(\pi)X_I^{-1}(\pi)X_I(\pi)X_S(\pi) \tag{7.214}$$

before $X_I(\pi/2)$ and its inverse after $X_I^{-1}(\pi/2)$ we then transform

$$\begin{aligned}&X_I(\pi)X_S(\pi)\exp\left(i(\Omega_I I_z + \Omega_S S_z - aI_z S_z)\tau\right)X_S^{-1}(\pi)X_I^{-1}(\pi)\\ &= \exp\left(-i(\Omega_I I_z + \Omega_S S_z)\tau\right)\exp\left(-iaI_z S_z \tau\right)\end{aligned} \tag{7.215}$$

and

$$X_I(\pi)X_S(\pi)I_y X_S^{-1}(\pi)X_I^{-1}(\pi) = X_I(\pi)I_y X_I^{-1}(\pi) = -I_y \quad . \tag{7.216}$$

Substituting, we then get

$$\begin{aligned}\langle I^+(t)\rangle = \frac{\hbar\omega_{0I}}{ZkT}\text{Tr}\,\{&I^+\exp\left(i(\Omega_I I_z + \Omega_S S_z - aI_z S_z)(t-\tau)\right)\\ &\times\exp\left(-i(\Omega_I I_z + \Omega_S S_z)\tau\right)\exp\left(-iaI_z S_z \tau\right)(-I_y)\\ &\times\exp\left(iaI_z S_z \tau\right)\exp\left(i(\Omega_I I_z + \Omega_S S_z)\tau\right)\\ &\times\exp\left(-i(\Omega_I I_z + \Omega_S S_z - aI_z S_z)(t-\tau)\right)\} \quad . \end{aligned} \tag{7.217a}$$

This expression can be simplified immediately since the terms $\exp\left(i\Omega_S S_z(t-\tau)\right)$ and $\exp\left(-i\Omega_S S_z \tau\right)$ on the left of $(-I_y)$ commute with everything between them and the terms $\exp\left(i\Omega_S S_z \tau\right)$ and $\exp\left(-i\Omega_S S_z(t-\tau)\right)$ to the right. Hence, they can be combined, giving a factor of unity. We then get

$$\begin{aligned}\langle I^+(t)\rangle = \frac{\hbar\omega_{0I}}{ZkT}\text{Tr}\,\{&I^+\exp\left(i(\Omega_I I_z - aI_z S_z)(t-\tau)\right)\\ &\times\exp\left(-i\Omega_I I_z \tau\right)\exp\left(-iaI_z S_z \tau\right)(-I_y)\exp\left(iaI_z S_z \tau\right)\\ &\times\exp\left(+i\Omega_I I_z \tau\right)\exp\left(-i(\Omega_I I_z - aI_z S_z)(t-\tau)\right)\} \quad . \end{aligned} \tag{7.217b}$$

At $t - \tau = \tau$ or $t = 2\tau$, the factors involving Ω_I combine to unity, leaving

$$\langle I^+(2\tau)\rangle = \frac{\hbar\omega_{0I}}{ZkT}\,\text{Tr}\,\{I^+e^{-iaI_zS_z2\tau}(-I_y)e^{iaI_zS_z2\tau}\} \quad . \tag{7.218}$$

The quantity in the curly bracket is just what we would have if $(h_0)_I$ and $(h_0)_S$ vanished, and the system developed solely under the influence of the spin-spin coupling, aI_zS_z. Thus, chemical shifts and field inhomogeneities are refocused, but not spin-spin couplings.

Using the m_Im_S representation to evaluate the trace, and recognizing that

$$(m_Im_S|I^+|m_I'm_S') = \langle\tfrac{1}{2}|I^+|-\tfrac{1}{2}\rangle\delta_{m_Sm_S'}\delta_{m_I,1/2}\delta_{m_I',-1/2} \tag{7.219}$$

we get

$$\text{Tr}\,\{I^+e^{-iaI_zS_z2\tau}I_ye^{iaI_zS_z2\tau}\}$$
$$= \sum_{m_S}\langle\tfrac{1}{2}|I^+|-\tfrac{1}{2}\rangle e^{iam_S\tau}\langle-\tfrac{1}{2}m_S|I_y|\tfrac{1}{2}m_S\rangle e^{iam_S\tau} \quad . \tag{7.220}$$

But

$$I_y = \frac{i}{2}(I^- - I^+) \quad \text{and} \tag{7.221}$$

$$\langle\tfrac{1}{2}|I^+|-\tfrac{1}{2}\rangle = \langle-\tfrac{1}{2}|I^-|\tfrac{1}{2}\rangle$$
$$= 1 \quad \text{so that} \tag{7.222}$$

$$\langle I^+(2\tau)\rangle = -\frac{\hbar\omega_{0I}}{ZkT}\frac{i}{2}\sum_{m_S}e^{i(am_S2\tau)} = -\frac{\hbar\omega_{0I}}{ZkT}i\cos a\tau \quad . \tag{7.223}$$

In the high temperature approximation, $Z = 4$ so that

$$\langle I^+(2\tau)\rangle = -\frac{\hbar\omega_{0I}}{4kT}i\cos a\tau \quad . \tag{7.224}$$

Using for thermal equilibrium

$$M_{0I} = \chi_0 H_0 = \gamma\hbar\langle I_z\rangle$$

$$\chi_0 = \frac{\gamma^2\hbar^2 I(I+1)}{3kT} \tag{7.225}$$

we get

$$\langle I_z\rangle = \frac{\hbar\omega_{0I}}{4kT} \equiv I_0 \quad \text{so that} \tag{7.226}$$

$$\langle I^+(2\tau)\rangle = -iI_0\cos a\tau \quad . \tag{7.227}$$

The $-i$ factor indicates that the net spin lies along the negative y-axis.

We can now utilize the formalism, in particular going back to (7.212), (7.213), and (7.215), to compare three pulse sequences, the S-flip-only echo of Sect. 7.19, the conventional spin echo and spin echo double resonance:

S-flip-only echo (Sect. 7.19):

Sequence: $X_I(\pi/2)\ldots\tau\ldots X_S(\pi)\ldots(t-\tau)$.

Conventional I-spin echo:

Sequence: $X_I(\pi/2)\ldots\tau\ldots X_I(\pi)\ldots(t-\tau)$.

SEDOR:

Sequence: $X_I(\pi/2)\ldots\tau\ldots[X_S(\pi), X_I(\pi)]\ldots(t-\tau)$.

For these we need to insert for (7.214) respectively

Basic S-flip only: $\quad X_S^{-1}(\pi)X_S(\pi)$

Conventional I echo: $\quad X_I^{-1}(\pi)X_I(\pi)$

SEDOR: $\quad X_S^{-1}(\pi)X_I^{-1}(\pi)X_I(\pi)X_S(\pi)$

so that at time 2τ we will have had for the transformation of the term $\exp[i(\Omega_I I_z + \Omega_S S_z - aI_z S_z)\tau]$

Basic S-flip only: $\quad \exp[i(\Omega_I I_z - \Omega_S S_z + aI_z S_z)\tau]$

Conventional I echo: $\quad \exp[i(-\Omega_I I_z + \Omega_S S_z + aI_z S_z)\tau]$

SEDOR: $\quad \exp[i(-\Omega_I I_z - \Omega_S S_z - aI_z S_z)\tau]$.

As in the transformation from (7.217a) to (7.217b), we can combine the terms involving exponentials of $\Omega_S S_z(t-\tau)$ or $\Omega_S S_z \tau$ on the left of $(-I_y)$ with their inverses on the right of $(-I_y)$. What counts are the terms involving I_z or $I_z S_z$. Thus we see that:

- The basic S-flip-only π pulse changes the sign of the a term but leaves the sign of the Ω_I alone. Therefore, it refocuses the spin-spin coupling at $t = 2\tau$ but leaves the I-spin chemical shift alone.
- The conventional echo π pulse changes the sign of both spin-spin and chemical shift terms. Therefore it refocuses both spin-spin and chemical shifts at $t = 2\tau$, eliminating both.
- The SEDOR π pulses change the sign of the chemical shift term only. Therefore they refocus the chemical shifts, at $t = 2\tau$, effectively eliminating them, but leaving the spin-spin coupling untouched.

A useful variant on the SEDOR pulse sequence was introduced by *Wang* et al. [7.53, 54]. The time of the I-spin π pulse is held fixed. The S-spin π pulse is applied at a time T after the I-spin $\pi/2$ pulse ($T \leq \tau$). This sequence gives, in the notation of (7.201),

$$\theta_1 = \theta_0 \mp aT/2 \pm a(\tau - T)/2 \quad \text{and} \tag{7.228a}$$

$$\theta_2 = \theta_0 \pm a\tau/2 \quad \text{so that} \tag{7.228b}$$

$$\Delta\theta = \theta_1 - \theta_2 = \mp aT \quad \text{and} \tag{7.229}$$

$$\langle I^+(2\tau) \rangle = -iI_0 \cos aT \quad . \tag{7.230}$$

This sequence has two advantages. The first is that one can work at very short times, T, while keeping the I-spin echo delayed a convenient time. This feature eliminates I-spin amplifier recovery problems. The second advantage is that the decay of the I-spin echo with pulse spacing (by T_2 processes) is held constant as T is varied, and thus eliminated from the result.

7.22 Two-Dimensional FT Spectra – The Basic Concept

We now arrive at two-dimensional Fourier transform magnetic resonance, one of the most exciting and powerful developments since the original discovery of magnetic resonance. In this section we introduce the concepts using a very simple example, the basic S-flip-only experiment discussed in Sect. 7.19. We shall find that it already contains all the essential ingredients of two-dimensional magnetic resonance.

The idea of two-dimensional Fourier transformation is due to *Jeener*. He first presented it in lectures at the Ampère International Summer School, Basko Potze, Yugoslavia, in 1971 [7.55]. An excellent review by *Freeman* and *Morris* [7.56] describes the explosive development of the field, and states and documents that "Ernst was the first to appreciate the great potential of the method". Indeed, both Freeman and Ernst with their collaborators provided many of the fundamental applications. *Aue* et al. gave a treatment of basic significance and broad generality [7.57] which also gives a detailed treatment of concrete applications. It is a fundamental reference for workers in this field. Recent reviews include those by *Bax* [7.58] and *Turner* [7.59].

We shall describe in later sections some of the particular forms of double resonance, but our goal in this section is to introduce the concepts using S-flip-only double resonance (Sect. 7.19).

We saw in Sect. 7.19, (7.177), that for a spin I off resonance in the frame rotating at ω_I by

$$\Omega_I = \omega_{0I} - \omega_I$$

the magnetization was

$$\langle M_I^+(t)\rangle = \frac{iC_0}{2}(e^{-i(\Omega_I+a/2)t} + e^{-i(\Omega_I-a/2)t}) \qquad (7.231)$$

(where the $-\Omega_I$ means the sense of rotation is negative) for $t < t_\pi$ and

$$\langle M_I^+(t)\rangle = \frac{iC_0}{2}e^{-i\Omega_I t}$$
$$\times (e^{+iat_\pi/2}e^{-ia(t-t_\pi)/2} + e^{-iat_\pi/2}e^{+ia(t-t_\pi)/2}) \text{ for } t > t_\pi. \quad (7.232)$$

Exactly at $t = 2t_\pi$, we then have

$$\langle M_I^+(t)\rangle = \frac{iC_0}{2}e^{-i\Omega_I(2t_\pi)} \quad .$$

This, as we remarked at the end of the previous section, means that at $t = 2t_\pi$, the magnetization is just what it would be if the spin-spin coupling were zero, so that only the chemical shift $\Omega_I = \omega_{0I} - \omega_I$ remains. [We here assume a uniform static field, so that $(h_0)_I$ arises because of chemical shifts.] The situation is illustrated in Fig. 7.33 where we show (Fig. 7.33a) that for times greater than $2t_\pi$, the phases ϕ_1 and ϕ_2 act as though from $0 < t < 2t_\pi$ there were only a chemical shift, but for $t > 2t_\pi$ both chemical shifts and spin-spin splittings exist. In Fig. 7.33b we have redrawn the situation to represent the situation of $a = 0$ for $0 < t < 2t_\pi$.

If we reset t_π to a new value, t'_π longer than t_π, the new situation is shown in Figs. 7.33c and d. We can plot phases for a sequence of values t_π, t'_π, t''_π, etc., as in Fig. 7.33e.

These plots can be summarized by defining two times, t_1 and t_2, by

$$\begin{aligned} t_1 &= 2t_\pi \quad (2t'_\pi, 2t''_\pi, \text{ etc.}) \\ t_2 &= t - 2t_\pi \quad (t - 2t'_\pi, \text{ etc.}) \end{aligned} \qquad (7.233)$$

In terms of t_1 and t_2, we have that for times after t_1,

$$\begin{aligned} \langle M_I^+(t)\rangle &= \frac{iC_0}{2}e^{-i\Omega_I t_1}e^{-i\Omega_I t_2}(e^{iat_2/2} + e^{-iat_2/2}) \\ &\equiv f(t_1, t_2) \quad . \end{aligned} \qquad (7.234)$$

This expression is derived for $t_1 \geq 0$, $t_2 \geq 0$. If at this point we treat this theoretical expression formally as existing for $-\infty < t_1 < +\infty$ and $-\infty < t_2 < +\infty$, we can take its Fourier transform with respect to the two time variables t_1 and t_2, to obtain

$$g(\omega_1, \omega_2) \equiv \frac{1}{(2\pi)^2} \int_{-\infty}^{+\infty} e^{-i\omega_1 t_1}e^{-i\omega_2 t_2} f(t_1, t_2) dt_1 dt_2 \quad . \qquad (7.235)$$

Utilizing the relation $\delta(x) = (1/2\pi)\int_{-\infty}^{+\infty} \exp(ixt)dt$, we get

$$g(\omega_1, \omega_2) = \frac{iC_0}{2}\delta(\omega_1 + \Omega_I)[\delta(\omega_2 + \Omega_I - a/2) + \delta(\omega_2 + \Omega_I + a/2)]. \quad (7.236)$$

These data can be represented as points in a two-dimensional plot (ω_1, ω_2) (Fig. 7.34a). If there were a second chemical shift and a second spin-spin splitting, of strength b, we could label the chemical shifts as Ω_{Ia}, Ω_{Ib}, and then

$$\langle M_I(t)\rangle = \frac{iC_{0a}}{2}e^{-i\Omega_{Ia}t_1}e^{-i\Omega_{Ia}t_2}(e^{iat_2/2}+e^{-iat_2/2})$$
$$+\frac{iC_{0b}}{2}e^{-i\Omega_{Ib}t_1}e^{-i\Omega_{Ib}t_2}(e^{ibt_2/2}+e^{-ibt_2/2}) \qquad (7.237)$$

and we would get

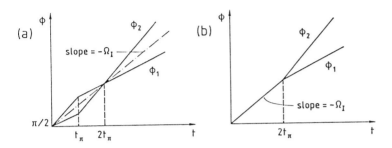

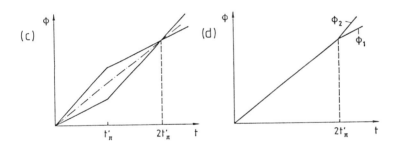

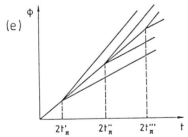

Fig. 7.33a–e. Phases of the two I-spin magnetization components versus time for the S-flip-only experiment. (a) The two phases $\phi_1(t)$ and $\phi_2(t)$, showing the time t_π at which the π pulse is applied to the S-spins, and the formation of the echo at $t=2t_\pi$ which refocuses the spin-spin splitting. If the spin-spin coupling were zero, the ϕ's would both follow the dashed curve of slope Ω_I representing a chemical shift only. (b) The behavior of ϕ equivalent, for $t>2t_\pi$, to that of (a). (c), (d) Replots of (a) and (b) respectively for a value $t'_\pi > t_\pi$. (e) A family of phase trajectories corresponding to a family of experiments with different times t_π (t'_π, t''_π, etc.)

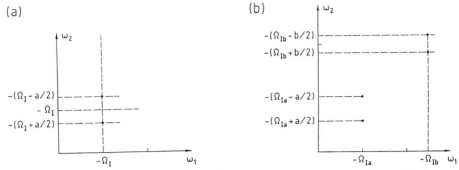

Fig. 7.34. (a) The two-dimensional plot of frequencies ω_1 and ω_2, showing points in ω_1 corresponding to the chemical shift $\omega_1 = -(\Omega_I \pm a/2)$. (b) A plot similar to (a) corresponding to two I-spins, each coupled to a single S-spin with corresponding chemical shifts Ω_{Ia} and Ω_{Ib} and spin-spin couplings a and b

$$g(\omega_1, \omega_2) = \frac{iC_{0a}}{2}\delta(\omega_1 + \Omega_{Ia})[\delta(\omega_2 + \Omega_{Ia} - a/2) + \delta(\omega_2 + \Omega_{Ia} + a/2)]$$
$$+ \frac{iC_{0b}}{2}\delta(\omega_1 + \Omega_{Ib})[\delta(\omega_2 + \Omega_{Ib} - b/2)$$
$$+ \delta(\omega_2 + \Omega_{Ib} + b/2)] \quad , \tag{7.238}$$

which would give the plot of Fig. 7.34b.

This plot allows us to go along the ω_1-axis to determine the values of the I-spin chemical shifts, then at each chemical shift position to observe in the ω_2 direction splitting of those particular I-spins by the S-spins to which they are coupled. That is, there is a *correlation* of the ω_1 frequencies with the ω_2 frequencies. The physical origin of this correlation can be traced back to the phase diagram Fig. 7.35, which shows that the single phase line over the time interval 0 to $2t_\pi$ splits into two lines for times after $2t_\pi$. For $t > 2t_\pi$, those

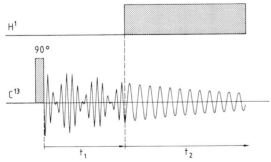

Fig. 7.35. Schematic representation of one form of C^{13} 2D-resolved spectroscopy after *Müller* et al. [7.60]. The C^{13} nuclei are the I-spins, H^1 the S-spins. For the interval t_1 the full Hamiltonian acts, but during t_2 the spin-spin interaction is turned off by broad-band decoupling. The situation is similar to S-flip-only resonance with the role of the times t_1 and t_2 interchanged

lines extrapolate back to an intersection at $t = 2t_\pi$. The two lines are diverging from their intersection since the chemical shift frequency Ω_I is split into two frequencies, $\Omega_I \pm a/2$, for times after $2t_\pi$.

The example illustrates the general feature of two-dimensional schemes that there is some initial preparation of the system [e.g. the $X_I(\pi/2)$ pulse], then the system evolves for a time interval t_1 under one effective Hamiltonian (in our case, one in which the spin-spin couplings are zero), then data are collected during a second time interval, t_2, for which the Hamiltonian takes on some new effective form (in our case the full Hamiltonian). The experiment must then be repeated for other values of t_1. These three intervals are often referred to as the preparation, the evolution, and the acquisition intervals, respectively.

The experiment we have just described is in fact essentially the famous experiment invented by *Müller* et al. [7.60] to resolve the C^{13} chemical shifts and C^{13}-H^1 spin-spin couplings. They demonstrated the technique using *n*-hexane (Figs. 7.35 and 36). They utilized broad-band (noise) decoupling of the protons instead of applying a π pulse as in our example, and they decoupled during the second time interval, t_2, instead of during the first time interval, t_1, as we did in our hypothetical example. They point out one could choose to decouple during either interval, and remark that since each point t_1 requires a separate experiment, one needs less data if the t_1 spectra is chosen to have the fewer spectral lines, i.e. is the time during which the protons are decoupled. In a later paper, they describe the sequence of our example [7.61]. This general method is referred to as *J*-resolved 2D NMR in the literature.

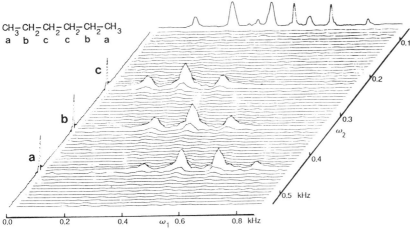

Fig. 7.36. Data of *Müller, Kumar,* and *Ernst* using the pulse sequence of Fig. 7.35 for the C^{13} spectrum of *n*-hexane (CH_3-CH_2-CH_2-CH_2-CH_2-CH_3). The ω_1-axis gives the combined chemical shift and spin-spin splitting, the ω_2-axis the chemical shifts. These data show how a 2D display enables one to greatly simplify unraveling a complex spectrum. Twenty-two experiments were coadded for each of the 64 values of t_1 between 0 and 35 ms. The authors state that the resolution is severely limited by the 64×64 data matrix used to represent the 2D Fourier transform. The absolute values of the Fourier coefficients are plotted. The undecoupled 1D spectrum is indicated along the ω_1-axis and the proton-decoupled spectrum is shown along the ω_2-axis

We shall discuss further examples in later sections. First, however, we need to take a side trip to deal briefly with a topic the reader will quickly encounter in studying the literature, the problem of 2D NMR line shapes.

7.23 Two-Dimensional FT Spectra – Line Shapes

In the previous section we developed the Fourier transform of the theoretical expression (7.234) by formally extending the meaning of t_1 and t_2 to negative times. This is easy to do given a theoretical expression. If, however, we have experimental data, we cannot in general guess the theoretical expression. One then encounters a problem which the reader will immediately come across if he reads the literature of 2D NMR. We therefore take it up briefly. A thorough treatment is given in the basic paper of *Aue* et al. [7.57].

Let us therefore reexamine how to treat experimental data considering positive times only. Since the problem is one of mathematics, we assume we have the simplest form of experimental data given by an exponentially decaying sine (cosine) wave. We put in the decay since such decays are always present, and to avoid difficulties at infinite times. Thus, suppose we have

$$f(t_1, t_2) = A e^{(i\Omega_1 - \alpha_1)t_1} e^{(i\Omega_2 - \alpha_2)t_2} \quad . \tag{7.239}$$

Then

$$g(\omega_1, \omega_2) = \frac{1}{(2\pi)^2} \int_0^\infty dt_1 \int_0^\infty dt_2 f(t_1, t_2) e^{-i\omega_1 t_1} e^{-i\omega_2 t_2} \tag{7.240a}$$

$$= \frac{1}{(2\pi)^2} \frac{A}{[\alpha_1 - i(\Omega_1 - \omega_1)][\alpha_2 - i(\Omega_2 - \omega_2)]} \tag{7.240b}$$

$$= \frac{A}{(2\pi)^2} \frac{\alpha_1 + i(\Omega_1 - \omega_1)}{\alpha_1^2 + (\Omega_1 - \omega_1)^2} \frac{\alpha_2 + i(\Omega_2 - \omega_2)}{\alpha_2^2 + (\Omega_2 - \omega_2)^2} \quad . \tag{7.240c}$$

In the corresponding one-dimensional case

$$f(t_1) = A e^{(i\Omega_1 - \alpha_1)t_1} \tag{7.241}$$

$$g(\omega_1) = \frac{1}{2\pi} \int_0^\infty e^{-i\omega_1 t_1} f(t_1) dt_1 \tag{7.242}$$

we get

$$g(\omega_1) = \frac{A}{2\pi} \frac{\alpha_1 + i(\Omega_1 - \omega_1)}{\alpha_1^2 + (\Omega_1 - \omega_1)^2} \quad . \tag{7.243}$$

We recognize that the real part is the Lorentzian line shape (the absorption), and the imaginary part is the dispersion. If we use quadrature detection (Sect. 5.8) we can select either the real or the imaginary part of $g(\omega_1)$ and thus either the absorption signal or the dispersion signal.

If, however, we take only the real part of $g(\omega_1, \omega_2)$, it involves (for A real)

$$\text{Re}\{g(\omega_1, \omega_2)\} = A \frac{\alpha_1 \alpha_2 - (\Omega_1 - \omega_1)(\Omega_2 - \omega_2)}{[\alpha_1^2 + (\Omega_1 - \omega_1)^2][\alpha_2^2 + (\Omega_2 - \omega_2)^2]} \quad , \tag{7.244}$$

which is a *mixture* of absorption and dispersion.

Of course, if one goes along the frequency contour $\omega_2 - \Omega_2$, one has

$$\text{Re}\{g(\omega_1, \omega_2)\} = \frac{A}{\alpha_2} \frac{\alpha_1}{\alpha_1^2 + (\Omega_1 - \omega_1)^2} \quad , \tag{7.245}$$

which is the ω_1 absorption line. However, the practical problem is that since dispersion extends out farther from the line-center than does absorption, the mixing of dispersion tends to decrease the resolution between adjacent lines. There are various methods of processing data to deal with these matters. In addition to the treatment of *Aue* [7.57], the review articles by *Freeman* and *Morris* [7.56], *Bax* [7.58], and *Turner* [7.59] discuss these issues.

7.24 Formal Theoretical Apparatus I – The Time Development of the Density Matrix

In discussing the S-flip-only experiment (Sect. 7.19), where we were applying π pulses to the S-spins, we assumed that the effect of the S-flips was simply to shift the precession frequency of the I-spins suddenly from one m_S multiplet to the other. An S-spin originally pointing up now points down, and vice versa. If, however, one applies a $\pi/2$ pulse to the S-spins, classically they are now *perpendicular* to H_0. How then do we treat what happens in a quantum mechanical system? Classically, if S_z is zero, the S-spin will not shift the I-spin precession. That would lead to a precession midway between the two I-spin lines at $\omega_{0I} + a/2$ and $\omega_{0I} - a/2$. But there is no transition at ω_{0I} among the eigenstates of the Hamiltonian, so this answer cannot be right. The next closest thing is for the spin to act as though it now has *both* frequencies. Thus, a spin which precessed at $\omega_{0I} + a/2$ before the S-pulse would have components precessing at $\omega_{0I} + a/2$ and $\omega_{0I} - a/2$ after the S-pulse. As we shall see, this latter answer turns out to be correct. We clearly need a systematic and rigorous derivation of the result.

Let us therefore treat carefully a concrete problem, a pulse sequence

$$X_I(\pi/2) \ldots \tau \ldots X_S(\theta) \ldots t - \tau \quad ,$$

where we will explore and contrast what happens when $\theta = \pi$ with the situation for $\theta = \pi/2$. We wish to calculate the expectation values of I_x and I_y in the I-spin rotating frame. We shall use the doubly rotating frame Hamiltonian.

We assume that the $X_I(\pi/2)$ pulse is applied at $t = 0$. At $t = 0^-$, just before, we assume the spin system is in thermal equilibrium. (We also treat the case of 100% polarization of the S-spins.) We employ the general formula

$$\langle I^+(t)\rangle = \text{Tr}\{I^+ \varrho_R(t)\} \quad , \tag{7.246}$$

where I^+ refers to axes in the rotating frame, as does ϱ_R. Since we will be working almost exclusively in the rotating frame, we drop the subscript R to simplify the notation.

Then, assuming $\mathcal{H}$ is given by (7.162) except during the $(H_1)_I$ and $(H_1)_S$ pulses, we characterize states by m_I and m_S, getting

$$\begin{aligned}\langle I^+(t)\rangle &= \sum_{m_I, m_S}(m_I m_S|I^+ \varrho(t)|m_I m_S)\\ &= \sum_{\substack{m_I,m_S\\m_I',m_S'}}(m_I m_S|I^+|m_I' m_S')(m_I' m_S'|\varrho(t)|m_I m_S) \quad .\end{aligned} \tag{7.247}$$

Making use of the properties of the raising operator and of the eigenstates $|m_I m_S)$, we have

$$\begin{aligned}(m_I m_S|I^+|m_I' m_S') &= \delta_{m_S m_S'}\delta_{m_I,1/2}\delta_{m_I',-1/2}(\tfrac{1}{2}|I^+|-\tfrac{1}{2})\\ &= \delta_{m_S,m_S'}\delta_{m_I,1/2}\delta_{m_I',-1/2} \quad .\end{aligned}$$

Abbreviating $+\tfrac{1}{2}$ as $+$, $-\tfrac{1}{2}$ as $-$ for m_I, we thus get

$$\langle I^+(t)\rangle = \sum_{m_S}(-m_S|\varrho(t)|+m_S) \quad . \tag{7.248}$$

Thus, there are only two matrix elements of $\varrho(t)$ needed, one for each value of m_S.

During those times when the rf pulses are not turned on, we have, using (5.133), that

$$\varrho(t_2) = \exp(-i\mathcal{H}(t_2-t_1)/\hbar)\varrho(t_1)\exp(i\mathcal{H}(t_2-t_1)/\hbar) \quad , \tag{7.249}$$

where

$$\mathcal{H} = -\hbar\Omega_I I_z - \hbar\Omega_S S_z + \hbar a I_z S_z \quad , \tag{7.250}$$

using the notation of (7.176). As a consequence

$$\begin{aligned}(-m_S|\varrho(t_2)|+m_S) &= \exp[-i(\Omega_I - am_S)(t_2-t_1)]\\ &\quad \times (-m_S|\varrho(t_1)|+m_S) \quad .\end{aligned} \tag{7.251}$$

Thus, if we know ϱ at time t_1, we can express it for later times as

$$\langle I^+(t)\rangle = \sum_{m_S}(-m_S|\varrho(t_1)|+m_S)\exp[-i(\Omega_I - am_S)(t-t_1)] \quad . \tag{7.252}$$

This expression, even without specifying $(-m_S|\varrho(t_1)|+m_S)$ already allows us to draw an important conclusion: $\langle I^+(t)\rangle$ contains just two frequencies $-(\Omega_I - a/2)$

and $-(\Omega_I + a/2)$. Therefore, no matter *what* rotation angle we give the S-spins, these are the only frequencies we observe. At most we can change the complex coefficients

$$(-m_S|\varrho(t_1)| + m_S)\exp[i(\Omega_I - am_S)t_1]$$

of the time-dependent terms

$$\exp[-i(\Omega_I - am_S)t] \quad .$$

Since the coefficients are complex, we have at our disposal the amplitude and phase of the two frequency components by manipulating $(-m_S|\varrho(t_1)| + m_S)$. To follow what happens in the actual pulse sequence, we use the method outlined in (5.167) in which we divide time into intervals during each of which the Hamiltonian can be taken as time independent.

In so doing, we will use the results of (5.253) and (5.254) for the time dependence of ϱ under the action of rf pulses. We must first, however, generalize these equations which apply to one spin for the case of two spins. In so doing, we consider the effect on ϱ of interactions in the rotating frame. Utilizing the concept that these interactions are large compared to the spin-spin coupling, we employ (7.206) and (7.207) to treat the interactions one at a time:

$$\mathcal{H}_I = -\gamma_I \hbar (H_1)_I I_x \equiv -\hbar \omega_{1I} I_x \quad \text{or} \tag{7.253a}$$

$$\mathcal{H}_S = -\gamma_S \hbar (H_1)_S S_x \equiv -\hbar \omega_{1S} S_x \quad . \tag{7.253b}$$

Thus, taking just $\mathcal{H}_S$

$$\frac{d}{dt}(m_I m_S|\varrho(t)|m_I' m_S') = \frac{i}{\hbar}(m_I m_S|\varrho \mathcal{H}_S - \mathcal{H}_S \varrho|m_I' m_S')$$

$$= \frac{i}{\hbar} \sum_{m_I'', m_S''} (m_I m_S|\varrho(t)|m_I'' m_S'')(m_I'' m_S''|\mathcal{H}_S|m_I' m_S')$$

$$- (m_I m_S|\mathcal{H}_S|m_I'' m_S'')(m_I'' m_S''|\varrho(t)|m_I' m_S') \quad . \tag{7.254}$$

Since

$$(m_I'' m_S''|\mathcal{H}_S|m_I' m_S') = \delta_{m_I'' m_I'}(m_S''|\mathcal{H}_S|m_S') \quad ,$$

we get

$$\frac{d}{dt}(m_I m_S|\varrho(t)|m_I' m_S') = \frac{i}{\hbar} \sum_{m_S''} [(m_I m_S|\varrho(t)|m_I' m_S'')(m_S''|\mathcal{H}_S|m_S')$$

$$- (m_S|\mathcal{H}_S|m_S'')(m_I m_S''|\varrho(t)|m_I' m_S')] \quad . \tag{7.255}$$

Equation (7.255) shows that if we are driving the S-spins only (i.e. $\mathcal{H}_I = 0$, $\mathcal{H}_S \neq 0$), the I-spin quantum numbers do not change. There is therefore no mixing of matrix elements $(m_I m_S|\varrho(t)|m_I' m_S')$ with matrix elements such as $(m_I m_S|\varrho(t)|m_I'' m_S'')$. The problem is thus transformed to a one-spin problem.

As a result, we can immediately utilize the results of (5.253) and (5.254). Note, however, that though matrix elements of $\mathcal{H}_S$ are necessarily diagonal in m_I, matrix elements of ϱ are not. This is a very important point to which we will return later in discussing coherence transfer and multiple quantum coherence.

Thus, under the action of $\mathcal{H}_S$, acting for a time t, we can write (using $\overline{m}_S$ for $-m_S$) for the matrix element diagonal in m_S:

$$(-m_S|\varrho(t)|+m_S) = \tfrac{1}{2}[(-m_S|\varrho(0)|+m_S) + (-\overline{m}_S|\varrho(0)|+\overline{m}_S)]$$
$$+\tfrac{1}{2}[(-m_S|\varrho(0)|+m_S) - (-\overline{m}_S|\varrho(0)|+\overline{m}_S)]\cos\omega_{1S}t$$
$$+\tfrac{i}{2}[(-\overline{m}_S|\varrho(0)|+m_S) - (-m_S|\varrho(0)|+\overline{m}_S)]\sin\omega_{1S}t \qquad (7.256)$$

and for the matrix element off diagonal in m_S

$$(-\overline{m}_S|\varrho(t)|+m_S) = \tfrac{1}{2}[(-m_S|\varrho(0)|+\overline{m}_S) + (-\overline{m}_S|\varrho(0)|+m_S)]$$
$$+\tfrac{1}{2}[(-\overline{m}_S|\varrho(0)|+m_S) - (-m_S|\varrho(0)|+\overline{m}_S)]\cos\omega_{1S}t$$
$$+\tfrac{i}{2}[(-m_S|\varrho(0)|+m_S) - (-\overline{m}_S|\varrho(0)|+\overline{m}_S)]\sin\omega_{1S}t \quad . \qquad (7.257)$$

An analogous pair of equations holds for the effect of $(H_1)_I$. Therefore, at $t = 0^-$, just before the pulse $X_I(\pi/2)$ we have

$$(m_I m_S|\varrho(0^-)|m'_I m'_S) = \delta_{m_I m'_I}\delta_{m_S m'_S}(m_I m_S|\varrho(\theta_l)|m_I m_S) \quad ,$$

where θ_l means the matrix at thermal equilibrium with the lattice temperature.

Utilizing (7.256) and (7.257), we see that $(H_1)_I$ will leave the matrix diagonal in m_S, but will generate elements off-diagonal in m_I. Since $(\omega_{1I}t) = \pi/2$, we get at time 0^+, just after the $(H_1)_I$ pulse, from the equation analogous to (7.256)

$$(-m_S|\varrho(0^+)|+m_S) = \tfrac{i}{2}[(+m_S|\varrho(0^-)|+m_S)$$
$$-(-m_S|\varrho(0^-)|-m_S)] \quad . \qquad (7.258)$$

But

$$\tfrac{1}{2}[(+m_S|\varrho(0^-)|+m_S) - (-m_S|\varrho(0^-)|-m_S)]$$

is just the net I-spin z-polarization at thermal equilibrium for those I-spins coupled to S-spins which have a particular m_S. Since to a very good approximation m_S is as likely to be $+\tfrac{1}{2}$ as $-\tfrac{1}{2}$, the $\langle I_z(0^-)\rangle$ is equally distributed between the two values of m_S.

Defining

$$I_0 \equiv \langle I_z(0^-)\rangle \quad , \qquad (7.259)$$

we get

$$(-m_S|\varrho(0^+)|+m_S) = \frac{iI_0}{2} \quad . \qquad (7.260)$$

We now let $\varrho(t)$ evolve for a time τ until time τ^- just before the S-spin pulse. Utilizing (7.250) we get

$$(-m_S|\varrho(\tau^-)| + m_S) = \exp[-i(\Omega_I - am_S)\tau](-m_S|\varrho(0^+)| + m_S)$$
$$= +\frac{iI_0}{2}\exp[-i(\Omega_I - am_S)\tau] \quad . \tag{7.261}$$

We now consider two cases for the $X_S(\theta)$ pulse, $\theta = \pi$ and $\theta = \pi/2$. Since $(H_1)_S$ will *not* change the m_I quantum numbers, and since $\langle I^+(t)\rangle$ arises *only* from terms off-diagonal in m_I, we need consider only the effect of $(H_1)_S$ on the matrix elements $(-m_S|\varrho(\tau^-)| + m_S)$. We have if $\theta = \pi$, $\sin\omega_{1S}t = 0$, $\cos\omega_{1S}t = -1$. For $\theta = \pi/2$, $\sin\omega_{1S}t = 1$, $\cos\omega_{1S}t = 0$. For $\theta = \pi$

$$(-m_S|\varrho(\tau^+)| + m_S) = (-\overline{m}_S|\varrho(\tau^-)| + \overline{m}_S) \quad . \tag{7.262}$$

For $\theta = \pi/2$

$$(-m_S|\varrho(\tau^+)| + m_S) = \tfrac{1}{2}[(-m_S|\varrho(\tau^-)| + m_S)$$
$$+ (-\overline{m}_S|\varrho(\tau^-)| + \overline{m}_S)] \quad , \tag{7.263a}$$

which also implies

$$(-\overline{m}_S|\varrho(\tau^+)| + \overline{m}_S) = \tfrac{1}{2}[(-m_S|\varrho(\tau^-)| + m_S)$$
$$+ (-\overline{m}_S|\varrho(\tau^-)| + \overline{m}_S)] \quad . \tag{7.263b}$$

These results have simple meanings. For the π pulse, (7.262) says that the π pulse interchanges the matrix elements of ϱ between m_S and $-m_S$. Thus, the matrix element develops under the action of m_S up to τ, then develops under the action of $-m_S$. This result is just what we assumed in Sect. 7.19, and is displayed in Fig. 7.18a,b, as well as Fig. 7.21. We will discuss the meaning of (7.263) below.

Denoting $\langle I^+(t)\rangle$ for a θ pulse as $\langle I^-(t)\rangle_\theta$, we get for $t > \tau$,

$$\langle I^+(t)\rangle_\pi = \frac{i}{2}\sum_{m_S}[(+\overline{m}_S|\varrho(0^-)| + \overline{m}_S) - (-\overline{m}_S|\varrho(0^-)| - \overline{m}_S)]$$
$$\times \exp[-i(\Omega_I - a\overline{m}_S)\tau]\exp[-i(\Omega_I - am_S)(t - \tau)] \quad . \tag{7.264}$$

Utilizing (7.260), this becomes

$$\langle I^+(t)\rangle_\pi = \frac{iI_0}{2}\sum_{m_S}e^{-i\Omega_I t}e^{iam_S(t-2\tau)} \quad . \tag{7.265}$$

This result is exactly what we deduced in Sect. 7.18 if one replaces τ by the notation t_π. For $t = 2\tau$ it displays the refocusing of the spin-spin coupling. The term involving Ω_I gives the chemical shift. If we set Ω_I equal to zero and utilize (7.260), (7.264) is identical to (7.171).

329

For the case $\theta = \pi/2$ we get

$$\langle I^+(t)\rangle_{\pi/2} = \frac{i}{2}\sum_{m_S}\frac{1}{2}\{[(+m_S|\varrho(0^-)|+m_S) - (-m_S|\varrho(0^-)|-m_S)]$$
$$\times \exp[-i(\Omega_I - am_S)\tau] + [(+\overline{m}_S|\varrho(0^-)|+\overline{m}_S)$$
$$- (-\overline{m}_S|\varrho(0^-)|-\overline{m}_S)]\exp[-i(\Omega_I - a\overline{m}_S)\tau]\}$$
$$\times \exp[-i(\Omega_I - am_S)(t-\tau)] \quad . \tag{7.266}$$

Again, utilizing (7.26), we get, for $t > \tau$,

$$\langle I^+(t)\rangle_{\pi/2} = \frac{iI_0}{2}e^{-i\Omega_I t}\left(\frac{e^{ia\tau/2} + e^{-ia\tau/2}}{2}\right)$$
$$\times (e^{ia(t-\tau)/2} + e^{-ia(t-\tau)/2}) \quad . \tag{7.267}$$

This expression has a simple graphical interpretation (Fig. 7.37). The $-(\Omega_I t)$ term gives the phase shift arising from chemical shifts. The two terms $\exp(ia\tau/2)$ and $\exp(-ia\tau/2)$ show that up until time τ there are two phases of magnetization, analogous to the ϕ_1 and ϕ_2 of Fig. 7.21. The *product* with the last bracket shows that at $t = \tau$, each of these phase lines splits into two lines.

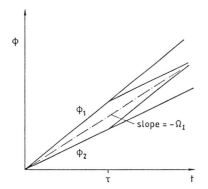

Fig. 7.37. The phases ϕ of the I-spins which result from application of a $\pi/2$ pulse to unpolarized S-spins at time τ. Up until $t = \tau$, there are two phases, one for each value of m_S. At $t = \tau$, the phases each split in two, but the *slopes* are related so that there are still only two precession frequencies for the I-spins, although there are now four phases

We can get added insight by taking the S-spins to be 100% polarized, so that at $t = 0^-$ only $m_S = +\frac{1}{2}$ is to be found. Then, for $t > \tau$

$$\langle I^+(t)\rangle_{\pi/2} = \frac{i}{2}[(++|\varrho(0^-)|++) - (-+|\varrho(0^-)|-+)]$$
$$\times \exp[-i(\Omega_I - a/2)\tau]$$
$$\times \left(\frac{\exp[-i(\Omega_I - a/2)(t-\tau)] + \exp[-i(\Omega_I + a/2)(t-\tau)]}{2}\right). \tag{7.268}$$

This shows that up until $t = \tau$, we get precession at a single angular frequency, $-(\Omega_I - a/2)$, corresponding to $m_S = +\frac{1}{2}$, but that after the $\pi/2$ pulse at time τ, the precession splits into two components of equal amplitude, one precessing at

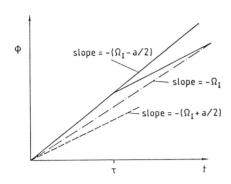

Fig. 7.38. The phases ϕ of the I-spins which result from application of a $\pi/2$ pulse to a 100% polarized S-spin sample ($m_S = \frac{1}{2}$ initially) at $t = \tau$. Before τ, only one ϕ is needed, corresponding to the single value of m_S. The $\pi/2$ pulse creates a situation in which both $m_S = +\frac{1}{2}$ and $-\frac{1}{2}$ are present, hence leads to two phases. Note that the same two precession frequencies are present as in Fig. 7.37, but the amplitudes of the components differ corresponding to the different initial m_S polarizations of the two cases

$-(\Omega_I - a/2)$, the other precessing at $-(\Omega_I + a/2)$ (Fig. 7.38). Thus the *average* phase angle advances at frequency $-\Omega_I$, as though there were zero coupling, but the effective *length* of the average magnetization goes as $\cos(a(t - \tau)/2)$ and hence is not constant. The fact that the *average* phase is independent of a for the $\pi/2$ pulse has as its classical analogy that the S-spins are pointing perpendicular to H_0, hence produce no shift in the I-spin precession. Indeed, this is the physical consequence of (7.263a) and (7.263b), which together show that even if only $m_S = \frac{1}{2}$ is present after the $\pi/2$ pulse, both $m_S = \frac{1}{2}$ and $m_S = -\frac{1}{2}$ are present after the $X_S(\pi/2)$ is applied.

The sorts of considerations we have been exploring are treated in a different manner in two illuminating papers by *Pegg* et al. [7.62], and applied to various double resonance methods.

7.25 Coherence Transfer

Now that we have employed formal methods to follow in detail the role of $\pi/2$ or π pulses, we return to one of the important applications of spin-coherence double resonance, the transfer of the magnetization of one set of spins to the other. The method of doing this transfer was discovered by *Feher* in 1956 [7.63, 64]. In 1962, *Baker* discovered a related form of nuclear-nuclear double resonance which he called INDOR. *Baker* pointed out that it was much like *Feher*'s ENDOR experiment [7.65]. In 1973, *Pachler* and *Wessels* [7.66] invented a pulsed version of the INDOR experiment, which they named selective population inversion. *Jeener*'s proposal [7.55] for two-dimensional NMR, as we shall see in Sect. 7.27 includes among other things a homonuclear pulsed version of the Feher experiment. Then a heteronuclear pulsed version, with all the benefits of Fourier transform, was proposed by *Maudsley* and *Ernst* [7.67], with modifications by *Bodenhausen* and *Freeman* [7.68], *Maudsley* et al. [7.69], and *Morris* and *Freeman* [7.70].

Feher was studying the electron spin resonance of donors in Si doped with phosphorous. His work was part of the research that led also to his discovery of ENDOR (Sect. 7.7). He showed that he could transfer the electron polarization to the P^{31} nuclei. Since his experiment is conceptually very simple and straightforward, we begin with it.

In keeping with the model system of Sect. 7.19, we assume the energy levels are those of Fig. 7.39a labeled by $|m_I m_S\rangle$, and also labeled 1, 2, 3, 4 as shown. In thermal equilibrium, the populations of each state are given by the Boltzmann factors, which, in the high temperature approximation, may be written as

$$p(m_I, m_S) = \frac{1}{4}\left[1 + \left(\frac{\hbar\omega_{0I} m_I + \hbar\omega_{0S} m_S}{kT}\right)\right] \quad , \tag{7.269}$$

where $p(m_I, m_S)$ is the probability of finding a given I-S pair in the state $|m_I, m_S\rangle$. In this expression we have neglected the small spin-spin coupling in computing the energies of the levels. The leading term, $\frac{1}{4}$, does not produce any net magnetization signals, so it is only the terms involving m_I and m_S which express net polarizations.

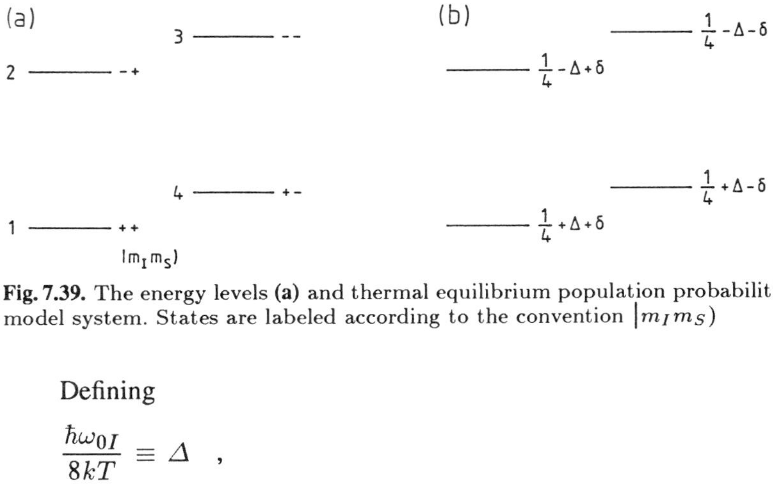

Fig. 7.39. The energy levels (a) and thermal equilibrium population probabilities (b) of the model system. States are labeled according to the convention $|m_I m_S\rangle$

Defining

$$\frac{\hbar\omega_{0I}}{8kT} \equiv \Delta \quad , \tag{7.270a}$$

$$\frac{\hbar\omega_{0S}}{8kT} \equiv \delta \quad , \tag{7.270b}$$

we have for $p(m_I, m_S)$

State 1 $p(++) = \frac{1}{4} + (\Delta + \delta)$ (7.271a)

State 2 $p(-+) = \frac{1}{4} - (\Delta - \delta)$ (7.271b)

State 3 $p(+-) = \frac{1}{4} + (\Delta - \delta)$ (7.271c)

State 4 $\quad p(--) = \frac{1}{4} - (\Delta + \delta)$. (7.271d)

The spin polarizations are

$$\langle I_z \rangle = \sum_{m_I, m_S} (m_I m_S | I_z | m_I m_S) p(m_I m_S)$$

$$= \sum_{m_I} m_I \left[\sum_{m_S} p(m_I, m_S) \right] \qquad (7.272)$$

$$\langle S_z \rangle = \sum_{m_S} m_S \left[\sum_{m_I} p(m_I, m_S) \right] . \qquad (7.273)$$

Now, though we have neglected the spin-spin coupling in computing the thermal equilibrium populations, we must *not* neglect them in calculating the spectrum since they make transition 1-2 have a different energy from that of transition 3-4.

Since the "$\frac{1}{4}$" in (7.271a) is the same in all energy levels, it has no effect on the intensity of any transition, and can thus be conveniently dropped in recording population probabilities, giving one the population excess above $\frac{1}{4}$. (A negative excess is a deficiency.) These are conveniently expressed in a matrix

	$m_S = \frac{1}{2}$	$m_S = -\frac{1}{2}$
$m_I = -\frac{1}{2}$	$-\Delta + \delta$	$-\Delta - \delta$
$m_I = +\frac{1}{2}$	$\Delta + \delta$	$\Delta - \delta$.	(7.274)

Thus, the resonances for both *I*-spins and *S*-spins are doublets, as shown in Fig. 7.40. For the sake of explanation, we assume the *I*-spins have a much larger population difference (Fig. 7.40a) than the *S*-spins ($\Delta \gg \delta$). So that we see a larger intensity in the *I*-spin absorption lines (Fig. 7.40b).

Feher's method consists then of two steps. In the first, he makes an adiabatic passage of frequency ω_I across one of the *I*-spin transitions (but not the other) such as the transition 1-2 for $m_S = +\frac{1}{2}$. Since the adiabatic passage inverts the magnetization, it interchanges the populations of the states involved, 1 and 2. This step leads to the population excesses shown in Fig. 7.41a, and listed in the following matrix:

	$m_S = \frac{1}{2}$	$m_S = -\frac{1}{2}$
$m_I = -\frac{1}{2}$	$\Delta + \delta$	$-\Delta - \delta$
$m_I = +\frac{1}{2}$	$-\Delta + \delta$	$\Delta - \delta$.	(7.275)

The *S*-spin population difference corresponding to a transition in which m_S changes, such as from $|-+\rangle$ to $|--\rangle$ (states 2 to 4), would be given by

$$p(-+) - p(--) = (\frac{1}{4} - \Delta + \delta) - (\frac{1}{4} - \Delta - \delta) = 2\delta \qquad (7.276a)$$

before the adiabatic passage, but

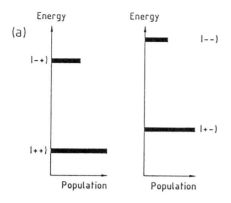

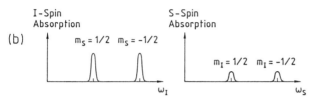

Fig. 7.40. (a) The energy level and thermal equilibrium populations (schematic) for a case in which $\gamma_I \gg \gamma_S$ so that $\Delta \gg \delta$. (b) The I-spin and S-spin resonance intensities showing the spin-spin splitting. The greater I-spin intensity arises because $\gamma_I \gg \gamma_S$

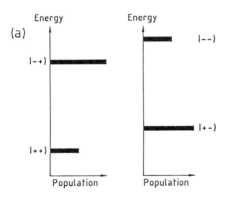

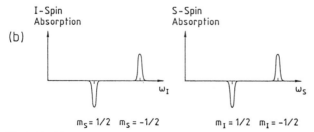

Fig. 7.41. (a) The energy level populations after an adiabatic passage through the I-spin line for $m_S = +\frac{1}{2}$, showing the population inversion in which the lower energy $|++\rangle$ level has *smaller* population than the higher energy $|-+\rangle$ level. (b) The I-spin and S-spin resonances showing the inversion of the I-spin line corresponding to $m_S = \frac{1}{2}$, and the greatly enhanced S-spin resonance intensities resulting from the populations of (a)

334

$$p(-+) - p(--) = (\tfrac{1}{4} + \Delta + \delta) - (\tfrac{1}{4} - \Delta - \delta) = 2\Delta + 2\delta \tag{7.276b}$$

after the adiabatic passage. Thus, for $m_I = -\tfrac{1}{2}$, the S-spins have an enhanced polarization.

On the other hand, for the S-spin transition involving $m_I = +\tfrac{1}{2}$ (the transition between states 1 and 3), we have a population difference before of

$$p(++) - p(+-) = 2\delta \tag{7.277a}$$

but after of

$$p(++) - p(+-) = -2\Delta - 2\delta \quad , \tag{7.277b}$$

which, for $\Delta \gg \delta$, is an inverted population. The population differences for both (7.276b) and (7.277b) are much bigger after this first step than before.

Since (7.277b) corresponds to a population inversion of the S-spins [the upper energy state $|m_I m_S = |+-\rangle$ has a larger population than the lower energy state $|m_I m_S\rangle = |+-\rangle$], there is no *net* polarization of the S-spins.

To achieve a *net* polarization, one may therefore invert the population difference of either (7.276b) or (7.277b). Feher achieved this by his second step, an adiabatic passage of $(H_1)_S$ across either (but not both) of the S-spin doublets. Thus, if the passage were across the 1-3 transitions, the 1-3 populations would be interchanged leading to

	$m_S = \tfrac{1}{2}$	$m_S = -\tfrac{1}{2}$
$m_I = -\tfrac{1}{2}$	$\Delta + \delta$	$-\Delta - \delta$
$m_I = +\tfrac{1}{2}$	$\Delta - \delta$	$-\Delta + \delta$

(7.278)

Using these populations, one gets readily

$$\langle S_z \rangle = \tfrac{1}{2}[\Delta + \delta - (-\Delta - \delta) + (\Delta - \delta) - (-\Delta + \delta)] = 2\Delta \tag{7.279}$$

compared to its value *before* the two steps

$$\langle S_z \rangle = 2\delta \quad . \tag{7.280}$$

Such a scheme of magnetization transfer can be used for transferring electron (I-spin) polarization to nuclei (S-spins), or for transferring large γ nuclear magnetization (protons) to lower γ nuclei (e.g. C^{13}).

In order to be able to carry out an adiabatic inversion of one multiplet, one must be able to pass through the multiplet with an H_1 sufficiently small to invert one but not the other transition. Thus γH_1 must be less than a. Another requirement is that spin-lattice or spin-spin relaxation times not prevent the adiabatic condition in this time interval. Thus, since it takes many precession periods $1/\gamma H_1$ to have the passage adiabatic, we have $\gamma H_1 \gg 1/T_1, 1/T_2$.

The *Feher* approach to coherence transfer using adiabatic passage can be replaced by a pulse technique, the method used by *Jeener* [7.55] and by *Maudsley* and *Ernst* [7.67]. Although superficially (7.279) and (7.280) suggest that one is performing an Overhauser effect, in fact, since the method does *not* use spin-lattice relaxation, it is clearly based on different principles.

We now show how to use pulses to invert one hyperfine line. Consider an ω_I tuned exactly to the unsplit I-spin resonance frequency ω_{0I},

$$\omega_I = \omega_{0I} \ . \tag{7.281}$$

At $t = 0$ we apply an $X_I(\pi/2)$ pulse (Fig. 7.42a). The magnetization vectors of the $m_S = \frac{1}{2}$ and $m_S = -\frac{1}{2}$ lines are shown in Fig. 7.42b at $t = 0^+$. We label them $M_I(\frac{1}{2})$ and $M_I(-\frac{1}{2})$ corresponding to $m_S = +\frac{1}{2}$ and $-\frac{1}{2}$ respectively. They now precess in the left-handed sense at angular frequencies $(\omega_{0I} - am_S)$ in the lab frame, or at $-am_S$ in the rotating frame. (This is equivalent to precession in the rotating frame frequency am_S in the right-handed sense, see Fig. 7.42c.) At a time, t_0, such that

$$at_0 = \pi \tag{7.282}$$

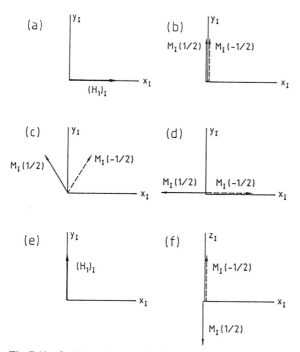

Fig. 7.42a–f. The effect of $(H_1)_I$ pulses tuned to resonance $(\omega_I = \omega_{0I})$ on the I-spin magnetization $M_I(m_S)$ for the two components $m_S = +\frac{1}{2}$ and $m_S = -\frac{1}{2}$. (a) At $t = 0$, $(H_1)_I$ is applied along the x-axis in the rotating frame giving an $X_I(\pi/2)$ pulse; (b) at $t = 0^+$, both $M_I(\frac{1}{2})$ and $M_I(-\frac{1}{2})$ lie along the y-axis; (c) the two components $M_I(m_S)$ precess in opposite directions at angular frequencies $\pm a/2$; (d) at a time $t = \pi/a$, $M_I(\frac{1}{2})$ and $M_I(-\frac{1}{2})$ point along the $-x$- and $+x$-axes respectively; (e) an $(H_1)_I$ lying along the y-axis gives a $Y_I(\pi/2)$ pulse; (f) the I-magnetization vectors after the $Y_I(\pi/2)$, with $M_I(-\frac{1}{2})$ pointing parallel to the z-direction, but $M_I(\frac{1}{2})$ pointing antiparallel

they are pointed opposite to one another as shown (Fig. 7.42d) along the x-axis. At this time (Fig. 7.42e), we apply a $\pi/2$ pulse with $(H_1)_I$ along the y-axis [a $Y_I(\pi/2)$ pulse]. This puts the $M_I(+\frac{1}{2})$ along the negative z-axis, and $M_I(-\frac{1}{2})$ along the positive z-axis (Fig. 7.42f). Thus, we have achieved a population inversion of the $m_S = +\frac{1}{2}$ line, while leaving the $m_S = -\frac{1}{2}$ line with a normal population difference. This is equivalent to an adiabatic passage across just the $m_S = \frac{1}{2}$ line. There are two important points to note. The first is that the splitting, a, is essential since it is responsible for producing the situation of Fig. 7.42d in which the two components of M_I are pointing opposite to one another. The second point is that the phase of the $(H_1)_I$ pulses (i.e. their orientation in the x-y plane) is important. If, for example, the second pulse had been another $X_I(\pi/2)$, it would have been parallel to the M_I vectors, and would have had no effect on their orientation. Likewise an $\overline{X}(\pi/2)$, where $\overline{X}$ means $(H_1)_I$ pointing along the negative x-axis, could be used to obtain a state in which the $M_I(-\frac{1}{2})$ would be inverted instead of $M_I(\frac{1}{2})$.

So far we have only inverted the $m_S = \frac{1}{2}$ transition, therefore we have achieved the effect of the first of *Feher's* two steps. To complete the coherence transfer, we must now do a similar thing for the S transitions. Thus, with ω_S tuned to ω_{0S}, we can apply the sequence $X_S(\pi/2)$ followed at a time delay $t_0 = \pi/a$ by a $Y_S(\pi/2)$ pulse to invert the $M_I = +\frac{1}{2}$ transition for the S-spins. This would have completed Feher's coherence transfer and produced the populations in the matrix of (7.278).

If instead of applying four pulses we apply the sequence

$$X_I(\pi/2)\ldots(\pi/a)\ldots(Y_I(\pi/2), X_S(\pi/2))\ldots \text{observe } S \quad , \tag{7.283}$$

the I pulses will have produced a situation in which the $M_I(\frac{1}{2})$ population is inverted, but not the $M_I(-\frac{1}{2})$, so the situation will look like Fig. 7.41a. Therefore the $X_S(\pi/2)$ pulse (Fig. 7.43a) will produce a situation in the rotating frame such

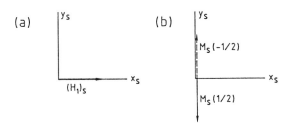

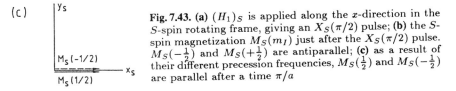

Fig. 7.43. (a) $(H_1)_S$ is applied along the x-direction in the S-spin rotating frame, giving an $X_S(\pi/2)$ pulse; (b) the S-spin magnetization $M_S(m_I)$ just after the $X_S(\pi/2)$ pulse. $M_S(-\frac{1}{2})$ and $M_S(+\frac{1}{2})$ are antiparallel; (c) as a result of their different precession frequencies, $M_S(\frac{1}{2})$ and $M_S(-\frac{1}{2})$ are parallel after a time π/a

as shown in Fig. 7.43b in which the two components of M_S point opposite to one another initially. At later times, their relative phase will have changed, making the magnetization components, in fact, parallel to each other at time π/a after the $X_S(\pi/2)$ pulse (Fig. 7.43c).

Indeed, following the $X_S(\pi/2)$ pulse, there will be an S-spin free induction signal. In the lab frame, it can be thought of as two signals, one at $\omega_{0S} - a/2$, the other at $\omega_{0S} + a/2$, or it can be viewed as a signal at ω_{0S} which is amplitude modulated as $\sin at/2$. This signal is of course similar to what would result from using *Feher*'s two adiabatic passages to transfer the I polarization to the S-spins, followed by a $\pi/2$ pulse to the S-spins to generate magnetization perpendicular to H_0. If one performed that total sequence using pulses instead of adiabatic passage, it would take a total of five $\pi/2$ pulses (two of the I-spins, three for the S-spins). However, the last two $\pi/2$ pulses can be omitted since the first of them rotates transverse S-magnetization to lie along the z-direction, and the second rotates the magnetization from the z-direction back to a transverse orientation. In their technique of selective population transfer, *Pachler* and *Wessels* [7.66] apply a π pulse to *one* of the I-spin transitions, inverting their population difference. They use an $(H_1)_I$ much smaller than the spin-spin splitting. Then they apply a strong $\pi/2$ pulse to the S-spins to observe the free induction decay. The Fourier transform of the S-spin signal shows one hyperfine line inverted with respect to the other, and a large signal increase if $\gamma_I \gg \gamma_S$. Thus their experiment is equivalent to (7.283).

For the sequence (7.283) we have required that $\omega_I = \omega_{0I}$, exact resonance for the I-spins. For the S-spins, $X_S(\pi/2)$ does not have to be exactly at resonance. It will still produce the situation of Fig. 7.43b. Then, those magnetization components will precess freely at $(\omega_{0S} - am_I)$, giving a signal at ω_{0I} which is amplitude modulated at angular frequency a.

The sequence (7.283) is almost the Maudsley-Ernst sequence. What they do is (i) remove the requirement that $\omega_I = \omega_{0I}$ and thereby (ii) remove the special phase requirement of the second I pulse. Then, taking a sequence

$$(\pi/2)_I \ldots t_1 \ldots ((\pi/2)_I, (\pi/2)_S) \ldots \text{observe } S\text{-spins}(t_2) \qquad (7.284)$$

in which t_2 is the observation time, and for which $(\pi/2)_I$ means an I-spin $\pi/2$ pulse without a special phase condition, they make a 2D Fourier transform. The fact that $\omega_I \neq \omega_{0I}$ produces a situation in which, following the initial $(\pi/2)_I$, the I-spin components precess both with respect to one another and also with respect to the rotating frame. Thus, if one is off resonance by an amount large compared to the spin-spin splitting, the pattern of Fig. 7.42f will rotate rapidly at $(\omega_{0I} - \omega_I)$ in the ω_I frame while the relative orientation of $M_I(\frac{1}{2})$ and $M_I(-\frac{1}{2})$ will change slowly. The situation is shown in Fig. 7.44. The initial $X_I(\pi/2)$ (Fig. 7.44a) puts $M_I(\frac{1}{2})$ and $M_I(-\frac{1}{2})$ along the y-axis in the frame rotating at ω_I (Fig. 7.44b). Under the combined influence of chemical shift (Ω_I) and spin-spin splittings the M_I vectors rotate and separate (Fig. 7.44c). A pattern of the M_I's at successive times is shown in Fig. 7.44d–h. If another $\pi/2$ pulse with

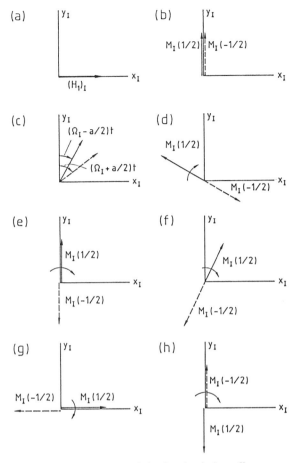

Fig. 7.44a–h. The effect of the I-spins being off resonance ($\omega_I \neq \omega_{0I}$). (a) At $t = 0$, $(H_1)_I$ is applied along the x-axis in the rotating frame; (b) the $X_I(\pi/2)$ pulse puts the two components of $M_I(m_S)$ along the y-axis at $t = 0^+$. $M_I(\frac{1}{2})$ is the solid arrow, $M_I(-\frac{1}{2})$ is the dashed arrow; (c) the components precess at $\Omega_I - am_S$, where $\Omega_I = \omega_{0I} - \omega_I$, under the combined effect of the chemical shift (Ω_I) and spin-spin coupling (am_S); (d)–(h) give $M_I(\frac{1}{2})$ and $M_I(-\frac{1}{2})$ at successively later times near to π/a under the assumption that $a \ll \Omega_I$ so that the relative alignment of the two vectors changes slowly compared to their joint rotation rate Ω_I. An $X_I(\pi/2)$ pulse would have no effect for situation (g), would make $M_I(\frac{1}{2})$ inverted relative to H_0 for situation (e), and make $M_I(-\frac{1}{2})$ inverted relative to H_0 for situation (h)

$(H_1)_I$ along the x-axis is applied (Fig. 7.44a), there will be complete inversion of $M_I(\frac{1}{2})$ for the situation (e), complete inversion of $M_I(-\frac{1}{2})$ for situation (h), *no* inversion of either component for situation (g), and partial inversion of both $M_I(\frac{1}{2})$ and $M_I(-\frac{1}{2})$ for situations (d) and (f). We see therefore that the chemical shift frequency Ω_I will determine the times at which the two M_I components are inverted, and thus they will determine the times at which an S-spin pulse will find transferred magnetization. Of course, a will also determine the general range of times for which $M_I(\frac{1}{2})$ and $M_I(-\frac{1}{2})$ point opposite to one another. Indeed, *Bodenhausen* and *Freeman* [7.68] analyze just this model to show that if the I-spins are protons and the S-spins are C^{13}, the resulting 2D spectrum has the proton spectrum in the ω_1 direction and the C^{13} spectrum in the ω_2 direction. Thus by observing the C^{13} spectrum one gets the indirect detection of the proton resonance frequencies, and the correlation of the proton and carbon chemical shifts.

It is useful to think of this experiment in terms of the density matrix analysis of Sect. 7.24. We have a pulse sequence

$$X_I(\pi/2)\dots\tau\dots(X_I(\pi/2),X_S(\pi/2))\dots\text{observe }S \tag{7.285}$$

in which we shall treat the $X_S(\pi/2)$ as having occurred just after the second $X_I(\pi/2)$ pulse.

Using our theorem of (7.207) we define τ^- as just before and τ^+ as just after the second $X_I(\pi/2)$ pulse, τ^{++} as just after the $X_S(\pi/2)$ pulse. Since eventually we will observe the S-spins, we seek $(m_{I-}|\varrho(t)|m_{I+})$ during the observation time $t > \tau^{++}$. Now $(m_{I-}|\varrho(\tau^{++})|m_{I+})$ is produced by $X_S(\pi/2)$, hence from elements of $\varrho(\tau^+)$ which are diagonal in m_I. At $t = 0^-$, before the first $X_I(\pi/2)$ pulse, we assume ϱ is in thermal equilibrium, hence diagonal in both m_I and m_S. Hence, under the action of the two $X_I(\pi/2)$ pulses, ϱ remains diagonal in m_S. Thus, $(m_{I-}|\varrho(\tau^{++})|m_{I+})$ arises solely from elements of $\varrho(\tau^+)$ just before the pulse $X_S(\pi/2)$ which are diagonal in both m_I and m_S. These matrix elements can be traced back to elements $(m_I m_S|\varrho(\tau^-)|m_I m_S)$ as well as to elements $(\mp m_S|\varrho(\tau^-)|\pm m_S)$ using (5.253):

$$(\pm m_S|\varrho(\tau^+)|\pm m_S) = \tfrac{1}{2}[(+m_S|\varrho(\tau^-)|+m_S) + (-m_S|\varrho(\tau^-)|-m_S)]$$
$$\pm \frac{i}{2}[(-m_S|\varrho(\tau^-)|+m_S)$$
$$- (+m_S|\varrho(\tau^-)|-m_S)] \quad . \tag{7.286}$$

These in turn arise from the same elements at $t = 0^+$, just after the first $X_I(\pi/2)$ pulse, using

$$(\mp m_S|\varrho(\tau^-)|\pm m_S) = e^{\mp i(\Omega_I - am_S)\tau}(\mp m_S|\varrho(0^+)|\pm m_S) \tag{7.287}$$

and

$$(\mp m_S|\varrho(0^+)|\pm m_S) = \pm\frac{i}{2}[(+m_S|\varrho(0^-)|+m_S)$$
$$-(-m_S|\varrho(0^-)|-m_S)] \quad \text{and} \tag{7.288}$$

$$(\pm m_S|\varrho(0^+)|\pm m_S) = \tfrac{1}{2}[(+m_S|\varrho(0^-)|+m_S)$$
$$+(-m_S|\varrho(0^-)|-m_S)] \quad . \tag{7.289}$$

We can thus relate the diagonal elements of $\varrho(\tau^+)$ to the elements of $\varrho(0^-)$, just before the I-spin pulse. The result is

$$(m_I m_S | \varrho(\tau^+) | m_I m_S)$$
$$= \tfrac{1}{2}[(m_I m_S | \varrho(0^-) | m_I m_S) + (\overline{m}_I m_S | \varrho(0^-) | \overline{m}_I m_S)]$$
$$+ \tfrac{1}{2}[(\overline{m}_I m_S | \varrho(0^-) | \overline{m}_I m_S) - (m_I m_S | \varrho(0^-) | m_I m_S)]$$
$$\times \cos(\Omega_I - am_S)\tau \tag{7.290}$$

for the diagonal elements just before the $X_S(\pi/2)$ pulse.

There are several interesting limiting cases. First, if $\Omega_I = 0$, the result depends on $\cos(am_S\tau)$, hence is the same for $m_S = \tfrac{1}{2}$ and $m_S = -\tfrac{1}{2}$. This result confirms our point earlier that in the absence of a chemical shift an $X_I(\pi/2)\ldots\tau\ldots X_I(\pi/2)$ will *not* invert only one multiplet. Instead, one needs an $X_I(\pi/2)\ldots\tau\ldots Y_I(\pi/2)$.

In fact, if $\Omega_I\tau = (\pi/2)$, one has the case of Fig. 7.44g, and indeed then $\cos(\Omega_I - am_S)\tau = \sin(am_S\tau)$ so the result differs between $m_S = +\tfrac{1}{2}$ and $m_S = -\tfrac{1}{2}$. Thus one can invert one multiplet only.

Note also that as τ goes to zero, $(m_I m_S | \varrho(\tau^+) | m_I m_S)$ goes to $(\overline{m}_I m_S | \varrho(0^-) | \overline{m}_I m_S)$, representing the fact that two closely spaced $\pi/2$ pulses are equivalent to a single π pulse. We can now use (5.254) to get the S-spin magnetization

$$\langle S^+(t_2) \rangle = \text{Tr}\{S^+ \varrho(t_2)\}$$
$$= \sum_{m_I}(m_I - | \varrho(t_2) | m_I +) \tag{7.291a}$$

$$= \sum_{m_I}(m_I - | \varrho(\tau^{++}) | m_I +) e^{-i(\Omega_S - am_I)t_2} \tag{7.291b}$$

$$= \frac{i}{2}\sum_{m_I}[(m_I + | \varrho(\tau^+) | m_I +) - (m_I - | \varrho(\tau^+) | m_I -)]$$
$$\times \exp[-i(\Omega_S - am_S)t_2] \tag{7.291c}$$

where the matrix elements of $\varrho(\tau^+)$ are given by (7.290). Equation (7.291c) is the mathematical expression of the fact that the S-spin signal is determined by the population differences between the states $m_S = \tfrac{1}{2}$ and $m_S = -\tfrac{1}{2}$, for each value of m_I, existing just prior to the $X_S(\pi/2)$ pulse. These population differences oscillate as a function of τ, the time difference between the $X_I(\pi/2)$ pulses, at the rates $\Omega_I - a/2$ and $\Omega_I + a/2$.

We can utilize (7.290) together with (7.275) to follow what the two $X_I(\pi/2)$ do to the population excesses. Thus we have for $t = 0^-$, the population excesses

	$m_S = \tfrac{1}{2}$	$m_S = -\tfrac{1}{2}$
$m_I = -\tfrac{1}{2}$	$-\Delta - \delta$	$-\Delta - \delta$
$m_I = \tfrac{1}{2}$	$\Delta + \delta$	$\Delta - \delta$.	(7.292)

which become, for $t = \tau^+$, just after the second I-spin pulse, using (7.290),

$$m_S = \tfrac{1}{2} \qquad\qquad\qquad m_S = -\tfrac{1}{2}$$

$$m_I = -\tfrac{1}{2} \quad \delta + \Delta \cos(\Omega_I - a/2)\tau \quad -\delta + \Delta \cos(\Omega_I + a/2)\tau$$
$$m_I = \tfrac{1}{2} \quad \delta - \Delta \cos(\Omega_I - a/2)\tau) \quad -\delta - \Delta \cos(\Omega_I + a/2)\tau \quad .$$
(7.293)

Therefore, from the point of view of the S-spins, the $m_I = \tfrac{1}{2}$ line has a population difference at $t = \tau^+$ of

$$(++|\varrho(\tau^+)|++) - (+-|\varrho(\tau^+)|+-)$$
$$= 2\delta - \Delta[\cos(\Omega_I - a/2)\tau - \cos(\Omega_I + a/2)\tau] \qquad (7.294a)$$

and the $m_I = -\tfrac{1}{2}$ has correspondingly

$$(-+|\varrho(\tau^+)|-+) - (--|\varrho(\tau^+)|--)$$
$$= 2\delta + \Delta[\cos(\Omega_I - a/2)\tau - \cos(\Omega_I + a/2)\tau] \quad . \qquad (7.294b)$$

Bodenhausen and *Freeman* derive the result of (7.290) by use of the classical spin picture (Fig. 7.45). The $X_I(\pi/2)$ pulse (Fig. 7.45a) puts the I-magnetization along the y-direction in the I rotating frame (Fig. 7.45b). This precesses at $(\Omega_I - am_S)$ (Fig. 7.45c). The second $X_I(\pi/2)$ pulse rotates this vector down into the x-z plane, with a z-component $M_{zI}(m_S)$ given by (Fig. 7.45d)

$$M_{zI}(m_S) = -M_I(m_S)\cos(\Omega_I - am_S)\tau \quad . \qquad (7.295)$$

This equation can be converted to an equation for elements of the density matrix since

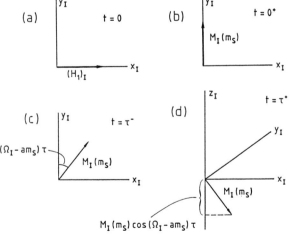

Fig. 7.45a–d. The effect of two $X_I(\pi/2)$ pulses, one at $t = 0$, the other at $t = \tau$ on the I-spin magnetization $M_I(m_S)$. (a) The direction of $(H_1)_I$ for the $X_I(\pi/2)$ pulses; (b) the magnetization $M_I(m_S)$ lies along the y-axis at $t = 0^+$; (c) $M_I(m_S)$ precesses through an angle $(\Omega_I - am_S)\tau$ at time τ^-; (d) an $X_I(\pi/2)$ pulse at $t = \tau$ puts $M_I(m_S)$ into the x-z plane, with a component along the z-axis of $-M_I \cos(\Omega_I - am_S)\tau$

$$M_I(m_S) = \frac{\gamma_I \hbar}{2}[(+m_S|\varrho(0^-)|+m_S) - (-m_S|\varrho(0^-)|-m_S)] \quad (7.296)$$

and since

$$M_{zI}(m_S) = \frac{\gamma_I \hbar}{2}[(+m_S|\varrho(\tau^+)|+m_S) - (-m_S|\varrho(\tau^+)|-m_S)] \quad . \quad (7.297)$$

Therefore, using (7.295)

$$(+m_S|\varrho(\tau^+)|+m_S) - (-m_S|\varrho(t^+)|-m_S)$$
$$= [(+m_S|\varrho(0^-)|+m_S) - (-m_S|\varrho(0^-)|-m_S)]\cos(\Omega_I - am_S)\tau \quad .$$
$$(7.298)$$

Moreover,

$$(+m_S|\varrho(0^-)|+m_S) + (-m_S|\varrho(0^-)|-m_S)$$
$$= (+m_S|\varrho(\tau^+)|+m_S) + (-m_S|\varrho(\tau^+)|-m_S) \quad . \quad (7.299)$$

Combining (7.298) and (7.299) we get (7.290). In summary, the pulses work by having the X pulses take $(m_I m_S|\varrho(0^-)|m_I m_S)$ into $(\pm m_S|\varrho(t)|\mp m_S)$, letting these elements evolve, then taking them back into diagonal elements $(m_I m_S|\varrho(\tau^+)|m_I m_S)$ with the second I-pulse. These modified populations are like those of Fig. 7.41a. Thus we have modulated the populations of a given m_I seen by the S-spins, and set up a case where the S-spin pulse can utilize an I-spin population excess. Indeed, the $X_S(\pi/2)$ pulse sets in motion magnetization oscillating at angular frequencies $(\Omega_S - am_I)$ but with amplitudes and phases dependent on $(\Omega_I - am_S)\tau$.

The situation we have just described shows that if one has an inhomogeneous magnetic field smearing out the chemical shifts, there is no time at which the second rf pulse will find all the spins properly oriented for inversion of the magnetization of one multiplet line. Certainly if two lines corresponding to $m_S = \pm\frac{1}{2}$ are not resolved, the method of adiabatic passage through one line will not work. But, since we can eliminate magnetic field inhomogeneity effects with a spin echo, we are tempted to look for a pulse solution to inversion of one multiplet. It is easy to find one. What we must do is refocus the field inhomogeneity without refocusing the spin-spin coupling. A

$$X_I(\pi/2)\ldots\tau\ldots X_I(\pi)\ldots\tau\ldots \quad (7.300)$$

would refocus both the field inhomogeneity and the spin-spin splitting. However, thinking back to spin-echo double resonance, we see that if we also flip the S-spin by a π pulse simultaneously with the I-spins, the spin-spin coupling will not refocus (see discussion at the end of Sect. 7.21). Thus, the sequence

$$X_I(\pi/2)\ldots\tau\ldots(X_I(\pi),X_S(\pi))\ldots\tau\ldots Y_I(\pi/2) \quad (7.301)$$

with

$$2\tau a = \pi \quad (7.302)$$

will produce a situation just after the last pulse in which the $M_I(\frac{1}{2})$ magnetization is inverted, but not the $M_I(-\frac{1}{2})$ magnetization.

Then, a $\pi/2$ pulse applied to the S-spins will generate transverse magnetization. It, too, will dephase if there is magnetic field inhomogeneity, but can be refocused by an $X_S(\pi)$ pulse accompanied by an $X_I(\pi)$ to prevent the spin-spin splitting from being refocused. The same condition $2\tau a = \pi$ of (7.287) will guarantee that at 2τ following the first $X_S(\pi/2)$ pulse the magnetization will look like Fig. 7.43c.

We can write this sequence in a slightly different format as

$$X_I(\pi/2)\ldots\tau\ldots X_I(\pi)\ldots\tau\ldots Y_I(\pi/2) \qquad X_I(\pi)$$
$$X_S(\pi) \qquad X_S(\pi/2)\ldots\tau\ldots X_S(\pi)\ldots\tau\ldots S-\text{echo}$$

with (7.303)

$$2a\tau = \pi \tag{7.304}$$

in which we utilized two lines, one for the I-spins, another for the S-spins to help keep track of the two spin systems.

If the "inhomogeneity" comes instead from the existence of a number of chemical shifts of the I-spins, then one can still transfer I-spin magnetization to enhance S-spin signals:

$$X_I(\pi/2)\ldots\tau\ldots X_I(\pi)\ldots\tau\ldots Y_I(\pi/2)$$
$$X_S(\pi) \qquad X_S(\pi/2)\ldots\text{acquire} \quad , \tag{7.305}$$

in which we assume the values of a are rather similar for all chemical shifts. This sequence was discovered by *Morris* and *Freeman* [7.70] who gave it a name "INEPT" to stand for "insensitive nuclei enhanced by polarization transfer".

7.26 Formal Theoretical Apparatus II – The Product Operator Method

In our example of coherence transfer, we saw that it was important to have a formal method to carry out a calculation in order to be sure one was dealing correctly with such things as the effect on one spin (e.g. the I-spin) of applying a $\pi/2$ pulse to the other (S) spin to which it is coupled. The differential equations for the density matrix gave us such a method. However, the density matrix for two spins $(m_I m_S | \varrho | m'_I m'_S)$ has sixteen elements, since m_I, m_S, m'_I and m'_S can each take on two values. Therefore, the differential equations are really quite complicated. We picked an example in which only a few elements were involved and in which we could see clearly just which elements we needed. Naturally one would like a formal method for calculating the time development of the density matrix which is straightforward to apply, and which is readily extendable to cases of more than two spins.

We turn now to such a method, which has been named the "product operator" formalism. This method was developed independently by at least three different groups: *Sorensen* et al. [7.71], *Van den Ven* and *Hilbers* [7.72] and *Wang* and *Slichter* [7.53,73]. We shall present the method in connection with the two-spin system. The generalization to more spins is straightforward [7.71]. We confine ourselves to spin $\frac{1}{2}$ nuclei.

The typical situation one encounters is illustrated by (7.212) which describes spin echo double resonance:

$$X_I(\pi/2)\ldots\tau\ldots(X_I(\pi), X_S(\pi))\ldots(t-\tau)$$

$$\langle I^+(t)\rangle = \frac{\hbar\omega_{0I}}{ZkT} \mathrm{Tr}\,\{I^+[\exp{(\mathrm{i}(\Omega_I I_z + \Omega_S S_z - aI_z S_z)(t-\tau))} X_I(\pi) X_S(\pi)$$
$$\times \exp{(\mathrm{i}(\Omega_I I_z + \Omega_S S_z - aI_z I_z)\tau)} X_I(\pi/2) I_z X_I^{-1}(\pi/2)$$
$$\times \exp{(-\mathrm{i}(\Omega_I I_z + \Omega_S S_z - aI_z S_z)\tau)} \qquad (7.212)$$
$$\times X_S^{-1}(\pi) X_I^{-1}(\pi) \exp{(-\mathrm{i}(\Omega_I I_z + \Omega_S S_z - aI_z S_z)(t-\tau))}]\} \;.$$

The system has an initial density matrix corresponding to thermal equilibrium given by (7.210):

$$\varrho(0^-) = \frac{1}{Z}\left(1 + \frac{\hbar\omega_{0I} I_z + \hbar\omega_{0S} S_z}{kT}\right) \;. \qquad (7.210)$$

The effect of the rf pulses is to produce rotations. We may symbolize these in general by unitary operators R. An example of such an R is $X_I(\theta)$:

$$X_I(\theta) = \mathrm{e}^{\mathrm{i}I_x \theta} \qquad (7.306\mathrm{a})$$

$$X_I^{-1}(\theta) = \mathrm{e}^{-\mathrm{i}I_x \theta} \;. \qquad (7.306\mathrm{b})$$

Likewise, between pulses the system evolves under the influence of the time development operator $T(t)$ given by

$$T(t) = \exp{(\mathrm{i}(\Omega_I I_z + \Omega_S S_z - aI_z S_z)t)} \qquad (7.307\mathrm{a})$$

$$= T_I(t) T_S(t) T_{IS}(t) \quad \text{where} \qquad (7.307\mathrm{b})$$

$$T_I \equiv \mathrm{e}^{\mathrm{i}\Omega_I I_z t} \qquad (7.308\mathrm{a})$$

$$T_S \equiv \mathrm{e}^{\mathrm{i}\Omega_S S_z t} \qquad (7.308\mathrm{b})$$

$$T_{IS} \equiv \mathrm{e}^{-\mathrm{i}aI_z S_z t} \;. \qquad (7.308\mathrm{c})$$

Since T_I, T_S, and T_{IS} commute with one another, they can be applied in any order in (7.307b). We have already worked a good deal with the R's and T's. We use the symbol U to stand for R or T and U^{-1} for their inverse. Then if $f_1(I, S)$ and $f_2(I, S)$ are two functions of the spin components of I and S,

$$U f_1(I, S) f_2(I, S) U^{-1} = U f_1(I, S) U^{-1} U f_2(I, S) U^{-1} \quad . \tag{7.309}$$

Hence, expanding the exponential in a power series, applying (7.309) term by term, and converting back to an exponential, we get

$$U e^{i f_1(I,S)} U^{-1} = e^{i U f_1(I,S) U^{-1}} \quad . \tag{7.310}$$

Now, the R's plus T_I and T_S are all simple rotation operators of the general form $\exp(iI_x\theta)$, $\exp(iI_y\theta)$, $\exp(iI_z\theta)$. Their effect on a density matrix $\varrho(0^-)$ such as (7.210) can be easily followed using equations such as

$$e^{iI_x\theta} I_z e^{-iI_x\theta} = I_z \cos\theta + I_y \sin\theta \tag{7.311}$$

based on (2.55).

We can construct formulas such as (7.311) by simply noting that when the exponential on the left has a positive exponent it generates a left-handed rotation θ of the vector, I_z, sandwiched between the two exponentials.

The problem arises from T_{IS} in expressions such as

$$T_{IS}(t_1) I_y T_{IS}^{-1}(t_1) = e^{-i a I_z S_z t_1} I_y e^{i a I_z S_z t_1} \tag{7.312}$$

which have bilinear forms $(I_z S_z)$ rather than linear forms (I_z) in the exponent. If we now consider

$$-i a S_z t_1 \equiv \theta \quad , \tag{7.313}$$

then (7.312) looks much like (7.311):

$$e^{-i a I_z S_z t_1} I_y e^{i a I_z S_z t_1} = I_y \cos(a S_z t_1) - I_x \sin(a S_z t_1) \quad . \tag{7.314}$$

This equation has an operator, S_z, as the argument of the cosine and sine, a situation which is quite acceptable if we recall that for operators the trigonometric and exponential functions are defined in terms of their power series expansions.

It turns out we can simplify these expressions by starting with their power series:

$$\cos(a S_z t_1) = 1 - \frac{(a S_z t_1)^2}{2!} + \frac{(a S_z t_1)^4}{4!} \cdots \tag{7.315}$$

$$\sin(a S_z t_1) = a S_z t_1 - \frac{(a S_z t_1)^3}{3!} + \frac{(a S_z t_1)^5}{5!} \cdots \quad . \tag{7.316}$$

We now make use of the spin $\frac{1}{2}$ property. If we choose as a basis set functions $|m_S\rangle$, $(m_S = +\frac{1}{2}$ or $-\frac{1}{2})$, the most general function is of the form

$$\psi(t) = C_+(t) |m_S = \tfrac{1}{2}\rangle + C_-(t) |m_S = -\tfrac{1}{2}\rangle \quad . \tag{7.317}$$

Explicit evaluation gives

$$\begin{aligned} S_z^2 \psi(t) &= C_+(t) S_z^2 |\tfrac{1}{2}\rangle + C_-(t) S_z^2 |-\tfrac{1}{2}\rangle \\ &= \tfrac{1}{4} C_+(t) |\tfrac{1}{2}\rangle + \tfrac{1}{4} C_-(t) |-\tfrac{1}{2}\rangle \\ &= \tfrac{1}{4} \psi(t) \quad . \end{aligned} \tag{7.318}$$

Since the same is true for S_x^2 or S_y^2, we can replace S_z^2 by $\frac{1}{4}$ [S_z^4 by $(\frac{1}{4})^2$, etc.] in (7.315), obtaining

$$\cos(aS_zt_1) = 1 - \frac{(at_1/2)^2}{2!} + \frac{(at_1/2)^4}{4!} \cdots \qquad (7.319)$$

$$\cos(aS_zt_1) = \cos(at_1/2) \quad .$$

For (7.316), we get

$$\sin(aS_zt_1) = S_z\left(at_1 - \frac{(at_1)^3}{3!}S_z^2 + \frac{(at_1)^5}{5!}S_z^4 + \cdots\right)$$

$$= S_z\left[at_1 - \frac{(at_1)^3}{3!}\left(\frac{1}{2}\right)^2 + \frac{(at_1)^5}{5!}\left(\frac{1}{2}\right)^4 + \cdots\right]$$

$$= 2S_z\left(\frac{(at_1)}{2} - \frac{(at_1/2)^3}{3!} + \cdots\right) \qquad (7.320)$$

$$\sin(aS_zt_1) = 2S_z \sin(at_1/2) \quad . \qquad (7.321)$$

Therefore, utilizing (7.319) and (7.321) in (7.314), we get

$$e^{-iaI_zS_zt_1} I_y e^{iaI_zS_zt_1} = I_y \cos(at_1/2) - I_x S_z 2\sin(at_1/2) \quad . \qquad (7.322)$$

It is useful in carrying out calculations to have these various results in a table (Table 7.1) in which we list $UI_\alpha U^{-1}$ for $\alpha = x, y, z$ for different operators U. The bottom line applies to $I = \frac{1}{2}$.

Table 7.1. The effect of $UI_\alpha U^{-1}$ for various U's and for $\alpha = x, y, z$. The bottom line applies for spin $\frac{1}{2}$ only

U \ I_α	I_x	I_y	I_z
$e^{i\theta I_x}$	I_x	$I_y \cos\theta - I_z \sin\theta$	$I_z \cos\theta + I_y \sin\theta$
$e^{i\theta I_y}$	$I_x \cos\theta + I_z \sin\theta$	I_y	$I_z \cos\theta - I_x \sin\theta$
$e^{i\theta I_z}$	$I_x \cos\theta - I_y \sin\theta$	$I_y \cos\theta + I_x \sin\theta$	I_z
$e^{-i\theta I_z S_z}$	$I_x \cos(\theta/2)$ $+I_y(2S_z \sin(\theta/2))$	$I_y \cos(\theta/2)$ $-I_x(2S_z \sin(\theta/2))$	I_z

We now see how to evaluate $T_{IS}I_\alpha T_{IS}^{-1}$ for $\alpha = x, y, z$. An interesting point is that it takes single spin operators, I_α, into a sum of a single spin operator and a product of spins $I_\alpha S_\beta$. We see therefore that if applied several times, it might develop terms like $I_\alpha I_\beta S_\gamma S_\delta$. But such terms can be simplified as we now explain.

Suppose we think of either the I-spins or the S-spins in terms of the 2×2 Pauli matrix representation. We recognize that spin products such as I_xI_y are products of one 2×2 matrix with another. The result is also a 2×2 matrix.

Now any 2 × 2 unitary matrix can be represented as a linear combination of the three Pauli matrices plus the 2 × 2 identity matrix. Thus, it must be possible to reduce products such as $I_x I_y$ to linear combinations of the identity operator with I_α's.

In fact, we have shown one such example of a product $I_\alpha I_\beta$ for the case that $\alpha = \beta$:

$$I_\alpha^2 = \tfrac{1}{4} \; , \quad \alpha = x, y, z \; . \tag{7.323}$$

We shall show that if $\alpha \neq \beta$, we can also easily reduce $I_\alpha I_\beta$ to a single spin operator. The result is

$$I_x I_y = \frac{iI_z}{2} \; , \quad I_y I_x = -\frac{iI_z}{2} \tag{7.324}$$

plus cyclic permutations.

To prove (7.324), we note that terms such as $I_x I_y$ are encountered in the commutation relation

$$I_x I_y - I_y I_x = iI_z \; . \tag{7.325}$$

This relationship, however, also involves $I_y I_x$, so we need one more relationship involving $I_x I_y$ and $I_y I_x$. Another such relationship arises from products such as $(I_x + I_y)^2$. Let us then consider the coordinate transformation from (I_x, I_y) to $(I_{x'}, I_{y'})$ with

$$I_{x'} = \frac{1}{\sqrt{2}}(I_x + I_y) \; , \tag{7.326}$$

a $\pi/4$ rotation. Then

$$I_{x'}^2 = \tfrac{1}{2}(I_x + I_y)(I_x + I_y) \tag{7.327}$$

$$\tfrac{1}{4} = \tfrac{1}{2}(I_x^2 + I_y^2) + \tfrac{1}{2}(I_x I_y + I_y I_x) = \tfrac{1}{2}(\tfrac{1}{4} + \tfrac{1}{4}) + \tfrac{1}{2}(I_x I_y + I_y I_x) \; .$$

Therefore

$$I_x I_y + I_y I_x = 0 \tag{7.328}$$

for a spin $\tfrac{1}{2}$ particle. Substituting (7.328) into (7.325) yields (7.324).

Let us now illustrate how these formulas are used. Suppose we analyze the S-flip-only double resonance for the case of a $\pi/2$ S pulse:

$$X_I(\pi/2) \ldots t_1 \ldots X_S(\pi/2) \ldots t_2 \; .$$

The initial density matrix at $t = 0$ is given by (7.210). Suppose we consider the portion of $\varrho(0^-)$ given by

$$\varrho(0^-) = \frac{\hbar \omega_{0I}}{ZkT} I_z \; . \tag{7.329}$$

Evaluation of the portion involving S_z is left as a homework problem.

For simplicity, we consider a case in which $\Omega_I = \Omega_S = 0$ (exact resonance for both spins). We apply an $X_I(\pi/2)$ pulse at $t = 0$, and an $X_S(\pi/2)$ pulse at time t_1, let the system evolve under T_{IS} from 0 to t_1, and after t_1 for a time t_2.

We define θ_1 and θ_2 by

$$\theta_1 = at_1 \quad , \quad \theta_2 = at_2 \quad . \tag{7.330}$$

Then

$$\varrho(t_1 + t_2) = \frac{\hbar\omega_{0I}}{ZkT} T_{IS}(\theta_2) X_S(\pi/2) T_{IS}(\theta_1) X_I(\pi/2) I_z X_I^{-1}(\pi/2) T_{IS}^{-1}(\theta_1)$$
$$\times X_S^{-1}(\pi/2) T_{IS}^{-1} \theta_2 \quad . \tag{7.331}$$

The effect of the $X_I(\pi/2)$ pulse is to rotate I_z into I_y:

$$X_I(\pi/2) I_z X_I^{-1}(\pi/2) = I_y \quad . \tag{7.332}$$

Next we apply T_{IS}. Utilizing Table 7.1 we get

$$T_{IS}(\theta_1) I_y T_{IS}^{-1}(\theta_1) = I_y \cos\theta_1/2 - 2I_x S_z \sin\theta_1/2 \quad . \tag{7.333}$$

Next we apply $X_S(\pi/2)$, a left-handed $\pi/2$ rotation of spin S, giving

$$X_S(\pi/2)(I_y \cos\theta_1/2 - 2I_x S_z \sin\theta_1/2) X_S^{-1}(\pi/2)$$
$$= I_y \cos\theta_1/2 - 2I_x S_y \sin\theta_1/2 \quad . \tag{7.334}$$

We now act on this with $T_{IS}(\theta_2)$:

$$T_{IS}(\theta_2) I_y T_{IS}^{-1}(\theta_2) \cos\theta_1/2 - 2T_{IS}(\theta_2) I_x T_{IS}^{-1}(\theta_2) T_{IS}(\theta_2) S_y T_{IS}^{-1}(\theta_2) \sin\theta_1/2$$
$$= (I_y \cos\theta_2/2 - 2I_x S_z \sin\theta_2/2) \cos\theta_1/2$$
$$- 2(I_x \cos\theta_2/2 + 2I_y S_z \sin\theta_2/2)(S_y \cos\theta_2/2 - 2I_z S_x \sin\theta_2/2) \sin\theta_1/2$$
$$= I_y \cos\theta_2/2 \cos\theta_1/2 - 2I_x S_z \sin\theta_2/2 \cos\theta_1/2$$
$$- 2I_x S_y \cos^2\theta_2/2 \sin\theta_1/2 + 4I_x I_z S_x \cos\theta_2/2 \sin\theta_2/2 \sin\theta_1/2$$
$$- 4I_y S_z S_y \sin\theta_2/2 \cos\theta_2/2 \sin\theta_1/2 + 8I_y S_z I_z S_x \sin^2\theta_2/2 \sin\theta_1/2 \quad . \tag{7.335}$$

We now have a number of terms involving spin products such as $I_\alpha I_\beta$ or $S_\alpha S_\beta$. Utilizing (7.324), we reduce the various terms as follows:

$$I_x I_z S_x = -\frac{i}{2} I_y S_x$$

$$I_y S_z S_y = -\frac{i}{2} I_y S_x \tag{7.336}$$

$$I_y I_z S_z S_x = \left(\frac{i}{2}\right) I_x \left(\frac{i}{2}\right) S_y = -\frac{I_x S_y}{4} \quad .$$

When these are substituted into (7.335), the two terms in $I_y S_x$ cancel, and the terms in $I_x S_y$ combine ($\cos^2\theta_2/2 + \sin^2\theta_2/2 = 1$) giving

$$T_{IS}(\theta_2)X_S(\pi/2)T_{IS}(\theta_1)X_I(\pi/2)I_zX_I^{-1}(\pi/2)T_{IS}^{-1}(\theta_1)X_S^{-1}(\pi/2)T_{IS}^{-1}(\theta_2)$$
$$= I_y \cos\theta_2/2 \cos\theta_1/2 - I_xS_z 2\sin\theta_2/2 \cos\theta_1/2 - I_xS_y 2\sin\theta_1/2 \quad .$$
(7.337)

Therefore, the contribution of (7.329) to ϱ is

$$\varrho(t_1+t_2) = \frac{\hbar\omega_{0I}}{ZkT}(I_x\cos\theta_2/2\cos\theta_1/2 - I_xS_z\sin\theta_2/2\cos\theta_1/2$$
$$- I_xS_y 2\sin\theta_1/2) \quad .$$
(7.338)

These various terms tell us what elements $(m_I m_S|\varrho(t)|m'_I m'_S)$ are nonzero at time t_1+t_2. For example, the first term involves I_x, hence gives nonvanishing elements

$$(\pm m_S|\varrho(t_1+t_2)|\mp m_S) \quad .$$
(7.339)

The term I_xS_z gives similar elements, weighted with m_S:

$$(m_I m_S|I_xS_z|m'_I m'_S) = m_S(\pm|I_x|\mp)\delta_{m_S,m'_S}\delta_{m_I,\pm}\delta_{m'_I,\mp} \quad .$$
(7.340)

Lastly I_xS_y gives nonvanishing terms

$$(\pm \pm|\varrho(t_1+t_2)|\mp \mp)$$
(7.341a)

$$(\pm \mp|\varrho(t_1+t_2)|\mp \pm)$$
(7.341b)

whose further significance we discuss in Sect. 9.1.

In Table 7.2 we list the other useful relationships employed in dealing with spin operators.

Table 7.2. Useful relations for unitary operators U. (U is a function of I, or S, or I and S)

$$Uf_1(I,S)f_2(I,S)U^{-1} = Uf_1(I,S)U^{-1}Uf_2(I,S)U^{-1}$$
$$U\exp[if(I,S)]U^{-1} = \exp[iUf(I,S)U^{-1}]$$

For spin $\frac{1}{2}$: $\cos(I_z\theta) = \cos(\theta/2)$
$\sin(I_z\theta) = 2I_z\sin(\theta/2)$

$I_\alpha^2 = \frac{1}{4}$ $\alpha = x,y,z$
$I_xI_y = \frac{i}{2}I_z$
$I_yI_x = -\frac{i}{2}I_z$ plus cyclic permutations

7.27 The Jeener Shift Correlation (COSY) Experiment

Since the early days of high resolution NMR, an important goal has been to tell which nuclear resonance lines arise from nuclei which are bonded to one another. The existence of bonding manifests itself in liquids through the indirect spin-

spin coupling [7.74, 75] (Sect. 4.9). The bonding might be direct, as in a $C^{13}H^1$ fragment, or remote via an electronic framework as in the H-H splittings in ethyl alcohol (CH_3CH_2OH) between the CH_3 protons and the CH_2 protons. In solids, the dipolar coupling proves proximity. In liquids, the dipolar coupling shows up through the nuclear Overhauser effect. One then distinguishes between bonding and proximity, an important distinction in large biomolecules [7.76] in which a long molecule may fold back on itself. A variety of methods such as spin echo double resonance, spin tickling [7.71], INDOR [7.65] and selective population transfer [7.66] have been employed, to utilize the fact that if I and S are coupled, perturbing the I-spin resonance will produce an effect on the S-spin resonance. All of those methods involve sitting on one line and point-by-point exploring the other lines. *Jeener*'s discovery of the two-dimensional Fourier transform method converts the approach to a Fourier transform method with all its advantages. In this section we analyze the original Jeener proposal, which now is commonly referred to as the COSY (correlated spectroscopy) method since it reveals which pairs of chemical shifts are correlated by a spin-spin coupling between them. It involves a single nuclear species (e.g. H^1) and a pulse sequence

$$X(\pi/2) \ldots t_1 \ldots X(\pi/2) \ldots \text{acquire}(t_2) \quad . \tag{7.342}$$

However, since two nuclei are involved, with $\gamma_I = \gamma_S$ but $\Omega_I \neq \Omega_S$ owing to their chemical shift difference, it is better to think of this as

$$X_I(\pi/2), X_S(\pi/2) \ldots t_1 \ldots X_I(\pi/2), X_S(\pi/2) \ldots \text{acquire } I(t_2) \text{ and } S(t_2) \quad , \tag{7.343}$$

where a single H_1 of sufficient strength flips both spins.

We have already discussed the basic principles of coherence transfer in Sect. 7.25. There we saw that if we are observing the S-spins in time interval two, the density matrix elements $(m_I - |\varrho(t_1+t_2)|m_I+)$ are responsible for the signal. These elements, however, have passed through both $(m_I m_S |\varrho(t)| m_I m_S)$ and $(\mp m_S |\varrho(t)| \pm m_S)$ during time interval t_1. Thus, the I-spin resonance frequencies modulate the S-spin signal during t_2 as a function of t_1. We can express these facts colloquially by saying the $(\mp m_S |\varrho(t)| \pm m_S)$ matrix elements during t_1 "feed" the $(m_I \mp |\varrho(t)| m_I \pm)$ elements during t_2.

Such a description almost suffices to describe the pulse sequence (7.343). All one need add is that at $t = 0^+$ *both* $(\mp m_S |\varrho(t)| \pm m_S)$ and $(m_I \mp |\varrho(t)| m_I \pm)$ are excited, so that during t_2 $(m_I - |\varrho(t)| m_I+)$ is fed by *both* $(\mp m_S |\varrho| \pm m_S)$ and $(m_I \mp |\varrho(t)| m_I \pm)$. [Note that $(m_I m_S |\varrho(t)| m_I m_S)$ during t_1 also "feeds" $(m_I - |\varrho(t)| m_I+)$ during t_2, as is shown by (7.291) and (7.290), however, since the diagonal elements of ϱ are independent of time they do not by themselves introduce any t_1 dependence to ϱ.]

We could try to carry out an analysis using ϱ similar to our approach to spin coherence. However, this method rapidly becomes quite cumbersome. Indeed, one finds that all 16 elements of ϱ are excited by this pulse sequence! We turn, therefore, to the method of spin operators.

Since the two nuclei are identical, we consider a Hamiltonian in the singly rotating frame, rotating at the angular frequency ω of the alternating field. The Hamiltonian includes a chemical shift difference so that Ω_I and Ω_S are not equal. Rather, they are given by

$$\Omega_I = \gamma H_0(1 - \sigma_I) - \omega_I \quad , \quad \Omega_S = \gamma H_0(1 - \sigma_S) - \omega_S \tag{7.344}$$

with σ_I and σ_S being the two chemical shifts. The Hamiltonian in the rotating frame is still

$$\mathcal{H} = -\hbar(\Omega_I I_z + \Omega_S S_z - aI_z S_z) \quad . \tag{7.345}$$

Both nuclei experience the same alternating field, H_1, so that for any pulse the two spins experience the same phase for H_1 and the same angle of rotation, θ. Thus, for example, if there is an $X_I(\theta)$ there is also an $X_S(\theta)$.

Since we observe *both* I and S spins, we seek

$$\langle I^+(t) + S^+(t)\rangle = \text{Tr}\{I^+ + S^+)\varrho(t)\} \tag{7.346}$$

with an initial density matrix prior to application of the first pulse given in the high temperature approximation as

$$\varrho(0^-) = \frac{1}{Z}\left(1 + \frac{\hbar\gamma H_0}{kT}(I_z + S_z)\right) \quad . \tag{7.347}$$

Utilizing the time development operators, T, T_I, T_S, T_{IS}, defined in (7.307), we readily get

$$\begin{aligned}\langle I^+(t) + S^+(t)\rangle = \frac{\hbar\gamma H_0}{ZkT}&\text{Tr}\,\{(I^+ + S^+)T(t_2)X_I(\pi/2)X_S(\pi/2)T(t_1)\\ &\times X_I(\pi/2)X_S(\pi/2)(I_z + S_z)X_S^{-1}(\pi/2)X_I^{-1}(\pi/2)T^{-1}(t_1)\\ &\times X_S^{-1}(\pi/2)X_I^{-1}(\pi/2)T^{-1}(t_2)\} \quad .\end{aligned} \tag{7.348}$$

The expression on the right can be written as the sum of two terms, one containing the I_z part of $\varrho(0^-)$, the other containing the S_z part. We define these as

$$\langle I^+(t) + S^+(t)\rangle_{I_z} \quad \text{and} \quad \langle I^+(t) + S^+(t)\rangle_{S_z} \quad . \tag{7.349}$$

Explicit examination of the two expressions shows that one can be converted to the other if every place one has an I_α one replaces it with S_α ($\alpha = x, y, z$) and every place one has S_α one replaces it with an I_α. Thus if one evaluates the term involving I_z only, one can get the contribution from S_z by interchanging in the first answer S_α's with I_α's, I_α's with S_α's, Ω_I with Ω_S, and Ω_S with Ω_I.

Therefore, we evaluate just the I_z term. Then

$$\begin{aligned}X_S(\pi/2)&X_I(\pi/2)I_z X_I^{-1}(\pi/2)X_S^{-1}(\pi/2)\\ &= X_S(\pi/2)I_y X_S^{-1}(\pi/2) = I_y \quad .\end{aligned} \tag{7.350}$$

Expressing $T(t_1)$ as

$$T(t_1) = T_I(t_1)T_S(t_1)T_{IS}(t_1) \tag{7.351}$$

we get

$$\begin{aligned}T(t_1)&I_y T^{-1}(t_1)\\&= T_I(t_1)T_{IS}(t_1)e^{i\Omega_S S_z t_1} I_y e^{-i\Omega_S S_z t_1} T_{IS}^{-1}(t_1)T_I^{-1}(t_1)\\&= T_I(t_1)T_{IS}(t_1)I_y T_{IS}^{-1}(t_1)T_I^{-1}(t_1) \quad .\end{aligned} \tag{7.352}$$

Utilizing the fact that

$$\mathrm{Tr}\{ABC\} = \mathrm{Tr}\{CAB\} \quad , \tag{7.353}$$

we pull the $T_I^{-1}(t_2)T_S^{-1}(t_2)$ from the far right hand end of (7.348), giving for the first part of the trace

$$T_S^{-1}(t_2)T_I^{-1}(t_2)(I^+ + S^+)T_I(t_2)T_S(t_2)T_{IS}(t_2)\ldots \quad . \tag{7.354}$$

Then, utilizing

$$e^{iI_z\theta}I^+ e^{-iI_z\theta} = I^+ e^{i\theta} \tag{7.355}$$

and the fact that I and S commute, we get instead of (7.354)

$$(I^+ e^{-i\Omega_I t_2} + S^+ e^{-i\Omega_S t_2})T_{IS}(t_2)\ldots \tag{7.356}$$

so that finally we have

$$\begin{aligned}\langle I^+(t) + S^+(t)\rangle_{I_z} = &\frac{\gamma\hbar H_0}{ZkT}\mathrm{Tr}\,\{I^+ e^{-i\Omega_I t_2} + S^+ e^{-i\Omega_S t_2})\\&\times [e^{-iaI_z S_z t_2} X_I(\pi/2) X_S(\pi/2) e^{i\Omega_I I_z t_1} e^{-iaI_z S_z t_1} I_y\\&\times e^{+iaI_z S_z t_1} e^{-i\Omega_I I_z t_1} X_S(\pi/2) X_I(\pi/2) e^{iaI_z S_z t_2}]\} \quad .\end{aligned} \tag{7.357}$$

We have collected inside the square brackets the set of operators which we need to evaluate using the spin-operator formalism. We saw in our example (Table 7.1) that the operators T_I, acting either on an I_α or on a product term $I_\alpha S_\beta$, at most double the number of terms, for example transforming an I_y to an $I_y \cos\theta + I_x \sin\theta$. An operator T_{IS} acting on a single term I_α or S_β will double the number of terms, but acting on a *product* will quadruple the number of terms. The operators $X_I(\pi/2)$ and $X_S(\pi/2)$ will change an I_z into an I_y, etc., but not increase the number of terms. Therefore T_{IS} acting on I_y will give two terms, which T_I will double to four terms (two single operators, two operator products), which T_{IS} will then convert to at most $2 \times 2 + 4 \times 4 = 12$ terms. (It turns out we get only 9.) They will be of the form I_α, or S_β, or $I_\alpha I_\beta$ after we have reduced any products such as $I_\alpha I_\beta$, or $I_\alpha I_\beta I_\gamma$, etc. to the appropriate $I_{\alpha'}$.

The only terms which will eventually contribute to $\langle I^+(t)+S^+(t)\rangle_{I_z}$ involve either

$$(-m_S|\varrho| + m_S) \quad \text{or} \quad (m_I - |\varrho|m_I+) \quad . \tag{7.358}$$

These are diagonal in one spin and off-diagonal in the other.

All possible spin operators can be expressed as linear combinations of the 15 spin functions I_α, S_α, $I_\alpha S_\beta$ ($\alpha = x, y, z$; $\beta = x, y, z$) plus the identity operator. But of these 16, only the eight operators

$$I_x, \ I_y, \ S_x, \ S_y \tag{7.359a}$$

$$I_x S_z, \ I_y S_z, \ I_z S_x, \ I_z S_y \tag{7.359b}$$

will give matrix elements of the form (7.358). In fact, using the symbol O to stand for these eight functions, (7.357) will consist of a sum of terms $\text{Tr}\{I^+O\}$ or $\text{Tr}\{S^+O\}$ which vanish if O is bilinear. Thus, only the terms (7.359a) of the form I_x, I_y, S_x and S_y contribute to $\langle I^+(t) + S^+(t)\rangle$.

As a result, of the 12 possible terms we will get from [] in (7.357), many will not be of interest. It is therefore useful, as shown by *Van den Ven* and *Hilbers* [7.72] to construct a table in which one keeps only the spin products, not all the other factors, to find the terms one needs eventually, then go back to get the coefficients. In so doing, we will designate what we are doing by writing a sequence of operators which act sequentially in the manner of *Sorensen* et al. [7.71].

Thus, to indicate that we wish to calculate the effect of the operators in the square brackets of (7.357) on an initial operator I_y, we write

$$(I_y); \ -aI_zS_zt_1, \ \Omega_I I_zt_1, \ X_S(\pi/2), \ X_I(\pi/2), \ -aI_zS_zt_2 \quad . \tag{7.360}$$

We put these results into a table (Table 7.3) in which over the dividing line between columns we list the operator which has transformed one column into the next column. If we make our table with 12 lines we should have all the space we need.

Comparing Table 7.3 with (7.359), we see that we will get contributions to $\langle I^+(t)\rangle_{I_z}$ from

line 3 I_x (7.361)

and to $\langle S^+(t)\rangle_{I_z}$ from

line 10 $S_x/4$. (7.362)

The contributions to I^+ can be seen to arise from terms which at all stages are either I_y or I_x, hence diagonal in m_S. Thus they are not fed by terms which are ever off-diagonal in m_S. Therefore, these terms do not involve transfer of coherence between the I-spins and the S-spins. On the other hand, (7.362) shows that the term $S_x/4$, which is off-diagonal in m_S but diagonal in m_I, evolved earlier from the initial I_z to I_y, to I_xS_z to I_yS_y to $-I_zS_y$. Hence the term I_y, which is off-diagonal in m_I but diagonal in m_S, feeds the term $S_x/4$, which is diagonal in m_I but off diagonal in m_S. This term involves transfer of I-spin

Table 7.3. Effects of operators from (7.357) on an initial operator I_y

Line	$-aI_zS_zt_1$	$\Omega_I I_z t_1$	$X_S(\pi/2)$	$X_I(\pi/2)$	$-aI_zS_zt_2$			
1	I_y	I_y	I_y	$-I_y$	$-I_z$	$-I_z$		1
2								2
3			I_x	I_x	I_x	I_x		3
4						I_yS_z		4
5	$-I_xS_z$	$-I_xS_z$	$-I_xS_y$	$-I_xS_y$	$\begin{bmatrix}-I_x\\-I_yS_z\end{bmatrix} \times \begin{bmatrix}S_y\\-S_xI_z\end{bmatrix} \Rightarrow$	$-I_xS_y$		5
6						$I_xI_zS_x = -(i/2)I_yS_x$		6
7						$-I_yS_zS_y = (i/2)I_yS_x$		7
8						$+I_yI_zS_zS_x = (i/2)^2I_xS_y$		8
9								9
10		I_yS_z	I_yS_y	$-I_zS_y$	$-I_zS_y$	$=$	$S_x/4$	10
11					$+I_zS_xI_z$			11
12								12

polarization to the S-spins. We see that by applying first $X_S(\pi/2)$, then second $X_I(\pi/2)$, we get from I_yS_z to $-I_zS_y$ via the state I_yS_y. This operator has matrix elements which are off-diagonal in *both* m_I and m_S. If instead we had first applied $X_I(\pi/2)$, then $X_S(\pi/2)$, we would have gone from $+I_yS_z$ to $-I_zS_z$ to $-I_zS_y$. Thus the intermediate state $-I_zS_z$ would have had matrix elements which are completely diagonal. This latter type of state is the one we encountered in our discussion of coherence transfer (7.286). The fact that we go from I_yS_z to $-I_zS_y$ no matter which order we apply $X_I(\pi/2)$ and $X_S(\pi/2)$ is an example of our theorem of (7.207) that one can interchange the order of the X_I and X_S pulses.

We now need to get the rest of the coefficients of I_x and S_x on lines 3 and 10 of Table 7.3. We readily find

$$\langle I^+(t) + S^+(t)\rangle_{I_z} = \frac{\gamma\hbar H_0}{ZkT}$$
$$\times (e^{-i\Omega_I t_2}\cos(at_1/2)\sin(\Omega_I t_1)\cos(at_2/2)\,\mathrm{Tr}\{I^+I_x\}$$
$$+ e^{-i\Omega_S t_2}2\sin(at_1/2)\sin(\Omega_I t_1)2\sin(at_2/2)\,\mathrm{Tr}\{S^+S_x/4\}). \quad (7.363)$$

Explicit evaluation gives

$$\mathrm{Tr}\{I^+I_x\} = \mathrm{Tr}\{I_x^2\}$$
$$= \sum_{m_I, m_S}(m_Im_S|I_x^2|m_Im_S)$$
$$= 4 \times \tfrac{1}{4} = 1 \quad \text{so that} \quad (7.364)$$

$$\langle I^+(t) + S^+(t)\rangle_{I_z} = \frac{\gamma\hbar H_0}{ZkT}[e^{-i\Omega_I t_2}\sin(\Omega_I t)\cos(at_1/2)\cos(at_2/2)$$
$$+ e^{-i\Omega_S t_2}\sin(\Omega_I t_1)\sin(at_1/2)\sin(at_2/2)] \,. \quad (7.365)$$

In writing expressions such as (7.365), *Van den Ven* and *Hilbers* [7.72] introduce a useful notation which greatly cuts down on the number of letters one needs to write down. They point out that using this formalism one encounters trigonometric factors such as

$$\cos(at_1/2), \quad \sin(at_2/2), \quad \sin(\Omega_I t_2), \quad \sin(\Omega_S t_1) \quad . \tag{7.366a}$$

They write these as

$$C_a^1 \equiv \cos(at_1/2) \,, \quad S_a^2 \equiv \sin(at_2/2),$$
$$S_I^2 \equiv \sin(\Omega_I t_2) \,, \quad S_S^1 \equiv \sin(\Omega_S t_1) \tag{7.366b}$$

in a notation which is self-evident. One merely has to be careful to remember the "$\frac{1}{2}$" in the arguments involving a which is not present in those involving Ω_I or Ω_S.

Returning to our discussion of (7.365), we now add to this the contribution from S_z to $\varrho(0^-)$, which we obtain simply by interchanging Ω_I and Ω_S in (7.365). The final result is

$$\langle I^+(t) + S^+(t) \rangle = \frac{\gamma \hbar H_0}{ZkT} \{ \cos(at_1/2) \cos(at_2/2)$$
$$\times [e^{-i\Omega_I t_2} \sin(\Omega_I t_1) + e^{-i\Omega_S t_2} \sin(\Omega_S t_1)]$$
$$+ \sin(at_1/2) \sin(at_2/2) [e^{-i\Omega_I t_2} \sin(\Omega_S t_1) + e^{-i\Omega_S t_2} \sin(\Omega_I t_1)] \} \quad . \tag{7.367}$$

The various terms in this expression have simple physical meanings. We note first that we have a term involving $\cos(at_1/2)\cos(at_2/2)$. This term is present even as t_1 and t_2 approach zero. It has the terms

$$e^{-i\Omega_I t_2} \sin(\Omega_I t_1) \quad \text{and} \tag{7.368a}$$

$$e^{-i\Omega_S t_2} \sin(\Omega_S t_1) \tag{7.368b}$$

which correspond to oscillation near Ω_I during both t_1 and t_2, or oscillation near Ω_S during both t_1 and t_2. Of course the actual eigenfrequencies are $\Omega_I \pm a/2$ and $\Omega_S \pm a/2$, but if $a \ll |\Omega_I - \Omega_S|$ the term "near Ω_I" or "near Ω_S" is well defined. Near $t_1 = t_2 = 0$, this term is independent of a. It therefore does not require a coupling to exist, and represents an effect of uncoupled spins, as is evident from the fact that the precession frequency is the same (Ω_I or Ω_S) during both time intervals.

On the other hand, the second term is proportional to $\sin(at_1/2)\sin(at_2/2)$. It reaches a maximum when $at_1/2 = at_2/2 = \pi/2$ or

$$t_1 = t_2 = \pi/a \quad . \tag{7.369}$$

This is the condition we encountered previously in (7.356) as the time which optimizes coherence transfer. This is the time, t_1, after the first $\pi/2$ pulse at which the spins have precessed to the condition of Fig. 7.42d. It is also the time, t_2, after the second pulse that the spins reach the condition of Fig. 7.43c. That coherence transfer is involved is also shown by the fact that this term also involves

$$e^{-i\Omega_I t_2} \sin(\Omega_S t_1) + e^{-i\Omega_S t_2} \sin(\Omega_I t_1) \quad . \tag{7.370}$$

The first term corresponds to a spin precessing near Ω_S during the first interval, and near Ω_I during the second. The second term represents precession near Ω_I during the first interval, and near Ω_S during the second.

If one makes a double Fourier transform of (7.366), one gets two types of peaks. From the term involving $\cos(at_1/2)\cos(at_2/2)$, one gets peaks near $\omega_1 = \omega_2 = \Omega_I$ or Ω_S, which lie close to the diagonal line $\omega_1 = \omega_2$ in a (ω_1, ω_2) plot. From the term involving $\sin(at_1/2)\sin(at_2/2)$, one gets peaks near the points $\omega_1 = \Omega_I$, $\omega_2 = \Omega_S$ or $\omega_1 = \Omega_S$, $\omega_2 = \Omega_I$. Such peaks immediately show that the spins with chemical shifts Ω_I and Ω_S are coupled. Since the frequencies are always $\Omega_I \pm a/2$, not Ω_I, and $\Omega_S \pm a/2$, not Ω_S, the peaks are actually clusters of four peaks. Thus, the diagonal peaks occur at

$$\begin{aligned} \omega_1 &= \Omega_I + a/2 & \omega_2 &= \Omega_I + a/2 \\ \omega_1 &= \Omega_I + a/2 & \omega_2 &= \Omega_I - a/2 \\ \omega_1 &= \Omega_I - a/2 & \omega_2 &= \Omega_I - a/2 \\ \omega_1 &= \Omega_I - a/2 & \omega_2 &= \Omega_I + a/2 \quad . \end{aligned} \tag{7.371}$$

Likewise, the off-diagonal peaks are actually a cluster of four peaks at

$$\omega_1 = \Omega_I \pm a/2 \quad , \quad \omega_2 = \Omega_S \pm a/2 \tag{7.372a}$$

and another cluster of four peaks at

$$\omega_1 = \Omega_S \pm a/2 \quad , \quad \omega_2 = \Omega_I \pm a/2 \quad . \tag{7.372b}$$

If we think of the factor $\sin(at_1/2)\sin(at_2/2)$ as modulating signals at Ω_I and Ω_S, we note that the buildup of the cross terms $\exp(-i\Omega_I t_2)\sin(\Omega_S t_1)$ or $\exp(-i\Omega_S t_2)\sin(\Omega_I t_1)$ requires a time $\sim \pi/a$. In a typical molecule, a spin I will have coupling constants to both nearby and more distant nuclei. The stronger coupling constants a will produce their cross peaks in a shorter time than the weaker coupling constants. Thus, by limiting the intervals t_1 and t_2, one can select large or small a's. If, indeed, two spins are not coupled at all, so that a is strictly zero, the cross peaks will *never* arise.

7.28 Magnetic Resonance Imaging

The use of NMR to produce two- and three-dimensional images has, since its invention in the early 1970s, increased at an explosive rate. We refer the reader to the books by *Mansfield* [7.77] and by *Ernst* et al. [7.78] for extended treatments of the many variants. We present a short treatment of the principles at this point in the text since it enables us to discuss the method of Kumar, Welti, and Ernst, which utilizes the two-dimensional Fourier transform approach.

Two pioneering papers, published within a few months of one another, are the proposals of *Lauterbur* [7.79] and of *Mansfield* and *Grannell* [7.80], which independently propose use of NMR to form images. The basic concept is straightforward. We start with a sample which has a narrow NMR line in a uniform static magnetic field. To the uniform field we add a second magnetic field which is nonuniform, ideally one with a constant gradient. The field gradient broadens the NMR line. For a constant gradient (e.g. in the z-direction) the magnetic field component, H_z, parallel to the uniform field is

$$H_z = H_0 + z\left(\frac{\partial H_z}{\partial z}\right) , \qquad (7.373)$$

so that planes of constant z correspond to planes of constant precession frequency. Thus, a given frequency interval is bounded by two frequencies which correspond to two planes (of constant z). The total NMR intensity in that frequency interval is proportional to the number of nuclei in the sample lying between those planes. Thus the NMR absorption spectrum provides a projection of the sample spin density integrated over planes perpendicular to the gradient direction. From a series of such projections for various gradient directions, one can reconstruct the object.

We are considering, then, a uniform field kH_0 to which there is added a small additional field $h(r)$ which is static in time:

$$\boldsymbol{H} = \boldsymbol{k}H_0 + \boldsymbol{h}(\boldsymbol{r}) \quad \text{with} \qquad (7.374\text{a})$$

$$\boldsymbol{h}(\boldsymbol{r}) = \boldsymbol{i}h_x(\boldsymbol{r}) + \boldsymbol{j}h_y(\boldsymbol{r}) + \boldsymbol{k}h_z(\boldsymbol{r}) \quad \text{and} \qquad (7.374\text{b})$$

$$|\boldsymbol{h}| \ll H_0 . \qquad (7.374\text{c})$$

We wish to show first that we can neglect the components h_x and h_y perpendicular to $\boldsymbol{k}H_0$.

The precession frequency depends on the magnitude of $\boldsymbol{H}$:

$$\omega(\boldsymbol{r}) = \gamma H(\boldsymbol{r}) . \qquad (7.375)$$

Since $H_0 \gg |\boldsymbol{h}(\boldsymbol{r})|$, the effects of $\boldsymbol{i}h_x$ and $\boldsymbol{j}h_y$ are merely to rotate $\boldsymbol{H}$ slightly without the first order changing its magnitude, whereas h_z changes $H(\boldsymbol{r})$ to first order:

$$\begin{aligned} H &= \sqrt{(H_0 + h_z)^2 + h_x^2 + h_y^2} \\ &= \sqrt{H_0^2 + 2H_0 h_z + h_x^2 + h_y^2 + h_z^2} . \end{aligned} \qquad (7.376)$$

Neglecting the terms quadratic in the components of $\boldsymbol{h}$, we get

$$H = H_0\sqrt{1 + 2h_z/H_0} \cong H_0(1 + h_z/H_0) = H_0 + h_z . \qquad (7.377)$$

What spatial forms can h_z take? If we simply consider it to be a general function of $\boldsymbol{r}$, we can expand it in a power series in x, y, z about a convenient origin. We assume we have shaped the field so that only the lowest terms are needed. Then

$$h_z(\boldsymbol{r}) = h_z(0) + x\left(\frac{\partial h_z}{\partial x}\right)_0 + y\left(\frac{\partial h_y}{\partial y}\right)_0 + z\left(\frac{\partial h_z}{\partial z}\right)_0 + \ldots$$
$$\equiv h_z(0) + xG_x + yG_y + zG_z \quad , \tag{7.378}$$

which defines the components G_x, G_y, G_z of the gradient $\boldsymbol{G}$ of $h_z(\boldsymbol{r})$ at the origin:

$$\boldsymbol{G} = \boldsymbol{i}G_x + \boldsymbol{j}G_y + \boldsymbol{k}G_z \quad . \tag{7.379}$$

The reader may wonder whether we can guarantee that it is possible to generate an h_z with the spatial dependence of (7.378). After all, $\boldsymbol{h}$ must obey the laws of physics ($\vec{\nabla} \cdot \boldsymbol{h} = 0$, $\vec{\nabla} \times \boldsymbol{h} = 0$). We explore these aspects in a homework problem. The answer is that we can achieve (7.378), but for there to be a term such as $\boldsymbol{k}xG_x$ for example, there must also be terms in the $\boldsymbol{i}$ and/or $\boldsymbol{j}$ directions. However, since as we have seen the transverse components do not affect $|\boldsymbol{H}|$ to first order, we can neglect them.

Let us then define two functions, a frequency distribution function, $f(\omega)$, which gives the NMR intensity at frequency ω and is normalized to satisfy

$$\int f(\omega)d\omega = 1 \quad , \tag{7.380}$$

and a spin density function $\varrho(\boldsymbol{r})$, which gives the number of nuclear spins in a unit volume at point $\boldsymbol{r}$, and which is likewise normalized so that

$$\int \varrho(\boldsymbol{r})d^3\boldsymbol{r} = 1 \quad . \tag{7.381}$$

If, then, the gradient $\boldsymbol{G}$ is oriented in the z'-direction,

$$\boldsymbol{G} = \hat{\boldsymbol{z}}'G \quad , \tag{7.382}$$

planes $z' = $ constant are planes of constant precession frequency. Since the magnetic field changes by $G\,dz'$ between the planes at $z' = $ constant and $(z' + dz') = $ constant, the precession frequency change, $d\omega$, is

$$d\omega = \gamma G\,dz' \quad . \tag{7.383}$$

Therefore

$$f(\omega)d\omega = dz' \iint_{z'=\text{const}} dx'dy'\varrho(x',y',z') \quad . \tag{7.384}$$

The integral in (7.384) is the projection of the spin density function $\varrho(x',y',z')$ on the z'-axis.

Thus, utilizing (7.383),

$$\gamma G f(\omega) = \iint_{z'=\text{const}} dx'dy'\varrho(x',y',z') \quad . \tag{7.385}$$

Since this projection depends on the direction of G, and since the magnitude of $f(\omega)$ depends on the magnitude of G, we put a subscript on $f(\omega)$:

$$\gamma G f_G(\omega) = \iint_{z'=\text{const}} dx' dy' \varrho(x', y', z') \quad . \tag{7.386}$$

Since $f_G(\omega)$ can be measured, we can view the integral on the right as known experimentally as a function of z'.

We can think of making a variety of measurements to determine the ω dependence of $f_G(\omega)$ for a variety of different G directions, thereby getting the projected spin density for any desired projection axis, z'.

It is now easy to show that such measurements enable one to reconstruct $\varrho(x, y, z)$. First we take the experimental data for a given z' and form the Fourier transform of the data:

$$\varrho_k = \int e^{-ikz'} dz' \iint dx' dy' \varrho(x', y', z') \quad , \tag{7.387a}$$

where

$$\boldsymbol{k} = k\hat{\boldsymbol{z}}' \quad . \tag{7.387b}$$

Utilizing (7.373) with $\omega_0 \equiv \gamma H_0$ and (7.386) we can express (7.387) as

$$\varrho_k = \exp(ik\omega_0/\gamma G) \int \exp(-ik\omega/\gamma G) f_G(\omega) d\omega \quad . \tag{7.388}$$

If we know ϱ_k for all $\boldsymbol{k}$, we can get $\varrho(\boldsymbol{r})$ by the relationship

$$\varrho(\boldsymbol{r}) = \frac{1}{(2\pi)^3} \int e^{i\boldsymbol{k}\cdot\boldsymbol{r}} \varrho_k d^3k \quad . \tag{7.389}$$

In practice, since each direction $\boldsymbol{k}$ is explored in a separate experiment, ϱ_k is never known experimentally for $\boldsymbol{k}$ as a continuous variable. For example, if one has a two-dimensional object, the various $\boldsymbol{k}$'s would be chosen to lie in the plane of the object. Then the $\boldsymbol{k}$ values explored could be represented as in Fig. 7.46. It is seen that they represent radial lines out to some maximum k, and $k_{\max}$. Since performing an integral such as (7.389) is a numerical operation, the

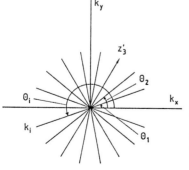

Fig. 7.46. The paths in k-space ($\boldsymbol{k} = k\hat{\boldsymbol{z}}'$) explored by applying the gradient $\boldsymbol{G} = G_{\hat{z}'}$ in a variety of directions $\hat{\boldsymbol{z}}'$ in a plane. The various directions θ_1, θ_2, ..., θ_i correspond to different directions $\hat{\boldsymbol{z}}'_1$, $\hat{\boldsymbol{z}}'_2$, ..., $\hat{\boldsymbol{z}}'_i$

problem then becomes one of finding a suitable numerical approach for getting the most accurate value of the integral from the data as given. Of course, in the process of selecting the approach, one gets guidance as to what data to collect.

The approach of reconstructing $\varrho(r)$ by Fourier transforming the spin density distributions is one approach to finding $\varrho(r)$ from data for $f_G(\omega)$. *Lauterbur* [7.79] demonstrated a simple graphical means, and pointed out that the solution to the problem of getting $\varrho(r)$ from spin-density projections was well known from other areas of science.

Another experimental approach was proposed by *Mansfield*, based on observing free induction decays. If one applies an $X(\pi/2)$ pulse to the spins and observes the free induction decay in the reference frame rotating at ω_0, the complex magnetization is

$$\begin{aligned}\langle M^+(t)\rangle &= \langle M_x(t)\rangle + \mathrm{i}\langle M_y(t)\rangle \\ &= \mathrm{i}M_0 \int \varrho(x',y',z')\mathrm{e}^{-\mathrm{i}\omega(x',y',z')t}dx'dy'dz' \\ &= \mathrm{i}M_0 \int \varrho(x',y',z')\mathrm{e}^{-\mathrm{i}\gamma Gz't}dx'dy'dz' \quad ,\end{aligned} \qquad (7.390\mathrm{a})$$

which can be rewritten to make its meaning clearer as

$$\langle M^+(t)\rangle = \mathrm{i}M_0 \int dz' \mathrm{e}^{-\mathrm{i}\gamma Gz't} \iint dx'dy'\varrho(x',y',z') \qquad (7.390\mathrm{b})$$

so that

$$\langle M^+(t)\rangle = \mathrm{i}M_0 \int dz' \mathrm{e}^{-\mathrm{i}\boldsymbol{q}\cdot\boldsymbol{r}} \iint dx'dy'\varrho(x',y',z') \qquad (7.391)$$

where

$$\boldsymbol{q} \equiv \gamma Gt\hat{z}' \quad . \qquad (7.392)$$

We can rewrite (7.391) to emphasize that on the left we are observing $\langle M^+(t)\rangle$ for a particular gradient direction $\hat{z}'$, and rewrite the right hand side to bring out its meaning to get

$$\frac{\langle M^+(t)\rangle_{\hat{z}'}}{\mathrm{i}M_0} = \int \mathrm{e}^{-\mathrm{i}\boldsymbol{q}\cdot\boldsymbol{r}}\varrho(\boldsymbol{r})d^3\boldsymbol{r} \quad . \qquad (7.393)$$

The expression on the right of (7.393) is identical to that on the right hand side of (7.387a) if one replaces q by k. We will henceforth replace q by k. Thus, in the presence of a static field gradient, the NMR signal following a $\pi/2$ pulse sweeps out in time the Fourier transform ϱ_k, where the direction of k is given by the direction of the field gradient, and each point in time corresponds to a magnitude of wave vector given by (7.392). Clearly, the free induction decay is also producing ϱ_k along radial lines in k-space as in Fig. 7.46.

In his initial paper, *Mansfield* [7.80] proposed using the free induction decay, giving (7.393). Since his initial idea related to crystallography, he explored the case that the spin density is given by a periodic lattice, and demonstrated the imaging scheme by making a macroscopic layered sample to simulate atomic periodicity.

In Fig. 7.46 we display a set of k trajectories for a planar sample. Even for nonplanar samples, it is convenient to collect data from planar slices in the sample, a sequence of slices thus providing the full three-dimensional object.

The methods of selecting well-defined slices are part of the present art of imaging. We refer the reader to [7.77] for example. A simple approach to generate a slice perpendicular to the z-direction is to start by applying a gradient in the z-direction. If one applies a pulse at a particular frequency ω_I, the spins at coordinate z such that

$$\gamma_I(H_0 + zG_z) = \omega_I \tag{7.394}$$

are perfectly at resonance. An $X_I(\pi/2)$ pulse will then rotate them into the x-y plane. Spins within a Δz given by

$$G_z \Delta z = (H_1)_I \tag{7.395}$$

will also be flipped, but spins farther away in z will be less effectively tilted into the x-y plane. By in fact shaping the time dependence of $(H_1)_I$ one can improve on the z-selection.

In any event, if one now turns off G_z, one has produced an initial transverse magnetization for all x, y within Δz. One can then apply gradients in the x-y plane to record the free induction decay for any direction $\hat{k} = \hat{G}$ in the x-y plane. All planar imaging methods must initially carry out some such precedure to define the slice Δz.

Kumar et al. [7.81] proposed a two-dimensional (or three-dimensional) Fourier transform approach to collecting data which has been adopted by many workers in the field.

We start by describing the two-dimensional version in which one first excites a slice in the z-direction. One then applies a gradient

$$\boldsymbol{G} = i G_x \tag{7.396a}$$

for a time called t_x, then a gradient

$$\boldsymbol{G} = j G_y \tag{7.396b}$$

for a time called t_y. During t_y, one records the free induction decay. Using (7.170) and (7.390) one has, neglecting relaxation, in the rotating frame

$$\langle M^+(t)\rangle = iM_0 \Delta z \int e^{-ix\gamma G_x t_x} e^{-iy G_y t_y} \varrho(x,y,z) dx\, dy \quad, \tag{7.397}$$

where $t = t_y + t_x$ and where M_0 is the total magnetization of the sample.

Defining

$$k_x = \gamma G_x t_x \quad, \quad k_y = \gamma G_y t_y \tag{7.398}$$

one can write

$$\frac{\langle M^+(t)\rangle}{iM_0} = \Delta z \int e^{-ik_x x} e^{-ik_y y} \varrho(x,y,z) dx\, dy \quad, \tag{7.399}$$

which is the Fourier transform in the plane z of the spin density. Since t_x is fixed while recording $\langle M^+(t)\rangle$, the data are for a fixed k_x, but for a continuous k_y. They may be represented as in Fig. 7.47. They thus sample ϱ_k for the z-slice along parallel lines in (k_x, k_y) space.

In a full three-dimensional version, one applies the initial $\pi/2$ pulse in the absence of gradients. Then, one applies

$$\begin{aligned} G &= iG_x \quad \text{for} \quad t_x \; , \\ G &= jG_y \quad \text{for} \quad t_y \; , \\ G &= kG_z \quad \text{for} \quad t_z \; , \end{aligned} \qquad (7.400)$$

recording the free induction decay during t_z. Then

$$\langle M^+(t)\rangle = iM_0 \int e^{-ixG_x t_x} e^{-iyG_y t_y} e^{-izG_z t_z} \varrho(x,y,z)dx\,dy\,dz \quad (7.401\text{a})$$

with

$$k_x \equiv \gamma G_x t_x \quad , \quad k_y \equiv \gamma G_y t_y \quad , \quad k_z \equiv \gamma G_z t_z \qquad (7.401\text{b})$$

and

$$t_z = t - (t_x + t_y) \; . \qquad (7.401\text{c})$$

This approach gives one the Fourier transform of $\varrho(r)$ lines parallel to the k_z axes formed by the intersection of the planes k_x = constant and k_y = constant (Fig. 7.48).

If one is interested in knowing the density pattern in a slice of known z, it is much faster to collect the data using a z-gradient to define the slice excitation and then work with two-dimensional transforms.

An exceedingly rapid way to sample all the points in one slice within a single sweep was proposed by *Mansfield* [7.82] and called by him the planar echo method. In the planar form of this method, one first selects a slice (say

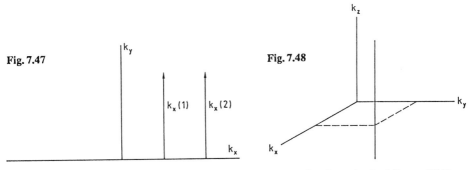

Fig. 7.47. The paths in k-space explored by the two-dimensional method of *Kumar, Welti, and Ernst*. For a particular t_x, k_x is fixed. As the time t_y grows (7.397), k_y grows as shown

Fig. 7.48. The path explored in k-space by the three-dimensional method of *Kumar, Welti, and Ernst*

at fixed z), then applies a weak steady gradient in the x-direction, and a strong gradient in the y-direction which is switched alternately between two values $+G_0$ and $-G_0$ (Fig. 7.49). Thus, following slice selection one has

$$G = iG_x + jG_0 f(t) \quad \text{where} \tag{7.402}$$

$$\begin{aligned} f(t) &= +1 && \text{for} \quad 0 < t < \tau_b \\ & -1 && \text{for} \quad \tau_b < t < 3\tau_b \\ & +1 && \text{for} \quad 3\tau_b < t < 5\tau_b \end{aligned} \tag{7.403}$$

and so on (Fig. 7.49).

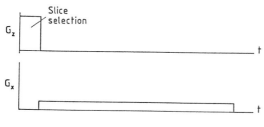

Fig. 7.49. The time dependence of the three gradients G_x, G_y, and G_z employed in the planar echo method of *Mansfield*

In order to grasp the effect of such a time-dependent gradient, it is useful to follow our earlier procedure of defining a wave vector k. The approach we then take was developed by *King* and *Moran* [7.83], *Brown* et al. [7.84], *Twieg* [7.85], and *Ljunggren* [7.86].

Since the precession frequency ω and the angle a spin has precessed ϕ, are related by

$$\omega = \frac{d\phi}{dt} \quad , \tag{7.404}$$

when the gradient is time dependent, we have

$$\langle M^+(t) \rangle = iM_0 \int \exp\left[-i \int_0^t \omega(t') dt'\right] \varrho(r) d^3r \tag{7.405a}$$

$$= iM_0 \int \exp\left[-i\gamma r \cdot \int_0^t G(t') dt'\right] \varrho(r) d^3r \quad . \tag{7.405b}$$

Therefore, defining

$$k(t) \equiv \gamma \int_0^t G(t')dt' \tag{7.406}$$

we get

$$\frac{\langle M^+(t)\rangle}{iM_0} = \int e^{-ik(t)\cdot r}\varrho(r)d^3r = \varrho_{k(t)} \quad . \tag{7.407}$$

Thus, as t develops, $k(t)$ moves through k space on the trajectory given by (7.406). $\langle M^+(t)\rangle/iM_0$ gives $\varrho_{k(t)}$ for the successive values of k through which (7.406) takes the system.

Examining the gradient of (7.403) or in Fig. 7.49, we see that the weak constant component G_x gives a component k_x which increases slowly proportional to time. The strong gradient G_y builds up a large phase angle $x\gamma G_x t$ from 0 to τ_b, but the sign reversal of G_x which then occurs leads to this phase accumulation unwinding during the next interval τ_b. Thus at $t = 2\tau_b$ the various planes x = constant get back in phase, producing an echo that is

$$k_x(t = 2\tau_b) = 0 \quad . \tag{7.408}$$

In k-space, the vector $ik_x(t)+jk_y(t)$ follows the trajectory shown in Fig. 7.50, in which we have assumed both G_x and G_0 are positive. As can be seen, this method explores positive and negative k_y for positive k_x. One can get the results for negative k_x by several methods. One is to apply a strong negative G_x for a short interval after slice selection, and before applying G_y. This process would move k to a point on the negative k_x-axis. Then one begins the sequence of (7.374a).

Another approach would be to utilize the mathematical result that

$$\frac{\langle M^+(-k)\rangle}{iM_0} = \left(\frac{\langle M^+(k)\rangle}{iM_0}\right)^* , \tag{7.409}$$

which can be seen to be true from (7.407). This is the statement that with quadrature detection the signal of one phase gives the cosine transform, the other gives the sine transform. The practical result is that one need only collect quadrature data in one half-plane of (k_x, k_y). This remark of course holds true for the schemes we described earlier.

We see, therefore, that the "planar echo" method samples all the half-plane in each cycle, and thus contains the information for total spin-density reconstruction in each sweep.

A more "physical" description of the planar echo method provided *Mansfield*'s original motivation. He thought in terms of a discrete lattice (as of atoms in a regular array). Suppose we imagine an apple orchard with trees planted in equally spaced rows and columns. Suppose some of the trees have died and been cut down. If one stands at the end of one row and looks down that row, one cannot tell whether any trees might be missing from it. However, since one sees the next row at an angle, one can readily spot any vacant site. This phenomenon is illustrated in Fig. 7.51 from *Mansfield*'s book [7.77]. To utilize this principle in imaging, one must produce a discrete lattice, and view it at an angle. The imag-

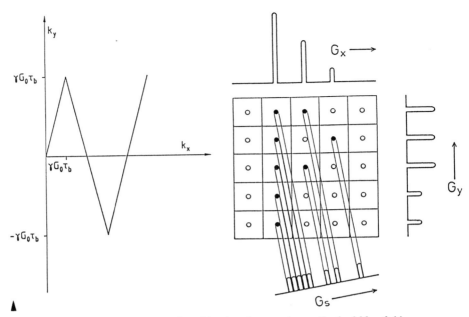

Fig. 7.50. The path in k-space explored in the planar echo method of *Mansfield*

Fig. 7.51. Projections of a discrete matrix filled to form the letter *P*. Projections in G_x or G_y alone do not contain sufficient information to uniquely determine the original object. Projection orthogonal to the special gradient direction G_s contains all information necessary to completely determine the original object distribution

ing scheme of Fig. 7.50 has parallel scanning traces $2\gamma G_y \tau_b$ apart in k-space, which corresponds to a spatial wavelength

$$\lambda = \frac{2\pi}{k_x} = \frac{2\pi}{2\gamma G_y \tau_b} \quad . \tag{7.410}$$

This wavelength imposes discreteness (i.e. "rows") on the spin density function $\varrho(\mathbf{r})$.

Another way of saying this is that the echo train produced by the oscillatory gradient G_y has a period $2\tau_b$, hence corresponds to observations at a spectrum of frequencies spaced apart by angular frequency $\Delta\omega$ of

$$\Delta\omega = \frac{1}{2\tau_b} \quad . \tag{7.411}$$

But $\Delta\omega$ implies a Δz of

$$\Delta z = \frac{\Delta\omega}{\gamma G_y} = \frac{1}{\gamma G_y} \frac{1}{2\tau_b} \quad . \tag{7.412}$$

Each frequency corresponds to a row of trees. The weak x-gradient produces the "view at an angle" shown in Fig. 7.51.

8. Advanced Concepts in Pulsed Magnetic Resonance

8.1 Introduction

Hahn's discovery of spin echoes [8.1] demonstrated that one could defeat the effect of magnet inhomogeneity in obscuring the true width of magnetic resonance lines because the echo amplitude $M(2\tau)$ decayed exponentially with $2\tau/T_2$, where τ is the time between pulses. However, if the magnet inhomogeneity is large, and if the diffusion rate of the nuclei sufficiently great,

$$M(2\tau) = M(0)\exp(-2\tau/T_2)\exp\left[-\left(\gamma\frac{\partial H}{\partial z}\right)^2 \frac{2}{3}D\tau^3\right] \tag{8.1}$$

where D is the diffusion constant and $\partial H/\partial z$ the gradient in static field. In some instances, then, the τ^3 term might obscure the T_2. Or, conversely, if $\partial H/\partial z$ is known, the τ^3 term gives one a method of measuring D. The derivation of the diffusion term is given in Appendix G.

Carr and *Purcell* [8.2] invented a clever scheme for eliminating the τ^3 term when desired. Their proposal initated a class of experiments in which a sequence of pulses is applied to the resonance. In this chapter we discuss first the Carr-Purcell concept, then an ingenious modification invented by *Meiboom* [8.3], and finally turn to some of the most intricate pulse methods yet invented which have the remarkable property of enabling one to eliminate the dipolar broadening of resonance lines in solids. These concepts were introduced by *Waugh* [8.4] and *Mansfield* [8.5], and have been extended by *Vaughan* [8.6], as well as a group of very talented resonators who have spread out from these initiators.

8.2 The Carr-Purcell Sequence

We begin a discussion of the Carr-Purcell sequence by assuming we can neglect the τ^3 diffusion term. To measure a T_2 by the conventional spin echo, one must make a sequence of measurements, each at a different value of τ. The envelope of the echo amplitude as a function of 2τ gives one T_2. *Carr* and *Purcell* pointed out that the entire envelope could be obtained at a shot if one applied the proper sequence of pulses.

Suppose at $t = 0$ one applies a $\pi/2$ pulse in which H_1 lies along the $+x$-axis in the rotating reference frame with the rf frequency ω exactly at the Larmor frequency ω_0. Such a pulse turns the magnetization M_0 to lie along the $+y$-axis. If one applies a π pulse at $t = \tau$, also with H_1 along the $+x$-axis, an echo is formed at 2τ with the magnetization along the $-y$-axis. If now one applies a π pulse at 3τ, another echo will form at 4τ, this time along the $+$-axis. In this manner, successive π pulses at $(2n+1)\tau (n = 0, 1, 2, \ldots)$ form echoes at $(2n+2)\tau$, the echoes forming along the $-y$-axis for odd n, and the $+y$-axis for even n. Since all components of the magnetization in the $x-y$ plane are decaying exponentially with time constant T_2, the echo sequence likewise decays exponentially with T_2.

A sequence of such pulses and echoes is shown in Fig. 8.1, with the sign, positive or negative, showing whether the echo forms along the positive or negative y-axis.

The behavior we describe enables one to obtain the entire echo envelope in a single pulse train, clearly a convenience. If one is studying a weak signal, so that the signal-to-noise ratio is important, the Carr-Purcell sequence is a tremendous advantage as we explain. The noise depends on the bandwidth of the apparatus. One limits the bandwidth so as to pass the echo signal without undue attenuation. Clearly the same bandwidth can be used for either a Carr-Purcell train or a conventional echo. After each single echo, one must wait several times T_1 for the system to recover before measuring the next echo. However, in the time it takes to get one echo, we can obtain the full envelope with a Carr-Purcell sequence. If we take N echo signals to define the echo envelope, we could obtain N Carr-Purcell pulse trains, so each Carr-Purcell echo is recorded N times. If the spin echoes are taken to correspond to the times of the Carr-Purcell echoes, the Carr-Purcell train will have a better signal-to-noise by $\sqrt{N}$.

We have remarked that the Carr-Purcell pulse sequence reduces the effect of diffusion on the echo decay. The reason can be seen as follows. If there is no diffusion, a spin dephases during the interval τ following each echo and rephases

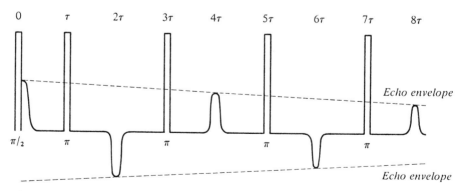

Fig. 8.1. The Carr-Purcell sequence and the resulting echoes. As described in the text, a positive echo forms along the $+y$-axis in the rotating frame, a negative echo forms on the $-y$-axis

during the interval after the π pulse. It is the fact that diffusion makes rephasing imperfect which causes the next echo to be smaller. In Appendix G we show that if a $\pi/2$ pulse at $t = 0$ tilts the magnetization M_0 into the $x - y$ plane, a π pulse τ later produces an echo at 2τ, $M(2\tau)$, smaller than M_0 owing to diffusion by

$$M(2\tau) = M_0 \exp\left[-D\left(\gamma\frac{\partial H}{\partial z}\right)^2 \frac{2}{3}\tau^3\right] \exp(-2\tau/T_2)$$

$$\equiv M_0 \alpha \quad . \tag{8.2}$$

When the echo is formed, we have recreated at 2τ the situation which existed at $t = 0$, except the magnetization is reduced by the factor α, and points along the $-y$-axis instead of the $+y$-axis. Therefore we can consider the magnetization at 2τ as another initial condition for applying the argument of Appendix G for the next interval from 2τ to 4τ with a pulse applied at 3τ.

The result is

$$M(4\tau) = M(2\tau)\alpha = M_0 \alpha^2 \quad . \tag{8.3}$$

If we have n cycles 2τ, we therefore get

$$M(n2\tau) = M_0 \alpha^n \quad . \tag{8.4}$$

Therefore

$$M(n2\tau) = M_0 \exp\left[-\left(\gamma\frac{\partial H}{\partial z}\right)^2 D(n2\tau)\frac{1}{3}\tau^2\right] \exp(-n2\tau/T_2) \quad . \tag{8.5}$$

Thus, if we change τ but hold $n2\tau$ constant so that we are comparing two different echoes of the Carr-Purcell train a fixed time apart, we can vary the factor involving diffusion but we do not vary the factor containing T_2. Hence we can make the effect of the diffusion term negligible.

It is clear that what is involved are two facts:

1. each echo cycle of the Carr-Purcell train reduces the magnetization by a factor α,
2. the *relative* contribution of the diffusion term compared to the T_2 term depends on τ.

8.3 The Phase Alternation and Meiboom-Gill Methods

A Carr-Purcell pulse sequence may contain many pulses. If the pulse rotations deviate slightly from π, there are cumulative effects which may become large. We illustrate in Fig. 8.2. (This figure is drawn for a negative γ, hence involves right hand rotations.) For simplicity we take the initial pulse to be a perfect $\pi/2$ pulse about the $+x$-axis, and we consider diffusion effects to be negligible. At

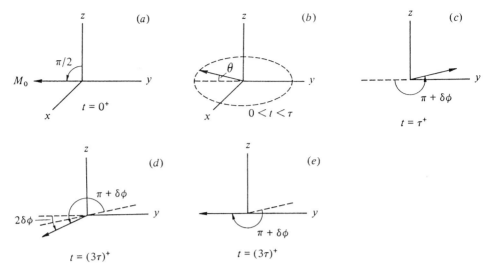

Fig. 8.2a-e. The effect of an imperfect π pulse. The initial $\pi/2$ pulse, assumed perfect, rotates M_0 to the $-y$-axis at $t = 0^+$. At $t = \tau^-$, just before the second pulse, a typical spin has precessed an amount θ in the $x - y$ plane away from the $-y$-direction, owing to static field inhomogeneities. We now apply a $\pi + \delta\phi$ rotation about the $+x$-axis. Fig. 8.2c shows the effect on a spin for which $\theta = 0$ lies in the $x - y$ plane. It no longer lies in the $x - y$ plane. Just after a second $\pi + \delta\phi$ pulse at $t = (3\tau)^+$, the same magnetization makes an angle $2\delta\phi$ with the $x - y$ plane. In Fig. 8.2e we see that if the H_1 of the second pulse lies along the -x-axis, the rotation direction is reversed, so the $-(\pi + \delta\phi)$ rotation restores the spin to its orientation at $t = 0^+$

$t = \tau^-$ the spins will have fanned out in the $x - y$ plane, a typical spin making an angle θ with the $-y$-axis (for negative γ). At $t = \tau$ we apply a $\pi + \delta\phi$ pulse about the $+x$-axis where $\delta\phi$ gives the deviation from being a perfect π pulse. For a spin having $\theta = \pi/2$, the spin lies along the x-axis. It is not affected by the pulse at $t = \tau$. A spin for which $\theta = 0$ would rotate as shown in Fig. 8.2c. It will lie an angle $\delta\phi$ above the $+y$-axis in the $y - z$ plane where it will sit until 2τ later when the next $\pi + \delta\phi$ pulse is applied. The result is shown in Fig. 8.2d. The spin now makes an angle $2\delta\phi$ with the $-y$-axis. Each successive imperfect π pulse adds another $\delta\phi$ to the deviation.

The cumulative effect becomes exceedingly serious in practice because the H_1 is never uniform through the sample. Consequently though part of the sample may have a π pulse, other parts do not.

These sorts of troubles have major implications for all multipulse sequences such as those described in the later portions of this chapter. One simple approach is to apply the H_1 for the π pulses alternating along the $+x$- and $-x$-axis. Thus the sense of rotation about the x-axis reverses in alternate pulses. The result is shown in Fig. 8.2e. Instead of getting a cumulative rotation error of $2\delta\phi$ as in Fig. 8.2d, the cumulative error is *zero*!

Meiboom and *Gill* [8.3] were the first people to find a solution to this problem. They likewise used a phase coherent method, but introduced a 90° phase shift into the radio frequency field between the initial $\pi/2$ pulse and the subsequent π pulses. Thus if H_1 for the $\pi/2$ pulse were along the $+y$-axis, H_1 for the π pulses would be applied along the $+x$-axis. As a result, the echoes are all formed along the $+x$-axis for negative γ. We leave it to the reader to show that if the pulse is $\pi - \delta$ instead of π, the error does not accumulate.

8.4 Refocusing Dipolar Coupling

The dipole-dipole coupling between neighbouring nuclei provides valuable information for many purposes, but on occasion it is a problem. For example, it may obscure chemical shifts, and it may cause free induction decays to be short-lived and thus difficult to see. We turn now to several interesting approaches to eliminating or effectively reducing dipolar coupling.

For a group of spins an ordinary spin echo [e.g. an $X(\pi/2)\ldots\tau\ldots X(\pi)$ sequence] refocuses the dephasing which arises because they are placed in an inhomogeneous magnetic field. Since the magnetic dipolar coupling among neighboring spins is in some way analogous to a magnetic field inhomogeneity, one might expect that the same spin echo pulse sequence would also refocus the dephasing resulting from magnetic dipolar coupling. If the dipolar coupling is between different nuclear species (e.g. H^1 and C^{13}), such an echo sequence does refocus the dipolar coupling. If, however, the coupling is between like nuclei (e.g. H^1 with H^1), the echo does not work. It is easy to see why. The π pulse inverts all the neighbors, so that a nucleus which was precessing more rapidly than average in the first time interval τ, finds its neighbors inverted, and thus precesses more slowly in the second interval. The usual echo works because the π pulse converts a precession phase lead into a precession phase lag, of equal size. If at the same time the rate of precession changes, the spins do not come back into phase.

Nevertheless, the fact that a well-defined relationship exists between the precession frequency when a neighbor points up versus down makes one feel that there should be some way to undo the dephasing which the dipolar coupling produces. In the following sections, we explore these ideas.

8.5 Solid Echoes

Powles and *Mansfield* [8.7] discovered that the free induction decay of a coupled pair of identical spin $\frac{1}{2}$ nuclei can be refocused by applying a pair of $\pi/2$ pulses shifted in phase by $\pi/2$ with respect to each other. Such a sequence might be denoted by

$$X(\pi/2)\ldots\tau\ldots Y(\pi/2)\ldots t_1 \tag{8.6}$$

with the echo occurring when $t_1 = \tau$. This sequence has come to be known as the solid echo. *Powles* and *Mansfield* demonstrated it experimentally for $CaSO_4 \cdot 2H_2O$, and showed that it followed theoretically. The pulse sequence perfectly refocuses spin $\frac{1}{2}$ nuclei which interact only in pairs, but does not refocus larger groupings of spins perfectly. In this section, we explain the echo for coupled pairs. This same pulse sequence refocuses the first order quadrupole coupling of a spin 1 nucleus for reasons explained in Appendix H. It is therefore important for deuterium and N^{14} NMR. The existence of echoes related to quadrupole splittings was first demonstrated by *Solomon* [8.8] for a case of $I = \frac{5}{2}$. It appears to have been *Davis* et al. [8.9] who recognized that for the $I = 1$ case sequence (8.6) gave perfect refocusing.

The fact that one needs exactly sequence (8.6) may seem surprising at first glance. Why does one need to phase shift the second pulse by $\pi/2$? Why does one use a $\pi/2$ pulse instead of a π pulse? To help give a feel for the answers to these questions, we shall derive the result three ways. The first method is a semigraphical one, utilizing a special value of τ. The second derivation generalizes the first for arbitrary τ. The third derivation utilizes the spin operator method.

We start with the simple physical picture. However, we must present it rather carefully. Consider a sample consisting of pairs of nuclei. That is, each nucleus has one and only one neighbor which is sufficiently close that only the dipolar coupling to it matters. In the presence of a static field, the nuclei are weakly polarized, there being a slight excess with magnetic moments parallel to the static field, H_0. If there are N nuclei and if we denote by p_+ and p_- the probabilities for a given nuclear moment to point either parallel (p_+) or antiparallel (p_-) to H_0, we have then that there are N_+ and N_- nuclei pointing parallel or antiparallel, respectively, given by

$$N_+ = Np_+ \quad , \quad N_- = Np_- \quad . \tag{8.7}$$

The population difference, n, then obeys

$$n = N(p_+ - p_-) \quad . \tag{8.8}$$

p_+ and p_- are related by the Boltzmann factor

$$p_+/p_- = e^{\gamma\hbar H_0/kT} \tag{8.9}$$

so that, if $\gamma\hbar H_0 \ll kT$,

$$p_+ = \tfrac{1}{2}(1 + \gamma\hbar H_0/2kT) \quad , \quad p_- = \tfrac{1}{2}(1 - \gamma\hbar H_0/2kT) \quad . \tag{8.10}$$

Note that the expressions for p_+ or p_- do *not* depend on the orientation of the other nucleus in the pair, because we have assumed that the only field acting on a nucleus is H_0. This assumption should be valid if $H_0 \gg \gamma\hbar/r^3$, where r is the distance between neighbors. Now, it is only the excess of nuclei N_+ over those N_- which can give rise to a NMR signal. To follow what happens, we need only follow what happens to the excess, since the remaining nuclei carry no net polarization.

Let us therefore in a thought experiment reach in at random to examine the pairs. We take $n/2$ pairs in which both nuclei point up, and $n/2$ pairs in which one nucleus points up, and the other points down. This will give us a net n nuclei pointing up. The other $N - 2n$ nuclei in the sample must then be equally pointing up and down, hence be unpolarized. We neglect them. Now, among the pairs with one nucleus pointing up, the other pointing down, we label the up nucleus blue, the down nucleus red. Among the pairs with both nuclei up, we also label one blue, the other red (it does not matter which of the pair we choose for which color). We thus have $n/2$ red nuclei with spin up, and $n/2$ red nuclei with spin down. Then the blue nuclei give the magnetization, since the red nuclei have no net magnetization.

We represent the spin-spin coupling by a term $\mathcal{H}_{\text{spin-spin}}$ of the form

$$\mathcal{H}_{\text{spin-spin}} = \hbar a I_z S_z \quad . \tag{8.11}$$

For pairs of identical particles of spin $\frac{1}{2}$ this is exactly equivalent to dipolar coupling, as explained in Appendix H. We assume that apart from $\mathcal{H}_{\text{spin-spin}}$ we are exactly at resonance for any pulses we apply. Then, following an $X(\pi/2)$ pulse, the blue spins lie along the $+y$-axis in the rotating frame.

Now, although we have flipped both red and blue spins by $\pi/2$, the red spins at this point have no net polarization and thus their density matrix remains diagonal with the two diagonal elements equal to each other. Thus, we can describe the situation at $t = 0^+$, just after the pulse, as in Fig. 8.3b, in which we represent the blue spins by vectors 1 and 3, lying along the y-axis, and the red spins by vectors 2 and 4, one pointing up, the other down, along the z-axis.

Spins 1 and 3 will now precess at rates $-a/2$ and $+a/2$ (in the left-handed sense) respectively about the $+z$-axis. The explanation of how the echo forms becomes very simple for a particular value of τ, the value which makes $a\tau/2 = \pi/2$ (i.e. $a\tau = \pi$). For that time, the spins will be aligned as in Fig. 8.3c.

At this time, the $Y(\pi/2)$ pulse will produce the situation of Fig. 8.3d. Now, during the next time interval τ, spins 1 and 3 will not precess (they lie along the z-axis), but spins 2 and 4 will precess at rates $+a/2$ and $-a/2$ respectively so that at $t = 2\tau$, 2 and 4 will lie along the $+y$-axis. An echo therefore occurs at $t = 2\tau$. It is interesting to note that in this process we have transferred the net polarization which was initially in spins 1 and 3 to spins 2 and 4. However, we have not made an unpolarized system become polarized. Thus, we can remain confident that the other $N - 2n$ spins, which we have said are initially unpolarized, do not become polarized by the echo sequence.

If one chooses a τ other than the one for which $a\tau/2 = \pi/2$, the explanation becomes more complicated. Then one needs to utilize the ideas of Sect. 7.24 in which we discuss the time development of one spin when it is coupled to another spin which is in a mixture of the spin-up and spin-down eigenstates. By choosing $a\tau/2 = \pi/2$ we have made this mixture be in fact pure eigenstates.

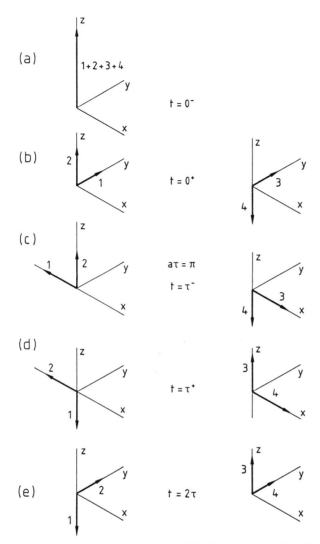

Fig. 8.3a-e. The formation of a solid echo for two pairs of spins (1 and 2, 3 and 4) for a particular value of the time τ at which the refocusing pulse is applied. (**a**) The thermal equilibrium magnetization at $t = 0^-$ showing the vector sum of the four magnetization vectors. Spins 1 and 3 are the blue spins, spins 2 and 4 are the red spins. (**b**) Just after the $X(\pi/2)$ pulse ($t = 0^+$), the blue spins (1 and 3) lie along the y-axis while the red spins (2 and 4) can be taken to point along the positive and negative z-axes respectively. (**c**) The magnetization vectors at times $t = \tau^-$ just before the $Y(\pi/2)$ pulse, for the particular τ such that $a\tau/2 = \pi/2$ (or $a\tau = \pi$). (**d**) The magnetization vectors at time $t = t^+$ immediately after the $Y(\pi/2)$ pulse. (**e**) The magnetization vectors at time $t = 2\tau$, showing that there is a net magnetization along the y-axis from spins 2 and 4. Note that initially spins 1 and 3 carried the magnetization. At $t = 2\tau$, the magnetization arises from spins 2 and 4

However, a true echo means that the refocusing condition is independent of the strength of the source of the dephasing. Thus, if spins are dephased by an inhomogeneous magnet, all the spins get back in phase at the echo, independent of how far the inhomogeneity has shifted their resonance from the average field. In our case, that means the existence of an echo should be independent of a, hence of the product $a\tau$, hence of τ.

Let us therefore reexamine the situation for a more general τ, using Fig. 8.4. We start (Fig. 8.4a) with the magnetization along the z-axis, and apply an $X(\pi/2)$ pulse to produce the situation in Fig. 8.4b. A time τ later the vectors 1 and 3 have rotated in the $x-y$ plane through angles $(a\tau/2)$ and $(-a\tau/2)$ to produce Fig. 8.4c. Then the $Y(\pi/2)$ pulse produces the situation of Fig. 8.4d in which magnetization vectors 1 and 3 are in the $y-z$ plane, making angles $-a\tau/2$ and $a\tau/2$ respectively with respect to the y-axis. In Fig. 8.4e we show the projections of the four vectors onto the $x-y$ plane. Spins 2 and 4 lie entirely in the plane. We define their lengths as M_0. Spins 1 and 3 have projections in the $x-y$ plane of $M_0 \cos(a\tau/2)$. Now, using the concepts of Sect. 7.22, we realize that each of the four magnetization vectors is coupled to a spin which is in a mixture of up and down states.

Since the quantum states of spins 2 and 4 are equal mixtures of up and down, spins 1 and 3 will each break into two counter-rotating components of equal amplitude (Fig. 8.4f). Spins 1 and 3 have z-components of $-M_0 \sin(a\tau/2)$ and $+M_0 \sin(a\tau/2)$ respectively. Thus they have an excess of $m = -\frac{1}{2}$ (for spin 1) and $m = +\frac{1}{2}$ (for spin 3). We need to calculate what fraction of their states correspond to spin up and spin down.

Denoting the occupation probability of state m for spin i as $(m|\varrho_i|m)$ we have that

$$\left(\tfrac{1}{2}|\varrho_1|\tfrac{1}{2}\right) + \left(-\tfrac{1}{2}|\varrho_1|-\tfrac{1}{2}\right) = 1 \tag{8.12}$$

from normalization. The z-component of magnetization of spin 1, M_{z1}, is

$$M_{z1} = M_0 \left(\tfrac{1}{2}|\varrho_1|\tfrac{1}{2}\right) - M_0 \left(-\tfrac{1}{2}|\varrho_1|-\tfrac{1}{2}\right) \quad, \tag{8.13}$$

expressing the fact that if the spin is entirely in $\frac{1}{2}$ or $-\frac{1}{2}$, M_{z1} is M_0 or $-M_0$ respectively. But

$$M_{z1} = -M_0 \sin(a\tau/2) \tag{8.14a}$$

so we get

$$\left(\tfrac{1}{2}|\varrho_1|\tfrac{1}{2}\right) - \left(-\tfrac{1}{2}|\varrho_1|-\tfrac{1}{2}\right) = -\sin(a\tau/2) \tag{8.14b}$$

Solving (8.12) and (8.13) we get

$$\begin{aligned}\left(\tfrac{1}{2}|\varrho_1|\tfrac{1}{2}\right) &= \tfrac{1}{2}[1 - \sin(a\tau/2)] \quad, \\ \left(-\tfrac{1}{2}|\varrho_1|-\tfrac{1}{2}\right) &= \tfrac{1}{2}[1 + \sin(a\tau/2)] \quad.\end{aligned} \tag{8.15}$$

In a similar way, we get

Fig. 8.4 Caption see opposite page

(a)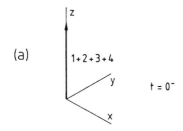

(b)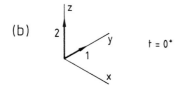

(c)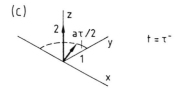

(d)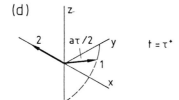

(e)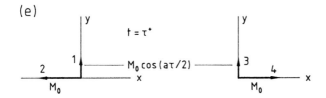

(f) (g)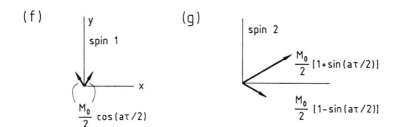

376

$$\left(\tfrac{1}{2}|\varrho 3|\tfrac{1}{2}\right) = \tfrac{1}{2}[1 + \sin(a\tau/2)]$$
$$\left(-\tfrac{1}{2}|\varrho 3|-\tfrac{1}{2}\right) = \tfrac{1}{2}[1 - \sin(a\tau/2)] \quad . \tag{8.16}$$

Thus, as spin 2 precesses, it will break into two counter-rotating components whose amplitudes are determined by the relative amounts of spin 1 in the up and down states. That is, the component of 2 which rotates clockwise (looking down on the $x-y$ plane) which comes from the down component of 1 will have amplitude

$$\frac{M_0}{2}[1 + \sin(a\tau/2)] \tag{8.17}$$

while the counterclockwise component of 2 will have amplitude

$$\frac{M_0}{2}[1 - \sin(a\tau/2)] \quad , \tag{8.18}$$

as shown in Fig. 8.4f.

We now add up the y-components, M_{yi} ($i = 1, 2, 3, 4$) of the spins at time τ after the $Y(\pi/2)$ pulse:

$$\begin{aligned}
M_{y1} &= \frac{M_0}{2}\cos^2(a\tau/2) + \frac{M_0}{2}\cos^2(a\tau/2) \\
M_{y2} &= \frac{M_0}{2}[1 + \sin(a\tau/2)]\sin(a\tau/2) \\
&\quad - \frac{M_0}{2}[1 - \sin(a\tau/2)]\sin(a\tau/2) \\
&= \frac{M_0}{2}[\sin^2(a\tau/2) + \sin^2(a\tau/2)] \quad .
\end{aligned} \tag{8.19}$$

Therefore

$$M_{y(1+2)} = M_0[\sin^2(a\tau/2) + \cos^2(a\tau/2)] = M_0 \quad . \tag{8.20a}$$

By a similar argument

$$M_{y(3+4)} = M_0 \quad . \tag{8.20b}$$

Fig. 8.4a-g. Formation of a solid echo for two pairs of spins (1 and 2, 3 and 4) for a general value of the spin-spin coupling constant a. (a) The thermal equilibrium magnetization at $t = 0^-$ showing the vector sum of the four magnetization vectors. Spins 1 and 3 are the blue spins, spins 2 and 4 are the red spins. (b) The magnetization vectors at $t = 0^+$ immediately following the $X(\pi/2)$ pulse. (c) The magnetization vectors at time $t = \tau^-$, just before the $Y(\pi/2)$ pulse. (d) The magnetization vectors at time $t = \tau^+$, immediately after the $Y(\pi/2)$ pulse. (e) The projections of the magnetization vectors on the $x - y$ plane at time $t = \tau^+$, immediately after the $Y(\pi/2)$ pulse. (f) The projection of the magnetization components of a spin 1 on the $x - y$ plane at later times, showing how its magnetization breaks into two counter-rotating components whose amplitude is determined by the orientation of spin 2. (g) The projection of the magnetization components of a spin 2 on the $x - y$ plane at later times, showing how its magnetization vector breaks into two counter-rotating components whose amplitude is determined by the z-components of spin 1

This result is independent of a, therefore corresponds to a real echo. Indeed, (8.20) shows that in a powder sample (in which a would take on different values for each crystal orientation if it were representing magnetic dipolar coupling), spins of all the crystallites would refocus at time 2τ.

This exercise has been long, laborious and exhausting, but it does serve to show how coupled precessing vectors behave in quantum mechanics. Now we turn to a much more elegant approach to analyze the same problem: use of the product operator method. It has the advantage of giving us a simple way of figuring out what the second pulse should be in order to produce the echo. (See Sect. 7.26. See also [8.10,11]).

We start with a system in thermal equilibrium and work in the rotating reference frame. The density matrix just before the first pulse ($t = 0^-$) is

$$\varrho(0^-) = A(I_z + S_z) \tag{8.21}$$

where A is a constant. Following the pulse $X(\pi/2)$, the density matrix becomes

$$\varrho(0^+) = A(I_y + S_y) \quad . \tag{8.22}$$

The density matrix at time $t = \tau^-$ just before the $Y(\pi/2)$ pulse is then

$$\varrho(\tau^-) = e^{-iaI_zS_z\tau}\varrho(0^+)e^{+iaI_zS_z\tau} \tag{8.23a}$$

which, utilizing Table 7.1, gives

$$\varrho(\tau^-) = (I_y + S_y)\cos(a\tau/2) - (I_xS_z + I_zS_x)2\sin(a\tau/2) \quad . \tag{8.23b}$$

Now, we wish to apply a pulse which produces some rotation R which will cause an echo. The effect of R is to change ϱ from its value at $t = \tau^-$ to a new value at $t = \tau^+$ given by

$$\varrho(\tau^+) = R\varrho(\tau^-)R^{-1} \quad . \tag{8.24}$$

We want $\varrho(\tau^+)$ to be such that during the next interval τ, ϱ returns to its value at $t = 0^+$, just after the first pulse. Thus, the condition for an echo is

$$\varrho(2\tau) = \varrho(0^+) \quad \text{but} \tag{8.25}$$

$$\varrho(2\tau) = e^{-iaI_zS_z\tau}\varrho(\tau^+)e^{+iaI_zS_z\tau} \quad . \tag{8.26}$$

If the effect of R on $\varrho(\tau^-)$ were equivalent to changing the sign of a during the interval between the X pulse and R, so that

$$R\varrho(\tau^-)R^{-1} = e^{+iaI_zS_z\tau}\varrho(0^+)e^{-iaI_zS_z\tau} \quad , \tag{8.27}$$

then we would satisfy (8.25) and produce an echo, since putting (8.27) in (8.26) we would have

$$\begin{aligned}\varrho(2\tau) &= e^{-iaI_zS_z\tau}\varrho(\tau^+)e^{+iaI_zS_z\tau} \\ &= e^{-iaI_zS_z\tau}R\varrho(0^+)R^{-1}e^{+iaI_zS_z\tau} \\ &= e^{-iaI_zS_z\tau}e^{+iaI_zS_z\tau}\varrho(0^+)e^{-iaI_zS_z\tau}e^{+iaI_zS_z\tau} \\ &= \varrho(0^+) \quad . \end{aligned} \tag{8.28}$$

Now, utilizing (8.23) we have

$$e^{+iaI_zS_z\tau}\varrho(0^+)e^{-iaI_zS_z\tau}$$
$$= (I_y + S_y)\cos(a\tau/2) + (I_zS_z + I_zS_x)2\sin(a\tau/2) \quad , \qquad (8.29)$$

which, using (8.24), is equal to

$$R\varrho(\tau^-)R^{-1} \qquad (8.30)$$

and using (8.23) gives

$$R[(I_y + S_y)\cos(a\tau/2) - (I_x + S_x)2\sin(a\tau/2)]R^{-1} \quad . \qquad (8.31)$$

Comparing (8.29) with (8.31) we see that R must satisfy the relations

$$I_y + S_y = R(I_y + S_y)R^{-1} \quad \text{and} \qquad (8.32a)$$

$$(I_xS_z + I_zS_x) = -R(I_xS_z + I_zS_x)R^{-1} \quad . \qquad (8.32b)$$

We automatically satisfy (8.32a) if we make R be a rotation about the y-axis. To satisfy (8.32b), the rotation must be $\pi/2$:

$$R = Y(\pi/2) \quad , \qquad (8.33)$$

a result which follows from the relationships (Table 7.1) such as

$$Y(\pi/2)I_yY^{-1}(\pi/2) = I_y$$
$$Y(\pi/2)I_xY^{-1}(\pi/2) = +I_z$$
$$Y(\pi/2)I_zY(\pi/2) = -I_y \quad \text{and} \qquad (8.34)$$

$$RI_xS_zR^{-1} = RI_xR^{-1}RS_zR^{-1} \quad . \qquad (8.35)$$

Before leaving the solid echo, we should remark on one more matter. In Fig. 8.3c we show the spins 1, 2, 3, and 4 at a time $a\tau/2 = \pi/2$. We want to refocus the spins to their condition in Fig. 8.3b where 1 and 3 lie along the $+y$-direction. Looking at Fig. 8.3c we note that if we simply reverse vectors 2 and 4, vectors 1 and 3 will reverse their precession directions, and thus refocus as desired a time τ later along the $+y$-direction. All that is needed is a π pulse about the x-axis. So why do we use a $Y(\pi/2)$ pulse instead of an $X(\pi)$ pulse? Clearly the idea will work if $a\tau/2 = \pi/2$. But to give a true echo, the scheme must work for other values of a, hence of $a\tau$. The easiest test is to look at (8.32). An $X(\pi)$ pulse will leave I_x and S_x alone, but will reverse I_y, S_y, I_z, S_z. It will therefore satisfy (8.32b), but it will *not* satisfy (8.32a). Note, however, in (8.29) that when $(a\tau/2) = \pi/2$, the coefficient of the term $(I_y + S_y)$ vanishes, so we will refocus this particular product $a\tau$. We will not, however, refocus for either a general value of a or of τ. Thus an $X(\pi)$ pulse will not produce a true echo.

8.6 The Jeener-Broekaert Sequence for Creating Dipolar Order

We have seen in Chap. 6 that it is possible to convert the order of a spin system from a net alignment of spins along an external magnetic field to an alignment of the spins in the local fields of their neighbors. That is, one converts Zeeman order to dipolar order. A standard method of transferring the order is by an adiabatic demagnetization either in the laboratory or the rotating frame. The time taken to demagnetize the sample must be sufficiently slow that the spin system remains in an internal equilibrium, otherwise one introduces irreversibility and a loss of order. *Jeener* and *Broekaert* [8.12] invented an ingenious method for transferring the order rapidly. Although there is some irreversible loss of order, their method is very useful in practice, and is based on a deep understanding of dipolar order which is of great fundamental significance.

The Jeener-Broekaert sequence consists of a $\pi/2$ pulse followed at a properly chosen time, τ, by a $\pi/4$ pulse phase shifted by $\pi/2$ from the first pulse. We might denote their sequence then as

$$X(\pi/2)\ldots\tau\ldots Y(\pi/4)\ldots t_1\ldots \quad .$$

The achievement of dipolar order means that a spin is aligned in the local field of its neighbors. Thus if the local field points up, the spin is more likely to be up than down, whereas if the local field is down, the spin is more likely to be down than up. This is the sort of effect we have neglected in writing expressions such as (8.10).

We can discuss this problem using the very same example used to explain solid echoes. That is, we consider a system in which the spins occur in pairs. We take the coupling between one pair and another to be weak, and we represent the spin-spin coupling by Hamiltonian (8.11)

$$\mathcal{H} = \hbar a I_z S_z \quad . \tag{8.11}$$

We then pick out just the subset of n spins which are polarized, as we did earlier, labeling spins as blue and red as before (Fig. 8.5a). Then, we apply an $X(\pi/2)$ pulse (Fig. 8.5b), and wait a time τ such that

$$a\tau/2 = \pi/2 \quad . \tag{8.36}$$

The resulting magnetization is shown in Fig. 8.5c. So far, everything we have done is similar to our first steps in discussing the echo using Fig. 8.3. Examination of the four magnetization vectors in Fig. 8.5c shows that their resultant is zero. We appear to have lost the magnetic order. However, that cannot be true since we know that a $Y(\pi/2)$ pulse will enable us to recover the magnetization. The secret of the Jeener-Broekaert pulse sequence is the recognition that the spins in Fig. 8.5c have spin-spin order. This can be understood by analyzing what happens if one applies a $Y(\pi/4)$ pulse, producing the result of Fig. 8.5d. If now

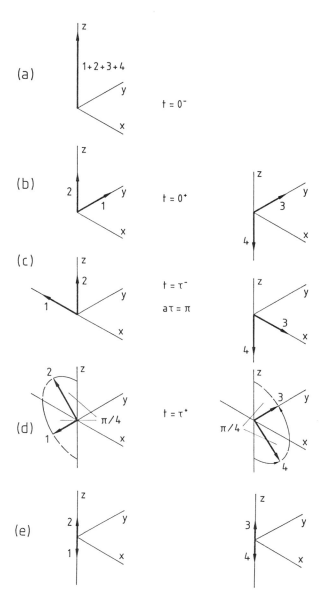

Fig. 8.5a-e. The Jeener-Broekaert method of creating dipolar order from pairs of spins (1 and 2, 3 and 4). (**a**) The thermal equilibrium magnetization at $t = 0^-$ showing the vector sum of the four magnetization vectors. Spins 1 and 3 are the blue spins, spins 2 and 4 the red spins. (**b**) The magnetization vectors at $t = 0^+$ immediately following the $X(\pi/2)$ pulse. (**c**) The magnetization vectors at time $t = \tau^-$, just before the $Y(\pi/4)$ pulse, for the case $a\tau/2 = \pi/2 (a\tau = \pi)$. (**d**) The magnetization vectors at $t = \tau^+$, immediately after the $Y(\pi/4)$ pulse. (**e**) The magnetization vectors a long time after $t = \tau$. The transverse components of magnetization have decayed to zero from spin-spin couplings to *other* pairs, leaving only the resultant components along the z-direction which are $1/\sqrt{2}$ shorter than the vectors of (**d**)

we have in our ensemble of spins a weak coupling of one pair with another, there will be a slight spread to the precession frequencies $\pm a/2$. Thus, at a much later time, the transverse components of the magnetization will decay, leading to the situation of Fig. 8.5e. The remarkable point about this state is that every spin has its neighbor pointing in the opposite direction. This is to be contrasted to the condition at $t = 0^-$, before the first pulse, when half the spins had a neighbor pointing parallel, half had their neighbor pointing antiparallel. Thus, the dipolar engergy $\hbar a I_z S_z$, which was initially zero, now takes on a nonzero value as in an adiabatic demagnetization experiment.

We employ the operator approach using a derivation due to *Wang* et al. [7.73]. The density matrix at time τ after the $X(\pi/2)$ pulse but before any other pulse is given by (8.23b)

$$\varrho(\tau) = A[(I_y + S_y)\cos(a\tau/2) - (I_x S_z + I_z S_x)2\sin(a\tau/2)] \quad . \tag{8.37}$$

The pulse $Y(\pi/4)$ then rotates the spin vectors to give

$$\begin{aligned}
\varrho(\tau^+) &= AY(\pi/4)(I_y + S_y)Y^{-1}(\pi/4) \\
&\quad - AY(\pi/4)2(I_x S_z + I_z S_x)\sin(a\tau/2)Y^{-1}(\pi/4) \\
&= A(I_y + S_y) - 2A\left(\frac{(I_x + I_z)}{\sqrt{2}}\frac{(S_z - S_x)}{\sqrt{2}} + \frac{(I_z - I_x)}{\sqrt{2}}\frac{(S_x + S_z)}{\sqrt{2}}\right) \\
&\quad \times \sin(a\tau/2) \\
&= A[(I_y + S_y) - 2(I_z S_z - I_x S_x)\sin(a\tau/2)] \quad .
\end{aligned} \tag{8.38}$$

We now invoke the argument that couplings to more distant spins, not included in the Hamiltonian (8.11), cause the off-diagonal components of ϱ, which arise from the terms $(I_y + S_y)$ and $I_x S_x$ in (8.38), to decay. Thus at a later time T we can drop the off-diagonal elements of ϱ, so that $\varrho(\tau + T)$ is then

$$\varrho(\tau + T) = -2A I_z S_z \sin(a\tau/2) \quad . \tag{8.39}$$

This Hamiltonian has zero net magnetization, as can be seen easily by explicit evaluation of

$$<I_\alpha>_{\alpha=x,y,z} = \text{Tr}\{I_\alpha \varrho(\tau + T)\} \quad . \tag{8.40}$$

However, the average dipolar energy

$$\begin{aligned}
<\mathcal{H}(\tau + T)> &= -\text{Tr}\{\hbar a I_z S_z \varrho(\tau + T)\} \\
&= -2A\hbar a \text{Tr}\{I_z^2 S_z^2 \sin(a\tau/2)\} \\
&= -\frac{A\hbar a}{2}\sin(a\tau/2) \quad .
\end{aligned} \tag{8.41}$$

This is maximized in magnitude for $(a\tau/2) = \pi/2$ at an energy of $(-A\hbar a/2)$. This result is to be contrasted to its value at $t = 0^-$ before the $X(\pi/2)$ pulse:

$$\begin{aligned}
<\mathcal{H}(0^-)> &= \text{Tr}\{\hbar a I_z S_z \varrho(0^-)\} \\
&= \text{Tr}\{\hbar a I_z S_z A(I_z + S_z)\} \\
&= 0 \quad .
\end{aligned} \tag{8.42}$$

Indeed, because $I_z S_z$ is, apart from a constant, the dipolar energy, we can see that the part of ϱ proportional to $I_z S_z$ is a measure of the dipolar order.

If the dipolar system now contains some order, one should be able to "read" the order by converting it from dipolar order to "Zeeman" order, i.e. to net magnetization. *Jeener* and *Broekaert* do this by applying yet one more $Y(\pi/4)$ pulse, which produces an echo a time t_1 later given by $t_1 = \tau$. It is not surprising that an echo results since the Jeener-Broekaert sequence then of

$$X(\pi/2)\ldots\tau\ldots Y(\pi/4)\ldots T\ldots Y(\pi/4)\ldots t_1 \tag{8.43}$$

can be seen, in the limit $T \to 0$, to be

$$\begin{aligned}X(\pi/2)\ldots\tau\ldots & Y(\pi/4)Y(\pi/4)\ldots t_1 \\ &= X(\pi/2)\ldots\tau\ldots Y(\pi/2)\ldots t_1 \quad ,\end{aligned} \tag{8.44}$$

which is just the Powles-Mansfield dipolar echo.

To show that this happens in the general case, we take ϱ just before the second pulse to be given by (8.39)

$$\varrho(\tau + T^-) = -2AI_z S_z \sin(a\tau/2) \quad . \tag{8.45}$$

So

$$\begin{aligned}\varrho(\tau + T^+) &= Y(\pi/4)(-2AI_z S_z)Y^{-1}\pi/4)\sin(a\tau/2) \\ &= -2A\frac{(I_z - I_x)}{\sqrt{2}}\frac{(S_z - S_x)}{\sqrt{2}}\sin(a\tau/2) \\ &= -A[I_z S_z - I_x S_z - I_z S_x + I_x S_x]\sin(a\tau/2) \quad .\end{aligned} \tag{8.46}$$

If this is acted on by $\mathcal{H}$ for a time t_1, then we get

$$\begin{aligned}\varrho(\tau + T + t_1) &= e^{-iaI_z S_z t_1}\varrho(\tau + T^+)e^{iaI_z S_z t_1} \\ &= -A[I_z S_z + I_x S_x - (I_x S_z + S_x I_z)\cos(at_1/2) \\ &\quad - (I_y S_z^2 + S_y I_z^2)2\sin(at_1/2)]\sin(a\tau/2) \quad .\end{aligned} \tag{8.47}$$

Utilizing the fact that I_z^2 and S_z^2 are $\frac{1}{4}$, we get

$$\varrho(\tau + T + t_1) = -A\left[I_z S_z + I_x S_x - (I_x S_z + S_x I_z)\cos(at_1/2)\right. \\ \left. - \left(\frac{I_y + S_y}{2}\right)\sin(at_1/2)\right]\sin(a\tau/2) \quad . \tag{8.48}$$

Of these terms, only the last one will contribute a signal:

$$\begin{aligned}\langle I^+ + S^+\rangle &= \text{Tr}\{(I^+ + S^+)\varrho(\tau + T + t_1)\} \\ &= \frac{iA}{2}\sin(a\tau/2)\sin(at_1/2)\text{Tr}\{(I_y + S_y)^2\} \quad .\end{aligned} \tag{8.49}$$

This signal is clearly maximized if we choose τ and t_1 equal to each other, and both of such duration that

$$a\tau/2 = at_1/2 = \pi/2 \quad . \tag{8.50}$$

Comparing this signal with that of the free induction decay following an $X(\pi/2)$ pulse,

$$\begin{aligned} \langle I^+ + S^+ \rangle &= \mathrm{Tr}\{(I^+ + S^+)A(I_y + S_y)\} \\ &= \mathrm{i}A\,\mathrm{Tr}\{(I_y + S_y)^2\} \quad , \end{aligned} \tag{8.51}$$

we see the maximum signal is one half that of a pure $X(\pi/2)$ pulse. If there is a distribution of values of a, there is still a signal at $t_1 = \tau$ since

$$\sin^2(a\tau/2) > 0 \quad . \tag{8.52}$$

Thus, we can create dipolar order by applying $X(\pi/2)\ldots\tau\ldots Y(\pi/4)$ with τ chosen such that a $\tau \approx \pi/2$, and we can inspect the dipolar order by later applying a $Y(\pi/4)$ pulse to produce an echo a time τ later. The inspection pulse can be used to study the decay of the dipolar order resulting from either conventional spin-lattice relaxation mechanisms (e.g. coupling to conduction electrons in a metal), or from motional effects which modulate the strength of the dipolar coupling.

We have treated the problem for coupled pairs. *Jeener* and *Broekaert* treat the general dipolar case in their classic paper [8.12].

8.7 The Magic Angle in the Rotating Frame – The Lee-Goldburg Experiment

Another important concept concerning dipolar order is connected with the famous Lee-Goldburg experiment [8.13] to which we now turn. Suppose one has a number of spins all with the same precession frequency, coupled together by dipolar coupling. We keep just the secular terms of the dipolar coupling, giving a Hamiltonian in the rotating frame of

$$\begin{aligned}\mathcal{H} =\ & -\gamma\hbar h_0 I_z - \gamma\hbar H_1 I_x \\ & + \sum_{i>j} \frac{\gamma^2\hbar^2}{r_{ij}^3}(1 - 3\cos^2\theta_{ij})(3I_{zi}I_{zj} - \boldsymbol{I}_i\cdot\boldsymbol{I}_j) \end{aligned} \tag{8.53a}$$

where as usual

$$I_z = \sum_j I_{zj}, \quad \text{etc.}$$

$$h_0 = H_0 - \omega/\gamma \quad . \tag{8.53b}$$

It is convenient to collect all the radial and angular terms of (8.53a) in a simple symbol by defining

$$B_{ij} \equiv \frac{1}{2}\frac{\gamma^2\hbar^2}{r_{ij}^3}(1 - 3\cos^2\theta_{ij}) \quad . \tag{8.54a}$$

Then
$$\mathcal{H} = -\gamma\hbar h_0 I_z - \gamma\hbar H_1 I_x + \sum_{i>j} B_{ij}(3I_{zi}I_{zj} - \boldsymbol{I}_i \cdot \boldsymbol{I}_j) \quad . \tag{8.54b}$$

We now define an effective field in the rotating frame
$$\boldsymbol{H}_{\text{eff}} = \boldsymbol{i}H_1 + \boldsymbol{k}h_0 \quad . \tag{8.55}$$

We shall concern ourselves with cases in which $\boldsymbol{H}_{\text{eff}}$ is much larger than the local fields. In that case, it is appropriate to quantize the spins along the effective field. Defining this direction to be the Z-direction, we make a coordinate transformation (Fig. 8.6) for all spins, getting
$$\begin{aligned} I_Z &= I_z \cos\theta + I_x \sin\theta \\ I_X &= I_x \cos\theta - I_y \sin\theta \\ I_Y &= I_y \quad , \end{aligned} \tag{8.56a}$$

giving the inverse transformations for an individual spin j
$$\begin{aligned} I_{zj} &= I_{Zj} \cos\theta - I_{Xj} \sin\theta \\ I_{xj} &= I_{Xj} \cos\theta + I_{Zj} \sin\theta \\ I_{yj} &= I_{Yj} \quad . \end{aligned} \tag{8.56b}$$

Substituting into $\mathcal{H}$, we get
$$\mathcal{H} = -\gamma\hbar H_{\text{eff}} I_Z + \sum_{M=-2}^{+2} \lambda_M(\theta)\mathcal{H}_M \quad \text{where} \tag{8.57}$$

$$\mathcal{H}_0 = \sum_{i>j} B_{ij}(3I_{zi}I_{zj} - \boldsymbol{I}_i \cdot \boldsymbol{I}_j) \tag{8.58}$$

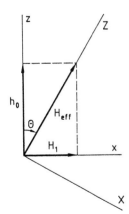

Fig. 8.6. The effective field in the rotating frame, showing the axes X, Z with respect to x, z

$$\mathcal{H}_{+1} = \sum_{i>j} 3B_{ij}(I_i^+ I_{zj} + I_{zi} I_j^+)$$

$$\mathcal{H}_{+2} = \sum_{i>j} 3B_{ij}(I_i^+ I_j^+) \quad,$$

$$\mathcal{H}_{-1} = (\mathcal{H}_{+1})^*, \mathcal{H}_{-2} = (\mathcal{H}_{+2})^* \quad \text{and} \tag{8.59}$$

$$\lambda_0(\theta) = \tfrac{1}{2}(3\cos^2\theta - 1)$$
$$\lambda_{\pm 1}(\theta) = -\tfrac{1}{2}\sin\theta\cos\theta \tag{8.60}$$
$$\lambda_{\pm 2}(\theta) = -\tfrac{1}{4}\sin^2\theta \quad.$$

The $\mathcal{H}_M$'s satisfy a commutation relation

$$[I_Z, \mathcal{H}_M] = M\mathcal{H}_M \tag{8.61}$$

as is easy to show by explicit calculations of matrix elements. From (8.61), or by examination of the explicit form of the $\mathcal{H}_M$'s, we see that only $\mathcal{H}_0$ commutes with the Zeeman interactions of the effective field. In the limit of large effective field, we can then as a first approximation drop all terms but the term involving $\mathcal{H}_0$. We get then a truncated Hamiltonian

$$\mathcal{H} = \gamma\hbar H_{\text{eff}} I_Z + \tfrac{1}{2}(3\cos^2\theta - 1)\sum_{i>j} B_{ij}(3I_{Zi}I_{Zj} - \bm{I}_i\cdot\bm{I}_j) \quad. \tag{8.62}$$

We can compare this with the secular part of the Hamiltonian of the spins in the lab frame, $\mathcal{H}_{\text{lab}}$, in the absence of an alternating field, H_1:

$$\mathcal{H}_{\text{lab}} = -\gamma\hbar H_0 I_z + \sum_{i>j} B_{ij}(3I_{zi}I_{zj} - \bm{I}_i\cdot\bm{I}_j)$$
$$\equiv \mathcal{H}_{\text{Zeeman}} + \mathcal{H}_{\text{dipolar}} \quad. \tag{8.63}$$

Equations (8.62) and (8.63) are identical in form, except in the rotating frame the dipolar term has been multiplied by the factor

$$\frac{3\cos^2\theta - 1}{2} \quad. \tag{8.64}$$

[This is just the term $\lambda_0(\theta)$]. θ is determined by the relative size of h_0, the amount one is off resonance, with H_1, the strength of the rotating field. *Lee* and *Goldburg* noted that if they chose H_1 and h_0 properly, they could make

$$\cos^2\theta = \tfrac{1}{3} \quad, \tag{8.65}$$

in which case the dipolar term in (8.62) vanished. In this manner they could effectively eliminate the dipolar broadening. An effective field for which (8.65) is satisfied is said to be at the *magic angle*.

We can rephrase their result by saying that if the effective field is at the magic angle, a spin will precess in the rotating frame solely under the influence of $\bm{H}_{\text{eff}}$, without suffering a dephasing and consequent decay of the components of magnetization perpendicular to H_{eff} as usually occurs when there is dipolar coupling.

To make a precise test of this concept, they did pulse experiments in which they observed the decay of the magnetization with time following a sudden turn-on of H_1 at a frequency ω somewhat off resonance. They varied the angle θ and the strength of the effective field. To measure the effective dipolar strength, they determined the second moment from the transform of the decay curves. They studied the F^{19} resonance of CaF_2. Figure 8.7a shows their measurements of the square root of the second moment versus $(3\cos^2\theta - 1)/2$. At the magic angle, the second moment should vanish. It does not quite do so owing to the nonsecular terms involving $\mathcal{H}_{\pm 1}$ and $\mathcal{H}_{\pm 2}$. The effect of these terms should vanish in the limit of infinite H_{eff}. In Fig. 8.7b we show their measurements of the fall-off of second moment, normalized to its value in the lab frame, versus $1/\omega_e^2$ where

$$\omega_e \equiv \gamma H_{\text{eff}} \quad . \tag{8.66}$$

The failure of the dashed theoretical curve to intercept the origin arises because of slight inhomogeneity of H_1. *Barnaal* and *Low* [8.14] made an extensive study of the systems in which H_1 was exactly at resonance. In this case

$$\frac{3\cos^2\theta - 1}{2} = -\frac{1}{2} \quad . \tag{8.67}$$

They did experiments in which they turned on H_1 for a time τ, then observed the free induction decay a time t after H_1 is turned off. They solved the problem of an interacting pair exactly. For an H_1 along the x-axis they found

$$M_y(t) = M(0)\frac{\omega_1}{\Omega} \sin(\Omega\tau) \cos\left[\left(\frac{3B_{12}}{2\hbar}\right)(t + \tau/2)\right] \tag{8.68}$$

where

$$\Omega^2 = \left(\frac{3B_{12}}{4\hbar}\right)^2 + \omega_1^2 \quad . \tag{8.69}$$

In the limit of large H_1, this expression agrees with the result of the truncated Hamiltonian:

$$M_y(t) = M(0) \sin(\omega_1\tau) \cos\left[\left(\frac{3B_{12}}{2\hbar}\right)(t + \tau/2)\right] \quad . \tag{8.70}$$

A striking feature of (8.68) is that the oscillation after the turn-off of H_1 is identical to what it would be following an $X(\pi/2)$ pulse except for (1) a slight amplitude correction and (2) a change of τ in the apparent zero of time t. In a beautiful set of experiments, *Barnaal* and *Lowe* showed that for $CaSO_4 \cdot 2H_2O$, CaF_2, and ice the first zero crossing of the free induction decay moves to later times as τ is increased. The delay in zero crossing is $\tau/2$, for values of τ which are up to about half of the normal free induction decay zero crossing.

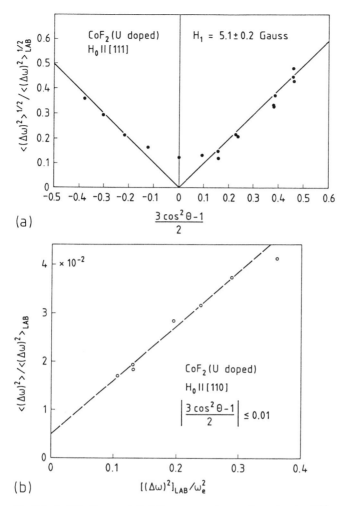

Fig. 8.7a,b. The Lee and Goldburg experimental results for F^{19} resonance in CaF_2 single crystals. (**a**) The normalized second moment of the F^{19} resonance for H_0 parallel to the [111] direction, as a function of $(3\cos^2\theta - 1)/2$, where θ specifies the orientation of $\boldsymbol{H}_{\text{eff}}$ in the rotating frame (see Fig. 8.6). (**b**) The normalized second moment as a function of the second moment in the lab frame $\langle(\Delta\omega^2)\rangle_{\text{LAB}}$ divided by ω_e^2, where $\omega_e = \gamma H_{\text{eff}}$. H_0 is parallel to the crystal [110] direction

8.8 Magic Echoes

In Sects. 2.10 and 5.6 we saw that one way of understanding how a spin echo comes about is to view the first pulse as initiating a time development of the magnetization under the influence of the Hamiltonian, and to view the second pulse a time τ later as suddenly transforming the Hamiltonian to be the negative of the actual Hamiltonian. Then, over the next time interval τ, the time development of the magnetization unwinds, returning the magnetization to its value just after the initial pulse.

In discussing spin temperature (Chap. 6), we talked about irreversible processes connected with the complexity of a system of many coupled dipoles. We now turn to a remarkable discovery which shows how one can run backwards the dephasing produced by dipolar coupling, thus showing that it is possible, even after the free induction decay is over, to recover the initial magnetization. This experiment shows that the spin temperature approach does not in all cases accurately describe the evolution of a spin system. In the process an echo is formed. This sort of echo has become known as a *magic echo*. The first intimation that one could refocus dephasing arising from dipolar coupling was noticed by *Rhim* and *Kessemeier* [8.15,16], who discovered the effect experimentally, and showed theoretically by an approximate method that an echo was formed in which the loss of signal owing to dipolar dephasing was recovered. Following this work, done at the University of North Carolina, *Rhim* went to MIT where, in collaboration with *Pines* and *Waugh*, he extended and perfected the experimental techniques and the theoretical analysis [8.17].

In order to refocus spins which are defocused by dipolar coupling, one needs to be able to reverse the sign of the dipolar Hamiltonian. Then, as we have seen, we can effectively run the system backwards in time to undo the dephasing. The trick is closely related to the Lee-Goldburg experiment. The critical equation is (8.62), the truncated Hamiltonian in the rotating frame:

$$\mathcal{H} = -\gamma\hbar H_{\text{eff}} I_Z + \frac{(3\cos^2\theta - 1)}{2}\sum_{i>j} B_{ij}(3I_{Zi}I_{Zj} - \boldsymbol{I}_i \cdot \boldsymbol{I}_j) \quad . \tag{8.62}$$

In this equation, θ is the angle between the static field $\boldsymbol{H}_0$ and the effective field, $\boldsymbol{H}_{\text{eff}}$, and the Z-direction is the direction of the effective field. If $\boldsymbol{H}_{\text{eff}}$ is nearly parallel to $\boldsymbol{H}_0$, $\cos\theta = 1$ and the angular factor, $(3\cos^2\theta - 1)/2$, in front of the dipolar term is $+1$. However, if $\theta = \pi/2$, as when H_1 is exactly at resonance, the angular factor becomes

$$\frac{3\cos^2\theta - 1}{2} = -\frac{1}{2} \quad . \tag{8.71}$$

Thus, it has a *negative* sign. The magic echo makes use of this negative sign to unwind the dipolar dephasing.

Let us then rewrite (8.62), the truncated Hamiltonian, exactly at resonance, using the coordinates x, y, z in which H_1 lies along the x-axis:

$$\mathcal{H} = -\gamma\hbar H_1 I_x - \frac{1}{2}\sum_{j>k} B_{jk}(3I_{xj}I_{xk} - \boldsymbol{I}_j \cdot \boldsymbol{I}_k) \quad . \tag{8.72}$$

In addition to these terms, there is the nonsecular term

$$\mathcal{H}_{\text{nonsec}} = \frac{1}{4} \sum_{j>k} 3B_{jk}(I_j^+ I_k^+ + I_j^- I_k^-) \quad , \tag{8.73}$$

which we are for the moment neglecting.

How can we utilize the negative sign to unwind dipolar dephasing? For the moment, consider an experiment in which we produce some transverse magnetization [for example, by a $Y(\pi/2)$ pulse]. Let it dephase under the action of the dipolar system, for a time τ, then turn on H_1 to produce a negative dipolar coupling. Will this refocus the dipolar dephasing? If the density matrix (in the rotating frame for our whole discussion) is initially ($t = 0^-$) given by

$$\varrho(0^-) = AI_z \quad , \tag{8.74}$$

where A is some constant which we set equal to 1 for convenience, the pulse sequence would produce at a time t a density matrix

$$\varrho(t) = \exp\left[-\frac{i}{\hbar}\left(-\gamma\hbar H_1 I_x - \frac{1}{2}\mathcal{H}_{xx}\right)(t-\tau)\right]$$
$$\times \exp\left(-\frac{i}{\hbar}\mathcal{H}_{zz}\tau\right) Y(\pi/2) I_z \{\text{inv}\} \tag{8.75}$$

where {inv} is the inverse of the operators to the left of I_z and where we define

$$\mathcal{H}_{xx} \equiv \sum_{j>k} B_{jk}(3I_{xj}I_{xk} - \mathbf{I}_j \cdot \mathbf{I}_k)$$
$$\mathcal{H}_{zz} \equiv \sum_{j>k} B_{jk}(3I_{zj}I_{zk} - \mathbf{I}_j \cdot \mathbf{I}_k) \quad . \tag{8.76}$$

This expression has two dipolar terms with opposite signs in front, but one is $\mathcal{H}_{xx}$, the other $\mathcal{H}_{zz}$, so they are not the negatives of each other. Moreover, there is also a term involving H_1 which induces precession around the x-axis, whereas the initial dipolar term involves only $\mathcal{H}_{zz}$, the dipolar dephasing.

It is easy to get rid of the precession, as was shown by *Solomon* [8.18] in his paper describing the *rotary echo*. He showed that he could get rid of dephasing from precessing in an inhomogenous H_1 by suddenly reversing the phase of H_1 so that the spins precess in the opposite sense. We employ the same idea here. We keep H_1 on for a time τ' then reverse it for an equal time τ'. Then we get

$$\varrho(\tau + 2\tau') = \exp\left[-\frac{i}{\hbar}\left(\gamma\hbar H_1 I_x - \frac{1}{2}\mathcal{H}_{xx}\right)\tau'\right]$$
$$\times \exp\left[-\frac{i}{\hbar}\left(-\gamma\hbar H_1 I_x - \frac{1}{2}\mathcal{H}_{xx}\right)\tau'\right]$$
$$\times \exp\left(-\frac{i}{\hbar}\mathcal{H}_{zz}\tau\right) Y(\pi/2) I_z \{\text{inv}\}$$
$$= \exp\left[-\frac{i}{\hbar}\left(-\frac{\mathcal{H}_{xx}}{2}2\tau'\right)\right]$$
$$\times \exp\left(-\frac{i}{\hbar}\mathcal{H}_{zz}\tau\right) Y(\pi/2) I_z \{\text{inv}\} \quad . \tag{8.77}$$

This expression is still not quite what we want since it still has two different dipolar Hamiltonians.

However, examination of (8.76) shows that $\mathcal{H}_{xx}$ and $\mathcal{H}_{zz}$ differ simply by a coordinate rotation of z into x or x into z. Such a transformation is produced by a rotation about the y-axis. In fact

$$Y(\pi/2)\mathcal{H}_{xx}Y^{-1}(\pi/2) = \mathcal{H}_{zz} \quad . \tag{8.78}$$

Therefore, we add two pulses, a $Y(\pi/2)$ and a $Y^{-1}(\pi/2)$, where the latter is simply a $Y(-\pi/2)$ before and after the time interval $2\tau'$ giving

$$\begin{aligned}\varrho(\tau + 2\tau') &= Y(\pi/2)\exp\left[-\frac{i}{\hbar}\left(-\frac{\mathcal{H}_{xx}}{2}2\tau'\right)\right]Y^{-1}(\pi/2) \\ &\quad \times \exp\left(-\frac{i}{\hbar}\mathcal{H}_{zz}\tau\right)Y(\pi/2)I_z\{\text{inv}\} \\ &= \exp\left(\frac{i}{\hbar}\mathcal{H}_{zz}\tau'\right)\exp\left(-\frac{i}{\hbar}\mathcal{H}_{zz}\tau\right)Y(\pi/2)I_z\{\text{inv}\} \quad . \end{aligned} \tag{8.79}$$

Clearly when $\tau' = \tau$, the dipolar dephasing has vanished. Our pulse sequence is thus (reading left to right)

$$Y(\pi/2)\ldots\tau\ldots Y^{-1}(\pi/2)\ldots\tau'^{H_1}\ldots\tau'^{-H_1}\ldots Y(\pi/2)\ldots t_1 \quad , \tag{8.80}$$

where t_1 is the observation period. Picking $\tau' = \tau$ will cause the signal at $t_1 = 0$ to correspond to the full magnetization.

It is easy to extend the discussion to show that if one were to hold τ' fixed and reduce τ, the dipolar refocusing would occur at a time t_1 given by

$$t_1 + \tau = \tau' \quad , \tag{8.81}$$

so that if τ goes to zero, there would be a dipolar echo at $t_1 = \tau'$. Note that if τ goes to zero, the $Y(\pi/2)$ and $Y^{-1}(\pi/2)$ pulses just undo one another, so both could be omitted. This is exactly what *Rhim* and *Kessemeier* did in their first experiments.

From (8.78), it is clear that

$$\begin{aligned}&Y^{-1}(\pi/2)\ldots\tau'^{H_1}\ldots\tau'^{-H_1}\ldots Y(\pi/2) \\ &= \exp\left(-\frac{i}{\hbar}(-\mathcal{H}_{zz}\tau')\right) \quad . \end{aligned} \tag{8.82}$$

Rhim, Pines, and *Waugh* label such a sequence a "burst". In fact, they argue that effects of the nonsecular term (8.73) can be reduced if one makes a single burst by putting together many pairs of (H_1, τ') and $(-H_1, \tau')$ using very short τ's instead of using a single $H_1, -H_1$ pair of longer duration. Their argument, the details of which they omit, appears to be related to their thinking about so-called average Hamiltonian theory. Of course, the most obvious way of reducing errors from neglect of the nonsecular term is to make H_1 very large. *Rhim, Pines* and *Waugh* used an H_1 of 100 Gauss for CaF_2 with τ's of the order of 1.25 μs.

In their work, they also implied that one should pick $\gamma H_1 \tau = n\pi$ where n is an integer, although in fact they demonstrate data which violates this condition. *Takegoshi* and *McDowell* [8.19] have shown experimentally that this is evidently not necessary.

It is worth noting that there is a strong experimental reason for reversing H_1 many times at short intervals rather than once at a long interval. The requirement of applying a large H_1 for a long time often places a great strain on the power supply of the rf power amplifier, causing the amplitude of H_1 to droop with time. For a "burst" involving only one phase reversal of the rf, there may be significantly lower H_1 during the second half (the phase reversed period) of the burst than was present in the first half. If, however, there are many short cycles $(H_1, -H_1)$, the fractional difference between H_1 and $-H_1$ will be greatly reduced.

It is interesting to note that if one has a group of nuclei with several chemical shifts the shifts can be revealed by a magic echo. During the time the H_1 is on, the chemical shift fields, which lie along the z-axis, can be neglected since they are perpendicular to the much larger H_1. During the time τ when H_1 is off, the full chemical shift acts. Since that is only one-third of the time, the precession frequency is displaced from ω by one-third of the chemical shift frequency.

Rhim, Pines, and *Waugh* demonstrated that one can apply a train of magic echoes which, similar to a Carr-Purcell sequence, will refocus the magnetization again and again.

8.9 Magic Angle Spinning

In Sect. 3.4 we described experiments by *Andrew* and *Eades* on the use of line width studies to reveal the presence of rapid molecular motions. This effect had actually been discovered by *Gutowsky* and *Pake* [8.20]. For the case of rapid motion, *Gutowsky* and *Pake* had shown tht the angular factor $(1 - 3\cos^2\theta_{jk})$ should be replaced by its time average as discussed in Chap. 3. That average is given by (3.61)

$$(1 - 3\cos^2\theta_{jk})_{\mathrm{avg}} = (1 - 3\cos^2\theta')\left(\frac{3\cos^2\gamma_{jk} - 1}{2}\right) \tag{3.61}$$

where θ' is the angle the molecular rotation axis makes with the static magnetic field, and γ_{jk} is the angle made by the internuclear vector r_{ij} with the rotation axis.

This expression suggested independently to *Lowe* [8.21] and to *Andrew* et al. [8.22] that one could produce the rotation artificially by turning the entire sample. In that case, the angle θ' would be the same for all pairs of nuclei throughout the sample. Then if one chose θ' to satisfy the condition

$$1 - 3\cos^2\theta' = 0 \tag{8.83}$$

the time averaged dipolar coupling would vanish [8.21, 22b]. This value of θ' soon became known as the *magic angle*, and such a method of line narrowing is called magic angle spinning (abbreviated as MAS) or magic angle sample spinning (MASS). It is historically the first of the methods for narrowing dipolar broadened lines.

The effect of spinning on dipolar coupling is only one of several important uses of spinning to eliminate unwanted couplings. It can also be used to eliminate chemical shift anisotropies and first order quadrupole splittings. These are all interactions which involve angular functions made up of the $l = 2$ spherical harmonics, Y_{lm}. To analyze these various methods, it is convenient to begin with a minor digression about spherical harmonics.

Table 8.1 lists the normalized Y_{lm}'s for $l = 0, 1,$ and 2 as well as unnormalized forms involving the coordinates x, y, z where

$$z = r\cos\theta \quad , \quad x = r\sin\theta\cos\phi \quad , \quad y = r\sin\theta\sin\phi \quad . \tag{8.84}$$

We recall that the Y_{lm}'s are related to solutions of Laplace's equation

$$\nabla^2(r^l Y_{lm}) = 0 \quad . \tag{8.85}$$

[The fact that the unnormalized column of $r^l Y_{lm}$'s satisfies (8.85) is readily verified by expressing ∇^2 in rectangular coordinates]. This is the reason that, for example, $l = 2$ functions are made up of linear combinations of x^2, y^2, z^2, xy, xz, yz, and do not include terms such as x or xy^2.

Table 8.1. Listing of the normalized spherical harmonics for $l = 0, 1, 2$ and the unnormalized forms

Normalized spherical harmonics Y_{lm}	Unnormalized $r^l Y_{lm}$ in rectangular coordinates
$Y_{0,0} = \frac{1}{\sqrt{4\pi}}$	constant
$Y_{1,1}(\theta,\phi) = -\sqrt{\frac{3}{8\pi}} \sin\theta\, e^{i\phi}$	$x + iy$
$Y_{1,0}(\theta,\phi) = \sqrt{\frac{3}{4\pi}} \cos\theta$	z
$Y_{1,-1}(\theta,\phi) = \sqrt{\frac{3}{8\pi}} \sin\theta\, e^{-i\phi}$	$x - iy$
$Y_{2,2}(\theta,\phi) = \sqrt{\frac{15}{32\pi}} \sin^2\theta\, e^{2i\phi}$	$(x + iy)^2 = x^2 - y^2 + 2ixy$
$Y_{2,1}(\theta,\phi) = -\sqrt{\frac{15}{8\pi}} \sin\theta\cos\theta\, e^{i\phi}$	$z(x + iy) = xz + iyz$
$Y_{2,0}(\theta,\phi) = \sqrt{\frac{5}{16\pi}}(3\cos^2\theta - 1)$	$3z^2 - r^2 = 2z^2 - x^2 - y^2$
$Y_{2,-1}(\theta,\phi) = \sqrt{\frac{15}{8\pi}} \sin\theta\cos\theta\, e^{-i\phi}$	$z(x - iy) = xz - iyz$
$Y_{2,-2}(\theta,\phi) = \sqrt{\frac{15}{32\pi}} \sin^2\theta\, e^{-2i\phi}$	$(x - iy)^2 = x^2 - y^2 - 2ixy$

Suppose now we consider two coordinate systems x, y, z and x', y', z', with corresponding angles θ, ϕ and θ', ϕ' defined by (8.84). The $Y_{lm}(\theta, \phi)$'s form a complete set for a given l, which means that we can always express $Y_{l\mu}(\theta', \phi')$ in terms of the $Y_{lm}(\theta, \phi)$'s:

$$Y_{l\mu}(\theta', \phi') = \sum_m a_{\mu m} Y_{lm}(\theta, \phi) \quad \text{and} \tag{8.86a}$$

$$Y_{lm}(\theta, \phi) = \sum_\mu c_{m\mu} Y_{l\mu}(\theta', \phi') \quad. \tag{8.86b}$$

It is then straightforward, utilizing the orthogonality properties of the Y_{lm}'s when integrated over 4π solid angle, to show that

$$a_{\mu m} = c^*_{m\mu} \quad. \tag{8.87}$$

Noting the forms of the Y_{lm}'s in rectangular coordinates, we realize that the coefficients $a_{\mu m}$ or $c_{m\mu}$ could be found by explicit substitution of the relationship between the primed and the unprimed coordinates:

$$\begin{aligned} x' &= b_{11}x + b_{12}y + b_{13}z \\ y' &= b_{21}x + b_{22}y + b_{23}z \\ z' &= b_{31}x + b_{32}y + b_{33}z \quad. \end{aligned} \tag{8.88}$$

They can also be found using the formalism of the Wigner rotation matrices (see for example the text [8.23]).

Let us now specify the orientation of the z-axis as being at an angle (θ'_0, ϕ'_0) in the primed system. Then there is a famous theorem, the addition theorem for spherical harmonics, which tells us that

$$Y^*_{l0}(0,0)Y_{l0}(\theta, \phi) = \sum_m Y^*_{lm}(\theta'_0, \phi'_0) Y_{lm}(\theta', \phi') \quad. \tag{8.89}$$

[This theorem can be derived by making an expansion of a δ-function lying along the z-axis. One utilizes (8.86), the fact that the δ-function is axially symmetric about the z-axis, and equates the expansion in the $Y_{lm}(\theta, \phi)$'s to that in the $Y_{lm}(\theta', \phi')$'s].

We now wish to consider a spinning sample. To do this, we wish to define some axes (Fig. 8.8). We define first the laboratory z-axis, z_L, along the static field, and an axis z_R about which we will eventually rotate the sample. The angle between z_L and z_R is θ_0 (Fig. 8.8a). We can then, without loss in generality, define the laboratory axis x_L to be perpendicular to both z_L and z_R (Fig. 8.8b). We define x_R to be coincident with x_L (Fig. 8.8c). The axes x_R, y_R, and z_R are fixed in the laboratory frame (Fig. 8.8d). We then define axes x_S, y_S, and z_S which are fixed in the sample with z_S coincident with z_R at all times, and x_S and y_S coincident with x_R and y_R respectively at time $t = 0$, but making an angle Ωt at later times (Fig. 8.8e). Ω is the angular velocity of sample spinning. Utilizing these definitions, we can specify the orientation of H_0 in the coordinate system x_R, y_R, z_R (at angle θ_{0R}, ϕ_{0R}) and in the system x_S, y_S, z_S (at angle θ_{0S}, ϕ_{0S}).

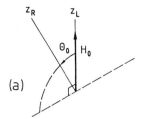
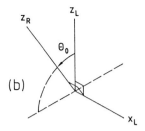
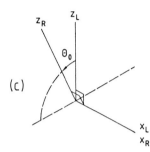
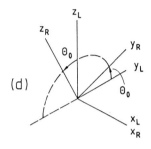
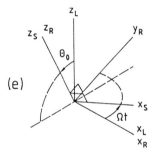

Fig. 8.8a-e. Axes important for sample spinning. (a) The laboratory z-axis, z_L, is chosen along the static field H_0. The axis z_R makes an angle θ_0. Together, z_L and z_R define a plane. (b) The lab axis x_L is defined as lying perpendicular to the (z_L, z_R) plane. (c) The axis x_R is coincident with x_L. (d) The axes y_L and y_R lie in the (z_R, z_L) plane. *All* the axes x_L, y_L, z_L and x_R, y_R, z_R are fixed in the laboratory frame. (e) The axes $x_S, y_S,$ and z_S are fixed in the sample and rotate with it, z_S coinciding with z_R. At $t = 0$, x_S is chosen to be coincident with x_L and x_R, so that it makes an angle Ωt with them at later times

From the figure we see

$$\theta_{0R} = \theta_0 , \qquad \phi_{0R} = \pi/2 \tag{8.90a}$$

$$\theta_{0S} = \theta_{0R} = \theta_0 , \qquad \phi_{0S} = \phi_{0R} - \Omega t = \pi/2 - \Omega t . \tag{8.90b}$$

We now consider the interaction between a pair of spins j, k which we have previously written as proportional to $3\cos^2\theta_{jk} - 1$ (3.7). We now, however, must distinguish between coordinate systems, hence we replace θ_{jk} by θ_{Ljk}, writing the coupling

$$3\cos^2\theta_{Ljk} - 1 = \sqrt{16\pi/5}\, Y_{20}(\theta_{Ljk}, \phi_{Ljk}) . \tag{8.91}$$

Then, we can utilize (8.89) to get

$$Y_{20}^*(0,0)Y_{20}(\theta_{Ljk}, \phi_{Ljk}) = \sum_m Y_{2m}^*(\theta_{0S}, \phi_{0S}) Y_{2m}(\theta_{Sjk}, \phi_{Sjk}) \tag{8.92}$$

395

where θ_{Sjk}, ϕ_{Sjk} give the orientation of the internuclear vector r_k in the coordinate system x_S, y_S, z_S which is fixed in the sample and where θ_{0S} and ϕ_{0S} are given by (8.90).

Substituting the explicit expressions for the Y_{lm}'s from Table 8.1, we get

$$3\cos^2\theta_{Ljk} - 1 = \left(\frac{3\cos^2\theta_{0S} - 1}{2}\right)(3\cos^2\theta_{Sjk} - 1)$$
$$+ 6\sin\theta_{0S}\cos\theta_{0S}\sin\theta_{Sjk}\cos\theta_{Sjk}\cos(\phi_{Sjk} - \phi_{0S})$$
$$+ \tfrac{3}{2}\sin^2\theta_{0S}\sin^2\theta_{Sjk}\cos 2(\phi_{Sjk} - \phi_{0S}) \quad . \tag{8.93}$$

Since ϕ_{0S} obeys (8.90b),

$$\phi_{0S} = \pi/2 - \Omega t \quad . \tag{8.90b}$$

We see that the second and third terms on the right vary as Ωt and $2\Omega t$ respectively, whereas the first term on the right is independent of time. Thus if one can average the time-dependent terms, we recover the result of (3.61) with the notation

$$\gamma_{jk} = \theta_{Sjk} \quad . \tag{8.94}$$

For steady-state experiments (i.e. if one talks about the "frequency" domain, perhaps with data which are the Fourier transform of data from a pulse experiment), the terms involving Ωt and $2\Omega t$ give rise to frequency modulation at frequencies Ω and 2Ω. Such a modulation gives rise to multiple sidebands spaced in frequency by Ω about the normal fequency ω_0. For frequencies Ω much bigger than the dipolar line width $\triangle\omega_d$, these sidebands will be displaced by an amount large compared to the width associated with the time-independent term. If we neglect the sidebands to focus our attention solely on the main transition, it will appear narrower. Indeed, since the time-independent term of the dipolar coupling for the spinning case is simply reduced from its value for the nonspinning case by the factor C_S where

$$C_S = \frac{3\cos^2\theta_{0S} - 1}{2} \quad , \tag{8.95}$$

the *form* of the absorption line should be identical in the two cases except for a scaling of the frequency. Thus, if $f(\omega - \omega_0)$ is the normalized intensity function without spinning, the normalized function with spinning $f_S(\omega - \omega_0)$ becomes

$$f_S(\omega - \omega_0) = \frac{1}{C_S}f(C_S(\omega - \omega_0)) \quad . \tag{8.96}$$

At the magic angle, the line theoretically becomes a δ-function. In practice, it is difficult to achieve the condition

$$\Omega \gg \triangle\omega_d \tag{8.97}$$

for large γ nuclei, since spinning speeds are typically a few kilohertz, comparable to typical dipolar line widths.

We turn now to a discussion of the effect of spinning on a pulse experiment in which we observe the free induction decay of a spinning sample following a $\pi/2$ pulse. Then we might express the dipolar Hamiltonian, $\mathcal{H}_d$ as

$$\mathcal{H}_d = \sum_{j<k} (1 - 3\cos^2\theta_{Sjk}) \frac{\gamma^2 \hbar^2}{r_{jk}^3} (3I_{zj}I_{zk} - \mathbf{I}_j \cdot \mathbf{I}_k)$$

$$= \sum_{j>k} [F_{0jk} + F_{1jk}(t) + F_{2jk}(t)] G_{jk}(\mathbf{I}_j, \mathbf{I}_k) \qquad (8.98a)$$

where

$$G_{jk}(\mathbf{I}_j, \mathbf{I}_k) \equiv \frac{\gamma^2 \hbar^2}{r_{jk}^3} (\mathbf{I}_j \cdot \mathbf{I}_k - 3I_{zj}I_{zk}) \qquad (8.98b)$$

is a function only of the x_L, y_L, and z_L components of the spin operators $\mathbf{I}_j$ and $\mathbf{I}_k$, and

$$\begin{aligned}
F_{0jk} &= \frac{(3\cos^2\theta_{0S} - 1)}{2} (3\cos^2\theta_{Sjk} - 1) \\
F_{1jk}(t) &= 6\sin\theta_{0S}\cos\theta_{0S}\sin\theta_{Sjk}\cos\theta_{Sjk}\cos(\phi_{Sjk} - \phi_{0S}) \\
F_{2jk}(t) &= \tfrac{3}{2}\sin^2\theta_{0S}\sin^2\theta_{Sjk}\cos^2(\phi_{Sjk} - \phi_{0S})
\end{aligned} \qquad (8.99)$$

Since $\phi_{0S} = \pi/2 - \Omega t$ from (8.90b), we see that F_{0jk} is independent of time, but F_{1jk} and F_{2jk} oscillate in time at Ω and 2Ω respectively. Explicit examination of the spin functions G_{jk} shows that two different functions which have one spin in common, for example the jth, such as G_{jk} and G_{jl} ($l \neq k$), do not commute. This makes it difficult in general to determine the time development. Such matters are discussed in Appendix J dealing with time-dependent Hamiltonians. Here, we are dealing with the case of a Hamilitonian composed of a sum of noncommuting parts each of which has a time-dependent coefficient. This is Case II of Appendix J. However, the time dependence is periodic, so we can consider what happens over a single period.

Suppose that the period $T (\equiv 2\pi/\Omega)$ is sufficiently short that not much happens during one cycle. Then, as we show in Appendix I, we can relate the wave function at the start of a period to its value at the end of that period by the unitary operator $U(T)$ given by

$$\psi(T) = U(T)\psi(0) = \exp\left(-\frac{i}{\hbar}\overline{\mathcal{H}}T\right)\psi(0) \quad \text{where} \qquad (8.100)$$

$$\overline{\mathcal{H}} = \frac{1}{T} \int_0^T \mathcal{H}(\tau)d\tau \quad . \qquad (8.101)$$

But

$$\int_0^T F_{1jk}(\tau)d\tau = 0 \quad , \quad \int_0^T F_{2jk}(\tau)d\tau = 0 \qquad (8.102)$$

so that it is only the term F_{0jk} which brings about a change in ψ over an integral period.

This is tantamount to saying that to the extent $F_{1jk}(t)$ and $F_{2jk}(t)$ produce changes during the time interval T, they must also undo these changes over a complete period T. The result is that the free induction decay has periodic maxima which *Lowe* called *rotational echoes*. Figure 8.9 shows the data from *Lowe*'s paper [8.21] for F^{19} in Teflon. The rotational echoes are clearly apparent, in Fig. 8.9a, as is the fact that at the magic angle (54.7°), the free induction decay lasts much longer. The Fourier transform of these data (Fig. 8.9b) shows the sideband structure. It illustrates the fact that even when spinning at over 6 kHz, the sidebands are still not separated from the central line for the high F^{19} nucleus.

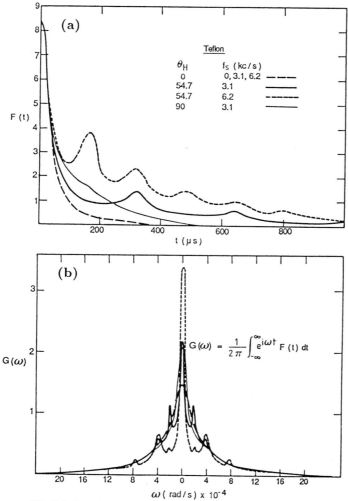

Fig. 8.9. Free induction decays (a) for spinning and nonspinning samples of Teflon, and their Fourier transforms (b). The curves are corrected for instrumental nonlinearities. The data are due to *I. Lowe, H. Kessemeier, W. Yen, G. Theiss*, and *R.E. Norberg* [8.20]

Andrew et al. [8.22] and *Kessemeier* and *Norberg* [8.24] demonstrated clear separations of the spinning sidebands for lower γ nuclei Na^{23} and P^{31} respectively. Indeed, removal of the dipolar coupling between like spins is done more readily by applying strong pulses to flip the spins, so-called spin-flip narrowing, which we discuss in the next section.

We can understand physically the significance of the requirement (8.97) that $\Omega \gg \Delta\omega_d$ and its relation to the fact that G_{ij} and G_{jl} do not commute as follows. For spinning to average out an interaction, the interaction must be constant in time over at least one cycle of rotation. To average the coupling G_{jk}, the local field that spin j produces at spin k must not change during one cycle. But the coupling G_{jl} permits spins j and l to undergo mutual flips, thereby interrupting the coherent averaging of G_{jk}. $\Delta\omega_d$ is a measure of τ_{jl}, how long spin j can maintain its orientation without such a mutual spin flip ($\tau_{jl} \approx 1/\Delta\omega_d$). For narrowing of the jk coupling to occur, the period of rotation ($1/\Omega$) must be less than τ_{jl},

$$\tau_{jl} > 1/\Omega \quad \text{where} \tag{8.103a}$$

$$\tau_{jl} \sim 1/\Delta\omega_d \quad . \tag{8.103b}$$

These can be rewritten to give (8.97)

$$\Omega > \Delta\omega_d \quad . \tag{8.97}$$

The case we have just discussed, which we call Case II in Appendix J, has been called an example of a "homogeneous" broadening mechanism by *Maricq* and *Waugh* [8.25]. (This term is closely related to an idea introduced by *Portis* [8.26] in electron spin resonance). Its essential feature is that the time-dependent parts of the Hamiltonian do not commute with one another. To produce line narrowing in this case, one needs $\Omega \gg \Delta\omega$ where $\Delta\omega$ is the line width from this mechanism. In the absence of spinning $\Delta\omega = \Delta\omega_d$ for dipolar broadening.

We now turn to a case which has a much less stringent condition, and thus is much easier to achieve experimentally, narrowing lines when the time-dependent parts of the Hamiltonian commute, Case I of Appendix J. We illustrate this case by line broadening from the anisotropy of chemical or Knight shifts. This case is called "inhomogeneous" since it is analogous to line broadening by inhomogeneous magnetic fields. Andrew recognized that magic angle spinning would remove line broadening from sources, other than dipolar coupling, which depend on the orientation of H_0 with respect to the crystal axes. The key point was that the broadening be described by appropriate angular factors arising from the orientation of the static magnetic field, H_0, with respect to the crystal axes. When the angular factors are proportional to Y_{lm}'s with $l = 2$, there is a magic angle. Other cases are line broadening from chemical and Knight shift anisotropies, and from electric quadrupole splittings. We treat the case of shift anisotropy.

Recalling (4.198), we get that, in the presence of anisotropic shifts,

$$\mathcal{H} = -\gamma\hbar H_0 \left[1 + (\overline{K} - \overline{\sigma}) + (K_{LO} - \sigma_{LO}) \left(\frac{3\cos^2\theta - 1}{2} \right) \right.$$
$$\left. + \left(\frac{K_{TR} - \sigma_{TR}}{2} \right) \sin^2\theta \cos 2\phi \right] I_z \quad . \tag{8.104}$$

We consider I_z to be the sum of the z components I_{zk} of N noninteracting spins ($k = 1$ to N). We shall analyze the effect of rotation on this Hamiltonian.

The angles θ and ϕ of (8.104) specify the orientation of $\boldsymbol{H}_0$ in the principal axis system of the shift tensors. We shall need to keep track of various reference frames. Therefore, we shall define the spherical angles θ_{0P} and ϕ_{0P} to designate the orientation of $\boldsymbol{H}_0$ in the principal axis system x_P, y_P, z_P. We keep the same definitions for the coordinate systems x_L, y_L, z_L ($\boldsymbol{H}_0$ lies along z_L), x_R, y_R, z_R, and x_S, y_S, z_S (Fig. 8.8).

We have then that at $t = 0$, z_L is at θ_{0P}, ϕ_{0P} or expressed in R or S at θ_{0R}, ϕ_{0R} or θ_{0S}, ϕ_{0S}, which in turn are described by (8.90a). Therefore, we can write at $t = 0$

$$\mathcal{H} = -\gamma\hbar H_0 \left[1 + (\overline{K} - \overline{\sigma}) + \left(\frac{K_{LO} - \sigma_{LO}}{2} \right) (3\cos^2\theta_{0P} - 1) \right.$$
$$\left. + \left(\frac{K_{TR} - \sigma_{TR}}{2} \right) \sin^2\theta_{0P} \cos 2\phi_{0P} \right] I_z \quad . \tag{8.105}$$

As time goes on, θ_{0P} and ϕ_{0P} change, owing to the sample rotation, hence should be viewed as time dependent.

Recognizing that the two angular functions are related to Y_{2m}'s apart from normalization constants, we define the normalizations $N_{2,0}, N_{2,2}, N_{2,-2}$ as

$$Y_{2,0} \equiv N_{2,0}(3\cos^2\theta - 1)$$
$$Y_{2,2} \equiv N_{2,2} \sin\theta\, e^{i\phi} \tag{8.106}$$
$$Y_{2,-2} \equiv N_{2,-2} \sin\theta\, e^{-i\phi}$$

and express

$$\mathcal{H} = -\gamma\hbar H_0 I_z \left[1 + (\overline{K} - \overline{\sigma}) + \left(\frac{K_{LO} - \sigma_{LO}}{2} \right) \frac{Y_{2,0}(\theta_{0P}, \phi_{0P})}{N_{2,0}} \right.$$
$$\left. + \left(\frac{K_{TR} - \sigma_{TR}}{2} \right) \frac{1}{2N_{2,2}} [Y_{2,2}(\theta_{0P}, \phi_{0P}) + Y_{2,-2}(\theta_{0P}, \phi_{0P})] \right] \quad .$$
$$\tag{8.107}$$

We can now express the angular functions in terms of the S-system, using (8.86),

$$Y_{2,0}(\theta_{0P}, \phi_{0P}) = \sum_{m_S} C_{0m_S}(P,S) Y_{2m_S}(\theta_{0S}, \phi_{0S})$$

$$Y_{2,2}(\theta_{0P}, \phi_{0P}) = \sum_{m_S} C_{2m_S}(P,S) Y_{2m_S}(\theta_{0S}, \phi_{0S}) \qquad (8.108)$$

$$Y_{2,-2}(\theta_{0P}, \phi_{0P}) = \sum_{m_S} C_{-2m_S}(P,S) Y_{2m_S}(\theta_{0S}, \phi_{0S}) \quad .$$

At later times, owing to the rotation, ϕ_{0S} will change with time, but since z_S is the rotation axis, θ_{0S} does not change, so

$$\theta_{0S}(t) = \theta_0 \quad , \quad \phi_{0S}(t) = \pi/2 - \Omega t \quad . \qquad (8.109)$$

Therefore the $m_S = \pm 2$ terms of (8.108) will oscillate at $\pm 2\Omega$, the $m_S = \pm 1$ terms will oscillate at $\pm \Omega$, and the $m_S = 0$ terms will be time independent. Collecting terms we get

$$\begin{aligned}\mathcal{H} = &-\gamma\hbar H_0 I_z \Bigg[1 + (\overline{K} - \overline{\sigma}) \\ &+ \left(\frac{K_{\text{LO}} - \sigma_{\text{LO}}}{2N_{2,0}}\right) \sum_{m_S} C_{0m_S}(P,S) Y_{2m_S}(\theta_0, \pi/2 - \Omega t) \\ &+ \left(\frac{K_{\text{TR}} - \sigma_{\text{TR}}}{4N_{2,2}}\right) \sum_{m_S} [C_{2m_S}(P,S) + C_{-2,m_S}(P,S)] \\ &\times [Y_{2m_S}(\theta_0, \pi/2 - \Omega\tau)] \Bigg] \quad . \end{aligned} \qquad (8.110)$$

This Hamiltonian consists of several time-dependent parts all multiplied by the same spin function, I_z. Thus the Hamiltonian at one time, t, commutes with the Hamiltonian at another time, t':

$$[\mathcal{H}(t), \mathcal{H}(t')] = 0 \quad . \qquad (8.111)$$

This Hamiltonian therefore belongs to Class I of Appendix J. Indeed, it is even described by (J.28)

$$\mathcal{H}(t) = a(t)\mathcal{H}_a \quad , \qquad (\text{J.28})$$

for which the formal solution is given in (J.30) as

$$\psi(t) = \exp\left(-\frac{i}{\hbar}\mathcal{H}_a \int_0^t a(\tau)d\tau\right) \psi(0) \quad . \qquad (\text{J.30})$$

Let us first consider just the time-independent term ($m_S = 0$). It will give a contribution to the Hamiltonian of

$$\mathcal{H}_{\text{time indep}} = -\gamma\hbar H_0 I_z \left\{ 1 + (\overline{K} - \overline{\sigma}) + \left[\left(\frac{K_{\text{LO}} - \sigma_{\text{LO}}}{2N_{2,0}} \right) C_{0,0}(\text{P},\text{S}) \right. \right.$$
$$+ \left(\frac{K_{\text{TR}} - \sigma_{\text{TR}}}{4N_{2,2}} \right) [C_{2,0}(\text{P},\text{S}) + C_{-2,0}(\text{P},\text{S})] \bigg]$$
$$\left. \times N_{2,0}(3\cos^2\theta_0 - 1) \right\} \ . \tag{8.112}$$

The coefficients $C_{2,0}$ (P, S) and $C_{-2,0}$ (P, S) can be obtained (if desired) as we explain below. But even before that, we note that when θ_0, the angle between the rotation axis and H_0, is at the magic angle, we get

$$\mathcal{H}_{\text{time indep}} = -\gamma\hbar H_0 I_z [1 + (\overline{K} - \overline{\sigma})] \ . \tag{8.113}$$

Thus, it is identical to the form taken in a liquid in which the anisotropic shift contributions average to zero.

If θ_0 is not at the magic angle we might wish to know the coefficients $C_{0,0}(\text{P},\text{S})$, $C_{2,0}(\text{P},\text{S})$, and $C_{-2,0}(\text{P},\text{S})$. These we get from the inverse relations (8.87)

$$\begin{aligned} C_{0,0}(\text{P},\text{S}) &= a^*_{0,0}(\text{S},\text{P}) \\ C_{2,0}(\text{P},\text{S}) &= a^*_{0,2}(\text{S},\text{P}) \\ C_{-2,0}(\text{P},\text{S}) &= a^*_{0,-2}(\text{S},\text{P}) \end{aligned} \tag{8.114}$$

and the spherical harmonic addition theorem (8.89)

$$a_{0,0} = \frac{Y^*_{2,0}(\alpha,\beta)}{Y^*_{2,0}(0,0)} \ , \quad a_{0,2} = \frac{Y^*_{2,2}(\alpha,\beta)}{Y^*_{2,0}(0,0)} \ , \quad a_{0,-2} = \frac{Y^*_{2,-2}(\alpha,\beta)}{Y^*_{2,0}(0,0)} \tag{8.115}$$

where (α, β) gives the orientation in spherical coordinates of the spinning axis z_S in the x_P, y_P, z_P (principal axis) coordinate system. The result is

$$\mathcal{H}_{\text{time indep}} = -\gamma\hbar H_0 I_z \left\{ 1 + (\overline{K} - \overline{\sigma}) + \left[(K_{\text{LO}} - \sigma_{\text{LO}}) \left(\frac{3\cos^2\alpha - 1}{2} \right) \right. \right.$$
$$\left. \left. + (K_{\text{TR}} - \sigma_{\text{TR}}) \left(\frac{\sin\alpha\cos 2\beta}{2} \right) \right] \left(\frac{3\cos^2\theta_0 - 1}{2} \right) \right\} \ . \tag{8.116}$$

(This expression can be checked in the limiting case that $\theta_0 = 0$, hence the spin axis lies along H_0. Then the spinning has no effect and the time-independent term is the entire Hamiltonian. Then $\alpha = 0$ or $\alpha = \pi/2$, $\beta = 0$ or $\pi/2$ correspond to H_0 lying along the three principal directions. The result is that for the three cases we get shifts respectively $K_3 - \sigma_3$, $K_1 - \sigma_1$, or $K_2 - \sigma_2$, where 3, 1, and 2 stand for the z_p, x_p, and y_p directions.)

The other very interesting point to consider is the time dependence when one is spinning at the magic angle. Examination of (8.110) shows that the time dependence of $\mathcal{H}$ comes from the factors $Y_{2m_S}(\theta_0, \pi/2 - \Omega t)$. Utilizing the properties of the Y_{2m_S}'s (Table 8.1) we see that $\mathcal{H}$ is a sum of terms which are either independent of time, or vary as $\exp(i\Omega t)$, $\exp(-i\Omega t)$, $\exp(2i\Omega t)$, $\exp(-2i\Omega t)$. Keeping in mind that the Hamiltonian must be real, and that we are at the magic angle, we can thus write

$$\mathcal{H} = -\gamma\hbar H_0 I_z \{1 + (\overline{K} - \overline{\sigma}) + A_1(\text{P,S})\cos[\Omega t - \Gamma_1(\text{P,S})] \\ + A_2(\text{P,S})\cos[2\Omega t - \Gamma_2(\text{P,S})]\} \quad , \tag{8.117}$$

where $A_1(\text{P,S})$ and $A_2(\text{P,S})$ are coefficients, and $\Gamma_1(\text{P,S})$ and $\Gamma_2(\text{P,S})$ are phase angles, all of which in general will depend on the orientation of z_S in the x_P, y_P, z_P system. Utilizing this form, we get

$$\int_0^t \mathcal{H}(\tau)d\tau = -\gamma\hbar H_0 I_z \Bigg([1 + (\overline{K} - \overline{\sigma})]t \\ + \frac{A_1(\text{P,S})}{\Omega}\{\sin[\Omega t - \Gamma_1(\text{P,S})] + \sin\Gamma_1(\text{P,S})\} \\ + \frac{A_2(\text{P,S})}{2\Omega}\{\sin[2\Omega t - \Gamma_2(\text{P,S})] + \sin\Gamma_2(\text{P,S})\}\Bigg) \quad . \tag{8.118}$$

Utilizing the fact that

$$\psi(t) = \exp\left(-\frac{i}{\hbar}\int_0^t \mathcal{H}(\tau)d\tau\right)\psi(0) \quad , \tag{8.119}$$

and recalling that $T = 2\pi/\Omega$, we see that when

$$t = nT \quad , \quad n = 0, 1, 2\ldots \quad , \tag{8.120a}$$

$$\psi(t) = \exp\left(-i\gamma H_0 I_z(1 + \overline{K} - \overline{\sigma})t\right)\psi(0) \quad , \tag{8.120b}$$

just the result we would have if the time-dependent terms were missing. Note that this result says that at the magic angle the time development over integral multiples of the rotation period is independent of the orientation of the crystal axes relative to H_0. Thus, if we applied a $\pi/2$ pulse to a spinning sample, the free induction decay at times given by (8.120a) would act as though there were no anisotropy to the shift tensor. If one had a powder sample which was not spinning, the spread in precession frequency arising from shift anisotropy, $\Delta\omega_{\text{shift}}$, would cause the free induction to decay in a time $1/\Delta\omega_{\text{shift}}$. If, however, the sample is spinning we can see from (8.120b) that the signals are rephased at times given by (8.120a), thereby producing a string of echoes.

Notice that we have not placed any requirement as yet on how fast the spinning must be, in contrast to the previous situation concerning narrowing of dipolar line broadening. For the dipolar case we compared the rigid lattice line width $\Delta\omega_{\text{d}}$ with Ω. What frequency should we compare with ω in the present case? Dimensionally, there is only one relevant frequency, the total excursion of the precession frequency over the spinning cycle. Note that all our equations so far imply there is a single crystal being rotated since there is a well-defined orientation of z_S in the x_P, y_P, z_P coordinate system. However, for a powder, all possible orientations occur. For the rest of our discussion we will focus on a powder. Then, the maximum frequency excursion one can have in a rotation cycle is less than

$$\gamma H_0[(K_{z_P z_P} - \sigma_{z_P z_P}) - [(K_{x_P x_P} - \sigma_{x_P x_P})] \equiv \triangle\omega_{\text{shift}} \quad , \tag{8.121}$$

where we are defining the z_P and x_P axes as having the maximum and minimum shifts respectively.

So we ask, what happens when we vary the relative sizes of Ω and $\triangle\omega_{\text{shift}}$? Suppose, first, that the powder sample is not spinning. Then, as we have noted, following a $\pi/2$ pulse, the transverse magnetization will decay to zero in a time $\approx 1/\triangle\omega_{\text{shift}}$. The Fourier transform of this free induction decay would give us the line shape. If we then applied a π pulse at time t_π, we could refocus the magnetization into an echo. If $t_\pi \gg 1/\triangle\omega_{\text{shift}}$, the echo and the free induction decay would be well separated in time. Indeed, we could apply a string of such π pulses (a Carr-Purcell sequence) producing a string of echoes. The Fourier transform of any one echo will give us the powder line shape in frequency space. If, instead, we took a Fourier transform of the *string* of echoes, we would now have introduced a periodicity. The resultant transform differs from that of a single echo in the same way that a Fourier series differs from a Fourier integral. The transform of the string of echoes would consist of δ-functions spaced apart in frequency by the angular frequency $1/t_{\text{rep}}$ where t_{rep} is the time between successive π pulses. The Fourier transform of a single echo would be the envelope function of the spikes. If now one shortened t_{rep}, one would get to a point where $t_{\text{rep}} \approx 1/\triangle\omega_{\text{shift}}$. Then the echo would not have decayed completely to zero when the π pulse is applied, so that over a single period one no longer would have the complete shape of a single free induction decay. The Fourier transform of the sequence of echoes would still have a center line plus sidebands spaced in angular frequency by $1/t_{\text{rep}}$, but the envelope of these sidebands would not be the powder line shape (the transform of a single free induction decay).

Stejskal et al. [8.27] realized that the same situation would arise from magic angle spinning. When $\Omega \ll \triangle\omega_{\text{shift}}$, the NMR signal following an initial $\pi/2$ pulse decays rapidly compared to the period of rotation, hence is essentially identical to what it would be with $\Omega = 0$. However, after one full rotation there is an echo, as described by (8.120). Indeed, a train of "spinning echoes" will be formed, analogous to a Carr-Purcell train. The Fourier transform of any echo will give the $\Omega = 0$ powder line shape. The Fourier transform of the train of echoes will give δ-function spikes spaced apart in angular frequency by Ω, with an intensity envelope versus frequency identical to the nonspinning powder pattern. This situation is illustrated beautifully by Fig. 8.10 showing the data of *Herzfeld* et al. for a P^{13} compound [8.28, 29].

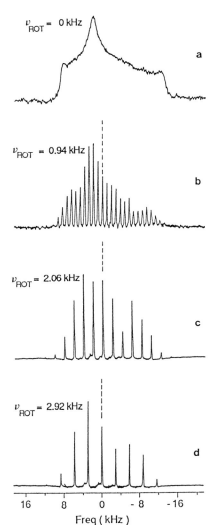

Fig. 8.10. Proton decoupled P^{31} spectra at 119.05 MHz of barium diethyl phosphate spinning at the magic angle. The figure illustrates how, in the limit of slow spinning, the spinning sidebands reproduce the shape of the nonspinning spectrum [8.28]

Once Ω becomes comparable to or larger than $\triangle\omega_{\text{shift}}$, the satellites occur outside the envelope of the nonspinning line. As can be seen from (8.117), the coefficients $A_1(P,S)$ and $A_2(P,S)$ are multiplied by $1/\Omega$, hence become progressively smaller than the larger Ω. This is part of a general theorem discussed by *Andrew* et al. [8.22] that the second moment is invariant under rotation. Thus, the intensity of lines spaced by multiples of Ω must fall off as $1/\Omega^2$ in the limit of large Ω.

When there are many spinning sidebands, how can one tell the *central* line? The easiest technique is to change Ω. One line will be undisplaced – the central line.

The beauty of the slow spinning regime of *Stejskal*, *Schaefer*, and *McKay* is that it gives one simultaneously a high precision determination of the isotropic average shift, as well as a picture of the associated powder pattern. It also removes the difficulty of spinning at a rate faster than $\triangle\omega_{\text{shift}}$, a genuine difficulty in the strong fields of superconducting magnets when one has large shift anisotropies.

A common use of the slow spinning regime is to study low abundance nuclei (e.g. C^{13}), so that there is no like-nuclei dipolar coupling with which to contend, combined with standard decoupling to any high abundance nuclei like H^1 whose mutual spin-flips would otherwise broaden the spinning side bands. In some instances it is desirable to remove sidebands in order to simplify the spectra. This can be done by applying π pulses synchronously with the rotation as shown by *Dixon* [8.30] and by *Raleigh* et al. [8.31].

8.10 The Relation of Spin-Flip Narrowing to Motional Narrowing

The strong spin-spin coupling characteristic of solids was essential for achieving the high sensitivity of double resonance. Its disadvantage is that it produces line broadening, and thus may obscure details of the resonance line such as the existence of anisotropic couplings (chemical shift, Knight shift, or quadrupole interactions). Within the past few years, a variety of clever techniques have been introduced which utilize strong rf pulses to eliminate much of the dipolar broadening. We now turn to the principles on which the ideas are based. The highly ingenious concepts owe their development to several groups. The pioneering experiments were performed by two groups, one headed by *John Waugh* [8.4], the other headed by *Peter Mansfield* [8.5]. Subsequently *Robert Vaughan* [8.6] and his colleagues have contributed importantly, as have many of the scientists [8.32,33] who worked with *Waugh, Mansfield* or *Vaughan*.

The essence of these schemes is to apply a repetitive set of rf pulses [8.34-39] which produce large spin rotations and which, by a process akin to motional narrowing, cause the dipolar coupling to average to zero. Each cycle consists of a small number of pulses (8 pulses per cycle is common). To achieve the effect, there need to be many cycles within the normal dephasing time of the rigid lattice line width. Since the rf pulses must produce large spin rotations (90° pulses are typically used), and since the rotations should occur within a small fraction of a cycle of pulses, one needs H_1's which are large compared to the rigid lattice line breadth.

We have remarked on the similarity between the multiple pulse schemes and motional narrowing. We start by making the analogy explicit. For convenience we shall name the multiple pulse schemes "spin-flip narrowing".

Consider two spins, I and S. Let r_{IS} be the vector from spin I to S. The magnetic field H_I at spin S due to I is then

$$H_I(r_{IS}) = \frac{3(M_I \cdot r_{IS})r_{IS}}{r_{IS}^5} - \frac{M_I}{r_{IS}^3} \quad . \tag{8.122}$$

We consider M_I to lie along the k-direction, and consider three cases in which S lies at a distance a respectively along the x-, y-, and z-axes (see Fig. 8.11). Application of the formula shows that when S is on the z-axis it experiences a field

$$H_I(0,0,a) = \frac{2M_I}{a^3} \tag{8.123}$$

whereas when S is on the x- or y-axes

$$H_I(a,0,0) = H_I(0,a,0) = -\frac{M_I}{a^3} \quad . \tag{8.124}$$

If, for some reason, S were to jump rapidly among the three positions, spending equal times in each on the average, it would experience a time-averaged field $<H_I>$ given by

$$\begin{aligned}<H_I> &= \frac{1}{3}[H_I(a,0,0) + H_I(0,a,0) + H_I(0,0,a)] \\ &= \frac{1}{3}\frac{1}{a^3}(2M_I - M_I - M_I) \\ &= 0 \quad . \end{aligned} \tag{8.125}$$

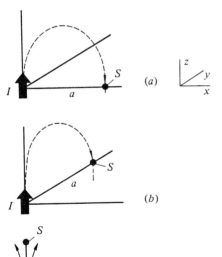

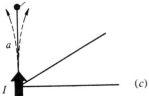

Fig. 8.11a-c. The magnetic field of I at S for three locations of S all the same distance, a, from spin I. The dashed line indicates a magnetic line of force

This result is the essence of motional narrowing. Though for our case we picked only discrete locations for spin S, the same result is found for continuously variable sites for which the strength of the interaction goes with angular position θ_{IS} of r_{IS} with respect to the z-axis as $3\cos^2\theta_{IS} - 1$. This averages to zero over a sphere.

Suppose now we consider a variation in the above picture. Suppose we position S at $(0,0,a)$, i.e. on the z-axis, and consider the orientation of both M_I and M_S (Fig. 8.12).

In terms of the orientation of the vector r_{IS} with respect to M_I, it is useful to introduce some names. Referring to Fig. 8.11, we denote the (c) configuration as the *on-axis position* of S, we denote the position of S in (a) and (b) as the *side position*. Note that for the *on-axis position* H_{IS} is parallel to M_I, whereas for the *side position* H_{IS} is anti-parallel to M_I and half of its on-axis magnitude.

Referring now to Fig. 8.12, we see that in (a) and (b) S occupies a *side position*, whereas in (c) it occupies an *on-axis position*. Moreover, since we have taken M_I and M_S parallel in all parts, for configurations (a) and (b) the magnetic energy $E_{\text{mag}} = -M_S \cdot H_{IS}$ is

$$E_{\text{mag})_{a,b}} = -M_S \cdot H_{IS} = \frac{M_I M_S}{a^3} \tag{8.126}$$

whereas for (c)

$$E_{\text{mag})_c} = -M_S \cdot H_{IS} = \frac{-2M_I M_S}{a^3} \ . \tag{8.127}$$

Suppose, then, we began with the configuration of Fig. 8.12c, the on-axis arrangement with M_I and M_S parallel. After time τ let us quickly rotate M_I and M_S by $\pi/2$ about the y-axis to the *side position* of Fig. 8.12a. After a τ give

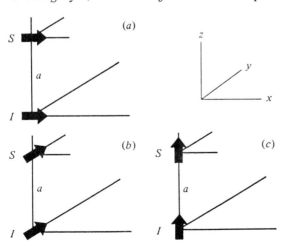

Fig. 8.12. The two spins M_I and M_s are oriented parallel to one another. In (**a**), (**b**), and (**c**) their magnetic moments are parallel respectively to the x-, y-, and z-directions

a quick rotation of both spins to the *side position* of Fig. 8.12b. Wait another time τ.

What is the average magnetic energy $<E_\mathrm{m}>$ over the interval 3τ? Using (8.126) and (8.127) we get

$$<E_\mathrm{m}> = \frac{1}{3\tau}\left(\frac{-2M_I M_S \tau}{a^3} + \frac{M_I M_S \tau}{a^3} + \frac{M_I M_S \tau}{a^3}\right) = 0 \quad . \tag{8.128}$$

That is, over such a cycle in which, at stated times, we apply selected $\pi/2$ rotations to *both* spins, we can make the magnetic energy average to zero. Thus by spin-flips we can cause the dipolar coupling to vanish just as we could for position jumps in Fig. 8.11. This is the principle of spin-flip line narrowing. *We make the dipolar energy vanish by flipping the spins among selected on-axis and side positions, spending twice as long in the side positions as in the on-axis ones.*

All of the complicated pulse cycles are based on exactly this principle.

8.11 The Formal Description of Spin-Flip Narrowing

From the previous section we see that it is possible to cause the dipolar interaction to average to zero if we cause the spins to flip between the "on-axis" and "side" arrangements. For the conventional motional narrowing, narrowing occurs when the correlation time τ_c and the rigid lattice line breadth (in frequency) $\delta\omega_\mathrm{RL}$ satisfy

$$\tau_\mathrm{c}\delta\omega_\mathrm{RL} \leq 1 \quad . \tag{8.129}$$

Without jumping, a set of spins precessing in phase initially get out of step in a time $\sim(1/\delta\omega_\mathrm{RL})$. The condition expresses the fact that for motional narrowing to occur, the jumping must occur before the dephasing can take place. In our example of motional narrowing, the correlation time would be on the order of the mean time spent in one of the three configurations of Fig. 8.11. If τ is the time in any one orientation, roughly

$$\tau_\mathrm{c} \approx 3\tau \quad . \tag{8.130}$$

In a similar way we expect that spin-flip narrowing will work only if the spins are flipped among the needed configurations before dephasing has occurred. Thus we get a condition on τ:

$$3\tau\delta\omega_\mathrm{RL} \leq 1 \quad . \tag{8.131}$$

The better the inequality is satisfied, the longer the spins will precess in phase.

We expect, therefore, that we shall wish to flip the spins again and again through the configuration of Fig. 8.12. This we can do by applying a given cycle of spin-flips repetitively. The basic cycle will bring the spins back to their starting point at the end of each cycle.

A formal description of what takes place begins with a specification of the Hamiltonian. We shall use it to compute the development of the wave function in time. We write the Hamiltonian as

$$\mathcal{H}(t) = \mathcal{H}_{\text{rf}}(t) + \mathcal{H}_{\text{int}} \quad . \tag{8.132}$$

We shall work in the rotating reference frame with the z-axis defined as the direction of the static field. $\mathcal{H}_{\text{rf}}(t)$ is the coupling to the applied rf pulses used to apply the spin rotations. It is time dependent because the pulses are switched on for very short intervals, t_w. In principle the corresponding H_1 can be oriented along the x-, y-, or z-axes. Selection of the x- versus the y-axis is a question of the phase of the rf pulse. Though pulses along the z-axis can be applied in principle, in practice they are not used since they would require addition of another coil to the rig. However, a z-axis rotation can be achieved by two successive rotations about the x- and y-axes. (Show how this is done!)

The term $\mathcal{H}_{\text{int}}$ consists of two terms,

$$\mathcal{H}_{\text{int}} = \mathcal{H}_0 + \mathcal{H}_{\text{d}}^0 \quad \text{where} \tag{8.133}$$

$$\mathcal{H}_0 = -\hbar(\omega_0 - \omega)I_z - \hbar\omega_0 \sum_i \sigma_{zzi} I_{zi} \quad \text{and} \tag{8.134}$$

$$\mathcal{H}_{\text{d}}^0 = \sum_{ij} B_{ij}(r_{ij})(\boldsymbol{I}_i \cdot \boldsymbol{I}_j - 3I_{zi}I_{zj}) \quad . \tag{8.135}$$

$\mathcal{H}_0$ includes the chemical and Knight shifts σ_{zzi}. $\mathcal{H}_{\text{d}}^0$ is the secular part of the dipolar coupling, the coefficients $B_{ij}(r_{ij})$ including the distance and angular factors [see (8.54a)].

The object of spin-flip narrowing is to cause $\mathcal{H}_{\text{d}}^0$ to vanish while maintaining $\mathcal{H}_0$ nonzero.

To illustrate the principles of spin-flip narrowing, it is useful to idealize the situation. Corrections to the idealization are important as a practical matter. We return to them in Sect. 8.13.

The idealization is to consider that $\mathcal{H}_{\text{rf}}$ is zero except for very short times during which it is so large that $\mathcal{H}_{\text{int}}$ can be neglected in comparison. This approximation enables us to say that $\mathcal{H}_{\text{rf}}$ produces spin rotations at the time a pulse is applied, but that between pulses the wave function $\psi(t)$ develops under the action of the time independent Hamiltonian $\mathcal{H}_{\text{int}}$.

Thus between pulses the wave function at time t can be related to its value at an earlier time t_1 by (2.49)

$$\begin{aligned}\psi(t) &= \exp[-(i/\hbar)\mathcal{H}_{\text{int}}(t-t_1)]\psi(t_1) \\ &= U_{\text{int}}(t-t_1)\psi(t_1)\end{aligned} \tag{8.136}$$

which defines U_{int}. The effect of pulse of amplitude H_1 and duration t_w along the α-axis ($\alpha = x, y, z$) at time t_1 is to produce a transformation by the unitary operator P_i for the ith pulse,

$$\psi(t_i^+) = e^{i\gamma H_1 t_w I_\alpha} \psi(t_i^-) \tag{8.137}$$

where, for example, $\gamma H_1 t_w$ is chosen as $\pi/2$ giving

$$P_i = e^{i(\pi/2)I_\alpha} \tag{8.138}$$

if the pulse produces a $\pi/2$ rotation about the $\alpha(= x, y, \text{or } z)$ axis.

Let us consider a three-pulse cycle for simplicity and concreteness. We always begin an experiment by having a sample which has reached thermal equilibrium in the magnet. We then tilt the magnetization into the $x-y$ plane at $t = 0$. We call that pulse the preparation pulse. After a time τ_0 we begin applying the repetitive pulse cycles. Choice of τ_0 and the phase of the preparation pulse is made in terms of producing a signal at some point in the pulse cycle which is convenient for observation (at a time which is called "the window" in the spin-flip literature).

Let $\psi(t_n)$ be the wave function after n cycles [i.e., just before pulse P_1 of the $(n + 1)$th cycle]. Then we can follow the wave function in time with Table 8.2.

Table 8.2. Variation of the wave function with time for a three-pulse sequence

Time	ψ
t_n^- (just before P_1)	$\psi = \psi(t_n)$
t_n^+ (just after P_1)	$\psi = P_1 \psi(t_n)$
$t_n + \tau_1^-$ (just before P_2)	$\psi = \exp(-i\mathcal{H}_{\text{int}} \tau_1/\hbar) P_1 \psi(t_n)$
	$= U_{\text{int}}(\tau_1) P_1 \psi(t_n)$
$t_n + \tau_1^+$ (just after P_2)	$\psi = P_2 U_{\text{int}}(\tau_1) P_1 \psi(t_n)$
$t_n + \tau_1 + \tau_2^-$ (just before P_3)	$\psi = U_{\text{int}}(\tau_2) P_2 U_{\text{int}}(\tau_1) P_1 \psi(t_n)$
$t_n + \tau_1 + \tau_2 + \tau_3^-$ [just before $(n+1)$th cycle]	$\psi(t_{n+1}) = U_{\text{int}}(\tau_3) P_3 U_{\text{int}}(\tau_2) P_2 U_{\text{int}}(\tau_1) P_1 \psi(t_n)$
	$\equiv U_{\text{T}} \psi(t_n)$ defining U_{T}

We can thus write

$$\psi(t_{n+1}) = U_{\text{T}} \psi(t_n) \tag{8.139}$$

where U_{T} is independent of n. Since U_{T} is a product of unitary operators, it is itself unitary. After N pulses, we have

$$\psi(t_N) = U_{\text{T}}^N \psi(t_0) \quad . \tag{8.140}$$

so that the problem can be considered solved if the effect of U_{T} can be deduced.

Let us examine one cycle, and introduce the unitary operators P_i^{-1} which are the inverses of the P_i's. For a unitary operator

$$P_i^{-1} = P_i^* \tag{8.141}$$

where the * stands for complex conjugate.

Then we write U_T as

$$\begin{aligned}U_T &= U_{\text{int}}(\tau_3)P_3 U_{\text{int}}(\tau_2)P_2 U_{\text{int}}(\tau_1)P_1 \\ &= P_3P_2P_1[P_1^{-1}P_2^{-1}P_3^{-1}U_{\text{int}}(\tau_3)P_3P_2P_1][P_1^{-1}P_2^{-1}U_{\text{int}}(\tau_2)P_2P_1] \\ &\quad \times [P_1^{-1}U_{\text{int}}(\tau_1)P_1] \quad .\end{aligned} \quad (8.142)$$

Now we showed in our example of Sect. 8.10 that a complete cycle of spin-flips should get us back to the starting point so that we can repetitively flip the spins among "on-axis" and "side" configurations. Therefore the cycle P_1, P_2, P_3 should get us back where we started. Hence

$$P_3P_2P_1 = 1 \quad (8.143)$$

giving us

$$\begin{aligned}U_T &= [P_1^{-1}P_2^{-1}P_3^{-1}U_{\text{int}}(\tau_3)P_3P_2P_1][P_1^{-1}P_2^{-1}U_{\text{int}}(\tau_2)P_2P_1] \\ &\quad \times [P_1^{-1}U_{\text{int}}(\tau_1)P_1] \end{aligned} \quad (8.144\text{a})$$

$$= U_{\text{int}}(\tau_3)[P_1^{-1}P_2^{-1}U_{\text{int}}(\tau_2)P_2P_1][P_1^{-1}U_{\text{int}}(\tau_1)P_1] \quad . \quad (8.144\text{b})$$

The meaning of the individual terms can be made evident by several transformations.

First, consider a unitary operator P and a Hamiltonian $\mathcal{H}$, and the corresponding U:

$$P^{-1}U(t-t_0)P = P^{-1}e^{-(\text{i}/\hbar)\mathcal{H}(t-t_0)}P \quad . \quad (8.145)$$

By expanding the exponential, inserting $P^{-1}P(=1)$ between factors, and regrouping, we find that

$$P^{-1}U(t-t_0)P = \exp\bigl[-(\text{i}/\hbar)(P^{-1}\mathcal{H}P)(t-t_0)\bigr] \quad (8.146)$$

so that the effect of $P^{-1}UP$ is to cause U to develop in time under a transformed Hamiltonian.

If we were to evaluate the expression in the exponent of the transformed $U_{\text{int}}(\tau_2)$,

$$P_1^{-1}P_2^{-1}\mathcal{H}_{\text{int}}(\tau_2)P_2P_1 \quad , \quad (8.147)$$

we could first do the transformation $P_2^{-1}\mathcal{H}_{\text{int}}P_2$, then sandwich the result between P_1^{-1} and P_1 and evaluate that, using the appropriate expressions for exponential operators. This order of application of the operators is the *reverse* of the order in time of application of the spin rotation operators. Why is that?

Consider a Schrödinger equation

$$-\frac{\hbar}{\text{i}}\frac{\partial \psi}{\partial t} = \mathcal{H}\psi \quad . \quad (8.148)$$

We can transform this equation with a unitary operator R, which is independent of time, to the problem

$$-\frac{\hbar}{i}\frac{\partial}{\partial t}(R\psi) = R\mathcal{H}R^{-1}(R\psi) \tag{8.149}$$

which leaves the *problem the same*.

Thus, if R were a spin rotation operator giving a $+\pi/2$ rotation of the spins about some axis, $R\mathcal{H}R^{-1}$ must transform the coordinates of the Hamiltonian corresponding to the same rotation.

If $\mathcal{H} = -\gamma\hbar H_0 I_z$ and the spin is in a state corresponding to the spin-up state, a transformation, R, which rotates the spin function into the up spin state along the $+y$-direction requires rotating $\mathcal{H}$ the same way, which is done by replacing I_z by I_y in $\mathcal{H}$.

We can therefore interpret

$$P^{-1}\mathcal{H}P = (P^{-1})\mathcal{H}(P^{-1})^{-1} \tag{8.150}$$

as the $\mathcal{H}$ corresponding to the rotation P^{-1}. We can consider the following two descriptions of the effect of $\mathcal{H}_d^0$ following a $\pi/2$ rotation:

i) Use $\mathcal{H}_d^0$ untransformed acting on a ψ which is rotated $+\pi/2$

or

ii) Leave the spin function alone but rotate the spin coordinates in $\mathcal{H}_d^0$ by $-\pi/2$, the inverse of the spin rotation in (i). Equation (8.144) corresponds to (ii).

If a rotation is made up of several rotations in succession, the inverse consists of the inverse rotations performed in the opposite order. Thus, if the spin is flipped by $P_2 P_1$, the inverse transformation Q is $(P_2 P_1)^{-1}$, so the $\mathcal{H}$ transformed by the inverse is

$$Q\mathcal{H}Q^{-1} = (P_2 P_1)^{-1}\mathcal{H}P_2 P_1 = P_1^{-1}P_2^{-1}\mathcal{H}P_2 P_1 \quad . \tag{8.151}$$

The prescription for finding the transformed $\mathcal{H}$ corresponding to the ith interval τ_i of an n pulse sequence is to take the spin rotations $P_1, P_2 \ldots \ldots P_i$, which preceded the interval and transform the coordinates in the Hamiltonian by applying the inverse rotations in the reverse sequence (e.g., first P_i^{-1}, then $P_{i-1}^{-1}, \ldots \ldots$, lastly P_1^{-1}).

We now define the three transformed Hamiltonians $\mathcal{H}_A$, $\mathcal{H}_B$, and $\mathcal{H}_C$ as

$$\begin{aligned}\mathcal{H}_A &= P_1^{-1}\mathcal{H}_{\text{int}}P_1 \\ \mathcal{H}_B &= P_1^{-1}P_2^{-1}\mathcal{H}_{\text{int}}P_2 P_1 \\ \mathcal{H}_C &= \mathcal{H}_{\text{int}} \quad . \end{aligned} \tag{8.152}$$

On expanding the exponentials we get

$$U_T = \exp\left(-\frac{i}{\hbar}\mathcal{H}_C\tau_3\right)\exp\left(-\frac{i}{\hbar}\mathcal{H}_B\tau_2\right)\exp\left(-\frac{i}{\hbar}\mathcal{H}_A\tau_1\right)$$

$$= 1 - \frac{i}{\hbar}(\mathcal{H}_C\tau_3 + \mathcal{H}_B\tau_2 + \mathcal{H}_A\tau_1)$$

$$- \left(\frac{i}{\hbar}\right)^2(\mathcal{H}_C\tau_3\mathcal{H}_B\tau_2 + \mathcal{H}_C\tau_3\mathcal{H}_A\tau_1 + \mathcal{H}_B\tau_2\mathcal{H}_A\tau_1)$$

$$+ \frac{1}{2}\left(\frac{i}{\hbar}\right)^2(\mathcal{H}_C^2\tau_3^2 + \mathcal{H}_B^2\tau_2^2 + \mathcal{H}_A^2\tau_1^2) + \cdots . \quad (8.153)$$

If the τ's are short,

$$\left|\left\langle\left|\frac{\mathcal{H}_{A,B,C}}{\hbar}\right|\right\rangle\right|\tau_i \ll 1 \quad (8.154)$$

where by $|\langle|\mathcal{H}_{A,B,C}/\hbar|\rangle|$ we mean a quantity of the magnitude of typical matrix elements of the transformed Hamiltonians. Under these circumstances, U_T is well approximated by keeping only the leading two terms on the right of (8.153). The condition on τ_i is similar to the requirement on the correlation time τ for there to be motional narrowing.

Introducing the period of a cycle, $t_c = \tau_1 + \tau_2 + \tau_3$, we get

$$U_T = 1 - \frac{i}{\hbar}\left(\mathcal{H}_C\frac{\tau_3}{t_c} + \mathcal{H}_B\frac{\tau_2}{t_c} + \mathcal{H}_A\frac{\tau_1}{t_c}\right)t_c \quad (8.155a)$$

$$= 1 - \frac{i}{\hbar}\overline{\mathcal{H}_{\text{int}}}t_c \quad (8.155b)$$

$$\cong \exp\left(-\frac{i}{\hbar}\overline{\mathcal{H}_{\text{int}}}t_c\right) \quad (8.155c)$$

where (b) defines $\overline{\mathcal{H}_{\text{int}}}$, the average $\mathcal{H}_{\text{int}}$, and (c) serves to remind us that over a cycle the system develops to a good approximation as though $\mathcal{H}_{\text{int}}$ were replaced by its average $\overline{\mathcal{H}_{\text{int}}}$, defined above.

Now we expand $\mathcal{H}_{\text{int}}$ into its elements

$$\mathcal{H}_{\text{int}} = -\hbar(\omega_0 - \omega)I_z - \hbar\omega_0\sum_i \sigma_{zzi}I_{zi} + \mathcal{H}_d^0$$

$$\equiv \mathcal{H}_0 + \mathcal{H}_d^0 . \quad (8.156)$$

The trick then is to choose the pulse cycle (P_1, P_2, P_3, etc.) so that we eliminate the dipolar coupling but maintain the chemical shift and Knight shift information:

$$\overline{\mathcal{H}_d^0} = 0 \quad (8.157a)$$

$$\overline{-\hbar(\omega_0 - \omega)I_z - \hbar\omega_0\sum_i \sigma_{zzi}I_{zi}} \neq 0 . \quad (8.157b)$$

Thus if

$$(\mathcal{H}_d^0)_A = \sum_{i<j} B_{ij}(\mathbf{I}_i \cdot \mathbf{I}_j - 3I_{yi}I_{yj})$$

$$(\mathcal{H}_d^0)_B = \sum_{i<j} B_{ij}(\mathbf{I}_i \cdot \mathbf{I}_j - 3I_{xi}I_{xj})$$

$$(\mathcal{H}_d^0)_C = \sum_{i<j} B(\mathbf{I}_i \cdot \mathbf{I}_j - 3I_{zi}I_{zj}) \tag{8.158}$$

and $\tau_1 = \tau_2 = \tau_3$,

$$\overline{\mathcal{H}_d^0} = \tfrac{1}{3}[(\mathcal{H}_d^0)_A + (\mathcal{H}_d^0)_B + (\mathcal{H}_d^0)_C] = 0 \quad . \tag{8.159}$$

These results could be achieved if P_1^{-1} were equivalent to a $\pi/2$ rotation about the x-axis which would transform I_{zi} into I_{yi} and if $P_1^{-1}P_2^{-1}$ were a $\pi/2$ rotation about the y-axis which would transform I_{zi} and I_{xi}.

These pulses would also transform $\mathcal{H}_0$

$$(\mathcal{H}_0)_A = -\hbar(\omega_0 - \omega)I_y - \hbar\omega_0 \sum_i \sigma_{zzi}I_{yi}$$

$$(\mathcal{H}_0)_B = -\hbar(\omega_0 - \omega)I_x - \hbar\omega_0 \sum_i \sigma_{zzi}I_{xi} \tag{8.160}$$

$$(\mathcal{H}_0)_C = -\hbar(\omega_0 - \omega)I_z - \hbar\omega_0 \sum_i \sigma_{zzi}I_{zi} \quad .$$

Setting $\omega = \omega_0$ for simplicity, we would then get

$$\mathcal{H}_0 = -\frac{\hbar\omega_0}{3}\sum_i \sigma_{zzi}(I_{xi} + I_{yi} + I_{zi})$$

$$= -\frac{\hbar\omega_0}{\sqrt{3}}\sum_i \sigma_{zzi}I_{z'i} \quad \text{where} \tag{8.161}$$

$$I_{z'i} = \frac{1}{\sqrt{3}}(I_{xi} + I_{yi} + I_{zi}) \quad . \tag{8.162}$$

Equation (8.162) shows that the chemical shift and Knight shift are reduced by $\sqrt{3}$ (i.e., multiplied by $1/\sqrt{3}$) from their value without averaging. We have succeeded in making the dipolar coupling vanish without eliminating the chemical shift. To the extent that the term in $(\omega - \omega_0)$ is included (it represents frequency offset or field inhomogeneity), it is also reduced in the same proportion.

According to (8.139) and (8.155c), $\psi(t_0)$ and $\psi(t)$ are given by

$$\begin{aligned}\psi(t) &= U_T^N \psi(t_0) \\ &= \exp\bigl(-(i/\hbar)\overline{\mathcal{H}_{\text{int}}}(t-t_0)\bigr)\psi(t_0)\end{aligned} \tag{8.163}$$

provided

$$t = t_0 + Nt_c \quad . \tag{8.164}$$

How about $\psi(t)$ at other times? Consider a time t in the Nth interval such that

$$t = t_0 + Nt_c + t_1 \quad , \quad 0 < t_1 < t_c \quad . \tag{8.165}$$

The time t_1 could fall into any one of the time intervals $\tau_1, \tau_2, \ldots \tau_i$ which make up the basic cycle t_c. Over any one *complete* cycle there is no large change in ψ if (8.152) is true. However, each large pulse produces a sudden big spin change – typically a $\pi/2$ rotation. These large pulses might, for example, progressively cycle the spin along the x-, y-, and z-axis in the rotating frame. If one always observes during the ith interval, however, the effect of the big pulses in successive cycles will always have returned the nuclear magnetization to the same direction in the rotating frame.

Thus we can write

$$\psi(t) = \exp\left(-(i/\hbar)\overline{\mathcal{H}_{\text{int}}}(t-t_0)\right)\psi(t_0) \tag{8.166}$$

when t and t_0 are both within the same subinterval τ_i of t_c.

If they are within different intervals τ_i and τ_j, with

$$Nt_c < t - t_0 < (N+1)t_c \tag{8.167}$$

we should properly go back to expressions like (8.136) and (8.137). If we consider that the individual τ's are so short that (8.154) is true, then $\mathcal{H}_{\text{int}}$ does not produce much of a change during t_c, but the pulses $i+1, i+2, i+3, \ldots, j-1, j$ do. Thus, defining

$$P \equiv P_j P_{j-1} \ldots, P_{i+2} P_{i+1} \quad , \tag{8.168}$$

there are two expressions which are nearly equal:

$$P \exp\left(-(i/\hbar)\overline{\mathcal{H}_{\text{int}}}Nt_c\right)\psi(t_0)$$
$$\cong P \exp\left(-(i/\hbar)\overline{\mathcal{H}_{\text{int}}}(N+1)t_c\right)\psi(t_0) \quad . \tag{8.169}$$

Since $Nt_c < t - t_0 < (N+1)t_c$ we can write

$$\psi(t) \cong P \exp\left(-(i/\hbar)\mathcal{H}_{\text{int}}(t-t_0)\right)\psi(t_0) \tag{8.170}$$

relating $\psi(t_0)$ in the ith interval to $\psi(t)$ in the jth interval of a pulse.

Equation (8.167) says in essence to compute the wave function evolution between t and t_0 as though $\mathcal{H}_{\text{int}}$ acted the whole time, and was followed by the rotations $P_{i+1}, \ldots, P_{j-1}, P_j$ in quick succession.

8.12 Observation of the Spin-Flip Narrowing

How can one experimentally observe the effect of spin-flip narrowing? Ordinary motional narrowing can be seen either by steady-state or pulsed apparatus as very narrow lines in the former case, and as long Bloch decays or slowly decaying echoes in the latter.

To produce spin-flip narrowing large rf pulses must be applied to the sample. Assuming one has the equipment to apply such pulses, one could then in principle add either a steady-state or a pulsed spectrometer to observe the character of the resulting resonance. In practice, the amplifiers of most steady-state apparatus would be blocked by a large rf pulse at their input, with a recovery from blocking which would be much slower than the short time τ between spin-flipping pulses. Consequently, to see spin-flip narrowing one goes to NMR apparatus designed to handle intense rf pulses—one uses a pulse rig.

With coherent pulse apparatus, one picks out a particular component (M_x or M_y) of the magnetization in the rotating frame. If one is adjusted to detect M_x, then one finds in general that it is largest in one particular interval τ_i of the cycle, perhaps zero in others. One could start with the system in thermal equilibrium, apply a $\pi/2$ pulse about the x-axis to rotate the magnetization to the $x - y$ plane, and then start the spin-flip cycles. For simplicity of the discussion we assume we can apply $\pi/2$ rotations about the x-, y-, and z-axes in the spin-flip cycle, though in practice z-axis rotations are not used. A possible set of pulses to initiate and carry on spin-flip narrowing is shown in Fig. 8.13. In Fig. 8.13a we see M along the direction of the static magnetic field (the $+z$-direction). At $t = 0$, P_0 rotates M to the $+y$-direction. This is called the preparing pulse. At time τ_0 we initiate the spin-flip pulses P_1, P_2, P_3 which successively rotate M by $-\pi/2$ about the z-axis, $+\pi/2$ about the y-axis, and lastly $+2\pi/3$ about the $-\boldsymbol{i}+\boldsymbol{j}+\boldsymbol{k}$ axis. The rotation P_3 is deduced as being the inverse of $P_2 P_1$ and can also be thought of simply as $P_1^{-1} P_2^{-1}$ mathematically, or physically as the successive steps $-\pi/2$ about the y-axis followed by $+\pi/2$ about the z-axis. Note

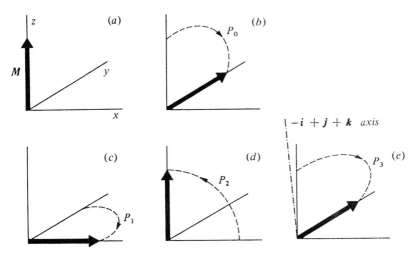

Fig. 8.13a-e. The effect of the various pulses as seen in the rotating frame. (a) The initial thermal equilibrium configuration with the magnetization M along the z-direction (the direction of the static field). (b) The rotation P_0 produced by the preparing pulse puts M along the y-axis. (c) The first pulse P of the spin-flip cycle rotates M to the $+x$-direction. (d) The second pulse P_2 rotates M to the $+z$-direction. (e) The third pulse P_3 must be the inverse of $P_2 P_1$. It is a rotation of $2\pi/3$ about the $(-\boldsymbol{i}+\boldsymbol{j}+\boldsymbol{k})$ axis. (Note this is *not* the same as a $-\pi/2$ rotation about the $+x$-direction)

that although P_3 looks superficially like a rotation of $-\pi/2$ about the x-axis, it is not, as can be verified if one considers rotating a three-dimensional object in P_1, P_2, P_3 instead of just a vector M pointing initially along the y-axis. This distinction is important because $\mathcal{H}_d$ is a second rank tensor; thus it should be thought of as an ellipsoid rather than as an arrow.

If the apparatus were adjusted to measure M_x, and if it were tuned exactly to resonance, including any possible chemical and Knight shifts, there would only be a nuclear resonance signal between pulses P_1 and P_2. If there were a resonance offset (as with a second class of nuclei chemically shifted from the first), M_x would gradually change in the P_1-to-P_2 interval for successive pulses, undergoing an oscillation (see Fig. 8.14) as we show below. It would appear during the P_3-to-P_1 window as well as time went on.

Let us turn to a mathematical formulation of the calculation of the observed NMR signal. For concreteness, we compute the time development of M_x in the rotating frame $<M_x(t)>$. We will use the language of wave functions rather than density matrix. Given the wave function $\psi(t)$, we have

$$<M_x(t)> \; = \; <\psi(t), M_x\psi(t)>$$
$$= \gamma\hbar <\psi(t), I_x\psi(\tau)> \quad . \tag{8.171}$$

We assume that we have a pulse sequence similar to that of Figs. 8.13 and 8.14. At $t = 0^-$ we let the wave function be $\psi(0^-)$; then the preparation pulse gives us

$$\psi(0^+) = P_0\psi(0^-) \quad . \tag{8.172}$$

The wave function then develops under $\overline{\mathcal{H}_{\text{int}}}$. If we ask for it in the interval P_3 to P_1, there will be an integral number of cycles P_1 followed by P_2 followed by P_3, so that, using (8.166),

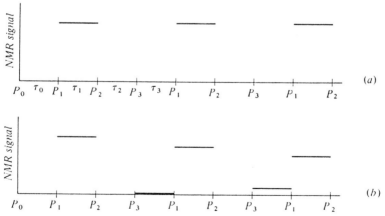

Fig. 8.14a,b. Graph of the oscilloscope signal of the resonance versus time for the pulse sequence when (**a**) apparatus is set exactly at resonance, or (**b**) a chemical shift causes the M_x signal to change with time appearing in the P_3-to-P_1 interval as it diminishes in the P_1-to-P_2 window

$$\psi(t) = e^{-(i/\hbar)\overline{\mathcal{H}_{\text{int}}}t} P_0 \psi(0^-) \quad . \tag{8.173}$$

If we ask for $\psi(t)$ later in the cycle, we use (8.170)

$$\psi(t) = P e^{-(i/\hbar)\overline{\mathcal{H}_{\text{int}}}t} P_0 \psi(0^-) \tag{8.174}$$

where $P = P_1$ or $P_1 P_2$ depending on whether t lies in the interval P_1-to-P_2 or P_2-to-P_3.

We now need to express the fact that the initial wave function $\psi(0^-)$ corresponds to the initial thermal equilibrium of the sample. To do so we express $\psi(0^-)$ in terms of the laboratory frame $\phi(0^-)$ instead of the rotating frame. In general the rotating frame wave function ψ and laboratory function ϕ are related by

$$\psi(t) = e^{-i\omega t I_z} \phi(t) \tag{8.175a}$$

so that at $t = 0^-$ we get

$$\psi(0^-) = \phi(0^-) \quad . \tag{8.175b}$$

Taking $|n\rangle$ to be the eigenstates of the secular Hamiltonian $\mathcal{H}$ in the laboratory frame, we then have

$$\phi(0^-) = \sum_n c_n(0^-) |n\rangle \quad . \tag{8.176}$$

Thus

$$\begin{aligned}
\langle M_x(t) \rangle &= \langle \psi(t), M_x \psi(t) \rangle \\
&= \gamma \hbar \sum_{n,m} c_n c_m^* \langle m | P_0^{-1} e^{(i/\hbar)\overline{\mathcal{H}_{\text{int}}}t} P^{-1} I_x P \\
&\quad \times e^{-(i/\hbar)\overline{\mathcal{H}_{\text{int}}}t} P_0 | n \rangle \quad .
\end{aligned} \tag{8.177}$$

We now make a statistical average over the coefficients $c_n c_m^*$ utilizing the fact that the states are occupied according to a Boltzmann distribution of the Hamiltonian $\mathcal{H}$, which describes the system before application of the pulses:

$$\overline{c_n c_m^*} = \delta_{nm} \frac{e^{-E_n/kT}}{Z} = \frac{\langle n | e^{-\mathcal{H}/kT} | m \rangle}{\text{Tr}\{e^{-\mathcal{H}/kT}\}} \quad . \tag{8.178}$$

Substituting (8.178) into (8.177) we get

$$\overline{\langle M_x(t) \rangle} = \frac{\gamma \hbar}{Z} \text{Tr}\left\{ P_0^{-1} e^{(i/\hbar)\overline{\mathcal{H}_{\text{int}}}t} P^{-1} I_x P e^{-(i/\hbar)\overline{\mathcal{H}_{\text{int}}}t} P_0 e^{-\mathcal{H}/kT} \right\} \quad . \tag{8.179}$$

In the high temperature approximation we can write

$$\frac{e^{-\mathcal{H}/kT}}{Z} = \frac{(1 - \mathcal{H}/kT)}{(2I+1)^{N_0}} \tag{8.180}$$

where there are N_0 spins of spin I. The leading term vanishes, so that the nonvanishing term gives a $1/kT$ contribution.

If we then write
$$\mathcal{H} = -\gamma \hbar H_0 I_z \quad , \tag{8.181}$$
transfer the P_0^{-1} from the left to the right in the trace, using
$$\mathrm{Tr}\{ABC\} = \mathrm{Tr}\{BCA\} \tag{8.182}$$
and if we pick P_0 to make
$$P_0 I_z P_0^{-1} = I_x \tag{8.183}$$
which rotates the spin from the z-axis to the x-axis, we get
$$\overline{<M_x(t)>} = \frac{\gamma^2 \hbar^2}{(2I+1)^{N_0}} \frac{H_0}{kT} \mathrm{Tr}\left\{ e^{(i/\hbar)\overline{\mathcal{H}_{\mathrm{int}}}t} P^{-1} I_x P e^{-(i/\hbar)\overline{\mathcal{H}_{\mathrm{int}}}t} I_x \right\} \quad . \tag{8.184}$$

A useful exercise for the reader is to derive (8.184) using the density matrix language. For simplicity we take $P = 1$ (i.e., an integral number of pulse cycles P_1, P_2, P_3 pass between $t = 0$ and the time of the measurement).

Now,
$$\overline{\mathcal{H}_{\mathrm{int}}} = -\frac{\hbar\omega_0}{3} \sum_i \sigma_{zzi}(I_{zi} + I_{xi} + I_{yi})$$
$$= -\frac{\hbar\omega_0}{3} \sum_i \sigma_{zzi} I_{z'i} \tag{8.185}$$

where $I_{z'i} = (1/\sqrt{3})(I_{xi} + I_{yi} + I_{zi})$. Suppose all spins have the same σ_{zz}. Then we get
$$\overline{\mathcal{H}_{\mathrm{int}}} = -\frac{\hbar\omega_0 \sigma_{zz} I_{z'}}{\sqrt{3}} \quad . \tag{8.186}$$

Now in terms of the prime coordinates we have, in an obvious notation,
$$I_x = I_{x'}\cos(x,x') + I_{y'}\cos(x,y') + I_{z'}\cos(x,z')$$
$$I_{x'} = I_x\cos(x,x') + I_y\cos(y,x') + I_z\cos(z,x') \tag{8.187}$$

so that
$$\exp\left(\frac{-i\omega_0 \sigma_{zz} t I_{z'}}{\sqrt{3}}\right) I_x \exp\left(\frac{i\omega_0 \sigma_{zz} t I_{z'}}{\sqrt{3}}\right)$$
$$= \cos(x,x')\left[I_{x'}\cos\left(\frac{\omega_0 \sigma_{zz} t}{\sqrt{3}}\right) + I_{y'}\sin\left(\frac{\omega_0 \sigma_{zz} t}{\sqrt{3}}\right)\right]$$
$$+ \cos(x,y')\left[-I_{x'}\sin\left(\frac{\omega_0 \sigma_{zz} t}{\sqrt{3}}\right) + I_{y'}\cos\left(\frac{\omega_0 \sigma_{zz} t}{\sqrt{3}}\right)\right]$$
$$+ \cos(x,z')I_{z'} \quad . \tag{8.188}$$

Using the fact that $\mathrm{Tr}\{I_{x'}I_{y'}\} = 0$, and $\mathrm{Tr}\{I_{x'}^2\} = \mathrm{Tr}\{I_{y'}^2\}$ we get, after evaluating the traces,

$$\overline{<M_x(t)>} = \frac{N_0\gamma^2\hbar^2 I(I+1)H_0}{3kT}\left[\frac{1}{3} + \frac{2}{3}\cos\left(\frac{\omega_0\sigma_{zz}t}{\sqrt{3}}\right)\right]$$
$$= M_0\left[\frac{1}{3} + \frac{2}{3}\cos\left(\frac{\omega_0\sigma_{zz}t}{\sqrt{3}}\right)\right] \quad (8.189)$$

where M_0 is the thermal equilibrium magnetization.

We see, therefore, that the chemical shift-Knight shift σ_{zz} shows up as an oscillatory contribution to the $\overline{<M_x(t)>}$, reduced in frequency by $1/\sqrt{3}$ from the normal frequency shift $\omega_0\sigma_{zz}$ one would see if one could resolve it in a steady-state resonance.

If more than one kind of chemical or Knight shift is present, the resultant $\overline{<M_x(t)>}$'s add, the respective strengths being in proportion to the number of nuclei possessing that chemical shift or Knight shift through the M_0 factor.

The form of (8.189) is that of a constant term plus an oscillatory term. By taking a Fourier transform one can deduce a "reduced chemical shift frequency" $(\omega_0\sigma_{zz}/\sqrt{3})$. It is possible to utilize Fourier transform NMR attachments therefore to decompose a transient signal when there exists more than one Knight shift-chemical shift to unscramble the discrete frequencies.

8.13 Real Pulses and Sequences

Our discussion up to now has been directed towards explaining the principles of spin-flip narrowing. Translating these notions into practice, we encounter certain problems. We discuss these briefly, then list a few of the pulse sequences which have been developed to cope with them. An excellent discussion of these problems has been given by *Haeberlen* in his NATO Summer School lectures.[1] *Vaughan* and colleagues [8.38, 8.39] give a thorough theoretical discussion including explicit formulas of use. See also the references in the Selected Bibliography.

8.13.1 Avoiding a z-Axis Rotation

Pulse rigs apply H_1's in the $x-y$ plane. By choosing the rf phase, a single one of these can produce rotations about any arbitrary axis lying in the $x-y$ plane, but not about the z-axis. To produce a z-axis rotation with a single pulse would require adding a coil with an axis parallel to the z-axis plus the electronics to energize it. In order to avoid so doing, one can go to a four-pulse technique. That in a sense is equivalent to composing a z-axis rotation from successive x-axis and y-axis rotations. Ordinarily, however, the two pulses are spaced apart in time so that the basic timing intervals can be maintained more simply.

---

[1] The 1974 NATO Advanced Study Institute held in Leuven, Belgium featured a number of useful lectures on NMR in solids. *Haeberlen* and *Van Hecke* in particular treat spin-flip narrowing in [8.40].

To write sequences, it is convenient to introduce a notation based on the fact that $\pi/2$ pulses are normally used. For example, consider the following expression which defines one four-pulse scheme.

$$X(\tau, \overline{X}, \tau, Y, 2\tau, \overline{Y}, \tau, X, \tau) \quad . \tag{8.190}$$

The first "X" represents the preparing pulse, a $\pi/2$ pulse about the $+x$-axis. The parentheses enclose the pulses of a single cycle, which in the example indicated is considered to start with a time interval τ prior to application of a $\pi/2$ pulse about the $-x$-axis (or a $-\pi/2$ pulse about the $+x$-axis). This sequence could also be written as

$$X, \tau(\overline{X}, \tau, Y, 2\tau, \overline{Y}, \tau, X, 2\tau) \quad ,$$

which simply chooses to call the starting point in the cycle as the time just prior to the first $-x$ pulse in the chain. This notation makes explicit the presence of two intervals of 2τ and two of τ. The first notation is useful in proving certain theorems which utilize the time symmetry of a single pulse cycle.

8.13.2 Nonideality of Pulses

Pulse Duration. We have utilized instantaneous rotations in the discussion up to now, which would require an infinite H_1 of zero duration. Actual pulses have nonzero duration. One's first guess might be that there would be major problems as a result of the nonzero duration. Theoretical investigation of this point reveals, however, that it is not serious [8.34, 35, 38, 39]. The point is that *slow* rotations can still maintain the property of averaging out the dipolar coupling. They are in this sense equivalent in motional narrowing to allowing $(3\cos^2\theta - 1)$ to average to zero over a *continuous* variation in θ as opposed to the discrete variation of Fig. 8.1. The important experimental consequence is that much lower rf power is needed that one would at first suppose.

Inhomogeneous H_1. As we discussed in Sect. 8.3 it is always difficult to make H_1 homogeneous over a sample, particularly when one is attempting to keep the rf coil small to cut down on the power requirements (for a given Q and H_1, the power needed is proportional to coil volume). This problem is a bad one in that it causes the rotation angles to vary throughout the sample. It can be removed to first order, however, by using the trick of *Meiboom* and *Gill* or of pairing positive and negative rotations about any given axis, thereby achieving a refocusing effect.

Phase Modulation. Turning on an H_1 which points along a given direction in the rotating frame sounds easy, but in practice is hard to achieve. Usually the phase of the H_1 swings during the buildup and during the decay. Such swings can be labeled as modulations of the phase angle ϕ of the rf. They are equivalent to being off-resonance $\Delta\omega = d\phi/dt$ during the time that $d\phi/dt \neq 0$ [8.41]. Their effect can be eliminated to first order by having $d\phi/dt$ during turn-on and $d\phi/dt$ during turn-off equal and opposite. Making $d\phi/dt$ zero is not practical, but the balancing during turn-on and turn-off is.

Among the various pulse sequences used or referred to in the literature, we shall list three. They are known as WAHUHA, HW-8, and REV-8. These mysterious symbols are made up from initials of their inventors plus numbers which tell how many pulses there are per cycle.

The WAHUHA is named after *J.S. Waugh, L.M. Huber,* and *U. Haeberlen* [8.34]. It is

$$X, \tau(\tau, \overline{X}, \tau, Y, 2\tau, \overline{Y}, \tau, X, \tau) \ .$$

This sequence can be sampled at the times corresponding to just before the parenthesis by measuring M_y. The name WAHUHA refers to the contents of the parentheses. The preparation pulse can be chosen to give convenient sets of timing. The particular sequences shown have the property that the X, τ, at the start are in fact like the last pulse and interval of the cycle; hence no special timing intervals are needed at the start. That is, the preparing pulse in this case is part of the cycle.

For the preparing pulse shown, the magnetization is initially turned to the y-direction. Thus, if one detects M_y at the inverval between the X and $-X$ pulses, one will get a signal.

We can write the HW-8 [8.35] as

$$(\tau, \overline{X}, 2\tau, X, \tau, \overline{Y}, 2\tau, Y, \tau, Y, 2\tau, \overline{Y}, \tau, X, 2\tau, \overline{X}, \tau) \ .$$

There are a number of equivalent variations.

The MREV-8 [8.38] sequence is

$$X, \tau(\tau, X, \tau, Y, 2\tau, \overline{Y}, \tau, \overline{X}, 2\tau, \overline{X}, \tau, Y, 2\tau, \overline{Y}, \tau, X, \tau) \ .$$

This can be viewed as pair of pulses X, Y separated by τ followed at 2τ later by their inverse $\overline{Y}, \overline{X}$.

MREV demonstrate the highest resolution obtained with their MREV-8.

8.14 Analysis of and More Uses for Pulse Sequence

The importance and utility of multipulse techniques goes far beyond spin-flip line narrowing. We turn now to several examples, as well as to illustrations of some calculations of the effects of the pulse techniques.

Before beginning our discussion, it is useful to introduce a notation and to demonstrate some useful relationships. We shall be following the effect of pulses on individual components of spins such as I_x, I_y, or I_z and on dipolar terms. For the dipolar term we introduce the notation $\mathcal{H}_{\alpha\alpha}(\alpha = x, y, z)$ defined as

$$\mathcal{H}_{\alpha\alpha} = \sum_{j>k} B_{ik}(\boldsymbol{I}_j \cdot \boldsymbol{I}_k - 3I_{\alpha j}I_{\alpha k}) \quad . \tag{8.191}$$

Explicit calculation then shows that

$$\mathcal{H}_{xx} + \mathcal{H}_{yy} + \mathcal{H}_{zz} = 0 \quad . \tag{8.192}$$

In general, the effect of pulses P_i will be rotations. To discuss the effects of nonzero pulse lengths, one must treat the P_i's as time-dependent operators such as

$$e^{-i\omega_1 t I_x} \tag{8.193}$$

where t is the time variable during the pulse. For our present purposes, however, we shall treat the pulses as negligible duration (so-called δ-function pulses) and moreover treat them as all being $\pi/2$ pulses.

We then have to evaluate expressions such as

$$P_1^{-1} I_z P_1 \tag{8.194a}$$

$$P_1^{-1} H_{zz} P_1 \quad . \tag{8.194b}$$

But once we know what (8.194a) is, it is easy to get (8.194b). For example, let

$$P_1 = X \quad \text{then} \tag{8.195a}$$

$$P_1^{-1} = \overline{X} \quad \text{then} \tag{8.195b}$$

$$\overline{X} I_z X = -I_y$$

$$\begin{aligned}
\overline{X} \mathcal{H}_{zz} X &= \sum_{j>k} B_{ij} (\overline{X} \boldsymbol{I}_j \cdot \boldsymbol{I}_k X - 3\overline{X} I_{zj} I_{zk} X) \\
&= \sum_{j>k} B_{ij} (\boldsymbol{I}_j \cdot \boldsymbol{I}_k - 3\overline{X} I_{zj} X \overline{X} I_{zk} X) \\
&= \sum_{j>k} B_{ij} (\boldsymbol{I}_j \cdot \boldsymbol{I}_k - 3 I_{yj} I_{yk}) \\
&= \mathcal{H}_{yy} \quad .
\end{aligned} \tag{8.196}$$

Thus, if

$$P_1^{-1} I_\alpha P_1 = k I_\beta \quad , \quad k = \pm 1 \quad , \quad \text{then} \tag{8.197}$$

$$P_1^{-1} H_{\alpha\alpha} P_1 = H_{\beta\beta}$$

whether $k = +1$ or -1.

The effect of a (left-handed) rotation P_1 in the expression

$$P_1^{-1} I_\alpha P_1 \tag{8.198}$$

can be worked out once one knows P_1 either by first computing P_1^{-1} (e.g. if

$P_1 = X$ then $P_1^{-1} = \overline{X}$) and then using the left-handed rotations *or* by instead pretending that one is calculating

$$P_1 I_\alpha P_1^{-1} \tag{8.199a}$$

using right-handed rotations. For a pair of pulses P_1 followed by P_2, then the left-handed rotations

$$P_1^{-1} P_2^{-1} I_\alpha P_2 P_1 \tag{8.199b}$$

would be replaced by the right-handed rotations P_1, P_2, etc.

$$P_1 P_2 I_\alpha P_2^{-1} P_1^{-1} \quad. \tag{8.199c}$$

Using either approach, one can work out a table for (8.198) which is then useful for all applications in which we list I_α across the top and in each row list first the operator P_1, then the effect of P_1 on I_α. The result in the table is the same whether we treat P_1 as a left-handed rotation and use (8.198), or a right-handed rotation and use (8.199a).

Rotation operator	Operator being rotated		
	I_x	I_y	I_z
X	I_x	I_z	$-I_y$
Y	$-I_z$	I_y	I_x
Z	I_y	$-I_x$	I_z

(8.200)

or expressed as coordinates x, y, z

	x	y	z
X	x	z	$\overline{y}$
Y	$\overline{z}$	y	x
Z	y	$\overline{x}$	z

(8.201)

where the bar indicates a negative. Suppose then that in the sense of (8.198) one has the operator H_{yy} acted on by $P_1 = X$. Then the table shows us that

$$\overline{X} H_{yy} X = H_{zz} \quad. \tag{8.202}$$

Let us now illustrate this with a calculation of the effect of a very useful pulse sequence utilized by *Warren* et al. [8.42] to refocus dipolar dephasing. They employ it as part of their method of generating multiple quantum coherence of a particular order, a topic we discuss in the next chapter. In discussing magic echoes, we showed that one could refocus the dephasing arising from dipolar coupling by creating a dipolar coupling of negative sign. In particular we showed how we could create a dipolar Hamiltonian $\mathcal{H}_d$ given by

$$\mathcal{H}_d = -\tfrac{1}{2}\mathcal{H}_{xx} \quad . \tag{8.203}$$

We now show how *Warren* et al. created such an $\mathcal{H}_d$ by pulses.
The basic approach is founded on (8.192)

$$\mathcal{H}_{xx} + \mathcal{H}_{yy} + \mathcal{H}_{zz} = 0 \quad .$$

Thus

$$-\mathcal{H}_{xx} = \mathcal{H}_{yy} + \mathcal{H}_{zz} \quad . \tag{8.204}$$

Therefore, if we create a Hamiltonian which spends equal time as $\mathcal{H}_{yy}$ and $\mathcal{H}_{zz}$, its average value will be $\overline{\mathcal{H}_{xx}}/2$.

Since without pulses $\mathcal{H}_d = \mathcal{H}_{zz}$, we simply need pulses to switch $\mathcal{H}_{zz}$ to $\mathcal{H}_{yy}$. Clearly, all we need is X's or $\overline{X}$'s for the P_i's. The simplest thing one can do is to switch $\mathcal{H}_{zz}$ to $\mathcal{H}_{yy}$ with $P_1 = X$, then switch it back, $P_2 = \overline{X}$, so that

$$P_2 P_1 = 1 \tag{8.205a}$$

for a sequence

$$(X, \tau, \overline{X}, \tau) \quad . \tag{8.205b}$$

To follow the action, we make a table. We need to follow two terms in the Hamiltonian: the dipolar coupling and the chemical shift. If we simply wish to create a negative $\mathcal{H}_d$ for all spins without regard to their chemical shifts, we need a pulse sequence which will make all chemical shifts zero. We will find that our simple $(X, \tau, \overline{X}, \tau)$ sequence has problems with chemical shifts. We construct a table in which time progresses downwards (Table 8.3).

Table 8.3. Sequence $(X, \tau, \overline{X}, \tau)$

	Duration	P_i	$P_1 \ldots P_i$	I_z	$\mathcal{H}_{zz}$
		$P_1 = X$	X		
	1			$\overline{y}$	yy
		$P_2 = \overline{X}$	$X\overline{X} = 1$		
	1			z	zz
Sum	2			$(-I_y + I_z)$	$\mathcal{H}_{yy} + \mathcal{H}_{zz}$
Average				$\tfrac{1}{2}(I_z - I_y)$	$\tfrac{1}{2}(\mathcal{H}_{yy} + \mathcal{H}_{zz})$

In Table 8.3, the "duration" column is measured in units of τ. The column $P_1 \ldots P_i$ lists $P_1, P_1P_2, P_1P_2P_3$, etc. as in (8.199c). Remember, the first operator to apply is the right hand one, but its effect is to be computed with right-handed rotations.

On line 1 we list P_1. Then on the second line under I_z and $\mathcal{H}_{zz}$ we list the operators produced by the action of P_1 on I_z and $\mathcal{H}_{zz}$, as well as the duration of the time between application of P_1 and the application of the next pulse P_2. We have omitted the "I" and the "$\mathcal{H}$" since only the subscript and sign is needed to specify the entry. On the third line we list P_2, then the product $P_2 P_1$, which we simplify to the extent possible. The last line labeled "sum" contains the sum of the time intervals, and the sums of the products of the operators with the time intervals for which each operator acts. The average values are then just the weighted sums of the operators divided by the duration. Using (8.204) we see that the pulse sequence does produce an average Hamiltonian

$$\overline{\mathcal{H}}_\mathrm{d} = -\tfrac{1}{2}\mathcal{H}_{xx} \quad . \tag{8.206}$$

Of course, the pulse sequence must be repeated continuously for as long as one wants to satisfy (8.206).

A pulse sequence which gives the same average dipolar Hamiltonian but eliminates the chemical shift term and has some other advantages is given by *Warren* et al. [8.42]:

$$X, \tau, X, \tau, \overline{X}, \tau, \overline{X}, \tau, \overline{X}, \tau, \overline{X}, \tau, X, \tau, X, \tau \quad .$$

We evaluate it in Table 8.4 utilizing (8.201).

Table 8.4. Sequence $(X, \tau, X, \tau, \overline{X}, \tau, \overline{X}, \tau, \overline{X}, \tau, \overline{X}, \tau, X, \tau, X, \tau)$

	Duration	P_i	$P_1 \ldots P_i$	I_z	$\mathcal{H}_{zz}$
		X	X		
	1			$\overline{y}$	yy
		X	X^2		
	1			$\overline{z}$	zz
		$\overline{X}$	X		
	1			$\overline{y}$	yy
		$\overline{X}$	1		
	1			z	zz
		$\overline{X}$	$\overline{X}$		
	1			y	yy
		$\overline{X}$	$\overline{X}^2$		
	1			$\overline{z}$	zz
		X	$\overline{X}$		
	1			y	yy
		X	1		
	1			z	zz
Sum	8			0	$4(\mathcal{H}_{zz} + \mathcal{H}_{yy})$
Average				0	$\tfrac{1}{2}(\mathcal{H}_{zz} + \mathcal{H}_{yy})$

The same approach can be used to analyze the spin-flip line narrowing sequences discussed in Sect. 8.13. For example, in Table 8.5 we analyze the four-pulse sequence $(\overline{X}, \tau, Y, 2\tau, \overline{Y}, \tau, X, 2\tau)$. As remarked in Sect. 8.13.1, this pulse sequence is equivalent to the three-pulse average we employed (rotations about the x-, y-, and z-axes) in Sect. 8.12. Utilizing Table 8.5 we see it produces a zero average dipolar Hamiltonian.

Table 8.5. Sequence $(\overline{X}, \tau, Y, 2\tau, \overline{Y}, \tau, X, 2\tau)$

	Duration	P_i	$P_1 \ldots P_i$	I_z	$\mathcal{H}_{zz}$
	1		$\overline{X}$	y	yy
	2	Y	$\overline{X}Y$	x	xx
	1	$\overline{Y}$	$\overline{X}Y\overline{Y} = \overline{X}$	y	yy
	2	X	$\overline{X}Y\overline{Y}X = 1$	z	zz
Sum	6			$2(I_x + I_y + I_z)$	$2(\mathcal{H}_{xx} + \mathcal{H}_{yy} + \mathcal{H}_{zz})$
Average				$\frac{1}{3}(I_x + I_y + I_z)$	$\frac{1}{3}(\mathcal{H}_{xx} + \mathcal{H}_{yy} + \mathcal{H}_{zz})$

For multiple quantum excitation, *Warren* et al. [8.42] show that it is useful to generate an effective operator proportional to $\mathcal{H}_{yy} - \mathcal{H}_{xx}$. In particular, they find

$$\tfrac{1}{2}(\mathcal{H}_{yy} - \mathcal{H}_{xx}) = \tfrac{1}{2}(2\mathcal{H}_{yy} + \mathcal{H}_{zz}) \tag{8.207a}$$

is generated by the eight-pulse sequence

$$(X, 2\tau, X, \tau, \overline{X}, 2\tau, \overline{X}, \tau, \overline{X}, 2\tau, \overline{X}, \tau, X, 2\tau, X, \tau) \tag{8.207b}.$$

After a bit of experience in calculating tables, one soon discovers a problem. Suppose one has just calculated the effect of the operator

$$P_1 P_2 \ldots P_n \equiv Q \tag{8.208}$$

acting to the right. Suppose it produces a result I_α in the I_z column. We next have to calculate the effect of the operator

$$P_1 P_2 \ldots P_n P_{n+1} \quad . \tag{8.209}$$

Now it would be *easy* if the operator we had to calculate was $P_{n+1}Q$ since then

$$P_{n+1} Q I_z = P_{n+1} I_\alpha \quad . \tag{8.210}$$

This situation would let us take the result of Q, which we already have, and apply P_{n+1} to it. However, what we must calculate is instead

$$Q P_{n+1} \quad ,$$

which has the P_{n+1} operate *before* Q operates. Therefore, we must go through all $n+1$ operator calculations, a tedious process for long pulse squences. Now suppose

$$P_{n+1} I_z = I_\beta \quad . \tag{8.211}$$

Then

$$Q P_{n+1} I_z = Q I_\beta \quad . \tag{8.212}$$

So if we knew what QI_β was, we could immediately evaluate (8.212). Since β will be x, y, or z (or $\bar{x}, \bar{y}, \bar{z}$), we can handle any eventuality if we know the results of QI_x, QI_y, and QI_z. A procedure which will provide us with these three results is to replace the single column I_z with three columns I_x, I_y, I_z. A portion of the table would then look like Table 8.6.

Table 8.6. Sample table to help in evaluating the effect of QP_{n+1}

Line		I_x	I_y	I_z
1	$P_1 \ldots P_{n-1}$			
2		$I_{\beta''}$	$I_{\beta'}$	I_β
3	$P_1 \ldots P_n (= Q)$			
4		$I_{\alpha''}$	$I_{\alpha'}$	I_α
5	$P_1 \ldots P_n P_{n+1}$			
6				

To compute line 6, we first find the effect of P_{n+1} on I_x, I_y, and I_z. There will be three results. We look for them on line 2. Thus, suppose

$$P_{n+1} I_z = I_{\beta'} \ .$$

Then

$$P_1 \ldots P_n I_{\beta'} = I_{\alpha'} \ .$$

So we would enter $I_{\alpha'}$ on line 6 under the I_z column.

This approach requires evaluating three columns, but it makes the step of going from one line to the next much simpler – one has only to evaluate two operator steps per column, hence six operator steps per line.

9. Multiple Quantum Coherence

9.1 Introduction

In Chapter 5 we introduced the density matrix as a powerful tool for the analysis of magnetic resonance experiments. In analyzing the case of spin $\frac{1}{2}$, we saw that the diagonal elements of ϱ were connected with the magnetization parallel to the static field, and the off-diagonal elements were related to the transverse components through the equations

$$\langle M^+(t)\rangle = \gamma\hbar\varrho_{-+}(t) \quad \text{and} \tag{5.257a}$$

$$\langle M_z(t)\rangle = \frac{\gamma\hbar}{2}[\varrho_{++}(t) - \varrho_{--}(t)] \quad . \tag{5.257b}$$

In a system with two spins I and S we saw that

$$\langle I^+(t)\rangle = \sum_{m_S}(-m_S|\varrho(t)|+m_S) \tag{7.248}$$

so that the transverse magnetization of the I-spins arises from elements of the density matrix which are off-diagonal in the I quantum numbers, but diagonal in the S quantum numbers. There are other off-diagonal elements of ϱ, for example

$$(-+|\varrho(t)|+-) \quad \text{or} \tag{9.1a}$$

$$(--|\varrho(t)|++) \quad . \tag{9.1b}$$

It is a fundamental tenet of statistical mechanics that for a system in thermal equilibrium such terms are zero, using the hypothesis of random phases described in the text just below (5.87). This hypothesis is based on two ideas. The first is that all observables of a system in thermal equilibrium must be time independent. This requirement really expresses the meaning of "equilibrium". It is a necessary (though not sufficient) condition for thermal equilibrium that all observables be time independent.

The second idea is that since the time dependence of ϱ comes only in the off-diagonal elements, there is some experimental means for probing every off-diagonal element of ϱ. If there were not, the existence of off-diagonal elements would not produce time-dependent observables, to conflict with the hypothesis of equilibrium. Since we have just seen that resonance experiments do not directly detect elements such as in (9.1), we naturally wonder if there is any experimental method for probing the existence of such matrix elements. A closely related question is whether there is any way to excite such matrix elements if they are initially zero.

We are already in a position to say that in certain cases we can excite these elements because indeed we have analyzed in detail an experiment which generates them. Referring to our analysis in Sect. 7.26 of the pulse sequence

$$X_I(\pi/2)\ldots t_1 \ldots X_S(\pi/2)\ldots t_2 \quad , \tag{9.2}$$

we see that in (7.338) we produced terms of ϱ which, expressing ϱ in operator form, are proportional to

$$I_x S_y \quad . \tag{9.3}$$

These gave nonvanishing elements of ϱ

$$(\pm\pm|\varrho(t_1+t_2)|\mp\mp) \quad \text{and} \tag{7.341a}$$

$$(\pm\mp|\varrho(t_1+t_2)|\pm\mp) \quad , \tag{7.341b}$$

where in each matrix element the upper signs go together, and the lower signs go together.

The matrix elements (7.341a) join states of energy difference, ΔE, given by

$$\Delta E = \pm \hbar(\omega_{0I} + \omega_{0S}) \tag{9.4}$$

and the matrix elements of (7.341b) join states differing in energy by

$$\Delta E = \pm \hbar(\omega_{0I} - \omega_{0S}) \quad . \tag{9.5}$$

If the two spins were identical, so that $\omega_{0I} \cong \omega_{0S}$, we would refer to these energy differences as two quantum and zero quantum and we would say that the pulse sequence (9.2) has generated 0- and 2-quantum matrix elements of ϱ in addition to the 1-quantum matrix which we observe directly.

A next question is, clearly, can we detect the zero- or two-quantum matrix elements given the fact that (7.248) shows that resonance directly sees only the one-quantum matrix elements? Before proceeding with that question, we discuss a bit more the concept of 0-, 1-, 2-, and in general n-quantum matrix elements.

In some cases, m_I or m_S are not individual constants of the motion. For example, consider two identical spins, with a chemical shift difference and a spin-spin coupling involving terms such as $\boldsymbol{I}\cdot\boldsymbol{S}$. If the spin-spin coupling is comparable in magnitude to the chemical shift difference, m_I or m_S are not constants of the motion. It is still true, however, that the operator, F_z, for the total z-component of angular momentum

$$F_z = I_z + S_z \tag{9.6}$$

commutes with the Zeeman energy and the secular part of the spin-spin coupling, and thus its eigenvalue, M, is a good quantum number.

In fact, for an N-spin system with total z-component of spin $I_{z\text{T}}$ given by

$$I_{z\text{T}} \equiv \sum_{k=1}^{N} I_{zk} \quad , \tag{9.7a}$$

the eigenvalue M of $I_{z\text{T}}$ is a good quantum number. Thus, if we have a set of quantum numbers M and α, where α are all the other quantum numbers needed, so that

$$I_{z\text{T}}|M\alpha) = M|M\alpha) \quad , \tag{9.7b}$$

we define matrix elements

$$(M'\alpha'|\varrho(t)|M\alpha) \tag{9.8}$$

as n-quantum matrix elements where

$$n \equiv |M - M'| \quad . \tag{9.9}$$

We follow the convention of calling n the *order* of the matrix element, and say that the existence of a nonvanishing matrix element (9.8) describes n-*order coherence*.

It is immediately evident that for a system of N spins of spin I, M can range from NI to $-NI$, so that the largest n one can have is $2NI$. Thus, for protons, a spin system involving an H_2 group (as in a CH_2 fragment) should permit generation of two-quantum matrix elements, a system involving three protons (as in a CH_3 group) should permit generation of both two- *and* three-quantum matrix elements.

The generation of a given order of coherence thus enables one to determine what spin groupings exist if one can detect the existence of orders different from one. Note, however, that a *pair* of CH_2 groups contains four protons, so one may ask, when should one consider a CH_2 fragment to be a pair of protons, and when should it be considered to be part of a larger group?

We thus see that there are exciting possibilities to use multiple quantum phenomena to classify spin groups, but that there are also important things we need to understand, such as, how do we create and how do we detect the various orders of coherence?

Multiple quantum phenomena are encountered in all branches of spectroscopy. *Bodenhausen* [9.1], in his excellent review of the field, includes a discussion of the manifestations of multiple quantum phenomena in cw magnetic resonance spectroscopy. It was the advent of pulse methods in multiple quantum spectroscopy which caused the technique to become of central importance as an almost routine spectroscopic tool.

In two fundamental papers, *Hatanaka* et al. [9.2, 3] showed how to generate and to detect multiple quantum coherence. They studied the Al^{27} resonance of Al_2O_3, which can produce multiple quantum transitions since the spin of Al^{27} is 5/2. *Aue* et al. [9.4], in their famous initial paper on two-dimensional Fourier transform NMR, show that, using two-dimensional Fourier transform spectroscopy, one can observe both zero- and double-quantum spectra. They describe several ways of producing the desired matrix elements.

Since the publication of the basic papers from the laboratories of Hashi and of Ernst, the field of multiquantum coherence has exploded. In addition to the review by *Bodenhausen,* there is an extensive review by *Weitekamp* [9.5]. There are useful reviews by *Emid* [9.6] and by *Munowitz* and *Pines* [9.7]. As in all aspects of pulsed NMR, the book by *Ernst* et al. [9.8] gives a thorough and detailed account of the principles and practice.

Our goal is to explain the physical principles. We turn next to showing that it is the nonlinearity in the equation of motion which makes possible excitation of multiple quantum coherence. We initially deal with cases in which we are tuned to discrete transitions, showing how we can then progressively pump higher and higher order coherences.

Next we deal with the much more common experimental case of applying pulses of sufficiently great amplitude to excite a number of transitions, discussing as well how to detect the fact that we have generated multiple quantum coherence. We then turn to methods of singling out in the detection coherence of a desired order, and lastly to the problem of how one can selectively excite a particular order of coherence.

9.2 The Feasibility of Generating Multiple Quantum Coherence – Frequency Selective Pumping

Consider for the moment a two-spin system. We have seen (Sect. 7.5) that we can produce two-quantum elements of ϱ by applying a driving signal at $\omega_I = \omega_{0I}$, followed by one at $\omega_S = \omega_{0S}$. Among the matrix elements produced is

$$(--|\varrho(t)|++) \quad , \tag{9.10}$$

which oscillates in time as

$$\exp[-i(\omega_{0I} + \omega_{0S})t] \quad . \tag{9.11}$$

We now proceed to show that this result is closely related to the idea of the beating of the frequencies of two oscillators.

Beats are produced by putting input signals into a nonlinear device so that the output signal contains terms proportional to the product of the input signals. Thus if we had a device whose electrical properties are characterized by the property that the current i through the device is proportional to the square of the voltage V applied across the device

$$i = AV^2 \quad , \tag{9.12}$$

application of a voltage

$$V = V_1 \cos \omega_1 t + V_2 \cos \omega_2 t \qquad (9.13)$$

would produce a current

$$i(t) = A(V_1^2 \cos^2 \omega_1 t + V_2^2 \cos^2 \omega_2 t + 2V_1 V_2 \cos \omega_1 t \cos \omega_2 t) \ . \qquad (9.14)$$

Utilizing the fact that

$$\cos \omega_1 t \cos \omega_2 t = \tfrac{1}{2}[\cos(\omega_1 + \omega_2)t + \cos(\omega_1 - \omega_2)t] \qquad (9.15)$$

we get

$$\begin{aligned} i = \frac{A}{2}\Big\{ &V_1^2[1 + \cos(2\omega_1 t)] + V_2^2[1 + \cos(2\omega_2 t)] \\ &+ 2V_1 V_2[\cos(\omega_1 + \omega_2)t + \cos(\omega_1 - \omega_2)t]\Big\} \ . \end{aligned} \qquad (9.16)$$

Thus the current i contains the frequencies $2\omega_1$, $2\omega_2$, $\omega_1 + \omega_2$, $\omega_1 - \omega_2$, and 0 (which is $\omega_1 - \omega_1$ and $\omega_2 - \omega_2$).

A similar thing happens for the density matrix equation

$$\frac{d\varrho}{dt} = \frac{i}{\hbar}(\varrho\mathcal{H} - \mathcal{H}\varrho) \ . \qquad (9.17)$$

Since this equation involves a product of ϱ with $\mathcal{H}$, if both ϱ and $\mathcal{H}$ are time dependent on the right hand side, they couple to terms on the left hand side which oscillate at the sum or difference frequencies. We want to show that this is just what we need to take a ϱ which has an n-quantum coherence and pump it either up to an $(n+1)$-quantum coherence or down to an $(n-1)$-quantum coherence by means of a time-dependent $\mathcal{H}$ which oscillates at the frequency difference between the n and $n+1$ (or $n-1$) quantum coherence.

To make those ideas more concrete, let us consider a system with a set of nondegenerate energy levels, a static Hamiltonian $\mathcal{H}_0$, and a time-dependent drive term $V(t)$ given by

$$V(t)/\hbar = Fe^{i\omega t} + F^* e^{-i\omega t} \qquad (9.18\text{a})$$

where the * refers to the complex conjugate.

For example, if we have a group of identical spins as in (9.7) acted on by a rotating magnetic field, we get

$$\begin{aligned} V(t) &= -\gamma\hbar H_1(I_{x\text{T}} \cos \omega t - I_{y\text{T}} \sin \omega t) \\ &= -\gamma\hbar H_1\left[I_{x\text{T}}\left(\frac{e^{i\omega t} + e^{-i\omega t}}{2}\right) + \frac{i}{2}I_{y\text{T}}(e^{i\omega t} - e^{-i\omega t})\right] \\ &= -\frac{\gamma\hbar H_1}{2}\left(I_\text{T}^+ e^{i\omega t} + I_\text{T}^- e^{-i\omega t}\right) \ , \end{aligned} \qquad (9.18\text{b})$$

so that, using the usual notation $\omega_1 = \gamma H_1$,

$$F = -\frac{\omega_1}{2}I_\text{T}^+ \ , \quad F^* = -\frac{\omega_1}{2}I_\text{T}^- \ . \qquad (9.18\text{c})$$

Note that a rotating field applied in the opposite sense is obtained by treating ω as negative.

Let the eigenstates of $\mathcal{H}_0$ be denoted by quantum numbers, j, k, l each of which stands for some set of quantum numbers such as M, α of (9.7b). Then we define

$$\omega_j \equiv E_j/\hbar \quad , \quad \omega_{jk} \equiv \omega_j - \omega_k \quad . \tag{9.19}$$

Then, denoting the interaction representation by a prime, we have

$$\varrho(t) = \mathrm{e}^{-(\mathrm{i}/\hbar)\mathcal{H}_0 t}\varrho'(t)\mathrm{e}^{(\mathrm{i}/\hbar)\mathcal{H}_0 t} \quad \text{or} \tag{9.20}$$

$$\varrho_{jk}(t) = \mathrm{e}^{-\mathrm{i}\omega_{jk}t}\varrho'_{jk}(t) \quad \text{and} \tag{9.21a}$$

$$\varrho'_{jk}(t) = \mathrm{e}^{\mathrm{i}\omega_{jk}t}\varrho_{jk}(t) \quad . \tag{9.21b}$$

The density matrix then obeys the matrix equation

$$\frac{d}{dt}\varrho_{jl} = \frac{\mathrm{i}}{\hbar}\sum_k [\varrho_{jk}(\mathcal{H}_0)_{kl} - (\mathcal{H}_0)_{jk}\varrho_{kl}]$$
$$+ \frac{\mathrm{i}}{\hbar}\sum_k [\varrho_{jk}V_{kl}(t) - V_{jk}(t)\varrho_{kl}] \quad . \tag{9.22}$$

If we had only the terms involving $\mathcal{H}_0$, ϱ_{jk} would oscillate at $\exp(-\mathrm{i}\omega_{jk}t)$. We can eliminate them from the right by substituting (9.21a) into (9.22). We also utilize the fact that

$$(\mathcal{H}_0)_{kl} = E_k \delta_{kl} \quad , \tag{9.23}$$

since the states, j, k, l are the eigenstates of $\mathcal{H}_0$. We thus get

$$\exp(-\mathrm{i}\omega_j l t)\frac{d\varrho'_{jl}}{dt} = \frac{\mathrm{i}}{\hbar}\sum_l [\varrho'_{jk}\exp(-\mathrm{i}\omega_{jk}t)V_{kl}(t)$$
$$- \frac{\mathrm{i}}{\hbar}V_{jk}(t)\varrho'_{kl}\exp(-\mathrm{i}\omega_{kl}t)] \quad . \tag{9.24}$$

This equation reminds us that the matrix elements of ϱ' would be independent of time (i.e. that $d\varrho'_{jl}/dt = 0$) if V were zero. Consequently, we expect that matrix elements of ϱ' to change slowly compared to the frequencies ω_{jl}, ω_{jk}, ω_{kl}. Therefore, to a good approximation the left hand side of (9.24) oscillates at ω_{jl}. This requires that the right hand side also oscillates at the same frequency if $(d\varrho'_{jl}/dt)$ is to be something other than zero.

Substituting (9.18) for $V/\hbar$ in (9.24), we thus get

$$\exp(-i\omega_{jl}t)\frac{d\varrho'_{jl}}{dt} = i\sum_k \varrho'_{jk}\{\exp[i(\omega - \omega_{jk})t]F_{kl} + \exp[-i(\omega + \omega_{jk})t]F^*_{kl}\}$$
$$- i\sum_k \{F_{jk}\exp[i(\omega - \omega_{kl})t]$$
$$+ F^*_{jk}\exp[-i(\omega + \omega_{kl})t]\}\varrho'_{kl} \quad . \tag{9.25}$$

The first two terms on the right give the sum and difference of ω and ω_{jk}, the third and fourth terms on the right have the sum and difference frequencies of ω with ω_{kl}. This is the same situation we got for our example of the nonlinear current/voltage relationship (9.12) and (9.16) in which the product of $V_1(t)$ with $V_2(t)$ gave the sum and difference frequencies. Here what is happening is that we consider the elements of ϱ' on the right hand (either ϱ'_{jk} or ϱ'_{kl}) to be the n-quantum coherence which we wish to pump by means of V up to the $(n+1)$-quantum coherence or down to the $(n-1)$-quantum coherence represented by ϱ'_{jl} on the left hand side. The point is that since ϱ'_{jl} [the $(n+1)$- or $(n-1)$-quantum coherence] is to be produced, its time derivative must be different from zero.

Let us suppose, then, that at $t = 0$ the matrix elements ϱ_{jl} and ϱ_{kl}, and thus ϱ'_{jl} and ϱ'_{kl}, are zero, but that we have previously excited the matrix element ϱ_{jk_0} for a particular value of k, labeled k_0. So, we have a nonzero ϱ_{jk_0}, hence ϱ'_{jk_0}. Let us suppose further that $V(t)$ possesses a nonzero matrix element between states k_0 and l. Then, we have

$$\exp(-i\omega_{jl}t)\frac{d\varrho'_{jl}}{dt} = i\varrho'_{jk_0}\{\exp[i(\omega - \omega_{jk_0})t]F_{k_0l}$$
$$+ \exp[-i(\omega + \omega_{jk_0})t]F^*_{k_0l}\} \quad . \tag{9.26}$$

Now, ordinarily, F is a raising operator and F^* a lowering operator. Thus, recalling that k_0 and l will in general be specified in part by the quantum number M, see (9.8b), we see that if $F_{k_0l} \neq 0$, then $F^*_{k_0l} = 0$, or vice versa. Let us assume

$$F_{k_0l} = 0 \quad , \quad F^*_{k_0l} \neq 0 \quad . \tag{9.27}$$

Then we would have

$$\exp(-i\omega_{jl}t)\frac{d\varrho'_{jl}}{dt} = i\varrho'_{jk_0}F^*_{k_0l}\exp[-i(\omega + \omega_{jk_0})t] \quad . \tag{9.28}$$

Then the condition that F pump ϱ_{jk_0} up to ϱ_{jl} is that

$$F^*_{k_0l} \neq 0 \quad \text{and} \tag{9.29a}$$

$$\omega_{jl} = \omega_{jk_0} + \omega \quad \text{or} \tag{9.29b}$$

$$\omega_j - \omega_l = \omega_j - \omega_{k_0} + \omega \quad . \tag{9.29c}$$

This equation may be interpreted as saying that the beat between the pumping term F (frequency ω), and the n-quantum coherence (frequency $\omega_j - \omega_{k_0}$) must add to give the frequency, $\omega_j - \omega_l$, of the $(n+1)$-quantum coherence. [If ω is negative, the left hand side is an $(n-1)$-quantum coherence.]

Equation (9.29c) can also be written as

$$\omega_{k_0} - \omega_l = \omega \quad . \tag{9.29d}$$

The meaning of this is that to excite ϱ_{jl} from ϱ_{jk_0} (or ϱ_{jk_0}), ω must match the resonance frequency of the transition k_0 to l, and V must have a matrix which joins k_0 to l.

If we start with a diagonal ϱ, we can first produce a one-quantum coherence with a $\pi/2$ pulse. The discussion above shows that we can pump this up to a two-quantum coherence. That in turn can be pumped to a three-quantum coherence, etc. Note that we are here assuming we can apply a rf which excites one transition at a time since we kept only one term on the right hand side of (9.25). Thus, if there are splittings, we are using an H_1 sufficiently weak to pick out the individual lines. It must therefore be smaller than the spacing between spectral lines. This mode of pumping is called *frequency selective excitation*. A strong H_1 covering all the splittings is called *nonselective excitation*. We treat it in the next section.

Let us follow our example in greater detail. We define three energy levels 1, 2, 3 (Fig. 9.1) with the order of energies such that

$$\omega_3 > \omega_2 > \omega_1 \quad . \tag{9.30}$$

We assume that

$$\omega_3 - \omega_2 \neq \omega_2 - \omega_1 \tag{9.31}$$

and that we have previously excited ϱ_{21} (hence ϱ_{12}), but not ϱ_{23} (hence not ϱ_{32}). An actual realization of such a system is shown for a pair of spins in Fig. 9.2. Assuming they have a coupling of the form

$$AI_z S_z \tag{9.32}$$

and are identical except for a chemical shift difference between spins I and S, the four allowed transitions occur at different frequencies, as we discussed in Chapter 7.

Let us rewrite (9.24) by multiplying both sides by $\exp(i\omega_{jl}t)$. Then

$$\frac{d\varrho'_{jl}}{dt} = \frac{i}{\hbar} \sum_k [\varrho'_{jk} e^{+i\omega_{kl}t} V_{kl}(t) - V_{jk}(t) e^{+i\omega_{jk}t} \varrho'_{kl}] \quad . \tag{9.33}$$

This is actually just what we would have gotten if we had started with the equation for ϱ in the interaction representation, with [following (9.21)]

$$V'_{kl}(t) = V_{kl}(t) e^{i\omega_{kl}t} \tag{9.34}$$

etc. Then,

Fig. 9.1

Fig. 9.2

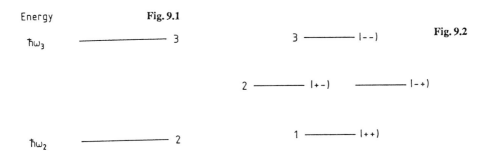

Fig. 9.1. Three energy levels labeled 1, 2, 3 with energies E_1, E_2, E_3 expressed as $\hbar\omega_1$, $\hbar\omega_2$, $\hbar\omega_3$ in the notation of (9.19). Note that $\omega_2 > \omega_2 > \omega_1$ and we assume $\omega_3 - \omega_2 \neq \omega_2 - \omega_1$

Fig. 9.2. The energy levels of a pair of spin-$\frac{1}{2}$ nuclei. We focus on the three levels labeled 1, 2, 3

$$\frac{d\varrho'_{31}}{dt} = -\frac{\mathrm{i}}{\hbar}\mathrm{e}^{\mathrm{i}\omega_{32}t}V_{32}\varrho'_{21} \quad \text{and} \quad \frac{d\varrho'_{21}}{dt} = -\frac{\mathrm{i}}{\hbar}\mathrm{e}^{\mathrm{i}\omega_{23}t}V_{23}\varrho'_{31} \quad . \tag{9.35}$$

Using (9.18c), we see that F involves the raising operator $I^+ + S^+$ and F^* the lowering operator $I^- + S^-$. Since state three is $|--)$ and state two is $|+-)$, we have

$$\begin{aligned} F^*_{32} &= -\frac{\omega_1}{2}(--|I^- + S^-|+-) \\ &= -\frac{\omega_1}{2}(--|I^-|+-) \\ &= -\frac{\omega_1}{2} \end{aligned} \tag{9.36a}$$

$$F_{32} = -\frac{\omega_1}{2}(--|I^+ + S^+|+-) = 0 \quad . \tag{9.36b}$$

Likewise

$$F^*_{23} = 0 \quad , \quad F_{23} = -\frac{\omega_1}{2} \quad . \tag{9.36c}$$

Therefore

$$\frac{d\varrho'_{31}}{dt} = i\frac{\omega_1}{2}\mathrm{e}^{\mathrm{i}(\omega_{32}-\omega)t}\varrho'_{21} \quad , \quad \frac{d\varrho'_{21}}{dt} = i\frac{\omega_1}{2}\mathrm{e}^{\mathrm{i}(\omega_{23}+\omega)t}\varrho'_{31} \quad . \tag{9.37}$$

The right hand side has an oscillation at

$$\omega_3 - \omega_2 - \omega \quad . \tag{9.38}$$

This causes the left hand sides to vanish unless

$$\omega_3 - \omega_2 \cong \omega \quad . \tag{9.39}$$

If we satisfy (9.39) with an equality, then we get

$$\frac{d^2 \varrho'_{31}}{dt^2} + \left(\frac{\omega_1}{2}\right)^2 \varrho'_{31} = 0 \quad , \quad \frac{d^2 \varrho'_{21}}{dt^2} + \left(\frac{\omega_1}{2}\right)^2 \varrho'_{21} = 0 \quad , \tag{9.40}$$

which we recognize as harmonic oscillator equations. Writing down the general solutions, then evaluating the constants of integration with the help of (9.37), we get

$$\varrho'_{31}(t) = \varrho'_{31}(0) \cos\left(\frac{\omega_1 t}{2}\right) + i\varrho'_{21}(0) \sin\left(\frac{\omega_1 t}{2}\right)$$

and

$$\varrho'_{21}(t) = \varrho'_{21}(0) \cos\left(\frac{\omega_1 t}{2}\right) + i\varrho'_{31}(0) \sin\left(\frac{\omega_1 t}{2}\right) \quad . \tag{9.41}$$

Thus, if we turn on H_1, ϱ'_{21} will transfer its amplitude to ϱ'_{31}, which will then transfer it back to ϱ'_{21} later and so on.

Note that if t is applied for a time t_w generating a π pulse ($\omega_1 t_w = \pi$) so that

$$\frac{\omega_1 t_w}{2} = \frac{\pi}{2} \quad , \tag{9.42}$$

$$\varrho'_{31}(t_w) = i\varrho'_{21}(0) \tag{9.43}$$

$$\varrho'_{21}(t_w) = 0 \quad [\text{since } \varrho'_{31}(0) = 0] \quad .$$

Thus, the full coherence has gone from one-quantum coherence to two-quantum coherence at time t_w.

These same equations show that if we start with pure two-quantum coherence $[\varrho'_{31}(0) \neq 0, \varrho'_{21}(0) = 0]$, we can transfer it to one-quantum coherence $[\varrho'_{21}(t_w) \neq 0, \varrho'_{31}(t_w) = 0]$ by a π pulse ($\omega_1 t_w = \pi$).

It is interesting to note that, if one starts with pure one-quantum coherence and applies a 2π pulse ($\omega_1 t_w = 2\pi$) to transition 2–3 (flipping the I-spins) $[\varrho'_{21}(0) \neq 0, \varrho'_{31}(0) = 0]$,

$$\varrho'_{31}(t_w) = 0 \quad , \quad \varrho'_{21}(t_w) = -\varrho'_{21}(0) \quad . \tag{9.44}$$

Now,

$$\varrho'_{21} = (+ - |\varrho'| + +) \tag{9.45}$$

gives the S-spin transverse magnetization. Thus, a 2π pulse applied to the I-spins reverses the sign of the S-spin transverse magnetization. This result is a manifestation of the spinor character of spin $\frac{1}{2}$ systems. It was employed by *Stoll* et al. [9.9, 10] and *Wolff* and *Mehring* [9.11] to demonstrate the spinor nature of a spin $\frac{1}{2}$ particle. *Mehring* et al. [9.12] have utilized this fact in an ingenious version of spin echo double resonance for electron spin echoes when the electron is coupled to a proton giving a resolved hyperfine splitting of the proton.

One can represent the basic equations (9.28) and (9.29), describing selective excitation of $(n \pm 1)$-quantum coherence from a state containing n-quantum coherence graphically, as shown in Fig. 9.3. We use a solid line between two energy levels to denote the matrix element of V joining those states. Dashed lines between states represent matrix elements of ϱ (or ϱ') between those states. Then, the requirement that to generate a "jl" matrix element on the left of (9.28) demands a jk_0 and a k_0l matrix element on the right simply means that the two dashed lines plus the solid line form a closed loop, as in Fig. 9.3a. Then one also automatically satisfies the energy conservation requirements (9.29c) or (9.29d).

Note that although V is in resonance with the 2–3 transition, we have assumed it is *not* in resonance with the 1–4 transition since we did not connect those states with a solid line. In Fig. 9.3b we show that if we tuned $V(\omega)$ to be resonant at the 1–4 transition, it cannot pump ϱ_{21} into ϱ_{31} since the lines do not form a closed figure. However, looking at Fig. 9.3c we see that V_{14} will pump ϱ_{21} into ϱ_{24}, the zero-quantum transition.

Note also that these figures show (Fig. 9.3a) that V_{23} can pump ϱ_{31}, the two-quantum coherence, into ϱ_{21}, a one-quantum coherence. Also (Fig. 9.3c) V_{14} can pump the zero-quantum coherence into the one-quantum coherence ϱ_{21}.

We began Sect. 9.1 with some general remarks about the density matrix of a system in thermal equilibrium, asking whether or not one could experimentally

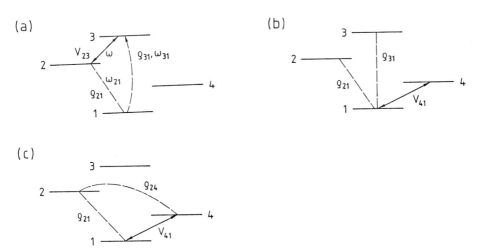

Fig. 9.3a–c. The four energy levels of a pair of spin-$\frac{1}{2}$ nuclei. (a) The density matrix elements ϱ_{21} and ϱ_{31}, which oscillate at ω_{21} and ω_{31} respectively, are coupled together by the time-dependent perturbation V oscillating at frequency ω provided (i) the matrix element V_{23} is nonvanishing and (ii) $\omega + \omega_{21} = \omega_{31}$. The solid line ($V_{23}$) and the two dashed lines (ϱ_{21}, ϱ_{31}) form a closed loop. (b) If ϱ_{21}, ϱ_{31}, and V do *not* form a closed loop, as in (b) where ω is tuned to the 4–1 transition making V_{41} the driving term, V does not couple ϱ_{21} to ϱ_{31}. Thus, V could not pump the one-quantum coherence ϱ_{31}. (c) However, V_{41} can pump ϱ_{21} into ϱ_{24} since the two dashed lines and the solid line form a closed figure

determine whether *all* ϱ_{ij} ($i \neq j$) were zero for a system in thermal equilibrium. The fact that we can pump n up or down gives us hope. Thus if we want to inspect a particular ϱ_{ij}, we look for a sequence of transitions to pump which eventually enable us to pump the ij coherence (if it is nonzero) into a pair of levels kl whose coherence, ϱ_{kl}, is directly observable. Alas, a simple example will suffice to show that the usual excitations will not do the job in all cases.

Consider two identical spin $\frac{1}{2}$ nuclei (e.g. a pair of protons) with *identical* chemical shifts. Their energy levels are the singlet and triplet states (Appendix H). Designating the total spin by F, we have

$$F = I_1 + I_2 \quad , \quad F_z = I_{1z} + I_{2z} \quad . \tag{9.46}$$

For the two spins, we have $F = 0$, $M_F = 0$ for the singlet, $F = 1$, $M_F = 1, 0, -1$ for the triplet.

We now demonstrate a type of matrix element of ϱ which we cannot probe. Since

$$F_x \equiv I_{1x} + I_{2x} \tag{9.47}$$

has no matrix elements connecting the singlet to the triplet, the only nonvanishing V_{ij}'s are entirely within the triplet. Thus, if we wish to inspect a matrix element

$$(FM_F|\varrho|F'M_{F'}) = (11|\varrho|00) \quad , \tag{9.48}$$

we could only pump it to states such as $(10|\varrho|00)$ or $(1-1|\varrho|00)$ in order to generate a closed set of lines. Figure 9.4 illustrates how we could use $(11|\varrho|00)$ to feed $(10|\varrho|00)$. But $(10|\varrho|00)$ is not observable either, nor is $(1-1|\varrho|00)$ which we could produce as in Fig. 9.4b. Thus, we can only succeed in pumping our original unobservable matrix element (9.48) into other nonobservable matrix elements!

In principle we could apply a spatially inhomogeneous alternating field of strength H_1 at nucleus 1 and H_2 at nucleus 2 so that we had a rotating frame Hamiltonian

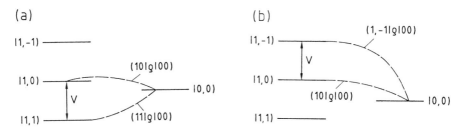

Fig. 9.4. (a) An alternating potential joining the states $|10\rangle$ and $|11\rangle$ can couple $(10|\varrho|00)$ to $(11|\varrho|00)$. If we wanted to inspect an element $(11|\varrho|00)$ which is itself not directly observable, we are unsuccessful since we can only pump it into an element $(10|\varrho|00)$, which is also unobservable. (b) If we produced $(10|\varrho|00)$, a further application of V tuned to the $|1-1\rangle - |10\rangle$ transition will produce $(1-1|\varrho|00)$ which is still not observable

$$\mathcal{H}_{\text{eff}} = -\gamma\hbar h_0 F_z + \hbar a I_{1z}I_{2z} - \gamma\hbar(H_1 I_{1x} + H_2 I_{2x})$$
$$= -\gamma\hbar h_0 F_z + \hbar a I_{1z}I_{2z} - \gamma\hbar\left(\frac{H_1 + H_2}{2}\right)F_x$$
$$- \gamma\hbar\left(\frac{H_1 - H_2}{2}\right)(I_{1x} - I_{2x}) \quad . \tag{9.49}$$

The term $I_{1x} - I_{2x}$ now connects singlet to triplet states, as can be seen by evaluating matrix elements between $|11\rangle$ and $|00\rangle$ using the spin-up and spin-down functions α and β to express the $|FM_F\rangle$ states:

$$|11\rangle = \alpha(1)\alpha(2) \quad \text{and} \tag{9.50a}$$

$$|00\rangle = \frac{1}{\sqrt{2}}[\alpha(1)\beta(2) - \beta(1)\alpha(2)] \quad . \tag{9.50b}$$

We find that

$$(11|I_{1x} - I_{2z}|00) = -\frac{1}{\sqrt{2}} \quad . \tag{9.50c}$$

Thus, if we had initially a nonzero matrix element $(10|\varrho(0)|00)$, we could use the coupling of (9.49) through its matrix element $(11|\mathcal{H}_{\text{eff}}|00)$ [or $(00|\mathcal{H}_{\text{eff}}|11)$] to produce a matrix element $(10|\varrho|11)$, which is one of the standard observable matrix elements of ϱ in the triplet ($F = 1$) state.

Another approach would be to apply a spatially inhomogeneous static field. This would mix the states $F = 1$, $M_F = 0$, and $F = 0$, $M_F = 0$. Such an effect is of course equivalent to having a chemical shift difference between the two nuclei. In practice, it would be difficult to make H_2 sufficiently different from H_1 for there to be an observable effect. A more practical approach would be to utilize the coupling to a third nucleus nearby, provided there is one which is closer to spin 1(2) than to spin 2(1). The third nucleus will cause the singlet and triplet states of the first nuclei to mix, making the rf singlet to triplet transition no longer forbidden, or one could drive the third nucleus with an rf field, to cause it to excite the singlet-triplet transition of the first pair.

The use of frequency selective pulses makes it easy to see how multiple quantum matrix elements can be generated. It also gives us great control in generating a particular element of ϱ which we might desire. However, it also requires that each time we pump n either up or down we be tuned precisely to the appropriate transition. Thus, referring to Fig. 9.3a, to produce the element ϱ_{31} starting from a system in thermal equilibrium we might first pump the 1–2 transition at frequency

$$\omega = \omega_{21} \tag{9.51}$$

to produce ϱ_{12}, then pump the 23 transition using

$$\omega = \omega_{32} \tag{9.52}$$

to produce ϱ_{13}. This would require shifting the frequency ω from ω_{21} to ω_{31}, something of a nuisance to do. In practice, one usually works with large H_1's which can flip spins corresponding to many multiplet lines. Such pulses are called nonselective. They cover all the lines within an angular frequency width

$$\Delta\omega = \gamma H_1 \tag{9.53}$$

of the oscillator frequency ω. We turn now to see how they work.

9.3 Nonselective Excitation

9.3.1 The Need for Nonselective Excitation

As we have seen, using frequency-selective pulses we can pump an n-quantum coherence either up or down. Thus, starting with the usual one-quantum coherence following a $\pi/2$ pulse, we can proceed in principle to produce a pure two-quantum coherence, then convert that to a pure three-quantum coherence, and so on. To do this, we would need to know the energy levels and the corresponding wave functions. Having produced some desired level of coherence, we could later inspect how it evolved in time by later reversing the process and pumping the n-quantum coherence to $n-1$, retuning the pump frequency to pump that to $n-2$, etc., eventually getting back to one-quantum (observable) coherence. If we studied how the observed coherence varied with the length of time the system was in the n-quantum state, we could learn about the evolution of the system in the n-quantum state. However, this approach implies we already know the energy levels of the system, and know therefore what spectral lines are connected with what pair of states. This situation would be satisfactory for some purposes – for example, if we wanted to verify that the two-quantum matrix elements of ϱ of a pair of spins vanishes when the system is in thermal equilibrium.

However, frequently one is attempting to utilize multiple quantum coherence to understand some unknown system for which the spectral lines may not yet be assigned. One then wishes some general approach to exciting multiple quantum coherences. One would then let them evolve, and lastly inspect what happened during the evolution, to try to learn about the system. For example, one might wish to know whether there are groups of two spins but not groups of three, as in the experiments of *Wang* et al. [9.13], who studied acetylene (C_2H_2) adsorbed on the surface of Pt metal and wished to determine whether or not any CCH_3 species were formed. They found they could generate both two- and three-quantum coherence. By comparing the intensity of two-quantum coherence with that of three-quantum coherence, they showed that groups of three or more protons were rare. By studying how the relative intensities of two- versus three-quantum coherence depended on the concentration of acetylene, they showed that the three-quantum coherence arose from coupling between different molecules, hence that there were no groups of three protons within a given molecule. They thereby showed that the species CCH$_3$ was not present.

9.3.2 Generating Multiple Quantum Coherence

To generate multiple quantum coherence in the absence of detailed spectral information we utilize an approach which is referred to as nonselective excitation. There are many variants. We start with the simplest. We will first describe without explanation the pulses we apply first to produce, then to detect the generation of multiple quantum coherence. Then we will explain how the sequence works.

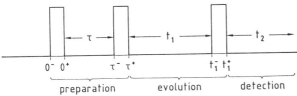

Fig. 9.5. A three-pulse sequence, showing the preparation, evolution, and detection periods, and defining the times $0^-, 0^+, \tau^-, \tau^+, t_1^-, t_1^+$

The simplest approach to generating multiple quantum coherence is illustrated in Fig. 9.5. A strong $\pi/2$ pulse is applied to a system in thermal equilibrium. "Strong" means γH_1 exceeds the spectral width, $\Delta\omega$, of the spectrum. That is, H_1 is larger than the spectral splittings arising from chemical shifts or spin-spin couplings. The system develops freely for a time τ. This period is called the preparation period. During this period, as we shall see, spin-spin coherences develop so that when a second $\pi/2$ pulse is applied at the end of τ, it converts the spin-spin coherence to a variety of multiple quantum coherences. These coherences then develop freely for a time t_1, called the development time. At the end of that time, one wishes to inspect what has happened. This is done by applying a $\pi/2$ pulse which transforms the multiple quantum coherence back to one-quantum coherence for detection. The signal is recorded as a function of the time t_2 after the detection pulse. Frequently t_1 is varied as a parameter to enable one to identify the contributions of the various multiple quantum orders. Thus, the signal, S, is a function of both t_1 and t_2. It will also depend on the preparation time, τ, giving us in general a complex signal, $S(\tau, t_1, t_2)$.

To explain how the above pulse sequence works, we consider a model system of two identical spin $\frac{1}{2}$ nuclei, spins I and S, possessing identical chemical shifts. We suppose that the rf pulses have a frequency ω and write the Hamiltonian in the coordinate system rotating at ω. In that frame, in the absence of the pulses, the Hamiltonian is then

$$\mathcal{H} = -\hbar\Omega(I_z + S_z) + \hbar a I_z S_z \quad \text{with} \tag{9.54a}$$

$$\Omega \equiv \gamma H_0 - \omega \tag{9.54b}$$

representing the possibility of being somewhat off resonance either from the existence of chemical shifts, or because the experimenter has deliberately chosen to be off resonance.

Recall that this form of coupling would give exact results for a pair of spins with dipolar coupling, but would not describe two spins with a pseudo-exchange coupling. See Appendix H. For more than two spins, the form of the interaction term $(I_z S_z)$ is only an approximation even if the coupling is dipolar. However, though this form of coupling gives only approximate results in that case, it makes possible simple explicit calculations, enabling us to get results whose physical significance we can explore.

We utilize the spin-operator method (Sect. 7.26) to follow the density matrix in time. Denoting times 0^-, 0^+, τ^-, τ^+, etc. as indicated in Fig. 9.5 to indicate times just before and just after the pulses at $t = 0$, τ, etc., we take the density matrix at $t = 0^-$ to be

$$\varrho(0^-) = (I_z + S_z) \quad , \tag{9.55}$$

omitting the constant of proportionality. Then the development of ϱ is given in Table 9.1 up to the time just after the second pulse. We utilize the shorthand notation

$$\cos(a\tau/2) \equiv C_a^\tau \quad , \quad \sin(\Omega\tau) \equiv S_\Omega^\tau \quad , \quad \text{etc.} \tag{9.56}$$

It is useful to express the spin components I_x, I_y, etc. in terms of raising and lowering operators such as

$$I_x = \frac{1}{2}(I^+ + I^-) \quad , \quad I_y = \frac{1}{2i}(I^+ - I^-) \quad .$$

We have collected the results for products of two spin components in Table 9.2. Utilizing these results we get the last column of Table 9.1, which expresses $\varrho(\tau^+)$ in terms of raising and lowering operators. Utilizing them, we see that line 1 of the table corresponds to elements $(m_I m_S |\varrho| m_I' m_S')$ for which

Table 9.1. The time dependence of ϱ during the preparation period

Column no.:	1	2	3	4	5	6	7
Driving terms:		$X(\pi/2)$		$a\tau I_z S_z$		$\Omega\tau(I_z + S_z)$	
Column labels:	$\varrho(0^-)$		$\varrho(0^+)$				$\varrho(\tau^-)$
	$I_z + S_z$		$I_y + S_y$		$(I_y + S_y)C_a^\tau$		$(I_y + S_y)C_\Omega^\tau C_a^\tau$
							$(I_x + S_x)S_\Omega^\tau C_a^\tau$
					$-(I_x S_z + I_z S_x)2S_a^\tau$		$-(I_x S_z + I_z S_x)C_\Omega^\tau 2S_a^\tau$
							$(I_y S_z + I_z S_y)S_\Omega^\tau 2S_a^\tau$

$$M \equiv m_I + m_S \quad , \quad M' \equiv m'_I + m'_S \tag{9.57}$$

are the same ($M' = M$), hence are zero-quantum matrix elements.

For line 2, the terms have

$$M' = M \pm 1 \quad , \tag{9.58}$$

hence are one-quantum matrix elements. Line 3 has products of two raising operators or two lowering operators, hence corresponds to

$$M' = M \pm 2 \quad , \tag{9.59}$$

which are two-quantum matrix elements. Lastly, line 4 has I_z or S_z multiplied by either a raising or a lowering operator, hence has

$$M' = M \pm 1 \quad , \tag{9.60}$$

again a one-quantum coherence.

Utilizing Table 9.2 we see that at $t = 0^-$, the density matrix consists entirely of zero-quantum coherence. It is converted to one-quantum coherence by the first $X(\pi/2)$ pulse. At $t = \tau^-$, just before the second $X(\pi/2)$ pulse, we note $\varrho(\tau^-)$ still consists entirely of one-quantum coherence. However, there are now four terms (lines 1 through 4) as opposed to the single term at $t = 0^+$. Half of these terms arise simply because we are off resonance. That is, they come from the effect of the off-resonance term $\Omega(I_z + S_z)$. The other half arise from the effect of the spin-spin interaction term aI_zS_z. For lines 1 and 2, the elements of ϱ are proportional to $\cos(a\tau/2)$. For short τ, those terms survive even if a goes to zero. Thus, they would be present even if I and S were not coupled together. In the limit of uncoupled spins, line 1 corresponds to magnetization which is flipped by the first pulse into the x-y plane, and then is flipped back onto the z-axis by the second pulse. Note that as τ goes to zero, the combined effect of the two $\pi/2$ pulses is to flip the magnetization onto the negative z-direction. They are thus equivalent to a single π pulse.

Table 9.1 (continued)

8 $X(\pi/2)$	9 $\varrho(\tau^+)$	10		Line
	$-(I_z+S_z)C_\Omega^\tau C_a^\tau$	$=$	$-(I_z+S_z)C_\Omega^\tau C_a^\tau$	1
	$(I_x+S_x)S_\Omega^\tau C_a^\tau$	$=$	$\frac{1}{2}(I^+ + S^+ + I^- + S^-)S_\Omega^\tau C_a^\tau$	2
	$-(I_xS_y+I_yS_x)C_\Omega^\tau 2S_a^\tau$	$=$	$-\frac{1}{4i}(I^+S^+ - I^-S^-)C_\Omega^\tau 2S_a^\tau$	3
	$-(I_zS_y+I_yS_z)S_\Omega^\tau 2S_a^\tau$	$=$	$-\frac{1}{2i}(I_zS^+ - I_zS^- + I^+S_z - I^-S_z)S_\Omega^\tau 2S_a^\tau$	4

Table 9.2. Products of spin components expressed in terms of raising and lowering operators

$$I_z S_x = \frac{1}{2} I_z (S^+ + S^-)$$

$$I_z S_y = \frac{1}{2i} I_z (S^+ - S^-)$$

$$S_z I_x = \frac{1}{2} S_z (I^+ + I^-)$$

$$S_z I_y = \frac{1}{2i} S_z (I^+ - I^-)$$

$$I_x S_x = \frac{1}{4} (I^+ S^+ + I^- S^- + I^+ S^- + I^- S^+)$$

$$I_y S_y = -\frac{1}{4} (I^+ S^+ - I^- S^+ - I^+ S^- + I^- S^+)$$

$$I_x S_y = \frac{1}{4i} (I^+ S^+ - I^- S^- - I^+ S^- + I^- S^+)$$

$$I_y S_x = \frac{1}{4i} (I^+ S^+ - I^- S^- + I^+ S^- - I^- S^+)$$

Line 2 for $a = 0$ corresponds to two uncoupled spins which are flipped into the x-y plane by the first pulse and left there by the second pulse (because they point parallel to H_1 of the second pulse).

On the other hand, lines 3 and 4 are proportional to $\sin(a\tau/2)$. Therefore, if one switches off the spin-spin coupling ($a \to 0$), these terms vanish. They do not therefore arise for uncoupled spins. In order to observe these terms, one must choose a time τ such that $\sin(a\tau/2)$ is appreciable, hence τ's for which

$$a\tau/2 \approx \pi/2, \ 3\pi/2, \ \text{etc.} \tag{9.61}$$

If a represents dipolar coupling as in solids, $a \approx 1/r^3$ where r is the distance between spins I and S. Thus, by selecting τ, one can pick out the values of a or of the distances r which will contribute. For example, the proton pairs in a CH$_2$ group have a characteristic H-H distance. The existence of such a pair will give rise to a characteristic splitting of the proton NMR line into a pair of lines whose separation is proportional to the coupling. This splitting, first observed by *Pake* [9.14] is often referred to as a *Pake doublet*. Our analysis shows that the existence of the coupling, hence of the doublet, will also give rise to the possibility of generating double-quantum coherence. The pulse spacing τ to produce the maximum coherence depends on the strength of the coupling, a, hence on the magnitude of the splitting. Note that one can distinguish coupling between protons in the same molecule from coupling of protons in different molecules either by setting τ to correspond to the desired distance, or by studying the effect on the two-quantum signal strength of diluting the molecules in a non-proton-containing matrix, as in the experiments of *Wang* et al. Extensive studies of this sort have been carried out by *Baum* et al. [9.15], demonstrating large clusters of protons in a given molecule.

We note that while line 3 gives double-quantum coherence, line 4 gives single-quantum coherence. The distinction between these two lines arises solely from the precession effect of being off resonance. The lines are identical except that the term $\cos \Omega \tau$ of line 3 is replaced by $\sin \Omega \tau$ for line 4. If one were exactly on resonance ($\Omega = 0$), line 4 would be zero. However, one could then excite this single-quantum term by making the second pulse be $Y(\pi/2)$ instead of $X(\pi/2)$. In the process, the double-quantum coherence would vanish. The point is that if one is exactly at resonance, the phase of the second rf pulse will determine whether one gets single- or double-quantum coherence. We discuss these points more in the next section.

9.3.3 Evolution, Mixing, and Detection of Multiple Quantum Coherence

We have now seen how the first two pulses produce zero-, one- and two-quantum coherences. We now wish to see how these develop in time. Moreover, since neither zero- nor double-quantum coherences are directly observable, we want to see how the third pulse converts them to something we can measure. The conversion is conventionally called "mixing". Since we observe nuclei by single-quantum coherence, we therefore wish to see how the third pulse converts zero- and double-quantum coherence to single-quantum coherence.

We therefore continue our study of how ϱ develops in time under the combined effect of the Hamiltonian of (9.54a) and a third (mixing) rf pulse at time t_1 after the second pulse. Using the spin-operator methods, we construct Table 9.3 to describe the evolution and Table 9.4 to describe the detection periods. From the tables we can follow the effect of the different interactions (spin-spin and Zeeman) on the time development of the density matrix. We now discuss some important features the tables reveal.

First, we back up a bit to express the time development of ϱ. During the time interval between pulses, the Hamiltonian (9.54a) is static, hence the time development of the density matrix obeys an equation similar to (5.85)

$$(\mu|\varrho(t)|\mu') = (\mu|\varrho(0)|\mu') \exp(i(\omega_{\mu'} - \omega_\mu)) \tag{9.62}$$

where

$$|\mu) = |m_I m_S) \quad , \quad |\mu') = |m_{I'} m_{S'}) \quad , \quad \omega_\mu = E_\mu/\hbar \quad . \tag{9.63}$$

Now

$$\omega_{m_I m_S} = -\Omega(m_I + m_S) + a m_I m_S = -\Omega M + a m_I m_S \tag{9.64}$$

where

$$M = m_I + m_S \tag{9.65}$$

gives the total z-component of spin angular momentum. Utilizing (9.62) we see then that the time dependence of ϱ between pulses is given by

Table 9.3. Time development of ϱ during the evolution and mixing period

Number of quanta in column 1	Column no.: Column label: Operator: Line number	1 $\varrho(\tau^+)$ 	2 $at_1\, I_z S_z$	3	4 $\Omega t_1(I_z+I_z)$
0	1	$-(I_z+S_z)C_\Omega^\tau C_a^\tau$		$-(I_z+S_z)C_\Omega^\tau C_a^\tau$	
1	2	$(I_x+S_x)S_\Omega^\tau C_a^\tau$		$(I_x+S_x)C_a^1 S_\Omega^\tau C_a^\tau$	
	3				
	4			$(I_y S_z + I_z S_y)2S_a^1 S_\Omega^\tau C_a^\tau$	
	5				
2	6	$-(I_x S_y + I_y S_x)C_\Omega^\tau 2S_a^\tau$		$-(I_x S_y + I_y S_x)C_\Omega^\tau 2S_a^\tau$	
	7				
1	8	$-(I_z S_y + I_y S_z)S_\Omega^\tau 2S_a^\tau$		$-(I_z S_y + I_y S_z)C_a^1 S_\Omega^\tau 2S_a^\tau$	
	9				
	10			$+(I_z^2 S_x + I_x S_z^2)2S_a^1 S_\Omega^\tau 2S_a^\tau$	
	11			$= (S_x+I_x)S_a^1 S_\Omega^\tau S_a^\tau$	

Table 9.4. Time development of ϱ during the detection period

Order quantum coherence during evolution period	Line no.	Column no.: 1 Column label: $\varrho(\tau+t_1^+)$ Operator:	2 $at_2(I_z S_z)$	3
0 Q	1	$+(I_y+S_y)C_\Omega^\tau C_a^\tau$	$+(I_y+S_y)C_a^2 C_\Omega^\tau C_a^\tau$	
	2			
	3		$-(I_x S_z + I_z S_x)2S_a^2 C_\Omega^\tau C_a^\tau$	
	4			
1 Q	5	$(I_x+S_x)C_\Omega^1 C_a^1 S_\Omega^\tau C_a^\tau$	$(I_x+S_x)C_a^2 C_\Omega^1 C_a^1 S_\Omega^\tau C_a^\tau$	
	6			
	7		$(I_y S_z + S_y I_z)2S_a^2 C_\Omega^1 C_a^1 S_\Omega^\tau C_a^\tau$	
	8			
1 Q	9	$-(I_z S_y + I_y S_z)C_\Omega^1 2S_a^1 S_\Omega^\tau C_a^\tau$	$-(I_z S_y + I_y S_z)C_a^2 C_\Omega^1 2S_a^1 S_\Omega^\tau C_a^\tau$	
	10			
	11		$+(I_z^2 S_x + I_x S_z^2)2S_a^2 C_\Omega^1 2S_a^1 S_\Omega^\tau C_a^\tau$	
	12			
2 Q	13	$-(I_x S_z + I_z S_x)C_{2\Omega}^1 C_\Omega^\tau 2S_a^\tau$	$-(I_x S_z + I_z S_x)C_a^2 C_{2\Omega}^1 C_\Omega^\tau 2S_a^\tau$	
	14			
	15		$-(I_y S_z^2 + I_z^2 S_y)2S_a^2 C_{2\Omega}^1 C_\Omega^\tau 2S_a^\tau$	
	16			
1 Q	17	$(I_y S_z + I_z S_y)C_\Omega^1 C_a^1 S_\Omega^\tau 2S_a^\tau$	$(I_y S_z + I_z S_y)C_a^2 C_\Omega^1 C_a^1 S_\Omega^\tau 2S_a^\tau$	
	18			
	19		$-(I_x S_z^2 + I_z^2 S_x)2S_a^2 C_\Omega^1 S_\Omega^\tau 2S_a^\tau$	
	20			
1 Q	21	$(S_x+I_x)C_\Omega^1 S_a^1 S_\Omega^\tau S_a^\tau$	$(S_x+I_x)C_a^2 C_\Omega^1 S_a^1 S_\Omega^\tau S_a^\tau$	
	22			
	23		$(S_y I_z + I_y S_z)2S_a^2 C_\Omega^1 S_a^1 S_\Omega^\tau S_a^\tau$	

Table 9.3 (continued)

5 $\varrho(\tau+t_1^-)$	6 $\bar{X}(\pi/2)$	7 $\varrho(\tau+t_1^+)$	8 Number of quanta represented by column 7
$-(I_z+S_z)C_\Omega^\tau C_a^\tau$		$+(I_y+S_y)C_\Omega^\tau C_a^\tau$	$1Q$
$(I_x+S_x)C_\Omega^1 C_a^1 S_\Omega^\tau C_a^\tau$		$(I_x+S_x)C_\Omega^1 C_a^1 S_\Omega^\tau C_a^\tau$	$1Q$
$-(I_y+S_y)S_\Omega^1 C_a^1 S_\Omega^\tau C_a^\tau$		$-(I_z+S_z)S_\Omega^1 C_a^1 S_\Omega^\tau C_a^\tau$	$0Q$
$(I_y S_z+I_z S_y)C_\Omega^1 2 S_a^1 S_\Omega^\tau C_a^\tau$		$-(I_z S_y+I_y S_z)C_\Omega^1 2 S_a^1 S_\Omega^\tau C_a^\tau$	$1Q$
$(I_x S_z+I_z S_x)S_\Omega^1 2 S_a^1 S_\Omega^\tau C_a^\tau$		$-(I_x S_y+I_y S_x)S_\Omega^1 2 S_a^1 S_\Omega^\tau C_a^\tau$	$2Q$
$-(I_x S_y+I_y S_x)[(C_\Omega^1)^2-(S_\Omega^1)^2]C_M^\tau 2 S_a^\tau$		$-(I_x S_z+I_z S_x)C_{2\Omega}^1 C_\Omega^1 2 S_a^\tau$	$1Q$
$+(I_y S_y-I_x S_x)2 C_\Omega^1 S_\Omega^1 C_\Omega^\tau 2 S_a^\tau$		$+(I_z S_z-I_x S_x)S_{2\Omega}^1 C_\Omega^1 2 S_a^\tau$	$0,2Q$
$-(I_z S_y+I_y S_z)C_\Omega^1 C_a^1 S_\Omega^\tau 2 S_a^\tau$		$(I_y S_z+I_z S_y)C_\Omega^1 C_a^1 S_\Omega^\tau 2 S_a^\tau$	$1Q$
$-(I_z S_x+I_x S_z)S_\Omega^1 C_a^1 S_\Omega^\tau 2 S_a^\tau$		$+(I_y S_x+I_x S_y)S_\Omega^1 C_a^1 S_\Omega^\tau 2 S_a^\tau$	$2Q$
$(S_x+I_x)C_\Omega^1 S_a^1 S_\Omega^\tau S_a^\tau$		$(S_x+I_x)C_\Omega^1 S_a^1 S_\Omega^\tau S_a^\tau$	$1Q$
$-(S_y+I_y)S_\Omega^1 S_a^1 S_\Omega^\tau S_a^\tau$		$-(S_z+I_z)S_\Omega^1 S_a^1 S_\Omega^\tau S_a^\tau$	$0Q$

Table 9.4 (continued)

4	5 $\varrho(\tau+t_1+t_2)$
$\Omega t_2(I_z+S_z)$	

$+(I_y+S_y)C_\Omega^2 C_a^2 C_\Omega^\tau C_a^\tau$

$+(I_x+S_x)S_\Omega^2 C_a^2 C_\Omega^\tau C_a^\tau$

$(I_x+S_x)C_\Omega^2 C_a^2 C_\Omega^1 C_a^1 S_\Omega^\tau C_a^\tau$

$-(I_y+S_y)S_\Omega^2 C_a^2 C_\Omega^1 C_a^1 S_\Omega^\tau C_a^\tau$

$(I_z^2 S_x+I_x S_z^2)C_\Omega^2 2 S_a^2 C_\Omega^1 2 S_a^1 S_\Omega^\tau C_a^\tau = (S_x+I_x)C_\Omega^2 S_a^2 C_\Omega^1 S_a^1 S_\Omega^\tau C_a^\tau$

$-(I_z^2 S_y+I_y S_z^2)S_\Omega^2 2 S_a^2 C_\Omega^1 2 S_a^1 S_\Omega^\tau C_a^\tau = -(S_y+I_y)S_\Omega^2 S_a^2 C_\Omega^1 S_a^1 S_\Omega^\tau C_a^\tau$

$-(I_y S_z^2+I_z^2 S_y)C_\Omega^2 2 S_a^2 C_{2\Omega}^1 C_\Omega^\tau 2 S_a^\tau = -(I_y+S_y)C_\Omega^2 S_a^2 C_{2\Omega}^1 C_\Omega^\tau S_a^\tau$

$-(I_x S_z^2+I_z^2 S_x)S_\Omega^2 2 S_a^2 C_{2\Omega}^1 2 S_a^\tau = -(I_x+S_x)S_\Omega^2 S_a^2 C_{2\Omega}^1 C_\Omega^\tau C_\Omega^\tau S_a^\tau$

$-(I_x S_z^2+I_z^2 S_x)C_\Omega^2 2 S_a^2 C_\Omega^1 C_a^1 S_\Omega^\tau 2 S_a^\tau = -(I_x+S_x)C_\Omega^2 S_a^2 C_\Omega^1 C_a^1 S_\Omega^\tau S_a^\tau$

$+(I_y S_z^2+I_z^2 S_y)S_\Omega^2 2 C_a^2 C_\Omega^1 C_a^1 S_\Omega^\tau 2 S_a^\tau = (I_y+S_y)S_\Omega^2 S_a^2 C_\Omega^1 C_a^1 S_\Omega^\tau S_a^\tau$

$(S_x+I_x)C_\Omega^2 C_a^2 C_\Omega^1 S_a^1 S_\Omega^\tau S_a^\tau$

$-(S_y+I_y)S_\Omega^2 C_a^2 C_\Omega^1 S_a^1 S_\Omega^\tau S_a^\tau$

$$\exp\left(\mathrm{i}[\Omega(M - M') + a(m'_I m'_S - m_I m_S)]t\right)$$
$$= \exp\left(\mathrm{i}\Omega(M - M')t\right) \exp\left(\mathrm{i}a(m'_I m'_S - m_I m_S)t\right) \quad , \tag{9.66}$$

where the first factor arises from the Zeeman offset, and the second from the spin-spin coupling. Recalling that $n = |M - M'|$ gives the order, n, of the coherence we see immediately that the time dependence of the frequency offset uniquely labels the order of coherence of the element. This fact is evident in Table 9.3. The time dependence of the Zeeman offset is given by the terms involving Ω and t_1. They are to be found in column 5, and are re-expressed in column 7. Since $M - M' = 0$ for zero-quantum terms, $(M - M')\Omega t_1 = 0$, hence the Zeeman offset term is merely a constant independent of Ω or t_1. This prediction agrees with the table where the zero-quantum terms (line 1) have *no* terms depending on Ωt_1. The one-quantum terms (lines 2, 3, 4, 5, 8, 9, 10, 11) all involve either C_Ω^1 or S_Ω^1. The two-quantum terms (lines 6 and 7) involve $(C_\Omega^1)^2 - (S_\Omega^1)^2 = C_{2\Omega}^1$ or $2C_\Omega^1 S_\Omega^1 = S_{2\Omega}^1$, hence are at frequency 2Ω.

A second important feature of the table can be seen by comparing column 3 with column 1, showing the effect of the spin-spin coupling (column 2) on ϱ. Examination of line 6, the matrix elements corresponding to two-quantum coherence, shows that the spin-spin coupling has no effect on ϱ. To get this result using the operator method, one gets *eight* terms. They can, however, be added together. Using relationships such as $\sin^2 + \cos^2 = 1$ and $I_x I_y = \mathrm{i}I_z/2$, they can be reduced to a single final term identical to the starting term. One naturally wonders if there is some deeper significance — indeed there is.

The effect of the spin-spin coupling is given by the second factor of (9.66)

$$\exp\left(\mathrm{i}a(m'_I m'_S - m_I m_S)t\right) \quad , \tag{9.67}$$

which, for the two-quantum term, has

$$m_I = m_S = \tfrac{1}{2} \quad \text{and} \quad m'_I = m'_S = -\tfrac{1}{2} \tag{9.68}$$

or vice versa. But

$$a\left(\tfrac{1}{2}\right)\left(\tfrac{1}{2}\right) = a\left(-\tfrac{1}{2}\right)\left(-\tfrac{1}{2}\right) \quad . \tag{9.69}$$

Therefore, the spin-spin factor becomes 1 and we get no difference on line 6 between columns 1 and 3.

If one has a system with more than two spins, a similar theorem is true. Consider a general system with N spins of $\tfrac{1}{2}$, and a Hamiltonian in the rotating frame

$$\mathcal{H} = -\sum_i \hbar \Omega_i I_{zi} + \sum_{ij}(A_{ij} \mathbf{I}_i \cdot \mathbf{I}_j + B_{ij} I_{zi} I_{zj}) \tag{9.70}$$

corresponding to various chemical shifts and a generalized secular portion of the spin-spin coupling. The total angular momentum operator

$$F_z \equiv \sum_i I_{zi} \tag{9.71}$$

with eigenvalues M_F commutes with $\mathcal{H}$.

Thus, we know that the state in which all the spins point up (with $M_F = N/2$) and the state in which all spins point down ($M = -N/2$) are both eigenstates. Using a notation $|m_1, m_2, m_3, \ldots, m_N\rangle$ for the m-values of spins 1, 2, 3, etc., we know that the state $|N/2\rangle$ in which all the spins point up (i.e. with $M = N/2$) is given in terms of individual spin m values by

$$|N/2\rangle = |\tfrac{1}{2}, \tfrac{1}{2}, \tfrac{1}{2}, \ldots, \tfrac{1}{2}\rangle \tag{9.72a}$$

and the state with all spins point down is

$$|-N/2\rangle = |-\tfrac{1}{2}, -\tfrac{1}{2}, \ldots, -\tfrac{1}{2}\rangle \quad . \tag{9.72b}$$

It is easy to show by explicit evaluation using (9.70) that these two states have Zeeman energies which are equal in magnitude but opposite in sign, whereas the dipolar energies are equal. Therefore, the energy difference between the states, which gives the time dependence of the N-quantum coherence, is independent of the strength of the dipolar coupling.

We turn now to the detection method period described by Tables 9.3 and 9.4. This requires converting the various orders of quantum coherence back to the observable one-quantum coherence, the step called mixing. Since a $\pi/2$ pulse initially converted one-quantum to n-quantum coherence, it is natural to ask if it would reverse the process. Indeed, in Table 9.3 we analyze the effect of an $\overline{X}(\pi/2)$ pulse, which is just the inverse of the original $X(\pi/2)$ pulse.

The result, in column 7 of Table 9.3, is labeled in column 8 according to the orders of the resulting quantum coherence. For example, line 3 represents zero-order quantum coherence. It would be undetectable.

In Table 9.4 we keep only the coherences which are single quantum during the detection interval, but label the rows by the order of the quantum coherence during the evolution period.

In column 3 there are terms such as $(I_x S_z + I_z S_x)$ which eventually give no signal because when we form $\mathrm{Tr}\{\varrho(I^+ + S^+)\}$ the I_z and S_z cause the trace to vanish. We do not therefore bother to follow these terms past column 3. In column 5, we include the effect of the Zeeman offset. The signal is observed as a function of t_2, and is given in terms of $\boldsymbol{F} = \boldsymbol{I} + \boldsymbol{S}$ as

$$\langle F^+ \rangle = \langle F_x \rangle + \mathrm{i}\langle F_y \rangle = \mathrm{Tr}\{(I^+ + S^+)\varrho\} \quad . \tag{9.73}$$

Utilizing (9.73), we can combine various lines of column 5 in Table 9.4 to get the contributions of the zero-, one-, and two-quantum coherences to $\langle F^+ \rangle$. Denoting by

$$\langle F^+ \rangle_n \tag{9.74}$$

the contribution of the nth order quantum coherence, we get for the zero-quantum coherence

$$\langle F^+ \rangle_0 = \frac{\mathrm{i}}{4}[e^{-\mathrm{i}(\Omega+a/2)t_2} + e^{-\mathrm{i}(\Omega-a/2)t_2}] \times [e^{\mathrm{i}(\Omega+a/2)\tau} + e^{\mathrm{i}(\Omega-a/2)\tau} + e^{-\mathrm{i}(\Omega+a/2)\tau} + e^{-\mathrm{i}(\Omega-a/2)\tau}] \quad . \tag{9.75}$$

For the one-quantum coherence

$$\langle F^+ \rangle_1 = 2e^{-i\Omega t_2} \cos[a(t_1 - \tau - t_2)/2] \cos(\Omega t_1) \sin(\Omega \tau) , \qquad (9.76)$$

which, in complex form, gives

$$\begin{aligned}\langle F^+ \rangle_1 = \frac{i}{4}[&e^{-i(\Omega - a/2)t_2}(e^{i(\Omega + a/2)t_1} + e^{-i(\Omega - a/2)t_1}) \\ &\times (e^{i(\Omega + a/2)\tau} + e^{-i(\Omega - a/2)\tau}) \\ &+ e^{-i(\Omega + a/2)t_2}(e^{i(\Omega - a/2)t_1} + e^{-i(\Omega + a/2)t_1}) \\ &\times (e^{i(\Omega + a/2)\tau} + e^{-i(\Omega - a/2)\tau})] .\end{aligned}$$

For two-quantum coherence we get

$$\begin{aligned}\langle F^+ \rangle_2 &= -2ie^{-i\Omega t_2} \sin(at_2/2) \cos(2\Omega t_1) \cos(\Omega \tau) \sin(a\tau/2) \\ &= \frac{i}{8}[e^{-i(\Omega + a/2)t_2} - e^{-i(\Omega - a/2)t_2}] \\ &\quad \times (e^{2i\Omega t_1} + e^{-2i\Omega t_1})(e^{i(\Omega + a/2)\tau} - e^{i(\Omega - a/2)\tau} \\ &\quad + e^{-i(\Omega - a/2)\tau} - e^{-i(\Omega + a/2)\tau}) .\end{aligned} \qquad (9.77)$$

Referring back to (9.62), we recognize that the elements of the density matrix will have time variations

$$e^{+i(\Omega \pm a/2)t} \qquad (9.78a)$$

for the states $M - M' = +1$, will have

$$e^{-i(\Omega \pm a/2)t} \qquad (9.78b)$$

for $M - M' = -1$, will have

$$e^{2i\Omega t} \qquad (9.78c)$$

for $M - M' = 2$, and

$$e^{-2i\Omega t} \qquad (9.78d)$$

for $M - M' = -2$. We know from the Hermitian property of ϱ that whenever we have generated an $M - M' = p$ matrix element, we have also generated an $M - M' = -p$ element. We note next that all three $\langle F^+ \rangle_n$'s have the factors $\exp[-i(\Omega + a/2)t_2]$ and $\exp[-i(\Omega - a/2)t_2]$. This corresponds to the fact that during t_2 the spins are precessing in a given sense.

We can identify the role of the preparation interval by looking at the τ-dependent terms. Note that these all involve single-quantum terms since their time dependence involves the frequencies $(\Omega \pm a/2)$. We note that the time dependence t_1 tags the evolution period, and the order of the coherence is immediately obvious from the Ω dependence which is $\exp(\pm 0i\Omega t_1)$ for zero order, $\exp(\pm 1i\Omega t_1)$ for single order, and $\exp(\pm 2i\Omega t_1)$ for double-order quantum coherence. Thus, if one studies how $\langle F^+ \rangle$ varies with t_1, one can isolate the zero-, one-, and two-quantum coherences. The expressions for $\langle F^+ \rangle_p$ show clearly the conversion of the initial z magnetization to one-quantum coherence for the preparation period, the fact that zero-, one-, and two-quantum coherences have been brought into existence for time interval t_1, and the conversion back to single-quantum coherence for observation.

We have seen that the pulse sequence $X(\pi/2)\ldots\tau\ldots X(\pi/2)$ will produce zero-, one-, and two-quantum coherence. As we remarked earlier, if we study column 10 of Table 9.1, we note that the zero- and two-quantum coherence terms (lines 1 and 3) are proportional to $\cos\Omega\tau$, whereas the one-quantum terms (lines 2 and 4) are proportional to $\sin\Omega\tau$. Therefore, if $\Omega = 0$ (exact resonance) we would not produce one-quantum coherence. The question, how, when we are tuned exactly to resonance, can we produce one-quantum coherence with a pair of spins, turns into the question, how do we produce odd-order quantum coherence for an arbitrary number of spins, when tuned exactly to resonance? We turn now to a simple approach.

The disappearance of the one-quantum coherence for a pair of spins can be traced back to column 5 of Table 9.1 where we see the effect of the spin-spin coupling alone [i.e. before the resonance offset term $\Omega\tau(I_z + S_z)$ acts]. It produces $(I_y + S_y)C_a^\tau$ on line 1 and $-(I_x S_z + I_z S_x)C_a^\tau$ on line 3. These terms, if acted on by $X(\pi/2)$, produce $(I_z + S_z)C_a^\tau$ (a zero-quantum coherence) and $-(I_x S_y + I_y S_x)2S_a^\tau$ (two-quantum coherence).

To get the one-quantum terms we allowed $\Omega\tau(I_z + S_z)$ to turn I_x into $-I_y$, I_y into I_x, then we applied an $X(\pi/2)$ pulse. Therefore, if we have $\Omega = 0$, so that I_x and I_y are not rotated by $\pi/2$, we should get one-quantum terms by rotating the pulse from being an $X(\pi/2)$ to being a $Y(\pi/2)$ pulse. Indeed, we see that a $Y(\pi/2)$ pulse would leave $(I_y + S_y)C_a^\tau$ alone as a one-quantum coherence, but turn $-(I_x S_z + I_z S_x)2S_a^\tau$ into $(I_z S_x + I_x S_z)2S_a^\tau$, a one-quantum term.

In general, of course, $\Omega \neq 0$. Then the points we have just discussed show that production of a particular order of quantum coherence depends on the amount one is off resonance. That is frequently a disadvantageous situation. It can be remedied by making Ω effectively zero by means of a spin echo, then choosing the phase of the second pulse to get the desired order of coherence. Thus, we can use a pulse sequence

$$X(\pi/2)\ldots\tau/2\ldots X(\pi)\ldots\tau/2\ldots [Y(\pi/2) \text{ or } X(\pi/2)]$$

to produce the multiple-quantum coherence. One can also then eliminate the offset effects during the detection period by introducing a π pulse (for example at $t_2 = \tau/2$) to give an echo (at $t_2 = \tau$).

9.3.4 Three or More Spins

We have seen that a pulse sequence $X(\pi/2)\ldots\tau\ldots X(\pi/2)$ will produce zero-, one-, and two-quantum coherence from a pair of spins if τ is sufficiently long. What happens if there are a larger number of coupled spins? Will this sequence produce higher orders, or must one add more pulses? We shall show that the sequence still works.

We approach this problem two ways, first by considering three spin $\frac{1}{2}$ nuclei with a coupling of the form $\hbar a_{ik} I_{jz} I_{kz}$, which we can solve exactly, then we will look at the general case which we will treat by an approximation method.

Let us start then with a Hamiltonian for three coupled spins which generalizes that of (9.54a):

$$\mathcal{H} = -\hbar\Omega I_{zT} + \hbar(a_{12} I_{z1} I_{z2} + a_{23} I_{z2} I_{z3} + a_{31} I_{z3} I_{z1})$$

where

$$I_{\alpha T} = I_{\alpha 1} + I_{\alpha 2} + I_{\alpha 3} \quad , \quad \alpha = x, y, z \quad . \tag{9.79}$$

(Note, we shall assume below that the coupling between any pair, such as a_{12}, can also be written as a_{21}.) As can be seen from the case of two spins, it is the spin-spin coupling terms which lead to multiple quantum coherence. Therefore we need to focus on the second term of (9.79). Then, as with (9.55), we take $\varrho(0^-)$, just before the first pulse, as

$$\varrho(0^-) = I_{zT} \quad , \tag{9.80}$$

so that at $t = 0^+$, just after the first $X(\pi/2)$ pulse,

$$\varrho(0^+) = I_{yT} \quad . \tag{9.81}$$

Then we wish to calculate

$$\exp\left(-\frac{i}{\hbar}\mathcal{H}_{ss}t\right) I_{xT} \exp\left(\frac{i}{\hbar}\mathcal{H}_{ss}t\right) \quad , \quad \text{where} \tag{9.82}$$

$$\mathcal{H}_{ss} = \hbar(a_{12} I_{z1} I_{z2} + a_{23} I_{z2} I_{z3} + a_{31} I_{z3} I_{z1}) \quad . \tag{9.83}$$

Since $I_{yT} = I_{y1} + I_{y2} + I_{y3}$, and since all three spin-spin terms commute with each other, we have to compute quantities such as

$$\exp(-ia_{12} I_{z1} I_{z2} t) \exp(-ia_{23} I_{z2} I_{z3} t) \exp(-ia_{31} I_{z3} I_{z1} t) I_{y1}$$
$$\times \exp(ia_{31} I_{z3} I_{z1}) \exp(ia_{23} I_{z2} I_{z3} t) \exp(ia_{12} I_{z1} I_{z2} t) \quad . \tag{9.84}$$

In (9.84) the exponentials involving $I_{z2} I_{z3}$ commute with I_{y1}, hence annihilate one another so that (9.34) reduces to

$$\exp(-ia_{12} I_{z1} I_{z2} t) \exp(-ia_{31} I_{z3} I_{z1} t) I_{y1}$$
$$\times \exp(ia_{31} I_{z3} I_{z1} t) \exp(ia_{12} I_{z1} I_{z2} t) \quad . \tag{9.85}$$

We can now apply either of the (1,3) operators to get a result similar to column 5 of Table 9.1. Then we apply the (1,2) operators to that result. The final result is

$$\exp\left(-\frac{i}{\hbar}\mathcal{H}_{ss}t\right) I_{x1} \exp\left(\frac{i}{\hbar}\mathcal{H}_{ss}t\right)$$
$$= \cos(a_{31}\tau/2) \cos(a_{12}\tau/2) I_{y1} - \cos(a_{31}\tau/2) 2 \sin(a_{12}\tau/2) I_{x1} I_{z2}$$
$$- 2 \sin(a_{31}\tau/2) \cos(a_{12}\tau/2) I_{x1} I_{z3}$$
$$- 2 \sin(a_{31}\tau/2) 2 \sin(a_{12}\tau/2) I_{y1} I_{z2} I_{z3} \quad . \tag{9.86}$$

We have written (9.86) in a manner intended to help indicate how, using the spin operator formalism, each term arises.

We note first of all, that only spin 1 has either x- or y-components. Secondly, we note that only coupling constants a_{12} and a_{31} involving spin 1 occur. This is because terms such as $a_{23}I_{z2}I_{z3}$ commute with I_{y1}. Thus, if we had four spins, we would get from I_{y1} terms involving a_{12}, a_{13}, and a_{14} but no terms involving a_{23}, a_{24}, a_{34}.

We note that we can easily find the contributions from I_{y2} and I_{y3} from (9.86) by cyclic permutation of the indices. Expressing I_{x1} and I_{y1} in terms of raising and lowering operators shows us that all terms of (9.86) connect states M and M' which satisfy

$$M = M' \pm 1 \quad . \tag{9.87}$$

Thus all are one-quantum operators.

The frequency off-set term in $\mathcal{H}$, $-\hbar\Omega I_z T$, will cause a precession of the x- and y-components of spin, as can be seen from column 7 of Table 9.1. It will leave the I_{zk} ($k = 2, 3$) terms alone, but will convert I_{x1} or I_{y1} into linear combinations of I_{x1} and I_{y1} so that we should replace I_{x1} by

$$I_{x1} \cos \Omega t - I_{y1} \sin \Omega t \tag{9.88}$$

and replace I_{y1} by

$$I_{y1} \cos \Omega t + I_{x1} \sin \Omega t \quad , \tag{9.89}$$

giving us for the contribution of I_{x1} to $\varrho(t)$

$$\begin{aligned}\varrho(t))_{I_{x1}} &= \cos(a_{31}\tau/2)\cos(a_{12}\tau/2)[\cos(\Omega t)I_{y1} + \sin(\Omega t)I_{x1}] \\&+ \cos(a_{31}\tau/2)2\sin(a_{12}\tau/2)[\cos(\Omega t)I_{x1}I_{z2} - \sin(\Omega t)I_{y1}I_{z2}] \\&+ 2\sin(a_{31}\tau/2)\cos(a_{12}\tau/2)[\cos(\Omega t)I_{x1}I_{z2} - \sin(\Omega t)I_{y1}I_{z3}] \\&- 2\sin(a_{31}\tau/2)2\sin(a_{12}\tau/2)[\cos(\Omega t)I_{y1}I_{z2}I_{z3} + \sin(\Omega t)I_{x1}I_{z2}I_{z3}] \\&+ \text{terms obtained by permuting the indices } 1, 2, 3 \quad . \end{aligned} \tag{9.90}$$

This Hamiltonian is now acted on by the $X(\pi/2)$ pulse. In order to see what spin coherences are thereby produced, we need only examine what happens to the spin factors. The result is a table of operators before and after, and the order of quantum coherences which can arise from each term. Denoting the values

$$M - M' = p$$

for nonvanishing $(M\alpha|\varrho|M'\alpha')$, we get Table 9.5. The various possible orders of quantum coherence can be deduced by expressing the spin operators as linear combinations of I_{zk}, I_k^+, or I_k^- ($k = 1, 2, 3$), then multiplying them to get terms such as for line 4

Table 9.5. Effect of $X(\pi/2)$ pulse on operators

Line	Operator before $X(\pi/2)$ pulse	Operator after $X(\pi/2)$ pulse	Orders of quantum coherence, p
1	I_{y1}	$-I_{z1}$	0
2	I_{x1}	$I_{x1} = \frac{1}{2}(I_1^+ + I_1^-)$	± 1
3	$I_{x1}I_{z2}$	$I_{x1}I_{y2} = \frac{1}{4i}(I_1^+ + I_1^-)(I_2^+ - I_2^-)$	$\pm 2, 0$
4	$I_{y1}I_{z2}$	$-I_{z1}I_{y2} = -\frac{1}{2i}I_{z1}(I_1^+ - I_1^-)$	± 1
5	$I_{x1}I_{z3}$	$I_{x1}I_{y3} = \frac{1}{4i}(I_1^+ + I_1^-)(I_3^+ - I_3^-)$	$\pm 2, 0$
6	$I_{y1}I_{z3}$	$-I_{z1}I_{y3} = -\frac{1}{2i}I_{z1}(I_3^+ - I_3^-)$	± 1
7	$I_{y1}I_{z2}I_{z3}$	$-I_{z1}I_{y2}I_{y3} = -\frac{1}{4i}I_{z1}(I_2^+ - I_2^-)(I_3^+ - I_3^-)$	$\pm 2, 0$
8	$I_{x1}I_{z2}I_{z3}$	$I_{x1}I_{y2}I_{y3} = \frac{1}{8}(I_1^+ + I_1^-)(I_2^+ - I_2^-)(I_3^+ - I_3^-)$	$\pm 3, \pm 1$

$$\frac{1}{4i}I_1^+ I_2^+ \quad p = +2 \quad \text{or}$$

$$\frac{1}{4i}I_1^+ I_2^- \quad p = 0 \quad \text{or}$$

$$\frac{1}{4i}I_1^- I_2^- \quad p = -2 \; . \tag{9.91}$$

Once one has made such a table, one can work out the result in one's head since an I_{xk} or an I_{yk} gives ± 1, so a product of two such operators gives p values of $1 + 1 = 2$, $1 - 1 = 0$, $-1 + 1 = 0$, $-1 - 1 = -2$.

We see that the three spins give us p values ranging from $+3$ to -3, as we proved earlier. Now, however, we can see how long τ must be to produce the coherence. Note that if $a_{12} \cong a_{13}$, that the three-quantum coherence will develop in a time when $\sin(a_{12}\tau/2) \cong 1$ or $a_{12}\tau/2 \cong \pi/2$. If $a_{13} \ll a_{12}$, then we must wait until $a_{13}\tau/2 \cong \pi/2$. In the process, as time develops, $\sin(a_{12}\tau/2)$ will have been through several oscillations. Thus, we must wait for the weakest coupling.

Suppose, however, we had three spins in a line so that spins 1 and 3, on opposite ends of the chain, are weakly coupled. Then we expect

$$a_{12} \cong a_{23} \gg a_{13} \; . \tag{9.92}$$

Then our expression (9.90) would imply a_{13} would control how long it would take to generate a three-quantum coherence. That conclusion would, however, be wrong since, had we analyzed the time dependence of I_{x2} instead of I_{x1}, we would have gotten a three-quantum coherence dependent on

$$\sin(a_{12}\tau/2)\sin(a_{23}\tau/2) \quad , \tag{9.93}$$

which, using (9.92), is produced more quickly.

Therefore, if *all* spins in the 3 system are close together, we can produce the one-, two-, and three-quantum coherences when the strong coupling a_{jk}'s satisfy

$$a_{jk}\tau/2 \cong \pi/2 \quad . \tag{9.94}$$

If, however, two spins are close, the third far away, two of the three coupling constants are small, so that while two-quantum coherence can be produced quickly, three-quantum coherence is produced slowly.

We can also analyze the two-pulse method for nuclei whose spin is not restricted to $I = \frac{1}{2}$, and for arbitrary forms of spin-spin coupling, such as dipolar coupling instead of coupling such as aI_zS_z, by using a power series expansion. Suppose then we have a Hamiltonian in the rotating frame

$$\mathcal{H} = -\hbar\Omega I_{z\mathrm{T}} + \sum_{j>k} \mathcal{H}_{jk} \equiv \mathcal{H}_z + \mathcal{H}_{ss} \quad , \tag{9.95a}$$

where the Ω represents being off resonance and $\mathcal{H}_{ss}$ and $\mathcal{H}_{jk}$'s are the secular part of the spin-spin couplings. That is

$$[\mathcal{H}_{jk}, I_{z\mathrm{T}}] = 0 \quad , \quad [\mathcal{H}_{ss}, I_{z\mathrm{T}}] = 0 \quad . \tag{9.95b}$$

For example, for dipolar coupling

$$\mathcal{H}_{jk} = B_{jk}(3I_{zj}I_{zk} - \boldsymbol{I}_j \cdot \boldsymbol{I}_k) \tag{9.96a}$$

$$= B_{jk}(2I_{zj}I_{zk} - I_{xj}I_{xk} - I_{yj}I_{yk}) \quad . \tag{9.96b}$$

As in (9.7b), we then have quantum numbers M and α where

$$I_{z\mathrm{T}}|M\alpha\rangle = M|M\alpha\rangle \quad \text{and} \quad \mathcal{H}_{ss}|M\alpha\rangle = E_\alpha|M\alpha\rangle \quad . \tag{9.97}$$

Then, the time development up to just before the second $X(\pi/2)$ pulse is given by

$$\begin{aligned}\varrho(\tau^-) &= \exp\left(-\frac{\mathrm{i}}{\hbar}\mathcal{H}\tau\right)X(\pi/2)I_{z\mathrm{T}}\overline{X}(\pi/2)\exp\left(\frac{\mathrm{i}}{\hbar}\mathcal{H}\tau\right) \\ &= \exp\left(-\frac{\mathrm{i}}{\hbar}\mathcal{H}\tau\right)I_{y\mathrm{T}}\exp\left(\frac{\mathrm{i}}{\hbar}\mathcal{H}\tau\right) \\ &= \exp(\mathrm{i}\Omega\tau I_z)\exp\left(-\frac{\mathrm{i}}{\hbar}\mathcal{H}_{ss}\tau\right)I_{y\mathrm{T}}\exp\left(\frac{\mathrm{i}}{\hbar}\mathcal{H}_{ss}\tau\right)\exp(-\mathrm{i}\Omega\tau I_z) \quad .\end{aligned} \tag{9.98}$$

The spin-spin exponential can be expanded using the theorem (see for example [9.16])

$$e^A B e^{-A} = 1 + [A,B] + \tfrac{1}{2}[A,[A,B]] + \ldots \tag{9.99}$$

to give

$$\varrho(\tau^-) = e^{+i\Omega\tau I_z}\left\{I_{y\text{T}} - \frac{i}{\hbar}\tau[\mathcal{H}_{ss}, I_{y\text{T}}]\right.$$
$$\left. + \left(\frac{i}{\hbar}\right)^2\frac{1}{2!}[\mathcal{H}_{ss},[\mathcal{H}_{ss}, I_{y\text{T}}]] + \ldots\right\}e^{-i\Omega\tau I_z} \quad . \tag{9.100}$$

Consider now the first commutator

$$[\mathcal{H}_{ss}, I_{y\text{T}}] = \sum_{\substack{j > k \\ l}}[\mathcal{H}_{jk}, I_{yl}] \quad . \tag{9.101}$$

If l is different from both j and k, the commutator vanishes. Thus, we have typical terms such as

$$[\mathcal{H}_{12}, I_{y1} + I_{y2}] \quad , \tag{9.102}$$

which are just what one would have for a pair of coupled spins. It is straightforward to evaluate the commutation for a specific choice of H_{12}. Thus, for dipolar coupling, we get

$$\begin{aligned}[\mathcal{H}_{12}, (I_{y1} + I_{y2})] &= B_{12}[2I_{z1}I_{z2} - I_{x1}I_{x2} - I_{y1}I_{y2}, (I_{y1} + I_{y2})] \\
&= B_{12}[2(I_{z1}, I_{y1})I_{z2} - (I_{x1}, I_{y1})I_{x2} \\
&\quad + 2(I_{z2}, I_{y2})I_{z1} - (I_{x2}, I_{y2})I_{x1}] \\
&= -3i(I_{x1}I_{z2} + I_{z1}I_{x2})B_{12} \\
&= -\frac{3i}{2}B_{12}[(I_1^+ + I_1^-)I_{z2} + I_{z1}(I_2^+ + I_2^-)] \quad . \end{aligned} \tag{9.103}$$

From (9.103) we see that this commutator at this stage has one-quantum coherence. Indeed, we can see this in general by calculating a matrix element of $[\mathcal{H}_{ss}, I_{y\text{T}}]$

$$\begin{aligned}(M\alpha|\mathcal{H}_{ss}I_{y\text{T}} &- I_{y\text{T}}\mathcal{H}_{ss}|M'\alpha') \\
&= \sum_{M'',\alpha''}(M\alpha|\mathcal{H}_{ss}|M''\alpha'')(M''\alpha''|I_{y\text{T}}|M'\alpha') \\
&\quad - (M\alpha|I_{y\text{T}}|M''\alpha'')(M''\alpha''|\mathcal{H}_{ss}|M'\alpha')\end{aligned} \tag{9.104}$$

which, using (9.97), which says $\mathcal{H}_{ss}$ is diagonal in M, gives

$$\begin{aligned}&= \sum_{\alpha''}(M\alpha|\mathcal{H}_{ss}|M\alpha'')(M\alpha''|I_{y\text{T}}|M'\alpha') \\
&\quad - (M\alpha|I_{y\text{T}}|M'\alpha'')(M'\alpha''|\mathcal{H}_s|M'\alpha') \quad .\end{aligned} \tag{9.105}$$

Thus M and M' are states which are joined by $I_{y\text{T}}$. But

$$I_{y\text{T}} = \frac{1}{2i}(I_\text{T}^+ - I_\text{T}^-) \quad \text{so that} \tag{9.106}$$

$$M = M' \pm 1 \quad . \tag{9.107}$$

Equation (9.107) is, however, the condition for one-quantum coherence. In a similar way, one can show that every term in the series (9.101) generates one-quantum coherence, a fact to which we will return.

Returning to the two-spin case, (9.100 and 103), we can now apply the Zeeman operators $\exp(i\Omega\tau I_{zT})$ to get for the density matrix at time τ^-

$$\varrho(\tau^-) = I_{yT}\cos\Omega\tau + I_{xT}\sin\Omega\tau - \frac{3i}{2}B_{12}[I_{x1}I_{z2}\cos\Omega\tau$$
$$- I_{y1}I_{z2}\sin\Omega\tau + I_{z1}I_{x2}\cos\Omega\tau - I_{z1}I_{y2}\sin\Omega\tau] \quad . \quad (9.108)$$

Then, the $X(\pi/2)$ pulse will produce

$$\varrho(\tau^+) = -I_{zT}\cos\Omega\tau + I_{xT}\sin\Omega\tau$$
$$-\frac{3i}{2}B_{12}[(I_{x1}I_{y2} + I_{y1}I_{x2})\cos\Omega\tau$$
$$+ (I_{z1}I_{y2} + I_{y1}I_{z2})\sin\Omega\tau]\ldots \quad . \quad (9.109)$$

Expressing the I_x and I_y operators in terms of raising and lowering operators, (or using Table 9.2) we have

$$\varrho(\tau^+) = -I_{zT}\cos\Omega\tau + I_{xT}\sin\Omega\tau$$
$$-\frac{3i}{2}B_{12}\frac{\tau}{2i}\{(I_1^+ I_2^+ - I_1^- I_2^-)\cos\Omega\tau$$
$$+ [I_{z1}(I_2^+ - I_2^-) + (I_1^+ - I_1^-)I_{z2}]\sin\Omega\tau\}\ldots \quad . \quad (9.110)$$

We see clearly the one- and two-quantum coherences. The fact that the B_{12} term is proportional to τ is analogous to the small τ behavior of the term $\sin(a\tau/2)$ of lines 3 and 4 of column 10 of Table 9.1.

Clearly, we can use a similar approach to the next term in the series of (9.100)

$$\frac{1}{2}\left(\frac{i\tau}{\hbar}\right)^2 [\mathcal{H}_{ss}, [\mathcal{H}_{ss}, I_{yT}]] \quad . \quad (9.111)$$

Here we will have terms in the double commutation

$$[\mathcal{H}_{jk}, [\mathcal{H}_{lm}, I_{yn}]] \quad , \quad (9.112)$$

but since either l or m must equal n, this will be of the form

$$[\mathcal{H}_{jk}, [\mathcal{H}_{lm}, (I_{ym} + I_{yl})]] \quad . \quad (9.113)$$

Now, if *neither* j nor k is l or m, the outer commutator will vanish. Thus, we get either two-spin terms such as

$$[\mathcal{H}_{12}, [\mathcal{H}_{12}, (I_{y1} + I_{y2})]] \quad (9.114)$$

or three-spin terms such as

$$[\mathcal{H}_{13}, [\mathcal{H}_{12}, (I_{y1} + I_{y2})]] \quad . \quad (9.115)$$

In this manner, we see that each higher order of τ in the series (9.100) adds one more possible spin to the couplings.

We can now see another useful point by looking in detail at (9.103). Although the top line involves products of three spin operators, the commutation step reduces the product to two operators (e.g. I_{x1} with I_{z2}). In a similar way, one finds that the three-spin term (9.115) involves three spin operators, such as $I_{x1}I_{y2}I_{x3}$. In general each higher term in the series involves products of the previous term with a pair of spin operators, but then the commutator reduces the number by one. Thus, each successive term has one more spin operator in the product. Since it takes three spin products to get a three-quantum coherence, the τ^2 term is the first one in the series which can give three-quantum coherence. The τ^3 term is the first one which could give four-quantum coherence. Note that we say "could", implying that it does not necessarily do so. Thus, if one has two spin-$\frac{1}{2}$ nuclei, the highest-order coherence one can produce is second, hence the τ^2, τ^3, and higher terms cannot produce more than second-order coherence (double-quantum coherence) in this case. To find what coherences are actually realized, one must look in detail at the terms or use the general rules about the maximum coherence $2NI$ of a group of N nuclei of spin I.

Suppose in (9.115) one has a three-spin product involving coordinates of spins 1, 2, and 3. Expressing the I_{xk} and I_{yk}'s in terms of raising and lowering operators, one might ask, will there be terms in $\varrho(\tau^-)$ [just before the second $X(\pi/2)$ pulse] such as

$$I_1^+ I_2^+ I_3^+ \ ? \tag{9.116}$$

The answer must be "no" since this is a three-quantum term whereas we have said each term in the infinite series *prior* to the second $X(\pi/2)$ pulse has one-quantum coherence. Permissible terms then would be of the general form

$$I_1^+ I_2^- I_3^+ \quad \text{or} \tag{9.117a}$$

$$I_1^+ I_{z2} I_{z3} \ . \tag{9.117b}$$

These may be thought of as grouping a raising (or lowering) operator (e.g. I_1^+) with either zero-quantum operators I_{zk} or with paired raising and lowering operators (e.g. $I_2^- I_3^+$).

When these are operated on by the Zeeman operators, they simply multiply the I_k^+ or I_k^- by a phase $\exp(i\Omega\tau)$ or $\exp(-i\Omega\tau)$ respectively, or leave the I_{zk}'s alone. Thus, the second $X(\pi/2)$ pulse still acts either on expressions such as (9.117a or b).

Now

$$X I_1^+ \overline{X} = X(I_{x1} + iI_{y1})\overline{X} = I_{x1} - iI_{z1}$$

$$X I_1^- \overline{X} = I_{x1} + iI_{z1} \quad \text{so}$$

$$X(I_1^+ I_2^- I_3^+)\overline{X}$$
$$= (I_{x1} - iI_{z1})(I_{x2} + iI_{z2})(I_{x3} - iI_{z3})$$
$$= I_{x1}I_{x2}I_{x3} - iI_{x1}I_{x2}I_{z3} + iI_{x1}I_{z2}I_{x3} + I_{x1}I_{z2}I_{z3}$$
$$- iI_{z1}I_{x2}I_{x3} - I_{z1}I_{x2}I_{z3} + I_{z1}I_{z2}I_{x3} - iI_{z1}I_{z2}I_{z3} \quad . \quad (9.118)$$

When we express I_{x1}, I_{x2}, and I_{x3} with raising and lowering operators, we see explicitly how the three-quantum coherence arises. The term (9.117b) will give

$$XI_1^+ I_{2z} I_{3z}\overline{X} = (I_{x1} - iI_{z1})I_{y2}I_{y3}$$
$$= I_{x1}I_{y2}I_{y3} - iI_{z1}I_{y2}I_{y3} \quad . \quad (9.119)$$

The term $I_{x1}I_{y2}I_{y3}$ contains three-quantum coherence.

From our analysis of the power series expansion, using (9.9a), our conclusion is, then, that the pair of pulses $X(\pi/2)\ldots\tau\ldots X(\pi/2)$ will generate *all* orders of quantum coherence permitted. While it is a question of proper choice of τ to achieve the desired orders, in general, all allowed orders will be present. We thus have two problems yet to consider. (1) How can we detect a particular desired order? (2) Is it possible to generate a particular order without generating other orders? We turn to these topics in the next two sections.

9.3.5 Selecting the Signal of a Particular Order of Coherence

We turn now to a discussion of how to select the signal arising from a particular order of coherence. We have already mentioned one approach at the end of Sect. 9.3.3, where we noted that if the signal is Fourier analyzed with respect to the evolution time t_1, the coherence of order p gives rise to lines at frequency $p\Omega$, where Ω is the amount one is off resonance. Thus, introducing a deliberate resonance offset by an amount Ω large compared to the spin-spin splittings produces groups of lines in which the spectra of the different orders are well separated. This approach, and a variant of it (time proportional phase incrementation, TPPI) in which the offset is actually zero but is made to appear nonzero, has been widely employed, especially by the Pines group [9.5, 7]. We discuss TPPI at the end of this section.

Another powerful method was introduced by *Wokaun* and *Ernst* [9.17]. We turn to it now. Their method is based on a recognition that the phase of the multiple quantum coherence depends on the phase of the exciting pulse in a manner which depends on the order of the multiple quantum coherence. Their concept also underlies the scheme for production of a selected order of coherence which we take up in the next section.

The basic Wokaun-Ernst scheme may be described as follows. Let us denote by X_ϕ, a $\pi/2$ pulse applied about an x'-axis in the rotating frame where the x'-axis lies in the x-y plane, making an angle ϕ with the x-direction according to the equations

$$x' = x\cos\phi - y\sin\phi \quad (9.120a)$$

$$I_{x'} = I_x \cos\phi - I_y \sin\phi \tag{9.120b}$$

$$= e^{iI_z\phi} I_x e^{-iI_z\phi} \quad . \tag{9.120c}$$

Note that x' is rotated from x in a left-handed sense about I_z by an amount ϕ. Wokaun and Ernst apply a sequence

$$X_\phi \ldots \tau \ldots X_\phi \ldots t_1 \ldots X \ldots t_2 \tag{9.121}$$

recording the complex signal $S(\tau, \phi, t_1, t_2)$ as a function of t_2, for fixed values of the parameters τ and t_1, and for a succession of appropriately chosen phases ϕ. Let us write the signal for $\phi = 0$ as a sum of contributions G_p from the various orders of coherence p ($p = -N$ to $+N$)

$$S(\phi = 0) = \sum_{p=-N}^{+N} G_p \tag{9.122}$$

and then consider $S(\phi)$ to be known as a continuous function of ϕ. The essence of the Wokaun-Ernst method is then that the G_p's can be found as transforms of $S(\phi)$

$$G_p = \frac{1}{2\pi} \int_0^{2\pi} S(\phi) e^{-ip\phi} d\phi \quad . \tag{9.123}$$

Since in practice there are $2N + 1$ values of p, measurement of $S(\phi)$ for $2N + 1$ values of ϕ should suffice, as we show below.

To understand the Wokaun-Ernst theorem, we need to

1. examine the general form of the contributions of a given order of coherence,
2. examine the effect of a phase shift on the different orders of coherence, and
3. show how to utilize this knowledge to pick out the contributions G_p of the particular order of coherence.

We consider a system characterized by a Hamiltonian in the rotating frame

$$\mathcal{H} = -\sum_k \hbar\Omega_k I_{zk} + \sum_{k>j} \mathcal{H}_{jk} \equiv \mathcal{H}_z + \mathcal{H}_{ss} \quad , \tag{9.124a}$$

where for the spin-spin interactions $\mathcal{H}_{jk}$ we keep only the secular terms so that

$$[\mathcal{H}_{jk}, I_{zT}] = 0 \quad . \tag{9.124b}$$

Then, we have the quantum numbers M, α such that

$$I_{zT}|M\alpha\rangle = M|M\alpha\rangle \tag{9.124c}$$

$$\mathcal{H}_{ss}|M\alpha\rangle = E_\alpha|M\alpha\rangle \quad .$$

Note that (9.124a) allows for a variety of chemical shifts.

Then, utilizing the fact that

$$X = e^{iI_xT\pi/2} \quad , \tag{9.125a}$$

we have

$$X_\phi = \exp\left[i\tfrac{\pi}{2}(I_xT\cos\phi - I_yT\sin\phi)\right] \quad . \tag{9.125b}$$

Defining

$$R = e^{iI_zT\phi} \quad , \tag{9.125c}$$

we get

$$\begin{aligned}X_\phi &= \exp(iRI_xTR^{-1}\pi/2) \\ &= Re^{iI_xT\pi/2}R^{-1} \\ &= e^{iI_zT\phi}Xe^{-iI_zT\phi} \quad .\end{aligned} \tag{9.125d}$$

We now wish to express the ϕ dependence of the signal, S. Abbreviating $S(\tau,\phi,t_1,t_2)$ as $S(\phi)$, we have

$$S(\phi) = \text{Tr}\left\{I_T^+\left[e^{-(i/\hbar)\mathcal{H}t_2}Xe^{-(i/\hbar)\mathcal{H}t_1}\right]\varrho(\tau^+,\phi)\;[\text{inverse}]\right\} \tag{9.126}$$

where $\varrho(\tau^+,\phi)$, defined as

$$\varrho(\tau^+,\phi) \equiv X_\phi e^{-(i/\hbar)\mathcal{H}\tau}X_\phi I_zTX_\phi^{-1}e^{(i/\hbar)\mathcal{H}\tau}X_\phi^{-1} \quad , \tag{9.127}$$

is the density matrix just after the second $\pi/2$ pulse, for the case that both π pulses have phase shifts ϕ. The ϕ dependence of $S(\phi)$ therefore arises in $\varrho(\tau^+,\phi)$. Expressing X_ϕ utilizing (9.125d) and utilizing the fact that I_zT, and thus $\exp(iI_zT\phi)$, commutes with $\mathcal{H}$, we get

$$\begin{aligned}\varrho(\tau^+,\phi) &= e^{iI_zT\phi}Xe^{-(i/\hbar)\mathcal{H}\tau}XI_zX^{-1}e^{(i/\hbar)\mathcal{H}\tau}X^{-1}e^{-iI_zT\phi} \\ &= e^{iI_zT\phi}\varrho(\tau^+,0)e^{-iI_zT\phi} \quad .\end{aligned} \tag{9.128}$$

We recall that $\varrho(\tau^+,0)$ is the density matrix just after the second X pulse, hence it contains all the multiple quantum coherence. In general, for a given spin system, $\varrho(\tau^+,0)$ consists of a number of terms representing not only the different possible orders of coherence, but also the various ways of generating a given order. Thus, for a three-spin system, of spins $\tfrac{1}{2}$, one can generate $p=0$, $\pm 1, \pm 2, \pm 3$. Consider $p=1$. It could arise from a variety of operators such as

$$I_1^+ \quad , \quad I_1^+ I_{2z} \quad , \quad I_1^+ I_{2z}I_{3z} \quad , \quad I_1^+ I_2^+ I_3^- \tag{9.129}$$

plus all the operators one can get by permuting the labels 1, 2, and 3. Utilizing a symbol β to designate all such various ways of generating a given order, we can clearly break $\varrho(\tau^+,0)$ into a set of components $g_{p\beta}$ of given p

$$\varrho(\tau^+,0) = \sum_{p,\beta} g_{p\beta} \quad . \tag{9.130}$$

Now, *a given order, p, is distinguished by having p more raising operators than lowering operators*. (If p is negative, there are more lowering operators than raising operators.)

We can then easily evaluate $\varrho(\tau^+, \phi)$ as

$$\varrho(\tau^+, \phi) = e^{iI_zT\phi}\varrho(\tau^+, 0)e^{-iI_zT\phi} = \sum_{p,\beta} e^{iI_zT\phi} g_{p\beta} e^{-iI_zT\phi} \quad . \tag{9.131}$$

But,

$$e^{iI_zT} I_k^\pm e^{iI_z\phi} = I_k^\pm e^{\pm i\phi} \quad , \quad e^{iI_zT} I_{zk} e^{-iI_zT} = I_{zk} \quad . \tag{9.132}$$

Taking expressions such as (9.129), inserting factors $\exp(-iI_zT\phi)\exp(iI_zT\phi)$ between each pair of spin operators, and using (9.132), one gets results such as

$$e^{iI_zT\phi} I_1^+ I_{2z} e^{-iI_zT\phi} = I_1^+ I_{2z} e^{i\phi}$$

$$e^{iI_zT\phi} I_1^+ I_2^+ I_3^- e^{-iI_zT\phi} = I_1^+ I_2^+ I_3^- e^{i\phi} \quad . \tag{9.133}$$

That is, the effect of the ϕ rotation for any operator with $p = 1$ is simply to multiply that operator by $\exp(i\phi)$. In a similar way, the rotation operators acting on a $p = 2$ operator merely multiply it by $\exp(2i\phi)$. In general, for the operators $g_{p\beta}$,

$$e^{iI_zT\phi} g_{p\beta} e^{-iI_zT\phi} = g_{p\beta} e^{ip\phi} \quad . \tag{9.134}$$

Note that this result is independent of β.

We can utilize (9.134) to get

$$\varrho(\tau^+, \phi) = \sum_{p=-N}^{N} \left(\sum_{\beta} g_{p\beta} \right) e^{ip\phi} \quad , \tag{9.135}$$

where N is the maximum allowed value of p for the particular spin system. Substituting this expression into (9.126) we get the signal as a function of ϕ to be

$$S(\phi) = \sum_{p=-N}^{N} e^{ip\phi} \text{Tr}\left\{ I_T^+ \left[e^{-(i/\hbar)\mathcal{H}t_2} X e^{-(i/\hbar)\mathcal{H}t_1} \right] \left(\sum_{\beta} g_{p\beta} \right) \text{[inverse]} \right\}$$

$$= \sum_{p=-N}^{N} G_p e^{ip\phi} \quad , \quad \text{where} \tag{9.136a}$$

$$G_p = \text{Tr}\left\{ I_T^+ \left[e^{-(i/\hbar)\mathcal{H}t_2} X e^{-(i/\hbar)\mathcal{H}t_1} \right] \left(\sum_{\beta} g_{p\beta} \right) \text{[inverse]} \right\} \quad . \tag{9.136b}$$

The individual terms $g_{p\beta}$ depend on τ, on the Ω_k's and the strength of the spin-spin couplings as illustrated in the tables we have worked out for $\varrho(\tau^+)$ for the two- and three-spin cases. Examination of (9.136a) shows that it says that $S(\phi)$ can be represented by a Fourier series in ϕ in which the harmonics are labeled by p, and in which the coefficients of the harmonics are the G_p's. Therefore, if one knew $S(\phi)$, one could deduce the G_p's by taking the Fourier transform of $S(\phi)$.

Thus, suppose we have deduced $S(\phi)$. (We can do this by measuring at a number of discrete values ϕ_i and using a smooth interpolation procedure to estimate S for values between adjacent pairs of values ϕ_i.) Then consider I, the transform of S,

$$I = \int_0^{2\pi} S(\phi) e^{-ip'\phi} d\phi \quad . \tag{9.137}$$

Substituting (9.136a) for $S(\phi)$ we get

$$I = \sum_p G_p \int_0^{2\pi} e^{i(p-p')\phi} d\phi$$

$$= 2\pi G_{p'} + \sum_{p \neq p'} \frac{G_p}{i(p'-p)} [e^{2\pi i(p-p')} - 1]$$

$$= 2\pi G_{p'} \tag{9.138}$$

or

$$G_p = \frac{1}{2\pi} \int_0^{2\pi} S(\phi) e^{-ip\phi} d\phi \quad . \tag{9.139}$$

In some systems, such as a dilute solution of some molecule of interest, there is an upper limit to the quantum order [the quantity N of (9.135)]. In others (for example an effectively infinite solid of dipolar coupled nuclei, such as CaF$_2$) N has no limit. Suppose we had the former case. Then, there are $2N+1$ values of p, consequently $2N+1$ G_p's. Clearly $2N+1$ measurements of $S(\phi)$ should suffice to deduce the G_p's.

Suppose, then, we pick the $2N+1$ values of ϕ, which we label ϕ_k, as

$$\phi_k = k \frac{2\pi}{2N+1} \tag{9.140a}$$

$$k = 0, 1, 2, \ldots, 2N \quad . \tag{9.140b}$$

Note if we chose $k = 2N+1$, we would have a value of $\phi = 2\pi$, which produces the same shift as $\phi = 0$, a value we have already obtained from $k = 0$. Then

$$S(\phi_k) \equiv S_k = \sum_p G_p \exp(ip\phi_k)$$

$$= \sum_{p=-N}^{+N} G_p \exp[ipk2\pi/(2N+1)] \quad . \tag{9.141}$$

This is a finite series. Consider then the transforms

$$\sum_k S(\phi_k)\exp(-ip'\phi_k) = \sum_{k,p} G_p \exp[i(p-p')\phi_k]$$
$$= \sum_p G_p \sum_k \exp[i(p-p')k2\pi/(2N+1)] \quad . \quad (9.142)$$

Now, defining
$$f = \exp[i(p-p')2\pi/(2N+1)] \quad (9.143)$$

we have
$$\sum_{k=0}^{2N} \exp[i(p-p')k2\pi/(2N+1)]$$
$$= \sum_{k=0}^{2N} f^k = \sum_{k=0}^{\infty} f^k - \sum_{k=2N+1}^{\infty} f^k$$
$$= \left(\sum_{k=0}^{\infty} f^k\right)(1-f^{2N+1}) = \frac{1-f^{2N+1}}{1-f} \quad \text{if} \quad f \neq 1 \quad . \quad (9.144)$$

Examination of (9.144) shows that as long as $p \neq p'$ the exponent in f is neither zero nor a multiple of 2π, hence $f \neq 1$. But
$$f^{2N+1} = e^{i(p-p')2\pi} = 1 \quad , \quad (9.145)$$

hence the sum over k in (9.144) vanishes.

If $p = p'$, $f = 1$, and the sum over k is trivially $2N+1$. Thus
$$G_{p'} = \left(\frac{1}{2N+1}\right)\sum_k S(\phi_k)e^{-ip'\phi_k} \quad \text{with} \quad (9.146)$$

$$\phi_k = k2\pi/(2N+1) \quad , \quad k = 0, 1, \ldots, 2N \quad .$$

Therefore, measurement of $S(\phi_k)$ for these $2N+1$ values of ϕ will enable one to decompose S into the contributions, G_p, from the various orders of coherence. Since G_p is a complex number, one must record signals $S(\phi)$ as complex numbers, i.e. in quadrature, to evaluate the formula of (9.146).

It is sometimes useful to carry out a simpler version of adding signals. One picks a set of phases (e.g. 0, $\pi/2$, π, $3\pi/2$) and then adds the signals together. Sometimes one adds signals of one phase but subtracts signals of another. To see how this works it is useful to keep in mind a graphical picture. From (9.136a) we have
$$S(\phi) = \sum_{p=-N}^{+N} G_p e^{ip\phi} \quad , \quad (9.147)$$

hence each order can be represented in the complex plane by vectors $\exp(ip\phi)$. Consider then an example with $\phi = 0$ or π. Then

$$S(0) + S(\pi) = \sum_p G_p(e^{ip0} + e^{ip\pi})$$
$$= \sum_{p \text{ even}} G_p(1+1) + \sum_{p \text{ odd}} G_p(1-1)$$
$$= 2 \sum_{p \text{ even}} G_p \quad . \tag{9.148}$$

Thus, all the odd-order coherences contribute nothing. If we took $\phi = 0$, $\pi/2$, π, $3\pi/2$ and added signals we would get

$$S(0) + S(\pi/2) + S(\pi) + S(3\pi/2)$$
$$= \sum_p G_p(e^{ip0} + e^{ip\pi/2} + e^{ip\pi} + e^{ip3\pi/2}) \quad . \tag{9.149}$$

Clearly for $p = 0$ this will give $4G_0$. For $p = 1$ this will give

$$G_1(1 + e^{i\pi/2} + e^{i\pi} + e^{i3\pi/2}) \quad . \tag{9.150}$$

The four complex numbers may be thought of as four vectors in the complex plane $(1, i, -1, -i)$ which, when added tail of one to head of the prior one, form the sides of a square, giving a zero result. Thus there is no signal from G_1, first-order coherence. Likewise, there is no signal from $p = 2$ or $p = 3$. But for $p = 4$ we get

$$G_4(e^{i0} + e^{i4\pi/2} + e^{i8\pi/2} + e^{i12\pi/2})$$
$$= G_4(1 + 1 + 1 + 1) = 4G_4 \quad . \tag{9.151}$$

That is, all four terms $\exp(i4\phi)$ are equal to 1.

Thus we will record G_0 and G_4, but not G_1, G_2, or G_3. In a similar way one can show that we will not record G_5, G_6, G_7, but will record G_8. In fact, as we see from (9.151), we will get a nonzero result whenever all the terms are of a given order are 1. Since for a phase shift increment of θ

$$\sum_p G_p[1 + e^{ip\theta} + (e^{ip\theta})^2 + \ldots] \quad , \tag{9.152}$$

we will get signals when

$$e^{ip\theta} = 1 \quad \text{or} \tag{9.153a}$$

$$p\theta = k2\pi \quad , \quad k = 0, 1, 2, 3 \quad . \tag{9.153b}$$

Thus, if we get a signal for $p = 2\pi/\theta$, we will also get a signal for $2p$, $3p$, etc. One sometimes speaks of a sequence as being "pk selective", since if it gives order p it will also give k times any integral value of k.

Note that as long as we add all the signals, we get G_0, the zero-quantum signal. To eliminate it we add half the signals and subtract the other half. Some cases given by *Wokaun* and *Ernst* are

- add $\phi = 0$, π gives $p = 0, 2, 4, 6, 8$
- add $\phi = 0$, subtract $\phi = \pi$, gives $p = 1, 3, 5, 7, 9$
- add $\phi = 0$, $\pi/2$, π, $3\pi/2$ gives $p = 0, 4, 8, \ldots$
- add $\phi = 0$, π, subtract $\phi = \pi/2$, $3\pi/2$, gives $p = 2, 6, \ldots$
- add $\phi = 0$, $2\pi/3$, $4\pi/3$, subtract $\phi = \pi/3 + (0, 2\pi/3, 4\pi/3)$, gives $p = 3, 9$.

These simple sequences are particularly useful under circumstances where the allowed values of p are limited. Thus, if $p = 9$ can be ruled out, the last sequence gives one just $p = 3$.

We remarked that one method of displaying spectra by quantum order is to tune off resonance by an amount Ω. Then lines associated with order p can be identified by Fourier transforming the signal with respect to the evolution time t_1, since they occur at a frequency offset of $p\Omega$. As we remarked, there are disadvantages with working off resonance. A solution is to use a technique called time proportional phase incrementation (TPPI). The idea is to work at $\Omega = 0$, exact resonance, but in collecting data for the variable t_1 (the evolution time) to introduce a phase shift which depends on t_1. If the phase and t_1 obey a linear relationship

$$\phi = At_1 \quad . \tag{9.154}$$

Then as t_1 is advanced the extra phase is just as though one were off resonance by an amount $\Delta\omega$ given by

$$\Delta\omega = A \quad . \tag{9.155}$$

Then, when one does the Fourier transform of the signal with respect to t_1, an effective frequency offset obeying (9.155) is introduced. This approach is unaffected by insertion of π pulses to refocus magnetic field inhomogeneities or real frequency offsets at any part of the cycle (i.e. at the midpoint of the preparation cycle, the evolution cycle, or the detection cycle).

9.4 High Orders of Coherence

We have seen some simple ways of producing and detecting multiple quantum coherence. For many purposes these methods suffice. For example, if one wished to distinguish protons in a CH_3 group from protons in a CH_2 group, one need only to be able to produce and detect two- or three-quantum coherence. What happens if one is concerned with much higher orders? It is easy to see that there is a fundamental problem. Each quantum order contains "information" – that is, it represents order in the statistical mechanical sense. But we are confident that the total statistical mechanical order will never exceed its initial value representing the sample's initial magnetization. If one pumps only low-order quantum coherences, the statistical order is shared by only a modest number of $(M\alpha|\varrho|M'\alpha')$'s. However, if one sets the time, τ, of the preparation to longer and longer times one finds one excites most of the low-p elements of ϱ, as well as matrix elements of progressively larger p. Thus, the statistical order is being shared by more and more elements of ϱ. As a result, individual elements get much smaller. One tends to lose in two ways. (1) During preparation one spreads the order among many different elements of ϱ, and (2) the detection pulse feeds back many elements $(M\alpha|\varrho|M'\alpha')$ in which though $M - M' = p$ may be the same, there is destructive interference from the range of values of the other quantum numbers, α.

It is thus clear that to get strong signals one should first of all restrict the elements of ϱ which one excites to just the desired ones. For example, if one is interested in the six-quantum spectrum, one would like to excite only the six-quantum coherences. Then, having excited the desired coherences, one would like to read them out at a later time in a manner which avoids any further loss of coherence. We turn now to these matters. Many of the ideas involved here are from the studies of *Pines* and his students. The two review articles by *Weitekamp* [9.5] and by *Munowitz* and *Pines* [9.7b] are especially useful for study, as are the basic papers by *Warren* et al. [9.18, 19] and *Yen* and *Pines* [9.20].

9.4.1 Generating a Desired Order of Coherence

Can one generate one and only one desired order of coherence? We have seen that it is possible to detect an order p and integral multiples of p by cycling the phase of the exciting pulses relative to the detection pulses. We will now show how to use phase cycling to generate coherence of a given p and multiples thereof.

First, however, it is useful to introduce another way of looking at generation of multiple quantum coherence. Utilizing two pulses, we have seen how the first $\pi/2$ pulse converts the density matrix from its initial value, I_z, to I_x, producing one-quantum coherence. As time develops, spin-spin correlations develop in the density matrix as is evident either from the spin operator formalism [e.g., (9.90)] or from the perturbation expansion (9.100). However, the density matrix still possesses only one-quantum coherence. The second $\pi/2$ pulse, by acting on one-quantum terms like $I_{z1}I_{z2}I_3^+$ which express spin-spin correlations (in this case of three spins) produces multiple quantum coherence through operators such as $I_1^+I_2^+I_3^+$. We can express this formally by writing the equation for $\varrho(\tau^+)$, where τ^+ is the time just after the second $\pi/2$ pulse. For convenience we will take the second pulse to be an $\overline{X}$, though the argument we present could be done, a bit less elegantly, with an X pulse. Since $\overline{X}$ undoes an X pulse, we have

$$\overline{X} = X^{-1} \quad \text{so} \tag{9.156}$$

$$\varrho(\tau^+) = X^{-1}\exp\left(-\frac{i}{\hbar}\mathcal{H}\tau\right)XI_zX^{-1}\exp\left(\frac{i}{\hbar}\mathcal{H}\tau\right)X \quad . \tag{9.157}$$

Suppose for concreteness we are exactly at resonance, with $\mathcal{H}$ being only the spin-spin coupling. We take

$$\mathcal{H} = \mathcal{H}_{zz} \quad , \tag{9.158}$$

where $\mathcal{H}_{zz}$ could be the usual secular dipolar coupling, such as

$$\mathcal{H}_{zz} = \sum_{j>k} B_{jk}(I_{xj}I_{xk} + I_{yj}I_{yk} - 2I_{zj}I_{zk}) \quad . \tag{9.159}$$

Then

$$X^{-1}\exp\left(-\frac{i}{\hbar}\mathcal{H}_{zz}\tau\right)X = \exp\left[-\frac{i}{\hbar}(X^{-1}\mathcal{H}_{zz}X)\tau\right]$$

$$= \exp\left(-\frac{i}{\hbar}\mathcal{H}_{yy}\tau\right) \tag{9.160}$$

so that

$$\varrho(\tau^+) = \exp\left(-\frac{i}{\hbar}\mathcal{H}_{yy}\tau\right)I_z \exp\left(\frac{i}{\hbar}\mathcal{H}_{yy}\tau\right) \quad , \tag{9.161}$$

where, for example, if $\mathcal{H}_{zz}$ were given by (9.159),

$$\mathcal{H}_{yy} = \sum_{j>k} B_{jk}(I_{xj}I_{xk} + I_{zj}I_{zk} - 2I_{yj}I_{yk}) \quad . \tag{9.162}$$

In this formulation, explicit mention of the pulses has disappeared, replaced by a transformed Hamiltonian, $\mathcal{H}_{yy}$, which starts to act at $\tau = 0$. Now, $\varrho(\tau^+)$ possesses, as we have seen, multiple quantum matrix elements. That is

$$(M'\alpha'|\varrho(\tau^+)|M\alpha) \neq 0 \quad \text{for} \quad M' \neq M \pm 1 \quad . \tag{9.163}$$

However, since the density matrix at $t = 0$, I_z, has no matrix elements between different states M and M', we may say that $\mathcal{H}_{yy}$ acting over time τ generates the multiple quantum coherence. Indeed, expanding the exponentials we have

$$\exp\left(-\frac{i}{\hbar}\mathcal{H}_{yy}\tau\right) = 1 - i\frac{\mathcal{H}_{yy}\tau}{\hbar} + \left(\frac{i}{2}\right)^2\frac{\mathcal{H}_{yy}\mathcal{H}_{yy}\tau^2}{\hbar^2} + \ldots \quad . \tag{9.164}$$

From (9.162), substituting raising and lowering operators (9.159),

$$\mathcal{H}_{yy} = \sum_{j>k} B_{jk}\bigl(\tfrac{3}{4}(I_j^+ I_k^+ + I_j^- I_k^-) + I_{zj}I_{zk}$$
$$- \tfrac{1}{4}(I_j^+ I_k^- + I_j^- I_k^+)\bigr) \quad , \tag{9.165}$$

which therefore contains zero- and two-quantum coherences.

We have seen numerous examples of ways in which the ingenious resonator can effectively modify a Hamiltonian: spin echoes, in which magnetic field inhomogeneities are effectively reversed; magic echoes, in which dipolar dephasing is undone by creating the negative of the dipolar Hamiltonian; and spin-flip line narrowing, in which the dipolar Hamiltonian $\mathcal{H}_{zz}$ is averaged to zero by being made to jump to $\mathcal{H}_{xx}$ and $\mathcal{H}_{yy}$. Therefore, we may rephrase our question "can we generate one and only one desired order of coherence" to "can we find a way to achieve a Hamiltonian which will generate one and only one desired order of coherence?" In fact, as we shall see, we can find a way to generate coherence of order pk, where $k = 0, 1, \ldots$. By special means, one can eliminate $k = 0$. Then, the means to distinguish among the other k values is by the duration of the excitation period.

How then do we limit generation to an order pk? We will find that there are two parts to the task. The first part is concerned with finding a way to limit the order, and the second part, with making the amount generated large. In order to keep the discussion sufficiently general, we will therefore not as yet specify the Hamiltonian. However, we suppose that as with $\mathcal{H}_{yy}$ it can generate various orders of coherence, p. That is

$$\mathcal{H} = \sum_p \mathcal{H}_p \quad . \tag{9.166}$$

We now want to see how to limit the orders generated. We already have a clue from the ideas of *Wokaun* and *Ernst*: we should utilize the phase shift properties. Let us think about what happens to $\mathcal{H}$ if we shift the phase of the rf pulses which generate it.

For the case of two $\pi/2$ pulses so that the coherence is generated by

$$\overline{X} \exp\left(-\frac{i}{\hbar}\mathcal{H}_{zz}\tau\right) X = \exp\left(-\frac{i}{\hbar}\mathcal{H}_{yy}\tau\right) \tag{9.167}$$

we have seen that

$$\overline{X}_\phi \exp\left(-\frac{i}{\hbar}\mathcal{H}_{zz}\tau\right) X_\phi$$
$$= \exp\left(\frac{i}{\hbar}I_z\mathrm{T}\phi\right) \overline{X} \exp\left(-\frac{i}{\hbar}I_z\mathrm{T}\phi\right) \exp\left(-\frac{i}{\hbar}\mathcal{H}_{zz}\tau\right)$$
$$\times \exp\left(\frac{i}{\hbar}I_z\mathrm{T}\phi\right) X \exp\left(-\frac{i}{\hbar}I_z\mathrm{T}\phi\right)$$
$$= \exp\left(\frac{i}{\hbar}I_z\mathrm{T}\phi\right) \overline{X} \exp\left(-\frac{i}{\hbar}\mathcal{H}_{zz}\tau\right) X \exp\left(-\frac{i}{\hbar}I_z\mathrm{T}\phi\right)$$
$$= \exp\left(\frac{i}{\hbar}I_z\mathrm{T}\phi\right) \exp\left(-\frac{i}{\hbar}\mathcal{H}_{yy}\tau\right) \exp\left(-\frac{i}{\hbar}I_z\mathrm{T}\phi\right) \quad . \tag{9.168}$$

Let us then think about a general Hamiltonian, which for zero phase, ϕ, we call $\mathcal{H}_0$. We decompose it into various orders of coherence

$$\mathcal{H}_0 = \sum_p \mathcal{H}_p \quad . \tag{9.169a}$$

Then

$$\mathcal{H}_\phi \equiv e^{iI_z\mathrm{T}\phi} \mathcal{H}_0 e^{-iI_z\mathrm{T}\phi} = \sum_p e^{iI_z\mathrm{T}\phi} \mathcal{H}_p e^{-iI_z\mathrm{T}\phi} \quad . \tag{9.169b}$$

In the spirit then of (9.133), we then get

$$\mathcal{H}_\phi = \sum_p \mathcal{H}_p e^{ip\phi} \quad . \tag{9.170}$$

This expression is reminiscent of (9.136)

$$S(\phi) = \sum_p G_p e^{ip\phi} \quad .$$

The way we got selectivity in signal detection was to form a sum of signals for a sequence of phase shifts. How can we use this idea for Hamiltonians? We have already found out how when we studied spin-flip line narrowing. What we must do is employ the conditions for a time average Hamiltonian. That is, we switch among the various values of the Hamiltonian, forming a cycle of desired values, then repeat the cycle again and again for some desired total time.

The development of the system for one cycle in which the Hamiltonian takes on n values

$$\mathcal{H}_l \quad , \quad l = 0 \text{ to } n-1 \quad , \tag{9.171}$$

each of duration τ_k, is given by the operator

$$\exp\left(-\frac{i}{\hbar}\mathcal{H}_{n-1}\tau_{n-1}\right)\ldots\exp\left(-\frac{i}{\hbar}\mathcal{H}_2\tau_2\right)\exp\left(-\frac{i}{\hbar}\mathcal{H}_0\tau_0\right)$$
$$\cong \exp\left(-\frac{i}{\hbar}\sum_l \mathcal{H}_l \tau_l\right) \quad , \tag{9.172}$$

where the equality holds provided the τ_k are sufficiently short. Then the average Hamiltonian

$$\mathcal{H}_{\text{av}} = \frac{1}{t_c}\sum_l \mathcal{H}_l \tau_l \quad \text{where} \tag{9.173a}$$

$$t_c = \sum_l \tau_l \quad . \tag{9.173b}$$

So, if we picked a set of n phases $\phi_0, \ldots, \phi_{n-1}$ each acting for the same duration $\Delta\tau$, we would have

$$\mathcal{H}_{\text{av}} = \frac{1}{t_c}\sum_l (\mathcal{H}_{\phi_l}\Delta\tau) = \frac{1}{n\Delta\tau}\sum_l \left(\sum_p \mathcal{H}_p e^{ip\phi_l}\Delta\tau\right) \tag{9.174a}$$

$$= \frac{1}{n}\sum_p \mathcal{H}_p \left(\sum_{l=0}^{n=1} e^{ipl\phi}\right) \quad , \tag{9.174b}$$

where in the last step we took $\phi_l = l\phi$. For a given p, the sum over l is nothing but

$$(1 + e^{ip\phi} + e^{i2p\phi} + \ldots e^{i(n-1)p\phi}) \quad . \tag{9.175}$$

We have seen in the previous section [note in particular (9.150)] that if

$$e^{inp\phi} = 1 \tag{9.176a}$$

the complex numbers of (9.175) form a closed polygon of exterior angle $p\phi$, and the sum in (9.175) vanishes unless

$$e^{ip\phi} = 1 \quad . \tag{9.176b}$$

Therefore, if we pick a p, which we call p_0, that we wish to generate, we just need to select ϕ to satisfy (9.176b). Therefore if

$$p_0\phi = 2\pi \quad \text{or} \quad \phi = 2\pi/p_0 \quad , \tag{9.177a}$$

which from (9.175) gives

$$n = p_0 \quad , \tag{9.177b}$$

we will have a nonvanishing $\mathcal{H}_{av}$. We will *also* have a nonvanishing $\mathcal{H}$ for p's which are integral multiples of p_0. Thus, we will select values p

$$\begin{aligned} p &= p_0 k \quad k = 0, 1, 2 \quad \text{using} \\ \phi &= 2\pi/p_0 \quad . \end{aligned} \tag{9.178}$$

For all other values of p, (9.175) and thus $\mathcal{H}_{av}$ vanish.

The theorem we have proven holds true for any Hamiltonian provided we satisfy the condition for average Hamiltonian theory:

$$|\mathcal{H}|\Delta\tau \ll 1 \quad . \tag{9.179}$$

The sorts of correction terms can be looked at using the Magnus expansion (see Appendix K, and references therein). Reference [9.20] discusses correction terms in great detail.

If we now look at the sequence of a pair of $\pi/2$ pulses which we have been considering, we have for $\mathcal{H}_0$

$$\mathcal{H}_0 = \mathcal{H}_{yy} \quad . \tag{9.180}$$

From (9.165) we see it contains only the terms $p = 0$ and ± 2. Now, we know that the two-pulse sequence can generate all values of p, but those larger than two require the τ^2 or higher terms in the expansion (9.164) of the exponential. These are the terms we do not wish to have since we require in (9.174) that the average Hamiltonian be valid.

The question is, then, what can we do to create an $\mathcal{H}_0$ which gives high orders of p *without* violating the conditions for average Hamiltonian theory to hold? The solution was found by *Warren* et al. [9.19, 20]. We turn now to an explanation of their approach.

The Hamiltonian $\mathcal{H}_{yy}$ resulted from the action of the pulses X and $\overline{X}(= X^{-1})$ acting on $\mathcal{H}_{zz}$. The crucial steps are (9.160). In particular,

$$\mathcal{H}_{yy} = X^{-1}\mathcal{H}_{zz}X \quad . \tag{9.181}$$

Of course, we start with $\mathcal{H}_{zz}$ — that is what nature gives us. We get $\mathcal{H}_{yy}$ by applying a transformation (physically realizable!) with the operators X^{-1} and X. Let us then look for some other operator – call it R – to use instead of X. That is, $\mathcal{H}_0$ is to be produced by applying R to $\mathcal{H}_{zz}$. Thus

$$\mathcal{H}_0 = R^{-1}\mathcal{H}_{zz}R \quad \text{so that} \tag{9.182}$$

$$\exp[-(i/\hbar)\mathcal{H}_0 t] = R^{-1}\left[\exp[-(i/\hbar)\mathcal{H}_{zz}t]\right]R \quad . \tag{9.183}$$

What properties should R possess? The trouble with X is that it produced solely zero- and two-quantum coherences. What we need is an $\mathcal{H}_0$ that is rich in the values of p which we desire to excite. That is, in (9.169a) we want large $\mathcal{H}_p$'s for the values of p we desire.

It is clear that no simple rotation operator like X or Y will do, since it simply rotates the basic dipolar coupling. However, we remember that actually

$$X = \exp[(i/\hbar)\gamma\hbar H_1 I_{x\mathrm{T}} t_\mathrm{p}] \tag{9.184}$$

where H_1 is the strength of the rf field, and t_p the pulse duration. This term is nothing but an approximation to the time development operator for the system while H_1 is on, an approximation because we neglected the dipolar coupling while H_1 was on. More generally, we should think that

$$R = \exp[-(i/\hbar)\mathcal{H}_R t_\mathrm{p}] \quad , \tag{9.185}$$

where $\mathcal{H}_R$ is the Hamiltonian which acts during the time t_p. Now, $\mathcal{H}_R$ can be a real Hamiltonian, such as is the case when $R = X$, *or* it can be some *effective* Hamiltonian created by the experimenter by manipulating a real Hamiltonian with pulses.

The game, then, is to try to guess an $\mathcal{H}_R$ which does what one wants, then, if it does, figure out how to create it. Note that in order to create both R and its inverse R^{-1}, we must be able to create both $\mathcal{H}_R$ and $-\mathcal{H}_R$. There are not many choices. We have already excluded $\mathcal{H}_R$'s which are proportional to $I_{x\mathrm{T}}$, $I_{y\mathrm{T}}$, or $I_{z\mathrm{T}}$. All that is left is

$$\mathcal{H}_R = \mathcal{H}_{xx} \tag{9.186a}$$

$$\mathcal{H}_R = \mathcal{H}_{yy} \tag{9.186b}$$

$$\mathcal{H}_R = \mathcal{H}_{zz} \quad \text{or} \tag{9.186c}$$

$$\mathcal{H}_R = \mathcal{H}_{z'z'} \quad , \tag{9.186d}$$

where z' is an axis other than x, y, or z. We could also take linear combinations of the operators, or take these operators multiplied by a constant. $\mathcal{H}_{zz}$ alone is not useful since in (9.183) R would commute with the exponential, leaving $\mathcal{H}_0 = \mathcal{H}_{zz}$. However, $\mathcal{H}_{xx}$, $\mathcal{H}_{yy}$, or $\mathcal{H}_{z'z'}$ all will do the job. Before showing this, we recall that we get $\mathcal{H}_{xx}$ or $\mathcal{H}_{yy}$ simply by sandwiching $\mathcal{H}_{zz}$ between two $\pi/2$ pulses (Y and $\overline{Y}$ for $\mathcal{H}_{xx}$, X and $\overline{X}$ for $\mathcal{H}_{yy}$). We have seen several ways of producing $-\mathcal{H}_{xx}$ (actually $-\mathcal{H}_{xx}/2$). The first method was used to produce magic echoes in Sect. 8.8. It involved producing a strong H_1 in the x-direction in the rotating frame which one phase shifted by π at periodic intervals. Several other methods involved trains of X and $\overline{X}$ pulses, as for example shown in Table 8.4.

Another useful sequence produces

$$\mathcal{H}_R = \tfrac{1}{3}(\mathcal{H}_{yy} - \mathcal{H}_{xx}) \quad , \tag{9.187}$$

which, for dipolar coupling given by (9.159), is

$$\begin{aligned}\mathcal{H}_R &= \sum_{j>k} B_{jk}(I_{xj}I_{xk} - I_{yj}I_{yk}) \\ &= \frac{1}{2}\sum_{j>k} B_{jk}(I_j^+ I_k^+ + I_j^- I_k^-) \quad .\end{aligned} \tag{9.188}$$

This sequence arises from the sequence of (8.207), which we rewrite as

$$(\Delta/2, X, \Delta', X, \Delta, \overline{X}, \Delta', \overline{X}, \Delta, \overline{X}, \Delta', \overline{X}, \Delta, X, \Delta', X, \Delta/2) \tag{9.189a}$$

where

$$\Delta' = 2\Delta + t_\mathrm{p} \tag{9.189b}$$

and t_p is the pulse duration [9.19]. The fact that $\Delta' \neq 2\Delta$ is to correct for finite pulse length. Note that since (9.188) is a two-quantum operator, we can easily create its negative by phase shifting the pulses by $\pi/2$, i.e. by replacing X and $\overline{X}$ by Y and $\overline{Y}$. In practice the schemes utilized are all based on sequences of $\pi/2$ pulses.

We now want to show that if we have such a dipolar $\mathcal{H}_R$ we can generate an $\mathcal{H}_0$ which is rich in large quantum orders $\mathcal{H}_p$. Referring to (9.182) and (9.183), we have

$$\begin{aligned}\mathcal{H}_0 &= R^{-1}\mathcal{H}_{zz}R \\ &= \exp[+(i/\hbar)\mathcal{H}_R T]\mathcal{H}_{zz}\exp[-(i/\hbar)\mathcal{H}_R T] \quad ,\end{aligned} \tag{9.190}$$

where T is the length of time we let the system evolve under $\mathcal{H}_R$ or under $-\mathcal{H}_R$. Note that in the case $\mathcal{H}_R = \mathcal{H}_{xx}$, we produce a negative $(-\mathcal{H}_R)$ of the form $-\mathcal{H}_{xx}/2$ which requires that we let it act for a time T' which is twice T. However, if we set $T' = 2T$, (9.190) still holds.

We now expand the products using the relationship [9.16]

$$e^A B e^{-A} = B + [A, B] + [A, [A, B]] + \ldots \quad . \tag{9.191}$$

Substituting this into (9.190) gives

$$\mathcal{H}_0 = \mathcal{H}_{zz} + \frac{iT}{\hbar}[\mathcal{H}_R, \mathcal{H}_{zz}] + \frac{1}{2}\frac{T^2}{2!}\frac{1}{\hbar^2}[\mathcal{H}_R, [\mathcal{H}_R, \mathcal{H}_{zz}]] + \ldots \quad . \tag{9.192}$$

Evaluation of the commutators is straightforward but tedious. The general character of the result is clear, however. Successive terms involve progressively larger numbers of spin products, hence include terms which correspond to progressively higher orders of quantum coherence. That is, they contain terms of progressively larger p. If T is very short, only the first few terms contribute, and only small values of p result. But if $|\mathcal{H}_R|T \approx 1$, many terms will be included, giving large p's, though each individual term is still of order $|\mathcal{H}_{zz}|$.

We now have the operator $\mathcal{H}_0$. If we define a time $\Delta\tau'$ for which $\mathcal{H}_{zz}$ alone acts, and a time $\Delta\tau$ for the time in the cycle for which we have a fixed phase, ϕ_l,

$$\Delta\tau = T + T' + \Delta\tau' \quad . \tag{9.193}$$

But the development of ϱ over the total time $\Delta\tau$ is given by

$$R^{-1}\exp\left(-\frac{i}{\hbar}\mathcal{H}_{zz}\Delta\tau'\right)R = \exp\left(-\frac{i}{\hbar}R^{-1}\mathcal{H}_{zz}R\Delta\tau'\right)$$
$$= \exp\left(-\frac{i}{\hbar}\mathcal{H}_0\Delta\tau'\right) \quad . \tag{9.194}$$

Suppose then that we wish to select $p = p_0$. From (9.178) we see we must pick a phase shift ϕ for (9.175) given by

$$\phi = \frac{2\pi}{p_0} \quad . \tag{9.195}$$

For one cycle, ϕ_l will then, according to (9.175) and (9.195), need to take on the different values

$$\phi_l = 0, \ (2\pi/p_0), \ 2(2\pi/p_0), \ 3(2p/p_0) \ \ldots \ (p_0 - 1)(2\pi/p_0) \quad . \tag{9.196}$$

This cycle lasts for a time t_c given by the number of phases n times the duration of each phase

$$t_c = n\Delta\tau = p_0\Delta\tau \quad . \tag{9.197}$$

However, in order for the average Hamiltonian theory to hold, we must be able to approximate

$$\exp\left(-\frac{i}{\hbar}\mathcal{H}_{\phi_{p_0-1}}\Delta\tau'\right)\ldots\exp\left(-\frac{i}{\hbar}\mathcal{H}_0\Delta\tau'\right) \quad \text{by}$$
$$\exp\left[-\frac{i}{\hbar}\left(\mathcal{H}_{\phi_{p_0-1}} + \ldots + \mathcal{H}_0\right)\Delta\tau'\right] \tag{9.198}$$

This condition is satisfied if

$$|\mathcal{H}_0|\Delta\tau' \ll 1 \quad . \tag{9.199}$$

Thus, it is not the long time $\Delta\tau$ which comes into satisfying the Magnus expansion, but the (shorter) time $\Delta\tau'$. Since $\mathcal{H}_0$ is independent of $\Delta\tau'$, we can set T and T' (if $T' \neq T$) to make $\mathcal{H}_0$ what we desire, then set $\Delta\tau'$ to satisfy the Magnus expansion.

We may describe the total preparation period to generate coherence p_0 (and integral multiples thereof) as follows:

1. We apply $\mathcal{H}_R$ for time T, $\mathcal{H}_{zz}$ for time $\Delta\tau'$, then $-\mathcal{H}_R$ for time T (or T'). This takes time $\Delta\tau$.
2. With the phase of $\mathcal{H}_R$ shifted by $2\pi/p_0$, we repeat step 1.
3. With the phase $\mathcal{H}_R$ shifted by $2(2\pi/p_0)$ relative to step 1, repeat step 1.
4. Continue the phase advances until $\phi = (p_0 - 1)(2\pi/p_0)$ relative to step one.

The total time to complete (1)–(4) is $p_0\Delta\tau$, and constitutes one cycle. The average Hamiltonian produced by this cycle is

$$\mathcal{H}_{av} = \frac{\Delta\tau'}{\Delta\tau} \sum_p{}' \mathcal{H}_p \tag{9.200a}$$

where the prime on the summation indicates including only those p's for which

$$p = kp_0 \quad , \quad k = 0, 1, 2, \ldots \quad . \tag{9.200b}$$

Note that we multiply the summation by $\Delta\tau'/\Delta\tau$ since during each interval $\Delta\tau$, $\mathcal{H}_\phi$ is acting only for the time $\Delta\tau'$. The rest of the time $(T + T')$ is spent generating R and R^{-1}.

The basic cycle is then repeated for as many times, m, as is necessary for the amplitude of the desired coherence to grow sufficiently. Thus the total preparation time τ_p is $m\Delta\tau$. To give an idea of what is involved, we may quote the numbers reported by *Warren* et al. [9.18] to produce a selected fourth-order coherence for protons of benzene molecules oriented in a liquid crystal. They use the sequence (9.189a) to produce

$$\mathcal{H}_0 = R^{-1}\mathcal{H}_{zz}R \tag{9.201}$$

with a 2.4 μs duration of X (or $\overline{X}$)

$$\tau = 3.0\,\mu s \quad , \quad 2\tau' = 8.8\,\mu s \quad , \quad T = 2\,\text{ms} \quad , \quad \Delta\tau' = 35\,\mu s \quad . \tag{9.202}$$

The entire cycle was repeated four times.

To get rid of zero-order terms, one must reverse the sign of $\mathcal{H}_\phi$ for half the values of ϕ in the cycle. For a Hamiltonian such as that of (9.201) this can be achieved by choice of the phase ϕ, since adding $\pi/2$ to ϕ changes the sign of $\mathcal{H}_\phi$.

We now have a situation in which many more than two pulses are used to generate the p_0 quantum coherence. We may then introduce the unitary operator $U(\tau_p)$ given by

$$U(\tau_p) = \exp\left[(i/\hbar)\mathcal{H}_{av}\tau_p\right] \tag{9.203}$$

to describe the development of ϱ during the preparation interval of duration τ_p:

$$\varrho(\tau_p) = U(\tau_p)\varrho(0)U^{-1}(\tau_p) \quad . \tag{9.204}$$

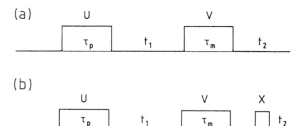

Fig. 9.6a,b. More general preparation and detection pulses represented by unitary operators U and V, preparation and mixing times τ_p and τ_m, evolution and detection times t_1 and t_2 respectively. In (b) we introduce explicitly an $X(\pi/2)$ pulse just before the detection period so that V is defined as all the mixing pulses which precede the $X(\pi/2)$ pulse

We then let the system evolve freely for a time t_1. To observe the effects of what happened during t_1, we need to mix down to one-quantum coherence. Previously we did this with a third $\pi/2$ pulse. As we shall see, we can do better by applying a fancy set of pulses, closely related to those which prepared the multiquantum coherence. Let us call the total mixing time τ_m. Then we represent mixing by a unitary operator $V(\tau_m)$, such that

$$\varrho(\tau_p + t_1 + \tau_m) = V(\tau_m)\varrho(\tau_p + t_1)V^{-1}(\tau_m) \quad . \tag{9.205}$$

The preparation, evolution, mixing, and detection periods may then be thought of as in Fig. 9.6.

We now need to investigate the optimum way to carry out the mixing so that we do not lose the advantages of confining the coherence we have generated to a small number of orders, p.

9.4.2 Mixing to Detect High Orders of Coherence

Our task now is to find an appropriate set of one or more mixing pulses to provide large detected signals when we have generated large-order coherences. The Pines group has done extensive work on this problem [9.20]. Specifically, given a preparation specified by $U(\tau_p)$, what should we construct for $V(\tau_m)$? We of course eventually detect the signal as a function of t_2, but in order to study the nature of V we shall focus on $t_2 = 0$. Thus, if we were going to detect $\langle I_y \rangle$ at $t_2 = 0$

$$\begin{aligned}\langle I_y \rangle &= \text{Tr}\left\{I_y \varrho(\tau_p + t_1 + \tau_m)\right\} \\ &= \text{Tr}\bigg\{I_y V(\tau_m) \exp\left(-\frac{i}{\hbar}\mathcal{H}_{zz}t_1\right) U(\tau_p) I_z U^{-1}(\tau_p) \\ &\quad \times \exp\left(\frac{i}{\hbar}\mathcal{H}_{zz}t_1\right) V^{-1}(\tau_m)\bigg\} \quad , \end{aligned} \tag{9.206}$$

where $\mathcal{H}_{zz}$ is taken to be the Hamiltonian in the absence of pulses. It turns out to be convenient to think of V as ending in an X pulse (Fig. 9.6b). We then redefine V to make that explicit, replacing $V(\tau_m)$ with $XV(\tau_m)$ so that

$$\begin{aligned}\langle I_y \rangle &= \text{Tr}\left\{ I_y XV(\tau_m) \exp\left(-\frac{i}{\hbar}\mathcal{H}_{zz}t_1\right) U(\tau_p) I_z U^{-1}(\tau_p) \right. \\ &\quad \left. \times \exp\left(\frac{i}{\hbar}\mathcal{H}_{zz}t_1\right) V^{-1}(\tau_m) X^{-1} \right\} \\ &= \text{Tr}\left\{ (X^{-1} I_y X) V(\tau_m) \exp\left(-\frac{i}{\hbar}\mathcal{H}_{zz}t_1\right) U(\tau_p) I_z U^{-1}(\tau_p) \right. \\ &\quad \left. \times \exp\left(\frac{i}{\hbar}\mathcal{H}_{zz}t_1\right) V^{-1}(\tau_m) \right\} \\ &= \text{Tr}\left\{ V^{-1} I_z V \exp\left(-\frac{i}{\hbar}\mathcal{H}_{zz}t_1\right) U I_z U^{-1} \exp\left(\frac{i}{\hbar}\mathcal{H}_{zz}t_1\right) \right\}. \end{aligned} \quad (9.207)$$

To understand the content of this equation, we evaluate the trace using the eigenstates of the system. Recall that we have M, the eigenvalue of $I_z T$, as one quantum number, and a family of others related to the dipolar energy abbreviated by α. We thus denote states by $|M\alpha\rangle$. It is sometimes convenient to use a single symbol r or s to stand for the pair. We define ω_r then by

$$E_r = E_{M\alpha} = \hbar\omega_r \quad \text{and} \quad (9.208a)$$

$$\omega_{rs} \equiv \omega_r - \omega_s \quad , \quad \text{then} \quad (9.208b)$$

$$\begin{aligned}\langle I_y \rangle &= \sum_{r,s} ((r|V^{-1}I_zV|s)e^{-i\omega_s t_1}(s|UI_zU^{-1}|r)e^{-i\omega_r t_1}) \\ &= \sum_{r,s}(r|V^{-1}I_zV|s)(s|UI_zU^{-1}|r)e^{i\omega_{rs}t_1} \quad . \end{aligned} \quad (9.209)$$

Let us first look at a simple limiting case: what we get if both U and V leave ϱ alone. This would be the case if there were no spin-spin coupling (isolated spins) and no rf pulses other than the final X pulse. Then $U = 1$, $V = 1$. Since the states $|r\rangle$, $|s\rangle$ are eigenstates of I_z, the matrix elements are diagonal, with the result ω_{rs} vanishes, giving

$$\langle I_y \rangle = \sum_{M\alpha} |(M\alpha|I_z|M\alpha)|^2 \quad (9.210a)$$

$$= \text{Tr}\{I_z^2\} \quad , \quad (9.210b)$$

which is just the well-known result of applying an X pulse to an initial density matrix $I_z T$.

This signal is, of course, the maximum we can get since it corresponds to rotating the total magnetization perpendicular to the static field. There is no loss in order in that process, hence this is the maximum signal we could get. It is conceivable that during the development period, t_1, there is some loss of order. Let us therefore next examine the general result (9.209) for the case that we eliminate such losses by making $t_1 = 0$. Then, going back to the last line of (9.207), we get $\langle I_y \rangle_{t_1=0}$ as

$$\langle I_y \rangle_{t_1=0} = \text{Tr}\{I_z(VUI_zU^{-1}V^{-1})\} \quad . \tag{9.211}$$

Comparing this result with (9.210b), which gives us the maximum possible $\langle I_y \rangle$, we see that if we make

$$VU = 1 \quad \text{i.e.} \tag{9.212a}$$

$$V = U^{-1} \tag{9.212b}$$

we will guarantee the maximum signal. Physically this condition states that what U does, V undoes. This statement is clearly a general kind of condition analogous to forming the conventional echo or the magic echo. We can compare this result with what we would get if we used U to create the coherence, but only a single $\pi/2$ pulse for the mixing. Since as we see in (9.207) we have already explicitly included a final pulse X, this situation corresponds to having $V = 1$:

$$\langle I_y \rangle_{t_1=0} = \text{Tr}\{I_z(UI_zU^{-1})\} \quad . \tag{9.213}$$

How does this compare with the result when $V = U^{-1}$? U is a unitary transformation. This expression reminds us of the dot product between a vector (I_z) and the same vector rotated (by U). It should be less than $\text{Tr}\{I_z^2\}$ or for that matter than $\text{Tr}\{(UI_zU^{-1})^2\}$. Indeed

$$\text{Tr}\{(UI_zU^{-1})^2\} = \text{Tr}\{UI_zU^{-1}UI_zU^{-1}\}$$
$$= \text{Tr}\{UI_z^2U^{-1}\} = \text{Tr}\{I_z^2U^{-1}U\} = \text{Tr}\{I_z^2\} \quad . \tag{9.214}$$

Consider then $\text{Tr}\{[I_z - (UI_zU^{-1})]^2\}$. Since this is the trace of the square of an Hermitian operator, it must be positive. Thus

$$0 \leq \text{Tr}\{(I_z - (UI_zU^{-1}))^2\} = \text{Tr}\{I_z^2\} + \text{Tr}\{(UI_zU^{-1})^2\}$$
$$- \text{Tr}\{I_zUI_zU^{-1}\} - \text{Tr}\{UI_zU^{-1}I_z\} \quad . \tag{9.215}$$

Permuting the order of the operators of the last trace, and utilizing (9.214), we thus get that

$$2\left(\text{Tr}\{I_z^2\} - \text{Tr}\{I_zUI_zU^{-1}\}\right) \geq 0 \quad . \tag{9.216}$$

In general, as in the Schwarz inequality, we expect the inequality to hold unless U is the identity operator. Thus, the signal of (9.213) will be less than the maximum.

We can think of the signal for $t_1 = 0$ along these lines: the expression

$$\langle I_y \rangle_{t_1=0} = \text{Tr}\left\{(V^{-1}I_zV)(UI_zU^{-1})\right\} \tag{9.217}$$

says that both V^{-1} and U "rotate" I_z. Unless V^{-1} and U produce the same "rotation" the signal is diminished from its maximum value. In general, the bigger the *relative* "rotation", the smaller the net result. In particular, suppose U generates particularly large p's, but V^{-1} does not. Then U will put the order of the system into large p, but V will not be able to bring it back down to zero order to permit the X pulse to convert it to transverse magnetization.

The Pines group has made extensive use of sequences satisfying (9.212)

$$V = U^{-1}$$

to generate and then observe large-order coherences.

It is the author's opinion that the *Warren, Weitekamp, Pines* method of selective excitation of multiple quantum coherence can be thought of as a kind of graduation exercise in spin Hamiltonian manipulation. It requires that one be completely at home with the idea that pulses manipulate the Hamiltonian, with average Hamiltonian theory, possess knowledge of the specific kinds of Hamiltonian one can generate and how to generate them. It illustrates then how once one has these details in mind, one can advance to a higher level using the ideas as building blocks, much as in electronics one thinks of oscillators, amplifiers, mixers, pulses and so on, without then being overwhelmed by the details of each circuit.

10. Electric Quadrupole Effects

10.1 Introduction

So far we have considered only the magnetic interactions of the nucleus with its surroundings. To be sure, by implication we have considered the effect of the nuclear charge, since it determines the electron orbits and where the nucleus sits in a molecule. However, we have not considered any electrical effects on the energy required to reorient the nucleus. That such effects do exist can be seen by considering a nonspherical nucleus. Suppose it is somewhat elongated and is acted on by the charges shown in Fig. 10.1. We see that Fig. 10.1b will correspond to a lower energy, since it has put the tips of the positive nuclear charge closer to the negative external charges. There is, therefore, an electrostatic energy that varies with the nuclear orientation. Of course,[1] turning the nucleus end for end does not affect the electrostatic energy. Consequently, for spin $\frac{1}{2}$ nuclei the electrostatic energy does not split the m_I degeneracy.

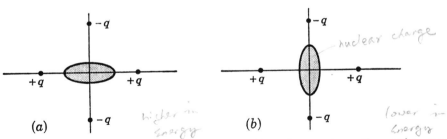

Fig. 10.1. (a) A cigar-shaped nucleus in the field of four charges, $+q$ on the x-axis; $-q$ on the y-axis. The configuration of (b) is energetically more favorable because it puts the positive charge of the ends of the cigar closer to the negative charges $-q$

---

[1] See references to "Quadrupole Effects" in the Bibliography and the articles by *Cohen* and *Reif* and by *Das* and *Hahn* under "Books, Monographs, or Review Articles" in the Bibliography.

10.2 Quadrupole Hamiltonian – Part 1

To develop a more quantitative theory, we begin with a description in terms of the classical charge density of the nucleus, ϱ. We shall obtain a quantum mechanical answer by replacing the classical ϱ by its quantum mechanical operator. Classically, the interaction energy E of a charge distribution of density ϱ with a potential V due to external sources is

$$E = \int \varrho(r) V(r) d\tau \quad . \tag{10.1}$$

We expand $V(r)$ in a Taylor's series about the origin:

$$V(r) = V(0) + \sum_\alpha x_\alpha \left(\frac{\partial V}{\partial x_\alpha}\right)_{r=0} + \frac{1}{2!} \sum_{\alpha,\beta} x_\alpha x_\beta \left(\frac{\partial^2 V}{\partial x_\alpha \partial x_\beta}\right)_{r=0} + \ldots \tag{10.2}$$

where x_α ($\alpha = 1,2,3$) stands for x, y, or z, respectively. Defining

$$V_\alpha \equiv \left(\frac{\partial V}{\partial x_\alpha}\right)_{r=0} \quad , \quad V_{\alpha\beta} \equiv \left(\frac{\partial^2 V}{\partial x_\alpha \partial x_\beta}\right)_{r=0} \tag{10.3}$$

we have

$$E = V(0) \int \varrho \, d\tau + \sum_\alpha V_\alpha \int x_\alpha \varrho \, d\tau + \frac{1}{2!} \sum_{\alpha,\beta} V_{\alpha\beta} \int x_\alpha x_\beta \varrho \, d\tau \ldots \quad . \tag{10.4}$$

Choosing the origin at the mass center of the nucleus, we have for the first term the electrostatic energy of the nucleus taken as a point charge. The second term involves the electrical dipole moment of the nucleus. It vanishes, since the center of mass and center of charge coincide. That they do coincide can be proved if the nuclear states possess a definite parity. All experimental evidence supports the contention that nuclei do have definite parity. Moreover, a nucleus in equilibrium experiences *zero* average electric field V_α. It is interesting to note that even if the dipole moment were not zero, the tendency of a nucleus to be at a point of vanishing electric field would make the dipole term hard to see. In fact it was for just this reason that *Purcell, Ramsey,* and *Smith* [10.1] looked for signs of a possible nuclear electrical dipole moment in neutrons rather than in charged nuclei.

The third term is the so-called electrical quadrupole term. We note at this point that one can always find principal axes of the potential V such that

$$V_{\alpha\beta} = 0 \quad \text{if} \quad \alpha \neq \beta \quad . \tag{10.5}$$

Moreover, V must satisfy Laplace's equation:

$$\nabla^2 V = 0 \quad . \tag{10.6}$$

This equation, evaluated at the origin, gives us

$$\sum_\alpha V_{\alpha\alpha} = 0 \quad . \tag{10.7}$$

(We note that sometimes Poisson's equation applies instead. Some care must then be exercised because we are, of course, interested only in the orientation dependent part of the potential, and must therefore subtract the spherically symmetric parts.) If one has a nucleus at a site of cubic symmetry,

$$V_{xx} = V_{yy} = V_{zz} \quad \text{(cubic symmetry)} \quad , \tag{10.8}$$

which, combined with (10.7), makes all three derivatives zero. The quadrupole coupling then vanishes. This situation arises, for example, with Na^{23} in Na metal. The face-centered cubic crystal structure puts each nucleus at a site of cubic symmetry.

It is convenient to consider the quantities $Q_{\alpha\beta}$ defined by the equation

$$Q_{\alpha\beta} = \int (3x_\alpha x_\beta - \delta_{\alpha\beta} r^2) \varrho \, d\tau \quad . \tag{10.9}$$

In terms of the $Q_{\alpha\beta}$'s, we have

$$\int x_\alpha x_\beta \varrho \, d\tau = \tfrac{1}{3}(Q_{\alpha\beta} + \int \delta_{\alpha\beta} r^2 \varrho \, d\tau) \quad . \tag{10.10}$$

As we shall see, the introduction of the $Q_{\alpha\beta}$'s amounts to our subtracting from the left side of (10.10) a term that does not depend on the orientation of the nucleus. We have, then, for the quadrupole energy $E^{(2)}$,

$$\begin{aligned}
E^{(2)} &= \frac{1}{2} \sum_{\alpha,\beta} V_{\alpha\beta} \int x_\alpha x_\beta \varrho \, d\tau \\
&= \frac{1}{6} \sum_{\alpha,\beta} (V_{\alpha\beta} Q_{\alpha\beta} + V_{\alpha\beta} \delta_{\alpha\beta} \int r^2 \varrho \, d\tau) \quad .
\end{aligned} \tag{10.11}$$

Since V satisfies Laplace's equation, the second term on the right of (10.11) vanishes, giving us

$$E^{(2)} = \frac{1}{6} \sum_{\alpha,\beta} V_{\alpha\beta} Q_{\alpha\beta} \quad . \tag{10.12}$$

Even if this term were not zero, we note that it would be independent of nuclear orientation.[2]

---

[2] If there is an electronic charge at the nucleus, we must use Poisson's equation. Then

$$\sum_\alpha V_{\alpha\alpha} = -4\pi e |\psi(0)|^2$$

where $|\psi(0)|^2$ is the electronic probability density at the nucleus. The orientation independent term, ΔE, of (10.11) becomes

$$\Delta E = \frac{1}{2} \sum_\alpha V_{\alpha\alpha} \int r^2 \varrho \, d\tau = -\frac{4\pi e}{6} |\psi(0)|^2 \int r^2 \varrho \, d\tau \quad .$$

This ΔE will be different for two nuclei of the same charge but different charge distributions (isotopes), or for two nuclei of the same mass and charge but different nuclear states (isomers). In an electronic transition between an s and a p-state, ΔE will make a contribution that will in general be different for different isotopes or isomers. Effects also show up in nuclear transition [10.2].

To obtain a quantum mechanical expression for the quadrupole coupling, we simply replace the classical ϱ by its quantum mechanical operator $\varrho^{(\mathrm{op})}$, given by

$$\varrho^{(\mathrm{op})}(\mathbf{r}) = \sum_k q_k \varrho(\mathbf{r} - \mathbf{r}_k) \quad, \tag{10.13}$$

where the sum runs over the nuclear particles, $1, 2, \ldots k \ldots N$, of charge q_k. Since the neutrons have zero charge, and the protons a charge e, we can simply sum over the protons:

$$\varrho^{(\mathrm{op})}(\mathbf{r}) = e \sum_{\mathrm{protons}} \delta(\mathbf{r} - \mathbf{r}_k) \quad. \tag{10.14}$$

By substituting (10.14) into the classical expression for $Q_{\alpha\beta}$, we obtain the quadrupole operator $Q_{\alpha\beta}^{(\mathrm{op})}$:

$$\begin{aligned} Q_{\alpha\beta}^{(\mathrm{op})} &= \int (3x_\alpha x_\beta - \delta_{\alpha\beta} r^2) \varrho^{(\mathrm{op})}(\mathbf{r}) d\tau \\ &= e \sum_{\mathrm{protons}} \int (3x_\alpha x_\beta - \delta_{\alpha\beta} r^2) \delta(\mathbf{r} - \mathbf{r}_k) d\tau \\ &= e \sum_{\mathrm{protons}} (3x_{\alpha k} x_{\beta k} - \delta_{\alpha\beta} r_k^2). \end{aligned} \tag{10.15}$$

We have, then, a quadrupole term for the Hamiltonian $\mathcal{H}_Q$, given by

$$\mathcal{H}_Q = \frac{1}{6} \sum_{\alpha,\beta} V_{\alpha\beta} Q_{\alpha\beta}^{(\mathrm{op})} \quad. \tag{10.16}$$

The expressions of (10.15) and (10.16) look exceedingly messy to handle because they involve all the nuclear particles. They appear to require us to treat the nucleus as a many-particle system, a complication we have avoided in discussing the magnetic couplings. Actually a similar problem is involved in both magnetic dipole and electric quadrupole cases, but we have simply avoided discussion in the magnetic case.

The quadrupole interaction represented by (10.15) enables us to treat problems of much greater complexity than those we encounter in a discussion of resonance phenomena. When performing resonances, we are in general concerned only with the ground state of a nucleus, or perhaps with an excited state when the excited state is sufficiently long-lived. The eigenstates of the nucleus are characterized by the total angular momentum I of each state, $2I + 1$ values of a component of angular momentum, and a set of other quantum numbers η, which we shall not bother to specify. Since we shall be concerned only with the spatial reorientation of the nucleus for a given nuclear energy state, we shall be concerned only with matrix elements diagonal in both I and η. Thus we shall need only matrix elements of the quadrupole operator, such as

$$(Im\eta | Q_{\alpha\beta}^{(\mathrm{op})} | Im'\eta) \quad .$$

These can be shown to obey the equation

$$(Im\eta|Q^{(op)}_{\alpha\beta}|Im'\eta) = C(Im|\tfrac{3}{2}(I_\alpha I_\beta + I_\beta I_\alpha) - \delta_{\alpha\beta}I^2|Im') \qquad (10.17)$$

where C is a constant, different for each set of the quantum numbers I and η. In order to justify (10.17), we need to digress to discuss the Clebsch-Gordan coefficients, the so-called irreducible tensor operators T_{LM}, and the Wigner-Eckart theorem.

10.3 Clebsch-Gordan Coefficients, Irreducible Tensor Operators, and the Wigner-Eckart Theorem

The Wigner-Eckart theorem is one of the most useful theorems in quantum mechanics. In order to state it, we must introduce the Clebsch-Gordan coefficients $C(LJ'J; MM_{J'}M_J)$, and the irreducible tensor operators T_{LM}. We shall first state the Wigner-Eckart theorem and then define the Clebsch-Gordan coefficients. Next we shall discuss the irreducible tensor operators, and lastly we shall indicate the derivation of the Wigner-Eckart theorem.

We consider a set of wave functions characterized by quantum numbers J and J' for the total angular momentum, M_J or $M_{J'}$ for the z-component of angular momentum, and as many other quantum numbers η or η' as are needed to specify the state. We are then concerned with calculating the matrix elements of the operators T_{LM}, using these functions as the basis functions. The Wigner-Eckart theorem states that all such matrix elements are related to the appropriate Clebsch-Gordan coefficients through a set of quantities $(J\eta \| T_L \| J'\eta')$ that depend on J, J', η, η' and L but which are independent of M_J, $M_{J'}$, and M. Stated mathematically, the Wigner-Eckart theorem is

$$(JM_J\eta|T_{LM}|J'M_{J'}\eta') = C(J'LJ; M_{J'}MM_J)(J\eta \| T_L \| J'\eta') \quad . \qquad (10.18)$$

Let us now define the Clebsch-Gordan coefficients. They are encountered when one discusses the addition of two angular momenta to form a resultant. We therefore consider a system made up of two parts. Let us describe one part of the system by the quantum numbers L and M, to describe the total angular momentum of that part and its z-component. Let us use the quantum numbers J' and $M_{J'}$ correspondingly for the second part of the system. For the system as a whole we introduce quantum numbers J and M_J. We have, then, wave functions ψ_{LM} and $\phi_{J'M_{J'}}$, to describe the two parts, and Ψ_{JM_J} for the whole system. The function Ψ_{JM_J} can be expressed as a linear combination of product functions of the two parts, since such products form a complete set:

$$\Psi_{JM_J} = \sum_{J'M_{J'};LM} C(J'LJ; M_{J'}MM_J)\phi_{J'M_{J'}}\psi_{LM} \quad . \qquad (10.19)$$

The coefficients $C(J'LJ; M_{J'}MM_J)$ are called the *Clebsch-Gordan coefficients*. Certain of their properties are very well known. For example, $C(J'LJ; M_{J'}MM_J)$ vanishes unless $M_J = M + M_{J'}$. A second property, often called the *triangle rule*, is that $C(J'LJ; M'_J MM_J)$ vanishes unless J equals one of the values $J' + L, J' + L - 1, \ldots |J' - L|$, a fact widely used in atomic physics.

Let us now define the irreducible tensor operators T_{LM}. Suppose we have a system whose angular momentum operators have components J_x, J_y, and J_z. We define the raising and lowering operators J^+ and J^- as usual by the relations

$$J^+ \equiv J_x + iJ_y \quad , \quad J^- \equiv J_x - iJ_y \quad . \tag{10.20}$$

One can construct functions G of the operators of the system and examine the commutators such as $[J^+, G]$, $[J^-, G]$, and $[J_z, G]$. It is often possible to define a family of $2L + 1$ operators (L is an integer) labeled by an integer M ($M = L, L-1, \ldots -L$) which we shall term *irreducible tensor operators* T_{LM}, which obey the commutation rules

$$[J^\pm, T_{LM}] = \sqrt{L(L+1) - M(M \pm 1)} T_{LM \pm 1} \tag{10.21}$$

$$[J_z, T_{LM}] = M T_{LM} \quad .$$

An example of such a set for $L = 1$ is

$$T_{11} = \frac{-1}{\sqrt{2}} J^+ \quad , \quad T_{10} = J_z \quad , \quad T_{1-1} = \frac{1}{\sqrt{2}} J^- \quad . \tag{10.22}$$

Another example of a T_{1M} can be constructed for an atom with spin and orbital angular momentum operators s and l, respectively, and total angular momentum J. Then we define the operators

$$l^+ = l_x + il_y \quad , \quad l^- = l_x - il_y \quad . \tag{10.23}$$

One can then verify the operators T_{1M} defined by

$$T_{11} = -\frac{1}{\sqrt{2}} l^+ \quad , \quad T_{10} = l_z \quad , \quad T_{1-1} = \frac{1}{\sqrt{2}} l^- \tag{10.24}$$

obey (10.21). (Actually the operators of (10.24) form components of an irreducible tensor T_{1M} with respect to the operators l^+, l^-, and l_z as well as J^+, J^-, and J_z.) We may write the T_{1M}'s of (10.22) as $T_{1M}(\boldsymbol{J})$, to signify that they are functions of the components J_x, J_y, and J_z of $\boldsymbol{J}$. The T_{1M}'s of (10.24) are in a similar manner signified as $T_{1M}(\boldsymbol{l})$.

It is helpful to have a more physical feeling for the definition of the operators T_{LM} by the commutation rules of (10.21). We realize that angular momentum operators can be used to generate rotations, as discussed in Chapter 2. It is not surprising, therefore, that (10.21) can be shown to guarantee that T_{LM} transforms under rotations of the coordinate axes into linear combinations $T_{LM'}$, in exactly the same way that the spherical harmonics Y_{LM} transform into linear combinations of $Y_{LM'}$'s. This theorem is shown in Chapter 5 of *Rose*'s excellent book [10.3].

We shall wish to compute matrix elements of the T_{LM}'s. We are familiar with the fact that it is possible to derive expressions for the matrix elements of angular momentum from the commutation rules among the components. It is possible to compute the matrix elements of the T_{LM}'s by means of (10.21) in a similar manner. Let us illustrate.

We have in mind a set of commuting operators J^2, J_z, plus others, with eigenvalues J, M_J, and η. We use η to stand for all other quantum numbers needed. We wish to compute matrix elements such as

$$(JM_J\eta|T_{LM}|J'M_{J'}\eta') \ . \tag{10.25}$$

By means of the commutation rule

$$[J_z, T_{LM}] = MT_{LM} \tag{10.26}$$

we have

$$(JM_J\eta|[J_z, T_{LM}]|J'M_{J'}\eta') = M(JM_J\eta|T_{LM}|J'M_{J'}\eta') \ . \tag{10.27a}$$

But

$$(JM_J\eta|[J_z, T_{LM}]|J'M_{J'}\eta') = \underbrace{(JM_J\eta|J_zT_{LM}|J'M_{J'}\eta')}_{1}$$
$$- \underbrace{(JM_J\eta|T_{LM}J_z|J'M_{J'}\eta')}_{2}$$
$$= (M_J - M_{J'})(JM_J\eta|T_{LM}|J'M_{J'}\eta')$$

where the last step follows from allowing the Hermitian operator J_z to operate on the function to its left in term 1 and to its right in term 2.

Therefore

$$(M_J - M_{J'})(JM_J\eta|T_{LM}|J'M_{J'}\eta') = M(JM_J\eta|T_{LM}|J'M_{J'}\eta') \ . \tag{10.27b}$$

Equation (10.27b) shows that

$$(JM_J\eta|T_{LM}|J'M_{J'}\eta') = 0 \quad \text{unless} \quad M_J - M_{J'} = M \ . \tag{10.28}$$

In a similar way we may find conditions on the matrix elements of the other terms of (10.21). Thus

$$(JM_J\eta|[J^{\pm}, T_{LM}]|J'M_{J'}\eta')$$
$$= \sqrt{L(L+1) - M(M \pm 1)}(JM_J\eta|T_{LM\pm 1}|J'M_{J'}\eta') \ . \tag{10.29}$$

But

$$(JM_J\eta|J^{\pm}T_{LM}|J'M_{J'}\eta')$$
$$= (JM_J\eta|J^{\pm}|JM_J \mp 1\eta)(JM_J \mp 1\eta|T_{LM}|J'M_{J'}\eta')$$
$$= \sqrt{J(J+1) - (M_J \mp 1)M_J}(JM_J \mp 1\eta|T_{LM}|J'M_{J'}\eta') \ . \tag{10.30}$$

By combining (10.29) and (10.30), we obtain the other recursion relations:

$$\sqrt{J(J+1) - (M_J \mp 1)M_J}(JM_J \mp 1\eta|T_{LM}|J'M_{J'}\eta')$$
$$- (JM_J\eta|T_{LM}|J'M_{J'} \pm 1\eta')\sqrt{J'(J'+1) - M_{J'}(M_{J'} \pm 1)}$$
$$= \sqrt{L(L+1) - M(M \pm 1)}(JM_J\eta|T_{LM \pm 1}|J'M_{J'}\eta') \quad . \tag{10.31}$$

We note that the only nonvanishing terms must satisfy (10.27b). However, if any one term in (10.31) satisfies this relation, all do. Equation (10.27b) and (10.31) constitute a set of recursion relations relating matrix elements for T_{LM} to one another and to those of $T_{LM'}$. These equations turn out to be sufficient to enable one to solve for all T_{LM} matrix elements for given J, J', η, η' in terms of any one matrix element.

A further insight into the significance of the recursion relations is shown by returning to the Clebsch-Gordan coefficients. In so doing, we shall sketch the proof of the Wigner-Eckart theorem.

As is shown by *Rose*, the C's obey recursion relations identical to those of the T_{LM}'s. We shall derive one: the selection rule on M, M_J, and $M_{J'}$. Consider the operator

$$J_z \equiv L_z + J'_z \quad \text{where} \tag{10.32}$$

$$J_z\Psi_{JM_J} = M_J\Psi_{JM_J}$$
$$L_z\psi_{LM} = M\psi_{LM}$$
$$J'_z\phi_{J'M_{J'}} = M_{J'}\phi_{J'M_{J'}} \quad . \tag{10.33}$$

Then, using (10.19), consider the following matrix element of the operator J_z:

$$(\psi_{LM}\phi_{J'M_{J'}}, J_z\Psi_{JM_J}) = M_J(\psi_{LM}\phi_{J'M_{J'}}, \Psi_{JM_J})$$
$$= M_J C(J'LJ; M_{J'}MM_J) \tag{10.34}$$

where we have let J_z operate to the right. But, writing J_z as $L_z + J'_z$ and operating on the functions to the left, we get

$$(\psi_{LM}\phi_{J'M_{J'}}, J_z\psi_{JM_J}) = (M + M_{J'})C(J'LJ; M_{J'}MM_J) \quad . \tag{10.35}$$

By equating (10.34) and (10.35), we find

$$(M + M_{J'} - M_J)C(J'LJ; M_{J'}MM_J) = 0 \quad . \tag{10.36}$$

This equation is quite analogous to (10.27b), provided we replace

$(JM_J\eta|T_{LM}|J'M_{J'}\eta')$ by

$C(J'LJ; M_{J'}MM_J) \quad .$

One can proceed in a similar manner to compute matrix elements of the raising and lowering operators, to get equations similar to (10.31). In fact the $C(J'LJ; M_{J'}MM_J)$'s obey recursion relations identical to those of the $(JM_J\eta|T_{LM}|J'M_{J'}\eta')$'s. As a result, one can say that the C's and the matrix elements of T_{LM}'s are related. The relationship is called the *Wigner-Eckart theorem*:

$$(JM_J\eta|T_{LM}|J'M_{J'}\eta') = C(J'LJ; M_{J'}MM_J)(J\eta \| T_L \| J'\eta') \quad (10.37)$$

where the notation $(J\eta \| T_L \| J'\eta')$ stands for a quantity that is a constant for a given J, L, J', η, η' independent of M_J, $M_{J'}$, and M.

As we can see specifically from (10.22) and (10.24), for a given L and M there may be a variety of functions, all of which are T_{LM}'s. The Clebsch-Gordan coefficient is the same for all functions T_{LM} that have the same L and M, but the constant $(J\eta \| T_L \| J'\eta')$ will depend on what variable is used to construct the T_{LM}'s.

To illustrate this point further, let us consider a particle with spin s and orbital angular momentum l and position r. The total angular momentum J is given by

$$J = s + l \quad \text{where} \quad (10.38)$$

$$l_x = \frac{1}{i}\left(y\frac{\partial}{\partial z} - z\frac{\partial}{\partial y}\right)$$

$$l_y = \frac{1}{i}\left(z\frac{\partial}{\partial x} - x\frac{\partial}{\partial z}\right) \quad (10.39)$$

$$l_z = \frac{1}{i}\left(x\frac{\partial}{\partial y} - y\frac{\partial}{\partial x}\right) .$$

We shall now list two T_{2M}'s: one a function of the angular momentum J; the other, of the coordinate r. One can verify that the functions of Table 10.1, which we shall call $T_{2M}(J)$ and $T_{2M}(r)$, indeed obey the commutation rules of (10.21) with respect to J^+, J^-, and J_z.

We have used the notation $T_{2M}(r)$ as shorthand for a T_{2M} constructed from the components x, y, and z of r. There is an obvious similarity between $T_{2M}(J)$ and $T_{2M}(r)$: Replacement of J^+ by $(x+iy)$, J^- by $(x-iy)$, J_z by z will convert $T_{2M}(J)$ into $T_{2M}(r)$. This similarity is a direct consequence of the similarity of the commutation relations of components of J and r with J_x, J_y, and J_z:

Table 10.1

	$T_{2M}(J)$	$T_{2M}(r)$
T_{22}	J^{+2}	$(x+iy)^2$
T_{21}	$-(J_z J^+ + J^+ J_z)$	$-2z(x+iy)$
T_{20}	$\sqrt{2/3}(3J_z^2 - J^2)$	$\sqrt{2/3}(3z^2 - r^2)$
T_{2-1}	$J_z J^- + J^- J_z$	$2z(x-iy)$
T_{2-2}	J^{-2}	$(x-iy)^2$

$$[J_x, y] = iz \tag{10.40a}$$

$$[J_x, J_y] = iJ_z \quad , \quad \text{etc.} \tag{10.40b}$$

where (10.40a) can be verified by means of (10.38) and (10.39). It is clear that any function $G(x, y, z)$ of x, y, z, constructed from a function $G(J_x, J_y, J_z)$ of J_x, J_y, J_z by direct substitution of x for J_x, and so on, will obey the same commutation rules with respect to J_x, J_y, and J_z. Thus, if a function of J_x, J_y, J_z is known to be a T_{LM}, the same will be true of the function formed by replacing J_x, J_y, J_z by x, y, z, respectively. The only caution we note in procedures such as this is that we must remember that the components of some operators do not commute among themselves; so that for example in $T_{21}(\boldsymbol{J})$ we have the symmetrized product $J^+J_z + J_zJ^+$, not $2J^+J_z$. The method of direct replacement will work for other variables as long as they obey commutation relations such as those of (10.40). For an excellent review article including tables of T_{LM}'s of various L and M, see [10.4].

Returning now to (10.37), let us consider two T_{LM}'s, one a function of variables q and the other a function of variables p. Then (10.37) tells us that

$$(JM_J\eta|T_{LM}(q)|J'M_{J'}\eta')$$
$$= (JM_J\eta|T_{LM}(p)|J'M_{J'}\eta')\frac{(J\eta\,\|\,T_L(q)\,\|\,J'\eta')}{(J\eta\,\|\,T_L(p)\,\|\,J'\eta')} \quad . \tag{10.41}$$

Since the factor $(J\eta\,\|\,T_L(q)\,\|\,J'\eta')/(J\eta\,\|\,T_L(p)\,\|\,J'\eta')$ is a constant (that is, independent of M, M_J, and $M_{J'}$), we see that we can compute all the matrix elements of $T_{LM}(q)$ of fixed J, J', η, and η' from knowledge of the constant and of the matrix elements $(JM_J\eta|T_{LM}(p)|J'M_{J'}\eta')$.

One word of caution is necessary. It may be that (10.41) is not meaningful, since for some operators p, the matrix element $(JM_J\eta|T_{LM}(p)|J'M_{J'}\eta')$ vanishes even though the matrix element $(JM_J\eta|T_{LM}(q)|J'M_{J'}\eta')$ does not. An example of such a case is when $T_{LM}(p)$ is made up of components of $\boldsymbol{J}$. Then all matrix elements in which $J' \neq J$ vanish. Of course $(J\,\|\,T_2(\boldsymbol{J})\,\|\,J')$ vanishes too, so that (10.41) becomes indeterminant.

10.4 Quadrupole Hamiltonian – Part 2

We now apply the Wigner-Eckart theorem to evaluate the matrix elements of $Q_{\alpha\beta}^{(\text{op})}$. Now

$$Q_{\alpha\beta}^{(\text{op})} = e \sum_{k(\text{protons})} (3x_{\alpha k}x_{\beta k} - \delta_{\alpha\beta}r_k^2) \quad . \tag{10.42}$$

By recalling that I_x, I_y, and I_z are the operators of the total angular momentum of the nucleus

$$I_x = \sum_k l_{xk} + s_{xk} \quad , \quad \text{etc., for } I_y \text{ and } I_z \quad , \tag{10.43}$$

where l_{xk} and s_{xk} are the x-components of the orbital and spin angular momenta of the kth nucleon; and by recalling that

$$[l_{xk}, y_k] = iz_k \quad [s_{xk}, y_k] = 0 \quad , \quad \text{etc.} \quad , \tag{10.44}$$

we see that

$$[I_x, y_k] = iz_k \quad , \quad \text{etc.} \tag{10.45}$$

The terms $3x_{\alpha k}x_{\beta k} - \delta_{\alpha\beta}r_k^2$ are linear combinations of $T_{2M}(r_k)$'s such as found in the right-hand column of Table 10.1.

Equation (10.41) applies in a somewhat more general form not only to T_{LM}'s but also to functions that are linear combinations of T_{LM}'s, all of the same L. Thus consider such a function $F(p)$, which is a function of the operators p:

$$F(p) = \sum_M a_M T_{LM}(p) \quad . \tag{10.46}$$

Let us define a function $G(q)$ of the operators q, using the same coefficients a_M:

$$G(q) \equiv \sum_M a_M T_{LM}(q) \quad . \tag{10.47}$$

Then one can easily verify, using (10.41, 46, 47), that

$$(JM_J\eta|G(q)|J'M_{J'}\eta') = (JM_J\eta|F(p)|J'M_{J'}\eta') \times \frac{(J\eta \| T_L(q) \| J'\eta')}{(J\eta \| T_L(p) \| J'\eta')} \quad . \tag{10.48}$$

We may apply this theorem to show that

$$(Im\eta|e \sum_{k(\text{protons})} (3x_{\alpha k}x_{\beta k} - \delta_{\alpha\beta}r_k^2)|Im'\eta)$$
$$= (Im\eta|3\frac{(I_\alpha I_\beta + I_\beta I_\alpha)}{2} - \delta_{\alpha\beta}I^2|Im'\eta)C \tag{10.49}$$

where C is a constant,[3] the same for all m, m', α, and β. We can express C in terms of the matrix element for which $m = m' = I$, $\alpha = \beta = z$ as follows:

$$(II\eta|e \sum_{k(\text{protons})} (3z_k^2 - r_k^2|II\eta) = C(II\eta|3I_z^2 - I^2|II\eta)$$
$$= CI(2I-1) \quad . \tag{10.50}$$

Since the quantum number η is assumed to be associated with a variable that commutes with I^2 and I_z, we can omit it in evaluating the right-hand side of (10.50). We shall also define a symbol eQ:

---

[3] Do not confuse C with the symbol for the Clebsch-Gordan coefficients.

$$eQ = (II\eta|e \sum_{k(\text{protons})} (3z_k^2 - r_k^2)|II\eta) \quad . \tag{10.51}$$

Q is called the *quadrupole moment* of the nucleus. We have, by combining (10.50) and (10.51),

$$C = \frac{eQ}{I(2I-1)} \quad . \tag{10.52}$$

The fact that we are concerned with matrix elements internal to one set of quantum numbers I, η enables us to use (10.49) and (10.52) to replace $Q_{\alpha\beta}^{(\text{op})}$ in the Hamiltonian. All matrix elements diagonal in I and η are just what we should calculate by adding an effective quadrupolar contribution $\mathcal{H}_Q$ to the Hamiltonian:

$$\mathcal{H}_Q = \frac{eQ}{6I(2I-1)} \sum_{\alpha,\beta} V_{\alpha\beta}[\tfrac{3}{2}(I_\alpha I_\beta + I_\beta I_\alpha) - \delta_{\alpha\beta} I^2] \quad . \tag{10.53}$$

It is interesting to note that of the nine components of $Q_{\alpha\beta}^{(\text{op})}$, only one nuclear constant, eQ, is needed. The reason is as follows: The fact that the nucleus is in a state of definite angular momentum is equivalent to the classical statement that the charge has cylindrical symmetry. Taking z as the symmetry axis, the energy change on reorientation depends, then, only on the *difference* between the charge distribution parallel and transverse to z:

$$\int z^2 \varrho \, d\tau \quad \text{and} \quad \int x^2 \varrho \, d\tau \quad .$$

This gives us the critical quantity

$$\int (z^2 - x^2)\varrho \, d\tau = \tfrac{1}{2} \int (2z^2 - x^2 - y^2)\varrho \, d\tau$$
$$= \tfrac{1}{2} \int (3z^2 - r^2)\varrho \, d\tau \quad . \tag{10.54}$$

The last integral, we see, is the classical equivalent of our eQ.

The effective quadrupole interaction of (10.53) applies for an arbitrary orientation of the rectangular coordinates $\alpha = x, y, z$. The tensor coupling to the symmetric (in x, y, z) tensor $V_{\alpha\beta}$ can be simplified by choice of a set of principal axes relative to which $V_{\alpha\beta} = 0$ for $\alpha \neq \beta$. In terms of these axes, we have

$$\mathcal{H}_Q = \frac{eQ}{6I(2I-1)}[V_{xx}(3I_x^2 - I^2) + V_{yy}(3I_y^2 - I^2)$$
$$+ V_{zz}(3I_z^2 - I^2)] \quad . \tag{10.55}$$

This expression can be rewritten, using Laplace's equation $\sum_\alpha V_{\alpha\alpha} = 0$, to give

$$\mathcal{H}_Q = \frac{eQ}{4I(2I-1)}[V_{zz}(3I_z^2 - I^2) + (V_{xx} - V_{yy})(I_x^2 - I_y^2)] \quad . \tag{10.56}$$

Equation (10.56) shows that only two parameters are needed to characterize the derivatives of the potential; V_{zz} and $V_{xx} - V_{yy}$. It is customary to define two symbols, η and q, called the *asymmetry parameter* and the *field gradient*, by the equations

$$eq = V_{zz}$$
$$\eta = \frac{V_{xx} - V_{yy}}{V_{zz}} \quad . \tag{10.57}$$

The case of axial symmetry, often a good approximation, is handled by taking the axis to be the z-direction, giving $\eta = 0$. ← Wrong. True for D

Since we have seen that the raising and lowering operators often provide particularly convenient selection rules, it is useful to write (10.53) in terms of I^+, I^-, and I_z for an arbitrary (that is, nonprincipal) set of axes. By defining

$$V_0 = V_{zz}$$
$$V_{\pm 1} = V_{zx} \pm i V_{zy} \tag{10.58}$$
$$V_{\pm 2} = \tfrac{1}{2}(V_{xx} - V_{yy}) \pm i V_{xy}$$

we find by straightforward algebraic manipulation that

$$\mathcal{H}_Q = \frac{eQ}{4I(2I-1)}[V_0(3I_z^2 - I^2) + V_{+1}(I^- I_z + I_z I^-)$$
$$+ V_{-1}(I^+ I_z + I_z I^+) + V_{+2}(I^-)^2 + V_{-2}(I^+)^2] \quad . \tag{10.59}$$

✗

Equation (10.59) gives a form of the quadrupole coupling that is particularly useful when considering relaxation for which the principal axes are not fixed in space but rather are functions of time. An attempt to use principal axes would then be exceedingly cumbersome. We shall not attempt to describe nuclear relaxation by the quadrupolar coupling, although it is a very important mechanism in insulating crystals, often dominant at room temperature.

10.5 Examples at Strong and Weak Magnetic Fields

In order to illustrate the use of the effective quadrupolar interaction, we shall make the simplifying assumption of a field with axial symmetry (or any other symmetry such that $V_{xx} = V_{yy}$ for a set of principal axes). Let us then consider a magnetic field applied along the z'-axis where in general the z- and z'-axes differ. Then we have, for our Hamiltonian,

$$\mathcal{H} = -\gamma_n \hbar H_0 I_{z'} + \frac{e^2 qQ}{4I(2I-1)}(3I_z^2 - I^2) \quad . \tag{10.60}$$

First we shall consider what happens when the quadrupole coupling is weak compared to the magnetic interaction. In this case we can consider the spin quantized along the z'-axis. We proceed to treat the quadrupolar coupling by

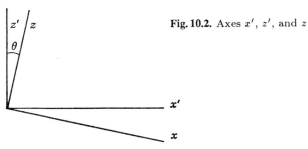

Fig. 10.2. Axes x', z', and z

perturbation theory. Defining the x'-axis to lie in the plane containing z' and z, we have (see Fig. 10.2)

$$I_z = I_{z'} \cos\theta + I_{x'} \sin\theta \quad . \tag{10.61}$$

By substituting in (10.60), we have

$$\mathcal{H} = -\gamma_n \hbar H_0 I_{z'} + \frac{e^2 qQ}{4I(2I-1)} [3I_{z'}^2 \cos^2\theta + 3I_{x'}^2 \sin^2\theta$$
$$+ 3(I_{z'}I_{x'} + I_{x'}I_{z'})\sin\theta\cos\theta - I^2] \quad . \tag{10.62}$$

In this equation, since $I_{z'}$ is diagonal in first order and $I_{x'}$ has vanishing diagonal elements, we have no contribution from terms such as $I_{z'}I_{x'}$ in first order. On the other hand, $I_{x'}^2$ has diagonal elements, since it involves the *product* of off-diagonal elements. By expressing $I_{x'} = \frac{1}{2}(I'^+ + I'^-)$ and $I_{y'} = (1/2i)(I'^+ - I'^-)$, it is straightforward to show that the diagonal elements of $I_{x'}^2$ and $I_{y'}^2$ are identical. We can therefore compute the diagonal matrix element

$$(m|I_{x'}^2|m) = (m|I_{y'}^2|m) = \tfrac{1}{2}(m|I^2 - I_{z'}^2|m)$$
$$= \tfrac{1}{2}[I(I+1) - m^2] \quad . \tag{10.63}$$

By collecting terms, we find

$$E_m = -\gamma_n \hbar H_0 m + \frac{e^2 qQ}{4I(2I-1)} \left(\frac{3\cos^2\theta - 1}{2}\right)[3m^2 - I(I+1)] \quad .$$

The effect of the quadrupole coupling is shown in Fig. 10.3 for the case of a spin $I = \frac{3}{2}$. It is helpful to note, since

$$\sum_m [3m^2 - I(I+1)] = \text{Tr}\{3I_z^2 - I^2\} = 0 \quad , \tag{10.64}$$

that the quadrupole coupling does not shift the center of gravity of the energy in first order. Moreover, the shifts of $+m$ and $-m$ are identical. With these points in mind we realize that the energy levels must look as shown.

One interesting result is that for a half-integral spin the $m = \pm\frac{1}{2}$ levels are shifted the same amount and the transition frequency between them is unaffected in first order by the quadrupole coupling. The $\frac{1}{2}$ to $-\frac{1}{2}$ transitions are quite insensitive to effects such as crystalline strains which may tend to shift

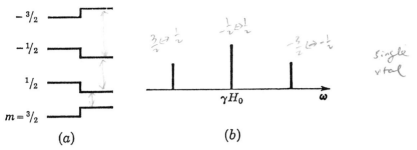

Fig. 10.3. (a) Effect of a quadrupole coupling in first order. The shifts of all levels for $I = \frac{3}{2}$ have the same magnitude. (b) Spectral absorption corresponding to the energy levels of (a). The central line is unaffected by the quadrupole coupling in first order

the frequency of the other transitions. When a nucleus has a particularly large quadrupole coupling the chance is great that even for well-annealed crystals one sees only the $+\frac{1}{2}$ to $-\frac{1}{2}$ transition.

If we carry the perturbation to the next higher order even the $\frac{1}{2}$ to $-\frac{1}{2}$ transition is shifted, the shift being of order $(e^2qQ)^2/\gamma_n \hbar H_0$.

A contrasting experimental situation arises when the quadrupole coupling is larger than that to the magnetic field H_0. Then it is appropriate to consider the quadrupole coupling as a first approximation. We have for the Hamiltonian in the absence of an external field (still assuming axial symmetry),

$$\mathcal{H} = \frac{e^2qQ}{4I(2I-1)}(3I_z^2 - I^2) \quad . \tag{10.65}$$

Clearly I^2 and I_z commute with $\mathcal{H}$, giving the quantum numbers I and m respectively. The energies are

$$E = \frac{e^2qQ}{4I(2I-1)}[3m^2 - I(I+1)] \quad . \tag{10.66}$$

A set of levels is shown for $I = \frac{5}{2}$ in Fig. 10.4.

We note that there is a degeneracy of $\pm m$ corresponding to the fact that turning the nucleus end for end does not affect the electrostatic energy. If an alternating magnetic field is applied with a nonvanishing component perpendicular to the z-axis it produces nonvanishing matrix elements of $|\Delta m| = 1$. It can therefore produce resonant transitions between the quadrupole levels. It is

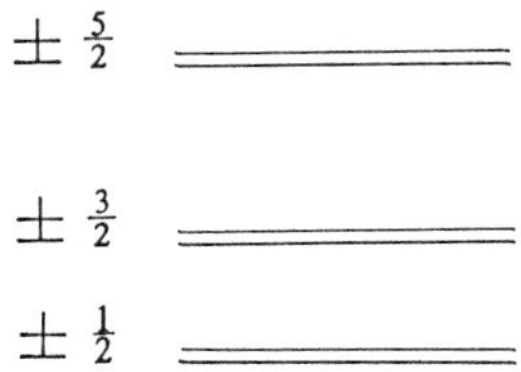

$\pm \frac{5}{2}$ ═══

$\pm \frac{3}{2}$ ═══

$\pm \frac{1}{2}$ ═══

Fig. 10.4. Energy levels of a quadrupole coupling when the Zeeman coupling is negligible

customary to speak then of "pure quadrupole resonance", although the transition is still induced by magnetic dipole coupling to the alternating field.

An important observation related to (10.66) is that when I is $\frac{1}{2}$, $\frac{3}{2}$, and so on (in general when $I = n + \frac{1}{2}$, where n is an integer) the energy levels are all doubly degenerate in the absence of a magnetic field but that for integral spin the degeneracy may be completely removed, as with the $m = 0$ state. This result is an example of an important theorem, due to Kramers, and applies to both electron and nuclear magnetic resonance. Kramers' theorem states:

For a system of angular momentum $I = n + \frac{1}{2}$, where n is 0, 1, 2, and so on, the degeneracy of any state can never be completely lifted by electric fields.

A corollary is that when a system is composed of an odd number of spin $\frac{1}{2}$ particles, electric fields can never completely lift the degeneracy.

The degeneracy is commonly called the *Kramers' degeneracy*. Proof of its existence actually depends on the properties of the system under time reversal.

10.6 Computation of Field Gradients

We have seen that the quadrupole coupling depends on the second derivatives $V_{\alpha\beta}$ of the potential, which reduce for the case of axes x, y, z, which are principal axes to V_{zz} and $V_{xx} - V_{yy}$. The potential V arises from external charges of either other nuclei or electrons. It is a straightforward matter of taking derivatives of the potential to show that a charge e at a point x, y, z produces a V_{zz} at the origin of

$$V_{zz} = e\frac{(3z^2 - r^2)}{r^5} \tag{10.67}$$

$$r^2 = x^2 + y^2 + z^2 \quad .$$

In terms of spherical coordinates we have (see Fig. 10.5)

$$V_{zz} = e\frac{(3\cos^2\theta - 1)}{r^3} \quad . \tag{10.68a}$$

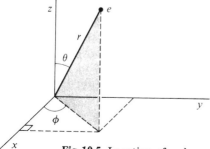

Fig. 10.5. Location of a charge e in terms of the spherical coordinates r, θ, ϕ

In general for α or $\beta = x, y, z$ we have

$$V_{\alpha\beta} = \frac{e}{r^3}\left(\frac{3x_\alpha x_\beta}{r^2} - \delta_{\alpha\beta}\right) \quad . \tag{10.68b}$$

Equation (10.67) and (10.68) emphasize through the $1/r^3$ dependence that charges close to the nucleus have the most important effect. We may suppose that the electrons belonging to the atom containing the nucleus would make major contributions to V_{zz}. Such is indeed the case. However, if we have a closed shell, the electronic charge is spherically symmetric, and there is no quadrupole coupling (see, however, some further remarks on closed shells below). The case of an incomplete shell is readily illustrated by an example of a single p-electron in an orbit $zf(r)$.

We wish, then, to compute the quadrupole operator for this example. Since the electronic motion is rapid, we shall average the expression of (10.68) over the electronic orbit. This procedure is equivalent to saying that, of the total Hamiltonian describing both the electron orbit and the nuclear spin, we shall compute only matrix elements that are diagonal in the electron orbital quantum numbers, and that we shall neglect the perturbation of the electron orbit by the nucleus.

We have, then,

$$\begin{aligned}
V_{zz} &= -e \int \psi_e^* \frac{(3\cos^2\theta - 1)}{r^3} \psi_e d\tau_e \\
&= -e \int \cos^2\theta \frac{(3\cos^2\theta - 1)}{r^3} \frac{r^2 f^2(r)}{4\pi} \sin\theta\, r^2\, dr\, d\phi\, d\theta \\
&= -e \frac{4}{15}\overline{\left(\frac{1}{r^3}\right)}
\end{aligned} \tag{10.69}$$

where we have designated the electronic charge as $-e$ and where, as usual, $\overline{(1/r^3)}$ is the average of $1/r^3$ for the p-orbit. We note that large-Z atoms, for which $\overline{(1/r^3)}$ is very large, will have large field gradients. This trend is shown by Table 10.2., which lists typical values of e^2qQ of halogen nuclei in covalently bonded crystals.

It is interesting to note that the values of e^2qQ of Table 10.2 put the frequency of pure quadrupole transitions for covalently bonded halogens at much higher frequencies than that of their Zeeman transitions $\gamma_n H_0$ for typical laboratory magnetic fields.

Table 10.2. Typical values of e^2qQ for halogen nuclei in covalently bonded crystals

Nucleus	e^2qQ [MHz]	Q [10^{-24} cm^2]
Cl^{35}	80	-7.97×10^{-2}
Br^{79}	500	0.30
I^{127}	2000	-0.59

When the electronic wave function contains a mixture of s- and p-states (a "hybridized bond") the s-part contributes nothing to the quadrupole coupling. A similar situation arises when a halogen atom is in a state corresponding to a mixture of a pure covalent bond (that is, a p-state) and an ionic bond (closed shell). The quadrupole coupling of the ionic bonding vanishes. One can therefore utilize quadrupole couplings to study bond hybridization, degree of covalency, double bonding, and so on.

The fact that the closed shell electrons are very close to the nucleus makes it important to consider their distortion from spherical symmetry. For example, a charge e will produce fields that will disturb the closed shell electrons. This effect has been studied extensively by various workers. It leads to a correction to the gradient V_{zz}^0 due to e alone. The actual field gradient V_{zz} is in fact given by

$$V_{zz} = V_{zz}^0[1 - \gamma(r)] \quad . \tag{10.70}$$

The quantity $\gamma(r)$ is called the *Sternheimer antishielding factor,* after one of the workers who has made some of the most important contributions to understanding of the phenomenon [10.5]. The fact that it is a function of the distance r from the charge e to the nucleus is emphasized by writing $\gamma(r)$. In general $\gamma(r) \ll 1$ as long as e is well inside the closed shell charge distribution. Once r is well outside, γ becomes independent of r. We shall denote this value by γ_∞. Some theoretical values for $1 - \gamma_\infty$ are shown in Table 10.3.

Table 10.3. Theoretical values of $1 - \gamma_\infty$

Ion	$1 - \gamma_\infty$
Cl$^-$	48
Cu$^+$	10
Rb$^+$	51
Cs$^+$	99

As we can see, the correction is enormous, amplifying the direct effect V_{zz}^0 by one or two orders of magnitude.

The existence of the Sternheimer effect greatly complicates the determination of nuclear quadrupole moments. It is difficult to know how accurate the theoretical γ's are. However, examination of (10.68) reminds us of the magnetic dipole coupling between a nucleus and an electron spin. The radial and angular terms are the same as the A and B terms of the dipolar coupling. Since the nuclear and electronic magnetic moments are known, it is therefore possible to use measured hyperfine couplings to get the average of $(3\cos^2\theta - 1)/r^3$. Since we are not using closed shells (the hyperfine coupling of a closed shell vanishes, since the electron spin is zero), the Sternheimer factor is only a small correction. This technique has been applied to the atomic beam experiments of halogens and is the basis of the most reliable experimental measurements of nuclear quadrupole moments.

11. Electron Spin Resonance

11.1 Introduction

So far we have confined our attention to nuclear magnetic resonance, although many of the basic principles apply to electron spin resonance. We have also considered questions concerning the electrons, such as the quenching of orbital angular momentum and the magnetic coupling of the nuclear spin to that of the electron. In this chapter we shall add a few more concepts that are important to the study of electron spin resonance[1] but which are not encountered in the study of nuclear resonance.

Probably the major difference between electron and nuclear magnetic resonance is the fact that the nuclear properties such as spin, magnetic moment, and quadrupole moment are to a very high degree of approximation unaffected by the surroundings, whereas for electronic systems, the relatively much greater physical size and the much smaller energy to excited states make the system strongly dependent on the surroundings. An atom, when placed in a crystal, may have angular momentum, magnetic moment, and quadrupole moment values entirely different from those of its free atom. It is as though in nuclear resonance we had to compute γ_n, I, and Q for each material in which the nucleus was to be studied.

The fact that the state of an atom in a solid or liquid is very different from that when it is free means that we cannot predict the properties or even the existence of a resonance from the free atom electronic angular momentum and magnetic moment.

For example, a sodium atom has zero orbital magnetic moment and angular momentum, but it has a spin of $\frac{1}{2}$ and a corresponding spin magnetic moment. The magnetic properties can be studied by the method of atomic beams. In sodium metal, the valence electrons form a conduction band, with substantial pairing of spins. However, there is a weak electronic spin magnetization whose spin resonance has been studied. In sodium chloride, the sodium gives up its outermost electron to complete the unfilled p-shell of the chlorines. The result is a zero spin magnetization and no electron spin resonance. Even if one has atoms whose bonding is covalent, as in molecular hydrogen, there is usually

---

[1] See references to "Electron Spin Resonance" listed in the Bibliography.

no net spin magnetization because the electron spins pair off into a spin singlet. There are exceptions, of course, such as the oxygen molecule. As we remarked in connection with chemical shifts, the orbital angular momentum is often quenched, so that there is no first-order orbital contribution to a resonance.

We see that most insulators will not exhibit a resonance, unless one takes special pains to unpair the spins. Some atoms, such as those in the iron group or rare earths, have incomplete inner shells. Even when ionized, they still possess a net moment. Thus neutral copper has a configuration $(3d)^{10}4s$. Cu^{++} has $(3d)^9$, which is paramagnetic. In an ionic substance such as $CuSO_4 \cdot 5H_2O$ (copper sulfate), the copper atoms are paramagnetic, and a resonance results.

We may list several classes of substances or circumstances in which one may expect to find resonances, although in individual cases the general rules may break down:

1. Materials containing atoms of the transition elements with incomplete inner shells; as, for example, the iron group or rare earths.
2. Ordinary metals, the conduction electrons.
3. Ferro- and ferrimagnets.
4. Imperfections in insulators, which may trap electrons or holes. For example, the F-center (electron trapped at the site of a missing halogen ion in an alkali halide) or donor and acceptor sites of semiconductors.

Treatment of all these situations on a unified basis is so hopelessly general that none of the interesting features emerge. The approximations important in one problem may not be at all justified in another. For example, if one is dealing with the resonance due to Cu^{++}, one knows already a great deal about the electronic wave function, since it will be closely related to that of a free Cu^{++} ion. One can therefore start by considering states of a free copper ion. On the other hand, there is no equivalent to the "free ion" if one is dealing with an F-center. We could not, therefore, define a set of "free ion" states.

What we shall do is list some of the more important interactions and then consider several examples that represent rather different physical situations but which involve the major phenomena.

The principal terms in the electron Hamiltonian will consist of:

1. The electron kinetic energy.
2. The electron potential energy. Often it is convenient to divide this into a "free ion" potential energy plus one due to the crystalline surroundings, the so-called crystalline potential. Such a decomposition makes sense provided there is such a thing as a "free ion", but, as remarked above, it would not have meaning for an F-center.
3. The spin-orbit coupling. An electron moving in an electric field E experiences a coupling of the spin to the orbital motion $\mathcal{H}_{SO}$:

$$\mathcal{H}_{SO} = \frac{e\hbar}{2m^2c^2} S \cdot (E \times p) \quad . \tag{11.1}$$

Often the electric field in an atom points radially outward and is a function of r only, so that

$$E(r) = \frac{r}{r} E(r) \quad .$$

Then $\mathbf{E} \times \mathbf{p}$ becomes $(1/r)E(r)\mathbf{r} \times \mathbf{p} = (\hbar/r)E(r)\mathbf{L}$. This circumstance leads to the well-known form of the spin-orbit coupling, utilizing the spin-orbit coupling constant λ:

$$\mathcal{H}_{SO} = \lambda \mathbf{L} \cdot \mathbf{S} \quad . \tag{11.2}$$

For free atoms that obey Russell-Saunders coupling, the spin-orbit coupling gives rise to the splitting of states of given L and S and their classification according to the total angular moment $J = L+S, L+S-1, \ldots, |L-S|$.

4. The coupling of the electron spin and orbital magnetic moments to an externally applied magnetic field.
5. The magnetic coupling of the nuclear spin to the electronic spin and orbital moments.
6. The coupling of the nuclear electrical quadrupole moment to the electronic charge.

Let us turn now to an example that will illustrate the role of some of the more important terms. We begin in the next section with a discussion of the role of the crystalline fields and spin-orbit coupling. In the Section 11.3 we shall consider the coupling to the nuclear magnetic moment.

11.2 Example of Spin-Orbit Coupling and Crystalline Fields

For our example we shall consider the case of an atom at the origin of a set of coordinates, possessing a single p-electron, acted on by four charges equidistant from the origin, two of the charges being positive, two negative, their magnitudes all being the same. The details of the arrangement (see Fig. 11.1) are seen to be identical to those that we discussed earlier when we considered the phenomenon of chemical shifts.

Neglecting nuclear coupling, we have for the Hamiltonian of the electron of charge q (q negative):

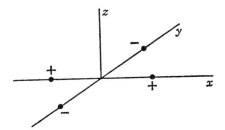

Fig. 11.1. Arrangement of two positive and two negative charges, all equidistant from the origin

$$\mathcal{H} = \frac{1}{2m}\left(p - \frac{q}{c}A\right)^2 + V_0 + V_1 + \lambda L \cdot S + 2\beta H \cdot S \tag{11.3}$$

where A is the vector potential associated with the applied static magnetic field H, V_0 is the potential of the "free atom", V_1 is the potential due to the four charges, and $2\beta H \cdot S$ represents the coupling of the electron spin moment to the external field. We are here using β, the Bohr magneton, to express the electron magnetic moment. It is related to γ_e, the electron gyromagnetic ratio, and μ_e, the spin magnetic moment, by the equation

$$\mu_e = -\gamma_e \hbar S = -2\beta S \quad \text{or} \tag{11.4}$$

$$\gamma_e \hbar = 2\beta \quad,$$

the negative sign representing the fact that the spin and moment are oppositely directed.

Expanding the first term on the right of (11.3) gives us

$$\frac{1}{2m}\left(p - \frac{q}{c}A\right)^2 = \frac{p^2}{2m} - \frac{q}{2mc}(p \cdot A + A \cdot p) + \frac{q^2}{2mc^2}A^2 \quad. \tag{11.5}$$

It is convenient to specify the vector potential as

$$A = \tfrac{1}{2} H \times r \quad, \tag{11.6}$$

which gives us

$$\frac{1}{2m}\left(p - \frac{q}{c}A\right)^2 = \frac{p^2}{2m} - \frac{q\hbar}{2mc} H \cdot L + \frac{q^2}{8mc^2}H^2(x'^2 + y'^2) \tag{11.7}$$

where, as usual, $L = (1/i)r \times \nabla$, and where the axes x' and y' are perpendicular to the field direction z'. (We distinguish between the field direction z' and the crystalline axis z). The term proportion to H^2 gives the usual diamagnetism. It turns out to be unimportant, compared to the term $H \cdot L$, in influencing the electron spin resonance. By utilizing the fact that $\beta = e\hbar/2mc$, we have, accordingly, as the Hamiltonian

$$\mathcal{H} = \frac{p^2}{2m} + \beta H \cdot L + V_0 + V_1 + \lambda L \cdot S + 2\beta H \cdot S \quad. \tag{11.8}$$

We shall consider that the principal energy terms are the kinetic energy and the "free atom" potential V_0. We shall treat the remaining terms by a perturbation method. For our example we shall think of the three degenerate p-states $xf(r)$, $yf(r)$, and $zf(x)$, which are solutions of the free atom potential V_0. We shall assume that the coupling to other free atom states is relatively unimportant, so that the effect of the remaining terms in the Hamiltonian can be found by considering only the submatrix of the Hamiltonian involving these three orbital states.

For practical laboratory fields the terms $H \cdot L$ and $H \cdot S$ are only about $1\,\text{cm}^{-1}$ whereas V_1 may be a substantial part of an electron volt (that is,

Table 11.1. Spin-orbit coupling constants per electron for several atoms

Atom	Coupling constant [cm^{-1}]
B	10
C	28
F	271
Cl	440
Br	1842

100 cm^{-1} to 10,000 cm^{-1}). The spin-orbit coupling constants vary substantially, some typical values of the coupling per electron being given in Table 11.1.

We see that, under some circumstances, V_1 will dominate; under other circumstances the spin-orbit coupling will be the major effect. The latter situation is, for example, typical of the rare earths, whereas the former is typical of the iron group.

Let us first consider the case that V_1 is much larger than λ. At first, we consider the effect of V_1 only. For the situation of Fig. 11.1, the effect of V_1 will be to lift the orbital degeneracy. The resultant energy levels, shown in Fig. 11.2, are all twofold degenerate because of the electron spin. We denote the wave functions as $xf(r)u_m$, and so on, where the function u_m is a spin function. If there were no spin-orbit coupling, the spin would be quantized independently of the orbital state so that the u_m's would be the usual eigenfunctions of $S_{z'}$, where z' is the magnetic field direction.

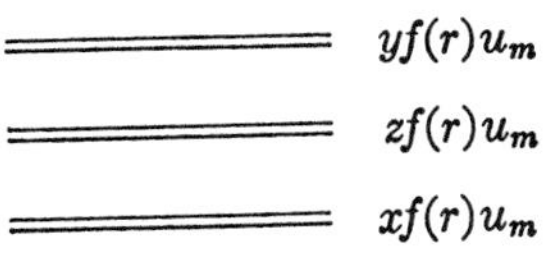

$$\begin{array}{l} = yf(r)u_m \\ = zf(r)u_m \\ = xf(r)u_m \end{array}$$

Fig. 11.2. Energy levels of the three p-states under the potential of Fig. 11.1

Let us consider, then, the effect of the two remaining terms:

$$\beta \boldsymbol{H} \cdot \boldsymbol{L} + \lambda \boldsymbol{L} \cdot \boldsymbol{S} \quad . \tag{11.9}$$

We examine the sorts of matrix elements these terms possess. There are two sorts: those that connect the same orbital state, and those that connect different orbital states. The former are clearly the more important, if they exist, since the orbital splittings are so large. We have, then, matrix elements such as

$$\int xf(r)u_m^* \beta H_z L_z xf(r)u_{m'} d\tau \, d\tau_s \quad \text{or} \tag{11.10}$$

$$\int xf(r)u_m^* \lambda L_z S_z xf(r)u_{m'} d\tau \, d\tau_s \tag{11.11}$$

where $d\tau$ stands for an integral over spatial coordinates, and $d\tau_\text{s}$, over spin variables. We see that both integrals of (11.10) and (11.11) involve

$$\int xf(r)L_z xf(r)d\tau \quad . \tag{11.12}$$

Recalling our earlier discussion about the quenching of orbital angular momentum, we realize that the integral of (11.12) vanishes. Therefore the only nonvanishing matrix elements of terms $\beta\boldsymbol{H}\cdot\boldsymbol{L}$ and $\lambda\boldsymbol{L}\cdot\boldsymbol{S}$ connect states differing in the orbital energy. They have, therefore, no effect in first order.

We have discussed this very problem in the absence of spin, noting that there was no first-order term $\beta\boldsymbol{H}\cdot\boldsymbol{L}$, since the states $xf(r)$ and so forth correspond to no net circulation of the electron. A similar remark applies to the spin-orbit coupling. The spin is coupled to states in which the electron has no preferential circulation. The average magnetic field due to orbital motion seen by the spin vanishes.

We know, however, from our discussion of chemical shifts that the term $\beta\boldsymbol{H}\cdot\boldsymbol{L}$ will induce some orbital circulation. The spin will not, therefore, experience a strictly zero field due to orbital motion.

We can think of solving exactly for the wave function under the influence of the applied field and of computing the matrix elements of $\lambda\boldsymbol{L}\cdot\boldsymbol{S}$, using the exact wave function. As a practical matter we use perturbation theory to compute the effect of $\beta\boldsymbol{H}\cdot\boldsymbol{L}$ on the wave function, keeping only the first term.[2] We have, then, for the modified wave function ψ_{xm}

$$\psi_{xm} = xf(r)u_m + \sum_{m'}\sum_{w=y,z}\frac{(wm'|\beta\boldsymbol{H}\cdot\boldsymbol{L}|xm)}{E_x - E_w}wf(r)u_{m'} \quad . \tag{11.13}$$

Since $\beta\boldsymbol{H}\cdot\boldsymbol{L}$ does not depend on spin, $m' = m$. By writing the interaction $\beta\boldsymbol{H}\cdot\boldsymbol{L}$ in component form, we find

$$\psi_{xm} = \left[xf(r) + \sum_{w=y,z}\sum_{q=x,y,z}\frac{(w|L_q|x)}{E_x - E_w}\beta H_q wf(r)\right]u_m \quad . \tag{11.14}$$

We now use this corrected function to compute matrix elements of $\lambda\boldsymbol{L}\cdot\boldsymbol{S}$ which involve the ground orbital state. There are actually matrix elements to the excited state, which in second order can couple back down to the ground state. However, they do not involve the applied field H. By neglecting them, we are finding the field dependent coupling energy. (The second-order terms in spin-orbit coupling produce no splitting when the spin is $\frac{1}{2}$.)

---

[2] We have two terms in the Hamiltonian, $\beta\boldsymbol{H}\cdot\boldsymbol{L}$ and $\lambda\boldsymbol{L}\cdot\boldsymbol{S}$, neither of which gives a first-order contribution. In second and higher orders, both terms perturb the wave function. Because of the similarity between the present problem and the chemical shift, first of all we treat the effect of the term $\beta\boldsymbol{H}\cdot\boldsymbol{L}$ on the wave function. It may seem more sensible to take the much larger $\lambda\boldsymbol{L}\cdot\boldsymbol{S}$ term first. As we shall see, however, our final answer involves an approximation that is proportional to the product $\beta\boldsymbol{H}\cdot\boldsymbol{L}$ and $\lambda\boldsymbol{L}\cdot\boldsymbol{S}$ (it is the interplay between the two energies that gives the effect). For this purpose it is immaterial which interaction we use to perturb the wave function.

$$\int \psi_{xm'}^* \lambda \mathbf{L} \cdot \mathbf{S} \psi_{xm} d\tau \, d\tau_s = \lambda\beta \sum_{w=y,z} \sum_{q,q'=x,y,z}$$

$$\times \frac{(m'|S_{q'}|m)[(x|L_{q'}|w)(w|L_q|x) + (x|L_q|w)(w|L_{q'}|x)]H_q}{E_x - E_w} \quad . \quad (11.15)$$

For computing matrix elements internal to the ground orbital state, (11.15) is equivalent to our replacing the terms $\beta \mathbf{H} \cdot \mathbf{L}$ and $\lambda \mathbf{L} \cdot \mathbf{S}$ by an effective term in the Hamiltonian ($\mathcal{H}_{\text{eff}}$):

$$\mathcal{H}_{\text{eff}} = \sum_{q,q'} S_q H_{q'} \lambda \beta \sum_w \frac{[(x|L_q|w)(w|L_{q'}|x) + (x|L_{q'}|w)(w|L_q|x)]}{E_x - E_w}$$

$$= \beta \sum_{q,q'} S_q a_{qq'} H_{q'} \quad . \quad (11.16)$$

Since the matrix elements that make up $a_{qq'}$ transform under coordinate rotations like L_q and $L_{q'}$, the $a_{qq'}$'s are components of a second-rank tensor. Examination of (11.16) shows that is a symmetric tensor ($a_{qq'} = a_{q'q}$).

For our particular case we can compute the matrix elements from the operators

$$L_x = \frac{1}{i}\left(y\frac{\partial}{\partial z} - z\frac{\partial}{\partial y}\right)$$

$$L_y = \frac{1}{i}\left(z\frac{\partial}{\partial x} - x\frac{\partial}{\partial z}\right) \quad (11.17)$$

$$L_z = \frac{1}{i}\left(x\frac{\partial}{\partial y} - y\frac{\partial}{\partial x}\right) \quad .$$

By using these expressions, we find that

$$L_x x f(r) = 0$$

$$L_y x f(r) = \frac{1}{i} z f(r) \quad (11.18)$$

$$L_z x f(r) = -\frac{1}{i} y f(r) \quad .$$

Thus the matrix elements of L_x vanish, those of L_y connect only to the state $zf(r)$, and those of L_z connect only to $yf(r)$. This gives us contributions only from terms with $q = q'$. That is, the x-, y-, z-axes are principal axes of the tensor $a_{qq'}$. Specifically we have, using (11.16) and (11.18),

$$\mathcal{H}_{\text{eff}} = 2\beta\left(\frac{\lambda}{E_x - E_z} S_y H_y + \frac{\lambda}{E_x - E_y} S_z H_z\right) \quad . \quad (11.19)$$

By combining this result with the Zeeman term, $2\beta \mathbf{H} \cdot \mathbf{S}$, we obtain a spin Hamiltonian for the ground orbital state

$$\mathcal{H} = \beta(g_{xx}H_x S_x + g_{yy}H_y S_y + g_{zz}H_z S_z) \tag{11.20}$$

where

$$g_{xx} = 2 \quad, \quad g_{yy} = 2\left(1 - \frac{\lambda}{E_z - E_x}\right) \quad, \quad g_{zz} = 2\left(1 - \frac{\lambda}{E_y - E_x}\right) \ . \tag{11.21}$$

We may employ the dyadic notation $\overleftrightarrow{g}$, defined by

$$\overleftrightarrow{g} = ig_{xx}\mathbf{i} + jg_{yy}\mathbf{j} + kg_{zz}\mathbf{k} \ , \tag{11.22}$$

to write the interaction as

$$\mathcal{H} = \beta \mathbf{H} \cdot \overleftrightarrow{g} \cdot \mathbf{S} \tag{11.23}$$

in place of

$$\mathcal{H} = 2\beta \mathbf{H} \cdot \mathbf{S} \ . \tag{11.24}$$

Comparison of (11.23) and (11.24) shows that the combined effect of the spin-orbit coupling and orbital Zeeman energy is as though the real field $\mathbf{H}$ were replaced by an effective field $\mathbf{H}_{\text{eff}}$, given by

$$\mathbf{H}_{\text{eff}} = \frac{\mathbf{H} \cdot \overleftrightarrow{g}}{2} = i H_x \frac{g_{xx}}{2} + j H_y \frac{g_{yy}}{2} + k H_z \frac{g_{zz}}{2} \tag{11.25}$$

with the resonance given by

$$\mathcal{H} = 2\beta \mathbf{H}_{\text{eff}} \cdot \mathbf{S} \ . \tag{11.26}$$

Since g_{xx}, g_{yy}, and g_{zz} are in general different, the effective field differs from the actual field in both magnitude and direction. If we denote by z'' the direction of the effective field, it is clear that a coordinate transformation will put (11.26) into the form

$$\mathcal{H} = 2\beta H_{\text{eff}} S_{z''} \tag{11.27}$$

where H_{eff} is the magnitude of $\mathbf{H}_{\text{eff}}$. The resonant frequency ω_0 therefore satisfies the condition

$$\begin{aligned}\hbar \omega_0 &= 2\beta H_{\text{eff}} \\ &= \beta \sqrt{H_x^2 g_{xx}^2 + H_y^2 g_{yy}^2 + H_z^2 g_{zz}^2} \\ &= \beta H \sqrt{\alpha_1^2 g_{xx}^2 + \alpha_2^2 g_{yy}^2 + \alpha_3^2 g_{zz}^2}\end{aligned} \tag{11.28}$$

where α_1, α_2 and α_3 are the cosines of the angle between $\mathbf{H}$ and the x-, y-, and z-axes. Often one writes (11.28) as

$$\hbar \omega_0 = g\beta H \tag{11.29}$$

where the "g-factor" is defined by the equation

$$g = \sqrt{g_{xx}^2 \alpha_1^2 + g_{yy}^2 \alpha_2^2 + g_{zz}^2 \alpha_3^2} \ . \tag{11.30}$$

Equations (11.29) and (11.30) emphasize the fact that for a given orientation of H, the splitting of the spin states is directly proportional to the magnitude of H. Frequently one talks about the g-shift, a term that refers to the difference between g and the free spin value of 2. From (11.21) and (11.30), recognizing that both $E_z - E_x$ and $E_y - E_x$ are positive, we see that positive values of λ make g less than or equal to 2, whereas negative λ's make g greater than or equal to 2. We associate positive λ's with atomic shells less than half-full, negative λ's with those more than half-full. Another terminology is to remark that electron resonances give positive λ and hole resonances give negative λ's. We shall return to this point in Sect. 11.4, where we shall find that a great deal of caution must be exercised in this simple interpretation as a general rule.

The size of the g-shift clearly increases with the nuclear charge, as we noted in Table 11.1. Its magnitude also depends on the magnitude of the splitting to the excited states to which the orbital angular momentum couples. Using an energy of about $1.3\,\text{eV}$ ($10{,}000\,\text{cm}^{-1}$) and a λ of $100\,\text{cm}^{-1}$, we see $2 - g \approx 0.02$, a readily observable effect.

We note that the g-shift arises because of the interplay between the spin-orbit and orbital Zeeman interactions. It is analogous to the chemical shift that arises from the interplay between nuclear spin-electron orbit coupling and the electron orbital Zeeman interaction. In both cases we say that the spin (electron or nuclear) experiences both the applied magnetic field and a sort of induced magnetic field. All such phenomena involving the interplay of two interactions can be viewed also as an application of a generalized form of second-order perturbation theory. This is in fact *Ramsey*'s method for deriving the chemical shift formulas. We shall illustrate by computing the g-shift.

The problem is treated in general in Appendix D. There it is shown that the perturbation effectively adds a term, $\mathcal{H}_{\text{new}}$, to the Hamiltonian, which has matrix elements between states $|0\rangle$ and $|0'\rangle$; for our example these elements have the same orbital part $xf(r)$ but may differ in spin function.

Defining a perturbation term $\mathcal{H}_{\text{pert}}$ by

$$\mathcal{H}_{\text{pert}} = \lambda \mathbf{L} \cdot \mathbf{S} + \beta \mathbf{H} \cdot \mathbf{L} \tag{11.31}$$

we find a matrix element $(0|\mathcal{H}_{\text{new}}|0')$ between $|0\rangle$ and $|0'\rangle$, as outlined in Appendix D, given by

$$(0|\mathcal{H}_{\text{new}}|0') = \sum_n \frac{(0|\mathcal{H}_{\text{pert}}|n)(n|\mathcal{H}_{\text{pert}}|0')}{E_0 - E_n} . \tag{11.32}$$

By substituting from (11.31), we get

$$(0|\mathcal{H}_{\text{new}}|0') = \sum_n \left[\frac{(0|\lambda \mathbf{L} \cdot \mathbf{S}|n)(n|\beta \mathbf{H} \cdot \mathbf{L}|0')}{E_0 - E_n} + \frac{(0|\beta \mathbf{H} \cdot \mathbf{L}|n)(n|\lambda \mathbf{L} \cdot \mathbf{S}|0')}{E_0 - E_n} \right.$$
$$\left. + \frac{(0|\beta \mathbf{H} \cdot \mathbf{L}|n)(n|\beta \mathbf{H} \cdot \mathbf{L}|0')}{E_0 - E_n} + \frac{(0|\lambda \mathbf{L} \cdot \mathbf{S}|n)(n|\lambda \mathbf{L} \cdot \mathbf{S}|0')}{E_0 - E_n} \right]. \tag{11.33}$$

The first two terms on the right give the g-shift we have calculated. The last two terms shift the two spin states equally. They do not, therefore, either produce a splitting of the doubly degenerate ground state or contribute to the g-shift. (If the spin were greater than $\frac{1}{2}$, however, such a term could give a splitting of the ground spin state even when $H = 0$.) The last two terms are just what we should have had if either perturbation were present by itself. Our previous calculation of the g-shift did not give them because it treated the effect of one term of the perturbation ($\beta H \cdot L$) on the other ($\lambda L \cdot S$). The method could be extended to find all the terms included in (11.33), but we see that direct application of (11.32) gives us a systematic method of getting all terms. On the other hand, the physical principles of the first calculation are somewhat more apparent.

In the example we have discussed so far, we have considered the crystalline potential V_1 to be much larger than the spin-orbit coupling constant λ. As a result, the orbital angular momentum is largely quenched, and the g-value is very close to the spin-only value of 2. This situation corresponds to the iron group atoms as well as to many electron and hole centers. We turn now to the opposite case, one of strong spin-orbit coupling and relatively much weaker crystalline fields, as encountered in the case of rare earth atoms.

If the spin-orbit coupling is dominant, the situation is in first approximation similar to that of a free atom. In fact the Hamiltonian

$$\mathcal{H} = \frac{p^2}{2m} + V_0 + \lambda L \cdot S + \beta H \cdot L + 2\beta H \cdot S + V_1 \tag{11.34}$$

is identical to that of a free atom except for the term V_1. As a first approximation we consider just the effect of the term $\lambda L \cdot S$ on the state formed from spin functions and the three p-states $xf(r)$, $yf(r)$, and $zf(r)$. The sum of the angular momenta L and S is the total angular momentum J:

$$J = L + S \quad . \tag{11.35}$$

By squaring, J, we have

$$\lambda L \cdot S = \frac{\lambda}{2}(J^2 - L^2 - S^2) \tag{11.36}$$

with eigenvalues E_{SO} given by

$$E_{SO} = \frac{\lambda}{2}[J(J+1) - L(L+1) - S(S+1)] \quad . \tag{11.37}$$

Since we are concerned with states all of which are characterized by an orbital quantum number L and spin quantum number S, the possible values of J are $L+S, L+S-1, \ldots |L-S|$. For our example, $L = 1$, $S = \frac{1}{2}$, so that $J = \frac{3}{2}$ or $\frac{1}{2}$. The $J = \frac{3}{2}$ and $\frac{1}{2}$ states are therefore split apart by an energy spacing

$$\Delta E = \frac{3}{2}\lambda \quad . \tag{11.38}$$

(More generally, the state J is λJ above the state $J - 1$.) The energy levels are shown in Fig. 11.3 for a positive λ.

$\Delta E = \tfrac{3}{2}\lambda$
$\overline{\overline{\overline{}}}\quad J = \tfrac{3}{2}$

$\overline{\overline{}}\quad J = \tfrac{1}{2}$

Fig. 11.3. Energy levels for $L = 1$, $S = \tfrac{1}{2}$ resulting from a spin orbit coupling $\lambda \mathbf{L} \cdot \mathbf{S}$

We consider next the effect of V_1. Here it becomes convenient to assume a specific form. Assuming that the potential arises from charges external to the atom, the potential in the region of the atom can be expressed as a sum of the form

$$V_1 = \sum_{l,m} C_{lm} r^l Y_{lm} \tag{11.39}$$

where the Y_{lm}'s are spherical harmonics and the C_{lm}'s are constants. If the potential is due to the charges of Fig. 11.1, it vanishes on the z-axis and changes sign if we replace x by y and y by $-x$ (a coordinate rotation). It is a maximum on the x-axis and a minimum on the y-axis for a given distance from the origin. The lowest l in the series of (11.39) is clearly $l = 2$. Of the five $l = 2$ functions, xy, xz, yz, $3z^2 - r^2$, $x^2 - y^2$, only the last is needed. We have, therefore, as an approximation, insofar as terms for $l > 2$ are not required,

$$V_1 = A(x^2 - y^2) \quad , \tag{11.40}$$

where A is a constant. (We shall see shortly that no higher terms are needed for an exact treatment.) We have, then, to consider the effect of V_1 on the states of Fig. 11.3. Two sorts of matrix elements will be important: those entirely within a given J, such as $(JM_J|V_1|JM'_J)$, and those connecting the different J states. The former will be the more important because they connect degenerate states. We can compute the matrix elements internal to a given J by means of the Wigner-Eckart theorem, for we notice that, with respect to L_x, L_y, and L_z (hence with respect to J_x, J_y, and J_z), V_1 is a linear combination of T_{2M}'s. That is, the commutation relations of V_1 with J_z, $J_x \pm iJ_y$ show V_1 to be a linear combination of T_{2M}'s. (In fact V_1 is proportional to $T_{22} + T_{2-2}$.) Thus we have

$$\begin{aligned}(JM_J|V_1|JM'_J) &= A(JM_J|x^2 - y^2|JM'_J) \\ &= C_J(JM_J|J_x^2 - J_y^2|JM'_J) \quad . \end{aligned} \tag{11.41}$$

This is equivalent to our replacing V_1 by the operator $\mathcal{H}_1$,

$$\mathcal{H}_1 = C_J(J_x^2 - J_y^2) \tag{11.42}$$

as long as we only compute matrix elements diagonal in J.

As an alternative, had the potential V_1 been $B(3z^2 - r^2)$, we should have used

$$\mathcal{H}_1 = C'_J(3J_z^2 - J^2) \quad . \tag{11.43}$$

This case of what is called the "axial field" is frequently encountered. We describe the calculation of C_J and C'_J below.

We have still to consider the effect of the magnetic field terms. Again let us consider only the matrix elements diagonal in J. This means that we wish to have

$$(JM_J|\beta \mathbf{H}\cdot\mathbf{L} + 2\beta\mathbf{H}\cdot\mathbf{S}|JM'_J) = \beta\mathbf{H}\cdot(JM_J|\mathbf{L}+2\mathbf{S}|JM'_J) \quad . \tag{11.44}$$

We can apply the Wigner-Eckart theorem to the matrix element, which, with respect to $\mathbf{J}$, is a linear combination of T_{1M}'s. We can therefore write

$$(JM_J|\mathbf{L}+2\mathbf{S}|JM'_J) = g_J(JM_J|\mathbf{J}|JM'_J) \tag{11.45}$$

where g_J is a constant for a given J, independent of M_J or M'_J. We recognize this problem to be the same as that of computing the Zeeman effect of free atoms, and the constant g_J is therefore the familiar Landé g-factor:

$$g_J = 1 + \frac{J(J+1) + S(S+1) - L(L+1)}{2J(J+1)} \quad . \tag{11.46}$$

Equation (11.45) is equivalent (as far as matrix elements diagonal in J are concerned) to our having a term $\mathcal{H}_Z$ replacing the two magnetic terms, where $\mathcal{H}_Z$ is given by

$$\mathcal{H}_Z = g_J\beta\mathbf{H}\cdot\mathbf{J} \quad . \tag{11.47}$$

By combining (11.42) and (11.47) with (11.36), we obtain an effective spin Hamiltonian $\mathcal{H}_{\text{eff}}$, which describes our problem accurately within a given J:

$$\mathcal{H}_{\text{eff}} = C_J(J_x^2 - J_y^2) + g_J\beta\mathbf{H}\cdot\mathbf{J} \tag{11.48}$$

or, for the axial field,

$$\mathcal{H}_{\text{eff}} = C'_J(3J_z^2 - J^2) + g_J\beta\mathbf{H}\cdot\mathbf{J} \quad . \tag{11.49}$$

The two terms on the right of (11.48) and (11.49) lift the $(2J+1)$-fold degeneracy of each J state. We shall not discuss the details of handling this problem other than to remark that clearly it is formally equivalent to the solution of the problem of a nucleus possessing a quadrupole moment acted on by an electric field gradient and a static magnetic field.

So far we have not computed the constants C_J or C'_J. We turn now to that task, illustrating the method by computing C'_J.

We have, using (11.43)

$$B(JM_J|3z^2 - r^2|JM'_J) = C'_J(JM_J|3J_z^2 - J^2|JM'_J) \quad . \tag{11.50}$$

By choosing $M_J = M'_J = J$, we have

$$B(JJ|3z^2 - r^2|JJ) = C'_J[3J^2 - J(J+1)] \tag{11.50a}$$

$$C'_J = \frac{B}{J(2J-1)}(JJ|3z^2 - r^2|JJ) \quad . \tag{11.51}$$

Now for $J = \frac{1}{2}$, all matrix elements of $3J_z^2 - J^2$ vanish (analogous to the fact that a quadrupole coupling cannot split a pair of spin $\frac{1}{2}$ levels). For $J = \frac{3}{2}$, we have

$$|JJ\rangle = \frac{1}{\sqrt{2}}(x + iy) \tag{11.52}$$

where we have denoted the spin function $S_z = +\frac{1}{2}$ by $u_{1/2}$. Therefore we find

$$C'_{3/2} = \frac{B}{3}\int \frac{1}{2}(x^2 + y^2)f^2(r)(3z^2 - r^2)d\tau \quad . \tag{11.53}$$

The angular portions of the integral can be carried through, using spherical coordinates, to give

$$C'_{3/2} = -\frac{2B}{15}\overline{r^2} \tag{11.53a}$$

where $\overline{r^2}$ is the average value of r^2 for the p-states.

The spin Hamiltonians of (11.48) and (11.49) do not include any matrix elements joining states of $J = \frac{3}{2}$ with those of $J = \frac{1}{2}$. If we wish to include such effects, we can actually apply the Wigner-Eckart theorem in an alternative manner. All matrix elements of V_1 are between states of $L = 1$. That is, we are concerned only with $(LM_L|V_1|LM'_L)$. However, the commutators of L_x, L_y, and L_z with V_1 show that V_1 is a linear combination of T_{2M}'s. Therefore all matrix elements are of the form

$$A(LM_L|x^2 - y^2|LM'_L) = C_L(LM_L|L_x^2 - L_y^2|LM'_L)$$
$$B(LM_L|3z^2 - r^2|LM'_L) = C'_L(LM_L|3L_z^2 - L^2|LM'_L) \quad . \tag{11.54}$$

These matrix elements are equivalent to those we should have had by replacing V_1 by an equivalent Hamiltonian $\mathcal{H}_1$:

$$\mathcal{H}_1 = C_L(L_x^2 - L_y^2) \quad \text{or} \quad \mathcal{H}_1 = C'_L(3L_z^2 - L^2) \quad . \tag{11.55}$$

In terms of $\mathcal{H}_1$, the effective Hamiltonian $\mathcal{H}_{\text{eff}}$, which will give all matrix elements formed from the three p-states $xf(r)$, $yf(r)$, and $zf(r)$, is

$$\mathcal{H}_{\text{eff}} = \lambda \mathbf{L} \cdot \mathbf{S} + C_L(L_x^2 - L_y^2) + \beta \mathbf{H} \cdot (2\mathbf{S} + \mathbf{L}) \quad \text{or}$$

$$\mathcal{H}_{\text{eff}} = \lambda(\mathbf{L} \cdot \mathbf{S}) + C'_L(3L_z^2 - L^2) + \beta \mathbf{H} \cdot (2\mathbf{S} + \mathbf{L}) \quad . \tag{11.56}$$

Equation (11.56) reduces to (11.48) or (11.49) for matrix elements diagonal in J.

In the absence of an external magnetic field, the energy levels of the Hamiltonian of (11.56) would remain at least doubly degenerate, according to *Kramer's* theorem, stated in Section 10.5 since $L = 1$, $S = \frac{1}{2}$.

It is interesting to note that the derivation of (11.45)

$$(JM_J|L + 2S|JM'_J) = g_J(JM_J|J|JM'_J) \tag{11.45}$$

using the Wigner-Eckart theorem, is quite analogous to the statement about the nuclear magnetic moment μ and spin I

$$\mu = \gamma \hbar I \quad , \tag{11.57}$$

which, stated more precisely, is

$$(IM|\mu|IM') = \gamma \hbar (IM|I|IM') \quad . \tag{11.58}$$

We expressed the potential of Fig. 11.1 by means of only *one* term in the expansion of the potential, that with $l = 2$. We might have supposed also that terms $l = 4$, 6, and so on, would have been needed. (The odd l's are not needed for this example, owing to the inversion symmetry of the charges.) If we had included an $l = 4$ term, we should have then needed to compute matrix elements such as

$$(LM_L|r^l Y_{lm}|LM_L') \tag{11.59}$$

with $l = 4$. However, with respect to L_x, L_y, and L_z, $r^l Y_{lm}$ is a T_{lm}. Therefore we can apply the Wigner-Eckart theorem to evaluate it. Recognizing that such an integral is closely related to the combination of angular momentum (by means of the Clebsch-Gordan coefficients), we note that it will vanish unless L and l can couple to form an angular momentum L (the triangle rule). For the case $L = 1$, $l = 4$, we can see that L and l could combine to give angular momenta of 5, 4, or 3, so that the integral must vanish. In fact, for $L = 1$, only $l = 2$ gives nonvanishing matrix elements. We need not, therefore, bother with $l = 4$, 6, and so on, in the expansion of the potential.

11.3 Hyperfine Structure

We have not as yet considered the magnetic coupling of the electron to the nearby nuclei. The basic form of the interaction has been discussed in Chapter 4. We distinguish between s-states and non-s-states.

s-states :

$$\mathcal{H}_{ISr} = \frac{8\pi}{3} \gamma_e \gamma_n \hbar^2 I \cdot S \delta(r) \quad . \tag{11.60}$$

Non-s-states :

$$\mathcal{H}_{ISr} = \frac{\gamma_e \gamma_n \hbar^2}{r^3} \left[\frac{3(I \cdot r)(S \cdot r)}{r^2} - I \cdot S \right] \quad . \tag{11.61}$$

The effect of the hyperfine coupling can be illustrated by considering the example of the preceding section in which the orbital angular momentum was quenched (corresponding to a crystalline potential V_1 much larger than the spin-orbit coupling λ). In the Hamiltonian matrix there will then be elements of the hyperfine coupling diagonal in the electron orbital energy as well as those connecting states of different crystalline energy such as $xf(r)$ and $yf(r)$. We shall neglect the elements that are off-diagonal. We have, then, only matrix elements such as

$$(xm_S m_I | \mathcal{H}_{ISr} | xm'_S m'_I) = \int \phi^*_{m_I} u^*_{m_S} xf(r) \mathcal{H}_{ISr}$$
$$\times \phi_{m'_I} u_{m'_S} xf(r) d\tau \, d\tau_S \, d\tau_I \quad (11.62)$$

where m_S and m_I stand for eigenvalues of S_z and I_z; ϕ_{m_I} and u_{m_S} are nuclear and electron spin functions; and where $d\tau$ and $d\tau_S$ stand for integration, respectively, over electron spatial and spin coordinates, and $d\tau_I$ over nuclear spin coordinate. It is, as usual, convenient to leave the specification of the quantization of both electron and nuclear spins until later, since the appropriate quantum states will depend on other parts of the Hamiltonian. We therefore omit the electron and nuclear spin functions and integrations, computing only $\int xf(r) \mathcal{H}_{ISr} xf(r) d\tau$. This integral leaves the nuclear and electron spin coordinates as operators. We shall therefore denote it by $\mathcal{H}_{IS}$:

$$\mathcal{H}_{IS} = \int xf(r) \mathcal{H}_{ISr} xf(r) d\tau \quad . \quad (11.63)$$

By substituting (11.61), we obtain

$$\mathcal{H}_{IS} = \gamma_e \gamma_n \hbar^2 \int \frac{1}{r^3} \left[3 \left(\frac{\boldsymbol{I} \cdot \boldsymbol{r}}{r^2} \right)(\boldsymbol{S} \cdot \boldsymbol{r}) - \boldsymbol{I} \cdot \boldsymbol{S} \right] x^2 f^2(r) d\tau \quad . \quad (11.64)$$

The terms such as $I_x x S_y y$ will contribute nothing, since for them the integrand is an odd function of x or y. The other terms can be expressed as the product of angular and radial integrals, giving

$$\mathcal{H}_{IS} = \gamma_e \gamma_n \hbar^2 \overline{\left(\frac{1}{r^3}\right)} \frac{2}{5}(3I_x S_x - \boldsymbol{I} \cdot \boldsymbol{S}) \quad (11.65)$$

where, as usual, $\overline{(1/r^3)}$ denotes the value of $1/r^3$ averaged over the state $xf(r)$.

If instead of a p-state we had an s-state, or more generally a wave function ψ containing some s-state, we could compute a corresponding term $\mathcal{H}_{IS}$ arising from the δ-function coupling:

$$\mathcal{H}_{IS} = \int \psi^*(r) \mathcal{H}_{ISr} \psi(r) d\tau = \frac{8\pi}{3} \gamma_e \gamma_n \hbar^2 |\psi(0)|^2 \boldsymbol{I} \cdot \boldsymbol{S} \quad . \quad (11.66)$$

The most general interaction is, of course, the sum of the couplings to the interactions of (11.60) and (11.61):

$$\mathcal{H}_{IS} = \int |\psi(r)|^2 \left\{ \frac{8\pi}{3} \gamma_e \gamma_n \hbar^2 \boldsymbol{I} \cdot \boldsymbol{S} \delta(r) \right.$$
$$\left. + \frac{\gamma_e \gamma_n}{r^3} \hbar^2 \left[3 \frac{(\boldsymbol{I} \cdot \boldsymbol{r})(\boldsymbol{S} \cdot \boldsymbol{r})}{r^2} \right] - \boldsymbol{I} \cdot \boldsymbol{S} \right\} d\tau \quad . \quad (11.67)$$

It will be linear in the spin variables I_x, I_y, I_z and S_x, S_y, and S_z, being of the general form

$$\mathcal{H}_{IS} = \sum_{\alpha,\alpha'=x,y,z} A_{\alpha\alpha'} S_\alpha I_{\alpha'} \tag{11.68}$$

but with the A's symmetric (that is, $A_{\alpha\alpha'} = A_{\alpha'\alpha}$). One can therefore always find the principal axes such that $A_{\alpha\alpha'}$ is diagonal, with values A_α and the hyperfine coupling given as

$$\mathcal{H}_{IS} = A_x I_x S_x + A_y I_y S_y + A_z I_z S_z = \boldsymbol{S} \cdot \overleftrightarrow{A} \cdot \boldsymbol{I} \tag{11.69}$$

where the dyadic $\overleftrightarrow{A}$ is given by

$$\overleftrightarrow{A} = \boldsymbol{i} A_x \boldsymbol{i} + \boldsymbol{j} A_y \boldsymbol{j} + \boldsymbol{k} A_z \boldsymbol{k} \quad . \tag{11.70}$$

If we did not have quenched orbital angular momentum, it would be necessary to include the coupling of the nuclear moment to the magnetic field arising from the orbital motion of the electron. We should also need to choose a new set of basic electronic states to obtain the "spin" Hamiltonian, such as, for example, states that are eigenfunctions of J^2.

We can combine (11.69) with (11.23) to obtain the spin Hamiltonian that includes both nucleus and electron for a case of quenched orbital angular momentum:

$$\mathcal{H} = \beta \boldsymbol{H} \cdot \overleftrightarrow{g} \cdot \boldsymbol{S} - \gamma_n \hbar \boldsymbol{H} \cdot \boldsymbol{I} + \boldsymbol{I} \cdot \overleftrightarrow{A} \cdot \boldsymbol{S} \quad . \tag{11.71}$$

(If the nucleus experiences a quadrupolar coupling, a term $\mathcal{H}_Q$ should be added.)

We shall examine the sorts of effects that the couplings of (11.71) produce by a simple example. We note first that there is no reason for the principal axes of $\overleftrightarrow{g}$ and $\overleftrightarrow{A}$ to coincide, although they do in fact for many simple cases. (Experimental situations have been reported in which they differ). We shall assume that they do, for our example. We begin, moreover, with the assumption that $\boldsymbol{H}$ is parallel to one of the principal axes, the z-axis, so that

$$\mathcal{H} = \beta H g_{zz} S_z - \gamma_n \hbar H I_z + A_x I_x S_x + A_y I_y S_y + A_z I_z S_z \quad . \tag{11.72}$$

We cannot solve this Hamiltonian in closed form without making some approximations. We shall assume that the electron spin Zeeman energy $\beta H g_{zz}$ is much bigger than the hyperfine coupling energies A_x, A_y, and A_z. This approximation is frequently good when one has strong magnetic fields ($2\beta H = 10^{10}$ Hz for $H = 3300$ Gauss, whereas A is often 10^9 Hz or less). If we take the electron Zeeman term to be large, we see that $\mathcal{H}$ commutes with S_z to a good approximation. We take the eigenfunctions to be eigenfunctions of S_z, with eigenvalue m_S. The terms $A_x S_x I_x$ and $A_y S_y I_y$ then have no matrix elements that are diagonal in m_S. We drop them from the Hamiltonian in first order. On the other hand, $A_z S_z I_z$ is diagonal in m_S and must be kept. This gives us an approximate Hamiltonian:

$$\mathcal{H} = \beta g_{zz} H S_z - \gamma_n \hbar H I_z + A_z S_z I_z \quad . \tag{11.73}$$

We see that I_z commutes with (11.73). We therefore take the states to be eigenfunctions of I_z with eigenvalues m_I. The first-order energy is therefore

$$E = \beta g_{zz} H m_S - \gamma_n \hbar H m_I + A_z m_S m_I \quad . \tag{11.74}$$

Since I^2, S^2, and I_z and S_z all commute with the Hamiltonian of (11.73) we can take the eigenfunctions to be a product of a nuclear spin and electron spin function:

$$\psi_{m_I m_S} = \phi_{I m_I} u_{S m_S} \quad . \tag{11.75}$$

The possible transitions produced by an alternating field are found by considering the matrix elements of the magnetic operator $\mathcal{H}_m(t)$:

$$\mathcal{H}_m(t) = (\gamma_e \hbar S_x - \gamma_n \hbar I_x) H_x \cos \omega t \tag{11.76}$$

between states such as those of (11.75). We find in this way that the S_x part of $\mathcal{H}_m(t)$ connects states with $\Delta m_S = \pm 1$, $\Delta m_I = 0$, whereas the I_x portion connects $\Delta m_S = 0$, $\Delta m_I = \pm 1$. We can consider these respectively to represent electron resonance and nuclear resonance. The transitions are allowed only if ω satisfies the conservation of energy, which, by using (11.74) and the selection rules, gives for electron resonance

$$\omega_e = \frac{g_{zz} \beta H_0 + A_z m_I}{\hbar} \tag{11.77}$$

and for nuclear resonance

$$\omega_n = \gamma_n H_0 + \frac{A_z m_S}{\hbar} \quad . \tag{11.78}$$

The effect of the hyperfine coupling on the electron resonance is seen to be equivalent to the addition of an extra magnetic field proportional to the z-component of the nuclear spin. Since the nucleus can take up only quantized orientation, the electron resonance is split into $2I + 1$ (equally spaced) lines. If the nuclei have no preferential orientation, the lines corresponding to various values of m_I occur with equal probability, and the resonance pattern looks like that of Fig. 11.4.

If one looks at the nuclear resonance, the frequencies are given by (11.78). To interpret the expression, we must know whether the nuclear Zeeman energy $\gamma_n \hbar H_0$ is larger or smaller than the hyperfine coupling $A_z m_S$, since (11.78) gives both positive and negative frequencies. In the former case (Fig. 11.5) the resonance is split into $2S+1$ lines spaced $A_z/\hbar$ apart and centered on the angular frequency $\gamma_n H_0$.

If the nuclear Zeeman energy is smaller than $A_z m_S$, the two values of m_S, m_S and $-m_S$, give rise to two lines occurring at

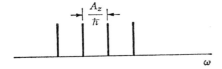

Fig. 11.4. Absorption versus frequency for the electron resonance for the case of a nucleus of spin $\frac{3}{2}$. The lines are, in first approximation, equally spaced

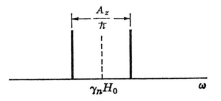

Fig. 11.5. Nuclear resonance when the nuclear Zeeman energy is larger than $|A_z\, m_S|$ drawn on the assumption of an electron spin of $\frac{1}{2}$

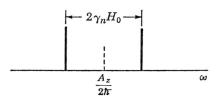

Fig. 11.6. Nuclear resonance for an electronic spin of $\frac{1}{2}$ when the hyperfine coupling is larger than the nuclear Zeeman coupling

$$\omega = \frac{A_z|m_S|}{\hbar} \pm \gamma_n H_0 \quad . \tag{11.79}$$

For an electron with spin $\frac{1}{2}$, the result is shown in Fig. 11.6.

An examination of Figs. 11.4, 11.5, and 11.6 shows that the electron resonance will enable one to measure the hyperfine coupling tensor and the nuclear spin, but will not by itself enable one to measure the nuclear moment. When combined with the results of a nuclear resonance, even the nuclear magnetic moment can be found. This latter feature is of particular importance in the study of color centers because it enables one to identify the nuclear species that gives rise to the hyperfine splitting, since the nuclear γ's are known from other experiments. The nuclear resonance, on the other hand, enables one to measure the spin of the electron system.

We have so far restricted ourselves to an orientation of the static field along one of the principal axes of the g- and hyperfine tensors. Interesting new effects arise when the field lies along other directions: The axis of quantization of the nucleus becomes different from that of the electron and in fact depends on whether the electron is oriented parallel or antiparallel to the field. For the sake of simplicity we shall illustrate these effects for the case where the electronic g-factor is isotropic. We shall take the field direction z' to lie in the x-z plane, where x, y, and z are the principal axes of the hyperfine coupling. The orientation is illustrated in Fig. 11.7.

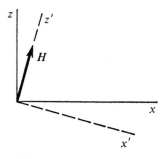

Fig. 11.7. Orientation of axes x, z, principal axes of the hyperfine tensor A, relative to the direction of the static field z'

The Hamiltonian becomes

$$\mathcal{H} = 2\beta H_{z'} S_{z'} - \gamma_n \hbar H_{z'} I_{z'} + A_x I_x S_x + A_y I_y S_y + A_z I_z S_z \quad . \quad (11.80)$$

We shall continue to assume $2\beta H_0 \gg A_x$, A_y, and A_z so that, to a good approximation, the Hamiltonian commutes with $S_{z'}$. We now seek to find those other parts of the Hamiltonian that will be diagonal in a representation in which $S_{z'}$ is diagonal. To do this, we express the spin components I_x, I_y, I_z, and S_x, S_y, and S_z in terms of the primed axes. (Actually it would be sufficient to express S_x, S_y and S_z in terms of $S_{x'}$, $S_{y'}$, and $S_{z'}$. However, the transformation of the nuclear axes enables us to see more readily that the nuclear quantization direction is not parallel to that of the electron). Noting that $y = y'$, we have

$$\begin{aligned}
S_x &= S_{x'} \cos\theta + S_{z'} \sin\theta \\
S_y &= S_{y'} \\
S_z &= S_{z'} \cos\theta - S_{x'} \sin\theta \\
I_x &= I_{x'} \cos\theta + I_{z'} \sin\theta \\
I_y &= I_{y'} \\
I_z &= I_{z'} \cos\theta - I_{x'} \sin\theta \quad .
\end{aligned} \quad (11.81)$$

By substituting these expressions into (11.80), we find

$$\begin{aligned}
\mathcal{H} = {} & 2\beta H_0 S_{z'} - \gamma_n \hbar H_0 I_{z'} + I_{x'} S_{x'}(A_x \cos^2\theta + A_z \sin^2\theta) \\
& + I_{y'} S_{y'} A_y + I_{z'} S_{z'}(A_x \sin^2\theta + A_z \cos^2\theta) \\
& + (I_{x'} S_{z'} + I_{z'} S_{x'})(A_x - A_z) \sin\theta \cos\theta \quad .
\end{aligned} \quad (11.82)$$

To first order, those terms in the hyperfine coupling involving $S_{x'}$ or $S_{y'}$ possess zero diagonal elements in the $S_{z'}$ scheme of quantization and therefore may be omitted. In second order, they will contribute energy shifts of the order of $A^2/2\beta H_0$, where by A^2 we mean the square of a matrix element of order A_x, A_y, or A_z, and the $2\beta H_0$ comes in because the "excited state" differs in electron spin orientation in the static field. The reduced Hamiltonian $\mathcal{H}_{\text{red}}$, which results from dropping all terms involving S_x or S_y, is then

$$\begin{aligned}
\mathcal{H}_{\text{red}} = {} & 2\beta H_0 S_{z'} - \gamma_n \hbar H_0 I_{z'} + [I_{z'}(A_x \sin^2\theta + A_z \cos^2\theta) \\
& + I_{x'}(A_x - A_z) \sin\theta \cos\theta] S_{z'} \quad .
\end{aligned} \quad (11.83)$$

Of course $S_{z'}$ commutes with $\mathcal{H}_{\text{red}}$. However, $I_{z'}$ does not. As far as the nucleus is concerned, $\mathcal{H}_{\text{red}}$ corresponds to a nucleus coupled to a magnetic field with components $H_{x'}$, $H_{y'}$, and $H_{z'}$ of

$$\begin{aligned}
H_{x'} &= \frac{1}{\gamma_n \hbar}(A_z - A_x) \sin\theta \cos\theta S_{z'} \\
H_{y'} &= 0 \\
H_{z'} &= H_0 - \left(\frac{A_x \sin^2\theta + A_z \cos^2\theta}{\gamma_n \hbar}\right) S_{z'} \quad .
\end{aligned} \quad (11.84)$$

These expressions involve the operator $S_{z'}$, by which we mean, of course, that the effective field depends on the electron quantum state. By denoting the eigenvalue of $S_{z'}$ by m_S, we have

$$H_{x'}(m_S) = 1\left(\frac{(A_z - A_x)\sin\theta\cos\theta}{\gamma_n \hbar}\right)m_S$$

$$H_{y'}(m_S) = 0 \qquad (11.85)$$

$$H_{z'}(m_S) = H_0 - \left(\frac{A_x \sin^2\theta + A_z \cos^2\theta}{\gamma_n \hbar}\right)m_S \quad.$$

The direction of quantization of the nucleus is clearly along the resultant of the effective field of (11.85) and *not* along the static field. The direction differs for the different m_S values. The magnitude of the nuclear energy separation $\Delta E_{\text{nuclear}}$ is as usual

$$\Delta E_{\text{nuclear}} = \gamma_n \hbar H_{\text{eff}}(m_S) \qquad (11.86)$$

where $H_{\text{eff}}(m_S)$ is the magnitude of the fields of (11.85) and depends on m_S. The total energy of electron and nucleus is therefore

$$E = 2\beta H_0 m_S - \gamma_n \hbar H_{\text{eff}}(m_S) m_I \quad. \qquad (11.87)$$

We may note another interesting side effect of the tilt in the nuclear quantization axis as one varies m_S. The nuclear functions u_{m_I} for one value of m_S are different from those for another value of m_S, owing to the change in the direction of nuclear quantization. We write them as $u_{m_I}(m_S)$ to emphasize this point. We can express the $2I+1$ functions $u_{m_I}(m_S+1)$ in terms of the $2I+1$ functions of $u_{m_I}(m_S)$ by the relationship

$$u_{m_I}(m_S + 1) = \sum_{m'_I} a_{m_S m_I m'_I} u_{m'_I}(m_S) \qquad (11.88)$$

where the a's are constants. We have, then, the possibility of transitions m_S, m_I to $m_S + 1$, m'_I where $m'_I \neq m_I$ (since $a_{m_S m_I m'_I}$ is in general not zero). Such a transition represents a simultaneous nuclear and electron spin transition.

Actually it does not really represent a nuclear orientation change. Equation (11.88) simply expresses the fact that a nucleus of given m_I relative to $H_{\text{eff}}(m_S)$ must (since its spatial orientation is fixed during the electron flip) go to a mixture of m'_I's when the electron orientation changes, the m'_I's referring as they do to a different quantization direction.

We can see this result explicitly by a simple example for a nucleus with spin $\frac{1}{2}$. Consider two different quantization directions z_1 and z_2 that make an angle ϕ with respect to each other; see Fig. 11.8. The z_2-axis is rotated relative to the z_1-axis by an angle ϕ (in the right-handed sense) about the $y_1 = y_2$ axis. Let u_+ and u_- be the eigenfunctions of I_{z_1} and v_+ and v_- be the eigenfunctions of I_{z_2}. Then the functions v and u are related through the Clebsch-Gordon coefficients [11.1]:

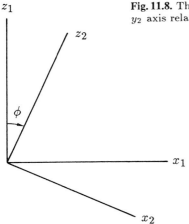

Fig. 11.8. The axes (x_2, z_2) are rotated an angle ϕ about the $y_1 = y_2$ axis relative to the axes (x_1, z_1)

$$v_+ = \cos(\phi/2)u_+ + \sin(\phi/2)u_- \quad ,$$
$$v_- = -\sin(\phi/2)u_+ + \cos(\phi/2)u_- \quad . \tag{11.89}$$

[Equation (11.89) can also be derived using the methods of Chap. 2. A rotation of ϕ about the y-axis can be treated by finding the solution of (2.73) for H_1 along the negative y-axis with $\gamma H_1 t_w = \phi$.]

Then, if the direction of the effective field acting on the nucleus when the electron spin points up is taken as the z_1 direction, and the direction when the electron spin is down is called the z_2 direction, and if α and β are the electron up and down functions (the eigenfunctions of $S_{z'}$), the complete eigenfunctions of the Hamiltonian are the four functions $|m_S m_I\rangle$ given by

$$\begin{aligned}
|++\rangle &= \alpha u_+ \\
|+-\rangle &= \alpha u_- \\
|-+\rangle &= \beta v_+ = \beta u_+ \cos(\phi/2) + \beta u_- \sin(\phi/2) \\
|--\rangle &= \beta v_- = -\beta u_+ \sin(\phi/2) + \beta u_- \cos(\phi/2) \quad .
\end{aligned} \tag{11.90}$$

[Explicit derivation of (11.90) plus determination of the directions z_1, z_2, and the angle ϕ are given in Sect. 11.4.] Clearly the operator $S_{x'}$ will have matrix elements which join either αu_- or αu_+ to *both* βv_+ and βv_-.

So far we have treated the electron as being quantized along the direction of the static field H_0. There are two other refinements worth mentioning. The first is concerned with anisotropy in the electron g-factor. If the electron g-factor is anisotropic, the electron will actually be quantized along the direction of the effective field acting on it, H_g, given by

$$H_g = H_0 \cdot \frac{\overleftrightarrow{g}}{g} = \frac{\hat{i}H_x g_x + \hat{j}H_y g_y + \hat{k}H_z g_z}{\sqrt{g_x^2 + g_y^2 + g_z^2}} \quad . \tag{11.91}$$

This refinement will slightly modify the treatment of (11.81–84).

The other refinement worth mentioning is to include the fact that the nucleus exerts a magnetic field on the electron. It therefore should play a role in the electron spin quantization. Thus, if the nucleus is flipped, the electron quantization direction changes slightly. This change is ordinarily small and can frequently be neglected. However, there are some situations in which it is important. *Waugh* and *Slichter* [11.2] have called this slight change in electron spin quantization "wobble." They point out that it gives the electron a degree of freedom much less costly in energy than flipping between the states α and β. They point out that it may play a role in nuclear relaxation by paramagnetic ions at very low temperatures since wobble can be excited by the small thermal energies, but electron spin flips between states α and β cannot.

11.4 Electron Spin Echoes

The use of pulse techniques, in particular of spin echoes and stimulated echoes, has become important in electron spin resonance. We turn now to some special features that arise for this case which are not encountered for spin echoes in nuclear magnetic resonance. As we have seen, if we have two coupled nuclei belonging to two different species (e.g. ^{13}C and ^{1}H), the spin-spin splitting of one species produced by the other does not affect the size of the echo. Indeed, it acts just like a static magnetic field inhomogeneity. Therefore, although it affects the free induction decay, it is refocused perfectly at time 2τ by application of the second pulse of the echo sequence. In order to observe the spin-spin coupling on an echo, one must do a double resonance experiment (Sects. 7.19–21).

For an electron (spin S) coupled to a nucleus (spin I) the situation is different owing to the fact, which we have just discussed, that in many cases the nuclear quantization direction depends on the electron spin orientation. As we shall see, the consequence of this coupling is that the amplitude of the electron spin echo is modulated as a function of the time τ between pulses by the $I - S$ spin-spin coupling. The modulation frequencies are the splitting frequencies of the nuclear magnetic resonance. These are the frequencies one observes in ENDOR. Therefore, the electron echo envelope in effect reveals the ENDOR frequencies. It is this special feature which has made spin echoes such an important technique for electron spin resonance. This phenomenon was first reported by *Rowan* et al. [11.3] and subsequently elucidated in greater generality by *Mims* [11.4].

Before providing a formal description, let us examine physically what is happening. In particular, why is the electron echo case different from the nuclear echo case? For a pair of coupled nuclei of spins I and S respectively, the quantization of one nucleus is to an excellent approximation independent of the orientation of the other nucleus. Consider, then, application of a $\pi/2 \cdots \tau \cdots \pi$ pulse sequence to the S-spins. If the I-spin has a particular orientation, it maintains this orientation despite the S-spin pulses. Therefore, whether the $I - S$ coupling enhances or diminishes the S-spin precession rate does not matter. The dephasing produced thereby between the first and second pulse is refocused between the second pulse and the echo.

Now consider the electron case. Suppose the electron (S) is initially pointing up. Let the nucleus (I) be pointing up along the effective field. That is, the nucleus is in a state u_+ of (11.90) and the total system is in the state $|++\rangle$. The $\pi/2$ pulse now tilts the electron. As a result, the nucleus suddenly finds the effective field acting on it tilted. The nucleus thus begins to precess about the new field direction. Consequently the field it produces on the electron changes in time. Clearly the magnetic field acting on the electron during the interval 0 to τ will, in general, differ from that between τ and 2τ. Therefore, the echo will not refocus the spin-spin coupling completely. If, however, τ corresponds exactly to a 2π precession of the nucleus, then the spin-spin coupling will have the same time average during the two intervals, hence the echo will be perfect. Thus, we expect the echo envelope to oscillate with τ such that maxima in the echo envelope correspond to τ's that are integral multiples of the nuclear precession period. We will see in our formal treatment of the problem that the description we have given is exact. In that connection, we can note that to think about the quantum treatment of the effect of a $\pi/2$ pulse on the electron system, the general concepts developed in Sect. 7.24 are helpful. In particular, they help us deal with the problem that there is not just one nuclear precession frequency. Rather, there are two, depending on whether the electron is in its spin-up or spin-down state.

We begin by expressing the reduced Hamiltonian (11.83) in different notation. First we define the frequencies a, b, ω_e, and ω_n:

$$\begin{aligned} \hbar 2a &= A_x \sin^2\theta + A_z \sin^2\theta \;, \\ \hbar 2b &= (A_x - A_z)\sin\theta \cos\theta \;, \\ \omega_n &= \gamma_n H_0 \;, \\ \hbar \omega_e &= 2\beta H_0 \;. \end{aligned} \tag{11.92}$$

Then

$$\mathcal{H}_{\text{red}} = \hbar\left[\omega_e S_{z'} - \omega_n I_{z'} + (aI_{z'} + bI_{x'})2S_{z'}\right] \;. \tag{11.93}$$

We now transform this Hamiltonian to the electron spin rotating frame at frequency ω

$$\Omega_e = \omega_e - \omega \tag{11.94}$$

using the operator $\exp(i\omega S_{z'} t)$ to transform the wave function

$$\psi' = e^{-i\omega t S_{z'}}\psi \quad \text{with} \tag{11.95}$$

$$\mathcal{H}'_{\text{red}} = \hbar\left[\Omega_e S_{z'} - \omega_n I_{z'} + (aI_{z'} + bI_{x'})2S_{z'}\right] \quad . \tag{11.96}$$

To save in writing symbols, we now drop the primes from x' and z' in (11.96), understanding that from now on the new unprimed axes are really the old primed axes.

To solve for the eigenstates we note first that S_z commutes with $\mathcal{H}'_{\text{red}}$. Therefore, the eigenstates, ψ', of $\mathcal{H}'_{\text{red}}$ can be labeled by m_S, the eigenvalue of S_z. Thus

$$\mathcal{H}'_{\text{red}}\psi'(m_S) = \hbar\left(\Omega_e m_S - \omega_n I_z + (aI_z + bI_x)2m_S\right]\psi'(m_S) \tag{11.97a}$$

$$= \left[\hbar\Omega_e m_S + \mathcal{H}_I(m_S)\right]\psi'(m_S) \quad , \tag{11.97b}$$

where $\mathcal{H}_I(m_S)$ is an operator in the components of $\boldsymbol{I}$, given by

$$\mathcal{H}_I(m_S) = -\hbar\left[\omega_n I_z - (aI_z + bI_x)2m_S\right] \quad . \tag{11.98}$$

For $m_S = +\tfrac{1}{2}$ we have

$$\mathcal{H}_I(\tfrac{1}{2}) = -\hbar\left[(\omega_n - a)I_z - bI_x\right] \tag{11.99}$$

and for $m_S = -\tfrac{1}{2}$

$$\mathcal{H}_I(-\tfrac{1}{2}) = -\hbar\left[(\omega_n + a)I_z + bI_x\right] \quad . \tag{11.100}$$

We now define two vectors $\boldsymbol{\omega}_+$ and $\boldsymbol{\omega}_-$ by

$$\boldsymbol{\omega}_+ = \hat{\boldsymbol{\imath}}(-b) + \hat{\boldsymbol{k}}(\omega_n - a) \quad , \quad \boldsymbol{\omega}_- = \hat{\boldsymbol{\imath}}b + \hat{\boldsymbol{k}}(\omega_n + a) \quad . \tag{11.101a}$$

They enable us to write

$$\mathcal{H}_I(\tfrac{1}{2}) = -\hbar\boldsymbol{\omega}_+ \cdot \boldsymbol{I} \quad , \quad \mathcal{H}_I(-\tfrac{1}{2}) = -\hbar\boldsymbol{\omega}_- \cdot \boldsymbol{I} \quad . \tag{11.101b}$$

The vectors $\boldsymbol{\omega}_+$ and $\boldsymbol{\omega}_-$ are shown in Fig. 11.9. Their magnitudes are given by

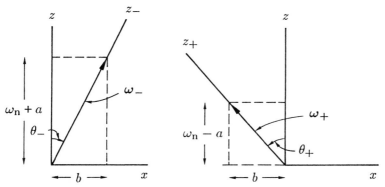

Fig. 11.9. The two angular momentum vectors $\boldsymbol{\omega}_+$ and $\boldsymbol{\omega}_-$ in the $x-z$ plane, defining the angles θ_+ and θ_- made by $\boldsymbol{\omega}_+$ and $\boldsymbol{\omega}_-$ respectively with respect to the z-axis. Note the sign convention (indicated by the arrows) for positive θ_+ and θ_-. The axes z_+ and z_- lie along $\boldsymbol{\omega}_+$ and $\boldsymbol{\omega}_-$ respectively

$$\omega_+ = \sqrt{b^2 + (\omega_n - a)^2} \quad , \quad \omega_- = \sqrt{b^2 + (\omega_n + a)^2} \quad . \tag{11.102}$$

The directions with respect to the z-axis are shown in Fig. 11.9. The angles θ_+ and θ_- shown in the figure obey the equations

$$\tan \theta_+ = \frac{b}{\omega_n - a} \quad , \quad \tan \theta_- = \frac{b}{\omega_n + a} \quad . \tag{11.103}$$

We then define the unit vectors $\boldsymbol{k}_+$ and $\boldsymbol{k}_-$ as

$$\boldsymbol{k}_+ = \boldsymbol{\omega}_+/\omega_+ \quad , \quad \boldsymbol{k}_- = \boldsymbol{\omega}_-/\omega_- \quad , \tag{11.104}$$

which give for the components of $\boldsymbol{I}$ along $\boldsymbol{\omega}_+$ and $\boldsymbol{\omega}_-$

$$I_{z+} = \boldsymbol{k}_+ \cdot \boldsymbol{I} \quad , \quad I_{z-} = \boldsymbol{k}_- \cdot \boldsymbol{I} \quad . \tag{11.105}$$

These enable us to write

$$\begin{aligned}\mathcal{H}_I(\tfrac{1}{2}) &= -\hbar \boldsymbol{\omega}_+ \cdot \boldsymbol{I} = -\hbar \omega_+ I_{z+} \quad , \\ \mathcal{H}_I(-\tfrac{1}{2}) &= -\hbar \boldsymbol{\omega}_- \cdot \boldsymbol{I} = -\hbar \omega_- I_{z-} \quad . \end{aligned} \tag{11.106}$$

Then if we define the eigenfunctions u_+, u_-, v_+, v_- by the equations

$$\begin{aligned} I_{z+} u_+ &= \tfrac{1}{2} u_+ \quad , \quad I_{z+} u_- = -\tfrac{1}{2} u_- \quad , \\ I_{z-} v_+ &= \tfrac{1}{2} v_+ \quad , \quad I_{z-} v_- = -\tfrac{1}{2} v_- \end{aligned} \tag{11.107}$$

we have that the u's and v's are the eigenfunctions of $\mathcal{H}_I(m_S)$ as follows:

$$\begin{aligned} \mathcal{H}_I(\tfrac{1}{2}) u_+ &= -\hbar(\omega_+/2) u_+ \quad , \\ \mathcal{H}_I(\tfrac{1}{2}) u_- &= \hbar(\omega_+/2) u_- \quad , \\ \mathcal{H}_I(-\tfrac{1}{2}) v_+ &= -\hbar(\omega_-/2) v_+ \quad , \\ \mathcal{H}_I(-\tfrac{1}{2}) v_- &= -\hbar(\omega_-/2) v_- \quad . \end{aligned} \tag{11.108}$$

Then, returning to (11.97) to express $\mathcal{H}'_{\text{red}}$, we find that when $\mathcal{H}'_{\text{red}}$ operates on any of the four functions

$$\begin{aligned} |++) &\equiv \alpha u_+ \quad , \quad |+-) \equiv \alpha u_- \quad , \\ |-+) &\equiv \beta u_+ \quad , \quad |--) \equiv \beta v_- \end{aligned} \tag{11.109}$$

that they are eigenfunctions of $\mathcal{H}'_{\text{red}}$ with eigenvalues

$$\begin{aligned} \hbar(\Omega_e - \omega_+)/2 \quad &, \quad \hbar(\Omega_e + \omega_+)/2 \quad , \\ \hbar(-\Omega_e - \omega_-)/2 \quad &, \quad \hbar(-\Omega_e + \omega_-)/2 \quad , \end{aligned} \tag{11.110}$$

respectively. The four functions of (11.109) therefore solve the reduced Hamiltonian.

Utilizing the definitions of θ_+ and θ_- given by (11.102) and shown in Fig. 11.9, we can calculate the angle θ_0 between z_+ and z_-, shown in Fig. 11.10,

$$\theta_0 = \theta_+ + \theta_- \quad . \tag{11.111}$$

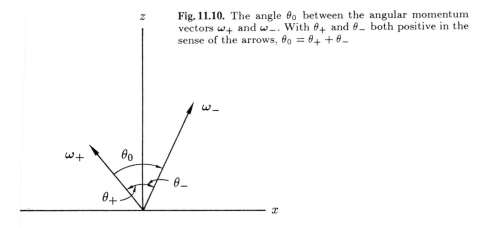

Fig. 11.10. The angle θ_0 between the angular momentum vectors ω_+ and ω_-. With θ_+ and θ_- both positive in the sense of the arrows, $\theta_0 = \theta_+ + \theta_-$

Referring to (11.89) and Fig. 11.8, we have that

$$\phi = -\theta_0 \quad \text{so that} \tag{11.112}$$

$$\begin{aligned} v_+ &= u_+ \cos(\theta_0/2) - u_- \sin(\theta_0/2) \ , \\ v_- &= u_+ \sin(\theta_0/2) + u_- \cos(\theta_0/2) \ . \end{aligned} \tag{11.113}$$

We are now ready to consider the electron echo signal. We begin by assuming that H_0 has some inhomogeneity. This may either result from actual inhomogeneity in the magnet or from some other source such as hyperfine coupling to nuclei other than the spin I. This inhomogeneity would cause the free induction signal of the electron spin to decay, and thus causes us to go to a spin echo to eliminate the effects of inhomogeneity. We therefore analyze for the signal at time $t = 2\tau$ for a pulse sequence

$$X_S(\pi/2) \cdots \tau \cdots X_S(\pi) \ . \tag{11.114}$$

Then defining the time development operation $T(t)$, in the electron spin rotating frame we have

$$\begin{aligned} T(t) &= \exp\left(-\frac{i}{\hbar}\mathcal{H}'_{\text{red}}t\right) \\ &= \exp\{-i[\Omega_e S_z + (aI_z + bI_x)2S_z - \omega_n I_z]t\} \ . \end{aligned} \tag{11.115}$$

We take the density matrix $\varrho(0^-)$ prior to the first pulse as $\varrho(0^-) = S_z$. Then the signal at time t (we take t to be after the π pulse), $\langle S^+(t)\rangle$, is given by

$$\begin{aligned} \langle S^+(t)\rangle &= \text{Tr}\{S^+ \varrho(t)\} \\ &= \text{Tr}\{S^+ \left[T(t-\tau)X_S(\pi)T(\tau)X_S(\pi/2)\right]_1 S_z[\ ]_1^{-1}\} \\ &= \text{Tr}\{S^+ \left[X_S(\pi)X_S^{-1}(\pi)T(t-\tau)X_S(\pi)T(\tau)\right]_2 S_y[\ ]_2^{-1}\} \ , \end{aligned} \tag{11.116}$$

where for simplicity we do not write out the contents of the inverse brackets and where the subscripts 1 and 2 on the brackets serve to distinguish the contents of the brackets.

Explicit evaluation of $X_S^{-1}(\pi)T(t-\tau)X_S(\pi)$ gives us

$$X_S^{-1}(\pi)T(\tau)X_S(\pi) = \exp\{i[\Omega_e S_z + (aI_z + bI_x)2S_z + \omega_n I_z](t-\tau)\} \quad . \tag{11.117}$$

We now note that we can commute the $\Omega_e S_z$ factors, to obtain

$$\langle S^+(t)\rangle = \text{Tr}\Big\{ S^+ \Big[X_S(\pi)\exp\{i[\Omega_e(t-\tau) - \Omega_e\tau]S_z\}$$
$$\times \exp\{i[(aI_z + bI_x)2S_z - \omega_n I_z](t-\tau)\}$$
$$\times \exp\{-i[(aI_z + bI_x)2S_z - \omega_n I_z]\tau\}\Big]_3 S_y [\ ]_3^{-1}\Big\} . \tag{11.118}$$

Now, owing to the spread in Ω_e due to magnetic field inhomogeneity, the term $\exp\{i[\Omega_e(t-\tau) - \Omega_e\tau]S_z\}$ makes the signal $\langle S^+(t)\rangle$ be small at most values of t. However, we note that when $t = 2t$ the factor

$$\exp\{i[\Omega_e(t-\tau) - \Omega_e\tau]S_z\} = \exp\{i\Omega_e(t-2\tau)S_z\} = 1 \quad . \tag{11.119}$$

Therefore, at $t = 2\tau$ the signal is independent of Ω_e, and an echo forms.

We now confine our attention to this time. We next factor out $X_S(\pi)$ from the left-hand side of $[\ ]_3$ and from the right-hand side of $[\ ]_3^{-1}$, and utilize the fact that $\text{Tr}\{ABC\} = \text{Tr}\{CAB\}$ to produce

$$X_S^{-1}(\pi)S^+ X_S(\pi) \quad , \tag{11.120a}$$

which we rewrite as

$$X_S^{-1}(\pi)S^+ X_S(\pi) = S^- \quad . \tag{11.120b}$$

Moreover, we get a more compact notation if we define the unitary operators $R(S_z)$ and $R(-S_z)$ by

$$R(S_z) \equiv \exp\{-i[(aI_z + bI_x)2S_z - \omega_n I_z]\tau\} \tag{11.121}$$

and

$$R(-S_z) \equiv \exp\{i[(aI_z + bI_x)2S_z + \omega_n I_z]\tau\} \quad , \tag{11.122}$$

where $R(-S_z)$ is obtained by replacing S_z of $R(S_z)$ by $-S_z$. Then we get

$$\langle S^+(2\tau)\rangle = \text{Tr}\{S_-[R(-S_z)R(S_z)]_4 S_y [\ ]_4^{-1}\} \quad .$$

We now evaluate the trace using a complete set of basis function. For the complete set we pick initially the two electron states α and β with quantum number m_S and the two nuclear states u_+ and u_-, which we denote by the quantum number μ, which can be $+\frac{1}{2}$ or $-\frac{1}{2}$. Thus, using the fact that $(m_S\mu|S_-|m_S'\mu') = 0$ unless $\mu = \mu'$, and so on,

529

$$\langle S^+(2\tau)\rangle = \sum_{\substack{m_S,\mu \\ m'_S,\mu',\mu''}} (m_S\mu|S_-|m'_S\mu)(m'_S\mu|R(-S_z)R(S_z)|m'_S\mu')$$
$$\times (m'_S\mu'|S'_y|m''_S\mu')(m''_S\mu'|R^{-1}(S_z)R^{-1}(-S_z)|m''_S\mu'') \quad. \tag{11.123}$$

In order to have the sums be *diagonal* sums, we get that $m''_S = m_S$ and $\mu'' = \mu$. Now the matrix element of S_- requires that

$$m_S = m'_S - 1 \quad. \tag{11.124}$$

Therefore, we get contributions only from

$$m_S = -\tfrac{1}{2} \quad, \quad m'_S = +\tfrac{1}{2} \tag{11.125}$$

and therefore $(m_S\mu|S_-|m'_S\mu) = \delta_{m_S,m'_S-1}$. Then we get for the matrix element of S_y

$$(m'_S\mu'|S_y|m''_S\mu') = (\tfrac{1}{2}|S_y|-\tfrac{1}{2}) = \frac{1}{2i}(\tfrac{1}{2}|S^+|-\tfrac{1}{2}) = \frac{1}{2i} \tag{11.126}$$

so that

$$\langle S^+(2\tau)\rangle = \frac{1}{2i}\sum_{\mu,\mu'} (\tfrac{1}{2}\mu|R(-S_z)R(S_z)|\tfrac{1}{2}\mu')$$
$$\times (-\tfrac{1}{2}\mu'|R^{-1}(S_z)R^{-1}(-S_z)|-\tfrac{1}{2}\mu) \quad. \tag{11.127}$$

Now, letting the operator S_z of $R(S_z)$ and $R(-S_z)$ operate, we replace S_z by m_S (and $-S_z$ by $-m_S$) in the matrix elements. Then, defining the operators

$$R_{-1/2} = \exp\{i(aI_z + bI_x + \omega_n I_z)\tau\} \quad,$$
$$R_{1/2} = \exp\{i(-aI_z - bI_x + \omega_n I_z)\tau\} \quad, \tag{11.128}$$

we get

$$\langle S^+(2\tau)\rangle = \frac{1}{2i}\sum_{\mu,\mu'} (\mu|R_{-1/2}R_{1/2}|\mu')(\mu'|R^{-1}_{-1/2}R^{-1}_{1/2}|\mu)$$
$$= \frac{1}{2i}\text{Tr}_I\{R_{-1/2}R_{1/2}R^{-1}_{-1/2}R^{-1}_{1/2}\} \quad, \tag{11.129}$$

where Tr_I means a trace over the nuclear spin eigenstates only. We can express the operators $R_{1/2}$ and $R_{-1/2}$ in a form that gives them a simple physical meaning now by utilizing (11.101 and 105)

$$R_{1/2} = \exp(i\omega_+\tau I_{z+}) \quad, \quad R_{-1/2} = \exp(i\omega_-\tau I_{z-}) \quad. \tag{11.130}$$

Therefore, $R_{1/2}$ generates rotations at angular velocity ω_+ about the z_+-axis, and $R_{-1/2}$ generates rotations about the z-axis at angular velocity ω_-.

Suppose now that τ corresponds to an integral number of 2π rotations about the z_+-axis:

$$\omega_+\tau = n2\pi \quad. \tag{11.131}$$

Then, for any function f expressible in powers of the components of I,

$$R_{1/2} f R_{-1/2} = \exp(i\omega_+ \tau I_z) f \exp(-i\omega_+ \tau I_z) = f \quad , \tag{11.132}$$

since every component of I is rotated onto itself for this value of $\omega_+ \tau$. Therefore

$$R_{-1/2} R_{1/2} R_{-1/2}^{-1} R_{1/2}^{-1} = R_{-1/2} R_{-1/2}^{-1} = 1 \quad . \tag{11.133}$$

Thus

$$\langle S^+(2\tau) \rangle = \frac{1}{2i} \text{Tr}_I \{1\} = \frac{1}{2i}(2I+1) \quad . \tag{11.134a}$$

But this is, apart from the usual sign change in a $\pi/2, \pi$ echo, just the signal we would have immediately after applying an $X_S(\pi/2)$ pulse:

$$\langle S^+(0^+) \rangle = \text{Tr}\{S^+ S_y\} = i \text{Tr}\{S_y^2\} = i(2I+1)\text{Tr}_S\{S_y^2\}$$

$$= i(2I+1)\frac{S(S+1)}{3}(2S+1) = \frac{i}{2}(2I+1) \quad . \tag{11.134b}$$

Therefore, if $\omega_+ \tau = n2\pi$ we get an echo corresponding to the full magnetization. A similar argument holds for the condition

$$\omega_- \tau = n2\pi \tag{11.135}$$

if we focus on the terms $R_{-1/2} R_{1/2} R_{-1/2}^{-1}$.

We can get further physical insight by writing

$$R_{-1/2} R_{1/2} R_{-1/2}^{-1} R_{1/2}^{-1} = \exp(i\omega_- \tau I_{z-}) R_{1/2} \exp(-i\omega_- \tau I_{z-}) R_{1/2}^{-1}$$

$$= \exp(i\omega_- \tau I_{z-}) \exp\left[-i\omega_- \tau \left(R_{1/2} I_z R_{1/2}^{-1}\right)\right] \quad . \tag{11.136}$$

Now $R_{1/2} I_{z-} R_{1/2}^{-1}$ provides a rotation of the vector $k_- I_{z-}$ about the vector ω_+. When this rotation is an integral number of cycles

$$R_{1/2} \exp(-i\omega_- \tau I_{z-}) R_{1/2}^{-1} = \exp(-i\omega_- \tau I_{z-}) \quad , \tag{11.137}$$

otherwise

$$R_{1/2} R_{-1/2}^{-1} R_{1/2}^{-1} \neq R_{1/2}^{-1} \quad , \tag{11.138}$$

and the product

$$R_{-1/2} R_{1/2} R_{-1/2}^{-1} R_{1/2}^{-1} \neq 1 \quad . \tag{11.139}$$

Clearly these equations express mathematically our initial physical description of why the spin echo envelope is modulated as τ is varied, returning to its full amplitude when either

$$\omega_+ \tau = n2\pi \quad \text{or} \quad \omega_- \tau = n2\pi \quad , \quad n = 1, 2, 3 \ldots \quad . \tag{11.140}$$

We now wish to evaluate the trace explicitly. Denoting the eigenvalues of I_{z+} by μ and of I_{z-} by σ so that

$$I_{z+}|\mu) = \mu|\mu) \quad \text{and} \quad I_{z-}|\sigma) = \sigma|\sigma) \quad , \tag{11.141}$$

we get

$$\text{Tr}_I\{R_{-1/2}R_{1/2}R_{-1/2}^{-1}R_{1/2}^{-1}\}$$
$$= \sum_{\substack{\sigma,\sigma' \\ \sigma'',\sigma'''}} (\sigma|R_{-1/2}|\sigma'')(\sigma'|R_{1/2}|\sigma'')\left(\sigma''|R_{-1/2}^{-1}|\sigma'''\right)\left(\sigma'''|R_{1/2}^{-1}|\sigma\right) \tag{11.142}$$

$$= \sum_{\sigma,\sigma''} e^{+i\sigma\omega_-\tau}(\sigma|R_{1/2}|\sigma'') e^{-i\sigma''\omega_-\tau}\left(\sigma''|R_{1/2}^{-1}|\sigma\right)$$

$$= \sum_{\sigma,\sigma''} e^{i\omega_-\tau(\sigma-\sigma'')}(\sigma|R_{1/2}|\sigma'')\left(\sigma''|R_{1/2}^{-1}|\sigma\right) \tag{11.143}$$

$$= \sum_{\substack{\sigma'\sigma'' \\ \mu,\mu',\mu'',\mu'''}} e^{i\omega_-\tau(\sigma-\sigma'')}(\sigma|\mu)\left(\mu|R_{1/2}|\mu'\right)(\mu'|\sigma'')(\sigma''|\mu'')$$
$$\times \left(\mu''|R_{1/2}^{-1}|\mu'''\right)(\mu'''|\sigma)$$

and relabeling σ'' as σ' and using the fact that the states $|\mu)$ are eigenstates of I_{z+}, we get

$$\text{Tr}_I\{R_{-1/2}R_{1/2}R_{-1/2}^{-1}R_{1/2}^{-1}\}$$
$$= \sum_{\substack{\sigma,\sigma' \\ \mu,\mu'}} e^{i\omega_-\tau(\sigma-\sigma')}(\sigma|\mu) e^{i\omega_+\tau\mu}(\mu|\sigma')(\sigma'|\mu') e^{-i\omega_+\tau\mu'}(\mu'|\sigma)$$

$$= \sum_{\substack{\sigma,\sigma' \\ \mu,\mu'}} e^{i\omega_-\tau(\sigma-\sigma')} e^{i\omega_+\tau(\mu-\mu')}(\sigma|\mu)(\mu|\sigma')(\sigma'|\mu')(\mu'|\sigma) \quad . \tag{11.144}$$

Since σ, σ', μ, and μ' can each take on two values ($+\frac{1}{2}$ or $-\frac{1}{2}$) there are 16 terms. The factors $(\sigma|\mu)$ are given by expressions such as, if

$$\sigma = \tfrac{1}{2} \quad , \quad \mu = -\tfrac{1}{2} \quad ,$$
$$(\sigma|\mu) = \int v_-^* u_+ d\tau \quad , \tag{11.145}$$

where the integral is easily evaluated using (11.109):

$$\left(\sigma = \tfrac{1}{2}|\mu = -\tfrac{1}{2}\right) = \sin(\theta_0/2) \tag{11.146}$$

and so on. It is then straightforward to show that

$$\text{Tr}_I\{R_{-1/2}R_{1/2}R_{-1/2}^{-1}R_{1/2}^{-1}\}$$
$$= 2\left[\cos^4(\theta_0/2) + \sin^4(\theta_0/2)\right] + 4\sin^2(\theta_0/2)\cos^2(\theta_0/2)$$
$$\times \left[\cos(\omega_+\tau) + \cos(\omega_-\tau) - \cos(\omega_+\tau)\cos(\omega_-\tau)\right] \quad . \tag{11.147}$$

If $\theta_0 = 0$, so that there is no change in the quantization direction, (11.147) reduces to

$$\text{Tr}_I\{R_{-1/2}R_{1/2}R_{-1/2}^{-1}R_{1/2}^{-1}\} = 2 \quad,$$

which agrees with (11.134). That is, it is independent of τ. Thus, the oscillation comes about because of the change in quantization direction of the nucleus when the electron spin flips.

Note also that if *either*

$$\cos(\omega_+\tau) = 1 \quad \text{or} \quad \cos(\omega_-\tau) = 1 \quad,$$

$$\begin{aligned}\text{Tr}_I\{R_{-1/2}R_{1/2}R_{-1/2}^{-1}R_{1/2}^{-1}\} \\ &= 2\left[\cos^4(\theta_0/2) + \sin^4(\theta_0/2) + 2\sin^2(\theta_0/2)\cos^2(\theta_0/2)\right] \\ &= 2\left[\cos^2(\theta_0 2) + \sin^2(\theta_0 2)\right]^2 \\ &= 2 \quad .\end{aligned}$$

Thus, we get the full signal, as we showed with (11.133).

11.5 V_k Center

A particularly interesting example of the application of the ideas of the previous sections is the discovery and identification of the so-called V_k center by *Känzig* and *Castner* [11.5]. The detailed analysis of the spectra will enable us to discuss the g-shift more thoroughly, including the effects of having several electrons and more than one force center.

Känzig and *Castner*'s first work in electron spin resonance was followed by a set of beautiful experiments by *Delbecq* [11.6], who combined optical techniques with electron resonance to determine (1) the optical absorptions associated with the center and (2) the energies of the excited states. A full account of all the work on V_k centers would take us too far afield. We shall remark on the method of identification of the center, and on certain features associated with the g-shift, which are not found in "one-atom" one-electron centers.

It is helpful to begin by, so to speak, giving the answer. The V_k center is formed in alkali halides by X-raying crystals at a temperature near that of liquid nitrogen. In this process, electrons are ejected from the negative halogen ion, changing it from a closed-shell configuration to one with one electron missing from the p-shell. The ejected electron may have a variety of fates. We shall simply assume that not all recombine with neutral halogens. As an example, consider Cl; the neutral chlorine atom is unstable and pulls together with a neighboring Cl$^-$ to form what may be conveniently called a Cl$_2^-$ molecule. The Cl-Cl axis turns out to lie in 110 or equivalent crystal directions, as illustrated in Fig. 11.11.

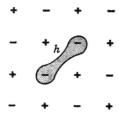

Fig. 11.11. A Cl_2^- molecule, or V_k center, in KCl. The center may be thought of as a hole, denoted by h, trapped on a pair of Cl^- ions

The electronic structure turns out to be very similar to that of the p-electron with quenched orbital angular momentum, discussed in Section 11.2, the electron with the unpaired spin being in an orbit whose axis is parallel to the bond direction of the Cl_2^- molecule. Coupling to excited states gives a g-shift that varies as the magnetic field orientation is varied with respect to the crystal axes, and the coupling of the unpaired spin with the nuclear moments of the *two* chlorines gives rise to a hyperfine coupling.

As we have remarked, this center was first discovered by electron spin resonance. A pattern observed for a case where the static magnetic field is along the 100 crystal direction is shown in Fig. 11.12.

At first sight the spectrum seems too hopelessly complicated to unravel, but fortunately a convenient starting point is the set of seven prominent lines. They are nearly equally spaced and are of intensities $1:2:3:4:3:2:1$. These were

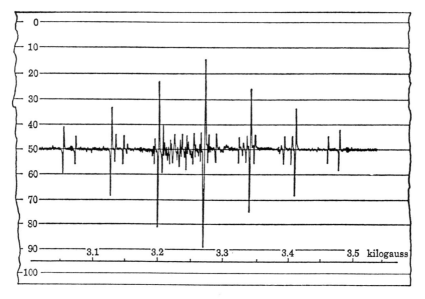

Fig. 11.12. V_k resonance in KCl for the static field parallel to a 100 crystallographic axis. (This figure kindly supplied by *Känzig* and *Castner*)

recognized to arise from the coupling of the unpaired electron spin with magnetic moments of a pair of chlorine nuclei.

Let us explain. There are two isotopes of chlorine: Cl^{35}, which is 75 percent abundant; and Cl^{37}, 25 percent abundant. Both are of spin $\frac{3}{2}$, but have slightly different magnetic moments ($\gamma_{37}/\gamma_{35} = 0.83$). As we shall show, the seven-line pattern arises from a pair of Cl^{35}'s. Let us assume that H_0 lies along a principal axis of the Cl_2^-; call it the z-axis. By recalling our discussion of the preceding section and generalizing it slightly, we see that the electron resonance condition will be

$$\omega = \frac{1}{\hbar}[g_{zz}\beta H_0 + A_z(m_1 + m_2)] \tag{11.148}$$

where m_1 and m_2 are the m-values of the two Cl nuclei. Therefore the frequency depends on m_1 and m_2 only through their sum $m_1 + m_2$. The largest $m_1 + m_2$ is $\frac{3}{2} + \frac{3}{2} = 3$. The next largest is $\frac{1}{2} + \frac{3}{2} = 2$. This value of $m_1 + m_2$ is also found if $m_1 = \frac{3}{2}$, $m_2 = \frac{1}{2}$. Since we assume the nuclei to be distributed at random among their m-states, the line at $m_1 + m_2 = 2$ will be twice as strong as that for $m_1 + m_2 = 3$. We indicate the possible combinations of m_1 and m_2 in Table 11.2.

The recognition that the seven main lines could arise from a scheme such as we have described was the first clue as to the nature of the resonance. If one accepts this clue, that the electron spends equal time on two Cl atoms, one must next consider what happens for Cl_2^- molecules in which the nuclei are both Cl^{37}'s or one is a Cl^{35} and the other is a Cl^{37}. The probability of finding a single atom of Cl^{35} is $\frac{3}{4}$; that of finding a Cl^{37} is $\frac{1}{4}$. Therefore the probability of finding pairs is as follows:

$$\begin{aligned}
Cl^{35} - Cl^{35}: &\quad \tfrac{3}{4} \times \tfrac{3}{4} = \tfrac{9}{16} \\
Cl^{35} - Cl^{37}: &\quad \tfrac{3}{4} \times \tfrac{1}{4} = \tfrac{3}{16} \\
Cl^{37} - Cl^{35}: &\quad \tfrac{1}{4} \times \tfrac{3}{4} = \tfrac{3}{16} \\
Cl^{37} - Cl^{37}: &\quad \tfrac{1}{4} \times \tfrac{1}{4} = \tfrac{1}{16}
\end{aligned} \Biggr\} \tfrac{6}{16} \; .$$

The generalization of (11.84) for unlike nuclei may be written as

$$\omega = \frac{1}{\hbar}[g_{zz}\beta H_0 + (A_z^{35}m_{35} + A_z^{37}m_{37})]] \tag{11.149}$$

in an obvious notation. The hyperfine coupling, as we have seen, is proportional to the nuclear γ's, so that

$$\frac{A^{35}}{A^{37}} = \frac{\gamma_{35}}{\gamma_{37}} \; . \tag{11.150}$$

The result of having nonequivalent nuclei is that configurations such as $(\frac{3}{2}, \frac{1}{2})$ and $(\frac{1}{2}, \frac{3}{2})$ no longer give the same frequency. If we call the $(\frac{3}{2}, \frac{3}{2})$ intensity

Table 11.2. Combinations (m_1, m_2) to make the same total value $m_1 + m_2$, and the corresponding frequency and statistical weight

(m_1, m_2)	$A(m_1 + m_2)$	Statistical weight
$(\frac{3}{2}, \frac{3}{2})$	$3A$	1
$(\frac{3}{2}, \frac{1}{2})(\frac{1}{2}, \frac{3}{2})$	$2A$	2
$(\frac{3}{2}, -\frac{1}{2})(\frac{1}{2}, \frac{1}{2})(-\frac{1}{2}, \frac{3}{2})$	$1A$	3
$(\frac{3}{2}, -\frac{3}{2})(\frac{1}{2}, -\frac{1}{2})(-\frac{1}{2}, \frac{1}{2})(-\frac{3}{2}, \frac{3}{2})$	$0A$	4
$(-\frac{3}{2}, \frac{1}{2})(-\frac{1}{2}, -\frac{1}{2})(\frac{1}{2}, -\frac{3}{2})$	$-1A$	3
$(-\frac{3}{2}, -\frac{1}{2})(-\frac{1}{2}, -\frac{3}{2})$	$-2A$	2
$(-\frac{3}{2}, -\frac{3}{2})$	$-3A$	1

"unity", the line that for like nuclei would have intensity 2 splits into two lines of unit intensity when the nuclei differ; that of intensity 3 splits into three lines of unit intensity, and soon. The positions are all predictable, using the measured A_z^{35} and (11.91). The intensity of the 35–37 lines is $\frac{6}{9}$ that of the outermost line of the 35–35 spectrum. There is in addition a set of seven lines from the Cl^{37} pairs, also having a predicted position, with an intensity $\frac{1}{9}$ that of the Cl^{35} pairs. All these lines are found at the proper positions and with the proper intensity. In this way many of the lines are accounted for.

An additional factor in determining the spectrum is the fact that the g-shift displaces the position of the center of the hyperfine patterns. The principal axes of the g-tensor with respect to the molecule are shown in Fig. 11.13. As we see in Fig. 11.13, if the magnetic field were perpendicular to the plane of the paper (pointing in a (001) direction), it would be parallel to the y-axis of the center. If it were along the 100 or 010 directions, it would make an angle of 45° with respect to the z-axis of the molecule in Fig. 11.13. For any given orientation of the magnetic field with respect to the crystal axes, there are in general several classes of V_k centers in terms of the angles made by H_0 with the principal axes of the center. If H_0 is parallel to a 111 direction, there are two classes of molecules. If H_0 is parallel to the 100 direction, $\frac{1}{3}$ of the centers have their bond axes perpendicular to H_0 and $\frac{2}{3}$ have a bond axis making a 45° angle with H_0.

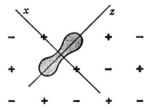

Fig. 11.13. Principal axes x, y, z of the g-tensor for the V_k center. The y-axis points out of the paper

We can see that in general there will be several hyperfine patterns whose centers of gravity are displaced because of the anisotropy of the g-factor. Moreover, the hyperfine splitting is itself strongly anisotropic, A_z being much bigger than either A_x or A_y (z being the bond axis). The anisotropy is interpreted as indicating that on each atom the individual bond function is a linear combination of an s-function $g(r)$ and a p-function $zf(r)$.[3]

The hyperfine coupling becomes, then, using (11.65) and (11.66),

$$\mathcal{H}_{IS} = \alpha^2 \gamma_e \gamma_n \hbar^2 \overline{\left(\frac{1}{r^3}\right)} \frac{3}{5}(I_{1z} + I_{2z})S_z$$

$$+ \left[(1-\alpha^2)\frac{4\pi}{3}|g(0)|^2 - \alpha^2 \frac{1}{5}\overline{\left(\frac{1}{r^3}\right)}\right] \gamma_e \gamma_n \hbar^2 (I_1 + I_2) \cdot S \quad (11.151)$$

where α^2 is the fraction of p-function, $\overline{(1/r^3)}$ the average of $(1/r^3)$ for the state $zf(r)$, and where a factor of $\frac{1}{2}$ multiplies the expression of (11.67), since the wave function spreads over two atoms. (We have neglected renormalization due to overlap of atomic functions.) There is a near cancellation of the two terms multiplying $(I_1 + I_2) \cdot S$, leading to a strong anisotropy.

The fact that the electron ranges over more than one atom presents a new problem in calculating the g-shift. In the example we studied, we represented the spin-orbit coupling by a form $\lambda L \cdot S$ appropriate to a free atom. The origin about which the angular momentum was measured was, of course, the nuclear charge, since it is motion of the spin with respect to this charge that produces the spin-orbit coupling. When there is more than one nucleus, it is not apparent which nucleus to choose as the origin. The dilemma is resolved by using the more basic form of the spin-orbit coupling:

$$\mathcal{H}_{SO} = \frac{e\hbar}{2m^2c^2} S \cdot (E \times p) \quad . \quad (11.152)$$

In this expression E is the electric field through which the electron is moving, and p is the momentum operator of the electron, $(\hbar/i)\nabla$. Since for an isolated atom E is directed along the radius vector from the origin, $E \times p \propto r \times p$, the angular momentum. The usual $\lambda L \cdot S$ expression therefore follows from (11.152). Since the electric field E is largest near a nucleus, the principal contribution to $\mathcal{H}_{SO}$ comes when the electron is near a nucleus.

In addition to the presence of two force centers, we have the complication that we must deal with more than one electron. In fact the V_k center lacks only one electron to fill the valence shells of its two chlorine atoms. In order to proceed further, it is helpful to describe the electronic states. We shall describe

---

[3] The electronic wave functions are discussed in the subsequent text. We may remark that the bond function referred to here is the function $z_1 + z_2$ mentioned in the later discussion.

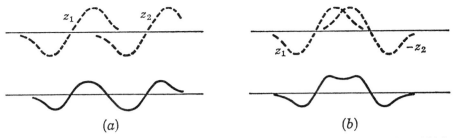

Fig. 11.14. Functions (a) $z_1 + z_2$ and (b) $z_1 - z_2$ shown schematically. The function of (a) is higher in energy, since 1) it has more nodes and 2) the node lies at a region of attractive potential for the electron

them in a molecular orbital scheme in which the molecular orbitals are made up as linear combinations of the free-atom p-states. We shall denote by x_1 an atomic p-function $xf(r)$ centered on atom 1 of the center. The atomic functions are thus x_1, y_1, z_1, x_2, y_2, and z_2, the z-axis lying along the bond. The functions $z_1 + z_2$ and $z_1 - z_2$ are shown schematically in Fig. 11.14. A study of the figure shows that $z_1 - z_2$ corresponds to a lower energy than does $z_1 + z_2$, since it has fewer nodes and tends to concentrate the electronic density between the two atoms where it can share their attractive potential. The states are in fact referred to as bonding ($z_1 - z_2$) and antibonding ($z_1 + z_2$). In a similar manner it turns out that $x_1 + x_2$ and $y_1 + y_2$ are bonding, and $x_1 - x_2$ and $y_1 - y_2$ are antibonding. (The z-states are the so-called σ-states and the x or y states are the π-states.) The energy levels of these states are shown schematically in Fig. 11.15. Actually the states $x_1 + x_2$ and $y_1 + y_2$, which are degenerate in the free molecules, are not degenerate in a crystal, but we neglect that splitting.

Since there are 6 orbital functions, there is room for 12 p-electrons. The V_k center, which has only 11, therefore has a hole in the $z_1 + z_2$ state. That is, there is an unpaired electron in that state. We have introduced, in Fig. 11.15, a labeling u or g (ungerade or gerade) that describes the parity of the orbital state.

One may expect to observe an optical absorption due to V_k centers. Since electrical dipole transitions are allowed only for transitions u to g or g to u, the optical absorption will arise from transitions of an electron in either the states $z_1 - z_2$, $x_1 - x_2$, or $y_1 - y_2$ to the empty $z_1 + z_2$. The strongest optical transition

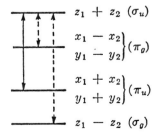

Fig. 11.15. Molecular orbitals formed from p-states in a halogen molecule ion. The allowed optical transitions into the unfilled σ_u orbit are shown by dashed lines. The g-shift "transitions" are shown by a solid line

is from the $z_1 + z_2$ to $z_1 - z_2$, since the electronic dipole matrix element here is the largest (in fact, corresponding to a dipole moment arm equal to the length of the molecule).

We see, therefore, that we must generalize our previous discussion to account for two new features: the lack of a single force center, and the fact that more than one electron is involved.

In order to illustrate the first point (more than one force center) without the complications of the second (more than one electron) let us consider an example in which we have only one electron occupying the V_k center orbitals. Then the ground orbital state function Ψ_0 is, neglecting overlap,

$$\Psi_0 = \frac{1}{\sqrt{2}}(z_1 - z_2) \quad , \tag{11.153}$$

which has quenched orbital angular momentum.

In our earlier discussion of the g-shift for a problem with only one force center we saw that the g-shift arose from the interplay between the spin-orbit coupling and the slight unquenching of orbital angular momentum produced by the orbital Zeeman energy. When more than one force center is present there is no unique point about which to measure angular momentum and it is natural to return to the more basic form of the spin-orbit coupling:

$$\mathcal{H}_{SO} = \frac{e\hbar}{2m^2c^2} \boldsymbol{S} \cdot (\boldsymbol{E} \times \boldsymbol{p}) \quad . \tag{11.154}$$

In the absence of an applied magnetic field this expression is correct but in the presence of a magnetic field described by a vector potential $\boldsymbol{A}$ we must modify it to obtain a gauge invariant result:

$$\mathcal{H}_{SO} = \frac{e\hbar}{2m^2c^2} \boldsymbol{S} \cdot \left[\boldsymbol{E} \times \left(\boldsymbol{p} + \frac{e}{c}\boldsymbol{A}\right)\right] \tag{11.155}$$

where $-e$ is the charge of the electron. Equation (11.155) follows directly from the Dirac equation but it is also intuitively obvious because, as we have discussed earlier, we always replace $\boldsymbol{p}$ by $\boldsymbol{p} - (q/c)\boldsymbol{A}$ in the presence of a magnetic field where, in our case $q = -e$.

The orbital Zeeman interaction $\mathcal{H}_{OZ}$ is

$$\mathcal{H}_{OZ} = \frac{2}{2mc}(\boldsymbol{p} \cdot \boldsymbol{A} + \boldsymbol{A} \cdot \boldsymbol{p}) + \frac{e^2}{2mc^2}A^2 \quad . \tag{11.156}$$

We can consider that $\mathcal{H}_{OZ}$ and $\mathcal{H}_{SO}$ together constitute a perturbation $\mathcal{H}_{\text{pert}}$:

$$\mathcal{H}_{\text{pert}} = \mathcal{H}_{SO} + \mathcal{H}_{OZ} \quad . \tag{11.157}$$

We are concerned with calculating matrix elements designated by an orbital quantum number n and a spin quantum number σ. In particular the effect of $\mathcal{H}_{\text{pert}}$ is, according to Appendix D, equivalent to our having an additional interaction $\mathcal{H}_{\text{new}}$ whose matrix elements diagonal in the ground orbital state $|0\rangle$ are

$$(0\sigma|\mathcal{H}_{\text{new}}|0\sigma') = (0\sigma|\mathcal{H}_{\text{pert}}|0\sigma')$$
$$+ {\sum_{n,\sigma''}}' \frac{(0\sigma|\mathcal{H}_{\text{pert}}|n\sigma'')(n\sigma''|\mathcal{H}_{\text{pert}}|0\sigma')}{E_0 - E_n} \quad (11.158)$$

where the prime on the summation means omitting $n = 0$ and where we have neglected the spin contributions to the energy denominators. The g-shift arises from keeping just those terms of (11.158) that are linear in the vector potential and the electron spin. Thus we get $(0\sigma|\mathcal{H}_{\Delta g}|0\sigma')$ as

$$(0\sigma|\mathcal{H}_{\Delta g}|0\sigma') = \frac{e\hbar}{2m^2 c^2}\left(0\sigma\left|\boldsymbol{S}\cdot\boldsymbol{E}\times\frac{c}{c}\boldsymbol{A}\right|0\sigma'\right)$$
$$+\frac{e^2\hbar}{4m^3 c^3}{\sum_{n\sigma''}}' \frac{(0\sigma|\boldsymbol{S}\cdot\boldsymbol{E}\times\boldsymbol{p}|n\sigma'')(n\sigma''|\boldsymbol{p}\cdot\boldsymbol{A}+\boldsymbol{A}\cdot\boldsymbol{p}|0\sigma')}{E_0 - E_n}$$
$$+\frac{(0\sigma|\boldsymbol{p}\cdot\boldsymbol{A}+\boldsymbol{A}\cdot\boldsymbol{p}|n\sigma'')(n\sigma''|\boldsymbol{S}\cdot\boldsymbol{E}\times\boldsymbol{p}|0\sigma')}{E_0 - E_n} \quad . \quad (11.159)$$

Since $\boldsymbol{p}\cdot\boldsymbol{A}+\boldsymbol{A}\cdot\boldsymbol{p}$ does not depend on spin

$$(n\sigma''|\boldsymbol{p}\cdot\boldsymbol{A}+\boldsymbol{A}\cdot\boldsymbol{p}|0\sigma') = (n|\boldsymbol{p}\cdot\boldsymbol{A}+\boldsymbol{A}\cdot\boldsymbol{p}|0)\delta_{\sigma'\sigma''} \quad . \quad (11.160)$$

Therefore we get

$$(0\sigma|\mathcal{H}_{\Delta g}|0\sigma') = \frac{e^2\hbar}{2m^2 c^3}(\sigma|\boldsymbol{S}|\sigma')\cdot\left\{(0|\boldsymbol{E}\times\boldsymbol{A}|0)\right.$$
$$\left.+\frac{1}{2m}{\sum_n}'\left[\frac{(0|\boldsymbol{E}\times\boldsymbol{p}|n)(n|\boldsymbol{p}\cdot\boldsymbol{A}+\boldsymbol{A}\cdot\boldsymbol{p}|0)+\text{c.c.}}{E_0 - E_n}\right]\right\} \quad .$$
$$(11.161)$$

This expression is the basis for a proper treatment of the problem of several force centers. However, in order to proceed with that problem, we should first understand some aspects of the single force center problem that we have not discussed. In particular, what is the best choice of the gauge for the vector potential, and what happens when we change gauge?

Suppose the atom in question is located at the origin. Then the wave functions $|n)$ are in general either classified by angular momentum about the origin or are perhaps linear combinations of such atomic orbitals. If we took as the vector potential $\boldsymbol{A}(\boldsymbol{R})$ defined as

$$\boldsymbol{A}(\boldsymbol{R}) \equiv \tfrac{1}{2}\boldsymbol{H}_0 \times (\boldsymbol{r} - \boldsymbol{R}) \quad (11.162)$$

where $\boldsymbol{R}$ is an arbitrary constant vector, then, using the fact that div $\boldsymbol{A}(\boldsymbol{R}) = 0$, we could write the matrix elements $(n|\boldsymbol{p}\cdot\boldsymbol{A}+\boldsymbol{A}\cdot\boldsymbol{p}|0)$ as

$$(n|\boldsymbol{p}\cdot\boldsymbol{A}+\boldsymbol{A}\cdot\boldsymbol{p}|0) = \int u_n^* \boldsymbol{H}_0\times(\boldsymbol{r}-\boldsymbol{R})\cdot\boldsymbol{p}u_0 d\tau$$
$$= \hbar\boldsymbol{H}_0 \cdot \int u_n^* \boldsymbol{L}(\boldsymbol{R})u_0 d\tau \quad (11.163)$$

where
$$L(R) \equiv \frac{1}{i}(r - R) \times \nabla \tag{11.164}$$

is the dimensionless angular momentum operator about the arbitrary point R. Integrals such as the lower one on the right of (11.163) are readily evaluated by the methods of (11.17) and (11.18) provided R is chosen as zero so that the angular momentum is measured about the natural origin of the atomic orbitals that make up the functions $|n\rangle$. We shall call this the "natural" gauge.

An even more important point is seen by examining the first-order term of (11.161):

$$\frac{e^2\hbar}{2m^2c^3}(\sigma|S|\sigma') \cdot (0|E \times A(R)|0)$$
$$= \frac{e^2\hbar}{4m^2c^3}(\sigma|S|\sigma') \cdot (0|E \times [H_0 \times (r - R)]|0) \quad . \tag{11.165}$$

By using the fact that the electric field E is large only near the nucleus where it is in fact to a good approximation radial we have

$$E(r) = \frac{r}{r}E(r) \quad . \tag{11.166}$$

Then taking the z-axis to a lie along the direction of the static field, we find

$$\frac{e^2\hbar}{2m^2c^3}(\sigma|S|\sigma') \cdot (0|E \times A|0)$$
$$= H_0 \frac{e^2\hbar}{4m^2c^3}(\sigma|S|\sigma') \cdot \left\{ \left(0\left|\frac{E(r)}{r}[k(x^2 + y^2) - ixz - jyz]\right|0\right) \right.$$
$$\left. - \left(0\left|\frac{E(r)}{r}[k(xX + yY) - iXz - jYz]\right|0\right) \right\} \tag{11.167}$$

where X and Y are two of the components of R.

If the wave function $|0\rangle$ has a definite parity the second term on the right vanishes. If $|0\rangle$ does not have a definite parity (as for example if it were an s-p hybrid) the second term does not vanish. Since this term depends on the choice of R it can in the latter case be made to take on any value. In order that the g-shift be independent of the gauge, there must be a compensating change in the terms of (11.159) that have the energy denominators. Such is in fact the case.

If we take the "natural" gauge for which $R = 0$, the order of magnitude of the right-hand side of (11.167) is approximately $\beta H_0 r_0/a_H$, where r_0 is the classical electron radius e^2/mc^2 ($\cong 10^{-13}$ cm) and a_H is the Bohr radius (0.5×10^{-8} cm). The matrix element is therefore $\cong 10^{-5}\beta H_0$ and is in general negligible. It is for this reason that one is justified in omitting the first-order term, as is ordinarily done.

We have seen that the "natural" gauge makes for simple evaluation of the matrix elements such as those in (11.163). When there is more than one force center no single gauge appears natural, and we should like, in fact, to be able to use a mixture of gauges: one gauge when in the vicinity of one nucleus, the other in the vicinity of a second nucleus. Such a trick is actually possible, provided we can neglect certain overlap integrals. Let us state a theorem; then we shall outline its proof and then show how the theorem enables us to use such a technique of several "natural" gauges to treat the problem of multiple force centers.

Let us therefore consider a system with two atoms. The ground state $|0)$ will be a linear combination

$$|0) = u_0 + v_0 \tag{11.168}$$

where u_0 is a linear combination of atomic orbitals on the first atom and v_0 is a linear combination of atomic orbitals on the second atom. The excited states $|n)$ are also linear combinations:

$$|n) = u_n + v_n \quad . \tag{11.169}$$

We shall neglect all contributions to matrix elements involving a product of a u and a v. This approximation is often good, but can lead to errors in some cases.

We have, then, as our theorem that the combined effect of the spin-orbit and orbital Zeeman coupling is to give a g-shift characterized by

$$(0\sigma|\mathcal{H}_{\Delta g}|0\sigma')$$
$$= \frac{e^2\hbar}{2m^2c^3}(\sigma|S|\sigma') \cdot \Big\{ (u_0|\boldsymbol{E} \times \boldsymbol{A}'|u_0) + (v_0|\boldsymbol{E} \times \boldsymbol{A}''|v_0)$$
$$+ \frac{1}{2m}\sum_n{}' \frac{(0|\boldsymbol{E} \times \boldsymbol{p}|n)[(u_n|\boldsymbol{A}'\cdot\boldsymbol{p} + \boldsymbol{p}\cdot\boldsymbol{A}'|u_0) + (v_n|\boldsymbol{A}''\cdot\boldsymbol{p} + \boldsymbol{p}\cdot\boldsymbol{A}''|v_0)] + \text{c.c.}}{E_0 - E_n} \Big\}$$
$$\tag{11.170}$$

where $\boldsymbol{A}'$ and $\boldsymbol{A}''$ are any vector potentials that give the static field $\boldsymbol{H}_0$ (they differ therefore, at most by a gauge transformation), and where

$$(u_n|\boldsymbol{A}'\cdot\boldsymbol{p} + \boldsymbol{p}\cdot\boldsymbol{A}'|u_0) \equiv \int u_n^*(\boldsymbol{A}'\cdot\boldsymbol{p} + \boldsymbol{p}\cdot\boldsymbol{A}')u_0 d\tau \quad . \tag{11.171}$$

The beauty of (11.170) is that it allows us to choose the vector potential $\boldsymbol{A}'$ used to evaluate the integrals with the u's independently of the vector potential $\boldsymbol{A}''$ used for integrals involving v's. (We shall discuss handling the matrix elements $(0|\boldsymbol{E} \times \boldsymbol{p}|n)$ shortly.)

In particular, we shall see, if the two nuclei are at $\boldsymbol{R}_1$ and $\boldsymbol{R}_2$, respectively, we can evaluate the matrix elements readily by choosing

$$\begin{aligned}\boldsymbol{A}' &= \boldsymbol{A}_1 \equiv \tfrac{1}{2}\boldsymbol{H}_0 \times (\boldsymbol{r} - \boldsymbol{R}_1) \\ \boldsymbol{A}'' &= \boldsymbol{A}_2 = \tfrac{1}{2}\boldsymbol{H}_0 \times (\boldsymbol{r} - \boldsymbol{R}_2) \quad . \end{aligned} \tag{11.172}$$

To prove the theorem of (11.170), we start with (11.161). We express the matrix elements involving $\boldsymbol{A}$ in terms of the u's and v's, and neglect overlap terms. For example,

$$(0|\boldsymbol{E}\times\boldsymbol{A}|0) = (u_0|\boldsymbol{E}\times\boldsymbol{A}|u_0) + (v_0|\boldsymbol{E}\times\boldsymbol{A}|v_0) \quad . \tag{11.173}$$

Then we introduce two vector potentials, $\boldsymbol{A}'$ and $\boldsymbol{A}''$, which differ by a gauge transformation:

$$\boldsymbol{A}' = \boldsymbol{A}'' + \nabla\phi \tag{11.174}$$

defining the function ϕ. (That (11.174) is simply a gauge transformation follows, of course, from the fact that it satisfies the requirement $\nabla\times\boldsymbol{A}' = \nabla\times\boldsymbol{A}''$). We then substitute $\boldsymbol{A}'$ for $\boldsymbol{A}$ in integrals involving u's, and $\boldsymbol{A}'' + \nabla\phi$ for $\boldsymbol{A}$ in integrals involving v's. By collecting terms, we get

$$(0\sigma|\mathcal{H}_{\Delta g}|0\sigma') = +\frac{e^2\hbar}{2m^2c^3}(\sigma|S|\sigma')\cdot\Big\{(u_0|\boldsymbol{E}\times\boldsymbol{A}'|u_0) + (v_0|\boldsymbol{E}\times\boldsymbol{A}''|v_0)$$
$$+\frac{1}{2m}\sum_n{}'\frac{(0|\boldsymbol{E}\times\boldsymbol{p}|n)[(u_n|\boldsymbol{p}\cdot\boldsymbol{A}' + \boldsymbol{A}'\cdot\boldsymbol{p}|u_0)}{E_0 - E_n}$$
$$+\frac{(u_n|\boldsymbol{p}\cdot\boldsymbol{A}'' + \boldsymbol{A}''\cdot\boldsymbol{p}|v_0)] + \text{c.c.}}{E_0 - E_n} + (v_0|\boldsymbol{E}\times\nabla\phi|v_0)$$
$$+\frac{1}{2m}\sum_n{}'\frac{(0|\boldsymbol{E}\times\boldsymbol{p}|n)(v_n|\boldsymbol{p}\cdot\nabla\phi + \nabla\phi\cdot\boldsymbol{p}|v_0) + \text{c.c.}}{E_0 - E_n}\Big\} \quad .$$
$$\tag{11.175}$$

To derive our theorem, (11.170), we must show that the terms involving ϕ add to zero. By making use of the fact that we are neglecting overlap terms, our proof is equivalent to showing that the quantity $(0\sigma|\mathcal{H}_{\Delta g}(\phi)|v_0\sigma')$, defined below, vanishes:

$$(0\sigma|\mathcal{H}_{\Delta g}(\phi)|v_0\sigma') \equiv \frac{e^2\hbar}{2m^2c^3}(\sigma|S|\sigma')\cdot\Big\{(0|\boldsymbol{E}\times\nabla\phi|v_0)$$
$$+\frac{1}{2m}\sum_n{}'\frac{(0|\boldsymbol{E}\times\boldsymbol{p}|n)(n|\boldsymbol{p}\cdot\nabla\phi + \nabla\phi\cdot\boldsymbol{p}|v_0) + \text{c.c.}}{E_0 - E_n}\Big\}$$
$$= 0 \quad . \tag{11.176}$$

The integral I, defined as

$$I \equiv \int \psi_n^*(\boldsymbol{p}\cdot\nabla\phi + \nabla\phi\cdot\boldsymbol{p})v_0 d\tau \quad , \tag{11.177}$$

can be transformed, making use of the fact that the wave functions are real and utilizing partial integrations, to be

$$I = \frac{\hbar}{i}\int(v_0\phi\nabla^2\psi_n - \psi_n\phi\nabla^2 v_0)d\tau \quad . \tag{11.178}$$

It is simple to re-express the first term on the right, since

$$\nabla^2\psi_n = \frac{2m}{\hbar^2}(V - E_n)\psi_n \tag{11.179}$$

where V is the potential acting on the electron. We evaluate the second term by noting that, neglecting overlap,

$$\int \psi_n \phi \nabla^2 v_0 d\tau = \int v_n \phi \nabla^2 v_0 d\tau = \int v_n \phi \nabla^2 \psi_0 d\tau \quad . \tag{11.180}$$

By utilizing (11.179) in (11.180) and again neglecting overlap, we obtain finally

$$(n|\mathbf{p}\cdot\nabla\phi + \nabla\phi\cdot\mathbf{p}|v_0) = \frac{\hbar}{i}\cdot\frac{2m}{\hbar^2}(E_0 - E_n)\int \psi_n \phi v_0 d\tau \quad . \tag{11.181}$$

We can substitute this expression into (11.176) and collect terms to obtain

$$(0\sigma|\mathcal{H}_{\Delta g}(\phi)|v_0\sigma') = \frac{e^2\hbar}{2m^2c^3}(\sigma|S|\sigma')\Big[(0|\mathbf{E}\times\nabla\phi|v_0)$$

$$- 2\sum_n{}'(0|\mathbf{E}\times\nabla|n)(n|\phi|v_0)\Big] \quad . \tag{11.182}$$

The prime can be removed from the summation, since the diagonal spin-orbit matrix elements vanish, giving

$$(0\sigma|\mathcal{H}_{\Delta g}(\phi)|v_0\sigma') = \frac{e^2\hbar}{2m^2c^3}(\sigma|S|\sigma')\cdot[(0|\mathbf{E}\times\nabla\phi - 2(\mathbf{E}\times\nabla\phi|v_0)]$$

$$= -\frac{e^2\hbar}{2m^2c^3}(\sigma|S|\sigma')\cdot\int \mathbf{E}\times\nabla(\phi v_0^2)d\tau \tag{11.183}$$

where $(\mathbf{E}\times\nabla)$ signifies that $\mathbf{E}\times\nabla$ is to operate on all functions to its right, that is, on both ϕ and v_0.

But the integral can be shown to vanish, utilizing the fact that $\nabla\times\mathbf{E} = 0$ and transforming the integral $\int \nabla\times(\mathbf{E}\phi v_0^2)d\tau$ into a surface integral. We omit these details since they are quite standard. Our theorem is thus proved.

We have not as yet said anything about the spin-orbit matrix elements to excited states. By utilizing the fact that the electric field $\mathbf{E}$ is large only near the nuclei, we are *always* able to neglect overlap when evaluating spin-orbit matrix elements. Thus

$$(0|\mathbf{E}\times\mathbf{p}|n) = (u_0|\mathbf{E}\times\mathbf{p}|u_n) + (v_0|\mathbf{E}\times\mathbf{p}|v_n) \quad . \tag{11.184}$$

To see the full import of (11.184) as well as to illustrate our theorem (11.170) concretely, we now turn to the evaluation of the specific problem of a molecular complex in which only one electron occupies the V_k center orbitals. The ground state is therefore given by (11.153), $\psi_0 = (1/\sqrt{2})(z_1 - z_2)$. We are concerned with excited states such as $(1/\sqrt{2})(x_1 \pm x_2)$. We have, using (11.184),

$$(0|\mathbf{E}\times\mathbf{p}|n) = \left(\frac{z_1 - z_2}{\sqrt{2}}\Big|\mathbf{E}\times\mathbf{p}\Big|\frac{x_1 \pm x_2}{\sqrt{2}}\right)$$

$$= \tfrac{1}{2}[(z_1|\mathbf{E}\times\mathbf{p}|x_1) \mp (z_2|\mathbf{E}\times\mathbf{p}|x_2)] \quad . \tag{11.185}$$

Since the two atoms are identical and since $\mathbf{E}$ is large only near a nucleus,

$$(z_1|\mathbf{E}\times\mathbf{p}|x_1) = (z_2|\mathbf{E}\times\mathbf{p}|x_2) \quad . \tag{11.186}$$

Therefore, when the upper sign applies, the two terms in the square brackets of (11.185) cancel, and $(0|E \times p|n)$ vanishes. On the other hand, for the lower sign, the terms add, giving twice either one. Thus the state $(x_1 + x_2)/\sqrt{2}$ does not contribute to the g-shift, but the state $(x_1 - x_2)/\sqrt{2}$ does. Of course a similar argument shows that $(y_1 + y_2)/\sqrt{2}$ also makes no contribution although $(y_1 - y_2)/\sqrt{2}$ does. The states involved in the g-shift are shown in Fig. 11.16 by the solid arrow.

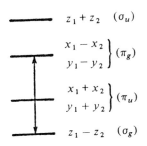

Fig. 11.16. The solid line indicates the states joined to the ground state $z_1 - z_2$ by the spin-orbit coupling

If we had a free atom, the spin-orbit matrix elements could be expressed in terms of the free-atom spin-orbit coupling constant λ according to the equation

$$\left(k\left|\frac{e\hbar}{2m^2c^2}S \cdot E \times p\right|l\right) = \lambda(k|L \cdot S|l) \tag{11.187}$$

where k and l denote free-atom states associated with the particular λ. For our example, x_1, y_1, and z_1 are being taken as free-atom p-states. Therefore we can write

$$\frac{e\hbar}{2m^2c^2}(\sigma|S|\sigma') \cdot (z_1|E \times p|x_1) = \lambda(\sigma|S|\sigma') \cdot (z_1|L_1|x_1) \tag{11.188}$$

where $\hbar L_1$ is the angular momentum about the nucleus of the first atom, and where λ is the spin-orbit coupling constant appropriate to the (np) electron configuration of the outer electron. Evaluation of the matrix element $(z_1|L_1|x_n)$ proceeds as in (11.17) and (11.18).

We now turn to evaluation of the matrix elements $(u_n|A' \cdot p + p \cdot A'|u_0)$ of (11.170). We have that $u_n = x_1/\sqrt{2}$ or $y_1/\sqrt{2}$, $u_0 = z_1/\sqrt{2}$. By utilizing (11.163) and the fact that the u's are real, we have

$$(u_n|A_1 \cdot p + p \cdot A_1|u_0) = \hbar H_0 \cdot \int u_n L_1 u_0 d\tau$$
$$= \frac{\hbar H_0}{2} \cdot (x_1|L_1|z_1) \quad \text{or} \quad \frac{\hbar H_0}{2} \cdot (y_1|L_1|z_1) . \tag{11.189}$$

But, by the symmetry of the atoms,

$$(u_n|A_1 \cdot p + p \cdot A_1|u_0) = (v_n|A_2 \cdot p + p \cdot A_2|v_0) \tag{11.190}$$

so that we have, neglecting the first-order terms such as given by (11.170),

$$(0\sigma|\mathcal{H}_{\Delta g}|0\sigma') = 2\beta\lambda(\sigma|S|\sigma') \cdot \left[\frac{(z_1|L_1|x_1)(x_1|L_1|z_1)}{E_{z_1-z_2} - E_{x_1-x_2}} \right.$$
$$\left. + \frac{(z_1|L_1|y_1)(y_1|L_1|z_1)}{E_{z_1-z_2} - E_{y_1-y_2}} \right] \cdot H_0 \quad . \tag{11.191}$$

This is equivalent to having

$$\mathcal{H}_{\Delta g} = \sum_{qq'=x,y,z} 2\beta S_q a_{qq'} H_{q'} \quad . \tag{11.192}$$

By evaluating the matrix elements, we get that $a_{qq'} = 0$ if $q \neq q'$, and

$$a_{xx} = -\frac{\lambda}{E_{y_1-y_2} - E_{z_1-z_2}}$$
$$a_{yy} = -\frac{\lambda}{E_{x_1-x_2} - E_{z_1-z_2}} \tag{11.193}$$
$$a_{zz} = 0 \quad .$$

It is interesting to comment a bit more on why the states $(x_1 + x_2)/\sqrt{2}$ and $(y_1 + y_2)/\sqrt{2}$ do not come into the g-shift. We note not only that the spin-orbit matrix elements to those states vanish, but also that the orbit-Zeeman terms cancel. Mixture of these excited states corresponds to production of a current flow in the ground state, shown in Fig. 11.17.

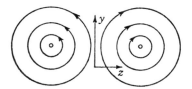

Fig. 11.17. Current flow produced by mixing some of the function $(x_1 + x_2)/\sqrt{2}$ into the ground state $(z_1 - z_2)/\sqrt{2}$

According to (11.155), the gauge-invariant spin-orbit interaction is between the spin and the gauge-invariant current density $j(r)$. With a current flow such as that given by Fig. 11.17, flowing in opposite senses on the two atoms, the net spin-orbit coupling vanishes. That is the significance of the vanishing of the spin-orbit matrix element. The vanishing of the orbit-Zeeman terms represents the fact that the applied field would never induce a current flow oppositely directed on the two atoms. Rather, it gives a flow such as that shown in Fig. 11.18.

We see that the technique of handling the problem of several force centers is to break the integrals into terms that are large only near the individual force centers, thereby converting the problem to a sum of single force center problems.

The second problem with which we must grapple in order to analyze the V_k center is how to compute the g-shift when we have a system with more than one electron. Since the spin and orbit are uncoupled in the absence of spin-orbit

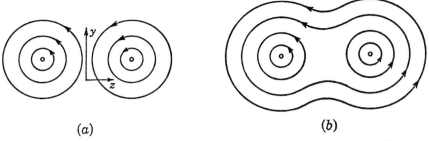

Fig. 11.18. (a) Current flow produced by the external field H_0 in the molecular complex. The fact that no current crosses the boundary between the two atoms results from the neglect of overlap. If overlap is included, the pattern is shown in (b)

coupling, let us assume that we can characterize the multi-electron states by a total spin quantum number S with eigenvalues M for some component. An extra quantum number n will also be needed to define the energy. We designate the ground state, then, as $|0SM\rangle$ and excited states by $|nS'M'\rangle$. We are concerned as before with the spin-orbit and the orbital Zeeman couplings $\mathcal{H}_{SO}$ and $\mathcal{H}_{OZ}$, respectively. The same expressions will apply as for the case of one electron, except that we must now label the coordinates by a symbol j, to specify which of the N electrons is involved.

We have, then,

$$\mathcal{H}_{SO} = \sum_j \mathcal{H}_{SO}^{(j)} \quad , \quad \mathcal{H}_{OZ} = \sum_j \mathcal{H}_{OZ}^{(j)} \tag{11.194}$$

where

$$\mathcal{H}_{SO}^{(j)} = \frac{e\hbar}{2m^2c^2} \mathbf{S}_j \cdot \left[\mathbf{E}_j \times \left(\mathbf{p}_j + \frac{e}{c}\mathbf{A}_j\right)\right] \tag{11.195}$$

$$\mathcal{H}_{OZ}^{(j)} = \frac{e}{2mc}(\mathbf{p}_j \cdot \mathbf{A}_j + \mathbf{A}_j \cdot \mathbf{p}_j)$$

and where we have neglected the term involving the square of the vector potential in the orbit-Zeeman coupling because we seek terms linear in H_0.

For simplicity we divide the spin-orbit coupling into two terms, one involving the vector potential $\mathbf{A}$; the other, not:

$$\mathcal{H}_{SOA}^{(j)} = \frac{e\hbar}{2m^2c^2} \mathbf{S}_j \cdot \left(\mathbf{E}_j \times \frac{e}{c}\mathbf{A}_j\right) \tag{11.196}$$

$$\mathcal{H}_{SO0}^{(j)} = \frac{e\hbar}{2m^2c^2} \mathbf{S}_j \cdot (\mathbf{E}_j \times \mathbf{p}_j) \ .$$

Therefore $\mathcal{H}_{SO0}$ is the spin-orbit operator in zero-applied field. Assuming that the orbital angular momentum is quenched, we have, therefore, that

$$(0SM|\mathcal{H}_{OZ}|0SM') = 0 \quad , \quad (0SM|\mathcal{H}_{SO0}|0SM') \ . \tag{11.197}$$

Then the spin-orbit and orbital Zeeman coupling combine to give matrix elements equivalent to our adding a term $\mathcal{H}_{\Delta g}$ to the Hamiltonian, where

$$(0SM|\mathcal{H}_{\Delta g}|0SM') = (0SM|\mathcal{H}_{SOA}|0SM')$$
$$+ \sum_{nS'M''} \frac{(0SM|\mathcal{H}_{SO0}|nS'M'')(nS'M''|\mathcal{H}_{OZ}|0SM')}{E_0 - E_n}$$
$$+ \frac{(0SM|\mathcal{H}_{OZ}|nS'M'')(nS'M''|\mathcal{H}_{SO0}|0SM')}{E_0 - E_n}, \quad (11.198)$$

where we have neglected the spin in the energy denominators, and where we are also keeping only the terms that give rise to a g-shift.

In the case of the V_k center, we may take the wave functions $|nSM)$ to be a product of one-electron molecular orbital states, properly antisymmetrized. The calculation proceeds along lines similar to that of the indirect nuclear coupling in Section 4.9. Let us denote the state $(z_1 - z_2)/\sqrt{2}$ containing electron number 1 with spin-up ($m = +\frac{1}{2}$) as

$$u_{z_1-z_2,+}(1) \quad . \quad (11.199)$$

Since the total spin of the V_k center is $\frac{1}{2}$, the state $|0SM) = |0\frac{1}{2}\frac{1}{2})$ is then

$$\left|0\frac{1}{2}\frac{1}{2}\right) = \frac{1}{\sqrt{11!}} \sum_p (-1)^P P u_{z_1-z_2,+}(1) u_{z_1-z_2,-}(2)$$
$$\times u_{x_1+x_2,+}(3) \cdots z_1+z_2,+(11) \quad . \quad (11.200)$$

That is, all the orbitals except $u_{z_1+z_2,-}$ are occupied by an electron.

It is convenient to denote functions such as $(z_1 - z_2)/\sqrt{2}$ by a symbol l and the spin quantum number by σ (since m is also used for the mass of an electron). In this notation the individual electron orbitals are denoted as $|l\sigma)$. As discussed in Section 4.9, all matrix elements of (11.198) arise from one-electron operators, so that the states joined by the operators can at most differ in the occupation of one orbital. Thus we find that we can express $(0SM|\mathcal{H}_{SOA}|0SM')$ in terms of the one-electron operator $\mathcal{H}_{SOA}^{(1)}$:

$$(0SM|\mathcal{H}_{SOA}|0SM') = \sum_{l\sigma,l\sigma'} (l\sigma|\mathcal{H}_{SOA}^{(1)}|l\sigma') \quad (11.201)$$

where $l\sigma$ goes over all values occupied in $|0SM)$ and where $l\sigma'$ goes over all values occupied in $|0SM')$. We do not include matrix elements $(l\sigma|\mathcal{H}_{SOA}|l'\sigma')$ where $l' \neq l$, since these states imply a change in the occupation of molecular orbitals. That would imply that the ground state possessed orbital degeneracy, a circumstance we do not wish to consider.

The second-order terms are handled in a similar manner. The sum over electrons can be converted to a sum over orbitals occupied in the ground state and sum over n to a sum over orbitals not occupied in the ground state. Therefore we get

$$(0SM|\mathcal{H}_{\Delta g}|0SM') = \sum_{\substack{l\sigma \text{ in } |0SM)\\ l\sigma' \text{ in } |0SM')}} (l\sigma|\mathcal{H}_{SOA}^{(1)}|l\sigma')$$

$$+ \sum_{\text{restriction } A}{}' \frac{(l\sigma|\mathcal{H}_{OZ}^{(1)}|l'\sigma)(l'\sigma|\mathcal{H}_{SO0}^{(1)}|l\sigma')}{E_l - E_{l'}}$$

$$+ \sum_{\text{restriction } B}{}' \frac{(l\sigma|\mathcal{H}_{SO0}^{(1)}|l'\sigma')(l'\sigma'|\mathcal{H}_{OZ}^{(1)}|l\sigma')}{E_l - E_{l'}} \qquad (11.202)$$

where by restriction A we mean:

$|l\sigma)$ is occupied in $|0SM)$

$|l\sigma')$ is occupied in $|0SM')$

$|l'\sigma)$ is occupied in neither $|0SM)$ nor $|0SM')$,

and by restriction B we mean:

$|l\sigma)$ is occupied in $|0SM)$

$|l\sigma')$ is occupied in $|0SM')$

$|l'\sigma')$ is occupied in neither $|0SM)$ nor $|0SM')$.

Equation (11.202) will hold for any system in which the wave function can be taken as a product of individual spin functions.

We wish, of course, to incorporate our earlier theorem, (11.170), to enable us to use "natural" gauges. This can be done readily by noting two things: The first is that, since $\mathcal{H}_{OZ}^{(1)}$ is independent of spin,

$$(l\sigma|\mathcal{H}_{OZ}^{(1)}|l'\sigma) = (l|\mathcal{H}_{OZ}^{(1)}|l') = (l\sigma'|\mathcal{H}_{OZ}^{(1)}l'\sigma') \quad . \qquad (11.203)$$

By utilizing this fact, we can see the second point. If we remove the condition that $|l'\sigma)$ be unoccupied in $|0SM)$ or $|0SM')$ from restriction A, and that $|l'\sigma')$ be unoccupied in $|0SM)$ or $|0SM')$ from restriction B, the extra terms we acquire will exactly cancel in pairs. We can therefore write

$$(0SM|\mathcal{H}_{\Delta g}|0SM')$$

$$= \sum_{l\sigma;l\sigma'} (l\sigma|\mathcal{H}_{SOA}^{(1)}|l\sigma')$$

$$+ \sum_{l\sigma;l\sigma';l'} \frac{(l\sigma|\mathcal{H}_{OZ}^{(1)}|l'\sigma)(l'\sigma|\mathcal{H}_{SO0}^{(1)}|l\sigma') + (l\sigma|\mathcal{H}_{SO0}^{(1)}|l'\sigma')(l'\sigma'|\mathcal{H}_{OZ}^{(1)}|l\sigma')}{E_l - E_{l'}} \qquad (11.204)$$

where now we require only that $|l\sigma)$ be occupied in $|0SM)$ and $|l\sigma')$ be occupied in $|0SM')$.

Consider now all the terms of fixed $l\sigma$ and $l\sigma'$:

$$(l\sigma|\mathcal{H}_{SOA}^{(1)}|l\sigma') + \sum_{l'}{}' \frac{(l\sigma|\mathcal{H}_{OZ}^{(1)}|l'\sigma)(l'\sigma|\mathcal{H}_{SO0}^{(1)}|l\sigma')}{E_l - E_{l'}}$$
$$+ \frac{(l\sigma|\mathcal{H}_{SO0}^{(1)}|l'\sigma')(l'\sigma'|\mathcal{H}_{OZ}^{(1)}|l\sigma')}{E_l - E_{l'}} \quad . \tag{11.205}$$

This is identical in form to (11.161). It can therefore be converted to the expression involving the mixed gauge. Let us therefore define

$$(l\sigma|V^{(1)}|l\sigma') = \frac{e^2\hbar}{2m^2c^3}(\sigma|\mathbf{S}_1|\sigma')\cdot[(u_l|\mathbf{E}_1\times\mathbf{A}_1'|u_l) + (v_l|\mathbf{E}_1\times\mathbf{A}_1''|v_l)]$$

and

$$(l\sigma|U^{(1)}|l'\sigma') = \frac{e^2\hbar}{2m^2c^3}(\sigma|\mathbf{S}_1|\sigma')\cdot[(u_l|\mathbf{A}_1'\cdot\mathbf{p}_1 + \mathbf{p}_1\cdot\mathbf{A}_1'|u_{l'})$$
$$+ (v_l|\mathbf{A}_1''\cdot\mathbf{p}_1 + \mathbf{p}_1\cdot\mathbf{A}_1''|v_{l'})] \quad . \tag{11.206}$$

In terms of these definitions we can rewrite the expression of (11.205) as

$$(l\sigma|V^{(1)}|l\sigma') + \sum_{l'}{}' \frac{(l\sigma|U^{(1)}|l'\sigma)(l'\sigma|\mathcal{H}_{SO0}^{(1)}|l\sigma')}{E_l - E_{l'}}$$
$$+ \frac{(l\sigma|\mathcal{H}_{SO0}^{(1)}|l'\sigma')(l'\sigma'|U^{(1)}|l\sigma')}{E_l - E_{l'}} \quad . \tag{11.207}$$

Therefore we find

$$(0SM|\mathcal{H}_{\Delta g}|0SM') = \sum_{l\sigma;l\sigma'} (l\sigma|V^{(1)}|l\sigma')$$
$$+ \sum_{l\sigma;l\sigma';l'} \frac{(l\sigma|U^{(1)}|l'\sigma)(l'\sigma|\mathcal{H}_{SO0}^{(1)}|l\sigma')}{E_l - E_{l'}}$$
$$+ \frac{(l\sigma|\mathcal{H}_{SO0}^{(1)}|l'\sigma')(l'\sigma'|U^{(1)}|l\sigma')}{E_l - E_{l'}} \tag{11.208}$$

where $l\sigma$ includes all values occupied in $|0SM\rangle$ and $l\sigma'$ includes all values occupied in $|0SM'\rangle$. As we have remarked, allowing $l'\sigma$ or $l'\sigma'$ of the excited states to include values occupied in either $|0SM\rangle$ or $|0SM'\rangle$ introduces pairs of terms that cancel. It is therefore simplest in practice to reimpose conditions A and B so that no superfluous terms arise in the summation.

By means of the Wigner-Eckart theorem, it is possible to show that (11.208) implies that all matrix elements $(0SM|\mathcal{H}_{\Delta g}|0SM')$ can be obtained from a Hamiltonian of the form

$$\mathcal{H}_{\Delta g} = 2\beta \sum_{\substack{q=x,y,z \\ q'=x,y,z}} H_q a_{qq'} S_{q'} \tag{11.209}$$

where the components S_q are, for example,

$$S_x = \sum_{i=1}^{N} S_{xi} \quad , \quad \text{etc.} \tag{11.210}$$

For the V_k center, however, rather than employ the Wigner-Eckart theorem, we shall simply evaluate (11.208). Symmetry tells us that the principal axes of the g-tensor are the x-, y-, z-axes of the molecule, where the z-axis lies along the bond. Suppose, therefore, that the static field lies along the x-axis and that M is taken as the eigenvalue of S_x. Since this is a principal axis, the only nonvanishing matrix elements have $M = M'$. Of course we can verify this by explicit evaluation of (11.208). S is, of course, $\frac{1}{2}$. Let us compute (11.208) for $M = M' = \frac{1}{2}$. The simplest way to discuss the matrix elements is, then, in terms of a diagram of the states. Since we are computing a diagonal term in $(0SM|\mathcal{H}_{\Delta g}|0SM')$, the states labeled $|l\sigma\rangle$ and $|l\sigma'\rangle$ of (11.208) must be identical (we must return the electron to the state from which it was virtually excited). Setting $\sigma' = \sigma$ and neglecting the terms involving $(l\sigma|V^{(1)}|l\sigma)$, we have

$$(0SM|\mathcal{H}_{\Delta g}|0SM)$$

$$= \sum_{\substack{|l\sigma\rangle \text{ occupied} \\ |l'\sigma\rangle \text{ unoccupied}}} \left[\frac{(l\sigma|U^{(1)}|l'\sigma)(l'\sigma|\mathcal{H}^{(1)}_{SO0}|l\sigma) + \text{c.c.}}{E_l - E_{l'}} \right] \tag{11.211}$$

where $|l\sigma\rangle$ occupied or $|l'\sigma\rangle$ unoccupied refer respectively to whether or not $|l\sigma\rangle$ is occupied or $|l'\sigma\rangle$ is unoccupied in the ground state. For $S = \frac{1}{2}$, $M = \frac{1}{2}$, the states $|l'\sigma\rangle$ and $|l\sigma\rangle$ can be summarized by a diagram in which solid arrows designate states occupied by electrons, an arrow pointing up, ↑, referring to $\sigma = +\frac{1}{2}$, and so on; and dashed arrows, ↓ or ↑, referring to unoccupied states. The ground state is shown in Fig. 11.19.

An excited state is obtained by transferring an electron from an occupied to a vacant orbital. For a field in the x-direction, the orbital Zeeman term couples only the state $y_1 + y_2$ to $z_1 + z_2$, as can be seen by an argument similar to that for deriving (11.162). The states joined in the g-shift are shown on Fig. 11.20.

The explicit evaluation of the matrix elements follows the discussion relating to (11.184) and (11.189) giving

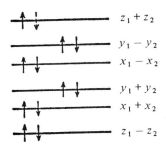

Fig. 11.19. Ground state $|0\frac{1}{2}\frac{1}{2}\rangle$ of the V_k center. We assume that the crystal field splits the states $x_1 \pm x_2$ slightly from $y_1 \pm y_2$. The solid arrows indicate an occupied state (↑, ↓); the dashed one (↓) is unoccupied

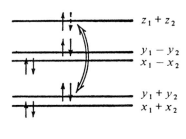

Fig. 11.20. The double arrow indicates states joined by the matrix elements of 11.120) for a field in the x-direction. The arrows indicate the electron spin quantization, ↑ signifying spin parallel to the static field and ↓ signifying spin antiparallel. The dashed arrow ↓ state is vacant in the ground state

$$\left(0\frac{1}{2}\frac{1}{2}|\mathcal{H}_{\Delta g}|0\frac{1}{2}\frac{1}{2}\right) = 2\beta H_x \frac{(y_1|L_x^{(1)}|z_1)(z_1 - \frac{1}{2}|\mathcal{H}_{SO0x}^{(1)}|y_1 - \frac{1}{2})}{E_{y_1+y_2} - E_{z_1+z_2}} \quad (11.212)$$

where $\mathcal{H}_{SO0x}$ is the one-electron, spin-orbit coupling associated with the x-component of spin and is given by

$$\left(z_1 - \frac{1}{2}|\mathcal{H}_{SO0x}^{(1)}|y_1 - \frac{1}{2}\right) = \frac{e\hbar}{2m^2c^2}\left[\int z_1^*(\boldsymbol{E}\times\boldsymbol{p})_x y_1 d\tau\right]$$
$$\times \left(-\frac{1}{2}|S_x^{(1)}|-\frac{1}{2}\right) \quad (11.213)$$

where $d\tau$ indicates integration over the electron position coordinate.

The expression of (11.213) involves functions on one atom only, and can therefore be related to the free-atom expression. In fact for a free atom, if one neglects the coupling of one spin to the orbit of another electron, one can write the spin-orbit coupling of N electrons as

$$\mathcal{H}_{SO} = \sum_{i=1}^{N} \zeta_i \boldsymbol{L}^i \cdot \boldsymbol{S}^i \quad . \quad (11.214)$$

For equivalent electrons, the ζ_i's are all equal. If one has Russell-Saunders coupling in the free atom, the total angular momentum quantum numbers L and S are good quantum numbers and for matrix elements internal to a given L and S, we have

$$\mathcal{H}_{SO} = \lambda \boldsymbol{L} \cdot \boldsymbol{S} \quad . \quad (11.215)$$

If $N = 1$, clearly $\lambda = \zeta$. If N represents a shell that has only one missing electron, $\lambda = -\zeta$. Since ζ is always positive, we obtain in this way the fact that holes have negative λ.

We can utilize the free-atom ζ's to evaluate (11.213) since it enables us to write

$$\frac{e\hbar}{2m^2c^2}\left[\int z_1^*(\boldsymbol{E}\times\boldsymbol{p})_x y_1 d\tau\right]\left(-\frac{1}{2}|S_x|-\frac{1}{2}\right)$$

$$= \zeta(z_1|l_x^{(1)}|y_1)\left(-\frac{1}{2}|S_x^{(1)}|-\frac{1}{2}\right) . \tag{11.216}$$

We obtain in this manner

$$\left(0\frac{1}{2}\frac{1}{2}|\mathcal{H}_{\Delta g}|0\frac{1}{2}\frac{1}{2}\right) = -\frac{2\beta H_x \zeta(-1/2|S_x^{(1)}|-1/2)}{E_{z_1+z_2} - E_{y_1-y_2}} . \tag{11.217}$$

Since the spin Zeeman energy $(0\frac{1}{2}\frac{1}{2}|\mathcal{H}_{SZ}|0\frac{1}{2}\frac{1}{2})$ is

$$(0\frac{1}{2}\frac{1}{2}|\mathcal{H}_{SZ}|0\frac{1}{2}\frac{1}{2}) = 2\beta H_x(\frac{1}{2}|S_x^{(1)}|\frac{1}{2}) , \tag{11.218}$$

being just that of the one unpaired spin, we have

$$g_{xx} = 2\left(1 + \frac{\zeta}{E_{z_1+z_2} - E_{y_1-y_2}}\right)$$
$$= 2\left(1 - \frac{\lambda}{E_{z_1+z_2} - E_{y_1-y_2}}\right) \tag{11.219}$$

where λ is the free Cl atom spin-orbit coupling constant. We note $g_{xx} > 2$.

In a similar manner we get

$$g_{yy} = 2\left(1 - \frac{\lambda}{E_{z_1+z_2} - E_{x_1-x_2}}\right) , \quad g_{zz} = 2 . \tag{11.220}$$

It is interesting to note in our calculation that the origin of the positive g-shift rather than the negative one that we would have for a single electron is the spin matrix element $(-\frac{1}{2}|S_x^{(1)}|-\frac{1}{2})$ of (11.217). The $|-\frac{1}{2}\rangle$ states come in because we have excited one of the paired spins into an originally unpaired state. We deal with a spin that points opposite to M.

We may contrast the situation with one in which only five electrons fill the states, as shown in Fig. 11.21. We shall assume the degeneracy of the states y_1+y_2 and x_1+x_2 is lifted, as shown, and likewise for y_1-y_2 and x_1-x_2. A field in the x-direction would join states $|y_1+y_2,\frac{1}{2}\rangle$ and $|z_1+z_2,\frac{1}{2}\rangle$ and would make $g<2$ (an "electron" shift). On the other hand, a field in the z-direction joins the nearly degenerate $|y_1+y_2,-\frac{1}{2}\rangle$ and $|x_1+x_2,-\frac{1}{2}\rangle$ states. It would have $g>2$ (a "hole"

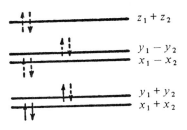

Fig. 11.21. Filled states when the V center orbitals contain only five electrons

shift). The close proximity of these two states would make $|\Delta g_{zz}| \gg |\Delta g_{xx}|$, and g_{yy}, of course, is still 2, since the pertinent matrix elements vanish.

It is clear that the 12 states are less than half full, yet the predominant g-shift is that of a "hole". We see, therefore, that we must use extreme caution in characterizing centers as "electron" or "hole" centers simply from the g-shift data.

We should also comment that we have assumed very simple functions with no overlap between atoms, to compute the matrix elements. In general, we would need to make corrections both for overlap and for the possibility that the functions x_1, y_1, z_1, and so on are linear combinations of atomic orbitals, as we did in discussing the hyperfine coupling. However, these corrections do not alter the principles, although they do complicate the numerical calculations.

12. Summary

We have considered a variety of effects — line widths, chemical shifts, Knight shifts, hyperfine splittings — a bewildering array of seemingly special cases. As we look back, we see some effects that occur in first-order perturbation theory, others that require a higher order. Since we have discussed the phenomena one by one, it is appropriate to summarize by writing a single Hamiltonian that includes everything. As we contemplate it, we should remind ourselves of the significance of each term. We write below the Hamiltonian describing a nucleus interacting with an electron in the presence of a magnetic field H_0. We define the vector potentials A_0, associated with the field H_0, and A_n, associated with the field at the electron owing to the nuclear moment ($A_n = \mu \times r/r^3$ normally). We also define the quantity

$$\pi = \frac{\hbar}{i}\nabla + \frac{e}{c}A_0 \quad . \tag{12.1}$$

Then we have the following Hamiltonian:

$$\mathcal{H} = -\frac{\hbar^2}{2m}\nabla^2 \quad + V_0 + V_{\text{cryst}}$$

electron kinetic energy   electron potential energy in the field of the nucleus and of other electrons   electron potential energy due to charges outside the atom

$$+ \frac{e\hbar}{2m^2c^2} S \cdot \left[E \times \left(p + \frac{e}{c}A_0\right)\right] + \gamma_e \hbar H_0 \cdot S$$

electron spin-orbit coupling   electron spin Zeeman energy

$$+ \frac{e}{2mc}(p \cdot A_0 + A_0 \cdot p) + \frac{e^2}{2mc^2}A_0^2$$

coupling of electron orbital motion to H_0

$$+ \frac{e}{2mc}(\pi \cdot A_n + A_n \cdot \pi) + \frac{\gamma_e \gamma_n \hbar^2}{r^3}\left[\frac{3(I \cdot r)(S \cdot r)}{r^2} - I \cdot S\right]$$

coupling of nuclear moment to electron orbital motion   coupling of nuclear moment with electron spin moment for non s-states

$$+ \frac{8\pi}{3}\gamma_e \gamma_n \hbar^2 I \cdot S\delta(r) + \mathcal{H}_Q \qquad\qquad - \gamma_n \hbar H_0 \cdot I .$$

coupling of nuclear moment with electron spin moment for s-states   coupling of nuclear quadrupole moment to field gradient due to electron and external charges   nuclear Zeeman energy

We can add to this the coupling of nuclei with each other and the magnetic coupling of electrons with each other.

Problems

Chapter 2

2.1 Consider an Hermitian operator F which is an explicit function of time. (For example, $F = -\gamma\hbar I_x H_x \cos \omega t$, the interaction energy of a spin with an alternating magnetic field in the x-direction.) Prove that

$$\frac{dF}{dt} = \frac{i}{\hbar}[\mathcal{H}, F] + \frac{\partial F}{\partial t}$$

where $\partial F/\partial t$ represents an actual derivative of $F(t)$ with respect to time.

2.2 Equation (2.22a) gives an expression for $\langle \mu_x(t) \rangle$ for a particle of spin $\frac{1}{2}$. Generalize the expression for a spin I.

2.3 A magnet has an inhomogeneous static magnetic field. The fraction of spins df that experience a magnetic field between H and $H + dH$ is

$$df = p(H)dH \quad \text{where} \quad \int_0^\infty p(H)dH = 1 \quad .$$

Assume the inhomogeneity is slight and that it simply gives a spread in field with no change in the direction. Take the field to be in the z direction.

Compute the magnetization in the x-direction, perpendicular to the static field, as a function of time, assuming that at $t = 0$ the total magnetization was M_0, pointing in the x-direction, for the three forms of $p(H)$:

a) $p(H)$ is a constant for $H_0 - a < H < H_0 + a$ and is zero for all other fields.
b) $p(H) \propto \exp[-(H - H_0)^2/a^2]$.
c) $p(H) \propto \frac{1}{1+(H-H_0)^2/a^2}$.

2.4 A nucleus of spin $\frac{1}{2}$ is quantized by a field H_0 in the z-direction. It is in the $m = +\frac{1}{2}$ state at $t = 0$ when a rotating magnetic field of amplitude H_1 is applied for a time t_w, producing a 90° pulse.

a) Compute the wave function of the spin in the rotating reference system as a function of time during and after the pulse.

b) Compute the wave function in the laboratory reference system during and after the pulse.
c) Compute the $\langle \mu_x(t) \rangle$ during and after the pulse.

2.5 A coil of length l, cross-sectional area A, and n turns is wound in the form of a solenoid. The axis of the coil is in the x-direction, and a static field H_0 is in the z-direction.

a) Assuming the nuclear moments are in thermal equilibrium, what is the nuclear magnetization per unit volume M_0 produced by H_0, in terms of H_0 and the static nuclear susceptibility χ_0?
b) Compute a numerical value of χ_0 for protons in water at room temperature, using the formula
$$\chi_0 = \frac{N\gamma^2 \hbar^2 I(I+1)}{3kT}$$
where N is the number of spins per unit volume. (For protons, γ can be found from the fact that the resonance occurs at 42 MHz for $H_0 = 10^4$ Gauss.)
c) The magnetization M_0 is turned by a 90° pulse. Derive an expression for amplitude V_0 of the voltage induced in the coil by the precessing M_0.
d) Make a numerical estimate of V_0 for protons, assuming a 10-turn coil 2 cm long, 1 cm in diameter, and an H_0 of 5000 Gauss.

2.6 Suppose the coil of Problem 2.5 has inductance L_0 and resistance R_0 and is in series resonance with a condenser C.

a) Derive an expression for the voltage across the condenser in terms of the induced voltage V_0 of Problem 2.5, L_0, R_0, C, and $Q(= L_0\omega/R)$.
b) Using the numerical estimate of Problem 2.5(d), and assuming the coil has a Q of 100, compute the size of the voltage across C.

2.7 *Rotation behavior of a spin 1 particle:* Consider a particle of spin 1, with eigenstates of I_z given as u_1, u_0, u_{-1} corresponding to the eigenvalues of 1, 0, and -1 respectively of I_z. Consider that it is acted on by a rotating field, H_1, exactly tuned to resonance as in the discussion of the spinor behavior of a spin $\frac{1}{2}$ particle in Sect. 2.6.

a) Taking the wave function in the rotating frame, ψ', to be
$$\psi'(t) = a(t)u_1 + b(t)u_0 + c(t)u_{-1}$$
and defining ω_1 as γH_1, show that a, b, and c obey the differential equations
$$\frac{da}{dt} = i\frac{\omega_1}{\sqrt{2}}b \quad , \quad \frac{db}{dt} = i\frac{\omega_1}{\sqrt{2}}(a+c) \quad , \quad \frac{dc}{dt} = i\frac{\omega_1}{\sqrt{2}}b \quad .$$

b) show that the solution of these equations is

$$a(t) = \tfrac{1}{2}[a(0)(1 + \cos \omega_1 t) - c(0)(1 - \cos \omega_1 t) + i\sqrt{2}b(0) \sin \omega_1 t]$$

$$b(t) = b(0) \cos \omega_1 t + \frac{i}{\sqrt{2}}[a(0) + c(0)] \sin \omega_1 t$$

$$c(t) = \tfrac{1}{2}[c(0)(1 + \cos \omega_1 t) - a(0)(1 - \cos \omega_1 t) + i\sqrt{2}b(0) \sin \omega_1 t] \quad .$$

c) Show that these equations give

$$\psi(t) = \psi(0) \quad \text{if} \quad \omega_1 t = 2\pi \quad ,$$

in contrast to a spin $\tfrac{1}{2}$ particle for which $\psi(t) = -\psi(0)$ if $\omega_1 t = 2\pi$.

2.8 In Sect. 2.9 the existence of spin echoes is derived assuming that the "$\pi/2$" pulse gives a $\pi/2$ rotation which is positive in the right-handed sense. This implies a negative γ. Carry through the derivation assuming the $\pi/2$ rotation is positive in the left-handed sense, as with a positive γ, showing that the echo is formed along the negative y-axis of Fig. 2.13.

2.9 Consider an experiment in which *initially* the magnetization is at its thermal equilibrium value, $\mathbf{k}M_0$, along a static field $\mathbf{k}H_0$. At $t = 0$ the magnetization is inverted by a π pulse. It then grows back towards its thermal equilibrium exponentially according to the equation

$$\frac{dM_z}{dt} = \frac{M_0 - M_z}{T_1} \quad .$$

At a time τ_1, M_z is inspected by applying a $\pi/2$ pulse and observing the initial amplitude of the free induction signal generated in an rf coil oriented transverse to $\mathbf{H}_0$.

a) Show that M_x, the value of the transverse magnetization immediately after the $\pi/2$ pulse, is

$$M_x = M_0(1 - 2e^{-\tau_1/T_1}) \quad .$$

b) Assume now that owing to natural line breadths, the transverse magnetization decays rapidly to zero. Suppose one now waits a time τ_2, long compared to the time for transverse magnetization to decay, then again applies a π pulse, and again applies a second $\pi/2$ pulse at time τ_1 later.

Suppose one repeats this pulse sequence many times:

$$\ldots \| \leftarrow \tau_1 \rightarrow \| \leftarrow \tau_2 \rightarrow \| \leftarrow \tau_1 \rightarrow \| \ldots \text{ etc.}$$
$$\quad \pi \qquad\qquad \pi/2 \qquad\qquad \pi \qquad\quad \pi/2$$

Show that the initial value of the free induction decay following the $\pi/2$ pulse becomes

$$M_x = M_0[1 - 2e^{-\tau_1/T_1} + e^{-(\tau_1+\tau_2)/T_1}] \quad .$$

2.10 This problem concerns the calculation of the time at which a spin echo occurs for a $(\pi/2, \pi)$ echo sequence, when the duration of the pulse is not taken as zero. Refer to the end of Sect. 2.9 and Figs. 2.15 and 2.16 for the notation. The goal is to derive the relationship of (2.73), $\tau' = \tau + 1/\gamma H_1$, in the approximation that we neglect corrections to τ' involving $(1/H_1)^2$. In all that follows, this approximation is assumed. Thus $\tan(\delta H/H_1) = \delta H/H_1$ etc.

a) Prove that the "$\pi/2$ pulse" rotates the magnetization, δM, from its initial direction along the z-axis to the x'-y' plane (where the y and y' axes coincide), and that it makes an angle $\Delta\theta = H_1/\delta H$ with the negative y' axis.

b) Prove that the projection of δM in the x-y plane makes the same angle, $\Delta\theta$, with the y-axis.

c) Show that the effect of the "π pulse" on the projection of δM in the x-y plane is a reflection in the x-z plane, as in the "infinite H_1" π pulse (i.e. there is *no* correction for the "π pulse" similar to the correction for the "$\pi/2$ pulse").

d) Derive (2.73)

$$\tau' = \tau + 1/\gamma H_1 \quad . \tag{2.73}$$

e) Show that (2.73) is *also* valid for an echo produced with a $(3\pi/2, 2\pi)$ pulse sequence, but that (2.74) becomes

$$\tau' = \tau + \frac{2}{3\pi}t_{3\pi/2} \quad . \tag{2.74}$$

2.11 Consider an operator R and its inverse R^{-1} (i.e. $RR^{-1} = R^{-1}R = 1$). An example might be

$$R = e^{i\theta I_z} \quad ,$$

which is associated with rotation about the z-axis.

Let G be some operator function which in general does not commute with R (i.e. $RG - GR = 0$). Prove that

$$R^{-1}e^{iG}R = e^{i(R^{-1}GR)} \quad .$$

2.12 This problem is concerned with the quantum mechanical treatment of the free induction decay and the spin echo from noninteracting spins subjected to a distribution of static field strength.

Consider the time development of the wave function in the rotating frame for a spin echo in which the H_1 is applied along the x-axis in the rotating frame. The first pulse is a $\pi/2$ pulse, and the second, applied time τ later, is a π pulse. Take the distribution of magnetic fields H to be symmetric about ω/γ, so that if $h_0 \equiv H - \omega/\gamma$, the distribution function $p(h_0)$ of h_0 is symmetrical about $h_0 = 0$.

a) Show that the expectation value of the total x-component of spin, obtained by adding up the contributions of all the spins (i.e. integrating over dh_0), vanishes at $t = 2\tau$.
b) Show that the time development of expectation value of the total y-component of spin $\langle I_{y,\text{tot}}(t)\rangle$ after the 90° pulse (but before the 180° pulse) is the *same* as $\langle I_{y,\text{tot}}(t - 2\tau)\rangle$, the behavior after the peak of the spin echo.

2.13 From (2.170) we have that

$$\chi_0 = \frac{2}{\pi} \frac{1}{\omega_0} \int_0^\infty \chi''(\omega')d\omega' \quad .$$

Show that if $\chi''_{\max}$ is the maximum value of χ'' in an absorption line,

$$\chi''_{\max} = \frac{\pi}{2}\chi_0 \frac{\omega_0}{\Delta\omega}$$

where $\Delta\omega$ is a suitably defined line breadth.

Assuming that the line width of the nuclear resonance of protons in water is 0.1 Gauss broad (because of magnet inhomogeneity) and that $H_0 = 10^4$ Gauss, compute $\chi''_{\max}$ for water, and also compute the maximum fractional change in coil resistance for a coil of $Q = 100$.

2.14 The response of a certain piece of material to a step of magnetic field of unit height, applied at $t = 0$, is

$$M_{\text{step}}(t) = \chi_0(1 - e^{-t/T}) \quad .$$

a) Compute $\chi'(\omega)$ and $\chi''(\omega)$.
b) Show that χ' and χ'' satisfy the Kramers-Kronig relations.

2.15 In Appendix F, an expression is developed for the complex magnetization $M_x + iM_y$

$$M_x + iM_y = \frac{i\gamma H_1 M_0 \tau [2 + \tau(\alpha_a + \alpha_b)/2]}{(1 + \alpha_a \tau)(1 + \alpha_b \tau) - 1} \quad .$$

a) Show that in the limit of very slow jumping ($\tau\delta\omega \gg 1$) the absorption line has the shape of two distinct resonances at frequencies

$$\omega = \gamma H_0 \pm \frac{\delta\omega}{2} \quad .$$

b) Show that in the limit of very fast jumping ($\tau\delta\omega \ll 1$) a single resonance results at frequency

$$\omega = \gamma H_0 \quad .$$

2.16 In Appendix F the Bloch equations are employed to analyze the case of a group of nuclei which spend equal times on the average in either of two sites at which the resonant frequency differs.

This equal time aspect appears in the statement that the two quantities C_1 and C_2 are equal, and in the use of $M_0/2$ in the term involving H_1. Suppose the equations were to be used to describe a problem in which the nucleus could jump between two sites which were not equally populated on the average. For example site "a" might be at a higher energy than site "b" so that the thermal equilibrium populations differed.

a) If M_a and M_b are the thermal equilibrium static magnetizations, what must the relationship be of C_1 to C_2?
b) Set up the appropriate form of (F.13) for this case and derive the expression for the resulting $M_x + iM_y$.

2.17 A static magnetic field $H(r)$ may be expressed as

$$H = \nabla \phi$$

where ϕ is the magnetic potential which, in free space, satisfies

$$\nabla^2 \phi = 0 \ .$$

If the magnetic field of a laboratory magnet is assumed axially symmetric about the z-axis, show that to lowest order the inhomogeneity can be expressed as

$$H - H_0 = z\left(\frac{\partial H}{\partial z}\right)$$

where H is the magnitude of $\mathbf{H}$, and H_0 is the magnitude of $\mathbf{H}$ at the origin and where $(\partial H/\partial z)$ is evaluated at the origin.

2.18 Show, following the methods of Problem 2.17, that the most general form of $H(x, y, z)$ based on including the spherical harmonics of $l = 2$ or less in the magnetic potential is

$$H(x, y, z) = H_0 + Ax + By + Cz \ .$$

2.19 Utilize the expression for $H(x, y, z)$ of Problem 2.18 and the methods of Appendix G to get a more general expression for the effect of diffusion on the echo envelope decay than that of (6.15).

2.20 Appendix F uses the Bloch equations to discuss the problem of a spin which can precess at either of two natural frequencies, depending on whether the spin is in site "a" or site "b".

a) Suppose that $T_2 = \infty$, $H_1 = 0$ (free precession), that the quantities C_1 and C_2 are zero, and that at $t = 0$, M_a and M_b lie along the x-axis in the frame which rotates at γH_0. Draw a picture of M_a and M_b in the x-y plane at a time at which M_a makes an angle θ with the x-direction.

b) Draw a picture of the vectors $\delta M_a = C_1 M_b \delta t$ and $\delta M_b = -C_1 M_b \delta t$ which would occur if C_1 is now switched on at the time corresponding to part (a).

c) Show by vector addition the new vectors

$$M'_a = M_a + \delta M_a \quad \text{and} \quad M'_b = M_b + \delta M_b \ .$$

Chapter 3

3.1 A pair of identical spins of $I_1 = I_2 = \frac{1}{2}$ is coupled by their magnetic dipole moments. Assuming zero external static magnetic field, show that the proper eigenstates of the spins are the singlet and triplet states, and then find the energies of the different states.

3.2 Suppose in Problem 3.1 that a static magnetic field H_0 is applied parallel to the internuclear axis.

a) Find the energy levels and eigenfunctions as a function of H_0.

b) An alternating magnetic field is applied perpendicular to the internuclear axis. Find the allowed transitions, their frequencies and relative intensities: (1) for H_0 much less than the dipolar coupling; (2) for H_0 much larger than the dipolar coupling.

3.3 Equation (3.43) involves $\text{Tr}\{[\mathcal{H}_d^0, \mu^-]\mu^+\}$. Prove that this trace vanishes.

3.4 Consider two identical spin $\frac{1}{2}$ nuclei. Let $I_z \equiv I_{1z} + I_{2z}$ be the total z-component of angular momentum. Evaluate $\text{Tr}\{I_z^2\}$ by explicit evaluation of the diagonal matrix elements for two schemes of quantization: (a) the m_1, m_2 scheme and (b) the I, M scheme. Show that the answers agree with each other.

3.5 Consider a nucleus of spin I. Compute $\text{Tr}\{I_z I_x\}$ and $\text{Tr}\{I^2 I_x^2\}$.

3.6 Consider a group of N noninteracting spins of spin I and gyromagnetic ratio γ. The total wave function can be taken as a product of the individual spin states, and the total energy as the sum of the individual eigenenergies. Evaluate the expression for $\chi''(\omega)$ of (2.190) to give the absorption. For simplicity, let $Z = (2I+1)^N$ and $\exp(-E_a/kT) = 1$, the expressions for the high-temperature limit.

3.7 The electrostatic exchange coupling between two electrons can be represented by adding a term $A S_1 \cdot S_2$ to the Hamiltonian.

563

a) Prove that this term commutes with the Zeeman energy.
b) Prove that addition of such a term to the direct dipole coupling does not affect the second moment computed when assuming dipolar coupling alone.

3.8 Consider three operators, A, B, and C. Prove that

$$\text{Tr}\{ABC\} = \text{Tr}\{CAB\} = \text{Tr}\{BCA\} \quad .$$

3.9 Consider a nucleus with spin $\frac{3}{2}$ whose Hamiltonian is $\mathcal{H} = \mathcal{H}_Z + \mathcal{H}_Q$, where

$$\mathcal{H}_Z = \gamma_n \hbar H_0 I_z \quad , \quad \mathcal{H}_Q = A(3I_z^2 - I^2) \quad .$$

The form of $\mathcal{H}_Q$ is similar to the one that sometimes arises when a nucleus has an electrical quadrupole moment. An alternating field is applied to the system to produce absorption.

a) Prove that $\mathcal{H}_Z$ and $\mathcal{H}_Q$ commute.
b) Treating $\mathcal{H}_Q$ as analogous to $\mathcal{H}_d^0$ in Section 3.3, and assuming $A \ll \gamma_n \hbar H_0$ prove that

$$\langle \omega \rangle = \gamma_n H_0$$

and find $\langle \Delta \omega^2 \rangle$.

3.10 Derive (3.61) for the case of a uniformly rotating pair of nuclei.

Chapter 4

4.1 In Section 4.4, gauge transformations are discussed.

a) Using (4.18), show that (4.19) is true.
b) Prove that the operator

$$\boldsymbol{r} \times \left(\frac{\hbar}{i} \nabla - \frac{q}{c} \boldsymbol{A} \right)$$

for the angular momentum is gauge invariant.

c) Consider an s-state $\psi(\boldsymbol{r}) = u(r)$ in the absence of a magnetic field. A uniform magnetic field with vector potential $\boldsymbol{A}_0 = \frac{1}{2} \boldsymbol{H}_0 \times \boldsymbol{r}$ is applied. Derive an expression for the resulting angular-momentum expectation value, and evaluate the answer in units of $\hbar$ for the ground state of hydrogen, assuming $H_0 = 10,000$ Gauss.

4.2 Calculate the numerical size of the magnetic field produced by the orbital motion of an electron in the $n = 2$, $l = 1$, $m = +1$ state at the nucleus of a hydrogen atom. (Neglect all effects associated with the electron spin.)

4.3 Consider the states $xf(r)$, $yf(r)$, and $zf(r)$ split by a crystal potential $A(x^2 - y^2)$. Let the system be started at time $t = 0$ in a state

$$\psi(0) = \frac{(x + iy)}{\sqrt{2}} f(r) \quad .$$

Compute the expectation value of the z-component of angular momentum, showing that it oscillates in time between the values $+1$ and -1 at the angular frequency of oscillation $A/\hbar$.

Note that your result corresponds to the classical picture that the effect of the crystal field is to cause the plane of the circular orbit to turn in such a way as to reverse the sense of circular motion periodically.

4.4 The hyperfine coupling of s-states may be found simply in the approximation that a nucleus is a uniformly magnetized sphere. In this problem, we derive the famous s-state formula for that model.

A uniformly magnetized sphere of magnetization M per unit volume can be represented by current distribution flowing on the surface, the current being proportional to $M \cdot n$, where n is the unit outer normal. We shall represent a nucleus by such a sphere of radius R. Consider the current to flow in circles about the z-axis; the surface current density $J(\theta)$ is then

$$J(\theta) = J_0 \sin \theta \quad .$$

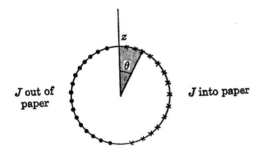

a) Show that the magnetic field H inside the sphere is uniform and that outside the sphere it is a pure dipole field.
b) Show that the field inside is

$$H = \frac{8\pi}{3} J_0 k \quad .$$

c) Show that the magnetic moment μ of the sphere is $(4\pi/3)R^3 J_0$.

d) Show that
$$\overline{H}_z = \int H_z |u^2(r)| d\tau = \frac{8\pi}{3} |u(0)|^2 \mu$$
where $u(r)$ is a spherically symmetric function that does not vary too rapidly within R of the origin and where H_z is the z-component of field due to the sphere.

4.5 An atom has a single valence electron in an s-state and a nucleus of spin I. The electron spin-lattice relaxation time is so short that the nucleus experiences only the time-average magnetic field of the electron. Derive an expression for the resonance frequency of the nucleus when a static field H_0 is applied, giving your answer in terms of the electron susceptibility χ_e. Discuss the temperature and field dependence: (a) at high temperature (where $kT \gg \gamma_e \hbar H_0$) and (b) at low temperature (where $\gamma_e \hbar H_0 \sim kT$).

4.6 In the text the Knight shift was calculated by first-order perturbation theory, using the fact that the static magnetic field H_0 causes a repopulation of the electrons among their spin states. It is possible to obtain an expression for the Knight shift by using second-order perturbation theory and by assuming that the applied field varies spatially in such a manner that there is no repopulation. Thus, suppose the applied field is in the z-direction, varying with x as

$$H_z = H_0 \cos qx \quad .$$

Consider a nucleus at $x = 0$. Using second-order perturbation, show that the electron wave function is perturbed in such a manner that a Knight shift is produced and that, in the limit of $q \approx 0$, the answer agrees with the result found in the text.

4.7 The electronic structure of the hydrogen molecule can be described in terms of the molecular orbital model, using molecular orbitals that are a linear combination of atomic orbitals.[1] The lowest molecular orbital is the bonding one formed by a linear combination of free-hydrogen $1s$-states. Compute an expression for the indirect coupling of the proton spins. As an approximation consider that the only excited state is the antibonding orbital formed from a linear combination of the free-hydrogen $1s$-states.

Chapter 5

5.1 From (5.39) and (5.40) verify that $W_{mn} = W_{nm} \exp\left[(E_m - E_n)\beta_L\right]$, where $\beta_L = 1/kT_L$ and T_L is the lattice temperature.

---

[1] See, for example, H. Eyring, J. Walter, G.E. Kimball: *Quantum Chemistry* (Wiley, New York 1944) Chaps. XI and XII.

5.2 Consider a system of N spins that interact with one another via a dipole-dipole coupling and with an external static field H in the z-direction. Assuming a density matrix ϱ given by

$$\varrho = \frac{e^{-\mathcal{H}/kT}}{Z}$$

corresponding to thermal equilibrium (where Z is the partition function), show that the thermal equilibrium expectation value of the total magnetization is

$$\overline{\langle M_x \rangle} = \overline{\langle M_y \rangle} = 0 \quad , \quad \overline{\langle M_z \rangle} = \frac{N\gamma_n^2 \hbar^2 I(I+1)}{3kT} H$$

in the high-temperature approximation.

It is interesting to note that these equations are of the form $M = CH/T$, Curie's law, and that the constant C does not depend on whether H is large or small compared with the local field due to neighboring dipoles – in contrast to one's naïve expectation.

5.3 Consider in a metal a system of nuclear spins that interact with a dipolar coupling only. By means of (5.49), prove that the spin-lattice relaxation time in zero static field is one-half its value in a strong field ($T_1 = 1/a_{00}$).

5.4 A nucleus of spin $\frac{3}{2}$ has a static Hamiltonian $\mathcal{H}_0 = -\gamma_n \hbar H_0 I_z$. It is acted on by a time-dependent interaction $\mathcal{H}_1(t)$, given by

$$\mathcal{H}_1(t) = A(t)(I_x^2 - I_y^2) \quad \text{quadrupole of}$$

fluctuating electric field gradient where $A(t)$ is a random function of time. Assume that the correlation function of $A(t)$ is

$$\overline{A(t)A(t+\tau)} = \overline{A(t)^2} e^{-|\tau|/\tau_0} \quad .$$

a) Express $\mathcal{H}_1(t)$ in terms of the raising and lower operators, I^+ and I^-.
b) Compute the probability per second of transitions from the $m = \frac{3}{2}$ state to the other three m-states induced by $\mathcal{H}_1(t)$.

5.5 Consider Problem 5.4. Assuming that the relative populations of the m-states always correspond to a spin temperature, compute the spin-lattice relaxation time due to the $\mathcal{H}_1(t)$.

5.6 In Section 5.13, the effect of an alternating field is included in the density matrix formalism.

a) Show that the solutions of (5.398) are correct for low V.
b) Carry out the solution for $\langle M_x(t) \rangle$, assuming large V, thereby obtaining the results for saturation.

5.7 Consider a system with three energy levels 1, 2, and 3. An alternating interaction $V(t) = V \cos \omega t$ is applied nearly at resonance with the transition between states 1 and 2.

a) Write down the differential equations for the density matrix analogous to (5.394) and (5.395).

b) In the limit of negligible saturation, compute $\langle M_x(t) \rangle$ and show that the width of the resonance is affected by the relaxation to level 3. (This is the phenomenon of lifetime broadening due to transitions to a level that is not directly involved in the spectral line.)

Chapter 6

6.1 Consider (6.3). Let $M_x = M_0$, $M_y = M_z = 0$ at $t = 0$. Show that for times less than approximately T_1, M_z and M_y will remain small, so that $\boldsymbol{M}$ will lie along H_1, and will decay to zero exponentially in T_2.

6.2 Show that the expression for the average energy given in Sect. 6.3

$$\overline{E} = \sum_n p_n E_n$$

can be found as a derivative of the partition function Z with respect to β ($\beta \equiv 1/kT$) thus confirming (6.1).

6.3 Show that the average energy $\overline{E}$ of a system of N nuclei acted on by an applied field H_0 and coupled together through a dipolar Hamiltonian $\mathcal{H}_d$ is

$$\overline{E} = -C \frac{(H_0^2 + H_L^2)}{\theta} \quad \text{where}$$

$$CH_L^2 = \frac{1}{k(2I+1)^N} \operatorname{Tr}\{\mathcal{H}_d^2\} \quad \text{and} \quad C = \frac{N\gamma^2 \hbar^2 I(I+1)}{3k} \quad .$$

6.4 The entropy σ is given by

$$\sigma = \frac{\overline{E} + k\theta \ln Z}{\theta} \quad .$$

Evaluate this expression in the high temperature approximation for a system of N nuclei to show that

$$\sigma = Nk \ln(2I+1) - \frac{C}{2} \frac{(H_0^2 + H_L^2)}{\theta^2} \quad .$$

6.5 Consider a Hamiltonian $\mathcal{H}$ which is independent of time. We know, then, that the most general solution is

$$\psi = \sum_n c_n u_n \exp\left(-\frac{i}{\hbar} E_n t\right)$$

where c_n are independent of time, and u_n are the eigenfunctions of the time-independent Schrödinger equation of energy eigenvalue E_n.

Prove that the expectation value of the energy for such a function is independent of time, i.e., that the energy of the system is conserved.

6.6 Derive (6.46)

$$W(\omega) = \frac{\pi}{2} \gamma^2 H_1^2 g(\omega)$$

from the formulas given in Sect. 2.12, using the high temperature approximation.

6.7 In Sect. 6.4, formulas are developed which show that for slow changes of the magnetic field for which the applied field H_i is always much larger than the local field,

$$\frac{H_f}{\theta_f} = \frac{H_i}{\theta_i} \quad \text{and} \quad M_f = M_i \quad .$$

Show that these results can also be obtained by considering what happens to the population of the $2I + 1$ Zeeman levels of the individual spins for an assembly of N noninteracting spins. (The lack of interaction corresponds to setting $H_L = 0$.)

6.8 Consider an adiabatic demagnetization experiment similar to that of Sect. 6.6. Show that the sign of the spin temperature in the rotating frame depends on whether H_1 is switched on when H_0 is above or below resonance.

For the case of negative spin temperature, draw a figure showing $\mathbf{M}$ and $\mathbf{H}_e$ in the x-z plane for several values of $\mathbf{H}_e$ as one approaches resonance.

6.9 Let $H_L = 3$ Gauss, Make a graph of M/M_0 versus h_0 for three cases

i) $H_1 = 1$ Gauss
ii) $H_1 = 3$ Gauss
iii) $H_1 = 9$ Gauss.

Assume that H_1 is switched on when $h_0 \gg H_0$ in each case.

Chapter 7

7.1 Show that (7.59), the exact expression for $\langle I_z \rangle$ corresponding to saturation of a forbidden transition, is equivalent to

$$\langle I_z \rangle = \frac{1}{2} \frac{\exp(A/2kT)[\exp(\gamma_e \hbar H_0/kT) - \exp(-\gamma_e \hbar H_0/kT)]}{2 + \exp(A/2kT)[\exp(\gamma_e \hbar H_0/kT) + \exp(-\gamma_e \hbar H_0/kT)]} \quad .$$

7.2 Show that the nuclear polarization $\langle I_z \rangle$ produced by the scheme of Fig. 7.8 is the negative of that produced by the scheme of Fig. 7.9 if $|A/2| \ll \gamma_e \hbar H_0$.

7.3 Consider the nuclear Overhauser effect for a pair of spin $\frac{1}{2}$ nuclei as in Sect. 7.5.

a) Show that if $(\omega_I \tau_c)^2 \ll 1$ and $(\omega_S \tau_c)^2 \ll 1$ in (7.47), that (7.61) becomes

$$\frac{d}{dt}(\langle I_z \rangle - I_0) = \delta((I_0 - \langle I_z \rangle) + \tfrac{1}{2}(S_0 - \langle S_z \rangle))$$

and

$$\frac{d}{dt}(\langle S_z \rangle - S_0) = \delta(\tfrac{1}{2}(I_0 - \langle I_z \rangle) + (S_0 - \langle S_z \rangle))$$

where $\delta = A_0^2 \tau_c / \hbar^2$.

b) Show that if the initial conditions on $\langle I_z \rangle - I_0$ and $\langle S_z \rangle - S_0$ are

$$\langle I_z(0) \rangle - I_0 = 0 \quad, \quad \langle S_z(0) \rangle - S_0 = S_i \quad \text{that}$$

$$\langle I_z(t) \rangle - I_0 = \tfrac{1}{2} S_i [\exp(-3\delta t/2) - \exp(-\delta t/2)] \quad \text{and}$$

$$\langle S_z(t) \rangle - S_0 = \tfrac{1}{2} S_i [\exp(-3\delta t/2) + \exp(-\delta t/2)] \quad .$$

7.4 Consider a system to be used to make a three-level maser with energies E_1, E_2, and E_3 (E_1 the highest; E_3 the lowest) (see Fig. 7.11). Suppose one saturates the transition between levels 1 and 3.

a) Set up the equations analogous to (7.13) for the case of a steady-state solution.
b) Solve the equations, expressing the populations p_2 and p_3.
c) What relationship between W_{12}, W_{23}, and the energy level spacing is necessary to achieve a population inversion between states 2 and 3? What is the corresponding relationship for a population inversion between states 1 and 2?

7.5 In Sect. 7.13, the double resonance method due to *Hahn* is discussed. Show that the operator T given in (7.91) transforms the Hamiltonian of (7.90) into the Hamiltonian of (7.94) of the double rotating frame.

7.6 *Decoupling.* Consider the spin-Hamiltonian of (7.190) which describes the Royden-Bloom-Shoolery-Bloch method of decoupling:

$$\mathcal{H}' = -\gamma_I \hbar H_0 I_z - \gamma_S \hbar[(h_0)_S S_z + (H_1)_S S_x] + A I_z S_z \quad . \tag{7.190}$$

Since $[I_z, \mathcal{H}'] = 0$, the eigenstates may be characterized by m_I, the eigenvalues of I_z.

a) Show that for a given m_I, the S-spins see an effective magnetic field $(H(m_I))_{\text{eff}}$ given by

$$(H(m_I))_{\text{eff}} = i(H_1)_S + k[(h_0)_S - (Am_I/\gamma_S\hbar)] \quad .$$

b) Draw a scale picture of $(H(m_I))_{\text{eff}}$ for the case $A = \gamma_S\hbar(h_0)_S$, $(h_0)_S = (H_1)_S$ for $m_I = \frac{1}{2}$ and $m_I = -\frac{1}{2}$.

c) Show that the energy eigenvalues of (7.190) are given by

$$E(m_I, m_{S'}) = -\gamma_I\hbar H_0 m_I - \gamma_S\hbar m_{S'} \\ \times \{[(h_0)_S - (Am_I/\gamma_S\hbar)]^2 + (H_1)_S^2\}^{1/2} \quad .$$

d) Explain in terms of the figure of part (b) why there are transitions of the I-spins in which $m_{S'}$ may change. These are called partially forbidden transitions.

e) For the numerical case of part (b), find the frequencies of the allowed and partially forbidden transitions of the I-spins.

7.7 Consider a spin echo double resonance experiment in which the echo of the I-spins is monitored, and in which the S-spins ($S = \frac{1}{2}$) produce a magnetic field $\pm h_{SI}$ on the I-spins. Let the I-spins be observed with a $\pi/2$-π pulse sequence, and the S-spins be flipped with a π pulse.

a) Show that the I echo, $M_I(2\tau)$, varies with τ according to the equation

$$M_I(2\tau) = M_0 \cos(2\gamma_I h_{IS}\tau) \quad .$$

7.8 Consider the energy levels shown in Fig. 7.4. Suppose one uses an adiabatic passage at the electron frequency $(E_1 - E_2)/\hbar$ to interchange the populations of levels 1 and 2 and then quickly observes nuclear resonance.

a) Show that the nuclear resonance transition $(E_2 - E_4)/\hbar$ has an increased absorption rate and compute the ratio of the rate of energy absorption to its normal value.

b) Show that the nuclear resonance transition $(E_1 - E_3)/\hbar$ has a stimulated emission, and compute the ratio of the rate of energy emission to the normal rate of energy absorption of this transition.

7.9 The density matrix of a pair of spins I and S in thermal equilibrium at temperature θ is, in the high temperature approximation,

$$\varrho(\theta) = \frac{1}{Z}\left(1 + \frac{\hbar H_0}{k\theta}(\gamma_I I_z + \gamma_S S_z)\right) \quad .$$

Consider the case of an $X_I(\pi/2)$ pulse. Then, the time-dependent expectation value of I^+ is

$$\langle I^+(t)\rangle = \text{Tr}\{I^+ \varrho(t)\}$$
$$= \text{Tr}\{I^+ \exp(i(\Omega_I I_z + \Omega_S S_z - aI_z S_z)t) X_I \varrho(\theta)$$
$$\times X_I^{-1} \exp(-i(\Omega_I I_z + \Omega_S S_z - aI_z S_z)t)\} \quad .$$

Show that the term in $\langle I^+(t)\rangle$ arising from S_z in $\varrho(\theta)$ can be written as $(\gamma_S \hbar H_0/Zk\theta) \times \text{Tr}\{I^+ S_z\}$, which vanishes.

7.10 Using the relationships for spin $\frac{1}{2}$ particles:

$$e^{iS_z\theta} = \cos(S_z\theta) + i\sin(S_z\theta)$$

$$\cos(S_z\theta) = \cos(\theta/2)$$

$$\sin(S_z\theta) = 2S_z \sin(\theta/2)$$

$$S_x S_y = \frac{i}{2} S_z \quad , \quad S_y S_x = -\frac{i}{2} S_z \quad , \quad \text{etc.}$$

Show that

$$e^{iS_z\theta} S_x e^{-iS_z\theta} = S_x \cos\theta - S_y \sin\theta \quad .$$

Chapter 8

8.1 Suppose that a spin echo is produced by applying the H_1 of the first pulse along the $+x$-axis in the rotating frame, rotating the magnetization to the $-y$-axis, and a π pulse a time τ later with H_1 along the $+y$-axis.

a) Show that an echo is formed at 2τ along the $-y$-axis.
b) Show that if H_1, when on, is produced by applying a linearly polarized alternating field in the sample coil along the laboratory frame x-axis

$$H_X(t) = 2H_1 \cos\omega t \quad , \quad t < \tau \quad ,$$

the situation described above would be produced by changing $H_X(t)$ to

$$H_X(t) = 2H_1 \cos\left(\omega t - \frac{\pi}{2}\right) \quad \text{for} \quad t > \tau \quad .$$

8.2 Draw vector diagrams for the rotating frame to prove that Fig. 8.1 correctly shows the echoes in a Carr-Purcell pulse sequence.

8.3 Apply the vector model and an argument similar to that in (8.12–20) to show that a pulse sequence

$$X(\pi/2)\ldots\tau\ldots X(\pi)\ldots\tau\ldots$$

will not produce an echo for a system of two identical spin $\frac{1}{2}$ particles coupled by a Hamiltonian

$$\mathcal{H} = AI_zS_z \quad .$$

8.4 In order to grasp the meaning of the various spin-flip narrowing pulse sequences, one can begin by assuming the dipolar broadening is negligible, that there is negligible magnetic field inhomogeneity, and follow what happens to the magnetization. Employ this method to the three-pulse cycle of Sect. 8.12 to verify the results of Fig. 8.14.

8.5 Employ the method of Problem 8.4 to describe the magnetization vector in the rotating frame through the first cycle for the four-pulse sequence $(\tau, \overline{X}, \tau, Y, 2\tau, \overline{Y}, \tau, X, \tau)$.

Chapter 10

10.1 It is stated in Sect. 10.3 that (10.27b) and (10.31) provide a set of recursion relations among the elements $(JM_J\eta|T_{LM}|J'M_{J'}\eta')$ for the various possible values of M_J, M', and $M_{J'}$, and a fixed set J, L, J', η, η'. For the case that $J = J'$ show that it is indeed true that specifying one matrix element (for example, that for which $M_J = M_{J'} = J$) enables all others to be computed by using the recursion relations.

10.2 Verify that the functions $T_{2M}(\boldsymbol{J})$ of Table 10.1 in Sect. 10.3 satisfy the commutation relations of a T_{2M} with respect to $\boldsymbol{J}$.

10.3 Consider an axially symmetric potential and a weak static field. The Hamiltonian is then

$$\mathcal{H} = \frac{e^2qQ}{4I(2I-1)}(3I_z^2 - I^2) - \gamma_n\hbar H I_{z'} \quad .$$

If $H = 0$, the spins are quantized by the quadrupole coupling as shown in Fig. 10.4. The states $m = \pm\frac{1}{2}$ are degenerate. Show that when H is weak, these states are split, the energy difference going from $\gamma_n\hbar H$ when z' is parallel to z to $(I + \frac{1}{2})\gamma_N\hbar H$ when z' is perpendicular to z.

10.4 Consider the Hamiltonian and energies given by (10.65) and (10.66). An alternating field $H_x \cos\omega t$ is applied perpendicular to the z-axis. Find the allowed transitions, their frequencies, and the relative intensities.
Work out numerical answers for the cases $I = \frac{3}{2}$ and $\frac{5}{2}$.

10.5 Show that a charge e located at a point x_0, y_0, z_0 produces a field gradient

$$V_{zz} \equiv \left(\frac{\partial^2 V}{\partial z^2}\right)_{x,y,z=0} \quad \text{of} \quad V_{zz} = e\frac{3z_0^2 - r_0^2}{r_0^5}.$$

10.6 A nucleus of spin $\frac{5}{2}$ experiences an electrical quadrupole coupling

$$\mathcal{H} = A(I_x^2 - I_y^2).$$

a) Show that the energy eigenvalues are 0, $\pm 2\sqrt{7}A$.
b) Show that the eigenfunctions are

$$E = 0 \begin{cases} \psi_1 = \left(\frac{9}{14}\right)^{1/2}\left[\phi_{5/2} - \left(\frac{5}{9}\right)^{1/2}\phi_{-3/2}\right] \\ \psi_2 = \left(\frac{9}{14}\right)^{1/2}\left[\phi_{-5/2} - \left(\frac{5}{9}\right)^{1/2}\phi_{3/2}\right] \end{cases}$$

$$E = 2(7)^{1/2}A \begin{cases} \psi_3 = \left(\frac{5}{28}\right)^{1/2}\left[\phi_{5/2} + \left(\frac{14}{5}\right)^{1/2}\phi_{1/2} + \left(\frac{9}{5}\right)^{1/2}\phi_{-3/2}\right] \\ \psi_4 = \left(\frac{5}{28}\right)^{1/2}\left[\phi_{-5/2} + \left(\frac{14}{5}\right)^{1/2}\phi_{-1/2} + \left(\frac{9}{5}\right)^{1/2}\phi_{3/2}\right] \end{cases}$$

$$E = -2(7)^{1/2}A \begin{cases} \psi_5 = \left(\frac{5}{28}\right)^{1/2}\left[\phi_{5/2} - \left(\frac{14}{5}\right)^{1/2}\phi_{1/2} + \left(\frac{9}{5}\right)^{1/2}\phi_{-3/2}\right] \\ \psi_6 = \left(\frac{5}{28}\right)^{1/2}\left[\phi_{-5/2} - \left(\frac{14}{5}\right)^{1/2}\phi_{-1/2} + \left(\frac{9}{5}\right)^{1/2}\phi_{3/2}\right] \end{cases}$$

where $\phi_{5/2}$ is an eigenfunction of I_z with $m = \frac{5}{2}$, and so on.

c) Show that application of a *small* static field H_0 in the z'-direction splits the degenerate $E = 0$ states, the splitting being $\gamma_n \hbar H_0(\frac{15}{7})$ independent of the orientation of z' with respect to x, y, or z.

10.7 Prove that the eigenvalues E_1, E_2, and so on of a Hamiltonian $\mathcal{H} = A(I_x^2 - I_y^2)$ come in pairs $\pm E_1$, $\pm E_2$, or else are zero. (*Hint:* Consider the effect of an operator R that changes x into y and y into $-x$).

Chapter 11

11.1 Evaluate the coefficient C_J, defined by (11.41), to obtain an answer analogous to (11.53a).

11.2 In the notation of Sect. 11.2 prove that, with respect to L_x, L_y, and L_z, the function $x^2 - y^2$ is a linear combination of T_{2M}'s.

11.3 Equation (11.65) says

$$\mathcal{H}_{IS} = \gamma_e \gamma_n \hbar^2 \left(\overline{\frac{1}{r^3}}\right) \frac{2}{5}(3I_z S_z - \mathbf{I} \cdot \mathbf{S}) \quad .$$

Show that this follows from (11.64):

$$\mathcal{H}_{IS} = \gamma_e \gamma_n \hbar^2 \int \frac{1}{r^3} \left[3\frac{(\mathbf{I} \cdot \mathbf{r})(\mathbf{S} \cdot \mathbf{r})}{r^2} - \mathbf{I} \cdot \mathbf{S}\right] x^2 f^2(r) d\tau \quad .$$

11.4 In Sect. 11.3 the hyperfine splitting was worked out for the case of an isotropic electron g-factor. Generalize the result to the case of a Hamiltonian for a field with x- and z-components only:

$$\mathcal{H} = \beta(g_{xx}H_x + g_{zz}H_z) + A_x I_x S_x + A_y I_y S_y + A_z I_z S_z$$
$$- \gamma_n \hbar (H_x I_x + H_z I_z) \quad .$$

11.5 Consider matrix elements $(\alpha|L_{\alpha'}|x)$, where $\alpha, \alpha' = x, y, z$, which are frequently encountered in problems for which $|x)$ is the ground electronic state.

Show that there are only two which are nonzero ($\alpha = y$, $\alpha' = z$ and $\alpha = z$, $\alpha' = y$) and evaluate the matrix elements for them. These results are useful for the following problems.

11.6 Consider an atom containing a single p-electron acted on by a crystal field giving an extra potential energy $A(x^2 - y^2)$. The resultant energy levels are shown in the figure (*neglect spin*). Assume $\Delta \gg kT$. Compute the principal values of the temperature independent susceptibility tensor.

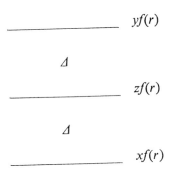

11.7 Consider the atom of Problem 11.6, but now include the spin. Let there be a spin-orbit interaction $\lambda(r)\mathbf{L} \cdot \mathbf{S}$ added to the Hamiltonian. Take as basis states the states $|\alpha m_S\rangle$ where $\alpha = x, y, z$ and $m_S = \pm\frac{1}{2}$, the spin quantum number.

a) Show that the matrix elements of the spin-orbit coupling $(xs|\lambda(r)\mathbf{L} \cdot \mathbf{S}|xs')$ which are internal to the ground orbital state vanish.

b) Using perturbation theory, show that the spin-orbit coupling mixes excited states into the original ground state wave function $|xs\rangle$ and find the corrected wave functions $|xs\rangle_{\text{corr}}$ correct to terms linear in λ.

c) Assuming an applied magnetic field along the z-direction, compute the total orbital plus spin Zeeman energy by first order perturbation theory,

  i) using functions $|xs\rangle$ and
  ii) using the corrected functions of (b).

d) Show that the Zeeman splitting of (c)–(i) is $\Delta E = 2\beta H$ but for (c)–(ii) it is of the form $\Delta E = g\beta H$, and find the expression for g correct to terms linear in λ.

11.8 An atom containing a single p-electron located at the origin is acted on by a set of 6 equal charges, q. The charges on the x- and y-axes are a distance a from the origin, those on the z-axis a distance b. Assume $b < a$. Neglect spin.

a) Show that the lowest nonvanishing term in the Hamiltonian which describes splitting of the p-states is $\mathcal{H} = A(3z^2 - r^2)$ and determine the sign of A.

b) Find the proper eigenstates and energies in terms of A and $\langle r^2 \rangle$, the mean square radius of the p-state orbit.

c) Suppose a magnetic field is applied along the z-direction; compute the 3×3 Hamiltonian matrix.

d) Find the eigenstates which resolve the degeneracy.

e) For which states is the angular momentum quenched?

11.9 Consider the system of Problem 11.8. Assume that the splitting of the p-states produced by the crystal field is large compared to kT, but that $\beta H \ll kT$.

a) Show that when $b > a$ the crystal splitting of the p-states has the opposite sign from its value for $b < a$.

b) Curie's law states that the magnetic susceptibility tensor per atom $\chi_{\alpha\alpha'}$ ($\alpha, \alpha' = x, y, z$) goes inversely with temperature

$$\chi_{\alpha\alpha'} = \frac{C_{\alpha\alpha'}}{T} \quad .$$

Show that for one sign of the crystal field Curie's law applies, show that for that sign the x-, y-, and z-axes are the principal axes, and evaluate the corresponding C_{xx}, C_{yy}, and C_{zz}.

11.10 Consider an atom with a single p-electron, the orbital angular momentum being quenched by a crystalline field such as that of Fig. 11.1.

By using second-order perturbation, show that interplay of the spin-orbit coupling $\lambda \mathbf{L} \cdot \mathbf{S}$ and the coupling $(e/2mc)(\mathbf{p} \cdot \mathbf{A}_n + \mathbf{A}_n \cdot \mathbf{p})$ between the nuclear moment and the electron orbit give an effective spin-spin coupling between the nucleus and electron.

11.11 Consider a single p-electron whose spin and orbit are strongly coupled so that the JM_J scheme of quantization applies. The nuclear moment and electron orbital motion are coupled by a Hamiltonian

$$\mathcal{H} = \frac{e}{2mc}(\boldsymbol{p}\cdot\boldsymbol{A}_n + \boldsymbol{A}_n\cdot\boldsymbol{p}) \tag{1}$$

where $\boldsymbol{A}_n = \gamma_n\hbar\boldsymbol{I}\times\boldsymbol{r}/r^3$ is the vector potential due to the nucleus.

a) Using the Wigner-Eckart theorem, show that for matrix elements diagonal in the electron quantum number, J, (1) is equivalent to an effective Hamiltonian

$$\mathcal{H}_{\text{eff}} = A_J\,\boldsymbol{J}\cdot\boldsymbol{I}$$

where A_J is a constant for a given J, independent of M_J.

b) Find A_J for the $J = \frac{3}{2}$ state.

Appendixes

A. A Theorem About Exponential Operators

We wish to prove a theorem about the exponential function of two operators, A and B, and their commutator C:

$$C \equiv [A, B] \; . \tag{A.1}$$

The theorem states that when both A and B commute with C, then

$$e^{A+B} = e^A e^B e^{-C/2} \quad \text{or} \tag{A.2}$$

$$e^{A+B} = e^B e^A e^{C/2} \; . \tag{A.3}$$

We shall prove (A.2).

The problem is most readily solved by considering the function $\exp[\lambda(A + B)]$. We seek the function $G(\lambda)$ such that

$$e^{\lambda(A+B)} = e^{\lambda A} e^{\lambda B} G(\lambda) \; . \tag{A.4}$$

To find $G(\lambda)$, we seek a differential equation in λ which it satisfies. In essence this amounts to finding the way in which the function $\exp[\lambda(A + B)]$ changes for small changes in λ, and then integrating from $\lambda = 0$ to $\lambda = 1$.

By taking the derivative of both sides of (A.4), we get

$$(A + B)e^{\lambda(A+B)} = e^{\lambda A}(A + B)e^{\lambda B} G(\lambda) + e^{\lambda A} e^{\lambda B} \frac{dG}{d\lambda} \; . \tag{A.5}$$

By utilizing (A.4) and multiplying from the left by $\exp(-\lambda B)\exp(-\lambda A)$, we can rewrite (A.5) as

$$e^{-\lambda B} e^{-\lambda A} B e^{\lambda A} e^{\lambda B} G - BG = \frac{dG}{d\lambda} \; . \tag{A.6}$$

We can evaluate the expression

$$e^{-\lambda A} B e^{\lambda A} \equiv R(\lambda) \tag{A.7}$$

as follows: by taking the derivative of both sides of (A.6) with respect to λ, we find

$$e^{-\lambda A}(BA - AB)e^{\lambda A} = \frac{dR}{d\lambda}$$
$$- C = \tag{A.8}$$

579

since $AB - BA \equiv C$ commutes with A.

Integrating (A.8) we have

$$R(\lambda) = -C\lambda + \text{constant} \quad . \tag{A.9}$$

We can evaluate the constant by setting $\lambda = 0$ and by noting from (A.7) that $R(0) = B$. Therefore

$$R(\lambda) = -C\lambda + B \quad . \tag{A.10}$$

By substituting (A.10) into (A.6) and using the fact that C commutes with B, we get

$$-\lambda C G = \frac{dG}{d\lambda} \tag{A.11}$$

which can be interpreted to give

$$G = \exp[-(\lambda^2 C/2 + \text{const})] \quad . \tag{A.12}$$

The constant must be zero because, from (A.4), $G(0) = 1$.

Therefore

$$e^{A+B} = e^A e^B e^{-C/2} \quad . \quad \text{Q.E.D.} \tag{A.13}$$

B. Some Further Expressions for the Susceptibility

(This appendix requires familiarity with Chaps. 2, 3, and 5.)

Equation (2.190) gives an expression for χ''. Another expression is frequently encountered in the literature. It provides an alternative derivation for the moments of the shape function. It can be obtained from (2.190),

$$\chi'' = \frac{\hbar\omega\pi}{kTZ} \sum_{a,b} e^{-E_a/kT} |(a|\mu_x|b)|^2 \delta(E_a - E_b - \hbar\omega) \tag{B.1}$$

by use of the integral representation of the δ-function:

$$\delta(x) = \frac{1}{2\pi} \int_{-\infty}^{+\infty} e^{-ix\tau} d\tau \quad . \tag{B.2}$$

By substituting into (B.1), we obtain

$$\chi''(\omega) = \frac{\hbar\omega}{2kTZ} \int_{-\infty}^{+\infty} \sum_{E_a, E_b} e^{-E_a/kT}$$
$$\times (a|\mu_x|b)(b|\mu_x|a) \exp[i(E_a - E_b - \hbar\omega)\tau] d\tau \quad , \tag{B.3}$$

and substituting for the variable τ a new variable t that has the dimensions of time,

$$\frac{t}{\hbar} = \tau \tag{B.4}$$

we get

$$\chi''(\omega) = \frac{\omega}{2kTZ} \int_{-\infty}^{+\infty} \sum_{a,b} e^{-E_a/kT}$$
$$\times (a|\mu_x|b)(b|\mu_x|a)\exp\{[i(E_a - E_b)t]/\hbar\}\exp(-i\omega t)dt \quad . \tag{B.5}$$

We can use the fact that the states $|a\rangle$ and $|b\rangle$ are eigenfunctions of the Hamiltonian $\mathcal{H}$ to express the expression more compactly as

$$\chi''(\omega) = \frac{\omega}{2kTZ} \int_{-\infty}^{+\infty} \sum_{a,b} (a|e^{-\mathcal{H}/kT}e^{i\mathcal{H}t/\hbar}\mu_x e^{-i\mathcal{H}t/\hbar}|b)$$
$$\times (b|\mu_x|a)e^{-i\omega t}dt \quad . \tag{B.6}$$

But the summation over a and b is clearly just a trace, so that

$$\chi''(\omega) = \frac{\omega}{2kTZ} \int_{-\infty}^{+\infty} \text{Tr}\{e^{-\mathcal{H}/kT}e^{i\mathcal{H}t/\hbar}\mu_x e^{-i\mathcal{H}t/\hbar}\mu_x\}e^{-i\omega t}dt \quad . \tag{B.7}$$

In the high-temperature approximation, we replace $\exp(-\mathcal{H}/kT)$ by unity. If we then define the operator $\mu_x(t)$ by

$$\mu_x(t) = e^{i\mathcal{H}t/\hbar}\mu_x e^{-i\mathcal{H}t/\hbar} \tag{B.8}$$

we can also express (B.7) as

$$\chi''(\omega) = \frac{\omega}{2kTZ} \int_{-\infty}^{+\infty} \text{Tr}\{\mu_x(t)\mu_x\}e^{-i\omega t}dt \quad . \tag{B.9}$$

The quantity $\text{Tr}\{\mu_x(t)\mu_x\}$ is a form of correlation function, and (B.9) states that $\chi''(\omega)$ is given by the Fourier transform of that correlation function.

By using this expression for $\chi''(\omega)$, it is easy to show that omission of the dipolar terms C, D, E, and F of (3.7) gives one absorption at the Larmor frequency only, but that their inclusion gives absorption at 0 and $2\omega_0$.

We get also a very compact expression for the shape function $f(\omega)$:

$$f(\omega) = \frac{\chi''(\omega)}{\omega} = \frac{1}{2kTZ} \int_{-\infty}^{+\infty} \text{Tr}\{\mu_x(t)\mu_x\}e^{-i\omega t}dt \quad . \tag{B.10}$$

We can prove another interesting theorem by taking Fourier transform of (B.10):

$$\frac{1}{2kTZ}\text{Tr}\{\mu_x(t)\mu_x\} = \frac{1}{2\pi} \int_{-\infty}^{+\infty} f(\omega)e^{i\omega t}d\omega \quad . \tag{B.11}$$

We see that, setting $t = 0$,

$$\frac{1}{2kTZ}\text{Tr}\{\mu_x(0)\mu_x\} = \frac{1}{2\pi}\int_{-\infty}^{+\infty} f(\omega)d\omega \quad . \tag{B.12}$$

By taking the nth derivative of (B.11) with respect to t, and evaluating at $t = 0$, we find

$$\frac{1}{2kTZ}\frac{d^n}{dt^n}\text{Tr}\{\mu_x(t)\mu_x\}\bigg|_{t=0} = \frac{(i)^n}{2\pi}\int_{-\infty}^{+\infty} \omega^n f(\omega)d\omega \quad . \tag{B.13}$$

We get, therefore, a compact expression for the nth moment of the shape function $f(\omega)$:

$$\langle \omega^n \rangle = \frac{\int_{-\infty}^{+\infty} \omega^n f(\omega)d\omega}{\int_{-\infty}^{+\infty} f(\omega)d\omega} = \frac{(i)^{-n}(d^n/dt^n)\text{Tr}\{\mu_x(t)\mu_x\}|_{t=0}}{\text{Tr}\{\mu_x(0)\mu_x\}} \quad . \tag{B.14}$$

As an illustration let us derive an expression for the second moment $\langle \omega^2 \rangle$. Taking the derivative of $\mu_x(t)$ gives

$$\frac{d^2}{dt^2}\text{Tr}\{e^{i\mathcal{H}t/\hbar}\mu_x e^{-i\mathcal{H}t/\hbar}\mu_x\} = \left(\frac{i}{\hbar}\right)^2 \text{Tr}\{e^{i\mathcal{H}t/\hbar}[\mathcal{H},[\mathcal{H},\mu_x]]e^{i\mathcal{H}t/\hbar}\mu_x\}$$

$$= -\frac{1}{\hbar^2}\text{Tr}\{e^{i\mathcal{H}t/\hbar}[\mathcal{H},\mu_x]e^{-i\mathcal{H}t/\hbar}[\mathcal{H},\mu_x]\} \quad . \tag{B.15}$$

Therefore

$$\langle \omega^2 \rangle = -\frac{1}{\hbar^2}\frac{\text{Tr}\{[\mathcal{H},\mu_x]^2\}}{\text{Tr}\{\mu_x^2\}} \quad . \tag{B.16}$$

This formalism provides a very simple way of generating expressions for the higher moments. Note, however, that the odd moments all vanish, since $f(\omega)$ is an even function of ω.

So far, apart from assuming the high-temperature approximation, we have left the specification of the Hamiltonian completely general. We can proceed further if we assume that it consists of the sum of a Zeeman term $\mathcal{H}_Z$ and a term $\mathcal{H}_p$, often a perturbation, which commutes with $\mathcal{H}_Z$. A typical $\mathcal{H}_p$ is the terms A and B of the dipolar coupling. Then, since $\mathcal{H}_p$ and $\mathcal{H}_Z$ commute,

$$\begin{aligned}\mu_x(t) &= e^{(i/\hbar)(\mathcal{H}_Z+\mathcal{H}_p)t}\mu_x e^{-(i/\hbar)(\mathcal{H}_Z+\mathcal{H}_p)t}\\ &= e^{(i/\hbar)\mathcal{H}_p t}e^{-i\omega_0 I_z t}\mu_x e^{\omega_0 I_z t}e^{-(i/\hbar)\mathcal{H}_p t}\\ &= e^{+(i/\hbar)\mathcal{H}_p t}(\mu_x \cos\omega_0 t + \mu_y \sin\omega_0 t)e^{-(i/\hbar)\mathcal{H}_p t}\end{aligned} \tag{B.17}$$

where we have used (2.55). We have, then,

$$\begin{aligned}\text{Tr}\{\mu_x(t)\mu_x\} &= \cos\omega_0 t\,\text{Tr}\{e^{(i/\hbar)\mathcal{H}_p t}\mu_x e^{-(i/\hbar)\mathcal{H}_p t}\mu_x\}\\ &+ \sin\omega_0 t\,\text{Tr}\{e^{(i/\hbar)\mathcal{H}_p t}\mu_y e^{-(i/\hbar)\mathcal{H}_p t}\mu_x\} \quad .\end{aligned} \tag{B.18}$$

In general, if $\mathcal{H}_\text{p}$ is invariant under a rotation of 180° about the x- or y-axes (as is usually the case), the second term vanishes. This can be shown by evaluating the trace, using coordinates $x' = x$, $y' = -y$, $z' = -z$, which differ only by a 180° rotation about x. Then $\mathcal{H}_\text{p} = \mathcal{H}'_\text{p}$ by our postulate, so that

$$\text{Tr}\{e^{i(\mathcal{H}_\text{p}/\hbar)t}\mu_y e^{-i(\mathcal{H}_\text{p}/\hbar)t}\mu_x\}$$
$$= \text{Tr}\{e^{i(\mathcal{H}'_\text{p}/\hbar)t}(-\mu_{y'})e^{-i(\mathcal{H}'_\text{p}/\hbar)t}\mu_{x'}\}$$
$$= -\text{Tr}\{e^{i(\mathcal{H}'_\text{p}/\hbar)t}\mu_{y'}e^{(i/\hbar)\mathcal{H}'_\text{p}t}\mu_{x'}\} \quad . \tag{B.19}$$

But the last trace is clearly the same as the first. Therefore the trace is equal to its own negative and must vanish.

We have, then, that the correlation function $\text{Tr}\{\mu_x(t)\mu_x\}$ is given as

$$\text{Tr}\{\mu_x(t)\mu_x\} = \cos\omega_0 t\,\text{Tr}\{e^{i(\mathcal{H}_\text{p}/\hbar)t}\mu_x e^{-(i/\hbar)\mathcal{H}_\text{p}t}\mu_x\} \quad . \tag{B.20}$$

Since this is the Fourier transform of the shape function $f(\omega)$, we see that the transient behavior consists of a term $\cos\omega_0 t$ multiplied by an envelope function.

If we define $\mu_x^*(t)$ as

$$\mu_x^*(t) = e^{(i/\hbar)\mathcal{H}_\text{p}t}\mu_x e^{-(i/\hbar)\mathcal{H}_\text{p}t} \tag{B.21}$$

we can say that the envelope function is $\text{Tr}\{\mu_x^*(t)\mu_x\}$.

By writing the $\cos\omega_0 t$ as

$$\cos\omega_0 t = \tfrac{1}{2}(e^{i\omega_0 t} + e^{-i\omega_0 t}) \tag{B.22}$$

we can say that the two exponentials correspond to lines at $+\omega_0$ and $-\omega_0$. If we wish to discuss only the line at $+\omega_0$, $f_+(\omega)$, we can therefore write

$$f_+(\omega) = \frac{1}{4kTZ}\int_{-\infty}^{+\infty}\text{Tr}\{\mu_x^*(t)\mu_x\}e^{+i\omega_0 t}e^{-i\omega t}dt \tag{B.23}$$

and obtain its transform:

$$\frac{1}{4kTZ}e^{+i\omega_0 t}\text{Tr}\{\mu_x^*(t)\mu_x\} = \frac{1}{2\pi}\int_{-\infty}^{+\infty}f_+(\omega)e^{+i\omega t}d\omega \quad . \tag{B.24}$$

This can be rewritten as

$$\frac{1}{4kTZ}\text{Tr}\{\mu_x^*(t)\mu_x\} = \frac{1}{2\pi}\int_{-\infty}^{+\infty}f_+(\omega)e^{i(\omega-\omega_0)t}d\omega \quad . \tag{B.25}$$

By taking derivatives as before, we now get

$$\frac{(d^n/dt^n)\text{Tr}\{[\mu_x^*(t)\mu_x]_{t=0}\}}{\text{Tr}\{\mu_x^*(0)\mu_x\}} = (i^n)\frac{\int_{-\infty}^{+\infty}(\omega-\omega_0)f_+(\omega)d\omega}{\int_{-\infty}^{+\infty}f_+(\omega)d\omega} = i^n\langle(\omega-\omega_0)^n\rangle. \tag{B.26}$$

This, then, gives the nth moment with respect to the frequency ω_0. This formalism has supressed the line at $-\omega_0$, which would, of course, make an inordinately large contribution to $\langle(\omega - \omega_0)^n\rangle$ were it included!

By following steps similar to those of (B.15), we now find

$$\langle(\omega - \omega_0)^2\rangle = \frac{1}{\hbar^2} \frac{\text{Tr}\{[\mathcal{H}_p, \mu_x]^2\}}{\text{Tr}\{\mu_x^2\}} . \tag{B.27}$$

We also see that

$$\langle\omega - \omega_0\rangle = \frac{1}{\hbar}\text{Tr}\{[\mathcal{H}_p, \mu_x]\mu_x\} , \tag{B.28}$$

which can be shown to vanish if $\mathcal{H}_p$ consists of the dipolar terms A and B, as shown in Chapter 3.

C. Derivation of the Correlation Function for a Field That Jumps Randomly Between $\pm h_0$

We shall assume the field jumps randomly between the two values $\pm h_0$, which we shall label as states 1 and 2. We shall call

$$+h_0 = H_1 , \quad -h_0 = H_2 . \tag{C.1}$$

Then we wish to know the correlation function $G(\tau)$:

$$G(\tau) = \overline{H(t)H(t+\tau)} \tag{C.2}$$

where the bar indicates an ensemble average.

If the field is H_1 at time $t = 0$, then we can write for a single member of the ensemble:

$$H(0)H(\tau) = H_1[P_1(\tau)H_1 + P_2(\tau)H_2] \tag{C.3}$$

where $P_1(\tau)$ and $P_2(\tau)$ are either zero or one, depending on whether at time τ the field is H_1 or H_2. We now perform an ensemble average of (C.3) over the various histories. This replaces quantities $P_1(\tau)$, $P_2(\tau)$ by their ensemble averages $p_1(\tau)$ and $p_2(\tau)$, which are the *probabilities* that in an ensemble in which the field was H_1 at $\tau = 0$, it will be H_1 or H_2 at time τ.

Thus we have

$$\overline{H(0)H(\tau)} = H_1[H_1 p_1(\tau) + H_2 p_2(\tau)]. \tag{C.4}$$

This equation, of course, assumes that at $\tau = 0$, $H(\tau) = H_1$, so that as $\tau \to 0$, $p_1(\tau) \to 1$, $p_2(\tau) \to 0$. Equally likely is the situation that the field is H_2 at $\tau = 0$, which will give a similar equation except that 1 and 2 are interchanged.

We shall assume the behavior of p_1 and p_2 as a function of τ to be given by a rate equation:

$$\frac{dp_1}{d\tau} = W(p_2 - p_1) \quad , \quad \frac{dp_2}{d\tau} = W(p_1 - p_2) \quad . \tag{C.5}$$

This is a "normal modes" problem, with solutions obtained by adding or subtracting:

$$\begin{aligned} p_1(\tau) + p_2(\tau) &= \text{const.} (= 1 \text{ from normalization}) \\ p_1(\tau) - p_2(\tau) &= Ce^{-2W\tau} \end{aligned} \tag{C.6}$$

where $C = p_1(0) - p_2(0) = p_1(0)$. Since $p_2(0)$ vanishes and since $p_1(0) = 1$, $C = 1$.

By making use of (C.1), (C.4), and (C.6), we have

$$\overline{H(0)H(\tau)} = H_1[H_1 p_1(\tau) + p_2(\tau)H_2] = h_0^2 e^{-2W\tau} \quad . \tag{C.7}$$

An identical answer is found for $H(0)H(\tau)$ if the field is assumed to be H_2 at $\tau = 0$. We must weigh these equally (that is, average the answers over the *initial* fields) to get the final ensemble average. We denote this by a double bar, to indicate the fact that we have averaged over an ensemble of initial conditions as well as a variety of histories for a given initial condition:

$$\overline{\overline{H(0)H(\tau)}} = h_0^2 e^{-2W\tau} = G(\tau) \quad .$$

This is the correlation time assumed in Chapter 5, with $2W \cong 1/\tau_0$.

D. A Theorem from Perturbation Theory

In this appendix we shall derive from perturbation theory a theorem that has wide utility in magnetic resonance. It is closely related to second-order perturbation theory but gives the results in a form particularly useful when there is degeneracy. A typical situation in which the theorem has great use is illustrated by the g-shift calculation of Section 11.2. We may divide the Hamiltonian into three terms;

$$\mathcal{H} = \mathcal{H}_0 + \mathcal{H}_1 + \mathcal{H}_2 \quad \text{where} \tag{D.1}$$

$$\begin{aligned} \mathcal{H}_0 &= \frac{p^2}{2m} + V_0 + V_1 \\ \mathcal{H}_1 &= 2\beta \boldsymbol{H} \cdot \boldsymbol{S} \\ \mathcal{H}_2 &= \lambda \boldsymbol{L} \cdot \boldsymbol{S} + \beta \boldsymbol{H} \cdot \boldsymbol{L} \quad . \end{aligned} \tag{D.2}$$

Since $\mathcal{H}_0$ does not depend on spin, its eigenstates may be taken as products of an orbital and a spin function. We denote the orbital quantum numbers by l and the spin quantum numbers by α. Then

$$\mathcal{H}_0|l\alpha\rangle = E_l|l\alpha\rangle \quad . \tag{D.3}$$

The states $|l\alpha\rangle$ are degenerate for a given l because of the spin quantum numbers.

The term $\mathcal{H}_1$ lifts the spin degeneracy. Since it depends on spin only, it has no matrix elements between different orbital states:

$$(l\alpha|\mathcal{H}_1|l'\alpha') = \delta_{ll'}(l\alpha|\mathcal{H}_1|l\alpha') \tag{D.4}$$

In general the matrix elements of $\mathcal{H}_1$ between states $|l\alpha)$ and $|l\alpha')$ where $\alpha \neq \alpha'$ will be nonzero. Therefore the presence of $\mathcal{H}_1$ still leaves us a group of submatrices $(l\alpha|\mathcal{H}_1|l\alpha')$ to diagonalize. For our example, since the spin was $\frac{1}{2}$, these submatrices are only 2×2 and are easily handled.

The presence of the term $\mathcal{H}_2$ spoils things, since $\mathcal{H}_2$ joins states of different l. However, as a result of the quenching of the orbital angular momentum, the matrix elements of $\mathcal{H}_2$ that are diagonal in l vanish:

$$(l\alpha|\mathcal{H}_2|l\alpha') = 0 \quad . \tag{D.5}$$

We may schematize things as shown in Fig. D.1, where the Hamiltonian matrix is illustrated and where we have labeled which terms $\mathcal{H}_1$ or $\mathcal{H}_2$ have nonvanishing matrix elements.

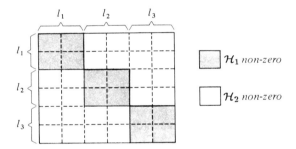

Fig. D.1. Hamiltonian matrix. The regions in which nonvanishing elements of $\mathcal{H}_1$ or $\mathcal{H}_2$ may be found are labeled by the shading. The quantum numbers l_1, l_2 and l_3 designate different eigenvalues of $\mathcal{H}_0$

The technique that we shall describe below in essence provides a transformation which reduces the size of the matrix elements of $\mathcal{H}_2$ joining states of different l. In the process, new elements are added which are diagonal in l. In this way states of different l are, so to speak, uncoupled, and we are once again faced with diagonalizing only the smaller submatrices diagonal in l.

The basic technique may be thought of formally as follows. The set of basis functions $\psi_{l\alpha}$ forms a complete set, but has the troublesome $\mathcal{H}_2$ matrix elements between states of different l. We seek a transformed set of functions $\phi_{l\alpha}$ given by

$$\phi_{l\alpha} = e^{iS}\psi_{l\alpha} \tag{D.6}$$

where S is a Hermitian operator that reduces the size of the troublesome matrix elements. In terms of the ϕ's, the Hamiltonian matrix elements are

$$\int \phi_{l\alpha}^* \mathcal{H} \phi_{l'\alpha'} d\tau \, d\tau_s \tag{D.7}$$

where $d\tau$ and $d\tau_s$ represent integration over spatial and spin variables, respectively. By utilizing (D.6) and the Hermitian property of S, we have

$$\int \phi_{l\alpha}^* \mathcal{H} \phi_{l'\alpha'} d\tau \, d\tau_s = \int \psi_{l\alpha}^* e^{-iS} \mathcal{H} e^{iS} \psi_{l'\alpha'} d\tau \, d\tau_s \tag{D.8a}$$

$$= (l\alpha | e^{-iS} \mathcal{H} e^{iS} | l'\alpha') \tag{D.8b}$$

where we have used the notation $|l\alpha)$ for matrix elements calculated using the ψ's. We may interpret (D.8) as saying that we can look either for transformed functions, $\phi_{l\alpha}$, or a transformed Hamiltonian, (D.8b).

If we define $\mathcal{H}'$ as

$$\mathcal{H}' = e^{-iS} \mathcal{H} e^{iS} \tag{D.9}$$

we may state as our goal the determination of a Hermitian operator S that generates a transformed Hamiltonian $\mathcal{H}'$ such that $\mathcal{H}'$ has no matrix elements between states of different l.

Presumably S must be small, since the original Hamiltonian $\mathcal{H}$ has small matrix elements off-diagonal in l. Therefore we may approximate by expanding the exponentials in (D.9):

$$\begin{aligned}\mathcal{H}' &= e^{-iS} \mathcal{H} e^{iS} \\ &= \left(1 - iS - \frac{S^2}{2!} + \ldots\right) \mathcal{H} \left(1 + iS - \frac{S^2}{2!} + \ldots\right) \\ &= \mathcal{H} + i[\mathcal{H}, S] + \left(S\mathcal{H}S - \frac{S^2}{2}\mathcal{H} - \frac{\mathcal{H}S^2}{2}\right) \\ &= \mathcal{H} + i[\mathcal{H}, S] - \tfrac{1}{2}[[\mathcal{H}, S], S] \quad . \end{aligned} \tag{D.10}$$

Writing $\mathcal{H} = \mathcal{H}_0 + \mathcal{H}_1 + \mathcal{H}_2$, we wish to choose S to eliminate $\mathcal{H}_2$.

By writing out (D.10), we have

$$\mathcal{H}' = \mathcal{H}_0 + \mathcal{H}_1 + \mathcal{H}_2 + i[\mathcal{H}_0 + \mathcal{H}_1, S] + i[\mathcal{H}_2, S] - \tfrac{1}{2}[S, [S, \mathcal{H}]] \quad . \tag{D.11}$$

We can eliminate the third term on the right by choosing

$$\mathcal{H}_2 + i[\mathcal{H}_0 + \mathcal{H}_1, S] = 0 \quad . \tag{D.12}$$

Then we have

$$\mathcal{H}' = \mathcal{H}_0 + \mathcal{H}_1 + i[\mathcal{H}_2, S] + \frac{i^2}{2}[[\mathcal{H}_0 + \mathcal{H}_1, S], S] + \frac{i^2}{2}[[\mathcal{H}_2, S], S] \quad . \tag{D.13}$$

If $\mathcal{H}_2$ were zero, S would vanish. Therefore, we expect S will be of order $\mathcal{H}_2$, and the last term of order $(\mathcal{H}_2)^3$. Neglecting it, and utilizing (D.12), we have

$$\mathcal{H}' = \mathcal{H}_0 + \mathcal{H}_1 + \frac{i}{2}[\mathcal{H}_2, S] \quad . \tag{D.14}$$

Equation (D.12) may be put in matrix form to obtain an explicit matrix for S. Using the facts that $\mathcal{H}_1$ has no matrix elements between states of different l and that $\mathcal{H}_2$ has none diagonal in l, we have

$$(l\alpha|\mathcal{H}_2|l'\alpha') + i\sum_{l''\alpha''}[(l\alpha|\mathcal{H}_0 + \mathcal{H}_1|l''\alpha'')(l''\alpha''|S|l'\alpha')$$
$$- (l\alpha|S|l''\alpha'')(l''\alpha''|\mathcal{H}_0 + \mathcal{H}_1|l'\alpha')] = 0 \quad . \tag{D.15}$$

Thus

$$(l\alpha|\mathcal{H}_2|l'\alpha') + i(E_l - E_{l'})(l\alpha|S|l'\alpha') + i\sum_{\alpha''}[(l\alpha|\mathcal{H}_1|l\alpha'')(l\alpha''|S|l'\alpha')$$
$$- (l\alpha|S|l'\alpha'')(l'\alpha''|\mathcal{H}_1|l'\alpha')] = 0 \quad . \tag{D.16}$$

If $l \neq l'$, we may neglect the terms in $\mathcal{H}_1$, as being small compared with those involving $E_l - E_{l'}$. Then

$$(l\alpha|S|l'\alpha') = \frac{1}{i}\frac{(l\alpha|\mathcal{H}_2|l'\alpha')}{(E_{l'} - E_l)} \quad . \tag{D.17}$$

If $l = l'$, we have, for (D.16)

$$\sum_{\alpha''}(l\alpha|\mathcal{H}_1|l\alpha'')(l\alpha''|S|l\alpha') = \sum_{\alpha''}(l\alpha|S|l\alpha'')(l\alpha''|\mathcal{H}_1|l\alpha') \quad . \tag{D.18}$$

This is readily satisfied by choosing (D.18a):

$$(l\alpha|S|l\alpha'') = 0 \quad . \tag{D.18a}$$

Therefore S does not join states of the same l. [Equation (D.17) and (D.18a) enable one to verify that S is Hermitian; that is, that $(l\alpha|S|l'\alpha') = (l'\alpha'|S|l\alpha)^*$, where the star indicates a complex conjugate.] By using (D.17) in (D.14), we may find the new matrix elements between $|l\alpha)$ and $|l'\alpha')$. First we note that the states off-diagonal in l are

$$(l\alpha|\mathcal{H}'|l'\alpha') = \frac{i}{2}(l\alpha|[\mathcal{H}_2, S]|l, \alpha') \tag{D.19}$$

since $\mathcal{H}_0$ and $\mathcal{H}_1$ are diagonal in l. Thus,

$$(l\alpha|\mathcal{H}'|l'\alpha') = \frac{i}{2}\sum_{l''\alpha''}[(l\alpha|\mathcal{H}_2|l''\alpha'')(l''\alpha''|S|l'\alpha')$$
$$- (l\alpha|S|l''\alpha'')(l''\alpha''|\mathcal{H}_2|l'\alpha')]$$
$$= \frac{1}{2}\sum_{l''\alpha''}(l\alpha|\mathcal{H}_2|l''\alpha'')(l''\alpha''|\mathcal{H}_2|l'\alpha')$$
$$\times \left[\frac{1}{E_{l'} - E_{l''}} + \frac{1}{E_l - E_{l''}}\right] \quad . \tag{D.20}$$

The off-diagonal matrix elements are therefore reduced in the ratio of $\mathcal{H}_2$ to the difference between eigenvalues of $\mathcal{H}_0$, and the states of different l are "uncoupled". The matrix elements diagonal in l are modified, too. They become, using (D.14) and (D.17),

$$(l\alpha|\mathcal{H}'|l\alpha') = (l\alpha|\mathcal{H}_0 + \mathcal{H}_1 + \frac{i}{2}[\mathcal{H}_2, S]|l\alpha')$$

$$= E_l \delta_{\alpha\alpha'} + (l\alpha|\mathcal{H}_1|l\alpha')$$

$$+ \frac{i}{2} \sum_{l'',\alpha''} [(l\alpha|\mathcal{H}_2|l''\alpha'')(l''\alpha''|S|l\alpha')$$

$$- (l\alpha|S|l''\alpha'')(l''\alpha''|\mathcal{H}_2|l\alpha')]$$

$$= E_l \delta_{\alpha\alpha'} + (l\alpha|\mathcal{H}_1|l\alpha') + \sum_{l'',\alpha''} \frac{(l\alpha|\mathcal{H}_2|l''\alpha'')(l''\alpha''|\mathcal{H}_2|l\alpha')}{E_l - E_{l''}}. \quad \text{(D.21)}$$

If $\alpha = \alpha'$, we recognize that the terms in $\mathcal{H}_2$ give the familiar expression for the energy shift in second-order perturbation theory. However, our expression also includes matrix elements for $\alpha \neq \alpha'$. In this connection we wish to emphasize that in degenerate perturbation theory, ordinarily one must find zero-order functions that have vanishing off-diagonal elements. The method we have described places no such restriction on the basis functions $|l\alpha)$. If the quantum numbers α lead to elements $(l\alpha|\mathcal{H}'|l\alpha')$ between states of different α, it means merely that we must still diagonalize the matrix $(l\alpha|\mathcal{H}'|l\alpha')$ of (D.21). We conclude that the presence of a term $\mathcal{H}_2$ is to a good approximation equivalent to adding to the Hamiltonian $\mathcal{H}_0 + \mathcal{H}_1$ matrix elements diagonal in l of

$$\sum_{l'',\alpha''} \frac{(l\alpha|\mathcal{H}_2|l''\alpha'')(l''\alpha''|\mathcal{H}_2|l\alpha')}{E_l - E_{l''}}$$

and neglecting the coupling between states of different l.

E. The High Temperature Approximation

In several places in the text we make use of the high temperature approximation. For example, on page 63 in Chapter 3 we replace the exponentials by unity in the expression for $\chi''(\omega)$:

$$\chi''(\omega) = \frac{\pi\hbar\omega}{kTZ} \sum_{a,b} e^{-E_a/kT} |(a|\mu_x|b)|^2 \delta(E_a - E_b - \hbar\omega) \quad \text{(E.1)}$$

where

$$Z = \sum_a e^{-E_a/kT}$$

is the partition function. Since the energies E_a are energies of the N-particle system, they may range from $-N\gamma\hbar H_0 I$ to $+N\gamma\hbar H_0 I$ as a result of the Zeeman energy alone. Of course, the energy $-N\gamma\hbar H_0 I$ would occur only if all N spins were in the state $m = I$, and is thus quite unlikely on a statistical basis. However, we expect to find typical values of $|E_a| \cong \sqrt{N}\gamma\hbar H_0 I$ combining the m values of the N spins at random. Since $N \cong 10^{23}$ in a typical sample, how can we approximate $E_a/kT \ll 1$?

It is clear that no one spin interacts with many others, so that in some sense we do not really need to consider 10^{23} spins to get a fair precision in computing χ''. That is, asserting that 10^{23} spins are involved is really a fiction. After all, in an applied field of reasonable strength we can predict the location of the absorption by considering only one spin.

We believe that the high temperature approximation will hold if the energy of a *single* spin is small compared to kT. We wish now to demonstrate how that comes about. To do so, we shall consider a simplified case, that of N noninteracting identical spins, and assert that a similar argument should hold for interacting spins provided that the effective interaction between pairs is still small compared to kT so that no drastic phenomena such as ferromagnetism results.

If the spins are noninteracting, we can choose as exact quantum numbers the individual spin quantum numbers $m_1, m_2, \ldots, m_N$. The energy E_a then becomes

$$E_a = -\gamma\hbar H_0 \sum_{j=1}^{N} m_j = -\hbar\omega_0 M \tag{E.2}$$

where ω_0 is the Larmor frequency, and $M = \sum_j m_j$. The wave function $|a\rangle$ and operator μ_x are

$$|a\rangle = |m_1, m_2, \ldots, m_N\rangle \tag{E.3}$$

$$\mu_x = \sum_j \mu_{xj} \quad . \tag{E.4}$$

Then

$$|\langle a|\mu_x|b\rangle|^2 = \langle a|\sum_j \mu_{xj}|b\rangle\langle b|\sum_k \mu_{xk}|a\rangle$$
$$= \sum_{j,k} \langle a|\mu_{xj}|b\rangle\langle b|\mu_{xk}|a\rangle \quad . \tag{E.5}$$

Since the μ_{xj}'s involve the coordinates of only one nucleus, we see from (E.3) that we only get terms in which $j = k$ so that

$$|\langle a|\mu_x|b\rangle|^2 = \sum_j |\langle m_1, m_2, \ldots, m_j, \ldots|\mu_{xj}|m_1, m_2, \ldots, m'_j, \ldots\rangle|^2 \quad, \tag{E.6}$$

giving us

$$\chi''(\omega) = \frac{\pi\hbar\omega}{kTZ} \sum_{m_1, m_2, \ldots, m_j, \ldots, m_N, m'_j} \exp(\hbar\omega_0 M/kT)$$
$$\times \sum_j |\langle m_1, m_2, \ldots, m_j, \ldots|\mu_{xj}|m_1, m_2, \ldots, m'_j, \ldots\rangle|^2$$
$$\times \delta[\hbar\omega_0(m_j - m'_j) - \hbar\omega] \quad . \tag{E.7}$$

If we define $m = M - m_j$, we can write

$$\exp(\hbar\omega_0 M/kT) = \exp(\hbar\omega_0 m/kT)\exp(\hbar\omega_0 m_j/kT) \quad . \tag{E.8}$$

Then, using the fact that $(m_1, m_2, \ldots, m_j, \ldots |\mu_{xj}| m_1, m_2, \ldots, m'_j, \ldots)$
$= (m_j|\mu_{xj}|m'_j|)$, (E.7) becomes

$$\chi''(\omega) = \frac{\pi\hbar\omega}{kTZ} \sum_{m_1, m_2, \ldots m_j-1, m_j+1, \ldots} \exp(m\hbar\omega_0/kT)$$

$$\times \sum_{j, m_j, m'_j} \exp(-m_j\hbar\omega_0/kT)|(m_j|\mu_{xj}|m'_j)|^2$$

$$\times \delta[\hbar\omega_0(m_j - m'_j) - \hbar\omega] \quad . \tag{E.9}$$

Here we have used the fact that the sum over the $N-1$ coordinates omitting m_j is independent of which j we omit since the spins are identical. But

$$Z = \sum_{m_1, m_2, \ldots, m_j-1, m_j+1 \ldots} \exp(m\hbar\omega_0/kT \sum_{m_j} \exp(m_j\hbar\omega_0/kT) \quad . \tag{E.10}$$

If, now, $I\hbar\omega_0 \ll kT$, we can replace the exponentials in the m_j sum by unity, and obtain

$$\sum_{m_j} \exp(m_j\hbar\omega_0/kT) = (2I+1) \tag{E.11}$$

giving

$$Z = (2I+1) \sum_{m_1, m_2, \ldots, m_j-1, m_j+1 \ldots} \exp(m\hbar\omega_0/kT) = (2I+1)Z(N-1)$$

where $Z(N-1)$ is the partition function of $N-1$ particles. The sum

$$\sum \exp(+m\hbar\omega_0/kT)$$

now factors out of the numerator of (E.9) and out of Z giving

$$\chi''(\omega) = \frac{\pi\hbar\omega}{(2I+1)kT} \sum_{j=1}^{N} \sum_{m_j, m'_j} |(m_j|\mu_{xj}|m'_j)|^2 \delta[\hbar\omega_0(m_j - m'_j) - \hbar\omega] \quad . \tag{E.12}$$

We wish to re-express this in terms of the states $|a\rangle$ and $|b\rangle$. To do so we note that

$$\sum_{m_1, m_2, \ldots, m_j \ldots m_N m'_j} |(m_1, m_2, \ldots m_j, \ldots |\mu_{xj}| m_1, m_2, \ldots, m'_j, \ldots)|^2$$

$$= (2I+1)^{N-1} \sum_{m_j, m'_j} |(m_j|\mu_{xj}|m'_j)|^2 \quad , \tag{E.13}$$

giving

$$\chi''(\omega) = \frac{\pi\hbar\omega}{(2I+1)^N kT} \sum_{a,b} |(a|\mu_x|b)|^2 \delta(E_a - E_b - \hbar\omega) \quad . \tag{E.14}$$

But this is just the result we would have had had we replaced all exponentials in (E.1) by unity.

We see that we have never asserted that $|E_a| \ll kT$. In fact, we have made no approximation on this score at all. Our only real approximation is that the partition function for N spins shall be $(2I+1)$ times that for $N-1$ spins.

A similar situation arises in numerous other places where we use the high temperature approximation. Essentially we are saying that although the energies correspond formally to a large N, in actual fact only a small number of spins is ever really important. Restrictions on the temperature which appear because $N \cong 10^{23}$ must therefore be fictions, and we need not worry unless the energy of a small number of spins becomes comparable to kT.

F. The Effects of Changing the Precession Frequency – Using NMR to Study Rate Phenomena

Hahn's observation that the bulk diffusion of nuclei in an inhomogeneous static field caused the spin echo to decay is an example of a general circumstance of great utility in applications of magnetic resonance in physics, chemistry, and biology. The existence of diffusion allows a nucleus to move from a place where its precession frequency has one value to another place where the different magnetic field strength produces a different precession frequency. At a later time the nucleus could move to still a third location, and so forth. There is therefore a frequency modulation associated with the motion.

Such effects are readily treated using the Bloch equations, but to do so we lay some background by treating the more elemental frequency modulation which arises when a nucleus possesses only two precession frequencies between which it can jump. For example, consider the molecule shown in Fig. F.1, an N,N-dimethyl amide. The two CH_3 groups have different electronic surroundings giving them different resonant frequencies in the same applied laboratory static field. This effect (called the chemical shift) is discussed in Chapter 4. As explained in the figure caption, the molecule is planar, but can jump between configurations (a) and (c) if given enough thermal energy to overcome the potential barrier which tends to keep the molecule planar. At low temperatures the molecule is therefore effectively rigid, giving rise to two resonances (one from each CH_3 group) displaced in frequency an amount $\delta\omega$. At high temperatures, the molecule makes frequent jumps between the two planar configurations. When that motion is sufficiently rapid, the protons in either CH_3 group respond only to their time average environment, the distinction between the two positions in the moleculer is lost, and only a single resonance is seen.

The situation is shown in Fig. F.2 which reproduces the data of *Gutowsky* and *Holm* [F.1] for a molecule in which the molecular fragment R is CCl_3. We note at $-8.5°$ C two lines of equal intensity which gradually broaden as the

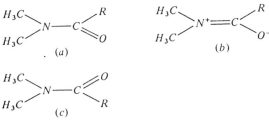

Fig. F.1a–c. The molecule N,N-dimethyl amide where R stands for either H or CH_3. The bonding structure on the left shows an N-C single bond, but admixture of states on the right gives a partial double bond, making the molecule assume a planar configuration, and giving a barrier to rotation about the N-C axis. The two CH_3 groups possess slightly different chemical shifts. However, if the RCO fragment jumps over the barrier, to achieve the configuration of (c), the CH_3 groups interchange chemical shifts

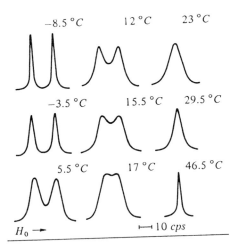

Fig. F.2. The proton magnetic resonance spectrum of N,N-dimethyltrichloroacetamide (DMTCA) as a function of temperature, at 60 MHz measured by *Gutowsky* and *Holm*. The frequency scale, but not the intensity scale, is the same for all temperatures

temperature is raised, coalesce into a single broad line, then eventually become a single narrow line at high temperatures.

This problem was first solved by *Slichter* [F.2] and independently by *Hahn* and *Maxwell* [F.3] using the Bloch equations. The method of *Hahn* and *Maxwell*, which we follow below, was also later rediscovered by *Van Vleck* [F.4] and by *McConnell* [F.5]. Using still different means, *Archer* [F.6] and *Anderson* [F.7] also discovered the same result. *Van Vleck* remarks, "It is rather remarkable that although calculations of line shape based on phase interruption are of very long standing, stemming mainly from early work by Lorentz a half a century ago, the formula ... based on the simplest example of frequency interruption was apparently not presented until 1953 although the case is one of considerable physical interest".

In order to analyze the situation, we consider that a weak rotating field H_1 is applied, we neglect saturation. Equation (2.87) described a single resonance

$$\frac{dM_+}{dt} = -M_+\alpha + i\gamma M_0 H_1 \quad \text{where} \tag{F.1}$$

$$M_+ = M_x + iM_y \tag{F.2}$$

$$\alpha = \frac{1}{T_2} + i\gamma h_0 \quad \text{where} \tag{F.3}$$

$$h_0 = H_0 - \omega/\gamma \quad . \tag{F.4}$$

We wish to describe nuclei at two sites, a and b. Let us denote the total $M_x + iM_y$ for nuclei at an a-type site as M_a and $M_x + iM_y$ for nuclei at a b-type site as M_b. M_a and M_b are then complex quantities. We define

$$\Delta\omega = \omega - \gamma H_0 \quad , \tag{F.5}$$

the displacement of the frequency from the resonance condition in H_0, and introduce the splitting $\delta\omega$ between the resonant frequencies of the two sites.

Then, were there no jumping, the nuclei at an a-site would obey one differential equation, those at a b-site would obey a second differential equation.

$$\frac{dM_a}{dt} = -\frac{M_a}{T_2} + i\left(\Delta\omega + \frac{\delta\omega}{2}\right)M_a + i\gamma\frac{M_0}{2}H_1 \tag{F.6a}$$

$$\frac{dM_b}{dt} = -\frac{M_b}{T_2} + i\left(\Delta\omega - \frac{\delta\omega}{2}\right)M_b + i\gamma\frac{M_0}{2}H_1 \tag{F.6b}$$

where M_0 is the thermal equilibrium static magnetization of the sum of the two sites.

The steady-state solution of (F.6), obtained by setting $dM_a/dt = dM_b/dt = 0$ goes through exactly as in Section 2.8, giving two Lorentzian lines of equal amplitude, one located at

$$\Delta\omega = -\frac{\delta\omega}{2} \quad \text{or} \quad \omega = \gamma H_0 - \frac{\delta\omega}{2} \tag{F.7a}$$

the other at

$$\Delta\omega = \frac{\delta\omega}{2} \quad \text{or} \quad \omega = \gamma H_0 + \frac{\delta\omega}{2} \quad . \tag{F.7b}$$

To include the molecular motion, we now assume that the molecule reorients, carrying spins from an a-site to a b-site, and vice versa. We assume that when the reorientation occurs, the actual process of going over the barrier is very rapid, so rapid in fact that the protons do not change their direction. Thus if we take a time interval δt sufficiently long that a number of spins at b-sites jump to a-sites, they will add to the a-magnetization an increment δM_a of the form

$$\delta M_a = C_1 M_b \delta t \tag{F.8}$$

expressing the facts that the extra magnetization δM_a has the orientation of M_b, and the amount is proportional to the small time interval δt. C_1 is a constant which depends on how often jumps occur. Of course, since M_a and M_b are complex quantities of the form $M_x + iM_y$, (F.8) is a vector relationship. The constant C_1 is real.

The process of (F.8) will diminish the b-magnetization by

$$\delta M_b = -C_1 M_b \delta t. \tag{F.9}$$

When a jump occurs, the a-spins in one CH_3 group leave the b-site, diminishing the b-magnetization, and go to an a-site, adding to the a-magnetization. The nature of the molecular formula for our example requires also that during the very same jump a-spins go to a b-site, but in a more general case (e.g., if we had a CD_3 group and a CH_3 group on the molecule and were considering only the proton resonance), the jump of a-sites to b-sites is independent of the jumps of b-sites to a-sites.

By the same token a-spins will jump to b-sites adding to M_b an amount

$$\delta M_b = C_2 M_a \delta t \tag{F.10}$$

and diminishing M_a by

$$\delta M_a = -C_2 M_a \delta t \; . \tag{F.11}$$

So we can say that as a result of jumps

$$\frac{dM_a}{dt} = C_1 M_b - C_2 M_a \quad \text{and} \tag{F.12a}$$

$$\frac{dM_b}{dt} = C_2 M_a - C_1 M_b \; . \tag{F.12b}$$

These rates can be added into (F.6a) and (F.6b), respectively.

$$\frac{dM_a}{dt} = -\frac{M_a}{T_2} + i\left(\Delta\omega + \frac{\delta\omega}{2}\right) M_a + C_1 M_b - C_2 M_a + i\gamma H_1 \frac{M_0}{2} \tag{F.13a}$$

$$\frac{dM_b}{dt} = -\frac{M_b}{T_2} + i\left(\Delta\omega - \frac{\delta\omega}{2}\right) M_b + C_2 M_a - C_1 M_b + i\gamma H_1 \frac{M_0}{2}. \tag{F.13b}$$

For our example, whenever the molecule reorients a-spins convert to b-spins and b-spins convert to a-spins, so that $C_1 = C_2 = C$.

These equations are easy to solve in the steady state, since then they are two simultaneous linear algebraic equations for the complex quantities M_a and M_b.

The result for the total complex magnetization, $M_x + iM_y$ is

$$M_x + iM_y = M_a + M_b = \frac{i\gamma H_1 M_0 \tau [2 + \tau(\alpha_a + \alpha_b)/2]}{(1 + \alpha_a \tau)(1 + \alpha_b \tau) - 1} \tag{F.14}$$

where

$$\alpha_a = \frac{1}{T_2} + i\left(\omega_0 - \omega - \frac{\delta\omega}{2}\right) \tag{F.15a}$$

$$\alpha_b = \frac{1}{T_2} + i\left(\omega_0 - \omega + \frac{\delta\omega}{2}\right) \quad \text{and} \tag{F.15b}$$

$$\frac{1}{\tau} = C \ . \tag{F.16}$$

The absorption signal is given by M_y, the imaginary part of (F.14). The results computed by *Gutowsky* and *Saika* [F.8] are displayed in Fig. F.3. The various curves labeled 1, 2, 3, and 4 correspond to various values of $\tau\delta\omega$. Note that the theoretical results have the same form as the experimental results of Fig. F.3. *From the curves we see that the transition from the case of infrequent jumping to the case of rapid jumping occurs when $\tau\delta\omega \approx 1$. This relationship is widely encountered and is of great importance for numerous applications of magnetic resonance. It describes motional narrowing.* Note though the total pattern changes character when $\tau\delta\omega \approx 1$, there are also changes apparent when $\tau\delta\omega \gg 1$ or $\tau\delta\omega \ll 1$. Thus, in Fig. F.2 the low temperature peaks begin to broaden long before they collapse and merge into a single peak. Likewise at high temperature the peak continues to narrow with increasing temperature long after it has become a single peak.

In general, associated with motion there is a splitting $\delta\omega$, characteristic of the zero-motion (long τ) limit, and a physical process which causes the nucleus to change resonant frequency, with an associated time τ describing how often on the average the frequency changes.

For example, in a solid, a typical nucleus experiences magnetic fields ΔH of order $\pm \mu/R^3$ where μ is the nuclear magnetic moment of the neighbors, and

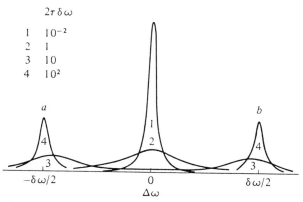

Fig. F.3. The averaging by molecular reorientation of a chemical shift in a high resolution NMR spectrum. It is assumed that mutual reorientation occurs between two sites (a and b) which are equally populated and which have resonance frequencies separated by $\delta\omega$ radians in the absence of the exchange

R the distance to the nearest neighbors and the $\pm$ signs represent the fact that the field due to the neighbor changes sign depending on the orientation of the neighbor. Consequently the resonance of the nucleus in question is spread above and below the value $\omega = \gamma H_0$. In a solid, a given nucleus has many neighbors, some aiding, some opposing the external field. Thus instead of having two peaks as in the case we analyzed, we get a smear with a width somewhat broader than would result from having only one neighbor. We thus make the rough identification

$$\frac{\delta\omega}{2} = \frac{Z\gamma\mu}{R^3} \qquad (F.17)$$

where Z is a small numerical factor representing the fact that there is more than a single neighbor. The process of self-diffusion enables the neighbor atoms to move, so that one neighbor is replaced by another. In the process, the resonance frequency of the nucleus under observation may change depending upon whether a neighbor is replaced by a nucleus whose moment is oriented in the same direction or in the opposite direction from the magnetic moment it replaced. If τ_m is the mean time a neighbor sits before jumping, we surmise

$$\tau = \tau_m \ . \qquad (F.18)$$

At low temperatures, then, we expect a nuclear resonance line width order $\delta\omega/2$ given by (F.17). As we increase the temperature, we expect τ_m to become shorter. When

$$\tau_m \delta\omega \approx 1 \qquad (F.19)$$

we expect the nuclear resonance line width to begin to narrow. Exactly such effects are seen, as is discussed in Chapter 5.

Chemical exchange can be studied by its effect on the nuclear resonance absorption lines. For example, the spin-spin coupling in a molecule gives rise to structure in liquids, as illustrated in Fig. 4.11. In liquid CH_3CH_2OH, the OH proton splits the CH_2 proton resonance structure if the OH proton remains on the molecule. But if there is rapid chemical exchange of the OH proton with other protons in the liquid, the structure may be washed out [F.9].

G. Diffusion in an Inhomogeneous Magnetic Field

In Section 2.9 we explained how diffusion in an inhomogeneous static magnetic field causes a spin echo to decay as the time between pulses is increased. The cause of the decay is the possibility for a nucleus to change its precession frequency by diffusing to a different point in the sample at which the static field is different owing to the inhomogeneity in the magnetic field. Such a situation is closely related to that treated in Appendix F in which the nucleus had two possible natural precession frequencies. For the case of diffusion in a magnetic field, there is a continuum of magnetic fields. For simplicity, we assume the magnetic field, though inhomogeneous, has axial symmetry so that

$$H(x, y, z,) = H_0 + z\frac{\partial H}{\partial z} \quad .\tag{G.1}$$

In Appendix F we described the process whereby spins switch between sites of different precession frequency by the rate term (F.12a) and (F.12b). When diffusion can take place, the possible precession frequencies form a continuum as expressed by (G.1).

The usual way to describe diffusion is by means of a diffusion equation. The use of a diffusion equation in conjunction with the Bloch equations was introduced by *Torrey* [G.1]. We shall follow a treatment which is a slight simplification of his.

Suppose we had a homogeneous static field in the z-direction, had no applied alternating field, and by some means had produced a nonuniform M_z, as in Fig. G.1a. Let us suppose T_1 is infinite. Then the total z-component of magnetization cannot change, but as a result of diffusion the region of magnetization will spread as in Fig. G.1b, eventually leading to a uniform M_z throughout the sample (Fig. G.1c). (Recall that the symbol M_z denotes the magnetization density.) The process of Fig. G.1 is described by the equation

$$\frac{\partial M_z}{\partial t} = D\nabla^2 M_z \tag{G.2}$$

where D is the diffusion constant.

If the magnetic field were uniform, and there were initially also x- and y-components to the magnetization density, there would also be diffusional effects for M_x and M_y. But since the static field causes a precession, two effects are present: (1) changes in M_x and M_y arising from diffusion, and (2) changes in

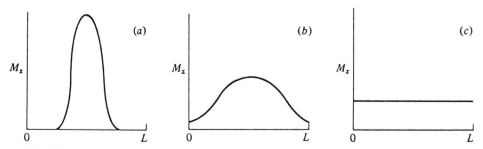

Fig. G.1a–c. A sample in a static field in the z-direction is magnetized in the z-direction. The magnetization density M_z is assumed initially to be nonuniform. Figure G.1a shows the initial M_z as a function of the x-coordinate, the sample extending from $x = 0$ to $x = L$. As a result of diffusion, the magnetization density at a later time (Fig. G.1b) has spread throughout the sample, with a diminished peak. Eventually, the M_z becomes uniformly spread as in Fig. G.1c. Note that if T_1 is infinite, the total z-magnetization, which is the area under the curve of $M_z(x)$, is the same in all three cases

M_x and M_y arising from precession. If we transform to the reference frame rotating at the precession frequency, we eliminate the precession and can write in this frame

$$\frac{\partial M_x}{\partial t} = D\nabla^2 M_x \tag{G.3a}$$

$$\frac{\partial M_y}{\partial t} = D\nabla^2 M_y \quad \text{or} \tag{G.3b}$$

$$\frac{\partial M^+}{\partial t} = D\nabla^2 M^+ \tag{G.4}$$

using the definition $M^+ \equiv M_x + iM_y$.

Since M^+ describes a two-dimensional effect, (G.4) treats vector effects. It is analogous to (F.12), if we bear in mind that M^+ is a function of position.

We now proceed as in Appendix F to recognize that when the static field is inhomogeneous, we must now include the effect of the spread in precession frequency.

We therefore add the precession driving terms giving

$$\frac{\partial M^+(x,y,z,t)}{\partial t} = -i\gamma h(x,y,z)M^+(x,y,z,t) - \frac{M^+(x,y,z,t)}{T_2} + D\nabla^2 M^+(x,y,z,t) \tag{G.5}$$

where

$$h(x,y,z) = H(x,y,z) - H_0 \tag{G.6}$$

and H_0 is the spatial average field over the sample. Substituting (G.1) for $h(x,y,z)$, we obtain an equation describing free precession in a static magnetic field which is inhomogeneous but possesses axial symmetry:

$$\frac{\partial M^+}{\partial t} = -i\gamma z\left(\frac{\partial H}{\partial z}\right)M^+ - \frac{M^+}{T_2} + D\nabla^2 M^+ \; . \tag{G.7}$$

Equation (G.7) includes the natural T_2 effects as well as the T_1 effect. [The fact that T_1 effects are included can be verified by examining (2.86) with H_1 set equal to zero]. Note that since $\partial H/\partial z$ is evaluated at the origin ($x = 0$, $y = 0$, $z = 0$), it is a constant. The only explicit dependence of (G.7) on position in the sample is therefore in the first term on the right-hand side.

If there were no diffusion ($D = 0$), (G.7) could be solved simply since over a layer of constant z the equation describes precession in the rotating frame in a constant field h, together with decay at the rate T_2. The result is

$$M^+(\mathbf{r},t) = M(\mathbf{r},0)e^{-t/T_2}e^{-i\gamma z(\partial H/\partial z)t} \tag{G.8}$$

where $M(\mathbf{r},0)$ is the complex magnetization density at $t = 0$. Suppose that $M(\mathbf{r},0)$ were uniform, as would be the case if it were prepared by applying a $\pi/2$ pulse to a sample initially magnetized to thermal equilibrium along the static field. (We here neglect the slight variation in thermal equilibrium static magnetization produced by the small inhomogeneity in a static field). Consider the magnetization density at $z - \Delta z$, z, and $z + \Delta z$, where Δz is a small distance. Since these three planes are equally spaced, their precession frequencies are equally spaced. Starting in phase at $t = 0$, the magnetization at $z + \Delta z$ develops a lead in angle over that at z, that at $z - \Delta z$ develops an angular lag relative to the magnetization at z, the lead angle and the lag angle growing progressively with time, but always remaining equal. When spins from both $z + \Delta z$ and $z - \Delta z$ diffuse to z, they will do so in equal amounts. Thus they will add in equal amounts increments which lead as increments which lag, giving no net phase change for the resultant magnetization at z. Therefore, we expect that diffusion will not affect the *phase* of the development of the magnetization at z, but will affect the *magnitude*. As long as we are not near a boundary, every plane z has planes symmetrically Δz above and below with phase lead and lag, respectively, from which magnetization diffuses to z, the phase lead or lag depending on Δz independent of z. Thus we expect that the diffusion-induced decay will be independent of z. We therefore try a solution

$$M^+(\mathbf{r}, t) = M_0 e^{-t/T_2} e^{-i\gamma(\partial H/\partial z)t} A(t) \quad . \tag{G.9}$$

When this solution is substituted in (G.7) we get a differential equation for $A(t)$

$$\frac{1}{A}\frac{dA}{dt} = -D\left(\gamma\frac{\partial H}{\partial z}\right)^2 t^2 \quad \text{or} \tag{G.10}$$

$$A = A(0)\exp\left[-D\left(\gamma\frac{\partial H}{\partial z}\right)^2 \frac{t^3}{3}\right] \quad . \tag{G.11}$$

The constant $A(0)$ we incorporate into M_0 giving

$$M^+(\mathbf{r}, t) = M_0 \exp(-t/T_2)$$
$$\times \exp\left[-D\left(\gamma\frac{\partial H}{\partial z}\right)^2 \frac{t^3}{3}\right]\exp\left[-i\gamma z\left(\frac{\partial H}{\partial z}\right)t\right] \tag{G.12}$$

for the magnetization following an initial $\pi/2$ pulse. This equation describes the development of magnetization with time from its initial value M_0 at $t = 0$.

We now need to consider what happens if we apply a π pulse at time τ producing a rotation about the y-axis. The magnetization density just prior to the pulse is

$$M^+(\mathbf{r}, \tau^-) = M_0 \exp\left(-\frac{\tau}{T_2}\right)\exp\left(-D\left(\gamma\frac{\partial H}{\partial z}\right)^2 \frac{\tau^3}{3}\right)$$
$$\times \exp\left(-i\gamma z \frac{\partial H}{\partial z}\tau\right) \quad . \tag{G.13}$$

The π pulse leaves M_y unchanged, and changes M_x into $-M_x$. This is equivalent to changing

$$-\gamma z \frac{\partial H}{\partial z}\tau \quad \text{into} \quad \pi + \gamma z \frac{\partial H}{\partial z}\tau \quad .$$

Thus at $t = \tau^+$

$$M^+(r, \tau^+) = M_0 \exp\left(-\frac{\tau}{T_2}\right)\exp\left(-D\left(\gamma\frac{\partial H}{\partial z}\right)^2 \frac{\tau^3}{3}\right)$$

$$\times \exp(i\pi)\exp\left(i\gamma z \frac{\partial H}{\partial z}\tau\right) \quad .$$

We now use (G.12) to describe the development of M^+ in time following τ:

$$M^+(r, t - \tau) = M^+(r, \tau)\exp\left(-\frac{(t-\tau)}{T_2}\right)\exp\left[-D\left(\gamma\frac{\partial H}{\partial z}\right)^2 \frac{(t-\tau)^3}{3}\right]$$

$$\times \exp\left(-i\gamma z \frac{\partial H}{\partial z}(t-\tau)\right) \quad . \tag{G.14}$$

Substituting (G.13) we see that at $t - \tau = \tau$ (or $t = 2\tau$), the complex phase factors cancel, giving

$$M^+(r, 2\tau) = -M_0 \exp\left(-\frac{2\tau}{T_2}\right)\exp\left[-D\left(\gamma\frac{\partial H}{\partial z}\right)^2 \frac{2\tau^3}{3}\right] \quad . \tag{G.15}$$

This is *Hahn*'s famous result.

It is important to note that the diffusion term after two intervals of τ has an exponent of $(\gamma\partial H/\partial z)^2 D(2\tau^3/3)$ rather than $(\gamma\partial H/\partial z)^2 D((2\tau)^3/3)$. That is, the diffusive phase loss takes place independently in each time interval τ. This fact is the basis of the technique of *Carr* and *Purcell* [G.2], discussed in Chapter 8, who note that if one applies a sequence of π pulses spaced 2τ apart, one obtains a sequence of echoes, and by making the pulse spacing sufficiently close, one can make the diffusive loss of magnetization as small as one wishes relative to the T_2 term.

H. The Equivalence of Three Quantum Mechanics Problems

In this appendix, we show the equivalence of three quantum mechanics problems:

1. A pair of identical spin $\frac{1}{2}$ nuclei coupled to a strong static field, interacting with each other by a coupling of the form $I_z S_z$, where z is the direction of H_0.
2. The same problem as (1) except the spin-spin coupling is dipolar.
3. A nucleus with spin $F = 1$, coupled to a strong static magnetic field, with a quadrupole coupling which is axially symmetric about the direction of $\mathcal{H}_0$.

Defining $\omega_0 = \gamma H_0$ the three Hamiltonians are then

(1) $\mathcal{H}_a = -\hbar\omega_0(I_z + S_z) + \hbar a I_z S_z$ (H.1)

(2) $\mathcal{H}_b = -\hbar\omega_0(I_z + S_z) + \hbar b(3I_z S_z - \boldsymbol{I}\cdot\boldsymbol{S})$ (H.2)

where we have kept just the secular part of the dipolar coupling

(3) $\mathcal{H}_0 = -\hbar\omega_0 F_z + \hbar c(3F_z^2 - F^2)$. (H.3)

For problems (1) and (2) we consider two basis sets

(1) $|m_I m_S\rangle$ where $m_I = \pm\tfrac{1}{2}$, $m_S = \pm\tfrac{1}{2}$. (H.4)

We use a notation $\alpha\beta$ to write wave functions for $m_I = +\tfrac{1}{2}, m_S = -\tfrac{1}{2}$

(2) The singlet and triplet states .

Defining $\boldsymbol{F} = \boldsymbol{I} + \boldsymbol{S}$ we have the commuting operators $|\boldsymbol{F}|^2$ and F_z with total angular momentum quantum numbers F associated with $|\boldsymbol{F}|^2$, and M, the eigenvalue of F_z giving four states

$F = 1$, $M = 1, 0, -1$ (the triplet states)
$F = 0$, $M = 0$. (H.5)

In terms of the states $|m_I m_S\rangle$ we have

$$\begin{aligned}\psi_{11} &= \alpha\alpha \\ \psi_{10} &= \frac{1}{\sqrt{2}}(\alpha\beta + \beta\alpha) \quad \psi_{00} = \frac{1}{\sqrt{2}}(\alpha\beta - \beta\alpha) \\ \psi_{1,-1} &= \beta\beta \end{aligned}$$ (H.6)

We note that the states ψ_{1M} are unchanged if we interchange the labeling of the two spins, whereas ψ_{00} changes sign. They are thus eigenstates of the permutation operator with different eigenvalues (+1 and −1 respectively). On the other hand, an operator such as $I_x + S_x$ is unchanged by the permutation ($S_x + I_x \to I_x + S_x$), hence commutes with it. Therefore, $I_x + S_x$ has nonzero matrix elements only between states with the same eigenvalue of the permutation operator. Thus, its matrix elements between the singlet and any of the triplet states vanish.

To solve the first Hamiltonian, $\mathcal{H}_a$, we note immediately that both I_z and S_z commute, hence we can take as exact eigenstates the individual states $|m_I m_S\rangle$. The energy levels are then immediately

$E_a = -\hbar\omega_0(m_I + m_S) + \hbar a m_I m_S$. (H.7)

It will be convenient for all three cases to express the energy eigenvalues in units of frequency, ω_i,

$\omega_i \equiv E_i/\hbar$. (H.8)

So
$\omega_{m_I m_S} = -\omega_0(m_I + m_S) + a m_I m_S$. (H.9)

See Table H.1. We note that the states $\alpha\beta$ and $\beta\alpha$ are degenerate. Thus any linear

602

Table H.1	
State	Energy
$\alpha\alpha$	$-\omega_0 + a/4$
$\alpha\beta$	$-a/4$
$\beta\alpha$	$-a/4$
$\beta\beta$	$\omega_0 + a/4$

Table H.2	
State	Energy
ψ_{11}	$-\omega_0 + a/4$
ψ_{10}	$-a/4$
$\psi_{1,-1}$	$\omega_0 + a/4$
ψ_{00}	$-a/4$

Table H.3	
State	Energy
ψ_{11}	$-\omega_0 + b/2$
ψ_{10}	$-b$
$\psi_{1,-1}$	$\omega_0 + b/2$
ψ_{00}	0

combination of them is also an eigenstate. In particular, the two unnormalized states $\alpha\beta + \beta\alpha$ and $\alpha\beta - \beta\alpha$ are also eigenstates. But when normalized, these are nothing but the two states ψ_{10} and ψ_{00} of (H.6) Therefore, *we can also take as the states* those shown in Table H.2.

We turn now to the second Hamiltonian

$$\mathcal{H}_b = -\hbar\omega_0(I_z + S_z) + \hbar b(3I_z S_z - \mathbf{I}\cdot\mathbf{S})$$

First, we note that the term $\mathbf{I}\cdot\mathbf{S}$ can be rewritten utilizing the fact that

$$|\mathbf{F}|^2 = (\mathbf{I}+\mathbf{S})^2 = I^2 + S^2 + 2\mathbf{I}\cdot\mathbf{S} \quad . \tag{H.10}$$

$$\mathcal{H}_b = -\hbar\omega_0(I_z + S_z) + \hbar b[3I_z S_z - \tfrac{1}{2}(|\mathbf{F}|^2 - |\mathbf{I}|^2 - |\mathbf{S}|^2)] \quad . \tag{H.11}$$

Now, up through the term $I_z S_z$ this Hamiltonian looks just like $\mathcal{H}_a$ with $3b$ replacing a. We could take either $\psi_{m_I m_S}$ or $\psi_{F,M}$ as eigenstates for this much of the Hamiltonian. For the last term, since $|\mathbf{I}|^2$ and $|\mathbf{S}|^2$ are the same for all states, they simply add constants to the energy. The term $|\mathbf{F}|^2$ is the significant difference from the Hamiltonian $\mathcal{H}_a$. However, if we take the states $\psi_{F,M}$, they are also eigenstates of $|\mathbf{F}|^2$.

Utilizing that

$$\begin{aligned}(FM||\mathbf{F}|^2|FM) &= F(F+1) \\ (FM||\mathbf{I}|^2|FM) &= \tfrac{3}{4} = (FM||\mathbf{S}^2||FM)\end{aligned} \tag{H.12}$$

or

$$\tfrac{1}{2}(|\mathbf{F}^2| - |\mathbf{I}|^2 - |\mathbf{S}^2|) = \tfrac{1}{2}F(F+1) - \tfrac{3}{4} \tag{H.13}$$

we get Table H.3. At this point it is useful to note that one can form a simple check on one's calculation of the individual eigenvalues from the observation, readily proven from the explicit forms of the Hamiltonians, that

$$\text{Tr}\{\mathcal{H}_a\} = 0 \quad , \quad \text{Tr}\{\mathcal{H}_b\} = 0 \quad , \quad \text{Tr}\{\mathcal{H}_c\} = 0 \quad . \tag{H.14}$$

Thus, adding the diagonal elements must give zero. The energy columns of Tables H.1–3 all add to zero.

The last Hamiltonian is

$$\mathcal{H}_c = -\hbar\omega_0 F_z + \hbar c(3F_z^2 - |\mathbf{F}|^2)$$

for $F = 1$. Obviously $|\boldsymbol{F}|^2$ and F_z commute with it so we immediately take the states ψ_{FM} as eigenstates, getting Table H.4.

For all three Hamiltonians, we could induce transitions by applying a transverse alternating field,

$$H_x(t) = H_{x0} \cos \omega t \tag{H.15}$$

producing a time-dependent perturbation, $\mathcal{H}_1(t)$, given by

$$\mathcal{H}_1(t) = -\gamma \hbar H_{x0} F_x \cos \omega t \tag{H.16}$$

where $F_x = I_x + S_x$ for cases (a) and (b). We have already shown that F_x has no matrix elements joining the $F = 1$ and $F = 0$ states of cases (1) and (2). Thus, the only transitions induced are between $|11\rangle$ and $|10\rangle$ and between $|10\rangle$ and $|1-1\rangle$. We can use the tables of energy levels to calculate these two frequencies. See Table H.5. These frequencies correspond to a pair of lines separated in frequency by

$$\Delta\omega = a \quad , \quad \Delta\omega = 3b \quad , \quad \Delta\omega = 6c \tag{H.17}$$

for the three cases.

Table H.4

State	Energy
ψ_{11}	$-\omega_0 + c$
ψ_{10}	$-2c$
$\psi_{1,-1}$	$+\omega_0 + c$

Table H.5. Frequencies of transitions

Hamiltonian	Frequency of the $\|11\rangle$ to $\|10\rangle$ transition	Frequency of the $\|10\rangle$ to $\|1-1\rangle$ transition
$\mathcal{H}_a$	$\omega_0 - a/2$	$\omega_0 + a/2$
$\mathcal{H}_b$	$\omega_0 - 3b/2$	$\omega_0 + 3b/2$
$\mathcal{H}_c$	$\omega_0 - 3c$	$\omega_0 + 3c$

Suppose we take the Hamiltonians $\mathcal{H}_a$, $\mathcal{H}_b$, or $\mathcal{H}_c$ as an unperturbed Hamiltonian $\mathcal{H}_0$, with eigenstates u_l and energies E_l. We can then consider the effect of any combination of pulses by adding their time-dependent perturbation $\mathcal{H}_p(t)$ to give

$$\mathcal{H}(t) = \mathcal{H}_0 + \mathcal{H}_p(t) \quad . \tag{H.18}$$

Then, the wave function $\psi(t)$ can be written as

$$\psi(t) = \sum_l a_l(t) u_l e^{-i\omega_l t} \quad . \tag{H.19}$$

Since the coefficients a_l are independent of time if $\mathcal{H}_p = 0$, the form (H.19) is the interaction representation. Substitution of $\psi(t)$ in Schrödinger's equation leads in a straightforward way [see (5.78,79)] to

$$\frac{da_{l'}}{dt} = \frac{i}{\hbar} \sum_l a_l \langle l' | \mathcal{H}_p | l \rangle e^{-i\omega_l t} \quad . \tag{H.20}$$

Since $(l'|\mathcal{H}_p|l)$ vanishes between $F = 1$ and $F = 0$ states, for cases (1) and (2) of coupled pairs of spins we get, using the notation a_{FM}

$$a_{00} = \text{constant} \quad . \tag{H.21}$$

The other a_{FM}'s obey the equations

$$\frac{da_{11}}{dt} = -\frac{i}{\hbar} a_{10}(11|\mathcal{H}_p|10) \exp[i(\omega_{11} - \omega_{10})t]$$

$$\frac{da_{10}}{dt} = -\frac{i}{\hbar}[a_{11}(10|\mathcal{H}_p|11) \exp[i(\omega_{10} - \omega_{11})t]$$
$$+ a_{1-1}(10|\mathcal{H}_p|1-1) \exp[i(\omega_{10} - \omega_{1-1})t]$$

$$\frac{da_{1-1}}{dt} = -\frac{i}{\hbar} a_{10}(1-1|\mathcal{H}_p|10) \exp[-i(\omega_{1-1} - \omega_{10})t] \quad .$$

These equations are identical for all three cases provided one notes the relation (H.17) between the three coupling constants:

a is equivalent to $3b$ or to $6c$.

Therefore there are no static or dynamic observations one could make which would enable one to tell which of the three systems one had.

One may ask, what if one had chosen to use instead the $|m_I m_S\rangle$ representation? Clearly the physical result should not change. All that will change is that one now has four instead of three "allowed" transitions ($\alpha\alpha$ to $\alpha\beta$ or $\beta\alpha$, and $\beta\beta$ to $\alpha\beta$ or $\beta\alpha$). However, if one were to do the problem explicitly one would need to be mindful of the *exact* degeneracy of the $\alpha\beta$ and $\beta\alpha$ states for case (1), which guarantees coherent effects. If one uses the product operator formalism all these things are automatically taken care of since we do not specify a representation when we use operator algebra. One could, of course, start with the representation $|m_I m_S\rangle$, formally express the states in terms of the states $|FM\rangle$, solve the problem using the $|FM\rangle$ states, then transform back to $|m_I m_S\rangle$.

An important practical consequence of the equivalence we have proved is that if an $X(\pi/2) \ldots \tau \ldots Y(\pi/2)$ pulse sequence refocuses the coupling $aI_z S_z$, it will also refocus the true dipolar coupling. As is shown by (10.59), the effect of a general quadrupole coupling *to first order* is always of the form of $\mathcal{H}_c$. Thus, for a spin I nucleus, $\mathcal{H}_c$ describes the first-order quadrupole effects. By analogy to the echoes described above, we can refocus the first-order quadrupole coupling of a spin 1 nucleus using the same $X(\pi/2) \ldots \tau \ldots Y(\pi/2)$ pulse sequence.

I. Powder Patterns

In Chapter 4 we saw that, in general, Knight shifts are anisotropic. The same thing is true for g-shifts encountered in electron spin resonance and discussed in

Chapter 11. These features give rise to a resonance frequency ω which, as shown in (4.187), goes as

$$\omega = \omega_0 + \omega_a \sin^2\theta \cos^2\phi + \omega_b \sin^2\theta \sin^2\phi + \omega_c \cos^2\theta \tag{I.1}$$

where (θ, ϕ) specify the orientation of the static field H_0 with respect to the crystal axes. For powder samples, all angles (θ, ϕ) are present randomly, giving rise to a spread in resonant frequencies with, however, quite characteristic singularities from which ω_a, ω_b, and ω_c can be deduced. Such spectra are called *powder patterns*. This problem was first treated by *Bloembergen* and *Rowland* [I.1]. We turn here to calculating the intensity patterns for such powders. In doing so, we are omitting discussion of fascinating systems such as liquid crystals or some form of oriented polymers which are conceptually in between single crystals and random powders.

We express the random orientation of the crystallites by saying that the probability dP that H_0 lies in any infinitesimal solid angle $d\Omega$ is given by

$$dP = \frac{d\Omega}{4\pi} = \sin\theta \, d\theta \, d\phi / 4\pi = d(-\cos\theta) d\phi / 4\pi \quad . \tag{I.2}$$

Introducing the variable z defined as

$$z = -\cos\theta \tag{I.3a}$$

we have that dP is

$$dP = \frac{d\phi dz}{4\pi} \quad . \tag{I.3b}$$

The advantage of the variables (ϕ, z) is that equal areas in the ϕ-z plane correspond to equal probabilities.

Introducing Ω defined as

$$\Omega \equiv \omega - \omega_0 \tag{I.4}$$

we have

$$\begin{aligned}\Omega &= \omega_a \sin^2\theta \cos^2\phi + \omega_b \sin^2\theta \sin^2\phi + \omega_c \cos^2\theta \\ &= \omega_a + (\omega_b - \omega_a)\sin^2\phi + [(\omega_c - \omega_a) - (\omega_b - \omega_a)\sin^2\phi]z^2 \quad .\end{aligned} \tag{I.5}$$

This expression can be thought of as determining lines of constant frequency, Ω, in the ϕ-z plane. Solving for z^2 we have

$$z^2 = \frac{\Omega - \omega_a - (\omega_b - \omega_a)\sin^2\phi}{(\omega_c - \omega_a) - (\omega_b - \omega_a)\sin^2\phi} \tag{I.6}$$

so that it is straightforward to calculate the curves of z versus ϕ at constant frequency in the ϕ-z plane. We illustrate such curves with Fig. I.1. Consider, then, two curves at frequencies Ω and $\Omega + \Delta\Omega$. As a result of (I.3), the probability of finding a nucleus whose frequency within this frequency range is just the area ΔA in the ϕ-z plane between these two curves. But the probability is also the normalized NMR line intensity $I(\Omega)\Delta\Omega$. Thus we have

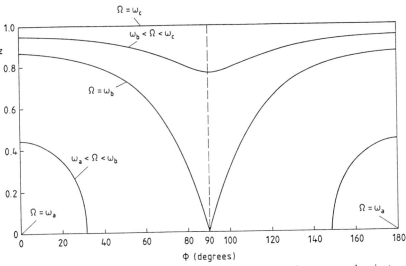

Fig. I.1. Curves of constant frequency in the ϕ-z plane for a general anisotropic chemical or Knight shift

$$I(\Omega)\Delta\Omega = \frac{\Delta A}{4\pi} \quad . \tag{I.7}$$

This relationship shows that if we divide the total frequency range into equal frequency intervals, the area between successive curves of constant frequency is proportional to the intensity at that frequency range, giving us a simple graphical picture of where the intensity is high, where it is low.

We now describe how to calculate $I(\Omega)$. We first treat the simplest case, axial symmetry, then do the general case.

Axial symmetry means that two of the principal frequencies are the same (e.g. $\omega_a = \omega_b \neq \omega_c$, or $\omega_a = \omega_c \neq \omega_b$, etc.). The calculation is easiest if we pick the $\theta = 0$ axis to be the symmetry axis ($\omega_a = \omega_b$) since then axial symmetry means that the ϕ-dependent terms of (I.5) vanish giving

$$\Omega = \omega_a + (\omega_c - \omega_a)z^2 \quad . \tag{I.8}$$

Therefore, curves of constant frequency in the ϕ-z plane are straight lines at constant z (Fig. I.2). The area, $\Delta A'$, between two curves differing in frequency by $\Delta\Omega$ is then

$$\Delta A' = 2\pi \Delta z \quad , \tag{I.9}$$

where the 2π comes from the total range of ϕ, and where Δz is the vertical spacing appropriate for the range $\Delta\Omega$:

$$\Delta z = \left(\frac{\partial z}{\partial \Omega}\right)\Delta\Omega \quad . \tag{I.10}$$

Such a strip occurs both for $0 < \theta < \pi/2$ and $\pi/2 < \theta < \pi$, since Ω is an even

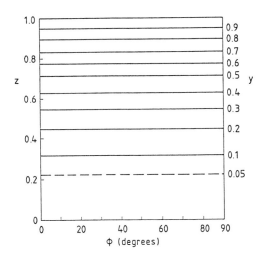

Fig. I.2. Curves of constant frequency in the ϕ-z plane for axially symmetric chemical or Knight shifts. y is the normalized frequency defined by (I.16)

function of $\cos\theta$ (i.e. z). Thus, the total area ΔA which has a given frequency range is twice $\Delta A'$:

$$\Delta A = 4\pi \Delta z \quad . \tag{I.11}$$

From (I.8) one readily finds

$$\frac{\partial z}{\partial \Omega} = \frac{1}{2\sqrt{(\Omega - \omega_a)(\omega_c - \omega_a)}} \quad , \tag{I.12}$$

giving

$$I(\Omega)\Delta\Omega = \Delta A/4\pi = \Delta z = \frac{\Delta\Omega}{2\sqrt{(\Omega - \omega_a)(\omega_c - \omega_a)}} \quad . \tag{I.13}$$

This formula holds for Ω between ω_a and ω_c. The intensity is zero for Ω outside this range. This famous line shape has a mild (square root) singularity (infinite but integrable intensity) for $\Omega = \omega_a$. This frequency corresponds to $z = 1$, or $\theta = \pi/2$, when $\boldsymbol{H}_0$ is perpendicular to the c (i.e. symmetry) axis. The infinity disappears if there is an additional broadening mechanism.

Note that this formula holds true whether $\omega_c > \omega_a$ or $\omega_c < \omega_a$. If $\omega_c > \omega_a$ the line extends from ω_a on the low end to ω_c on the high end, with

$$\omega_a \leq \Omega \leq \omega_c \quad . \tag{I.14}$$

If, however, $\omega_c < \omega_a$ (the symmetry axis is at the high frequency end of the spectrum),

$$\omega_c \leq \Omega \leq \omega_a \quad . \tag{I.15}$$

In this case, both $(\Omega - \omega_a)$ and $(\omega_c - \omega_a)$ are negative, so that the argument of the square root in (I.12) is still positive.

It is useful to introduce a dimensionless, variable y in place of Ω. We define y as

$$y = \frac{\Omega - \omega_a}{\omega_c - \omega_a} \quad . \tag{I.16}$$

From (I.13) and (I.14) we see that y is always positive whether $\omega_c > \omega_a$ or $\omega_c < \omega_a$, and extends from 0 to 1:

$$0 \leq y \leq 1 \quad . \tag{I.17}$$

The value $y = 0$ corresponds to $\Omega = \omega_a$, so y is a dimensionless frequency coordinate giving the fraction of the distance that Ω is from the ω_a end of the frequency interval. Then, dividing numerator and denominator of (I.12) by $\omega_c - \omega_a$ we get

$$I(y)\Delta y = \frac{1}{2\sqrt{y}} \Delta y \quad \text{or} \tag{I.18}$$

$$I(y) = \frac{1}{2\sqrt{y}} \quad .$$

Note that if we compute the total area of the line by integrating $I(y)$, the result is unity,

$$\int_0^1 I(y)dy = 1 \quad , \tag{I.19}$$

showing that $I(y)$ of (I.18) is properly normalized. This line shape is shown in Fig. I.3.

To deal with the more general case that $\omega_a \neq \omega_b \neq \omega_c$, we use two dimensionless variables y and f. We first rewrite (I.5) using (I.16) and (I.17):

$$y = \frac{\Omega - \omega_a}{\omega_c - \omega_a} \quad , \quad 0 \leq y \leq 1 \quad ,$$

for the fraction that $\Omega - \omega_a$ is of the total frequency interval, $\omega_c - \omega_a$. The second variable f is defined as

$$f = \frac{\omega_b - \omega_a}{\omega_c - \omega_a} \quad . \tag{I.20a}$$

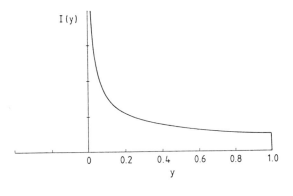

Fig. I.3. NMR line shape for an axially symmetric chemical or Knight shift tensor. NMR intensity $I(y)$ versus normalized frequency y

We adopt the convention that ω_b lies between ω_a and ω_c, but allow either $\omega_a > \omega_c$ or $\omega_a < \omega_c$. Thus ω_a can be either the low or the high end of the frequency spectrum. With these conventions we have

$$0 \leq f \leq 1 \quad . \tag{I.20b}$$

Note that f tells what fraction the total frequency interval $\omega_c - \omega_a$ lies between ω_a and ω_b. In other words, ω_b divides the interval into two fractions f, between ω_a and ω_b, and $1 - f$, between ω_b and ω_c. Then we get from (I.5)

$$y = f \sin^2 \phi + (1 - f \sin^2 \phi) z^2 \tag{I.21}$$

and from (I.6) [or by rearranging (I.21)]

$$z^2 = \frac{y - f \sin^2 \phi}{1 - f \sin^2 \phi} \quad . \tag{I.22}$$

From equation (I.22) we can draw some simple conclusions about the nature of the curves $z(\phi)$ at fixed frequency y. Since

$$\sin^2 \phi = \sin^2 (\phi + \pi) \tag{I.23}$$

the curves for $\pi \leq \phi \leq 2\pi$ simply repeat the curves for $0 \leq \phi \leq \pi$. Moreover, since

$$\sin \phi = \sin (\pi - \phi) \tag{I.24}$$

the curves for $\pi/2 \leq \phi \leq \pi$ are mirror images about $\phi = \pi/2$ of those for $0 \leq \phi \leq \pi/2$. Therefore, we need only consider the curves for $0 \leq \phi \leq \pi/2$.

Since (I.22) involves z^2, the curves are an even function of z. Thus, we need consider only positive values of z. The net result is that if we calculate any areas for the octant $0 \leq z \leq 1$, $0 \leq \phi \leq \pi/2$, we must multiply the net result by a factor of eight to get the full area. To get the axially symmetric case with $\omega_a = \omega_b$, we simply set $f = 0$ which gives then

$$z^2 = y \quad . \tag{I.25}$$

Figure I.4 shows the lines of fixed frequency y for the first octant ($0 \leq \phi \leq \pi/2$, $0 \leq z \leq 1$) for the case $f = 0.75$. For some values of frequency y the $z(\phi)$ curves go from $\phi = 0$ to $\pi/2$. Others intersect the $z = 0$ axis at smaller values of ϕ. The curve which divides these two classes has a y which has $z = 0$ when $\phi = \pi/2$. From (I.22) that is

$$y = f \quad , \tag{I.26a}$$

which means

$$\Omega = \omega_b \quad . \tag{I.26b}$$

Equation (I.22) also shows us that at the origin ($\phi = 0$, $z = 0$)

$$y = 0 \quad \text{or} \tag{I.27}$$

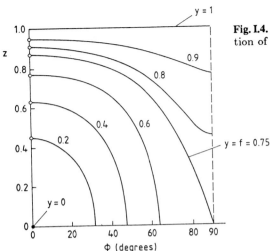

Fig. I.4. Lines of constant frequency y in one portion of the ϕ-z plane for $f = 0.75$

$$\Omega = \omega_a \quad .$$

For the frequency $y = 1$ ($\Omega = \omega_c$)

$$z = 1 \quad , \tag{I.28}$$

independent of ϕ. Therefore $\phi = \omega_a$ is the origin, $\phi = \omega_b$ ($y = f$) is the line which intersects the $z = 0$ axis at $\phi = \pi/2$, and $\omega = \omega_c$ is the line $z = 1$. We have marked these curves on Fig. I.4.

To calculate $I(y)$ we now consider two lines in the first octant of the ϕ-z plane, one corresponding to frequency Ω, the other to $\Omega + \Delta\Omega$. Then, recalling $I(y)\,\Delta y$ is given by eight times the area, $\Delta A'$, between the two lines at positive z,

$$I(y)\Delta y = 8\Delta A' \quad . \tag{I.29}$$

Focusing on two such curves, we examine Fig. I.5. We look first at the small region bounded by lines at $y = $ constant and $y + \Delta y = $ constant. The area $\Delta^2 A'$ of the small shaded region, essentially a parallelogram, is

$$\Delta^2 A' = \Delta z\, \Delta\phi \quad . \tag{I.30}$$

The spacing $\Delta\phi$ is arbitrary, but the spacing Δz must correspond to the frequency interval, hence is given by (I.10). Therefore

$$\Delta^2 A' = \frac{\partial z}{\partial y}\Delta y\, \Delta\phi \quad . \tag{I.31}$$

From (I.22) we have

$$\frac{\partial z}{\partial y} = \frac{1}{2z}\frac{1}{1 - f\sin^2\phi} = \frac{1}{2}\frac{1}{\sqrt{1 - f\sin^2\phi}\sqrt{y - f\sin^2\phi}}$$

$$= \frac{1}{2\sqrt{y}}\frac{1}{\sqrt{1 - f\sin^2\phi}\sqrt{1 - (f/y)\sin^2\phi}} \quad . \tag{I.32}$$

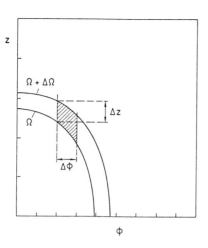

Fig. I.5. Curves of constant frequency in the ϕ-z plane for two nearby frequences Ω and $\Omega + \Delta\Omega$

Therefore from (I.29–31) we get

$$I(y) = \frac{8}{2\sqrt{y}} \int_0^{\phi_{\max}} \frac{d\phi}{\sqrt{1 - f \sin^2 \phi}\sqrt{1 - (f/y) \sin^2 \phi}} \quad . \tag{I.33}$$

In this equation, $\phi_{\max}$ is the maximum value of ϕ for the particular frequency, y. Thus, as can be seen from Fig. I.4,

$$\text{if } y > f \, , \quad \phi_{\max} = \pi/2 \quad (\Omega > \omega_b) \quad . \tag{I.34}$$

If $y < f$, $\phi_{\max}$ is easily found by setting $z = 0$ (I.22):

$$\sin^2 \phi_{\max} = \frac{y}{f} \quad . \tag{I.35}$$

If $\phi_{\max} = \pi/2$, the integral turns out to be a complete elliptic integral of the first kind. *Gradshteyn* and *Ryzhik* [I.2] show that if

$$0 < p^2 < q^2 < 1 \quad , \quad 0 < x \le \pi/2 \tag{I.36}$$

$$\int_0^\alpha \frac{dx}{\sqrt{1 - p^2 \sin^2 x}\sqrt{1 - q^2 \sin^2 x}} = \frac{1}{\sqrt{1 - p^2}} F\left(\alpha, \sqrt{\frac{q^2 - p^2}{1 - p^2}}\right)$$

where

$$F(a, k) = \int_0^\alpha \frac{d\alpha}{\sqrt{1 - k^2 \sin^2 \alpha}} \tag{I.37}$$

is an elliptic integral of the first kind. If $\alpha = \pi/2$, it is also called complete.

Thus if $y > f$ ($\Omega > \omega_b$) we have

$$I(y) = \frac{4}{\sqrt{y}} \int_0^{\pi/2} \frac{d\phi}{\sqrt{1 - f \sin^2 \phi}\sqrt{1 - (f/y) \sin^2 \phi}} \quad . \tag{I.38}$$

Since $y \leq 1$, $f < f/y$, and we can identify

$$p^2 = f \quad , \quad q^2 = f/y \tag{I.39}$$

giving

$$I(y) = \frac{4}{y\sqrt{(1-f)}} \int_0^{\pi/2} dx \bigg/ \sqrt{1 - \frac{(f/y) - f}{1 - f} \sin^2 x} \tag{I.40a}$$

$$= \frac{4}{y\sqrt{(1-f)}} \int_0^{\pi/2} dx \bigg/ \sqrt{1 - \frac{f(1-y)}{(1-f)y} \sin^2 x} \tag{I.40b}$$

$$= \frac{4(\omega_c - \omega_a)}{\sqrt{(\Omega - \omega_a)(\omega_c - \omega_b)}} \int_0^{\pi/2} dx \bigg/ \sqrt{1 - \frac{(\omega_b - \omega_a)(\omega_c - \Omega)}{(\omega_c - \omega_b)(\Omega - \omega_a)} \sin^2 x} \; . \tag{I.40c}$$

For the case of $y < f$ ($\Omega < \omega_b$), where the maximum value of ϕ, ϕ_{max}, is less than $\pi/2$, we can transform the integral by a substitution of a new variable ϕ' defined by

$$\sin^2 \phi' = \frac{f}{y} \sin^2 \phi \; . \tag{I.41}$$

ϕ' is chosen so that when $\phi = 0$, $\phi' = 0$, but when $\phi = \phi_{max}$, $\phi' = \pi/2$. With this substitution, it is straightforward to show that

$$\frac{1}{\sqrt{y}} \int_0^{\phi_{max}} \frac{d\phi}{\sqrt{1 - \sin^2 \phi}\sqrt{1 - (y/f)\sin^2 \phi}}$$

$$= \frac{1}{\sqrt{f}} \int_0^{\pi/2} \frac{d\phi'}{\sqrt{1 - y\sin^2 \phi'}\sqrt{1 - (y/f)\sin^2 \phi'}} \; . \tag{I.42}$$

Apart from the change in the integration limit, the right-hand side differs from the left solely in interchanging y and f. At all places, therefore, for $y < f$ ($\Omega < \omega_b$), we can transform (I.40b) to get

$$I(y) = \frac{4}{\sqrt{(1-y)}} \int_0^{\pi/2} d\phi' \bigg/ \sqrt{1 - \frac{y(1-f)}{(1-y)f} \sin^2 \phi'}$$

$$= \frac{4(\omega_c - \omega_a)}{\sqrt{(\omega_b - \omega_a)(\omega_c - \Omega)}} \int_0^{\pi/2} d\phi' \bigg/ \sqrt{1 - \frac{(\omega - \omega_a)(\omega_c - \omega_b)}{(\omega_c - \Omega)(\omega_b - \omega_a)} \sin^2 \phi'} \; . \tag{I.43}$$

An alternative notation commonly encountered for the complete elliptic integral of the first kind is $K(m)$ defined as

$$K(m) = \int_0^{\pi/2} \frac{dx'}{\sqrt{1 - m \sin^2 x'}} \quad . \tag{I.44}$$

Using it we can write for $y < f$ ($\Omega < \omega_b$)

$$I(y) = \frac{4}{\sqrt{f(1-y)}} K\left(\frac{y(1-f)}{(1-y)f}\right) \tag{I.45a}$$

and for $y > f$ ($\Omega > \omega_b$)

$$I(y) = \frac{4}{\sqrt{y(1-f)}} K\left(\frac{(1-y)f}{y(1-f)}\right) \quad , \tag{I.45b}$$

which brings out clearly the relationship between the frequency regions on the two sides of ω_b. Note that y is the distance in units of normalized frequency from the "lower" edge ($y = 0$), whereas $1 - y$ is the distance from the upper edge. A similar relation holds for f and $1 - f$. The arguments m of the two portions of the spectrum both approach $m = 1$ at $\Omega = \omega_b$. As m goes from 0 to 1, $K(m)$ goes from about 1.57079 to ∞, blowing up at $m = 1$. Thus, the line has a singularity at $\Omega = \omega_b$, but is finite everywhere else. The line shape is shown in Fig. I.6. Note from (I.45) at $y = 0$ and $y = 1$ the intensity is

$$I(0) = \frac{4}{\sqrt{f}} K(0) \quad , \quad I(1) = \frac{4}{\sqrt{1-f}} K(0) \tag{I.46}$$

so that

$$\frac{I(1)}{I(0)} = \sqrt{\frac{f}{1-f}} \tag{I.47a}$$

$$\frac{I(\omega_c)}{I(\omega_a)} = \sqrt{\frac{\omega_b - \omega_a}{\omega_c - \omega_b}} \quad . \tag{I.47b}$$

It is clear that from such a powder pattern one can very directly determine ω_a, ω_b, ω_c since they occur at the edges and at the peak.

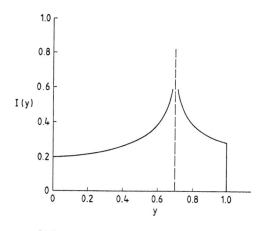

Fig. I.6. NMR intensity $I(y)$ versus normalized frequency y, for $f = 0.70$

Another important class of line-broadening mechanisms is that involving electric quadrupole coupling, a subject first treated by *Feld* and *Lamb* [I.3]. As we saw in Sect. 10.5, if the electric field gradient is axial, then in a strong field the energy levels obey (10.63)

$$E_m = -\gamma\hbar H_0 m + \frac{e^2qQ}{4I(2I-1)} \frac{(3\cos^2\theta - 1)}{2}[3m^2 - I(I+1)]$$

with transition frequencies $\omega_{m,m-1}$

$$\omega_{m,m-1} = (E_m - E_{m-1})/\hbar$$
$$= \omega_0 + \frac{e^2qQ}{4I(2I-1)\hbar} \frac{3\cos^2\theta - 1}{2} 3(2m+1) \quad . \tag{I.48}$$

This leads to the angular pattern for any one transition which is similar to that of Fig. I.3, with a singularity at $\theta = \pi/2$. However, for each transition $(m, m-1)$ there is a corresponding one between levels with the m values reversed [e.g. in addition to $(\frac{3}{2}, \frac{1}{2})$ there is $(-\frac{1}{2}, -\frac{3}{2})$]. This occurs at $(m'+1, m')$, where $m' = -(m+1)$, leading to replacing the factor $(2m+1)$ by

$$(2m'+1) = -2(m+1) + 1 = -(2m+1) \quad . \tag{I.49}$$

This reverses the sign of the coefficient of $(3\cos^2\theta - 1)$ in (I.48), so that the pattern is reversed about $\omega = \omega_0$ (Fig. I.7).

Bloembergen [I.4] noted that if I is $\frac{3}{2}$, $\frac{5}{2}$, etc., the term $(2m+1)$ vanishes when $m = -\frac{1}{2}$, so that the $(\frac{1}{2}, -\frac{1}{2})$ transition has zero first-order quadrupole splitting. He shows that it is shifted in second order by an amount

$$\hbar\Delta\omega = \frac{9}{64} \frac{(2I+3)}{4I^2(2I-1)} \frac{e^4Q^2q^2}{\hbar\omega_0}(1 - 9\cos^2\theta)(1 - \cos^2\theta)$$
$$= A(1 - 9\cos^2\theta)(1 - \cos^2\theta) \quad . \tag{I.50}$$

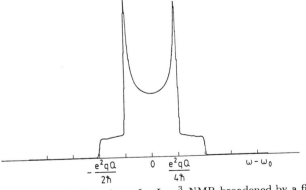

Fig. I.7. Powder line shape for $I = \frac{3}{2}$ NMR broadened by a first-order, axially symmetric electric field gradient. The singularities at $\omega - \omega_0 = \pm e^2qQ/4\hbar$ have been made finite by convolution with a narrow broadening function. This shape curve also arises in the powder line shape of a pair of dipolar-coupled identical spin $\frac{1}{2}$ nuclei, see (I.51)

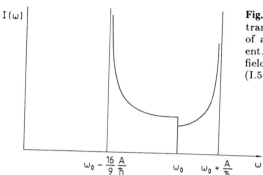

Fig. I.8. The line shape of the $\frac{1}{2}$ to $-\frac{1}{2}$ NMR transition resulting from the powder average of an axially symmetric electric field gradient. The broadening arises from treating the field gradient to second order. A is defined by (I.50)

If one expresses this relationship in terms of the variable z of (I.3a), and utilizes (I.3b), (I.7), (I.10), and (I.11), one obtains the angular pattern of Fig. I.8. The case of the second-order quadrupole broadening of the $(\frac{1}{2}, -\frac{1}{2})$ transition has been treated by various authors, and is summarized by *Gerstein* and *Dybowski* [I.5].

The first NMR paper to analyze a powder pattern line shape was *Pake*'s treatment [I.6] of a pair of spin $\frac{1}{2}$ nuclei coupled by the dipolar magnetic fields. This pattern is the same in form as that of the first-order quadrupole splitting for a spin $\frac{3}{2}$ nucleus except that the singularities occur at

$$|\omega - \omega_0| = \frac{3}{4} \frac{\gamma^2 \hbar}{r^3} \quad . \tag{I.51}$$

J. Time-Dependent Hamiltonians

We have seen that when the Hamiltonian is independent of time, we can formally solve the Schrödinger equation

$$-\frac{\hbar}{i} \frac{\partial \psi}{\partial t} = \mathcal{H}\psi \tag{J.1}$$

by use of exponential operators

$$\psi(t) = \exp\left(-\frac{i}{\hbar}\mathcal{H}t\right)\psi(0) \quad . \tag{J.2}$$

There are many circumstances however in which $\mathcal{H}$ depends on time. A very common example arises when one is applying pulses of alternating magnetic field, H_1, so that the Hamiltonian in the rotating frame makes sudden jumps from one value to another. Then during each time interval it is independent of time. During the turn-on or turn-off of the H_1, which we approximate as instantaneous, the wave function does not change (a finite $\mathcal{H}$ acting for an infinitesimal time produces an infinitesimal change in ψ — hence as the infinitesimal time goes to zero, so does the change in ψ). Thus the wave function just after the sudden change is identical to its value just before. So if we divide time into intervals $t_1, t_2, \ldots, t_N$ with time-dependent Hamiltonians in each interval $\mathcal{H}_1, \mathcal{H}_2$, etc., we immediately get that at the end of the nth interval:

$$\psi(t) = \exp\left(-\frac{i}{\hbar}\mathcal{H}_n t_n\right) \ldots \exp\left(-\frac{i}{\hbar}\mathcal{H}_2 t_2\right) \exp\left(-\frac{i}{\hbar}\mathcal{H}_1 t_1\right) \psi(0) \qquad (J.3)$$

If the $\mathcal{H}_k$'s commuted, we could write this answer as

$$\psi(t) = \exp\left(-\frac{i}{\hbar}\sum_k \mathcal{H}_k t_k\right) \psi(0) \quad \text{or as} \qquad (J.4)$$

$$\psi(t) = \exp\left(-\frac{i}{\hbar}\int \mathcal{H}(t')dt'\right) \psi(0) \quad . \qquad (J.5)$$

However, in general

$$[\mathcal{H}_i, \mathcal{H}_j] \neq 0 \quad , \qquad (J.6)$$

so one is left with (J.3) as the best one can do.

It is useful to have a more compact way of writing this result. This can be done by a method due to *Dyson* [J.1]. He introduced the concept of a time ordering operator, which we will denote by T_D (D for Dyson, T for time). Using it

$$T_D \exp\left(-\frac{i}{\hbar}\mathcal{H}_1 t_1\right) \exp\left(-\frac{i}{\hbar}\mathcal{H}_2 t_2\right) = \exp\left(-\frac{i}{\hbar}\mathcal{H}_2 t_2\right) \exp\left(-\frac{i}{\hbar}\mathcal{H}_1 t_1\right) \qquad (J.7)$$

or

$$T_D \exp\left(-\frac{i}{\hbar}\mathcal{H}_1 t_1\right) \exp\left(-\frac{i}{\hbar}\mathcal{H}_3 t_3\right) \exp\left(-\frac{i}{\hbar}\mathcal{H}_2 t_2\right)$$
$$= \exp\left(-\frac{i}{\hbar}\mathcal{H}_3 t_3\right) \exp\left(-\frac{i}{\hbar}\mathcal{H}_2 t_2\right) \exp\left(-\frac{i}{\hbar}\mathcal{H}_1 t_1\right) \quad . \qquad (J.8)$$

In other words, it simply tells one to rearrange the order of operators to a standard form, the time-ordered form.

A useful symbol in mathematics is the product symbol $\prod$. Thus, a product of n functions $F_1, F_2, \ldots, F_n$ is often written

$$F_1 F_2 \ldots F_n = \prod_{k=1}^{N} F_k \quad . \qquad (J.9)$$

Implicit in such a notation is the idea that the order of the functions, F_k, is not important, i.e. that the F_k's commute.

Thus, if we were careless we might write (J.3) as

$$\psi(t) = \prod_{k=1}^{n} \exp\left(-\frac{i}{\hbar}\mathcal{H}_k t_k\right) \psi(0) \quad . \qquad (J.10)$$

Then, if later we came to evaluate the product and continued our careless ways, we might write out the product as

$$\psi(t) = \exp(-i\mathcal{H}_1 t_1)\exp(-i\mathcal{H}_2 t_2)\ldots\exp(-i\mathcal{H}_n t_n)\psi(0) \quad . \tag{J.11}$$

If the $\mathcal{H}_k$'s did not commute, we clearly would not have rewritten the correct result (J.3)! The Dyson operator allows us to deal with this problem because if we wrote (J.11) as

$$\psi(t) = T_D \exp\left(-\frac{i}{\hbar}\mathcal{H}_1 t_1\right)\ldots\exp\left(\frac{i}{\hbar}\mathcal{H}_n t_n\right)\psi(0) \quad , \tag{J.12}$$

we immediately get

$$\psi(t) = \exp\left(-\frac{i}{\hbar}\mathcal{H}_n t_n\right)\exp\left(-\frac{i}{\hbar}\mathcal{H}_{n-1} t_{n-1}\right)\ldots$$
$$\times \exp\left(-\frac{i}{\hbar}\mathcal{H}_2 t_2\right)\exp\left(-\frac{i}{\hbar}\mathcal{H}_1 t_1\right)\psi(0) \quad . \tag{J.13}$$

Therefore, we can write (J.3) compactly as

$$\psi(t) = T_D \prod_{k=1}^{n} \exp\left(-\frac{i}{\hbar}\mathcal{H}_k t_k\right)\psi(0) \quad , \tag{J.14}$$

since no matter what order we use to write out the product of the exponentials, T_D tells us to reorder them to get the order of (J.3).

We can derive another useful theorem. Consider a product of two exponentials

$$f(A_2, A_1) \equiv e^{A_2} e^{A_1} \quad , \tag{J.15}$$

where A_1 refers to some operator at an earlier time t_1, and A_2 an operator at a later time t_2. We can use the Dyson operator to write this as

$$f(A_2, A_1) = T_D e^{A_2 + A_1} \quad \text{or} \tag{J.16a}$$

$$e^{A_2} e^{A_1} = T_D e^{A_2 + A_1} \quad . \tag{J.16b}$$

At first sight this result is surprising since (J.16) would not be correct without the T_D present. However, the theorem is easy to prove by considering the related functions as power series

$$e^{A_2 \tau} e^{A_1 \tau} = \left(\sum_l \frac{1}{l!} A_2^l \tau^l\right)\left(\sum_m \frac{1}{m!} A_1^m \tau^m\right) \quad . \tag{J.17}$$

The coefficient of τ^n in the product is then

$$\sum_{k=0}^{n} \frac{1}{(n-k)!} \frac{1}{k!} A_1^{n-k} A_2^k \quad . \tag{J.18}$$

On the other hand, expanding

$$e^{(A_2+A_1)\tau} = \sum_n \frac{(A_2+A_1)^n}{n!}\tau^n \tag{J.19}$$

we get the coefficient of τ^n as

$$\frac{(A_2+A_1)^n}{n!} . \tag{J.20}$$

Now, if A_1 and A_2 commuted, we could write (J.20) as

$$\frac{(A_2+A_1)^n}{n!} = \frac{1}{n!}\sum_{k=0}^n \frac{n!}{(n-k)!k!}A_2^{n-k}A_1^k$$

$$= \sum_{k_0}^n \frac{1}{(n-k)!k!}A_2^{n-k}A_1^k , \tag{J.21}$$

which is identical to (J.18). If we expand the product of (J.20), we will get

$$A_2^n + A_2^{n-1}A_1 + A_2^{n-2}A_1A_2 + A_2^{n-3}A_1A_2^2 \ldots A_2A_1A_2^{n-1} + A_1 + A_2^{n-1}$$
$$+ \text{terms involving } A_2 \text{ twice} + \text{etc} . \tag{J.22}$$

Operating on this expression by T_D will then give us

$$A_2^n + nA_2^{n-1}A_1 + \frac{n(n-1)}{2}A_2^{n-2}A_1^2 + \ldots , \tag{J.23}$$

which is just (J.18). That is

$$T_D(A_2+A_1)^n = \sum_{k=0}^n \frac{1}{(n-k)!}\frac{1}{k!}A_2^{n-k}A_1^k . \tag{J.24}$$

We can therefore write

$$T_D \prod_k \exp\left(\frac{i}{\hbar}\mathcal{H}_k t_k\right) = T_D \exp\left(\frac{i}{\hbar}\sum_k \mathcal{H}_k t_k\right) . \tag{J.25}$$

We can now deal with a general time-dependent interaction $\mathcal{H}(t)$ represented schematically by Fig. J.1a. First we approximate it by a series of short steps of duration $\tau_1, \tau_2, \ldots$ (Fig. J.1b) during which we approximate $\mathcal{H}(t)$ by $\mathcal{H}_1, \mathcal{H}_2$, etc. We now use this in (J.12) to write

$$\psi(t) = T_D \exp\left(\frac{i}{\hbar}\sum_{k=0}^n \mathcal{H}_k \tau_k\right)\psi(0) . \tag{J.26}$$

Next we let the τ_k's get smaller, thereby increasing n, and replace the summation sign by an integral sign to get

$$\psi(t) = T_D \exp\left(\frac{i}{\hbar}\int_0^t \mathcal{H}(\tau)d\tau\right)\psi(0) . \tag{J.27}$$

It is useful to keep in mind that to evaluate such an expression one can always go back to either (J.26) or indeed to (J.3) whenever doubts arise.

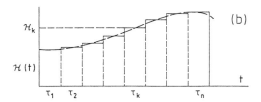

Fig. J.1. (a) A time-dependent Hamiltonian $\mathcal{H}(t)$ versus t. (b) The Hamiltonian of (a) approximated by a series of time-independent Hamiltonians of amplitude $\mathcal{H}_k$ ($k = 1$ to n) and duration τ_k

Sometimes the Hamiltonian has a particularly simple time dependence

$$\mathcal{H}(t) = a(t)\mathcal{H}_a \quad . \tag{J.28}$$

Then $\mathcal{H}(t_1)$ and $\mathcal{H}(t_2)$, the Hamiltonians at different times, commute

$$[\mathcal{H}(t_1), \mathcal{H}(t_2)] = a(t_1)a(t_2)[\mathcal{H}_a, \mathcal{H}_a] = 0 \quad . \tag{J.29}$$

In this case

$$\psi(t) = \exp\left(-\frac{i}{\hbar}\mathcal{H}_a \int_0^t a(\tau)d\tau\right)\psi(0) \quad . \tag{J.30}$$

An interesting situation arises when the Hamiltonian contains several parts with different time dependences. We shall distinguish between two cases.

Case I:

A Hamiltonian composed of several time-dependent parts which commute.

We can do a similar thing if the Hamiltonian contains several parts which commute but with differing time dependences. Let $a(t)$ and $b(t)$ be two different functions of time, and let $\mathcal{H}_a$ and $\mathcal{H}_b$ commute:

$$[\mathcal{H}_a, \mathcal{H}_b] = 0 \quad . \tag{J.31}$$

Then if

$$\mathcal{H}(t) = a(t)\mathcal{H}_a + b(t)\mathcal{H}_b \tag{J.32}$$

it is still true that that Hamiltonian at different times commutes since

$$[\mathcal{H}(t_1), \mathcal{H}(t_2)] = a(t_1)a(t_2)[\mathcal{H}_a, \mathcal{H}_a] + a(t_1)b(t_2)[\mathcal{H}_a, \mathcal{H}_b]$$
$$+ b(t_1)a(t_2)[\mathcal{H}_b, \mathcal{H}_a] + b(t_1)b(t_2)[\mathcal{H}_b, \mathcal{H}_b] = 0 \quad . \tag{J.33}$$

Therefore, we can write

$$\psi(t) = T_D \exp\left(-\frac{i}{\hbar}\int_0^t [a(\tau)\mathcal{H}_a + b(\tau)\mathcal{H}_b]d\tau\right)\psi(0)$$

$$= \exp\left(-\frac{i}{\hbar}\int_0^t [a(\tau)\mathcal{H}_a + b(\tau)\mathcal{H}_b]d\tau\right)\psi(0)$$

$$= \exp\left\{-\frac{i}{\hbar}\left[\mathcal{H}_a\left(\int_0^t a(\tau)d\tau\right) + \mathcal{H}_b\left(\int_0^t b(\tau)d\tau\right)\right]\right\}\psi(0) \quad . \tag{J.34}$$

If we define the average value of a or b over the time interval t as

$$\overline{a(t)} = \frac{1}{t}\int_0^t a(\tau)d\tau \quad , \quad \overline{b(t)} = \frac{1}{t}\int_0^t b(\tau)d\tau \tag{J.35}$$

then we can write (J.34) as

$$\psi(t) = \exp\left(-\frac{i}{\hbar}[\overline{a(t)}\mathcal{H}_a + \overline{b(t)}\mathcal{H}_b]t\right)\psi(0) \quad . \tag{J.36}$$

In general the average values $\overline{a(t)}$ and $\overline{b(t)}$ will vary with t. However, sometimes the time dependence is periodic, in which case if one chooses just those times t which are an integral multiple of the basic period T,

$$t = nT \quad , \quad n = 0, 1, 2, \text{ etc.} \quad , \tag{J.37}$$

then $\overline{a(t)}$ and $\overline{b(t)}$ are independent of time and

$$\psi(t) = \exp\left(-\frac{i}{\hbar}[\overline{a(T)}\mathcal{H}_a + \overline{b(T)}\mathcal{H}_b]nT\right)\psi(0) \quad . \tag{J.38}$$

This is a form of *stroboscopic observation*. Exactly this case arises in magic angle spinning. These formulas are easily generalized to a Hamiltonian containing more than two terms a and b.

Case II:

A Hamiltonian composed of several time-dependent parts which do *not* commute.

If we have two functions $a(t)$ and $b(t)$ with different time dependences and a Hamiltonian composed of two *noncommuting* parts

$$\mathcal{H}(t) = a(t)\mathcal{H}_a + b(t)\mathcal{H}_b \quad \text{with} \tag{J.39a}$$

$$[\mathcal{H}_a, \mathcal{H}_b] \neq 0 \quad , \tag{J.39b}$$

then we have to use the completely general relationship (J.27) or its equivalent forms.

If $a(t)$ and $b(t)$ are periodic, one can consider each period to define

$$U(T) = T_D \exp\left(-\frac{i}{\hbar}\int_0^T \mathcal{H}(\tau)d\tau\right)\psi(0) \quad . \tag{J.40}$$

Then for stroboscopic observation

$$\psi(nT) = T_D \exp\left(-\frac{i}{\hbar}\int_0^{nT} \mathcal{H}(\tau)d\tau\right)\psi(0)$$

$$= T_D \exp\left[-\frac{i}{\hbar}\sum_{k=1}^{n}\left(\int_{(k-1)T}^{kT} \mathcal{H}(\tau)d\tau\right)\right]\psi(0)$$

$$= T_D \prod_{k=1}^{n}\exp\left(-\frac{i}{\hbar}\int_{(k-1)T}^{kT} \mathcal{H}(\tau)d\tau\right)\psi(0) \quad . \tag{J.41}$$

But

$$\int_{(k-1)T}^{kT} \mathcal{H}(\tau)d\tau \tag{J.42}$$

is independent of k, so

$$\psi(nT) = \prod\left[T_D \exp\left(-\frac{i}{\hbar}\int_0^{T} \mathcal{H}(\tau)d\tau\right)\right]\psi(0) = U^n(T)\psi(0) \quad . \tag{J.43}$$

Thus, we need only work out $U(T)$ to find $\psi(nT)$.

Suppose, then, that the time interval T is very short so that $U(T)$ differs only slightly from the identity operator. In this case, we can keep as our lowest order approximation the first two terms in the exponential expansion of $U(T)$, getting

$$U(T) \cong 1 - T_D\frac{i}{\hbar}\int_0^{T}\mathcal{H}(\tau)d\tau \quad . \tag{J.44}$$

However, thinking of the integral as a sum of infinitesimal terms, we see that we never have a problem of commuting operators since we have no *products* of $\mathcal{H}(\tau)$ at one time with $\mathcal{H}(\tau)$ at another time. Consequently we can omit the time ordering operation, getting

$$U(T) = 1 - \frac{i}{\hbar}\int_0^{T}\mathcal{H}(\tau)d\tau \cong \exp\left(-\frac{i}{\hbar}\int_0^{T}\mathcal{H}(\tau)d\tau\right) \quad . \tag{J.45}$$

Defining the time average of the Hamiltonian $\overline{\mathcal{H}}$ as

$$\overline{\mathcal{H}} = \frac{1}{T}\int_0^{T}\mathcal{H}(\tau)d\tau \tag{J.46}$$

we can write

$$U(T) = \exp\left(-\frac{i}{\hbar}\overline{\mathcal{H}}T\right) \quad . \tag{J.47}$$

The theoretical formulas we have just given are sometimes referred to as *average Hamiltonian theory*. For it to hold true, typical matrix elements of $\mathcal{H}$, $(\alpha|\mathcal{H}|\alpha')$, must satisfy

$$\frac{(\alpha|\mathcal{H}|\alpha')}{\hbar}T \ll 1 \quad . \tag{J.48}$$

One useful way to think of this approximation is in terms of rotations. Rotations in three dimension do not commute, but infinitesimal ones do. Thus, if we think of $\mathcal{H}_a$ and $\mathcal{H}_b$ as two rotations, their total effect is changed if we reverse the order in which we perform the rotations. If, however, the rotations are infinitesimal, corresponding to satisfying (J.48), the order of performing the rotations is irrelevant.

In Appendix K we take up the corrections to the formula of (J.47).

K. Correction Terms in Average Hamiltonian Theory – The Magnus Expansion

In the theory of spin-flip line narrowing as well as in multiquantum excitation the effect of applications of sequences of pulses is analyzed using a time-averaged Hamiltonian. Specifically, in Sect. 8.11 we analyze a three-pulse sequence which causes the Hamiltonian to be effectively transformed as described in (8.152):

$$\begin{aligned}\mathcal{H}_A &= P_1^{-1}\mathcal{H}_{\text{int}}P_1 \\ \mathcal{H}_B &= P_1^{-1}P_2^{-1}\mathcal{H}_{\text{int}}P_2P_1 \\ \mathcal{H}_C &= \mathcal{H}_{\text{int}}\end{aligned} \tag{8.152}$$

where $\mathcal{H}_{\text{int}}$ is the sum of the Zeeman and dipolar Hamiltonians in the rotating reference frame. Then, over one cycle, the effect of the Hamiltonian is given by the unitary operator U_T of (8.153)

$$U_T = \exp\left(-\frac{i}{\hbar}\mathcal{H}_C\tau_3\right)\exp\left(-\frac{i}{\hbar}\mathcal{H}_B\tau_2\right)\exp\left(-\frac{i}{\hbar}\mathcal{H}_A\tau_1\right) \quad . \tag{8.153}$$

We expand the exponentials and keep just the leading terms, to write

$$\begin{aligned}U_T &= 1 - \frac{i}{\hbar}(\mathcal{H}_C\tau_3 + \mathcal{H}_B\tau_2\mathcal{H}_A\tau_1) \\ &\cong \exp\left(-\frac{i}{\hbar}(\mathcal{H}_C\tau_3 + \mathcal{H}_B\tau_2 + \mathcal{H}_A\tau_1)\right) \quad .\end{aligned}$$

The approximation is equivalent to writing

$$e^c e^b e^a = e^{c+b+a} \tag{K.1}$$

for three noncommuting operators a, b, c. Since clearly the expression (K.1) is only approximate, one would like to know how big the errors are. This problem has been treated by many authors. A pioneering paper in this regard is the treatment of *Haeberlen* and *Waugh* [K.1]. A useful mathematical treatment is that of *Wilcox* [K.2]. The basic paper is that of *Magnus* [K.3], who derived the first terms of a series expansion for a time-dependent Hamiltonian $\mathcal{H}(t)$ as being equivalent to a unitary operator $U(t)$ given by

$$U(t) = \exp\left(-\frac{i}{\hbar}(\overline{\mathcal{H}} + \mathcal{H}^{(1)} + \mathcal{H}^{(2)})t\right) \tag{K.2}$$

where

$$\overline{\mathcal{H}} = \frac{1}{t}\int_0^t \mathcal{H}(t_1)dt_1 \tag{K.3a}$$

$$\mathcal{H}^{(1)} = -\frac{i}{2t\hbar}\int_0^t dt_2 \int_0^{t_2} dt_1 [\mathcal{H}(t_2), \mathcal{H}(t_1)] \tag{K.3b}$$

$$\mathcal{H}^{(2)} = \frac{1}{6t\hbar^2}\int_0^t dt_3 \int_0^{t_3} dt_2 \int_0^{t_2} dt_1 \{[\mathcal{H}(t_3), [\mathcal{H}(t_2), \mathcal{H}(t_1)]] + [\mathcal{H}(t_1), [\mathcal{H}(t_2), \mathcal{H}(t_3)]]\} \quad . \tag{K.3c}$$

Our goal is to give the reader a simple idea of where these expressions come from, and what they say in concrete terms. To do so, let us consider the three pulse example.

Let us therefore examine the two sides of (K.1), utilizing power series expansion of the exponentials. Thus

$$\begin{aligned} e^{c+b+a} &= 1 + (c+b+a) + \frac{1}{2!}(c+b+a)^2 + \ldots \\ &= 1 + (c+b+a) + \frac{1}{2!}(c^2 + b^2 + a^2 + cb + bc + ca + ac + ba + ab) + \ldots \end{aligned}$$

whereas (K.4)

$$\begin{aligned} e^c e^b e^a &= \left(1 + c + \frac{c^2}{2!} + \ldots\right)\left(1 + b + \frac{b^2}{2!} + \ldots\right)\left(1 + a + \frac{a^2}{2!} + \ldots\right) \\ &= 1 + (c+b+a) + \frac{1}{2!}(c^2 + b^2 + a^2 + 2cb + 2ca + 2ba) + \ldots \end{aligned} \tag{K.5}$$

where in (K.4) and (K.5) we have collected all the terms involving products of 0, 1, or 2 of the operators but have not listed those involving 3 or more. Comparing, we see that (K.4) contains terms such as $bc + cb$, whereas (K.5) has only $2cb$, i.e. c only to the left of b. We can therefore write

$$\begin{aligned} e^c e^b e^a &\cong \{\exp(c+b+a)\} + \tfrac{1}{2}(cb - bc + ca - ac + ba - ab) + \ldots \\ &\cong \exp(c+b-a)\{1 + \tfrac{1}{2}([c,b] + [c,a] + [b,a])\} \end{aligned}$$

$$\cong \exp(c+b+a)\exp\left[\tfrac{1}{2}([c,b]+[c,a]+[b,a])\right]$$
$$\cong \exp\left[(c+b+a)+\tfrac{1}{2}([c,b]+[c,a]+[b,a])\right] \quad (K.6)$$

where in these expressions there is implicit the idea that the expression is to be thought of as an expansion, and is accurate only up to terms involving products of pairs of the operators.

As we show below, the expressions (K.6) is an example of including just the first two terms, $\overline{\mathcal{H}}$ and $\mathcal{H}^{(1)}$, of the Magnus expansion. Deriving the result (K.6) was easy. It is obvious now how to get the next term for our simple example of (K.1): merely by continuing the expansion of the exponentials. We leave that as an exercise for the reader.

We now want to show that if we evaluate (K.3), we get the same result as (K.6). First let us define times t_A, t_B, and t_C as follows. The time-dependent Hamiltonian exists as $\mathcal{H}_A$ from $t=0$ to $t=t_A$, as $\mathcal{H}_B$ from t_A to t_B, and as $\mathcal{H}_C$ from t_B to t_C. We further define τ_A, τ_B, and τ_C as the durations of the three intervals. Then we have, for one cycle, from (K.3)

$$\overline{\mathcal{H}} = \frac{1}{t_C}\int_0^{t_C} dt_1 \mathcal{H}(t_1) = \frac{1}{t_C}\left[\int_0^{t_A} dt_1 \mathcal{H}(t_1) + \int_{t_A}^{t_B} dt_1 \mathcal{H}(t_1) + \int_{t_B}^{t_C} dt_1 \mathcal{H}(t_1)\right]$$
$$= \frac{1}{t_C}[\mathcal{H}_A \tau_A + \mathcal{H}_B \tau_B + \mathcal{H}_C \tau_C] \quad . \quad (K.7)$$

This is simply the average Hamiltonian we have used previously.

We turn now to the next term, $\mathcal{H}^{(1)}$, given by (K.3b):

$$\mathcal{H}^{(1)} = -\frac{1}{2t\hbar}\int_0^t dt_2 \int_0^{t_2} dt_1 [\mathcal{H}(t_2), \mathcal{H}(t_1)] \quad . \quad (K.3b)$$

Now t_2 is always later than or equal to t_1. Thus, they could both fall into the same time interval, such as between 0 and t_A, or t_2 could be in one interval (e.g. between t_A and t_B) and t_1 in another (in this case necessarily 0 to t_A). Since $\mathcal{H}(t)$ is a constant during any one interval, it is useful to break the integrals into these time domains. We therefore adopt a notation for specifying the limits of integration:

$$\int_0^{t_c} dt_2 \int_0^{t_2} dt_1 F(t_2, t_1)$$
$$= \int_0^{t_A} dt_2 \int_0^{t_2 \le t_A} dt_1 F(t_2, t_1) + \int_{t_A}^{t_B} dt_2 \left(\int_0^{t_A} dt_1 F + \int_{t_A}^{t_2 \le t_B} dt_1 F\right)$$
$$+ \int_{t_B}^{t_c} dt_2 \left(\int_0^{t_A} dt_1 F + \int_{t_A}^{t_B} dt_1 F + \int_{t_B}^{t \le t_C} dt_1 F\right) \quad . \quad (K.8)$$

Consider the first integral. Since t_1 and t_2 are both between 0 and t_A,

$$\int_0^{t_A} dt_2 \int_0^{t_2 \leq t_A} dt_1 [\mathcal{H}(t_2), \mathcal{H}(t_1)] = \int_0^{t_A} dt_2 \int_0^{t_2 \leq t_A} dt_1 [\mathcal{H}_A, \mathcal{H}_A]$$
$$= 0 \ . \tag{K.9}$$

Consider then the second term on the right side of (K.8)

$$\int_{t_A}^{t_B} dt_2 \int_0^{t_A} dt_1 [\mathcal{H}(t_2), \mathcal{H}(t_1)] = \int_{t_A}^{t_B} dt_2 \int_0^{t_A} dt_1 [\mathcal{H}_B, \mathcal{H}_A]$$
$$= [\mathcal{H}_B, \mathcal{H}_A] \tau_B \tau_A \ . \tag{K.10}$$

We therefore get

$$\mathcal{H}^{(1)} = -\frac{i}{2t_C \hbar} \int_0^{t_C} dt_2 \int_0^{t_2} dt_1 [\mathcal{H}(t_2), \mathcal{H}(t_1)]$$
$$= -\frac{i}{2t_C \hbar} ([\mathcal{H}_B, \mathcal{H}_A] \tau_B \tau_A + [\mathcal{H}_C, \mathcal{H}_A] \tau_C \tau_A + [\mathcal{H}_C, \mathcal{H}_B] \tau_C \tau_B) \ , \tag{K.11}$$

so that using (K.11) the term of (K.2) involving $\mathcal{H}^{(1)}$, $-i\mathcal{H}^{(1)} t_C/\hbar$, is

$$-\frac{i\mathcal{H}^{(1)} t_C}{\hbar} = -\frac{1}{2\hbar^2}([\mathcal{H}_B, \mathcal{H}_A]\tau_A \tau_B$$
$$+ [\mathcal{H}_C, \mathcal{H}_A]\tau_C \tau_A + [\mathcal{H}_C, \mathcal{H}_B]\tau_C \tau_B) \ . \tag{K.12}$$

Now, identifying a, b, and c as

$$a = -\frac{i}{\hbar}\mathcal{H}_A \tau_A \ , \quad b = -\frac{i}{\hbar}\mathcal{H}_B \tau_B \ , \quad c = -\frac{i}{\hbar}\mathcal{H}_C \tau_C \tag{K.13}$$

we get that the correction term of (K.6)

$$\frac{1}{2}([c, b] + [c, a] + [b, a])$$
$$= -\frac{1}{2\hbar^2}([\mathcal{H}_C, \mathcal{H}_B]\tau_C \tau_B + [\mathcal{H}_C, \mathcal{H}_A]\tau_C \tau_A + [\mathcal{H}_B, \mathcal{H}_A]\tau_B \tau_A) \ , \tag{K.14}$$

which agrees with (K.12).

It is clear that if one has a succession of intervals $j = 1$ to N, of duration τ_j and Hamiltonians $\mathcal{H}_j$, the general form of $\mathcal{H}^{(1)}$ is

$$\mathcal{H}^{(1)} = -\frac{i}{2T\hbar} \sum_{j>k} [\mathcal{H}_j, \mathcal{H}_k]\tau_j \tau_k \quad \text{where} \tag{K.15a}$$

$$T = \sum_{j=1}^{N} \tau_j \ . \tag{K.15b}$$

An important application of these equations is to find pulse sequences which cause $\mathcal{H}^{(1)}$ to vanish. Here is a simple example. Suppose that one considers a pulse sequence of equal τ's. Then

$$\mathcal{H}^{(1)} = -\frac{i\tau^2}{2T\hbar} \sum_{j>k} [\mathcal{H}_j, \mathcal{H}_k] \tag{K.16}$$

so that the vanishing of $\mathcal{H}^{(1)}$ arises from the vanishing of the sum S defined as

$$S = \sum_{j>k} [\mathcal{H}_j, \mathcal{H}_k] \quad . \tag{K.17}$$

For a three-pulse sequence this can be written

$$S = [\mathcal{H}_3, \mathcal{H}_2] + [\mathcal{H}_3, \mathcal{H}_1] + [\mathcal{H}_2, \mathcal{H}_1] \quad . \tag{K.18}$$

Now suppose

$$\mathcal{H}_1 + \mathcal{H}_2 + \mathcal{H}_3 = 0 \tag{K.19}$$

as was true for the Hamiltonian of (8.158). Then, since

$$[\mathcal{H}_1, \mathcal{H}_1] = 0 \quad ,$$

we can write

$$S = [\mathcal{H}_3, \mathcal{H}_2] + [(\mathcal{H}_3 + \mathcal{H}_2 + \mathcal{H}_1), \mathcal{H}_1] = [\mathcal{H}_3, \mathcal{H}_2] \quad . \tag{K.20}$$

Alas, this does not vanish. How can we construct a pulse sequence which vanishes? *Haeberlen* and *Waugh* point out that one can readily find a six-pulse sequence which vanishes.

Introducing the notation

$$[\mathcal{H}_j, \mathcal{H}_k] \equiv j, k \tag{K.21}$$

we have

$$\begin{aligned} S = \ & 6,5 + 6,4 + 6,3 + 6,2 + 6,1 \\ & + 5,4 + 5,3 + 5,2 + 5,1 \\ & + 4,3 + 4,2 + 4,1 \\ & + 3,2 + 3,1 \\ & + 2,1 \quad . \end{aligned} \tag{K.22}$$

Now, previously we had

$$\mathcal{H}_1 + \mathcal{H}_2 + \mathcal{H}_3 = 0 \quad . \tag{K.23}$$

Suppose we keep this condition, and add to it the condition

$$\mathcal{H}_4 + \mathcal{H}_5 + \mathcal{H}_6 = 0 \quad . \tag{K.24}$$

Then, adding the columns of (K.22) vertically, and inserting items such as 4,4 in the second column, etc. we get that the second, third, and fifth columns add to zero, leaving solely contributions from the first and fourth columns:

$$S = 6,5 + 3,2 = [\mathcal{H}_6, \mathcal{H}_5] + [\mathcal{H}_3, \mathcal{H}_2] \quad .$$

It is now obvious that if we pick

$$\mathcal{H}_6 = \mathcal{H}_2 \quad , \quad \mathcal{H}_5 = \mathcal{H}_3$$

the commutators will be each other's negative, with the result that S will vanish. Thus, a cycle in which (K.23) is true and in addition

$$\mathcal{H}_6 = \mathcal{H}_A = \mathcal{H}_2 \quad , \quad \mathcal{H}_5 = \mathcal{H}_B = \mathcal{H}_3 \quad , \quad \mathcal{H}_4 = \mathcal{H}_A = \mathcal{H}_1$$

will have a vanishing $\mathcal{H}^{(1)}$.

There is an extensive literature on methods of finding such cycles. See for example [K.4, 5].

Selected Bibliography

The problem of preparing a complete bibliography of magnetic resonance is hopeless, there have been so many papers. Such a bibliography would not even be useful for a student, since he would not know where to begin. Therefore, a short list of articles has been selected which touches on a number of the most important ideas in resonance. In some instances, papers were chosen because they are basic references; others because they were representative of a class of papers. In some cases an attempt has been made to augment the treatment of the text. An important technique for searching the literature is use of the Citation Index, which lists all the papers in any given year which refer to a particular article or book (papers in the Citation Index are listed by the first name among the authors, e.g. by Bloembergen for the paper by Bloembergen, Purcell, and Pound). Thus to search for current work in a given area, one can look up a basic earlier paper to which later authors are likely to refer. The use of the Citation Index in connection with the following bibliography is the best method of obtaining an up-to-date picture of work in the areas listed.

Basic Papers

E.M. Purcell, H.C. Torrey, R.V. Pound: Resonance absorption by nuclear magnetic moments in a solid. Phys. Rev. **69**, 37 (1946)

F. Bloch, W.W. Hansen, M. Packard: Nuclear induction. Phys. Rev. **69**, 127 (1946)

F. Bloch, W.W. Hansen, M. Packard: The nuclear induction experiment. Phys. Rev. **70**, 474-485 (1946)

F. Bloch: Nuclear induction. Phys. Rev. **70**, 460-474 (1946)

N. Bloembergen, E.M. Purcell, R.V. Pound: Relaxation effects in nuclear magnetic resonance absorption. Phys. Rev. **73**, 679-712 (1948)

Books, Monographs, and Review Articles

G.E. Pake: "Nuclear Magnetic Resonance", in *Solid State Physics*, Vol. 2, ed. by F. Seitz, D. Turnbull (Academic, New York 1956) pp. 1-91

Collected articles. Nuovo cimento Suppl., Vol. VI, Ser X, p 808 ff. (1957)

E.R. Andrew: *Nuclear Magnetic Resonance* (Cambridge University Press, Cambridge 1955)

J.A. Pople, W.G. Schneider, H.J. Bernstein: *High-Resolution Nuclear Magnetic Resonance* (McGraw-Hill, New York 1959)

C.J. Gorter *Paramagnetic Relaxation* (Elsevier, New York 1947)

D.J.E. Ingram: *Spectroscopy at Radio and Microwave Frequencies* (Butterworth, London 1955)

A.K. Saha, T.P. Das: *Theory and Applications of Nuclear Induction* (Saha Institute of Nuclear Physics, Calcutta 1957)

M.H. Cohen. F. Reif: "Quadrupole Effects in Nuclear Magnetic Resonance Studies of Solids", in *Solid State Physics*, Vol. 5, ed. by F. Seitz, D. Turnbull (Academic, New York 1957) pp. 321-438

T.P. Das, E.L. Hahn: "Nuclear Quadrupole Resonance Spectroscopy", in *Solid State Physics*. Supplement 1, ed. by F. Seitz, D. Turnbull (Academic, New York 1958)

T.J. Rowland: Nuclear magnetic resonance in metals. *Prog. Mater. Sci.* **9**, 1-91 (1961)

R.V. Pound: *Prog. Nucl. Phys.* **2**, 21-50 (1952)
William Low: "Paramagnetic Resonance in Solids", in *Solid State Physics,* Supplement 2, ed. by F. Seitz, D. Turnbull (Academic, New York 1960)
J.S. Griffith: *The Theory of Transition-Metal Ions* (Cambridge University Press, Cambridge 1961)
N.F. Ramsey: *Nuclear Moments* (Wiley, New York 1953)
N.F. Ramsey: *Molecular Beams* (Clarendon, Oxford 1956)
A. Abragam: *The Principles of Nuclear Magnetism* (Clarendon, Oxford 1961)
N. Bloembergen: *Nuclear Magnetic Relaxation* (W.A. Benjamin, New York 1961)
J.D. Roberts: *Nuclear Magnetic Resonance* (McGraw-Hill, New York 1959)
R.T. Schumacher: *Introduction to Magnetic Resonance* (Benjamin-Cummings, Menlo Park, CA 1970)
K.A. McLauchlan: *Magnetic Resonance* (Clarendon, Oxford 1972)
G.E. Pake, T.L. Estle: *The Physical Principles of Electron Paramagnetic Resonance* (Benjamin-Cummings, Menlo Park, CA 1973)
A. Abragam, B. Bleaney: *Electron Paramagnetic Resonance of Transition Ions* (Clarendon, Oxford 1970)
Maurice Goldman: *Spin Temperature and Nuclear Magnetic Resonance in Solids* (Clarendon, Oxford 1970)
C. Kittel: *Introduction to Solid State Physics*, 5th ed. (Wiley, New York 1976) Chap. 16
C.P. Poole, Jr., H.A. Farach: *The Theory of Magnetic Resonance* (Wiley, New York 1972)

General Theory of Resonance

D. Pines, C.P. Slichter: Relaxation times in magnetic resonance. Phys. Rev. **100**, 1014-1020 (1955)
H.C. Torrey: Bloch equations with diffusion terms. Phys. Rev. **104**, 563-565 (1956)
R. Kubo, K. Tomita: A general theory of magnetic resonance absorption. J. Phys. Soc. Jpn. **9**, 888-919 (1954)
P.W. Anderson, P.R. Weiss: Exchange narrowing in paramagnetic resonance. Rev. Mod. Phys. **25**, 269-276 (1953)
P.W. Anderson: A mathematical model for the narrowing of spectral lines by exchange or motion. J. Phys. Soc. Jpn. **9**, 316-339 (1954)
R.K. Wangness, F. Bloch: The dynamical theory of nuclear induction. Phys. Rev. **89**, 728-739 (1953)
F. Bloch: Dynamical theory of nuclear induction II. Phys. Rev. **102**, 104-135 (1956)
A.G. Redfield: On the theory of relaxation processes. IBM J. **1**, 19-31 (1957)
H.C. Torrey: Nuclear spin relaxation by translational diffusion. Phys. Rev. **92**, 962-969 (1953)

Nuclear Magnetic Resonance in Metals

C.H. Townes, C. Herring, W.D. Knight: The effect of electronic paramagnetism on nuclear magnetic resonance frequencies in metals. Phys. Rev. **77**, 852-853 (1950) (letter)
W.D. Knight: "Electron Paramagnetism and Nuclear Magnetic Resonance in Metals", in *Solid State Physics*, Vol. 2, ed. by F. Seitz, D. Turnbull (Academic, New York 1956) pp. 93-136
J. Korringa: Nuclear magnetic relaxation and resonance line shift in metals. Physica **16**, 601-610 (1950)
D.F. Holcomb, R.E. Norberg: Nuclear spin relaxation in alkali metals. Phys. Rev. **98**, 1074-1091 (1955)
G. Benedek, T. Kushida: The pressure dependence of the Knight shift in the alkali metals and copper. J. Phys. Chem. Solids **5**, 241 (1958)

$I_1 \cdot I_2$ Coupling

H.S. Gutowsky, D.W. McCall, C.P. Slichter: Nuclear magnetic resonance multiplets in liquids. J. Chem. Phys. **21**, 279-292 (1953)
E.L. Hahn, D.E. Maxwell: Spin echo measurements of nuclear spin coupling in molecules. Phys. Rev. **88**, 1070-1084 (1952)
N.F. Ramsey, E.M. Purcell: Interactions between nuclear spins in molecules. Phys. Rev. **85**, 143-144 (1952) (letter)
N.F. Ramsey: Electron coupled interactions between nuclear spins in molecules. Phys. Rev. **91**, 303-307 (1953)
N. Bloembergen, T.J. Rowland: Nuclear spin exchange in solids: Tl^{203} and Tl^{205} magnetic resonance in thallium and thallic oxide. Phys. Rev. **97**, 1679-1698 (1955)
M.A. Ruderman, C. Kittel: Indirect exchange coupling of nuclear magnetic moments by conduction electrons. Phys. Rev. **96**, 99-102 (1954)
K. Yosida: Magnetic properties of Cu-Mn alloys. Phys. Rev. **106**, 893-898 (1957)
H.M. McConnell, A.D. McLean, C.A. Reilly: Analysis of spin-spin multiplets in nuclear magnetic resonance spectra. J. Chem. Phys. **23**, 1152-1159 (1955)
H.M. McConnell: Molecular orbital approximation to electron coupled interaction between nuclear spins. J. Chem. Phys. **24**, 460-467 (1956)
W.A. Anderson: Nuclear magnetic resonance spectra of some hydrocarbons. Phys. Rev. **102**, 151-167 (1956)

Pulse Methods

E.L. Hahn: Spin echoes. Phys. Rev. **80**, 580-594 (1950)
H.Y. Carr, E.M. Purcell: Effects of diffusion on free precession in nuclear magnetic resonance experiments. Phys. Rev. **94**, 630-638 (1954)

Second Moment

L.J.F. Broer: On the theory of paramagnetic relaxation. Physica **10**, 801-816 (1943)
J.H. Van Vleck: The dipolar broadening of magnetic resonance lines in crystals. Phys. Rev. **74**, 1168-1183 (1948)
G.E. Pake: Nuclear resonance absorption in hydrated crystals: Fine structure of the proton line. J. Chem. Phys. **16**, 327-336 (1948)
H.S. Gutowsky, G.E. Pake: Nuclear magnetism in studies of molecular structure and rotation in solids: ammonium salts. J. Chem. Phys. **16**, 1164-1165 (1948) (letter)
E.R. Andrew, R.G. Eades: A nuclear magnetic resonance investigation of three solid benzenes. Proc. R. Soc. London **A218**, 537-552 (1953)
H.S. Gutowsky, G.E. Pake: Structural investigations by means of nuclear magnetism II – Hindered rotation in solids. J. Chem. Phys. **18**, 162-170 (1950)

Nuclear Polarization

A.W. Overhauser: Polarization of nuclei in metals. Phys. Rev. **92**, 411-415 (1953)
T.R. Carver, C.P. Slichter: Experimental verification of the Overhauser nuclear polarization effect. Phys. Rev. **102**, 975-980 (1956)
A. Abragam: Overhauser effect in nonmetals. Phys. Rev. **98**, 1729-1735 (1955)
C.D. Jeffries: Polarization of nuclei by resonance saturation in paramagnetic crystals. Phys. Rev. **106**, 164-165 (1957) (letter)
J. Uebersfeld, J.L. Motchane, E.Erb: Augmentation de la polarisation nucléaire dans les liquides et gaz adsorbés sur un charbon. Extension aux solides contenant des impuretés paramagnétiques. J. Phys. Radium **19**, 843-844 (1958)
A. Abragam, W.G. Proctor: Une nouvelle méthode de polarisation dynamique des noyaux atomique dans les solides. C.R. Acad. Sci **246**, 2253-2256 (1958)

C.D. Jeffries: "Dynamic Nuclear Polarization" in *Progress in Cryogenics* (Heywood, London 1961)
G.R. Khutsishvili: The Overhauser effect and related phenomena. Soviet Phys. – Usp. **3**, 285-319 (1960)
R.H. Webb: Steady-state nuclear polarizations via electronic transitions. Am. J. Phys. **29**, 428-444 (1961)

Quadrupole Effects

R.V. Pound: Nuclear electric quadrupole interactions in crystals. Phys. Rev. **79**, 685-702 (1950)
N. Bloembergen: "Nuclear Magnetic Resonance in Imperfect Crystals", in Report of the Bristol Conference on Defects in Crystalline Solids (Physical Society, London 1955) pp. 1–32
T.J. Rowland: Nuclear magnetic resonance in copper alloys. Electron distribution around solute atoms. Phys. Rev. **119**, 900-912 (1960)
W. Kohn, S.H. Vosko: Theory of nuclear resonance intensity in dilute alloys. Phys. Rev. **119**, 912-918 (1960)
T.P. Das. M. Pomerantz: Nuclear quadrupole interaction in pure metals. Phys. Rev. **123**, 2070 (1961)
T. Kushida, G. Benedek, N. Bloembergen: Dependence of pure quadrupole resonance frequency on pressure and temperature. Phys. Rev. **104**, 1364 (1956)

Chemical Shifts

W.G. Proctor, F.C. Yu: The dependence of a nuclear magnetic resonance frequency upon chemical compound. Phys. Rev. **77**, 717 (1950)
W.C. Dickinson: Dependence of the F^{19} nuclear resonance position on chemical compound. Rev. **77**, 736 (1950)
H.S. Gutowsky, C.J. Hoffmann: Chemical shifts in the magnetic resonance of F^{19}. Phys. Rev. **80**, 110-111 (1950) (letter)
N.F. Ramsey: Magnetic shielding of nuclei in molecules. Phys. Rev. **78**, 699-703 (1950)
N.F. Ramsey: Chemical effects in nuclear magnetic resonance and in diamagnetic susceptibility. Phys. Rev. **86**, 243-246 (1952)
A. Saika, C.P. Slichter: A note on the fluorine resonance shifts. J. Chem. Phys. **22**, 26-28 (1954)
J.A. Pople: The theory of chemical shifts in nuclear magnetic resonance I – Induced current densities. Proc. R. Soc. London **A239**, 541-549 (1957)
J.A. Pople: The theory of chemical shifts in nuclear magnetic resonance II – Interpretation of proton shifts. Proc. R. Soc. London **A239**, 550-556 (1957)
H.M. McConnell: Theory of nuclear magnetic shielding in molecules I – Long range dipolar shielding of protons. J. Chem. Phys. **27**, 226-229 (1957)
R. Freeman, G. Murray, R. Richards: Cobalt nuclear resonance spectra. Proc. R. Soc. London **242A**, 455 (1957)

Spin Temperature

N. Bloembergen: On the interaction of nuclear spins in a crystalline lattice. Physica **15**, 386-426 (1949)
E.M. Purcell, R.V. Pound: A nuclear spin system at negative temperature. Phys. Rev. **81**, 279-280 (1951) (letter)
A. Abragam, W.G. Proctor: Experiments on spin temperature. Phys. Rev. **106**, 160-161 (1957) (letter)
A. Abragam, W.G. Proctor: Spin temperature. Phys. Rev. **109**, 1441-1458 (1958)

A.G. Redfield: Nuclear magnetic resonance saturation and rotary saturation in solids. Phys. Rev. **98**, 1787-1809 (1955)

A.G. Redfield: Nuclear spin-lattice relaxation time in copper and aluminium. Phys. Rev. **101**, 67-68 (1956)

J.H. Van Vleck: The physical meaning of adiabatic magnetic susceptibilities. Z. Phys. Chem. Neue Folge **16**, 358 (1958)

J.H. Van Vleck: The concept of temperature in magnetism. Il Nouvo Cimento, Suppl. **6**, Serie X, 1081 (1957)

C.P. Slichter, W.C. Holton: Adiabatic demagnetization in a rotating reference system. Phys. Rev. **122**, 1701-1708 (1961)

A.G. Anderson, A.G. Redfield: Nuclear spin-lattice relaxation in metals. Phys. Rev. **116**, 583-591 (1959)

L.C. Hebel, C.P. Slichter: Nuclear spin relaxation in Normal and Superconducting aluminium. Phys. Rev. **113**, 1504-1519 (1959)

A. Anderson: Nonresonant nuclear spin absorption in Li, Na, and Al. Phys. Rev. **115**, 863 (1959)

L.C. Hebel, Jr.: Spin Temperature and Nuclear Relaxation in Solids, Solid State Physics, Vol. 15 (Academic, New York 1963)

Rate Effects

H.S. Gutowsky, A. Saika: Dissociation, chemical exchange, and the proton magnetic resonance in some aqueous electrolytes. J. Chem. Phys. **21**, 1688-1694 (1953)

J.T. Arnold: Magnetic resonances of protons in ethyl alcohol. Phys. Rev. **102**, 136-150 (1956)

R. Kubo: Note on the stochastic theory of resonance absorption. J. Phys. Soc. Jpn. **9**, 935-944 (1954)

H.M. McConnell: Reaction rates by nuclear magnetic resonance. J. Chem. Phys. **28**, 430-431 (1958)

S. Meiboom. Z. Luz, D. Gill: Proton relaxation in water. J. Chem. Phys. **27**, 1411-1412 (1957) (letter)

Cross-Relaxation

N. Bloembergen, S. Shapiro, P.S. Pershan, J.O. Artman: Cross-relaxation in spin systems. Phys. Rev. **114**, 445-459 (1959)

P.S. Pershan: Cross relaxation in LiF. Phys. Rev. **117**, 109-116 (1960)

Electron Spin Resonance in Paramagnetic Systems

B. Bleaney, K.W.H. Stevens: Paramagnetic resonance. Rep. Prog. Phys. **XVI**, 108-159 (1953)

T.G. Castner, W. Känzig: The electronic structure of V-centers. J. Phys. Chem. Solids **3**, 178-195 (1957)

G.D. Watkins: Electron spin resonance of Mn^{++} in alkali chlorides: association with vacancies and impurities. Phys. Rev. **113**, 79-90 (1959)

G.D. Watkins: Motion of Mn^{++}-cation vacancy pairs in NaCl: study by electron spin resonance and dielectric loss. Phys. Rev. **113**, 91-97 (1959)

G. Feher: Observation of nuclear magnetic resonances via the electron spin resonance line. Phys. Rev. **103**, 834-835 (1956)

G. Feher: Electronic structure of F centers in KCl by the electron spin double resonance technique. Phys. Rev. **105**, 1122-1123 (1957)

G. Feher: Electron spin resonance experiments on donors in silicon – Electronic structure of donors by the electron nuclear double resonance technique. Phys. Rev. **114**, 1219-1244 (1959)

H.H. Woodbury, G.W. Ludwig: Spin resonance of transition metals in silicon. Phys. Rev. **117**, 102-108 (1960)

A.F. Kip, C. Kittel, R.A. Levy, A.M. Portis: Electronic structure of F centers: hyperfine interactions in electron spin resonance. Phys. Rev. **91**, 1066-1071 (1953)

C.J. Delbecq, B. Smaller, P.H. Yuster: Optical absorption of Cl_2 molecule-ions in irradiated potassium chloride. Phys. Rev. **111**, 1235-1240 (1958)

M. Weger: Passage effects in paramagnetic resonance experiments. Bell Syst. Tech. J. **39**, 1013-1112 (1960) (Monograph 3663)

George Feher, A.F. Kip: Electron spin resonance absorption in metals I – Experimental. Phys. Rev. **98**, 337-348 (1955)

Nuclear Resonance in Ferromagnets

A.M. Portis, A.C. Gossard: Nuclear resonance in ferromagnetic cobalt. J. Appl. Phys. **31**, 205S-213S (1960)

W. Marshall: Orientation of nuclei in ferromagnets. Phys. Rev. **110**, 1280-1285 (1958)

R.E. Watson, A.J. Freeman: Origin of effective fields in magnetic materials. Phys. Rev. **123**, 2027-2047 (1961)

G. Benedek, J. Armstrong: The pressure and temperature dependence of the Fe^{57} nuclear magnetic resonance frequency in ferromagnetic iron. J. Appl. Phys. **32**, 1065 (1961)

P. Heller, G. Benedek: Nuclear magnetic resonance in MnF_2 near the critical point. Phys. Rev. Lett. **8**, 428 (1962)

Nuclear Resonance in Paramagnetic and Antiferromagnetic Substances

N. Bloembergen: Fine structure of the magnetic resonance line of protons in $CuSO_4 \cdot 5H_2O$. Physica **16**, 95 (1950)

N.J. Poulis, G.E.G. Handeman: The temperature dependence of the spontaneous magnetization in an antiferromagnetic single crystal. Physica **19**, 391 (1953)

R.G. Shulman, V. Jaccarino: Nuclear magnetic resonance in paramagnetic MnF. Phys. Rev. **108**, 1219 (1957)

N. Jaccarino, R.G. Shulman: Observation of nuclear magnetic resonance in antiferromagnetic MnF, Phys. Rev. **107**, 1196 (1957)

G. Benedek, T. Kushida: Nuclear magnetic resonance in antiferromagnetic MnF_2 under hydrostatic pressure, Phys. Rev. **118**, 46 (1960)

W. Marshall, R.N. Stuart: Theory of transition ion complexes. Phys. Rev. **123**, 2048 (1961)

Ferromagnetic Resonance

B. Lax, K. Button: *Microwave Ferrites and Ferrimagnets* (McGraw-Hill, New York 1962)

Double Resonance in Solids

S.R. Hartmann, E.L. Hahn: Nuclear double resonance in the rotating frame. Phys. Rev. **128**, 2042 (1962)

F.M. Lurie, C.P. Slichter: Spin temperature in nuclear double resonance. Phys. Rev. **133**, A1108 (1964)

R.E. Slusher, E.L. Hahn: Sensitive detection of nuclear quadrupole interactions in solids. Phys. Rev. **166**, 332 (1968)

A.G. Redfield: Pure nuclear electric quadrupole resonance in impure copper. Phys. Rev. **130**, 589 (1963)

D.E. Kaplan, E.L. Hahn: Expérience de double irradiation en résonance magnétique par la méthode d'impulsions. J. Phys. Radium **19**, 821 (1958)

Double Resonance in Liquids

F. Bloch: Recent developments in nuclear induction. Phys. Rev. **93**, 944 (1954)
V. Royden: Measurement of the spin and gyromagnetic ratio of C^{13} by the collapse of spin-spin splitting. Phys. Rev. **96**, 543 (1954)
A.L. Bloom. J.N. Schoolery: Effects of perturbing radiofrequency fields on nuclear spin coupling. Phys. Rev. **97**, 1261 (1955)
W.P. Aue, E. Bartholdi, R.R. Ernst: Two-dimensional spectroscopy: application to nuclear magnetic resonance. J. Chem. Phys. **64**, 2299 (1976)

Masers and Lasers

J.P. Gordon. H.J. Zeiger, C.H. Townes: Molecular microwave oscillator and new hyperfine structure in the microwave spectrum of NH_3. Phys. Rev. **95**, 282 (1954)
N. Bloembergen: Proposal for a new type solid state maser. Phys. Rev. **104**, 324 (1956)
See also the 1964 Nobel Prize Lectures by Townes, Basov and Prokhorov, *Nobel Lectures – Physics* **1963-1970** (Elsevier, Amsterdam 1972). The lectures by Basov and Prokhorov give references to the Soviet literature

Slow Motion

C.P. Slichter, D.C. Ailion: Low-field relaxation and the study of ultraslow atomic motions by magnetic resonance. Phys. Rev. **135**, A1099 (1964)
D.C. Ailion, C.P. Slichter: Observation of ultra-slow translational diffusion in metallic lithium by magnetic resonance. Phys. Rev. **137**, A235 (1965)
D.C. Look, I.J. Lowe: Nuclear magnetic dipole-dipole relaxation along the static and rotating magnetic fields: application to gypsum. J. Chem. Phys **44**, 2995 (1966)

ENDOR

G. Feher: Electronic structure of F centers in KCl by the electron spin double resonance technique. Phys. Rev. **105**, 1122 (1957)
H. Seidel, H.C. Wolf: In *Physics of Color Centers*, ed. by W. Beall Fowler (Academic, New York 1968)

Spin-Flip Narrowing

J.S. Waugh, L.M. Huber, U. Haeberlen: Approach to high-resolution NMR in solids. Phys. Rev. **20**, 180 (1968)
U. Haeberlen, J.S. Waugh: Coherent averaging effects in magnetic resonance. Phys. Rev. **175**, 453 (1968)
P. Mansfield: Symmetrized pulse sequences in high resolution NMR in solids. J. Phys. **C4**, 1444 (1971)
P. Mansfield, M.J. Orchard, D.C. Stalker, K.H.B. Richards: Symmetrized multipulse nuclear-magnetic-resonance experiments in solids: measurement of the chemical shift shielding tensor in some compounds. Phys. Rev. **B7**, 90 (1973)
W.K. Rhim, D.D. Elleman, R.W. Vaughan: Analysis of multiple pulse NMR in solids. J. Chem. Phys. **59**, 3740 (1973)
W.K. Rhim, D.D. Elleman, L.B. Schreiber, R.W. Vaughan: Analysis of multiple pulse NMR in solids. II. J. Chem. Phys. **60**, 4595 (1974)
M. Mehring: *High Resolution NMR Spectroscopy in Solids*, NMR: Basic Principles and Progress, Vol. 11, ed. by P. Diehl, E. Fluck, P. Kosfeld (Springer, Berlin, Heidelberg 1976)

U. Haeberlen: *High Resolution NMR in Solids: Selective Averaging;* Supplement 1 to *Advances in Magnetic Resonance* (Academic, New York 1976)

Chemically Induced Nuclear Polarization (CIDNP)

G.L. Closs: A mechanism explaining nuclear spin polarizations in radical combination reactions. J. Am. Chem. Soc. **91**, 4552 (1969)

G.L. Closs, A.D. Trifunac: Chemically induced nuclear spin polarization as a tool for determination of spin multiplicities of radical-pair precursors. J. Am. Chem. Soc. **91**, 4554 (1969)

J.H. Freed, J.B. Pederson: The theory of chemically induced dynamic spin polarization. Adv. Magn. Reson. **8**, 2 (1976)

G.L. Closs: Chemically induced nuclear polarization. Adv. Magn. Reson. **7**, 157 (1974)

Composite Pulses

M.H. Levitt, R. Freeman: NMR population inversion using a composite pulse. J. Magn. Reson. **33**, 473 (1979)

M.H. Levitt: Composite pulses. Prog. NMR Spectros. **18**, 61 (1986)

A.J. Shaka, J. Keeler: Broadband spin decoupling in isotropic liquids. Prog. NMR Spectros. **19**, 47 (1987)

M.H. Levitt, R. Freeman, T. Frankiel: Broad band decoupling in high resolution nuclear magnetic resonance spectroscopy. Adv. Magn. Reson. **11**, 48 (1983)

Electron Spin Echo Modulation

L.G. Rowan, E.L. Hahn, W.B. Mims: Electron spin-echo envelope modulation. Phys. Rev. **137**, A61 (1965), errata Phys. Rev. **138**, AB4 (1965)

W.B. Mims: Envelope modulation in spin-echo experiments. Phys. Rev. B **5**, 2409 (1972)

Fourier Transform NMR

I.J. Lowe, R.E. Norberg: Free-induction decays in solids. Phys. Rev. **107**, 46 (1957)

C.R. Bruce: F^{19} nuclear magnetic resonance line shapes in CaF_2. Phys. Rev. **107**, 43 (1957)

R.R. Ernst, W.A. Anderson: Applications of Fourier transform spectroscopy to magnetic resonance. Rev. Sci. Instrum. **37**, 93 (1966)

T.C. Farrar, E.D. Becker: *Pulse and Fourier Transform NMR* (Academic, New York 1987)

R.R. Ernst, G. Bodenhausen, A. Wokaun: *Principles of Nuclear Magnetic Resonance in One and Two Dimensions* (Clarendon, Oxford 1987)

Magic Angle Spinning

I. Lowe: Free induction decays of rotating solids. Phys. Rev. Lett. **2**, 285 (1959)

E.R. Andrew, A. Bradbury, R.G. Eades: Nuclear magnetic resonance spectra from a crystal rotated at high speed. Nature **182**, 1659 (1958)

E.R. Andrew, A. Bradbury, R.G. Eades: Removal of dipolar broadening of nuclear magnetic resonance spectra of solids by specimen rotation. Nature **183**, 1802 (1959)

H. Kessmeier, R.E. Norberg: Pulsed nuclear magnetic resonance in rotating solids. Phys. Rev. **155**, 321 (1967)

M.M. Maricq and J.S. Waugh: NMR in rotating solids. J. Chem. Phys. **70**, 3300-3316 (1979)

E.O. Stejskal, J. Schaefer, R.A. McKay: High-resolution, slow-spinning magic-angle carbon-13 NMR. J. Magn. Reson. **25**, 569 (1977)

W.T. Dixon: Spinning-sideband-free and spinning-sideband-only NMR spectra in spinning samples. J. Chem. Phys. **77**, 1800 (1982)

Magnetic Resonance Imaging

P.C. Lauterbur: Image formation by induced local interactions: Examples employing nuclear magnetic resonance. Nature **242**, 190 (1973)

P. Mansfield, P.K. Granell: NMR "diffraction" in solids? J. Phys. **C6**, L422 (1973)

A. Kumar, D. Welti, R.R. Ernst: NMR Fourier zeugmatography. J. Magn. Reson. **18**, 69 (1975)

K.K. King. P.R. Moran: A unified description of NMR imaging data-collection strategies, and reconstruction. Med. Phys. **11**, 1 (1984)

P. Mansfield, P.G. Morris: *NMR Imaging in Biomedicine*, ed. by J.S. Waugh (Academic, New York 1982)

Superoperators

J. Jeener: Superoperators in magnetic resonance. Adv. Magn. Reson. **10**, 2 (1982)

R. Zwanzig: Ensemble method in theory of irreversibility. J. Chem. Phys. **33**, 1338 (1960)

Some Recent Books of Special Interest

F.R. Bovey: *Nuclear Magnetic Resonance Spectroscopy* (Academic, New York 1988)

W.S. Brey: *Pulse Methods in 1D and 2D Liquid-Phase NMR* (Academic, New York 1988)

C. Corso: *Physics of NMR Spectroscopy in Biology and Medicine*, Proc. Int. School of Physics, ed. by B. Maraviglia (North-Holland, Amsterdam 1988)

R.R. Ernst, G. Bodenhausen, A. Wokaun: *Principles of Nuclear Magnetic Resonance in One and Two Dimensions* (Clarendon, Oxford 1987)

R. Freeman: *A Handbook of Nuclear Magnetic Resonance* (Wiley, New York 1987)

E. Fukushima, S.B.W. Roeder: *Experimental Pulse NMR – A Nuts and Bolts Approach* (Addison-Wesley, Reading, MA 1981)

B.C. Gerstein, C.R. Dybowski: *Transient Techniques in NMR of Solids: An Introduction to Theory and Practice* (Academic, Orlando, FL 1985)

U. Haeberlen: *High Resolution NMR in Solids: Selective Averaging* (Academic, New York 1976)

P. Mansfield, P.G. Morris: *NMR Imaging in Biomedicine*, ed. by J.S. Waugh (Academic, New York 1982)

M. Mehring: *Principles of High Resolution NMR in Solids, 2nd ed.* (Springer, Berlin 1983)

M. Munowitz: *Coherence and NMR* (Wiley, New York 1988)

K. Wüthrich: *NMR of Proteins and Nucleic Acids* (Wiley, New York 1986)

R.N. Zare: *Angular Momentum: Understanding Spatial Aspects in Chemistry and Physics* (Wiley, New York 1988)

References

Chapter 1

1.1 C.J. Gorter, L.J.F. Broer: Physica **9**, 591 (1942)
1.2 E.M. Purcell, H.C. Torrey, R.V. Pound: Phys. Rev. **69**, 37 (1946)
1.3 F. Bloch, W.W. Hansen, M. Packard: Phys. Rev. **69**, 127 (1946)

Chapter 2

2.1 J.H. Van Vleck: Phys. Rev. **74**, 1168 (1948)
2.2 M. Tinkham: *Group Theory and Quantum Mechanics* (McGraw-Hill, New York 1964)
2.3 H. Rauch, A. Zeilinger, G. Badurek, A. Wilfing, W. Bauspiess, V. Bonse: Phys. Lett. **54A**, 425 (1975)
2.4 S.A. Werner, R. Colella, A.W. Overhauser, C.F. Eagen: Phys. Rev. Lett. **35**, 1053 (1975)
2.5 H.J. Bernstein: Phys. Rev. Lett. **18**, 1102 (1967)
2.6 Y. Aharanov, L. Susskind: Phys. Rev. **158**, 1237 (1967)
2.7 M.E. Stoll, A.J. Vega, R.W. Vaughan: Phys. Rev. **A16**, 1521 (1977)
2.8 M.E. Stoll, E.K. Wolff, M. Mehring: Phys. Rev. **A17**, 1561 (1978)
2.9 E.K. Wolff, M. Mehring: Phys. Lett. **70A**, 125 (1979)
2.10 E.L. Hahn: Phys. Rev. **80**, 580 (1950)
2.11 H.Y. Carr: Current Comments **20**, 24 (1983)
2.12 H.Y. Carr and E.M. Purcell: Phys. Rev. **94**, 630 (1954)
2.13 I. Solomon: Phys. Rev. **110**, 61 (1958)
2.14 J. Spokas: Thesis, University of Illinois (1957) (unpublished)
2.15 J.J. Spokas, C.P. Slichter: Phys. Rev. **113**, 1462 (1959)
2.16 D.F. Holcomb, R.E. Norberg: Phys. Rev. **98**, 1074 (1955)
2.17 Z. Wang: Private communication
2.18 C.J. Gorter: *Paramagnetic Relaxation* (Elsevier, New York 1947) p. 127
2.19 R. Kubo, K. Tomita: J. Phys. Soc. Japan **9**, 888 (1954)
2.20 E.W. Hobson: *The Theory of Functions of a Real Variable and the Theory of Fourier's Series* (Cambridge University Press, Cambridge 1926) p. 353 ff.
2.21 P.W. Anderson: J. Phys. Soc. Japan **9**, 316 (1954)
2.22 H.S. Gutowsky, G.E. Pake: J. Chem. Phys. **16**, 1164 (1948); ibid. **18**, 162 (1950) for example

Chapter 3

3.1 E.R. Andrew, R.G. Eades: Proc. R. Soc., London **A218**, 537 (1953)

Chapter 4

4.1 N.F. Ramsey: Phys. Rev. **78**, 699 (1950); Phys. Rev. **86**, 213 (1952)
4.2 W. Lamb: Phys. Rev. **60**, 817 (1941)

4.3 J.A. Pople: Proc. R. Soc., London **A239**, 541, 550 (1957)
4.4 N.F. Ramsey: Phys. Rev. **86**, 243 (1952)
4.5 D. Pines. *Solid State Physics*, Vol. 1, ed. by F. Seitz, D. Turnbull (Academic, New York 1955)
4.6 L. Pauling, E.B. Wilson, Jr.: *Introduction to Quantum Mechanics* (McGraw-Hill, New York 1935) p. 232
4.7 R.T. Schumacher, C.P. Slichter: Phys. Rev. **101**, 58 (1956);
B.R. Whiting, N.S. Van der Ven, R.T. Schumacher: Phys. Rev. **B18**, 5413 (1978)
4.8 Ch. Ryter: Phys. Rev. Lett. **5**, 10 (1960)
4.9 W. Kohn: Phys. Rev. **96**, 590 (1954)
4.10 T. Kjeldaas, Jr., W. Kohn: Phys. Rev. **101**, 66 (1956)
4.11 S.H. Vosko, J.P. Peradew, A.H. MacDonald: Phys. Rev. Lett. **35**, 1725 (1975)
4.12 A.C. Gossard, A.M. Portis: Phys. Rev. Lett. **3**, 164 (1959); J. Appl. Phys. Suppl. **31**, 2055 (1960)
4.13 W. Marshall: Phys. Rev. **110**, 1280 (1958)
4.14 R. Watson, A. Freeman: Phys. Rev. **123**, 2027 (1961)
4.15 N. Bloembergen, T.J. Rowland: Acta Metall. **1**, 731 (1953)
4.16 E.L. Hahn, D.E. Maxwell: Phys. Rev. **88**, 1070 (1952)
4.17 H.S. Gutowsky, D.W. McCall, C.P. Slichter: J. Chem. Phys. **21**, 279 (1953)
4.18 N.F. Ramsey, E.M. Purcell: Phys. Rev. **85**, 143-144 (1952) (letter)
4.19 N. Bloembergen, T.J. Rowland: Phys. Rev. **97**, 1679-1698 (1955)
4.20 M.A. Ruderman, C. Kittel: Phys. Rev. **96**, 99 (1954)
4.21 F. Fröhlich, F.R.N. Nabarro: Proc. R. Soc. London **A175**, 382 (1940)
4.22 K. Yoshida: Phys. Rev. **106**, 893 (1957)
4.23 J.H. Van Vleck: Rev. Phys. **34**, 681 (1962)

Chapter 5

5.1 C.J. Gorter: *Parametric Relaxation* (Elsevier, New York 1947)
5.2 L.C. Hebel, C.P. Slichter: Phys. Rev. **113**, 1504 (1959)
5.3 J. Korringa: Physica **16**, 601 (1950)
5.4 D. Pines: *Solid State Physics*, Vol. 1, ed. by F. Seitz, D. Turnbull (Academic, New York 1955)
5.5 R.C. Tolman: *The Principles of Statistical Mechanics* (Oxford University Press, New York 1946)
5.6 I.J. Lowe, R.E. Norberg: Phys. Rev. **107**, 46 (1957)
5.7 C.R. Bruce: Phys. Rev. **107**, 43 (1957)
5.8 R.R. Ernst, W.A. Anderson: Rev. Sci. Instrum. **37**, 93 (1966)
5.9 R.R. Ernst, G. Bodenhausen, A. Wokaun: *Principles of Nuclear Magnetic Resonance in One and Two Dimensions* (Clarendon, Oxford 1987)
5.10 A.G. Redfield: IBM J. Res. Dev. **1**, 19 (1957)
5.11 R.K. Wangsness, F. Bloch: Phys. Rev. **89**, 728 (1953);
F. Bloch: Phys. Rev. **102**, 104 (1956)
5.12 N. Bloembergen, E.M. Purcell, R.V. Pound: Phys. Rev. **73**, 679 (1948)
5.13 E.R. Andrew, R.G. Eades: Proc. R. Soc. London **A218**, 537 (1953)
5.14 D.F. Holcomb, R.E. Norberg: Phys. Rev. **98**, 1074 (1955)
5.15 E.F.W. Seymour: Proc. Phys. Soc., London **A66**, 85 (1953)
5.16 J.J. Spokas, C.P. Slichter: Phys. Rev. **113**, 1462 (1959)

Chapter 6

6.1 H.B.G. Casimir, F.K. duPré: Physica **5**, 507 (1938)
6.2 J.H. Van Vleck: Phys. Rev. **57**, 426 (1940)
6.3 J.H. Van Vleck: J. Chem. Phys. **5**, 320 (1937)
6.4 I. Waller: Z. Phys. **79**, 370 (1932)
6.5 A.G. Redfield: Phys. Rev. **98**, 1787 (1955)

6.6 C.P. Slichter, W.C. Holton: Phys. Rev. **122**, 1701 (1961)
6.7 E.H. Turner, A.M. Sachs, E.M. Purcell: Phys. Rev. **76**, 465 (A) (1949)
6.8 R.V. Pound: Phys. Rev. **81**, 156 (1951)
6.9 N.F. Ramsey, R.V. Pound: Phys. Rev. **81**, 278 (1951)
6.10 R.V. Pound, E.M. Purcell: Phys. Rev. **81**, 279 (1951)
6.11 J.H. Van Vleck: Nuovo Cimento Suppl. **6**, Serie X, 1081 (1957)
6.12 L.C. Hebel, C.P. Slichter: Phys. Rev. **113**, 1504 (1959)
6.13 A.G. Redfield: Phys. Rev. Lett. **3**, 85 (1959)
6.14 A.G. Redfield, A.G. Anderson: Phys. Rev. **116**, 583 (1959)
6.15 R.W. Morse, H.V. Bohm: Phys. Rev. **108**, 1094 (1957)
6.16 L. Cooper: Phys. Today **26**, (31 July 1973)
6.17 A.G. Anderson: Phys. Rev. **115**, 863 (1959)
6.18 A. Abragam, W.G. Proctor: Phys. Rev. **106**, 160 (1957)
6.19 A.G. Redfield: Phys. Rev. **98**, 787 (1955)
6.20 C.P. Slichter, W.C. Holton: Phys. Rev. **122**, 1701 (1961)
6.21 B.N. Provotorov: Soviet Phys. – JETP **14**, 1126 (1962)
6.22 R.T. Schumacher: Phys. Rev. **112**, 837 (1958)
6.23 C.P. Slichter, D. Ailion: Phys. Rev. **135**, A1099 (1964)
6.24 D. Ailion, C.P. Slichter: Phys. Rev. **137**, A235 (1965)
6.25 D.C. Look, I.J. Lowe: J. Chem. Phys. **44**, 2995 (1966)

Chapter 7

7.1 R.V. Pound: Phys. Rev. **79**, 685 (1950)
7.2 T.R. Carver, C.P. Slichter: Phys. Rev. **92**, 212 (1953); ibid. **102**, 975 (1956)
7.3 A.W. Overhauser: Phys Rev. **91**, 476 (1953); ibid. **92**, 411 (1953)
7.4 C.D. Jeffries (ed): *Dynamic Nuclear Orientation* (Interscience, New York 1963)
7.5 D.F. Holcomb: Thesis, University of Illinois (1954) unpublished
7.6 F. Bloch: Phys. Rev. **102**, 104 (1956)
7.7 I. Solomon: Phys. Rev. **99**, 559 (1955)
7.8 C.D. Jeffries: Phys. Rev. **106**, 164 (1957); ibid. **117**, 1056 (1960)
7.9 A. Abragam, J. Combrisson, I. Solomon: C.R. Acad. Sci. **247**, 2237 (1958)
7.10 E. Erb, J.L. Montchane, J. Uebersfeld: C.R. Acad. Sci. **246**, 2237 (1958)
7.11 G. Feher: Phys. Rev. **105**, 1122 (1957)
7.12 H. Seidel, H.C. Wolf: In *Physics of Color Centers*, ed. by W. Beall Fowler (Academic, New York 1968)
7.13 J.P. Gordon, H.J. Zeiger. C.H. Townes: Phys. Rev. **95**, 2821 (1954)
7.14 *Nobel Lectures–Physics 1963-1970* (Elsevier, Amsterdam 1972)
7.15 N.G. Basov, A.M. Prokhorov: Sov. Phys. – JETP **27**, 431 (1954)
7.16 N. Bloembergen: Phys. Rev. **104**, 324 (1956)
7.17 A. Abragam, W.G. Proctor: Phys. Rev. **106**, 160 (1957)
7.18 N. Bloembergen, S. Shapiro, P.S. Pershan, J.O. Artman: Phys. Rev. **114**, 445 (1959)
7.19 P.S. Pershan: Phys. Rev. **117**, 109 (1960)
7.20 A.G. Anderson: Phys. Rev. **115**, 863 (1959)
7.21 A.G. Redfield: Phys. Rev. **130**, 589 (1963)
7.22 N.C. Fernelius: Proc. of the XIV Colloque Ampere (1966) p. 497
7.23 R.E. Slusher, E.L. Hahn: Phys. Rev. **166**, 332 (1968)
7.24 M. Minier: Phys. Rev. **182**, 437 (1969)
7.25 D.P. Weitekamp, A. Bielecki, D. Zax, K. Zilm, A. Pines: Phys. Rev. Lett. **50**, 1807 (1983)
7.26 D.T. Edmonds: Phys. Rep. **29**, 233 (1977)
7.27 N. Bloembergen, P. Sorokin: Phys. Rev. **100**, 865 (1958)
7.28 S.R. Hartmann, E.L. Hahn: Phys. Rev. **128**, 2042 (1962)
7.29 F.M. Lurie, C.P. Slichter: Phys. Rev. **133**, A1108 (1964)
7.30 C.P. Slichter, W.C. Holton: Phys. Rev. **122**, 1701 (1961)
7.31 P.R. Spencer, N.D. Schmid, C.P. Slichter: Phys. Rev. **B1**, 2989 (1970)
7.32 D.V. Lang, P.R. Moran: Phys. Rev. **B1**, 53 (1970)

7.33 A. Pines, M.G. Gibby, J.S. Waugh: J. Chem. Phys. **56**, 1776 (1972); ibid. **59**, 569 (1973)
7.34 V. Royden: Phys. Rev. **96**, 543 (1954)
7.35 A.L. Bloom, J.N. Shoolery: Phys. Rev. **97**, 1261 (1955)
7.36 F. Bloch: Phys. Rev. **93**, 944 (1954)
7.37 R. Freeman, S.P. Kempsell, M.H. Levitt: J. Magn. Reson. **35**, 447 (1979)
7.38 W.A. Anderson, F.A. Nelson: J. Chem. Phys. **39**, 183 (1963)
7.39 R. Freeman, W.A. Anderson: J. Chem. Phys. **42**, 1199 (1965)
7.40 R. Ernst: J. Chem. Phys. **45**, 3845 (1966)
7.41 J.B. Grutzner, R.E. Santini: J. Magn. Reson. **19**, 173 (1975)
7.42 V.J. Basus, P.D. Ellis, H.D.W. Hill, J.S.Waugh: J. Magn. Reson. **35**, 19 (1979)
7.43 M.H. Levitt, R. Freeman: J. Magn. Reson. **33**, 473 (1979)
7.44 M.H. Levitt: Prog. Nucl. Magn. Reson. Spectrosc. **18**, 16 (1986)
7.45 A.J. Shaka, J. Keeler: Prog. Nucl. Magn. Reson. Spectrosc. **19**, 47 (1987)
7.46 R. Freeman: Bull. Magn. Reson. **8**, 120 (1986)
7.47 D.E. Kaplan, E.L. Hahn: J. Phys. Radium **19**, 821 (1958)
7.48 M. Emshwiller, E.L. Hahn, D. Kaplan: Phys. Rev. **118**, 414 (1960)
7.49 J.B. Boyce: Thesis, University of Illinois (1972)
7.50 D.V. Lang, J.B. Boyce, D.C. Lo, C.P. Slichter: Phys. Rev. Lett. **29**, 776 (1972)
7.51 D.V. Lang, D.C. Lo, J.B. Boyce, C.P. Slichter: Phys. Rev. **B 9**, 3077 (1974)
7.52 C.D. Makowka, C.P. Slichter, J.H. Sinfelt: Phys. Rev. Lett. **49**, 379 (1982); Phys. Rev. **B 31**, 5663 (1985)
7.53 P.K. Wang: Thesis, University of Illinois (1984) unpublished
7.54 P.K. Wang, C.P. Slichter, J.H. Sinfelt: Phys. Rev. Lett. **53**, 82 (1984)
7.55 J. Jeener: Lectures, Ampere International Summer School, Basko Potze, Yugoslavia (1971) unpublished
7.56 R. Freeman, G.A. Morris: Bull. Magn. Reson. **1**, 5 (1979)
7.57 W.P. Aue, E. Bartholdi, R.R. Ernst: J. Chem. Phys. **64**, 2229 (1976)
7.58 A. Bax: Bull. Magn. Reson. **78**, 167 (1985)
7.59 D.L. Turner: Prog. Nucl. Magn. Reson. Spectrosc. **17**, 281 (1985)
7.60 L. Müller, A. Kumar, R.R. Ernst: J. Chem. Phys. **63**, 5490 (1975)
7.61 L. Müller, A. Kumar, R.R. Ernst: J. Magn. Reson. **25**, 383 (1977)
7.62 D.T. Pegg, M.R. Bendall, D.D. Doddrell: J. Magn. Reson. **44**, 238 (1981); ibid. **45**, 8 (1981)
7.63 G. Feher: Phys. Rev. **103**, 500 (1956)
7.64 G. Feher, E.A. Gere: Phys. Rev. **103**, 501 (1956)
7.65 E.B. Baker: J. Chem. Phys. **37**, 911 (1962)
7.66 K.G.R. Pachler, P.L. Wessels: J. Magn. Reson. **12**, 337 (1973)
7.67 A.A. Maudsley, R.R. Ernst: Chem. Phys. Lett. **50**, 368 (1977)
7.68 G. Bodenhausen, R. Freeman: J. Magn. Reson. **28**, 463 (1977)
7.69 A.A. Maudsley, L. Müller, R.R. Ernst: J. Magn. Reson. **28**, 463 (1977)
7.70 G.A. Morris, R. Freeman: J. Am. Chem. Soc. **101**, 760 (1979)
7.71 O.W. Sorensen, G.W. Eich, M.H. Levitt, G. Bodenhausen, R.R. Ernst: Prog. Nucl. Magn. Reson. Spectrosc. **16**, 163 (1983)
7.72 F.J.M. Van den Ven, C.W. Hilbers: J. Magn. Reson. **54**, 512 (1983)
7.73 P.K. Wang, C.P. Slichter: Bull. Magn. Reson. **8**, 3 (1986)
7.74 E.L. Hahn, D.E. Maxwell: Phys. Rev. **88**, 1070 (1952)
7.75 H.S. Gutowsky, D.W. McCall, C.P. Slichter: J. Chem. Phys. **21**, 279 (1953)
7.76 K. Wüthrich: *NMR of Proteins and Nucleic Acids* (Wiley-Interscience, New York 1986)
7.77 P. Mansfield: *NMR Imaging in Biomedicine* (Academic, New York 1982)
7.78 R.R. Ernst, G. Bodenhausen, A. Wokaun: Principles of Nuclear Magnetic Resonance in One and Two Dimensions (Clarendon, Oxford 1987)
7.79 P.C. Lauterbur: Nature **242**, 190 (1973)
7.80 P. Mansfield, P.K. Grannell: J. Phys. **C 6**, L422 (1973)
7.81 A. Kumar, D. Welti, R.R. Ernst: J. Magn. Reson. **18**, 69 (1975)
7.82 P. Mansfield: J. Phys. C **10**, L55 (1977)
7.83 K.K. King, P.R. Moran: Med. Phys. **11**, 1 (1984)
7.84 T.R. Brown, B.M. Kincaid, K. Ugurbil: Proc. Natl. Acad. Sci. USA **79**, 3523 (1982)

7.85   D.B. Twieg: Med. Phys. **10**, 610 (1983)
7.86   S.J. Ljunggren: J. Magn. Reson. **54**, 38 (1983)

Chapter 8

8.1   E.L. Hahn: Phys. Rev. **80**, 580 (1950)
8.2   H.Y. Carr, E.M. Purcell: Phys. Rev. **94**, 630 (1954)
8.3   S. Meiboom, D. Gill: Rev. Sci. Instrum. **29**, 6881 (1958)
8.4   E.D. Ostroff, J.S. Waugh: Phys. Rev. Lett **16**, 1097 (1966)
8.5   P. Mansfield, D. Ware: Phys. Rev. Lett. **22**, 133 (1966)
8.6   L.M. Stacey, R.W. Vaughan, D.D. Elleman: Phys. Rev. Lett. **26**, 1153 (1971)
8.7   J.G. Powles, P. Mansfield: Phys. Lett. **2**, 58 (1962)
8.8   I. Solomon: Phys. Rev. **110**, 61 (1958)
8.9   J.H. Davis, K.R. Jeffrey, M. Bloom, M.I. Valic, T.P. Higgs: Chem. Phys. Lett. **42**, 390 (1976)
8.10   P.K. Wang: Thesis, University of Illinois (1984) unpublished
8.11   P.K. Wang, C.P. Slichter: Bull. Magn. Reson. **8**, 3 (1986)
8.12   J. Jeener, P. Broekaert: Phys. Rev. **157**, 232 (1967)
8.13   M. Lee, W.I. Goldburg: Phys. Rev. Lett. **11**, 255 (1963); Phys. Rev. **140**, A1261 (1965)
8.14   D. Barnaal, I.J. Lowe: Phys. Rev. Lett. **11**, 258 (1963)
8.15   W.-K. Rhim, H. Kessemeier: Phys. Rev. **B 3**, 3655 (1971)
8.16   H. Kessemeier, W.-K. Rhim: Phys. Rev. **B 5**, 761 (1972)
8.17   W.-K. Rhim, A. Pines, J.S. Waugh: Phys. Rev. Lett. **25**, 218 (1970); Phys. Rev. **B 3**, 684 (1971)
8.18   I. Solomon: Phys. Rev. Lett. **2**, 301 (1959)
8.19   K. Takegoshi, C.A. McDowell: Chem. Phys. Lett. **116**, 100 (1985)
8.20   H.S. Gutowsky, G.E. Pake: J. Chem. Phys. **16**, 1164 (1948); ibid. **18**, 162 (1950)
8.21   I. Lowe: Phys. Rev. Lett. **2**, 285 (1959)
8.22   E.R. Andrew, A. Bradbury, R.G. Eades: Nature **182**, 1659 (1958); ibid. **183**, 1802 (1959)
8.23   A.R. Edmonds: *Angular Momentum in Quantum Mechanics* (Princeton University Press, Princeton, NJ 1957)
8.24   H. Kessemeier, R.E. Norberg: Phys. Rev. **155**, 321 (1967)
8.25   M.M. Maricq, J.S. Waugh: J. Chem. Phys. **70**, 3300 (1979)
8.26   A.M. Portis: Phys. Rev. **91**, 1971 (1953)
8.27   E.O. Stejskal, J. Schaefer, R.A. McKay: J. Magn. Reson. **25**, 569 (1977)
8.28   J. Herzfeld, A. Roufosse, R.A. Haberkorn, R.G. Griffin, M.J. Glimcher: Philos. Trans. R. Soc. London, Series B **289**, 459 (1980)
8.29   J. Herzfeld, A.E. Berger: J. Chem. Phys. **73**, 6021 (1980)
8.30   W.T. Dixon: J. Chem. Phys. **77**, 1800 (1982)
8.31   D.P. Raleigh, E.T. Olejniczak, S. Vega, R.G. Griffin: J. Am. Chem. Soc. **106**, 8302 (1984); J. Magn. Reson. **72**, 238 (1987)
8.32   M. Mehring: *High Resolution NMR Spectroscopy in Solids*, NMR: Basic Principles and Progress, Vol. 11, ed. by P. Diehl, E. Fluck, R. Kosfeld (Springer, Berlin, Heidelberg 1976)
8.33   U. Haeberlen: *High Resolution NMR in Solids: Selective Averaging*; Supplement No. 1 to *Advances in Magnetic Resonance* (Academic, New York 1976)
8.34   J.S. Waugh, L.M. Huber, U. Haeberlen: Phys. Rev. Lett. **20**, 180 (1968)
8.35   U. Haeberlen, J.S. Waugh: Phys. Rev. **175**, 453 (1968)
8.36   P. Mansfield: J. Phys. **C 4**, 1444 (1971)
8.37   P. Mansfield, M.J. Orchard, D.C. Stalker, K.H.B. Richards: Phys. Rev. **B 7**, 90 (1973)
8.38   W.K. Rhim, D.D. Elleman, R.W. Vaughan: J. Chem. Phys. **59**, 3740 (1973)
8.39   W.K. Rhim, D.D. Elleman, L.B. Schreiber, R.W. Vaughan: J. Chem. Phys. **60**, 4595 (1974)
8.40   L. Van Gerven (ed.): *Nuclear Magnetic Resonance in Solids* (Plenum, New York 1977)

8.41 P.R. Moran: J. Phys. Chem. Solids **30**, 297 (1969)
8.42 W.S. Warren, S. Sinton, D.P. Weitekamp, A. Pines: Phys. Rev. Lett. **43**, 1791 (1979)

Chapter 9

9.1 G. Bodenhausen: Prog. Nucl. Magn. Reson. Spectrosc. **14**, 137 (1981)
9.2 H. Hatanaka, T. Terao, T. Hashi: J. Phys. Soc. Japan **39**, 835 (1975)
9.3 H. Hatanaka, T. Hashi: J. Phys. Soc. Japan **39**, 1139 (1975)
9.4 W.P. Aue, E. Bartholdi, R.R. Ernst: J. Chem. Phys. **64**, 2229 (1976)
9.5 D. Weitekamp: Adv. Magn. Reson **11**, 111 (1983)
9.6 S. Emid: Bull Magn. Reson. **4**, 99 (1982)
9.7 M. Munowitz, A. Pines: Science **233**, 525 (1986); Adv. Chem. Phys. **LXVI**, 1 (1987)
9.8 R.R. Ernst, G. Bodenhausen, A. Wokaun: *Principles of Nuclear Magnetic Resonance in One and Two Dimensions* (Clarendon, Oxford 1987)
9.9 M.E. Stoll, A.J. Vega, R.W. Vaughan: Phys. Rev. **A16**, 1521 (1977)
9.10 M.E. Stoll, E.K. Wolff, M. Mehring: Phys. Rev. **A17**, 1561 (1978)
9.11 E.K. Wolff, M. Mehring: Phys. Lett. **70A**, 125 (1979)
9.12 M. Mehring, P. Höfer, A. Grupp: Phys. Rev. **A33**, 3523 (1988)
9.13 P.K. Wang, C.P. Slichter, J.H. Sinfelt: Phys. Rev. Lett. **53**, 82 (1984)
9.14 G.E. Pake: J. Chem. Phys. **16**, 327 (1948)
9.15 J. Baum, M. Munowitz, A.N. Garroway, and A. Pines: J. Chem. Phys. **83**, 2015 (1985)
9.16 E. Merzbacher: *Quantum Mechanics*, 2nd ed. (Wiley, New York 1970) p. 167
9.17 A. Wokaun, R.R. Ernst: Chem. Phys. Lett. **52**, 407 (1977)
9.18 W.S. Warren, S. Sinton, D.P. Weitekamp, A. Pines: Phys. Rev. Lett. **43**, 1791 (1979)
9.19 W.S. Warren, D.P. Weitekamp, A. Pines: J. Chem. Phys. **73**, 2084 (1980)
9.20 Y.-S. Yen, A. Pines: J. Chem. Phys. **78**, 3579 (1983)

Chapter 10

10.1 J.H. Smith, E.M. Purcell, N.F. Ramsey: Phys. Rev. **108**, 120 (1957)
10.2 L.R. Walker, G.K. Wertheim, V. Jaccarino: Phys. Rev. Lett **6**, 98 (1961)
10.3 M.E. Rose: *Elementary Theory of Angular Momentum* (Wiley, New York 1957)
10.4 E. Ambler, J.C. Eisenstein, J.F. Schooley: J. Math. Phys. **3**, 118, 760 (1962)
10.5 R.M. Sternheimer: Phys. Rev. **84**, 244 (1951); ibid. **86**, 316 (1952); ibid. **95**, 736 (1954)

Chapter 11

11.1 A.R. Edmonds: *Angular Momentum in Quantum Mechanics* (Princeton University Press, Princeton, NJ 1957)
11.2 J.S. Waugh, C.P. Slichter: Phys. Rev. **B37**, 4337 (1988)
11.3 L.G. Rowan, E.L. Hahn, W.B. Mims: Phys. Rev. **137**, 61 (1965)
11.4 W.B. Mims: Phys. Rev. **B5**, 2409 (1972)
11.5 T.G. Castner, W. Känzig: J. Phys. Chem. Solids. **3**, 178 (1957)
11.6 C.J. Delbecq, B. Smaller, P.H. Yuster: Phys. Rev. **111**, 1235 (1958)

Appendix

F.1   H.S. Gutowsky, C. Holm: J. Chem. Phys. **25**, 1228 (1956)
F.2   H.S. Gutowsky, D.W. McCall, C.P. Slichter: J. Chem. Phys. **21**, 279 (1953)
F.3   E.L. Hahn, D.E. Maxwell: Phys. Rev. **88**, 1070 (1952)
F.4   J.H. Van Vleck: Ned. Tijdschr. Natuurkd. **27**, 1 (1961)
F.5   H.M. McConnell: J. Chem. Phys. **28**, 430 (1958)
F.6   D.H. Archer: Thesis, Harvard University (1953)
F.7   P.W. Anderson: J. Phys. Soc. Jpn. **9**, 316 (1954)
F.8   H.S. Gutowsky, A. Saika: J. Chem. Phys. **21**, 1688 (1953)
F.9   J.T. Arnold: Phys. Rev. **102**, 136 (1956)

G.1   H.C. Torrey: Phys. Rev. **104**, 563 (1956)
G.2   H.Y. Carr, E.M. Purcell: Phys. Rev. **94**, 630 (1954)

I.1   N. Bloembergen, T.J. Rowland: Acta Metall. **1**, 731 (1953)
I.2   I.S. Gradshteyn, I.M. Ryzhik: *Table of Integrals, Sums, Series, and Products*, 4th ed. (Academic, New York 1980) pp. 177, 904
I.3   B.T. Feld and W.E. Lamb: Phys. Rev. **67**, 15 (1945)
I.4   N. Bloembergen: In *Defects in Crystalline Solids* (Physical Society, London 1954) pp. 1-32
I.5   D. Gerstein, C. Dybowski: *Transient Techniques in NMR of Solids* (Academic, New York 1985)
I.6   G.E. Pake: J. Chem. Phys. **16**, 327 (1948)

J.1   F.J. Dyson: Phys. Rev. **75**, 486 (1949)

K.1   U. Haeberlen, J.S. Waugh: Phys. Rev. **125**, 453 (1968)
K.2   R.M. Wilcox: J. Math. Phys. **8**, 962 (1967)
K.3   W. Magnus: Commun. Pure Appl. Math. **7**, 649 (1954)
K.4   P. Mansfield: J. Phys. **C 4**, 1444 (1971)
K.5   D.P. Burum, M. Linder, R.R. Ernst: J. Magn. Reson. **44**, 173 (1981)

Author Index

Abragam, A. 231, 264, 271
Aharanov, Y. 32
Ailion, D. 245
Ambler, E. 494
Anderson, A.G. 230, 274
Anderson, P.W. 63
Anderson, W.A. 179, 183, 310
Andrew, E.R. 80, 82, 85, 215, 392, 399, 405
Ansermet, J.-P. 301
Archer, D.H. 593
Arnold, J.T. 597
Artman, J.O. 272
Aue, W.P. 319, 324, 325, 433

Badurek, B. 32
Baker, E.B. 331, 351
Bardeen, J. 231
Barnaal, D. 387
Bartholdi, E. 319, 324, 325, 433
Basov, N.G. 269, 270
Basus, V.J. 310
Baum, J. 448
Bauspiess, W. 32
Bax, A. 319, 325
Bendall, M.R. 331
Bernstein, H.J. 32
Bielecki, A. 274
Bloch, F. 9, 35, 199, 258, 297, 309
Bloembergen, N. 127, 133, 141, 143, 199, 206, 231, 269, 270, 275, 276, 281, 284, 289, 290, 292, 294, 606, 615
Bloom, A.L. 297, 307, 309, 310
Bloom, M. 372
Bodenhausen, G. 179, 331, 340, 342, 345, 350, 354, 357, 433, 434
Bohm, D. 115, 124
Bohm, H.V. 231
Bonse, V. 32

Boyce, J.B. 312
Bradbury, A. 392, 399, 405
Broekaert, P. 380, 381, 383, 384
Broer, L.J.F. 9
Brown, T.R. 364
Bruce, C.R. 179
Burum, D.P. 628

Carr, H.Y. 40, 44, 367–369, 392, 404, 601
Carver, T.R. 249, 250, 257
Casimir, H.B.G. 219
Castner, T.G. 533, 534
Cohen, M.H. 485
Colella, R. 32
Combrisson, J. 264
Cooper, L.N. 231

Das, T.P. 485
Davis, J.H. 372
DelBecq, C.J. 533
Dixon, W.T. 406
Doddrell, D.D. 331
du Pré, F.K. 219
Durand, D.J. 301
Dybowski, C. 616
Dyson, F. 617, 618

Eades, R.G. 80, 82, 85, 215, 392, 405
Edmonds, A.R. 522
Edmonds, D.T. 275
Eich, G.W. 345, 350, 354
Eisenstein, J.C. 494
Elleman, D.D. 367, 406, 422
Ellis, P.D. 310
Emid, S. 434
Emshwiller, M. 311
Erb, E. 264
Ernst, R.E. 179, 183, 310, 319, 322, 323–325, 331, 335, 338, 345, 350, 354, 357, 362, 363, 433, 434, 463, 464, 469, 473, 628

Feher, G. 266–268, 331, 332, 335, 337
Feld, B.T. 615
Fernelius, N. 274
Freeman, A. 126
Freeman, R. 307, 310, 311, 319, 325, 331, 340, 342
Fröhlich, F. 133, 143

Garroway, A.N. 448
Gerstein, D. 616
Gibby, M.G. 247, 293
Gill, D. 367, 369, 371, 422
Glimcher, M.J. 405
Goldburg, W.I. 384, 386, 388, 389
Gordon, J.P. 269
Gorter, C.J. 9, 52, 151, 219
Gossard, A.C. 126
Gradshteyn, I.S. 612
Granell, P.K. 358, 361
Griffin, R.G. 405, 406
Grupp, A. 440
Grutzner, J.B. 310
Gutowsky, H.S. 63, 80, 127, 132, 141, 351, 392, 592, 593, 596

Haberkorn, R.A. 405
Haeberlen, U. 421, 422, 423, 624, 627
Hahn, E.L. 39, 40, 43, 127, 132, 247, 274–277, 283, 287, 289, 293, 311, 351, 367, 485, 524, 593, 601
Hansen, W.W. 9
Hartmann, S. 247, 275–277, 283, 289, 293
Hashi, T. 433
Hatanaka, H. 433
Hebel, L.C. 155, 230
Herzfeld, J. 405
Higgs, T.P. 372
Hilbers, C.W. 345, 354, 356

647

Hill, H.D.W. 310
Hobson, E.W. 56
Höfer, P. 440
Holcomb, D.F. 42, 44, 215, 258
Holm, C. 592, 593
Holton, W.C. 220, 235
Huber, L.M. 422, 423

Jaccarino, V. 487
Jeener, J. 319, 331, 335, 351, 380, 381, 383, 384
Jeffrey, K.R. 372
Jeffries, C.D. 264

Känzig, W. 533, 534
Kaplan, D.E. 311
Keeler, J. 311
Kempsell, S.P. 307, 310
Kessemeier, H. 389, 391, 398, 399
Kincaid, B.M. 364
King, K.K. 364
Kittel, C. 133, 143
Kjeldaas, T. Jr. 123, 124
Knight, W. 113
Kohn, W. 123, 124
Korringa, J. 156, 157
Kubo, R. 55
Kumar, A. 322, 323, 362, 363

Lamb, W. 103, 615
Lang, D.V. 287
Lauterbur, P.C. 357, 358, 361
Lee, M. 384, 386, 388, 389
Levitt, M.H. 307, 310, 311, 345, 350, 354
Linder, M. 628
Ljunggren, S.J. 364
Look, D.C. 245
Lowe, I.J. 179, 183, 245, 387, 392, 398
Lurie, F. 275, 281, 283, 284, 286, 287, 289, 293

Magnus, W. 475, 478, 624, 625
Makowka, C.D. 312, 313
Mansfield, P. 357, 358, 361, 362, 365–367, 371, 372, 406, 628
Maricq, M.M. 399
Marshall, W. 126
Maudsley, A.A. 331, 335, 338
Maxwell, D.E. 127, 132, 351, 593

McCall, D.W. 127, 132, 141, 351, 593
McConnell, H.M. 593
McDowell, C.A. 392
McKay, R.A. 404, 406
Mehring, M. 33, 440
Meiboom, S. 367, 369, 371, 422
Merzbacher, E. 459
Mims, W.B. 524
Minier, M. 274
Montchane, J.L. 264
Moran, P.R. 287, 364, 422
Morris, G.A. 319, 325, 331
Morse, R.W. 231
Müller, L. 322, 323, 331
Munowitz, M. 434, 448, 463, 471

Nabarro, F.R.N. 133, 143
Nelson, F.A. 310
Norberg, R.E. 42, 179, 183, 215, 398, 399

Olejniczak, E.T. 406
Orchard, M.J. 406
Ostroff, E.D. 367
Overhauser, A.W. 32, 247, 249, 250, 254, 256, 257, 293–295

Pachler, K.G.R. 331, 338, 351
Packard, M. 9
Pake, G.E. 63, 80, 392, 448, 616
Pauling, L. 117
Pegg, D.T. 331
Pennington, C.H. 301
Pershan, P. 272–274
Pines, A. 247, 293, 389, 391, 392, 434, 448, 463, 471, 475, 477, 479, 480, 483
Pines, D. 115, 124, 157
Pople, J.A. 105
Portis, A.M. 126, 399
Pound, R.V. 9, 199, 206, 229–231, 247–250, 295
Powles, J.G. 371, 372
Proctor, W.G. 231, 271
Prokhorov, A.M. 269, 270
Provotorov, B.N. 239, 240
Purcell, E.M. 9, 40, 44, 132, 199, 206, 230, 231, 367–369, 392, 404, 486, 661

Raleigh, D.P. 406
Ramsey, N.F. 92, 107, 132, 229, 486, 511
Rauch, H. 32
Redfield, A.G. 34, 199, 200, 201, 204–206, 219–221, 230, 232, 234–236, 241, 242, 244, 274, 276, 279, 281, 289
Reif, F. 485
Rhim, W.-K. 389, 391, 392, 406, 422
Richards, K.H.B. 406
Rose, M.E. 490, 492
Roufosse, A. 405
Rowan, L.G. 524
Rowland, T.J. 127, 133, 141, 143, 606, 615
Royden, V. 297, 307, 309
Ruderman, M. 133, 143
Ryter, Ch. 122–124
Ryzhik, I.M. 612

Sachs, A.G. 229
Saika, A. 596
Santini, R.E. 310
Schaefer, J. 404, 406
Schmid, N.D. 287
Schoolery, J.N. 297, 307, 309, 310
Schooley, J.F. 494
Schreiber, L.B. 406, 422
Schrieffer, J.R. 231
Schumacher, R.T. 122–124, 240, 241
Seidel, H. 268
Seymour, E.F.W. 215
Shaka, A.J. 311
Shapiro, S. 272
Sinfelt, J.H. 312, 313, 318, 444, 448
Sinton, S. 428, 471, 479
Slichter, C.P. 42, 122–124, 127, 141, 155, 215, 220, 230, 235, 245, 249, 250, 257, 275, 281, 283, 284, 286, 287, 289, 293, 312, 313, 318, 345, 351, 378, 378, 382, 444, 448, 524, 593
Slusher, R.E. 274, 287
Smith, J.H. 486
Solomon, I. 42, 258, 262, 264, 372, 390
Sorensen, O.W. 345, 350, 354

Sorokin, P. 275, 276, 281, 284, 289, 290, 292, 294
Spencer, P.R. 287
Spokas, J. 42, 215
Stacey, L.M. 367, 406
Stalker, D.C. 406
Stejskal, E.O. 404, 406
Sternheimer, R.M. 502
Stoll, M.E. 33, 440
Susskind, L. 32

Takegoshi, K. 392
Terao, T. 433
Theiss, G. 398
Tinkham, M. 32
Tolman, R.C. 157
Tomita, K. 55
Torrey, H.C. 9, 598
Townes, C.H. 269
Turner, D.L. 319, 325
Turner, E.H. 229
Twieg, D.B. 364

Uebersfeld, J. 264
Ugurbil, K. 364

Valic, M.I. 372
Van der Ven, F.J.M. 345, 354, 356
Van Gerven, L. 421
Van Hecke, P. 421
Van Vleck, J.H. 14, 79, 133, 142, 219, 221, 230, 593
Vaughan, R.W. 406, 422
Vega, A.J. 33
Vega, S. 406

Walker, L.R. 487
Waller, I. 219
Wang, P.-K. 318, 345, 378, 382, 444, 448
Wang, Z. 45
Wangsness, R. 199
Ware, D. 367, 406
Warren, W.S. 428, 471, 475, 477, 479, 483
Watson, R. 126
Waugh, J.S. 247, 293, 310, 367, 389, 391, 392, 399, 422, 423, 524, 624, 627

Weitekamp, D.P. 274, 428, 434, 463, 471, 475, 477, 483
Welti, D. 362, 363
Werner, S.A. 32
Wertheim, G.K. 487
Wessels, P.L. 331, 338, 351
Wilcox, R.M. 624
Wilfing, A. 32
Wilson, E.B., Jr. 117
Wokaun, A. 179, 357, 434, 463, 464, 469, 473
Wolf, H.C. 268
Wolff, E.K. 33, 440
Wüthrich, K. 351

Yen, W. 398
Yen, Y.-S. 471, 475, 480
Yosida, K. 133, 142
Yuster, P. 533

Zax, D. 274, 301
Zeiger, H.J. 269
Zeilinger, A. 32
Zilm, K. 274

Subject Index

Absorption
— elementary theory 5ff
— expression for power absorbed 38
— general atomic theory 59ff
— introduced 5
Adiabatic changes in the magnetic field 24, 223ff
Adiabatic demagnetization
— in the lab frame 223ff
— in the rotating frame (ADRF) 235ff
Adiabatic inversion 24
Alternating magnetic field, *see* H_1
Anisotropy (crystalline)
— of chemical and Knight shifts 127ff
— effect on powder patterns 605ff
Antiferromagnets
— NMR in 124f
Antishielding factor 502
Asymmetry parameter, defined 497
Average Hamiltonian
— calculated for various pulse sequences 423ff
— corrections to the theory of 623ff
— introduced 414ff

Benzene, structure studied by second moments 80ff
Bloch equations
— applied to diffusion 597ff
— applied to rate processes 592ff
— failure under some conditions 220
— introduced 33
— modified for relaxations in the instantaneous field 217
— solution in low H_1 35ff
Bloch functions, defined 117
Bloch-Wangsness-Redfield theory 199ff
Bloembergen-Sorokin experiment
— described 275ff
— explained using spin temperature 289ff
Boltzmann ratio B_{ij} 255
Bond length, determined by magnetic resonance 82

Carr-Purcell sequence
— explained 367ff
— Meiboom-Gill variety 369ff
Chemical exchange, effect on NMR 592ff
Chemical shifts
— effect of electron spin 143f
— experimental facts 88
— formal theory 92
Clebsch-Gordon coefficients 489
Coherence transfer, explained 331ff
Collision broadening 213
Commutation relations for spins 18
Correlation function
— defined 193
— derived for a special case 584ff
— general properties 192ff
Correlation time 193
COSY 350ff
Cross-polarization (CP)
— Hartmann-Hahn method 277ff
— Pines-Gibby-Waugh method 293ff
Cross-relaxation
— and establishment of spin-temperature 147
— between unlike nuclei 271ff
Crystal fields
— role in chemical shifts 90, 101
— role in electron spin resonance 505ff
Current density $j(r)$, quantum mechanical expression for 93ff
Curie's law, derived 163

Decoupling of spins 303ff
Density matrix, *see also* Redfield theory of density matrix
— definition 159
— general treatment 157ff
— interaction representation 163ff
— perturbation theory 164ff
— product operator method applied 344ff
— response to a δ-function 174ff
— in the rotating reference frame 165ff
— for thermal equilibrium 162
— time dependence using eigenfunctions 161

651

— time-dependent equation 160
— time development for two spins 325ff
— of a two-level system 186ff
— used to derive transition probability 190ff
Detailed balance 148
Diffusion
— effect of Carr-Purcell pulse sequence 367ff
— effect on resonance 367, 597ff
Dirac equation, derivation of s-state coupling 111ff
Double resonance 247ff
— classification of three kinds 247, 248
— — cross-relaxation family 271ff
— — Pound-Overhauser family 247ff
— — spin-coherence family 295ff
— S-flip only 296ff
— spin-echo variety (SEDOR) 311

Effective magnetic field, H_{eff}, defined 21
Electric field gradient
— asymmetry parameter, defined 495
— calculated 500ff
— introduced 486
— from valence electrons 501
Electric quadrupole (Chap. 10) 485ff
— computation 500ff
— examples in strong and weak magnetic fields 497ff
— Hamiltonian
— — for general axes 496
— — general treatment 486ff, 494ff
— — for principal axes 496
— using raising and lowering operators 497
— Sternheimer antishielding factor 502
Electric quadrupole moment, defined 496
Electron-nuclear double resonance (ENDOR) 266ff
Electron spin
— coupling to nucleus 108ff
— role in chemical shift 143ff
— role in Knight shift 113ff
Electron spin echoes 524ff
Electron spin resonance (ESR) (Chap. 11) 503ff
Electrons
— electric field gradient produced by 500
— interaction of electron orbit with nuclei 89
— interaction of electron spin with nuclei 108ff
— magnetic interaction with nuclei (Chap. 4) 87ff
Equation of motion of a spin
— classical, in a rotating reference frame 12

— classical theory 11
— effect of alternating fields, classical 20ff
— expectation value 17
— quantum mechanical 13
— — in a rotating reference frame 29ff
Ethyl alcohol 88
Exponential operators
— defined 25
— expansion for $e^A B e^{-A}$ 459
— a theorem proved 579, 580
— use in solving time-independent wave equation 26

Fermi function, defined 118
Ferromagnets, NMR in 124ff
Fourier transform (FT) NMR
— derived 179ff
— experimental advantages 183, 184
— J-resolved 2D 323
— line shapes in 2D 324–5
— two-dimensional (2D), introduced 319ff
Free induction decay, defined 23
$f(\omega)$, defined 71

Gauge transformation,
— defined 92
— invariance under 92f
g-factor
— defined 510
— expression for 510
— Landé 514
— theory 505ff
— — for V_k center 533ff
g-shift see g-factor
Gyromagnetic ratio γ, defined 2

H_1
— defined 20
— effect of finite H_1 on time of the echo 49
— effect of inhomogeneity on pulses 422
— related to power level 38, 39
— relation to linearly polarized field 36
— rotating 20
Hartmann-Hahn condition 277
Hebel-Slichter equation for T_1 150
High-temperature approximation
— general discussion 589f
— introduced and defined 63
Hyperfine structure (in ESR)
— general theory 516f
— in the V_k center 533f

Impedance, effect of χ on 38
Indirect nuclear coupling 131f

Inductance, effect of χ on 37
Interaction representation 163f
Irreducible tensor operators T_{LM}
— applied to electron spin resonance 513f
— construction from Y_{LM}'s 493
— defined 490
— T_{1M}'s 490
— T_{2M}'s 493
— use in electric quadrupole problems 489f
Isomer shift 487

J-coupling 131f
Jeener-Brockaert pulse sequence 380f

Knight shift
— experimental facts 113
— formula 122
— general theory 114f
— relation to magnetization density 116
Korringa relation 156f

Larmor frequency 13
Laser 269f
Line shapes
— effect of crystalline anisotropy 605ff
— theory of dipolar effect on 127ff
Linear response theory
— applied to pulse NMR 174ff
— as basis for Fourier transform NMR 179ff
— general theory 51ff
— introduced 51
Local field
— in double resonance 282
— in laboratory frame 222
— in rotating frame 235
Lorentz local field 125, 182
Lorentzian line, defined 39

Magic angle
— in the rotating frame (Lee-Goldburg experiment) 384ff
— spinning 392ff
Magic echoes 388ff
Magnetic dipolar broadening of resonance lines 65ff
Magnetic dipolar coupling reduced by spin-flip method 406ff
Magnetic dipolar interaction between pairs of nuclei 66
Magnetic dipolar order
— produced by Jeener-Broekaert sequence 380ff
— produced by adiabatic demagnetization 227, 245
Magnetic dipolar refocusing dipolar coupling by echoes 371ff

Magnetic dipolar terms A, B, C, D, E, F, defined 66
Magnetic field gradient
— defined 358f
— effect when diffusion is present 367, 369, 597ff
Magnetic field at nucleus
— from electron orbit (classical) 89
— from electron spin 108ff
— quantum mechanical 95
Magnetic resonance imaging (MRI) 357ff
Magnetic susceptibility, $\chi = \chi' - i\chi''$
— atomic theory of χ' and χ'' 59ff
— complex χ, introduced 37
— effect on coil impedance 37
— further quantum mechanical expressions for 580ff
— measurement of spin contribution of conduction electrons 122ff
— orbital in terms of current density 104
— relation of χ'' to $f(\omega)$ 71
Magnetization, relation of rotating to linear polarizations 36
Magnus expansion 623f
Maser 269f
Master equation 149
Matrix elements of many-electron product functions 117
Meiboom-Gill method 369ff
Mixer 184f
Moments
— effect of molecular rotation 84
— further expressions for 582
— method of 71
— nth moment, $\langle \omega^n \rangle$, defined 71
— second moment formulas 79, 80
— zeroth, first, and second calculated 73ff
— $\langle \Delta\omega^n \rangle$ defined 72
Motional narrowing
— analyzed using the Bloch equations 592f
— defined 213
— simple derivation 213f
Multiple quantum coherence (Chap. 9) 431ff
— basic explanation 434ff
— evolution, mixing, and detection 449–455
— frequency selective pumping 434ff
— generating a desired order 471ff
— nonselective excitation 444ff
— observing a desired order 463ff
— preparation 445–449
— three or more spins 455f
$m(\tau)$
— defined 51
— derived by quantum mechanics 174ff

653

Nuclear Overhauser effect (NOE) 257ff
Nuclear polarization see Overhauser effect, Solid effect

Operators see Exponential operators, Irreducible tensor operators, Raising and lowering operators
Overhauser effect
— general theory 248ff
— nuclear (NOE) 257ff

Perturbation theory, a useful theorem 585ff
Phase coherent detection
— apparatus 42
— and mixers 184f
Powder line shapes, general theory 605ff
Product operator method 344ff
Pseudodipolar coupling 141
Pulse, 90° and 180°, defined 22

Quadrature detection 185
Quadrupole see Electric quadrupole
Quenching of orbital angular momentum 89ff

Raising and lowering operators, introduced 14
Random phases, hypothesis applied to density matrix in thermal equilibrium 161
Rate phenomena, studied by NMR 592ff
Redfield theory of the density matrix
— condition of validity 199
— including applied alternating fields 215
— including thermal equilibrium 204ff
— used to derive the Bloch equations 206f
Redfield theory of saturation, see Spin temperature in the rotating frame
Resonance, elementary quantum description 5ff
Rotating reference frame
— classical 12
— contrasted with interaction representation 168
— doubly rotating frame 279ff
— quantum mechanical 29

Saturation
— definition 8
— failure of classical saturation theory 231
Secular broadening 214
Sensitivity, the problem discussed 270

Signal-to-noise ratio
— and Carr-Purcell pulse sequences 367f
— and double resonance 270
— and phase coherent detection 42
Single crystal spectra 127ff
Slow motion, theory of effect on $T_{1\varrho}$ 244ff
Solid effect (in double resonance) 265
Spectral density 194
Spin echoes
— derivation using density matrix 169ff
— derivation using wave functions 46ff
— effect of diffusion on 44, 597ff
— in ESR 524ff
— introduced (classically) 39ff
— in solids 371ff
— for spin 1 nuclei 601ff
Spin-flip narrowing
— explained 406-409
— general theory 409ff
— observation 416ff
— real pulses and sequences 421ff
Spin-lattice relaxation see also $T_{1\varrho}$
— in the Bloch equations 34
— general theory 145ff
— the Korringa law 156f
— in a metal 151ff
— when a spin temperature applies 146ff
— time, T_1, introduced 8
— T_1 minimum 199
— using the density matrix 197
Spin locking 245
Spinor 31ff
Spin-orbit coupling
— defined 504, 505
— role in g-shift 505ff
— typical values 507
Spin susceptibility of conduction electrons
— defined 119
— and Knight shift 119ff
— measured and theoretical values 122ff
Spin temperature see also High-temperature approximation
— to analyze adiabatic and sudden changes 223ff
— in the laboratory frame 221ff
— in magnetism and magnetic resonance 219ff
— in the rotating frame (Redfield theory thereof)
— — approach to equilibrium 239ff
— — condition for validity 241
— — general theory 234ff
— — inclusion of spin-lattice effects 242ff

— — used to analyze Bloembergen-Sorokin experiments
— used in T_1 theory of solids 146ff
Sternheimer antishielding factor 502

T_1 *see* Spin-lattice relaxation
$T_{1\varrho}$
— defined 243
— use for study of slow motions 244ff
T_2, introduced 34
Thermal equilibrium
— density matrix corresponding to 163
— relation to transition rates 6ff
Time-dependent Hamiltonians
— corrections to approximate formulas 623ff
— general theory 616ff
Time development operator, defined 173
Time-ordering operator T_D (of Dyson), defined 617
Transition probability
— formula and conditions of validity 61
— introduced 5f
— role of the lattice in thermal processes 7ff
Triangle rule 490
Two-dimensional Fourier transform method (2D FT) 319ff
— *see also* Fourier transform NMR

V_k center 533ff

Wigner-Eckart theorem 489ff
Wobble (in electron systems) 524

$X(\theta)$, defined 47

Springer Series in Solid-State Sciences
Editors: M. Cardona P. Fulde K. von Klitzing H.-J. Queisser

1 **Principles of Magnetic Resonance**
   3rd Edition By C. P. Slichter
2 **Introduction to Solid-State Theory**
   2nd Printing. By O. Madelung
3 **Dynamical Scattering of X-Rays in Crystals** By Z. G. Pinsker
4 **Inelastic Electron Tunneling Spectroscopy**
   Editor: T. Wolfram
5 **Fundamentals of Crystal Growth I**
   Macroscopic Equilibrium and Transport Concepts. 2nd Printing
   By F. Rosenberger
6 **Magnetic Flux Structures in Superconductors** By R. P. Huebener
7 **Green's Functions in Quantum Physics**
   2nd Edition By E. N. Economou
8 **Solitons and Condensed Matter Physics**
   2nd Printing
   Editors: A. R. Bishop and T. Schneider
9 **Photoferroelectrics** By V. M. Fridkin
10 **Phonon Dispersion Relations in Insulators** By H. Bilz and W. Kress
11 **Electron Transport in Compound Semiconductors** By B. R. Nag
12 **The Physics of Elementary Excitations**
   By S. Nakajima, Y. Toyozawa, and R. Abe
13 **The Physics of Selenium and Tellurium**
   Editors: E. Gerlach and P. Grosse
14 **Magnetic Bubble Technology** 2nd Edition
   By A. H. Eschenfelder
15 **Modern Crystallography I**
   Symmetry of Crystals. Methods of Structural Crystallography By B. K. Vainshtein
16 **Organic Molecular Crystals**
   Their Electronic States. By E. A. Silinsh
17 **The Theory of Magnetism I**
   Statics and Dynamics. 2nd Edition
   By D. C. Mattis
18 **Relaxation of Elementary Excitations**
   Editors: R. Kubo and E. Hanamura
19 **Solitons,** Mathematical Methods for Physicists. 2nd Printing
   By G. Eilenberger
20 **Theory of Nonlinear Lattices**
   2nd Edition By M. Toda
21 **Modern Crystallography II** Structure of Crystals. By B. K. Vainshtein,
   V. M. Fridkin, and V. L. Indenbom
22 **Point Defects in Semiconductors I**
   Theoretical Aspects
   By M. Lannoo and J. Bourgoin
23 **Physics in One Dimension**
   Editors: J. Bernasconi, T. Schneider
24 **Physics in High Magnetic Fields**
   Editors: S. Chikazumi and N. Miura
25 **Fundamental Physics of Amorphous Semiconductors** Editor: F. Yonezawa

26 **Elastic Media with Microstructure I**
   One-Dimensional Models. By I. A. Kunin
27 **Superconductivity of Transition Metals**
   Their Alloys and Compounds.
   By S. V. Vonsovsky, Yu. A. Izyumov, and E. Z. Kurmaev
28 **The Structure and Properties of Matter**
   Editor: T. Matsubara
29 **Electron Correlation and Magnetism in Narrow-Band Systems** Editor: T. Moriya
30 **Statistical Physics I** 2nd Edition
   By M. Toda, R. Kubo, N. Saito
31 **Statistical Physics II**
   By R. Kubo, M. Toda, N. Hashitsume
32 **Quantum Theory of Magnetism**
   By R. M. White
33 **Mixed Crystals** By A. I. Kitaigorodsky
34 **Phonons: Theory and Experiments I**
   Lattice Dynamics and Models of Interatomic Forces. By P. Brüesch
35 **Point Defects in Semiconductors II**
   Experimental Aspects
   By J. Bourgoin and M. Lannoo
36 **Modern Crystallography III**
   Crystal Growth 2nd Edition
   By A. A. Chernov
37 **Modern Crystallography IV**
   Physical Properties of Crystals
   By L. A. Shuvalov
38 **Physics of Intercalation Compounds**
   Editors: L. Pietronero and E. Tosatti
39 **Anderson Localization**
   Editors: Y. Nagaoka and H. Fukuyama
40 **Semiconductor Physics** An Introduction
   3rd Edition By K. Seeger
41 **The LMTO Method**
   Muffin-Tin Orbitals and Electronic Structure
   By H. L. Skriver
42 **Crystal Optics with Spatial Dispersion, and Excitons**
   By V. M. Agranovich and V. L. Ginzburg
43 **Resonant Nonlinear Interactions of Light with Matter**
   By V. S. Butylkin, A. E. Kaplan,
   Yu. G. Khronopulo, and E. I. Yakubovich
44 **Elastic Media with Microstructure II**
   Three-Dimensional Models By I. A. Kunin
45 **Electronic Properties of Doped Semiconductors**
   By B. I. Shklovskii and A. L. Efros
46 **Topological Disorder in Condensed Matter**
   Editors: F. Yonezawa and T. Ninomiya
47 **Statics and Dynamics of Nonlinear Systems**
   Editors: G. Benedek, H. Bilz, and R. Zeyher
48 **Magnetic Phase Transitions**
   Editors: M. Ausloos and R. J. Elliott
49 **Organic Molecular Aggregates,** Electronic Excitation and Interaction Processes
   Editors: P. Reineker, H. Haken, and H. C. Wolf